Lecture Notes in Computer Science 5444

Commenced Publication in 1973
Founding and Former Series Editors:
Gerhard Goos, Juris Hartmanis, and Jan van Leeuwen

T0205306

Omer Reingold (Ed.)

Theory of Cryptography

6th Theory of Cryptography Conference, TCC 2009
San Francisco, CA, USA, March 15-17, 2009
Proceedings

 Springer

Volume Editor

Omer Reingold
The Weizmann Institute of Science
Faculty of Mathematics and Computer Science
Rehovot 76100 , Israel
E-mail: omer.reingold@weizmann.ac.il

Library of Congress Control Number: Applied for

CR Subject Classification (1998): E.3, F.2.1-2, C.2.0, G, D.4.6, K.4.1, K.4.3, K.6.5

LNCS Sublibrary: SL 4 – Security and Cryptology

ISSN 0302-9743

ISBN 978-3-642-00456-8 Springer Berlin Heidelberg New York

springer.com

© International Association for Cryptologic Research 2009

Typesetting: Camera-ready by author, data conversion by Scientific Publishing Services, Chennai, India
Printed on acid-free paper   SPIN: 12625369   06/3180   5 4 3 2 1 0

Preface

TCC 2009, the 6th Theory of Cryptography Conference, was held in San Francisco, CA, USA, March 15–17, 2009. TCC 2009 was sponsored by the International Association for Cryptologic Research (IACR) and was organized in cooperation with the Applied Crypto Group at Stanford University. The General Chair of the conference was Dan Boneh.

The conference received 109 submissions, of which the Program Committee selected 33 for presentation at the conference. These proceedings consist of revised versions of those 33 papers. The revisions were not reviewed, and the authors bear full responsibility for the contents of their papers. The conference program also included two invited talks: "The Differential Privacy Frontier," given by Cynthia Dwork and "Some Recent Progress in Lattice-Based Cryptography," given by Chris Peikert.

I thank the Steering Committee of TCC for entrusting me with the responsibility for the TCC 2009 program. I thank the authors of submitted papers for their contributions. The general impression of the Program Committee is that the submissions were of very high quality, and there were many more papers we wanted to accept than we could. The review process was therefore very rewarding but the selection was very delicate and challenging. I am grateful for the dedication, thoroughness, and expertise of the Program Committee. Observing the way the members of the committee operated makes me as confident as possible of the outcome of our selection process. I also thank the many external reviewers who assisted the Program Committee in its work. I have benefited from the experience and advice of past TCC Chairs, Ran Canetti, Moni Naor, and Salil Vadhan. I am indebted to Shai Halevi, who wrote a wonderful software package to facilitate all aspects of the PC work. Shai made his software available to us and provided rapid technical support. I am very grateful to TCC 2007 General Chair, Dan Boneh, who anticipated my requests before they were made. Thanks to our corporate Sponsors, Voltage Security, Google, Microsoft Research, the D. E. Shaw group, and IBM Research. I appreciate the assistance provided by the Springer LNCS editorial staff, including Ursula Barth, Alfred Hofmann, Anna Kramer, and Nicole Sator, and the assistance provided by IACR Director, Christian Cachin.

December 2008   Omer Reingold

TCC 2009

6th IACR Theory of Cryptography Conference

San Francisco, California, USA
March 15–17, 2009

Sponsored by *The International Association for Cryptologic Research*

With Financial Support from

Voltage Security
Google
Microsoft Research
The D. E. Shaw group
IBM Research

General Chair

Dan Boneh   Stanford University

Program Chair

Omer Reingold   Weizmann Institute

Program Committee

Ivan Damgård	University of Aarhus
Stefan Dziembowski	University of Rome
Marc Fischlin	Darmstadt University
Matthew Franklin	UC Davis
Jens Groth	University College London
Thomas Holenstein	Princeton University
Nicholas J. Hopper	University of Minnesota
Yuval Ishai	Technion and UC Los Angeles
Charanjit Jutla	IBM T.J. Watson Research Center
Daniele Micciancio	UC San Diego
Kobbi Nissim	Ben-Gurion University
Adriana M. Palacio	Bowdoin
Rafael Pass	Cornell
Manoj M Prabhakaran	Urbana-Champaign
Yael Tauman Kalai	Microsoft Research
Brent Waters	UT Austin
John Watrous	University of Waterloo

TCC Steering Committee

Mihir Bellare	UC San Diego
Ivan Damgård	University of Aarhus
Oded Goldreich (Chair)	Weizmann Institute
Shafi Goldwasser	MIT and Weizmann Institute
Russell Impagliazzo	UC San Diego and IAS
Johan Hastad	KTH
Ueli Maurer	ETH Zurich
Silvio Micali	MIT
Moni Naor	Weizmann Institute

External Reviewers

Ittai Abraham	Aggelos Kiayias	Amit Sahai
Adi Akavia	Robert König	Louis Salvail
Joel Alwen	Vladimir Kolesnikov	Eric Schost
Amos Beimel	Chiu-Yuen Koo	Dominique Schröder
Tor E. Bjorstad	Hugo Krawczyk	Gil Segev
Dario Catalano	Mikkel Kroigaard	Hovav Shacham
Yan Zong Ding	Eyal Kushilevitz	Abhi Shelat
Yevgeniy Dodis	Anja Lehmann	Elaine Shi
Serge Fehr	Huija (Rachel) Lin	Michael Steiner
Anna Lisa Ferrara	Noam Livne	Alain Tapp
Dario Fiore	Vadim Lyubashevsky	Stefano Tessaro
Matthias Fitzi	Yury Makarychev	Nikos Triandopoulos
David Freeman	Tal Malkin	Wei-lung (Dustin) Tseng
Juan Garay	Payman Mohassel	Dominique Unruh
Martin Geisler	Petros Mol	Salil Vadhan
Craig Gentry	Steven Myers	Vinod Vaikuntanathan
Clint Givens	Jesper Buus Nielsen	Jorge L. Villar
Dana Glasner	Alina Oprea	Ivan Visconti
Sharon Goldberg	Claudio Orlandi	Hoeteck Wee
Mark Gondree	Carles Padro	Stephanie Wehner
Dov Gordon	Omkant Pandey	Enav Weinreb
Ronen Gradwohl	Anindya Patthak	Daniel Wichs
Iftach Haitner	Chris Peikert	Severin Winkler
Danny Harnik	Krzysztof Pietrzak	Stefan Wolf
Carmit Hazay	Benny Pinkas	Jürg Wullschleger
Martin Hirt	Bartosz Przydatek	Scott Yilek
Dennis Hofheinz	Tal Rabin	Aaram Yun
Russell Impagliazzo	Renato Renner	Rui Zhang
Stanislaw Jarecki	Thomas Ristenpart	Yunlei Zhao
Ayman Jarrous	Alon Rosen	Hong-Sheng Zhou
Bhavana Kanukurthi	Mike Rosulek	Vassilis Zikas
Jonathan Katz	Guy Rothblum	

Table of Contents

An Optimally Fair Coin Toss

Tal Moran[*], Moni Naor[*,**], and Gil Segev[*]

Department of Computer Science and Applied Mathematics,
Weizmann Institute of Science, Rehovot 76100, Israel
talm@seas.harvard.edu, {moni.naor,gil.segev}@weizmann.ac.il

Abstract. We address one of the foundational problems in cryptography: the bias of coin-flipping protocols. Coin-flipping protocols allow mutually distrustful parties to generate a common unbiased random bit, guaranteeing that even if one of the parties is malicious, it cannot significantly bias the output of the honest party. A classical result by Cleve [STOC '86] showed that for any two-party r-round coin-flipping protocol there exists an efficient adversary that can bias the output of the honest party by $\Omega(1/r)$. However, the best previously known protocol only guarantees $O(1/\sqrt{r})$ bias, and the question of whether Cleve's bound is tight has remained open for more than twenty years.

In this paper we establish the optimal trade-off between the round complexity and the bias of two-party coin-flipping protocols. Under standard assumptions (the existence of oblivious transfer), we show that Cleve's lower bound is tight: we construct an r-round protocol with bias $O(1/r)$.

1 Introduction

A coin-flipping protocol allows mutually distrustful parties to generate a common unbiased random bit. Such a protocol should satisfy two properties. First, when all parties are honest and follow the instructions of the protocol, their common output is a uniformly distributed bit. Second, even if some of the parties collude and deviate from the protocol's instructions, they should not be able to significantly bias the common output of the honest parties.

When a majority of the parties are honest, efficient and completely fair coin-flipping protocols are known as a special case of general multiparty computation with an honest majority [1] (assuming a broadcast channel). When an honest majority is not available, and in particular when there are only two parties, the situation is more complex. Blum's two-party coin-flipping protocol [2] guarantees that the output of the honest party is unbiased only if the malicious party does not abort prematurely (note that the malicious party can decide to abort *after* learning the result of the coin flip). This satisfies a rather weak notion of fairness in which once the malicious party is labeled as a "cheater" the honest party is allowed to halt without outputting any value. Blum's protocol can rely on the

---

[*] Research supported in part by a grant from the Israel Science Foundation.
[**] Incumbent of the Judith Kleeman Professorial Chair.

O. Reingold (Ed.): TCC 2009, LNCS 5444, pp. 1–18, 2009.

existence of any one-way function [3, 4], and Impagliazzo and Luby [5] showed that one-way functions are in fact essential even for such a seemingly weak notion. While this notion suffices for some applications, in many cases fairness is required to hold even if one of the parties aborts prematurely (consider, for example, an adversary that controls the communication channel and can prevent communication between the parties). In this paper we consider a stronger notion: even when the malicious party is labeled as a cheater, we require that the honest party outputs a bit.

Cleve's impossibility result. The latter notion of fairness turns out to be impossible to achieve in general. Specifically, Cleve [6] showed that for any two-party r-round coin-flipping protocol there exists an efficient adversary that can bias the output of the honest party by $\Omega(1/r)$. Cleve's lower bound holds even under arbitrary computational assumptions: the adversary only needs to simulate an honest party, and decide whether or not to abort early depending on the output of the simulation. However, the best previously known protocol (with respect to bias) only guaranteed $O(1/\sqrt{r})$ bias [7, 6], and the question of whether Cleve's bound was tight has remained open for over twenty years.

Fairness in secure computation. The bias of coin-flipping protocols can be viewed as a particular case of the more general framework of fairness in secure computation. Typically, the security of protocols is formalized by comparing their execution in the *real model* to an execution in an *ideal model* where a trusted party receives the inputs of the parties, performs the computation on their behalf, and then sends *all* parties their respective outputs. Executions in the ideal model guarantee *complete fairness*: either all parties learn the output, or neither party does. Cleve's result, however, shows that without an honest majority complete fairness is generally impossible to achieve, and therefore the formulation of secure computation (see [8]) weakens the ideal model to one in which fairness is not guaranteed. Informally, a protocol is "secure-with-abort" if its execution in the real model is indistinguishable from an execution in the ideal model allowing the ideal-model adversary to chose whether the honest parties receive their outputs (this is the notion of security satisfied by Blum's coin-flipping protocol).

Recently, Katz [9] suggested an alternate relaxation: keep the ideal model unchanged (i.e., all parties always receive their outputs), but relax the notion of indistinguishability by asking that the real model and ideal model are distinguishable with probability at most $1/p(n) + \nu(n)$, for a polynomial $p(n)$ and a negligible function $\nu(n)$ (we refer the reader to Section 2 for a formal definition). Protocols satisfying this requirement are said to be $1/p$-secure, and intuitively, such protocols guarantee complete fairness in the real model except with probability $1/p$. In the context of coin-flipping protocols, any $1/p$-secure protocol has bias at most $1/p$. However, the definition of $1/p$-security is more general and applies to a larger class of functionalities.

1.1   Our Contributions

In this paper we establish the optimal trade-off between the round complexity and the bias of two-party coin-flipping protocols. We prove the following theorem:

Theorem 1.1. *Assuming the existence of oblivious transfer, for any polynomial $r = r(n)$ there exists an r-round two-party coin-flipping protocol that is $1/(4r-c)$-secure, for some constant $c > 0$.*

We prove the security of our protocol under the simulation-based definition of $1/p$-security[1], which for coin-flipping protocols implies, in particular, that the bias is at most $1/p$. We note that our result not only identifies the optimal trade-off asymptotically, but almost pins down the exact leading constant: Cleve showed that any r-round two-party coin-flipping protocol has bias at least $1/(8r+2)$, and we manage to achieve bias of at most $1/(4r-c)$ for some constant $c > 0$.

Our approach holds in fact for a larger class of functionalities. We consider the more general task of sampling from a distribution $\mathcal{D} = (\mathcal{D}_1, \mathcal{D}_2)$: party P_1 receives a sample from $\mathcal{D}_1$ and party P_2 receives a correlated sample from $\mathcal{D}_2$ (in coin-flipping, for example, the joint distribution $\mathcal{D}$ produces the values $(0, 0)$ and $(1, 1)$ each with probability $1/2$). Before stating our result in this setting we introduce a standard notation: we denote by $\mathrm{SD}(\mathcal{D}, \mathcal{D}_1 \otimes \mathcal{D}_2)$ the statistical distance between the joint distribution $\mathcal{D} = (\mathcal{D}_1, \mathcal{D}_2)$ and the direct-product of the two marginal distributions $\mathcal{D}_1$ and $\mathcal{D}_2$. We prove the following theorem which generalizes Theorem 1.1:

Theorem 1.2. *Assuming the existence of oblivious transfer, for any efficiently-sampleable distribution $\mathcal{D} = (\mathcal{D}_1, \mathcal{D}_2)$ and polynomial $r = r(n)$ there exists an r-round two-party protocol for sampling from $\mathcal{D}$ that is $\frac{\mathrm{SD}(\mathcal{D}, \mathcal{D}_1 \otimes \mathcal{D}_2)}{2r-c}$-secure, for some constant $c > 0$.*

Our approach raises several open questions that are fundamental to the understanding of coin-flipping protocols. These questions include identifying the minimal computational assumptions that are essential for reaching the optimal trade-off (i.e., one-way functions vs. oblivious transfer), extending our approach to the multiparty setting, and constructing a more efficient variant of our protocol that can result in a practical implementation. We elaborate on these questions in Section 5, and hope that our approach and the questions it raises can make progress towards resolving the complexity of coin-flipping protocols.

1.2   Related Work

Coin-flipping protocols. When security with abort is sufficient, simple variations of Blum's protocol are the most commonly used coin-flipping protocols.

---

[1] In a very preliminary version of this work we proved our results with respect to the definition of bias (see Section 2), and motivated by [10, 9] we switch to the more general framework of $1/p$-secure computation.

For example, an r-round protocol with bias $O(1/\sqrt{r})$ can be constructed by sequentially executing Blum's protocol $O(r)$ times, and outputting the majority of the intermediate output values [7,6]. We note that in this protocol an adversary can indeed bias the output by $\Omega(1/\sqrt{r})$ by aborting prematurely. One of the most significant results on the bias of coin-flipping protocols gave reason to believe that the optimal trade-off between the round complexity and the bias is in fact $\Theta(1/\sqrt{r})$ (as provided by the latter variant of Blum's protocol): Cleve and Impagliazzo [11] showed that in the *fail-stop model*, any two-party r-round coin-flipping protocol has bias $\Omega(1/\sqrt{r})$. In the fail-stop model adversaries are computationally unbounded, but they must follow the instructions of the protocol except for being allowed to abort prematurely. In this model commitment schemes exist in a trivial fashion[2], and therefore the Cleve–Impagliazzo bound also applies to any protocol whose security relies on commitment schemes in a black-box manner, such as Blum's protocol and its variants.

Coin-flipping protocols were also studied in a variety of other models. Among those are collective coin-flipping in the "perfect information model" in which parties are computationally unbounded and all communication is public [12, 13, 14, 15, 16], and protocols based on physical assumptions, such as quantum computation [17, 18, 19] and tamper-evident seals [20].

Fair computation. Some of the techniques underlying our protocols found their origins in a recent line of research devoted for achieving various forms of fairness in secure computation. The technique of choosing a secret "threshold round", before which no information is learned, and after which aborting the protocol is essentially useless was suggested by Moran and Naor [20] as part of a coin-flipping protocol based on tamper-evident seals. It was later also used by Katz [9] for partially-fair protocols using a simultaneous broadcast channel, and by Gordon et al. [21] for completely-fair protocols for a restricted (but yet rather surprising) class of functionalities. Various techniques for hiding a meaningful round in game-theoretic settings were suggested by Halpern and Teague [22], Gordon and Katz [23], and Kol and Naor [24]. Katz [9] also introduced the technique of distributing shares to the parties in an initial setup phase (which is only secure-with-abort), and these shares are then exchanged by the parties in each round of the protocol.

Subsequent work. Our results were very recently generalized by Gordon and Katz [10] to deal with the more general case of randomized functions, and not only distributions. Gordon and Katz showed that any efficiently-computable randomized function $f : X \times Y \to Z$ where at least one of X and Y is of polynomial size has an r-round protocol that is $O\left(\frac{\min\{|X|,|Y|\}}{r}\right)$-secure. In addition, they showed that even if both domains are of super-polynomial size but the range Z is of polynomial size, the f has an r-round protocol that is $O\left(\frac{|Z|}{\sqrt{r}}\right)$-secure. Gordon and Katz also showed a specific function $f : X \times Y \to Z$ where X, Y,

---

[2] The protocol for commitment in the fail-stop model is simply to privately decide on the committed value and send the message "I am committed" to the other party.

and Z are of size super-polynomial which cannot be $1/p$-securely computed for any $p > 2$ assuming the existence of exponentially-hard one-way functions.

1.3   Paper Organization

The remainder of this paper is organized as follows. In Section 2 we review several notions and definitions that are used in the paper (most notably, the definition of $1/p$-secure computation). In Section 3 we describe a simplified variant of our protocol and prove its security. In Section 4 we sketch a more refined and general variant of our protocol (due to space limitations we refer the reader to the full version for its complete specification and proof of security). Finally, in Section 5 we discuss several open problems.

2   Preliminaries

In this section we review the definitions of coin-flipping protocols, $1/p$-secure computation (taken almost verbatim from [10, 9]), security with abort, and one-time message authentication.

2.1   Coin-Flipping Protocols

A two-party coin-flipping protocol is defined via two probabilistic polynomial-time Turing machines (P_1, P_2), referred to as parties, that receive as input a security parameter 1^n. The parties exchange messages in a sequence of rounds, where in every round each party both sends and receives a message (i.e., a round consists of two moves). At the end of the protocol, P_1 and P_2 produce outputs bits c_1 and c_2, respectively. We denote by $(c_1|c_2) \leftarrow \langle P_1(1^n), P_2(1^n) \rangle$ the experiment in which P_1 and P_2 interact (using uniformly chosen random coins), and then P_1 outputs c_1 and P_2 outputs c_2. It is required that for all sufficiently large n, and every possible pair (c_1, c_2) that may be output by $\langle P_1(1^n), P_2(1^n) \rangle$, it holds that $c_1 = c_2$ (i.e., P_1 and P_2 agree on a common value). This requirement can be relaxed by asking that the parties agree on a common value with sufficiently high probability[3].

The security requirement of a coin-flipping protocol is that even if one of P_1 and P_2 is corrupted and arbitrarily deviates from the protocol's instructions, the bias of the honest party's output remains bounded. Specifically, we emphasize that a malicious party is allowed to abort prematurely, and in this case it is assumed that the honest party is notified on the early termination of the protocol. In addition, we emphasize that even when the malicious party is labeled as a cheater, the honest party must output a bit. For simplicity, the following definition considers only the case in which P_1 is corrupted, and an analogous definition holds for the case that P_2 is corrupted:

---

[3] Cleve's lower bound [6] holds under this relaxation as well. Specifically, if the parties agree on a common value with probability $1/2 + \epsilon$, then Cleve's proof shows that the protocol has bias at least $\epsilon/(4r + 1)$.

Definition 2.1. *A coin-flipping protocol (P_1, P_2) has bias at most $\epsilon(n)$ if for every probabilistic polynomial-time Turing machine P_1^* it holds that*

$$\left| \Pr\left[(c_1 | c_2) \leftarrow \langle P_1^*(1^n), P_2(1^n) \rangle : c_2 = 1 \right] - \frac{1}{2} \right| \leq \epsilon(n) + \nu(n) \ ,$$

for some negligible function $\nu(n)$ and for all sufficiently large n.

2.2 $1/p$-Indistinguishability and $1/p$-Secure Computation

$1/p$-Indistinguishability. A distribution ensemble $X = \{X(a, n)\}_{a \in \mathcal{D}_n, n \in \mathbb{N}}$ is an infinite sequence of random variables indexed by $a \in \mathcal{D}_n$ and $n \in \mathbb{N}$, where $\mathcal{D}_n$ is a set that may depend on n. For a fixed polynomial $p(n)$, two distribution ensembles $X = \{X(a, n)\}_{a \in \mathcal{D}_n, n \in \mathbb{N}}$ and $Y = \{Y(a, n)\}_{a \in \mathcal{D}_n, n \in \mathbb{N}}$ are *computationally $1/p$-indistinguishable*, denoted $X \overset{1/p}{\approx} Y$, if for every non-uniform polynomial-time algorithm D there exists a negligible function $\nu(n)$ such that for all sufficiently large $n \in \mathbb{N}$ and for all $a \in \mathcal{D}_n$ it holds that

$$|\Pr\left[D(X(a, n)) = 1 \right] - \Pr\left[D(Y(a, n)) = 1 \right]| \leq \frac{1}{p(n)} + \nu(n) \ .$$

$1/p$-Secure computation. A two-party protocol for computing a functionality $\mathcal{F} = \{(f^1, f^2)\}$ is a protocol running in polynomial time and satisfying the following functional requirement: if party P_1 holds input $(1^n, x)$, and party P_2 holds input $(1^n, y)$, then the joint distribution of the outputs of the parties is statistically close to $(f^1(x, y), f^2(x, y))$. In what follows we define the notion of $1/p$-secure computation [10,9]. The definition uses the standard real/ideal paradigm [25,8], except that we consider a completely fair ideal model (as typically considered in the setting of honest majority), and require only $1/p$-indistinguishability rather than indistinguishability (we note that, in general, the notions of $1/p$-security and security-with-abort are incomparable). We consider active adversaries, who may deviate from the protocol in an arbitrary manner, and static corruptions.

Security of protocols (informal). The security of a protocol is analyzed by comparing what an adversary can do in a real protocol execution to what it can do in an ideal scenario that is secure by definition. This is formalized by considering an ideal computation involving an incorruptible trusted party to whom the parties send their inputs. The trusted party computes the functionality on the inputs and returns to each party its respective output. Loosely speaking, a protocol is secure if any adversary interacting in the real protocol (where no trusted party exists) can do no more harm than if it was involved in the above-described ideal computation.

Execution in the ideal model. The parties are P_1 and P_2, and there is an adversary $\mathcal{A}$ who has corrupted one of them. An ideal execution for the computation of $\mathcal{F} = \{f_n\}$ proceeds as follows:

Inputs: P_1 and P_2 hold the security parameter 1^n and inputs $x \in X_n$ and $y \in Y_n$, respectively. The adversary $\mathcal{A}$ receives an auxiliary input aux.

Send inputs to trusted party: The honest party sends its actual input to the trusted party. The corrupted party may send an arbitrary value (chosen by $\mathcal{A}$) to the trusted party. Denote the pair of inputs sent to the trusted party by (x', y').

Trusted party sends outputs: If $x' \notin X_n$ the trusted party sets x' to some default element $x_0 \in X_n$ (and likewise if $y' \notin Y_n$). Then, the trusted party chooses r uniformly at random and sends $f_n^1(x', y'; r)$ to P_1 and $f_n^2(x', y'; r)$ to P_2.

Outputs: The honest party outputs whatever it was sent by the trusted party, the corrupted party outputs nothing, and $\mathcal{A}$ outputs any arbitrary (probabilistic polynomial-time computable) function of its view.

We denote by $\mathsf{IDEAL}_{\mathcal{F}, \mathcal{A}(\mathsf{aux})}(x, y, n)$ the random variable consisting of the view of the adversary and the output of the honest party following an execution in the ideal model as described above.

Execution in the real model. We now consider the real model in which a two-party protocol π is executed by P_1 and P_2 (and there is no trusted party). The protocol execution is divided into rounds; in each round one of the parties sends a message. The honest party computes its messages as specified by π. The messages sent by the corrupted party are chosen by the adversary, $\mathcal{A}$, and can be an arbitrary (polynomial-time) function of the corrupted party's inputs, random coins, and the messages received from the honest party in previous rounds. If the corrupted party aborts in one of the protocol rounds, the honest party behaves as if it had received a special $\perp$ symbol in that round.

Let π be a two-party protocol computing the functionality $\mathcal{F}$. Let $\mathcal{A}$ be a non-uniform probabilistic polynomial-time machine with auxiliary input aux. We denote by $\mathsf{REAL}_{\pi, \mathcal{A}(\mathsf{aux})}(x, y, n)$ the random variable consisting of the view of the adversary and the output of the honest party, following an execution of π where P_1 begins by holding input $(1^n, x)$, and P_2 begins by holding input $(1^n, y)$.

Security as emulation of an ideal execution in the real model. Having defined the ideal and real models, we can now define security of a protocol. Loosely speaking, the definition asserts that a secure protocol (in the real model) emulates the ideal model (in which a trusted party exists). This is formulated as follows:

Definition 2.2 (1/p-secure computation). *Let $\mathcal{F}$ and π be as above, and fix a function $p = p(n)$. Protocol π is said to 1/p-securely compute $\mathcal{F}$ if for every non-uniform probabilistic polynomial-time adversary $\mathcal{A}$ in the real model, there exists a non-uniform probabilistic polynomial-time adversary $\mathcal{S}$ in the ideal model such that*

$$\{\mathsf{IDEAL}_{\mathcal{F}, \mathcal{S}(\mathsf{aux})}(x, y, n)\}_{(x,y) \in X \times Y, \mathsf{aux}} \overset{1/p}{\approx} \{\mathsf{REAL}_{\pi, \mathcal{A}(\mathsf{aux})}(x, y, n)\}_{(x,y) \in X \times Y, \mathsf{aux}}$$

and the same party is corrupted in both the real and ideal models.

2.3   Security with Abort

In what follows we use the standard notion of computational indistinguishability. That is, two distribution ensembles $X = \{X(a,n)\}_{a \in \mathcal{D}_n, n \in \mathbb{N}}$ and $Y = \{Y(a,n)\}_{a \in \mathcal{D}_n, n \in \mathbb{N}}$ are *computationally indistinguishable*, denoted $X \stackrel{c}{=} Y$, if for every non-uniform polynomial-time algorithm D there exists a negligible function $\nu(n)$ such that for all sufficiently large $n \in \mathbb{N}$ and for all $a \in \mathcal{D}_n$ it holds that

$$|\Pr\left[D(X(a,n)) = 1\right] - \Pr\left[D(Y(a,n)) = 1\right]| \leq \nu(n) \ .$$

Security with abort is the standard notion for secure computation where an honest majority is not available. The definition is similar to the definition of $1/p$-security presented in Section 2.2, with the following two exceptions: (1) the ideal-model adversary is allowed to choose whether the honest parties receive their outputs (i.e., fairness is not guaranteed), and (2) the ideal model and real model are required to be computationally indistinguishable.

Specifically, the execution in the real model is as described in Section 2.2, and the execution in the ideal model is modified as follows:

Inputs: P_1 and P_2 hold the security parameter 1^n and inputs $x \in X_n$ and $y \in Y_n$, respectively. The adversary $\mathcal{A}$ receives an auxiliary input aux.

Send inputs to trusted party: The honest party sends its actual input to the trusted party. The corrupted party controlled by $\mathcal{A}$ may send any value of its choice. Denote the pair of inputs sent to the trusted party by (x', y').

Trusted party sends output to corrupted party: If $x' \notin X_n$ the trusted party sets x' to some default element $x_0 \in X_n$ (and likewise if $y' \notin Y_n$). Then, the trusted party chooses r uniformly at random, computes $z_1 = f_n^1(x', y'; r)$ and $z_2 = f_n^2(x', y'; r)$ to P_2, and sends z_i to the corrupted party P_i (i.e., to the adversary $\mathcal{A}$).

Adversary decides whether to abort: After receiving its output the adversary sends either "abort" of "continue" to the trusted party. In the former case the trusted party sends $\perp$ to the honest party P_j, and in the latter case the trusted party sends z_j to P_j.

Outputs: The honest party outputs whatever it was sent by the trusted party, the corrupted party outputs nothing, and $\mathcal{A}$ outputs any arbitrary (probabilistic polynomial-time computable) function of its view.

We denote by $\mathsf{IDEAL}^{\mathsf{abort}}_{\mathcal{F}, \mathcal{A}(\mathsf{aux})}(x, y, n)$ the random variable consisting of the view of the adversary and the output of the honest party following an execution in the ideal model as described above.

Definition 2.3 (security with abort). *Let $\mathcal{F}$ and π be as above. Protocol π is said to securely compute $\mathcal{F}$ with abort if for every non-uniform probabilistic polynomial-time adversary $\mathcal{A}$ in the real model, there exists a non-uniform probabilistic polynomial-time adversary $\mathcal{S}$ in the ideal model such that*

$$\{\mathsf{IDEAL}^{\mathsf{abort}}_{\mathcal{F}, \mathcal{S}(\mathsf{aux})}(x, y, n)\}_{(x,y) \in X \times Y, \mathsf{aux}} \stackrel{c}{=} \{\mathsf{REAL}_{\pi, \mathcal{A}(\mathsf{aux})}(x, y, n)\}_{(x,y) \in X \times Y, \mathsf{aux}} \ .$$

2.4   One-Time Message Authentication

Message authentication codes provide assurance to the receiver of a message that it was sent by a specified legitimate sender, even in the presence of an active adversary who controls the communication channel. A message authentication code is defined via triplet (Gen, Mac, Vrfy) of probabilistic polynomial-time Turing machines such that:

1. The key generation algorithm Gen receives as input a security parameter 1^n, and outputs an authentication key k.
2. The authentication algorithm Mac receives as input an authentication key k and a message m, and outputs a tag t.
3. The verification algorithm Vrfy receives as input an authentication key k, a message m, and a tag t, and outputs a bit $b \in \{0, 1\}$.

The functionality guarantee of a message authentication code is that for any message m it holds that $\mathsf{Vrfy}(k, m, \mathsf{Mac}(k, m)) = 1$ with overwhelming probability over the internal coin tosses of Gen, Mac and Vrfy. In this paper we rely on message authentication codes that are one-time secure. That is, an authentication key is used to authenticate a single message. We consider an adversary that queries the authentication algorithm on a single message m of her choice, and then outputs a pair (m', t'). We say that the adversary forges an authentication tag if $m' \neq m$ and $\mathsf{Vrfy}(k, m', t') = 1$. Message authentication codes that are one-time secure exist in the information-theoretic setting, that is, even an unbounded adversary has only a negligible probability of forging an authentication tag. Constructions of such codes can be based, for example, on pair-wise independent hash functions [26].

3   A Simplified Protocol

In order to demonstrate the main ideas underlying our approach, in this section we present a simplified protocol. The simplification is two-fold: First, we consider the specific coin-flipping functionality (as in Theorem 1.1), and not the more general functionality of sampling from an arbitrary distribution $\mathcal{D} = (\mathcal{D}_1, \mathcal{D}_2)$ (as in Theorem 1.2). Second, the coin-flipping protocol will only be $1/(2r)$-secure and not $1/(4r)$-secure.

We describe the protocol in a sequence of refinements. We first informally describe the protocol assuming the existence of a trusted third party. The trusted third party acts as a "dealer" in a pre-processing phase, sending each party an input that it uses in the protocol. In the protocol we make no assumptions about the computational power of the parties. We then eliminate the need for the trusted third party by having the parties execute a secure-with-abort protocol that implements its functionality (this can be done in a constant number of rounds).

The protocol. The joint input of the parties, P_1 and P_2, is the security parameter 1^n and a polynomial $r = r(n)$ indicating the number of rounds in the protocol. In the pre-processing phase the trusted third party chooses uniformly at random a value $i^* \in \{1, \ldots, r\}$, that corresponds to the round in which the parties learn their outputs. In every round $i \in \{1, \ldots, r\}$ each party learns one bit of information: P_1 learns a bit a_i, and P_2 learns a bit b_i. In every round $i \in \{1, \ldots, i^* - 1\}$ (these are the "dummy" rounds) the values a_i and b_i are independently and uniformly chosen. In every round $i \in \{i^*, \ldots, r\}$ the parties learn the same uniformly distributed bit $c = a_i = b_i$ which is their output in the protocol. If the parties complete all r rounds of the protocol, then P_1 and P_2 output a_r and b_r, respectively[4]. Otherwise, if a party aborts prematurely, the other party outputs the value of the previous round and halts. That is, if P_1 aborts in round $i \in \{1, \ldots, r\}$ then P_2 outputs the value b_{i-1} and halts. Similarly, if P_2 aborts in round i then P_1 outputs the value a_{i-1} and halts.

More specifically, in the pre-processing phase the trusted third party chooses $i^* \in \{1, \ldots, r\}$ uniformly at random and defines $a_1, \ldots, a_r$ and $b_1, \ldots, b_r$ as follows: First, it choose $a_1, \ldots, a_{i^*-1} \in \{0, 1\}$ and $b_1, \ldots, b_{i^*-1} \in \{0, 1\}$ independently and uniformly at random. Then, it chooses $c \in \{0, 1\}$ uniformly at random and lets $a_{i^*} = \cdots = a_r = b_{i^*} = \cdots = b_r = c$. The trusted third party creates secret shares of the values $a_1, \ldots, a_r$ and $b_1, \ldots, b_r$ using an information-theoretically-secure 2-out-of-2 secret sharing scheme, and these shares are given to the parties. For concreteness, we use the specific secret-sharing scheme that splits a bit x into $(x^{(1)}, x^{(2)})$ by choosing $x^{(1)} \in \{0, 1\}$ uniformly at random and letting $x^{(2)} = x \oplus x^{(1)}$. In every round $i \in \{1, \ldots, r\}$ the parties exchange their shares for the current round, which enables P_1 to reconstruct a_i, and P_2 to reconstruct b_i. Clearly, when both parties are honest, the parties produce the same output bit which is uniformly distributed.

Eliminating the trusted third party. We eliminate the need for the trusted third party by relying on a *possibly unfair* sub-protocol that securely computes with abort the functionality $\mathsf{ShareGen}_r$, formally described in Figure 1. Such a protocol with a constant number of rounds can be constructed assuming the existence of oblivious transfer (see, for example, [27]). In addition, our protocol also relies on a one-time message authentication code $(\mathsf{Gen}, \mathsf{Mac}, \mathsf{Vrfy})$ that is information-theoretically secure. The functionality $\mathsf{ShareGen}_r$ provides the parties with authentication keys and authentication tags so each party can verify that the shares received from the other party were the ones generated by $\mathsf{ShareGen}_r$ in the pre-processing phase. A formal description of the protocol is provided in Figure 2.

Proof of security. The following theorem states that the protocol is $1/(2r)$-secure. We then conclude the section by showing the our analysis is in fact tight:

---

[4] An alternative approach that reduces the *expected* number of rounds from r to $r/2+1$ is as follows. In round i^* the parties learn their output $c = a_{i^*} = b_{i^*}$, and in round $i^* + 1$ they learn a special value $a_{i^*+1} = b_{i^*+1} = \mathsf{NULL}$ indicating that they should output the value from the previous round and halt. For simplicity (both in the presentation of the protocol and in the proof of security) we chose to present the protocol as always having r rounds, but this is not essential for our results.

Functionality ShareGen$_r$

Input: Security parameter 1^n.

Computation:

1. Choose $i^* \in \{1, \ldots, r\}$ uniformly at random.
2. Define values $a_1, \ldots, a_r$ and $b_1, \ldots, b_r$ as follows:
   - For $1 \leq i \leq i^* - 1$ choose $a_i, b_i \in \{0, 1\}$ independently and uniformly at random.
   - Choose $c \in \{0, 1\}$ uniformly at random, and for $i^* \leq i \leq r$ let $a_i = b_i = c$.
3. For $1 \leq i \leq r$, choose $\left(a_i^{(1)}, a_i^{(2)}\right)$ and $\left(b_i^{(1)}, b_i^{(2)}\right)$ as random secret shares of a_i and b_i, respectively.
4. Compute $k_1^a, \ldots, k_r^a, k_1^b, \ldots, k_r^b \leftarrow \mathsf{Gen}(1^n)$. For $1 \leq i \leq r$, let $t_i^a = \mathsf{Mac}_{k_i^a}\left(i\|a_i^{(2)}\right)$ and $t_i^b = \mathsf{Mac}_{k_i^b}\left(i\|b_i^{(1)}\right)$.

Output:

1. Party P_1 receives the values $a_1^{(1)}, \ldots, a_r^{(1)}$, $\left(b_1^{(1)}, t_1^b\right), \ldots, \left(b_r^{(1)}, t_r^b\right)$, and $k^a = (k_1^a, \ldots, k_r^a)$.
2. Party P_2 receives the values $\left(a_1^{(2)}, t_1^a\right), \ldots, \left(a_r^{(2)}, t_r^a\right)$, $b_1^{(2)}, \ldots, b_r^{(2)}$, and $k^b = (k_1^b, \ldots, k_r^b)$.

Fig. 1. The ideal functionality ShareGen$_r$

there exists an efficient adversary that can bias the output of the honest party by essentially $1/(2r)$.

Theorem 3.1. *For any polynomial $r = r(n)$, if protocol π securely computes* ShareGen$_r$ *with abort, then protocol* CoinFlip$_r$ *is $1/(2r)$-secure.*

Proof. We prove the $(1/2r)$-security of protocol CoinFlip$_r$ in a hybrid model where a trusted party for computing the ShareGen$_r$ functionality with abort is available. Using standard techniques (see [25]), it then follows that when replacing the trusted party computing ShareGen$_r$ with a sub-protocol that security computes ShareGen$_r$ with abort, the resulting protocol is $1/(2r)$-secure.

Specifically, for every polynomial-time hybrid-model adversary $\mathcal{A}$ corrupting P_1 and running CoinFlip$_r$ in the hybrid model, we show that there exists a polynomial-time ideal-model adversary $\mathcal{S}$ corrupting P_1 in the ideal model with access to a trusted party computing the coin-flipping functionality such that the statistical distance between these two executions is at most $1/(2r) + \nu(n)$, for some negligible function $\nu(n)$. For simplicity, in the remainder of the proof we ignore the aspect of message authentication in the protocol, and assume that the only malicious behavior of the adversary $\mathcal{A}$ is early abort. This does not result in any loss of generality, since there is only a negligible probably of forging an authentication tag.

Protocol CoinFlip$_r$

Joint input: Security parameter 1^n.

Preliminary phase:

1. Parties P_1 and P_2 run protocol π for computing $\mathsf{ShareGen}_r(1^n)$ (see Figure 1).
2. If P_1 receives $\perp$ from the above computation, it outputs a uniformly chosen bit and halts. Likewise, if P_2 receives $\perp$ it outputs a uniformly chosen bit and halts. Otherwise, the parties proceed.
3. Denote the output of P_1 from π by $a_1^{(1)}, \ldots, a_r^{(1)}, \left(b_1^{(1)}, t_1^b\right), \ldots, \left(b_r^{(1)}, t_r^b\right)$, and $k^a = (k_1^a, \ldots, k_r^a)$.
4. Denote the output of P_2 from π by $\left(a_1^{(2)}, t_1^a\right), \ldots, \left(a_r^{(2)}, t_r^a\right), b_1^{(2)}, \ldots, b_r^{(2)}$, and $k^b = (k_1^b, \ldots, k_r^b)$.

In each round $i = 1, \ldots, r$ do:

1. P_2 sends the next share to P_1:
   (a) P_2 sends $\left(a_i^{(2)}, t_i^a\right)$ to P_1.
   (b) P_1 receives $\left(\hat{a}_i^{(2)}, \hat{t}_i^a\right)$ from P_2. If $\mathsf{Vrfy}_{k_i^a}\left(i || \hat{a}_i^{(2)}, \hat{t}_i^a\right) = 0$ (or if P_1 received an invalid message or no message), then P_1 outputs a_{i-1} and halts (if $i = 1$ it outputs a uniformly chosen bit).
   (c) If $\mathsf{Vrfy}_{k_i^a}\left(i || \hat{a}_i^{(2)}, \hat{t}_i^a\right) = 1$ then P_1 reconstructs a_i using the shares $a_i^{(1)}$ and $\hat{a}_i^{(2)}$.
2. P_1 sends the next share to P_2:
   (a) P_1 sends $\left(b_i^{(1)}, t_i^b\right)$ to P_2.
   (b) P_2 receives $\left(\hat{b}_i^{(1)}, \hat{t}_i^b\right)$ from P_1. If $\mathsf{Vrfy}_{k_i^b}\left(i || \hat{b}_i^{(1)}, \hat{t}_i^b\right) = 0$ (or if P_2 received an invalid message or no message), then P_2 outputs b_{i-1} and halts (if $i = 1$ it outputs a uniformly chosen bit).
   (c) If $\mathsf{Vrfy}_{k_i^b}\left(i || \hat{b}_i^{(1)}, \hat{t}_i^b\right) = 1$ then P_2 reconstructs b_i using the shares $b_i^{(1)}$ and $b_i^{(2)}$

Output: P_1 and P_2 output a_r and b_r, respectively.

Fig. 2. The coin-flipping protocol CoinFlip$_r$

On input $(1^n, \mathsf{aux})$ the ideal-model adversary $\mathcal{S}$ invokes the hybrid-model adversary $\mathcal{A}$ on $(1^n, \mathsf{aux})$ and queries the trusted party computing the coin-flipping functionality to obtain a bit c. The ideal-model adversary $\mathcal{S}$ proceeds as follows:

1. $\mathcal{S}$ simulates the trusted party computing the $\mathsf{ShareGen}_r$ functionality by sending $\mathcal{A}$ shares $a_1^{(1)}, \ldots, a_r^{(1)}, b_1^{(1)}, \ldots, b_r^{(1)}$ that are chosen independently and uniformly at random. If $\mathcal{A}$ aborts (i.e., if $\mathcal{A}$ sends abort to the simulated $\mathsf{ShareGen}_r$ after receiving the shares), then $\mathcal{S}$ outputs $\mathcal{A}$'s output and halts.
2. $\mathcal{S}$ chooses $i^* \in \{1, \ldots, r\}$ uniformly at random.

3. In every round $i \in \{1, \ldots, i^* - 1\}$, $\mathcal{S}$ chooses a random bit a_i, and sends $\mathcal{A}$ the share $a_i^{(2)} = a_i^{(1)} \oplus a_i$. If $\mathcal{A}$ aborts then $\mathcal{S}$ outputs $\mathcal{A}$'s output and halts.
4. In every round $i \in \{i^*, \ldots, r\}$, $\mathcal{S}$ sends $\mathcal{A}$ the share $a_{i^*+1}^{(2)} = a_{i^*+1}^{(1)} \oplus c$ (recall that c is the value received from the trusted party computing the coin-flipping functionality). If $\mathcal{A}$ aborts then $\mathcal{S}$ outputs $\mathcal{A}$'s output and halts.
5. At the end of the protocol $\mathcal{S}$ outputs $\mathcal{A}$'s output and halts.

We now consider the joint distribution of $\mathcal{A}$'s view and the output of the honest party P_2 in the ideal model and in the hybrid model. There are three cases to consider:

1. $\mathcal{A}$ aborts before round i^*. In this case the distributions are identical: in both models the view of the adversary is the sequence of shares, and the sequence of messages up to the round in which $\mathcal{A}$ aborted, and the output of P_2 is a uniformly distributed bit which is independent of $\mathcal{A}$'s view.
2. $\mathcal{A}$ aborts in round i^*. In this case $\mathcal{A}$'s view is identical in both models, but the distributions of P_2's output given $\mathcal{A}$'s view are not identical. In the ideal model, P_2 outputs the random bit c that was revealed to $\mathcal{A}$ by $\mathcal{S}$ in round i^* (recall that c is the bit received from the trusted party computing the coin-flipping functionality). In the hybrid model, however, the output of P_2 is the value b_{i^*-1} which is a random bit that is independent of $\mathcal{A}$'s view. Thus, in this case the statistical distance between the two distributions is $1/2$. However, this case occurs with probability at most $1/r$ since in both models i^* is independent of $\mathcal{A}$'s view until this round (that is, the probability that $\mathcal{A}$ aborts in round i^* is at most $1/r$).
3. $\mathcal{A}$ aborts after round i^* or does not abort. In this case the distributions are identical: the output of P_2 is the same random bit that was revealed to $\mathcal{A}$ in round i^*.

Note that $\mathcal{A}$'s view in the hybrid and ideal models is always identically distributed (no matter what strategy $\mathcal{A}$ uses to decide when to abort). The only difference is in the joint distribution of $\mathcal{A}$'s view and the honest party's output. Thus, conditioning on the round at which $\mathcal{A}$ aborts will have the same effect in the hybrid and ideal models; in particular, conditioned on case 1 or case 3 occurring, the joint distribution of $\mathcal{A}$'s view and the honest party's output will be identical in both models. We state this explicitly because in similar (yet inherently different) settings, using conditional probabilities in such a way might be problematic (see, for example, [28], Sect. 2).

The above three cases imply that the statistical distance between the two distributions is at most $1/(2r)$, and this concludes the proof.

Claim 3.2. *In protocol* $\mathsf{CoinFlip}_r$ *there exists an efficient adversarial party* P_1^* *that can bias the output of* P_2 *by* $\frac{1-2^{-r}}{2r}$.

Proof. Consider the adversarial party P_1^* that completes the pre-processing phase, and then halts in the first round $i \in \{1, \ldots, r\}$ for which $a_i = 0$. We denote by Abort the random variable corresponding to the round in which P_1^*

aborts, where $\mathsf{Abort} = \bot$ if P_1^* does not abort. In addition, we denote by c_2 the random variable corresponding to the output bit of P_2. Notice that if P_1^* aborts in round $j \le i^*$ then P_2 outputs a random bit, and if P_1^* does not abort then P_2 always outputs 1. Therefore, for every $i \in \{1, \ldots, r\}$ it holds that

$$
\begin{aligned}
&\Pr\left[c_2 = 1 \mid i^* = i\right] \\
&= \sum_{j=1}^{i} \Pr\left[\mathsf{Abort} = j \mid i^* = i\right] \Pr\left[c_2 = 1 \mid \mathsf{Abort} = j \wedge i^* = i\right] \\
&\qquad + \Pr\left[\mathsf{Abort} = \bot \mid i^* = i\right] \Pr\left[c_2 = 1 \mid \mathsf{Abort} = \bot \wedge i^* = i\right] \\
&= \sum_{j=1}^{i} \Pr\left[a_1 = \cdots = a_{j-1} = 1, a_j = 0\right] \Pr\left[c_2 = 1 \mid \mathsf{Abort} = j \wedge i^* = i\right] \\
&\qquad + \Pr\left[a_1 = \cdots = a_i = 1\right] \Pr\left[c_2 = 1 \mid \mathsf{Abort} = \bot \wedge i^* = i\right] \\
&= \sum_{j=1}^{i} \frac{1}{2^j} \cdot \frac{1}{2} + \frac{1}{2^i} \cdot 1 \\
&= \frac{1}{2} + \frac{1}{2^{i+1}} \ .
\end{aligned}
$$

This implies that

$$
\begin{aligned}
\Pr\left[c_2 = 1\right] &= \sum_{i=1}^{r} \Pr\left[i^* = i\right] \Pr\left[c_2 = 1 \mid i^* = i\right] \\
&= \sum_{i=1}^{r} \frac{1}{r}\left(\frac{1}{2} + \frac{1}{2^{i+1}}\right) \\
&= \frac{1}{2} + \frac{1 - 2^{-r}}{2r} \ .
\end{aligned}
$$

4   The Generalized Protocol

In this section we sketch a more refined and generalized protocol that settles Theorems 1.1 and 1.2 (due to space limitations, we defer the formal description of the protocol and its proof of security to the full version of the paper). The improvements over the protocol presented in Section 3 are as follows:

Improved security guarantee: In the simplified protocol party P_1 can bias the output of party P_2 (by aborting in round i^*), but party P_2 cannot not bias the output of party P_1. This is due to the fact that party P_1 always learns the output before party P_2 does. In the generalized protocol the party that learns the output before the other party is chosen uniformly at random (i.e., party P_1 learns the output before party P_2 with probability $1/2$). This is achieved by having the parties exchange a sequence of $2r$ values $(a_1, b_1), \ldots, (a_{2r}, b_{2r})$ (using the same secret-sharing exchange technique as in the simplified protocol) with the following property: for odd values of i,

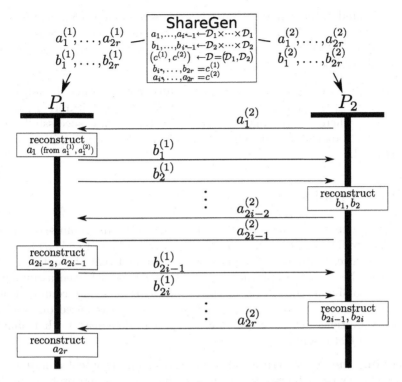

Fig. 3. Overview of the generalized protocol

party P_1 learns a_i before party P_2 learns b_i, and for even values of i party P_2 learns b_i before party P_1 learns a_i. Thus, party P_1 can bias the result only when i^* is odd, and party P_2 can bias the result only when i^* is even. The key point is that the parties can exchange the sequence of $2r$ shares in only $r + 1$ rounds by combining some of their messages[5].

Note that modifying the original protocol by having ShareGen randomly choose which player starts would also halve the bias (since with probability $\frac{1}{2}$ the adversary chooses a player that cannot bias the outcome at all). However, this is vulnerable to a trivial dynamic attack: the adversary decides which party to corrupt after seeing which party was chosen to start.

A larger class of functionalities: We consider the more general task of sampling from a distribution $\mathcal{D} = (\mathcal{D}_1, \mathcal{D}_2)$: party P_1 receives a sample from $\mathcal{D}_1$ and party P_2 receives a correlated sample from $\mathcal{D}_2$ (in coin-flipping, for example, the joint distribution $\mathcal{D}$ produces the values $(0,0)$ and $(1,1)$ each with probability $1/2$). Our generalized protocol can handle any polynomially-sampleable distribution $\mathcal{D}$. The basic idea here is ShareGen can be modified to output shares of samples for arbitrary (efficiently sampleable) distributions.

---

[5] Recall that each round consists of two moves: a message from P_2 to P_1 followed by a message from P_1 to P_2.

Up to round i^* the values each party receives are independent samples from the marginal distributions (i.e., P_1 receives independent samples from $\mathcal{D}_1$, and P_2 from $\mathcal{D}_1$). From round i^*, the values are the "real" output from the joint distribution.

Figure 3 gives a graphic overview of the protocol (ignoring the authentication tags). As in the simplified protocol, if P_2 aborts prematurely, P_1 outputs the value a_i, where i is the highest index such that a_i was successfully reconstructed. If P_1 aborts prematurely, P_2 outputs the last b_i value it successfully reconstructed.

5   Open Problems

Identifying the minimal computational assumptions. Blum's coin-flipping protocol, as well as its generalization that guarantees bias of $O(1/\sqrt{r})$, can rely on the existence of any one-way function. We showed that the optimal trade-off between the round complexity and the bias can be achieved assuming the existence of oblivious transfer, a complete primitive for secure computation. A challenging problem is to either achieve the optimal bias based on seemingly weaker assumptions (e.g., one-way functions), or to demonstrate that oblivious transfer is in fact essential.

Identifying the exact trade-off. The bias of our protocol almost exactly matches Cleve's lower bound: Cleve showed that any r-round protocol has bias at least $1/(8r+2)$, and we manage to achieve bias of at most $1/(4r-c)$ for some constant $c > 0$. It will be interesting to eliminate the multiplicative gap of $1/2$ by either improving Cleve's lower bound or by improving our upper bound. We note, however, that this cannot be resolved by improving the security analysis of our protocol since there exists an efficient adversary that can bias our protocol by essentially $1/(4r)$ (see Section 4), and therefore our analysis is tight.

Efficient implementation. Our protocol uses a general secure computation step in the preprocessing phase. Although asymptotically optimal, the techniques used in general secure computation often have a large overhead. Hence, it would be helpful to find an efficient sub-protocol to compute the $\mathsf{ShareGen}_r$ functionality that can lead to a practical implementation.

The multiparty setting. Blum's coin-flipping protocol can be extended to an m-party r-round protocol that has bias $O(m/\sqrt{r})$. An interesting problem is to identify the optimal trade-off between the number of parties, the round complexity, and the bias. Unfortunately, it seems that several natural variations of our approach fail to extend to the case of more than two parties. Informally, the main reason is that a coalition of malicious parties can guess the threshold round with a pretty good probability by simulating the protocol among themselves for any possible subset.

References

1. Ben-Or, M., Goldwasser, S., Wigderson, A.: Completeness theorems for non-cryptographic fault-tolerant distributed computation. In: Proceedings of the 20th Annual ACM Symposium on Theory of Computing, pp. 1–10 (1988)
2. Blum, M.: Coin flipping by telephone - A protocol for solving impossible problems. In: Proceedings of the 25th IEEE Computer Society International Conference, pp. 133–137 (1982)
3. Håstad, J., Impagliazzo, R., Levin, L.A., Luby, M.: A pseudorandom generator from any one-way function. SIAM Journal on Computing 28(4), 1364–1396 (1999)
4. Naor, M.: Bit commitment using pseudorandomness. Journal of Cryptology 4(2), 151–158 (1991)
5. Impagliazzo, R., Luby, M.: One-way functions are essential for complexity based cryptography. In: Proceedings of the 30th Annual IEEE Symposium on Foundations of Computer Science, pp. 230–235 (1989)
6. Cleve, R.: Limits on the security of coin flips when half the processors are faulty. In: Proceedings of the 18th Annual ACM Symposium on Theory of Computing, pp. 364–369 (1986)
7. Averbuch, B., Blum, M., Chor, B., Silvio Micali, S.G.: How to implement Bracha's O(log n) byzantine agreement algorithm (manuscript, 1985)
8. Goldreich, O.: Foundations of Cryptography: Basic Applications, vol. 2. Cambridge University Press, Cambridge (2004)
9. Katz, J.: On achieving the "best of both worlds" in secure multiparty computation. In: Proceedings of the 39th Annual ACM Symposium on Theory of computing, pp. 11–20 (2007)
10. Gordon, D., Katz, J.: Partial fairness in secure two-party computation. Cryptology ePrint Archive, Report 2008/206 (2008)
11. Cleve, R., Impagliazzo, R.: Martingales, collective coin flipping and discrete control processes (1993), http://www.cpsc.ucalgary.ca/~cleve/pubs/martingales.ps
12. Alon, N., Naor, M.: Coin-flipping games immune against linear-sized coalitions. SIAM Journal on Computing 22(2), 403–417 (1993)
13. Ben-Or, M., Linial, N.: Collective coin flipping. Advances in Computing Research: Randomness and Computation 5, 91–115 (1989)
14. Feige, U.: Noncryptographic selection protocols. In: Proceedings of the 40th Annual IEEE Symposium on Foundations of Computer Science, pp. 142–153 (1999)
15. Russell, A., Zuckerman, D.: Perfect information leader election in $\log^* n + O(1)$ rounds. Journal of Computer and System Sciences 63(4), 612–626 (2001)
16. Saks, M.: A robust noncryptographic protocol for collective coin flipping. SIAM Journal on Discrete Mathematics 2(2), 240–244 (1989)
17. Aharonov, D., Ta-Shma, A., Vazirani, U.V., Yao, A.C.: Quantum bit escrow. In: Proceedings of the 32nd Annual ACM Symposium on Theory of Computing, pp. 705–714 (2000)
18. Ambainis, A.: A new protocol and lower bounds for quantum coin flipping. Journal of Computer and System Sciences 68(2), 398–416 (2004)
19. Ambainis, A., Buhrman, H., Dodis, Y., Rohrig, H.: Multiparty quantum coin flipping. In: Proceedings of the 19th Annual IEEE Conference on Computational Complexity, pp. 250–259 (2004)
20. Moran, T., Naor, M.: Basing cryptographic protocols on tamper-evident seals. In: Caires, L., Italiano, G.F., Monteiro, L., Palamidessi, C., Yung, M. (eds.) ICALP 2005. LNCS, vol. 3580, pp. 285–297. Springer, Heidelberg (2005)

21. Gordon, S.D., Hazay, C., Katz, J., Lindell, Y.: Complete fairness in secure two-party computation. In: Proceedings of the 40th Annual ACM Symposium on Theory of Computing, pp. 413–422 (2008)
22. Halpern, J.Y., Teague, V.: Rational secret sharing and multiparty computation. In: Proceedings of the 36th Annual ACM Symposium on Theory of Computing, pp. 623–632 (2004)
23. Gordon, S.D., Katz, J.: Rational secret sharing, revisited. In: Proceedings on the 5th International Conference on Security and Cryptographyfor Networks, pp. 229–241 (2006)
24. Kol, G., Naor, M.: Cryptography and game theory: Designing protocols for exchanging information. In: Proceedings of the 5th Theory of Cryptography Conference, pp. 320–339 (2008)
25. Canetti, R.: Security and composition of multiparty cryptographic protocols. Journal of Cryptology 13(1), 143–202 (2000)
26. Wegman, M.N., Carter, L.: New hash functions and their use in authentication and set equality. Journal of Computer and System Sciences 22(3), 265–279 (1981)
27. Lindell, Y.: Parallel coin-tossing and constant-round secure two-party computation. Journal of Cryptology 16(3), 143–184 (2003)
28. Bellare, M., Rogaway, P.: Code-based game-playing proofs and the security of triple encryption. Cryptology ePrint Archive, Report 2004/331 (2004), http://eprint.iacr.org/2004/331.pdf

Complete Fairness in Multi-party Computation without an Honest Majority

S. Dov Gordon and Jonathan Katz*

Dept. of Computer Science, University of Maryland

Abstract. Gordon et al. recently showed that certain (non-trivial) functions can be computed with complete fairness in the *two-party* setting. Motivated by their results, we initiate a study of complete fairness in the *multi-party* case and demonstrate the first completely-fair protocols for non-trivial functions in this setting. We also provide evidence that achieving fairness is "harder" in the multi-party setting, at least with regard to round complexity.

1 Introduction

In the setting of secure computation, a group of parties wish to run a protocol for computing some function of their inputs while preserving, to the extent possible, security properties such as privacy, correctness, input independence and others. These requirements are formalized by comparing a real-world execution of the protocol to an *ideal world* where there is a trusted entity who performs the computation on behalf of the parties. Informally, a protocol is "secure" if for any real-world adversary $\mathcal{A}$ there exists a corresponding ideal-world adversary $\mathcal{S}$ (corrupting the same parties as $\mathcal{A}$) such that the result of executing the protocol in the real world with $\mathcal{A}$ is computationally indistinguishable from the result of computing the function in the ideal world with $\mathcal{S}$.

One desirable property is *fairness* which, intuitively, means that either *everyone* receives the output, or else *no one* does. Unfortunately, it has been shown by Cleve [1] that complete fairness is impossible *in general* without a majority of honest parties. Until recently, Cleve's result was interpreted to mean that *no* non-trivial functions could be computed with complete fairness without an honest majority. A recent result of Gordon et al. [2], however, shows that this folklore is wrong; there exist non-trivial functions than *can* be computed with complete fairness in the two-party setting. Their work demands that we re-evaluate our current understanding of fairness.

Gordon et al. [2] deal exclusively with the case of two-party computation, and leave open the question of fairness in the multi-party setting. Their work does *not* immediately extend to the case of more than two parties. (See also the discussion in the section that follows.) An additional difficulty that arises in the

---

* This work was supported by NSF CNS-0447075, NSF CCF-0830464, and US-Israel Binational Science Foundation grant #2004240.

O. Reingold (Ed.): TCC 2009, LNCS 5444, pp. 19–35, 2009.

multi-party setting is the need to ensure *consistency* between the outputs of the honest parties, even after a malicious abort. In more detail: in the two-party setting, it suffices for an honest party's output (following an abort by the other party) to be consistent only with its own local input. But in the multi-party setting, the honest parties' outputs must each be consistent with *all* of their inputs. This issue is compounded by the adversary's ability to adaptively abort the t malicious players in any order and at any time, making fairness in the multi-party setting even harder to achieve.

We initiate a study of complete fairness in the multi-party setting. We focus on the case when a private[1] broadcast channel (or, equivalently, a PKI) is available to the parties; note that Cleve's impossibility result applies in this case as well. Although one can meaningfully study fairness in the absence of broadcast, we have chosen to assume broadcast so as to separate the question of *fairness* from the question of *agreement* (which has already been well studied in the distributed systems literature). Also, although the question of fairness becomes interesting as soon as an honest majority is not assumed, here we only consider the case of completely-fair protocols tolerating an *arbitrary* number of corrupted parties. We emphasize that, as in [2], we are interested in obtaining *complete* fairness rather than some notion of *partial* fairness.

1.1   Our Results

A natural first question is whether two-party feasibility results [2] can be extended "easily" to the multi-party setting. More formally, say we have a function $f : \{0,1\} \times \cdots \times \{0,1\} \to \{0,1\}$ taking n boolean inputs. (We restrict to boolean inputs/outputs for simplicity only.) For any non-trivial subset $I \subset [n]$, define the *partition* f_I of f as the two-input function $f_I : \{0,1\}^{|I|} \times \{0,1\}^{n-|I|} \to \{0,1\}$ given by

$$f_I(y, z) = f(x),$$

where $x \in \{0,1\}^n$ is such that $x_I = y$ and $x_{\bar{I}} = z$. It is not hard to see that if there exists an I for which f_I *cannot* be computed with complete fairness in the two-party setting, then f cannot be computed with complete fairness in the multi-party setting. Similarly, the round complexity for computing f with complete fairness in the multi-party case must be at least the round complexity of fairly computing each f_I. What about the converse? We show the following negative result regarding such a "partition-based" approach to the problem:

Theorem 1. *(Under suitable cryptographic assumptions) there exists a 3-party function f all of whose partitions can be computed with complete fairness in $\mathcal{O}(1)$ rounds, but for which any multi-party protocol computing f with complete fairness requires $\omega(\log k)$ rounds, where k is the security parameter.*

This indicates that fairness in the multi-party setting is qualitatively harder than fairness in the two-party setting. (A somewhat analogous result in a different context was shown by Chor et al. [3].)

---

[1] We assume private broadcast so as to ensure security against passive eavesdroppers (who do not corrupt any parties).

The function f for which we prove the above theorem is interesting in its own right: it is the 3-party *majority* function (i.e., voting). Although the $\omega(\log k)$-round lower bound may seem discouraging, we are able to show a positive result for this function; to the best of our knowledge, this represents the first non-trivial feasibility result for complete fairness in the multi-party setting.

Theorem 2. *(Under suitable cryptographic assumptions) there exists an $\omega(\log k)$-round protocol for securely computing 3-party majority with complete fairness.*

Our efforts to extend the above result to n-party majority have been unsuccessful. One may therefore wonder whether there exists *any* (non-trivial) function that can be computed with complete fairness for general n. Indeed, there is:

Theorem 3. *(Under suitable cryptographic assumptions) for any number of parties n, there exists an $\Theta(n)$-round protocol for securely computing boolean OR with complete fairness.*

OR is non-trivial in our context: OR is complete for multi-party computation (without fairness) [4], and cannot be computed with information-theoretic privacy even in the two-party setting [5].

Relation to prior work. At a superficial level, the proof of the $\omega(\log k)$-round lower bound of Theorem 1 uses an approach similar to that used to prove an analogous lower bound in [2]. We stress, however, that our theorem does not follow as a corollary of that work (indeed, it cannot since each of the partitions of f can be computed with complete fairness in $\mathcal{O}(1)$ rounds). We introduce new ideas to prove the result in our setting; in particular, we rely in an essential way on the fact that the output of any two honest parties must agree (whereas this issue does not arise in the two-party setting considered in [2]).

Ishai et al. [6] propose a protocol that is resilient to a dishonest majority in a weaker sense than that considered here. Specifically, their protocol achieves the following guarantee (informally): when $t < n$ parties are corrupted then a real execution of the protocol is as secure as an execution in the ideal world with complete fairness *where the adversary can query the ideal functionality $\mathcal{O}(t)$ times* (using different inputs each time). While this definition may guarantee privacy for certain functions (e.g., for the sum function), it does not prevent the malicious parties from *biasing* the output of the honest parties. We refer the reader to their work for further discussion.

1.2   Outline of the Paper

We include the standard definitions of secure multi-party computation in the full version of this paper [7]. We stress that although the definitions are standard, what is *not* standard is that we are interested in attaining complete fairness even though we do not have an honest majority.

We begin with our negative result, showing that any completely-fair protocol for 3-party majority requires $\omega(\log k)$ rounds. Recall that what is especially interesting about this result is that it demonstrates a gap between the round complexities required for completely-fair computation of a function and its (two-party) partitions. In Section 3, we show an $\omega(\log k)$-round protocol for completely-fair computation of 3-party majority. In Section 4 we describe our feasibility result for the case of boolean OR.

2 A Lower Bound on the Round Complexity of Majority

2.1 Proof Overview

In this section, we prove Theorem 1 taking as our function f the three-party majority function maj. That is, $\mathsf{maj}(x_1, x_2, x_3) = 0$ if at least two of the three values $\{x_1, x_2, x_3\}$ are 0, and is 1 otherwise. Note that any partition of maj is just (isomorphic to) the greater-than-or-equal-to function, where the domain of one input can be viewed as $\{0, 1, 2\}$ and the domain of the other input can be viewed as $\{0, 1\}$ (in each case, representing the number of '1' inputs held). Gordon et al. [2] show that, under suitable cryptographic assumptions, the greater-than-or-equal-to function on constant-size domains can be securely computed with complete fairness in $\mathcal{O}(1)$ rounds.

We prove Theorem 1 by showing that any completely-fair 3-party protocol for maj requires $\omega(\log k)$ rounds. The basic approach is to argue that if Π is any protocol for securely computing maj, then eliminating the last round of Π results in a protocol Π' that still computes maj correctly "with high probability". Specifically, if the error probability in Π is at most μ (that we will eventually set to some negligible function of k), then the error probability in Π' is at most $c \cdot \mu$ for some constant c. If the original protocol Π has $r = \mathcal{O}(\log k)$ rounds, then applying this argument inductively r times gives a protocol that computes maj correctly on all inputs with probability significantly better than guessing *without any interaction at all*. This gives the desired contradiction.

To prove that eliminating the last round of Π cannot affect correctness "too much", we consider a constraint that holds for the ideal-world evaluation of maj. (Recall, we are working in the ideal world where complete fairness holds.) Consider an adversary who corrupts two parties, and let the input of the honest party P be chosen uniformly at random. The adversary can learn P's input by submitting $(0, 1)$ or $(1, 0)$ to the trusted party. The adversary can also try to bias the output of maj to be the opposite of P's choice by submitting $(0, 0)$ or $(1, 1)$; this will succeed in biasing the result half the time. But the adversary cannot both learn P's input and *simultaneously* bias the result. (If the adversary submits $(0, 1)$ or $(1, 0)$, the output of maj is always equal to P's input; if the adversary submits $(0, 0)$ or $(1, 1)$ the the output of maj reveals nothing about P's input.) Concretely, for any ideal-world adversary the *sum* of the probability that the adversary guesses P's input and the probability that the output of maj is not equal to P's input is at most 1. In our proof, we show that if correctness holds with significantly lower probability when the last round of Π is eliminated, then there exists a real-world adversary violating this constraint.

2.2   Proof Details

We number the parties P_1, P_2, P_3, and work modulo 3 in the subscript. The input of P_j is denoted by x_j. The following claim formalizes the ideal-world constraint described informally above.

Claim 4. *For all $j \in \{1, 2, 3\}$ and any adversary $\mathcal{A}$ corrupting P_{j-1} and P_{j+1} in an ideal-world computation of* maj, *we have*

$$\Pr\left[\mathcal{A} \text{ correctly guesses } x_j\right] + \Pr\left[\text{OUTPUT}_j \neq x_j\right] \leq 1,$$

where the probabilities are taken over the random coins of $\mathcal{A}$ and random choice of $x_j \in \{0, 1\}$.

Proof. Consider an execution in the ideal world, where P_j's input x_j is chosen uniformly at random. Let EQUAL be the event that $\mathcal{A}$ submits two equal inputs (i.e., $x_{j-1} = x_{j+1}$) to the trusted party. In this case, $\mathcal{A}$ learns nothing about P_j's input and so can guess x_j with probability at most $1/2$. It follows that:

$$\Pr\left[\mathcal{A} \text{ correctly guesses } x_j\right] \leq \frac{1}{2} \Pr\left[\text{EQUAL}\right] + \Pr\left[\overline{\text{EQUAL}}\right].$$

Moreover, $\Pr\left[\text{OUTPUT}_j \neq x_j\right] = \frac{1}{2} \Pr\left[\text{EQUAL}\right]$ since $\text{OUTPUT}_j \neq x_j$ occurs only if $\mathcal{A}$ submits $x_{j-1} = x_{j+1} = \bar{x}_j$ to the trusted party. Therefore:

$$\Pr\left[\mathcal{A} \text{ correctly guesses } x_j\right] + \Pr\left[\text{OUTPUT}_j \neq x_j\right]$$
$$\leq \frac{1}{2} \Pr\left[\text{EQUAL}\right] + \Pr\left[\overline{\text{EQUAL}}\right] + \frac{1}{2} \Pr\left[\text{EQUAL}\right]$$
$$= \Pr\left[\text{EQUAL}\right] + \Pr\left[\overline{\text{EQUAL}}\right] = 1,$$

proving the claim.

Let Π be a protocol that securely computes maj using $r = r(k)$ rounds. Consider an execution of Π in which all parties run the protocol honestly except for possibly aborting in some round. We denote by $b_j^{(i)}$ the value that P_{j-1} and P_{j+1} both[2] output if P_j aborts the protocol after sending its round-i message (and then P_{j-1} and P_{j+1} honestly run the protocol to completion). Similarly, we denote by $b_{j-1}^{(i)}$ (resp., $b_{j+1}^{(i)}$) the value output by P_j and P_{j+1} (resp., P_j and P_{j-1}) when P_{j-1} (resp., P_{j+1}) aborts after sending its round-i message. Note that an adversary who corrupts, e.g., both P_{j-1} and P_{j+1} can compute $b_j^{(i)}$ immediately after receiving the round-i message of P_j.

---

[2] Security of Π implies that the outputs of P_{j-1} and P_{j+1} in this case must be equal with all but negligible probability. For simplicity we assume this to hold with probability 1 but our proof can be modified easily to remove this assumption.

Since Π securely computes maj with complete fairness, the ideal-world constraint from the previous claim implies that for all $j \in \{1, 2, 3\}$, any inverse polynomial $\mu(k)$, and any poly-time adversary $\mathcal{A}$ controlling players P_{j-1} and P_{j+1}, we have:

$$\Pr_{x_j \leftarrow \{0,1\}} [\mathcal{A} \text{ correctly guesses } x_j] + \Pr_{x_j \leftarrow \{0,1\}} [\text{OUTPUT}_j \neq x_j] \leq 1 + \mu(k) \quad (1)$$

for k sufficiently large. Security of Π also guarantees that if the inputs of the honest parties agree, then with all but negligible probability their output must be their common input regardless of when a malicious P_j aborts. That is, for k large enough we have

$$x_{j+1} = x_{j-1} \Rightarrow \Pr\left[b_j^{(i)} = x_{j+1} = x_{j-1}\right] \geq 1 - \mu(k) \quad (2)$$

for all $j \in \{1, 2, 3\}$ and all $i \in \{0, \ldots, r(k)\}$.

The following claim represents the key step in our lower bound.

Claim 5. *Fix a protocol Π, a function μ, and a value k such that Equations (1) and (2) hold, and let $\mu = \mu(k)$. Say there exists an i, with $1 \leq i \leq r(k)$, such that for all $j \in \{1, 2, 3\}$ and all $c_1, c_2, c_3 \in \{0, 1\}$ it holds that:*

$$\Pr\left[b_j^{(i)} = \mathsf{maj}(c_1, c_2, c_3) \mid (x_1, x_2, x_3) = (c_1, c_2, c_3)\right] \geq 1 - \mu. \quad (3)$$

Then for all $j \in \{1, 2, 3\}$ and all $c_1, c_2, c_3 \in \{0, 1\}$ it holds that:

$$\Pr\left[b_j^{(i-1)} = \mathsf{maj}(c_1, c_2, c_3) \mid (x_1, x_2, x_3) = (c_1, c_2, c_3)\right] \geq 1 - 5\mu. \quad (4)$$

Proof. When $j = 1$ and $c_2 = c_3$, the desired result follows from Equation (2); this is similarly true for $j = 2$, $c_1 = c_3$ as well as $j = 3$, $c_1 = c_2$.

Consider the real-world adversary $\mathcal{A}$ that corrupts P_1 and P_3 and sets $x_1 = 0$ and $x_3 = 1$. Then:

- $\mathcal{A}$ runs the protocol honestly until it receives the round-i message from P_2.
- $\mathcal{A}$ then locally computes the value of $b_2^{(i)}$.
  - If $b_2^{(i)} = 0$, then $\mathcal{A}$ aborts P_1 *without sending its round-i message* and runs the protocol (honestly) on behalf of P_3 until the end. By definition, the output of P_2 will be $b_1^{(i-1)}$.
  - If $b_2^{(i)} = 1$, then $\mathcal{A}$ aborts P_3 *without sending its round-i message* and runs the protocol (honestly) on behalf of P_1 until the end. By definition, the output of P_2 will be $b_3^{(i-1)}$.
- After completion of the protocol, $\mathcal{A}$ outputs $b_2^{(i)}$ as its guess for the input of P_2.

Consider an experiment in which the input x_2 of P_2 is chosen uniformly at random, and then $\mathcal{A}$ runs protocol Π with P_2. Using Equation (3), we have:

$$\Pr\left[\mathcal{A} \text{ correctly guesses } x_2\right] = \Pr\left[b_2^{(i)} = x_2\right]$$
$$= \Pr\left[b_2^{(i)} = f(0, x_2, 1)\right] \geq 1 - \mu. \quad (5)$$

We also have:

$$\Pr\left[\text{OUTPUT}_2 \neq x_2\right] = \frac{1}{2} \cdot \Pr\left[\text{OUTPUT}_2 = 1 \mid (x_1, x_2, x_3) = (0, 0, 1)\right] \qquad (6)$$

$$+ \frac{1}{2} \cdot \Pr\left[\text{OUTPUT}_2 = 0 \mid (x_1, x_2, x_3) = (0, 1, 1)\right]$$

$$= \frac{1}{2} \left(\Pr\left[b_1^{(i-1)} = 1 \wedge b_2^{(i)} = 0 \mid (x_1, x_2, x_3) = (0, 0, 1)\right] \right.$$

$$+ \Pr\left[b_3^{(i-1)} = 1 \wedge b_2^{(i)} = 1 \mid (x_1, x_2, x_3) = (0, 0, 1)\right]$$

$$+ \Pr\left[b_3^{(i-1)} = 0 \wedge b_2^{(i)} = 1 \mid (x_1, x_2, x_3) = (0, 1, 1)\right]$$

$$\left. + \Pr\left[b_1^{(i-1)} = 0 \wedge b_2^{(i)} = 0 \mid (x_1, x_2, x_3) = (0, 1, 1)\right] \right).$$

From Equation (1), we know that the sum of Equations (5) and (6) is upperbounded by $1 + \mu$. Looking at the first summand in Equation (6), this implies that

$$\Pr\left[b_1^{(i-1)} = 1 \wedge b_2^{(i)} = 0 \mid (x_1, x_2, x_3) = (0, 0, 1)\right] \leq 4\mu. \qquad (7)$$

Probabilistic manipulation gives

$$\Pr\left[b_1^{(i-1)} = 1 \wedge b_2^{(i)} = 0 \mid (x_1, x_2, x_3) = (0, 0, 1)\right]$$

$$= 1 - \Pr\left[b_1^{(i-1)} = 0 \vee b_2^{(i)} = 1 \mid (x_1, x_2, x_3) = (0, 0, 1)\right]$$

$$\geq 1 - \Pr\left[b_1^{(i-1)} = 0 \mid (x_1, x_2, x_3) = (0, 0, 1)\right] - \Pr\left[b_2^{(i)} = 1 \mid (x_1, x_2, x_3) = (0, 0, 1)\right]$$

$$\geq 1 - \Pr\left[b_1^{(i-1)} = 0 \mid (x_1, x_2, x_3) = (0, 0, 1)\right] - \mu,$$

where the last inequality is due to the assumption of the claim. Combined with Equation (7), this implies:

$$\Pr\left[b_1^{(i-1)} = 0 \mid (x_1, x_2, x_3) = (0, 0, 1)\right] \geq 1 - 5\mu.$$

Applying an analogous argument starting with the third summand in Equation (6) gives

$$\Pr\left[b_3^{(i-1)} = 1 \mid (x_1, x_2, x_3) = (0, 1, 1)\right] \geq 1 - 5\mu.$$

Repeating the entire argument, but modifying the adversary to consider all possible pairs of corrupted parties and all possible settings of their inputs, completes the proof of the claim.

Theorem 6. *Any protocol Π that securely computes* maj *with complete fairness (assuming one exists[3] at all) requires $\omega(\log k)$ rounds.*

---

[3] In the following section we show that such a protocol does, indeed, exist.

Proof. Assume there exists a protocol Π that securely computes maj with complete fairness using $r = \mathcal{O}(\log k)$ rounds. Let $\mu(k) = \frac{1}{4 \cdot 5^{r(k)}}$, and note that μ is noticeable. By the assumed security of Π, the conditions of Claim 5 hold for k large enough; Equation (3), in particular, holds for $i = r(k)$. Fixing this k and applying the claim iteratively $r(k)$ times, we conclude that P_{j-1} and P_{j+1} can correctly compute the value of the function, on all inputs, with probability at least $3/4$ *without interacting with P_j at all.* This is clearly impossible.

3   Fair Computation of Majority for Three Players

In this section we describe a completely-fair protocol for computing maj for the case of $n = 3$ parties. The high-level structure of our protocol is as follows: the protocol consists of two phases. In the first phase, the parties run a secure-with-abort protocol to generate (authenticated) shares of certain values; in the second phase some of these shares are exchanged, round-by-round, for a total of m iterations. A more detailed description of the protocol follows.

In the first phase of the protocol the parties run a protocol π implementing a functionality ShareGen that computes certain values and then distributes authenticated 3-out-of-3 shares of these values to the parties. (See Fig. 1.) Three sets of values $\{b_1^{(i)}\}_{i=0}^m$, $\{b_2^{(i)}\}_{i=0}^m$, and $\{b_3^{(i)}\}_{i=0}^m$ are computed; looking ahead, $b_j^{(i)}$ denotes the value that parties P_{j-1} and P_{j+1} are supposed to output in case party P_j aborts after iteration i of the second phase; see below. The values $b_j^{(i)}$ are computed probabilistically, in the same manner as in [2]. That is, a round i^* is first chosen according to a geometric distribution with parameter $\alpha = 1/5$.[4] (We will set m so that $i^* \leq m$ with all but negligible probability.) Then, for $i < i^*$ the value of $b_j^{(i)}$ is computed using the true inputs of P_{j-1} and P_{j+1} but a random input for P_j; for $i \geq i^*$ the value $b_j^{(i)}$ is set equal to the correct output (i.e., it is computed using the true inputs of all parties). Note that even an adversary who knows all the parties' inputs and learns, sequentially, the values (say) $b_1^{(1)}, b_1^{(2)}, \ldots$ cannot determine definitively when round i^* occurs.

We choose the protocol π computing ShareGen to be secure-with-designated-abort [8] for P_1. Roughly speaking, this means privacy and correctness are ensured no matter what, and output delivery and (complete) fairness are guaranteed unless P_1 is corrupted; a formal definition in given in the full version [7].

The second phase of the protocol proceeds in a sequence of $m = \omega(\log n)$ iterations. (See Fig. 2.) In each iteration i, each party P_j broadcasts its share of $b_j^{(i)}$. (We stress that we allow rushing, and do not assume synchronous broadcast.) Observe that, after this is done, parties P_{j-1} and P_{j+1} *jointly* have enough information to reconstruct $b_j^{(i)}$, but neither party has any information about $b_j^{(i)}$ on its own. If all parties behave honestly until the end of the protocol, then in the final iteration all parties reconstruct $b_1^{(m)}$ and output this value.

---

[4] This is the distribution on $\mathbb{N} = \{1, 2, \ldots\}$ given by flipping a biased coin (that is heads with probability α) until the first head appears.

ShareGen

Inputs: Let the inputs to ShareGen be $x_1, x_2, x_3 \in \{0, 1\}$. (If one of the received inputs is not in the correct domain, then a default value of 1 is used for that player.) The security parameter is k.

Computation:

1. Define values $b_1^{(1)}, \ldots, b_1^{(m)}, b_2^{(1)}, \ldots, b_2^{(m)}$ and $b_3^{(1)}, \ldots, b_3^{(m)}$ in the following way:
   - Choose $i^* \geq 1$ according to a geometric distribution with parameter $\alpha = 1/5$ (see text).
   - For $i = 0$ to $i^* - 1$ and $j \in \{1, 2, 3\}$ do:
     - Choose $\hat{x}_j \leftarrow \{0, 1\}$ at random.
     - Set $b_j^{(i)} = \mathsf{maj}(x_{j-1}, \hat{x}_j, x_{j+1})$.
   - For $i = i^*$ to m and $j \in \{1, 2, 3\}$, set $b_j^{(i)} = \mathsf{maj}(x_1, x_2, x_3)$.
2. For $0 \leq i \leq m$ and $j \in \{1, 2, 3\}$, choose $b_{j|1}^{(i)}, b_{j|2}^{(i)}$ and $b_{j|3}^{(i)}$ as random three-way shares of $b_j^{(i)}$. (E.g., $b_{j|1}^{(i)}$ and $b_{j|2}^{(i)}$ are random and $b_{j|3}^{(i)} = b_{j|1}^{(i)} \oplus b_{j|2}^{(i)} \oplus b_j^{(i)}$.)
3. Let $(pk, sk) \leftarrow \mathsf{Gen}(1^k)$. For $0 \leq i \leq m$, and $j, j' \in \{1, 2, 3\}$, let $\sigma_{j|j'}^{(i)} = \mathsf{Sign}_{sk}(i\|j\|j'\|b_{j|j'}^{(i)})$.

Output:

1. Send to each P_j the public key pk and the values $\left\{ (b_{1|j}^{(i)}, \sigma_{1|j}^{(i)}), (b_{2|j}^{(i)}, \sigma_{2|j}^{(i)}), (b_{3|j}^{(i)}, \sigma_{3|j}^{(i)}) \right\}_{i=0}^{m}$. Additionally, for each $j \in \{1, 2, 3\}$ parties P_{j-1} and P_{j+1} receive the value $b_{j|j}^{(0)}$.

Fig. 1. Functionality ShareGen

If a single party P_j aborts in some iteration i, then the remaining players P_{j-1} and P_{j+1} jointly reconstruct the value $b_j^{(i-1)}$ and output this value. (These two parties jointly have enough information to do this.) If two parties abort in some iteration i (whether at the same time, or one after the other) then the remaining party simply outputs its own input.

We refer to Fig. 1 and Fig. 2 for the formal specification of the protocol. We now prove that this protocol securely computes maj with complete fairness.

Theorem 7. *Assume that* $(\mathsf{Gen}, \mathsf{Sign}, \mathsf{Vrfy})$ *is a secure signature scheme, that* π *securely computes* ShareGen *with designated abort, and that* π_{OR} *securely computes* OR *with complete fairness.[5] Then the protocol in Figure 2 securely computes* maj *with complete fairness.*

Proof. Let Π denote the protocol of Figure 2. Observe that Π yields the correct output with all but negligible probability when all players are honest. This is because, with all but negligible probability, $i^* \leq m$, and then $b_j^{(m)} = \mathsf{maj}(x_1, x_2, x_3)$. We thus focus on security of Π.

---

[5] It is shown in [2] that such a protocol exists under standard assumptions.

Protocol 1

Inputs: Party P_i has input $x_i \in \{0, 1\}$. The security parameter is k.

The protocol:

1. **Preliminary phase:**
   (a) Parties P_1, P_2 and P_3 run a protocol π for computing ShareGen. Each player uses their respective inputs, x_1, x_2 and x_3, and security parameter k.
   (b) If P_2 and P_3 receive $\perp$ from this execution, then P_2 and P_3 run a two-party protocol π_{OR} to compute the logical-or of their inputs. Otherwise, continue to the next stage.
   In what follows, parties always verify signatures; invalid signatures are treated as an abort.

2. **For $i = 1, \ldots, m - 1$ do:**
   Broadcast shares:
   (a) Each P_j broadcasts $(b_{j|j}^{(i)}, \sigma_{j|j}^{(i)})$.
   (b) If (only) P_j aborts:
       i. P_{j-1} and P_{j+1} broadcast $(b_{j|j-1}^{(i-1)}, \sigma_{j|j-1}^{(i-1)})$ and $(b_{j|j+1}^{(i-1)}, \sigma_{j|j+1}^{(i-1)})$, respectively.
       ii. If one of P_{j-1}, P_{j+1} aborts in the previous step, the remaining player outputs its own input value. Otherwise, P_{j-1} and P_{j+1} both output $b_j^{(i-1)} = b_{j|1}^{(i-1)} \oplus b_{j|2}^{(i-1)} \oplus b_{j|3}^{(i-1)}$. (Recall that if $i = 1$, parties P_{j-1} and P_{j+1} received $b_{j|j}^{(0)}$ as output from π.)
   (c) If two parties abort, the remaining player outputs its own input value.

3. **In round $i = m$ do:**
   (a) Each P_j broadcasts $b_{1|j}^{(m)}, \sigma_{1|j}^{(m)}$.
   (b) If no one aborts, then all players output $b_1^{(m)} = b_{1|1}^{(m)} \oplus b_{1|2}^{(m)} \oplus b_{1|3}^{(m)}$. If (only) P_j aborts, then P_{j-1} and P_{j+1} proceed as in step 2b. If two players abort, the remaining player outputs its own input as in step 2c.

Fig. 2. A protocol for computing majority

When no parties are corrupt, security is straightforward since we assume the existence of a private broadcast channel. We therefore consider separately the cases when a single party is corrupted and when two parties are corrupted. Since the entire protocol is symmetric except for the fact that P_1 may choose to abort π, without loss of generality we may analyze the case when the adversary corrupts P_1 and the case when the adversary corrupts $\{P_1, P_2\}$. In each case, we prove security of Π in a hybrid world where there is an ideal functionality computing ShareGen (with abort) as well as an ideal functionality computing OR (with complete fairness). Applying the composition theorem of [9] then gives the desired result. A proof for the case where only P_1 is corrupted turns out to be fairly straightforward, and is given in Appendix A.1.

Claim 8. *For every non-uniform, poly-time adversary $\mathcal{A}$ corrupting P_1 and P_2 and running Π in a hybrid model with access to ideal functionalities* *computing* ShareGen *(with abort) and* OR *(with completes fairness), there exists a non-uniform, poly-time adversary* $\mathcal{S}$ *corrupting* P_1 *and* P_2 *and running in the ideal world with access to an ideal functionality computing* maj *(with complete fairness), such that*

$$\left\{\text{IDEAL}_{\text{maj},\mathcal{S}}(x_1, x_2, x_3, k)\right\}_{x_i \in \{0,1\}, k \in \mathbb{N}}$$
$$\overset{\text{s}}{\equiv} \left\{\text{HYBRID}_{\Pi,\mathcal{A}}^{\text{ShareGen},\text{OR}}(x_1, x_2, x_3, k)\right\}_{x_i \in \{0,1\}, k \in \mathbb{N}}.$$

Proof. This case is significantly more complex than the case when only a single party is corrupted, since here $\mathcal{A}$ learns $b_3^{(i)}$ in each iteration i of the second phase. As in [2], we must deal with the fact that $\mathcal{A}$ might abort exactly in iteration i^*, after learning the correct output but before P_3 has enough information to compute the correct output.

We describe a simulator $\mathcal{S}$ who corrupts P_1 and P_2 and runs $\mathcal{A}$ as a black-box. For ease of exposition in what follows, we refer to the actions of P_1 and P_2 when more formally we mean the action of $\mathcal{A}$ on behalf of those parties.

1. $\mathcal{S}$ invokes $\mathcal{A}$ on the inputs x_1 and x_2, the auxiliary input z, and the security parameter k.

2. $\mathcal{S}$ receives x_1' and x_2' from P_1 and P_2, respectively, as input to ShareGen. If $x_1' \notin \{0,1\}$ (resp., $x_2' \notin \{0,1\}$), then $\mathcal{S}$ sets $x_1' = 1$ (resp., $x_2' = 1$).

3. $\mathcal{S}$ computes $(sk, pk) \leftarrow \text{Gen}(1^k)$, and then generates shares as follows:

   (a) Choose $\left\{b_{1|1}^{(i)}, b_{2|1}^{(i)}, b_{3|1}^{(i)}, b_{1|2}^{(i)}, b_{2|2}^{(i)}, b_{3|2}^{(i)}\right\}_{i=0}^{m}$ uniformly at random.

   (b) Choose $\hat{x}_3 \leftarrow \{0,1\}$ and set $b_3^{(0)} = \text{maj}(x_1', x_2', \hat{x}_3)$. Set $b_{3|3}^{(0)} = b_3^{(0)} \oplus b_{3|1}^{(0)} \oplus b_{3|2}^{(0)}$.

   $\mathcal{S}$ then hands $\mathcal{A}$ the public key pk, the values $\left\{b_{1|1}^{(i)}, b_{2|1}^{(i)}, b_{3|1}^{(i)}, b_{1|2}^{(i)}, b_{2|2}^{(i)}, b_{3|2}^{(i)}\right\}_{i=0}^{m}$ (along with their appropriate signatures), and the value $b_{3|3}^{(0)}$ as the outputs of P_1 and P_2 from ShareGen.

4. If P_1 aborts execution of ShareGen, then $\mathcal{S}$ extracts x_2'' from P_2 as its input to OR. It then sends $(1, x_2'')$ to the trusted party computing maj, outputs whatever $\mathcal{A}$ outputs, and halts.

5. Otherwise, if P_1 does not abort, then $\mathcal{S}$ picks a value i^* according to a geometric distribution with parameter $\alpha = \frac{1}{5}$.

   In what follows, for ease of description, we will use x_1 and x_2 in place of x_1' and x_2', keeping in mind that that $\mathcal{A}$ could of course have used substituted inputs. We also ignore the presence of signatures from now on, and leave the following implicit in what follows: (1) $\mathcal{S}$ always computes an appropriate signature when sending any value to $\mathcal{A}$; (2) $\mathcal{S}$ treats an incorrect signature as an abort; and (3) if $\mathcal{S}$ ever receives a valid signature on a previously unsigned message (i.e., a *forgery*), then $\mathcal{S}$ outputs fail and halts.

   Also, from here on we will say that $\mathcal{S}$ *sends* b *to* $\mathcal{A}$ *in round* i if $\mathcal{S}$ sends a value $b_{3|3}^{(i)}$ such that $b_{3|3}^{(i)} \oplus b_{3|1}^{(i)} \oplus b_{3|2}^{(i)} = b_3^{(i)} = b$.

6. For round $i = 1, \ldots, i^* - 1$, the simulator S computes and then sends $b_3^{(i)}$ as follows:
   (a) Select $\hat{x}_3 \leftarrow \{0, 1\}$ at random.
   (b) $b_3^{(i)} = \mathsf{maj}(x_1, x_2, \hat{x}_3)$.
7. If P_1 aborts in round $i < i^*$, then S sets $\hat{x}_2 = x_2$ and assigns a value to $\hat{x}_1$ according to the following rules that depend on the values of (x_1, x_2) and on the value of $b_3^{(i)}$:
   (a) If $x_1 = x_2$, then S sets $\hat{x}_1 = x_1$ with probability $\frac{3}{8}$ (and sets $\hat{x}_1 = \bar{x}_1$ otherwise).
   (b) If $x_1 \neq x_2$ and $b_3^{(i)} = x_1$, then S sets $\hat{x}_1 = x_1$ with probability $\frac{1}{4}$ (and sets $\hat{x}_1 = \bar{x}_1$ otherwise).
   (c) If $x_1 \neq x_2$ and $b_3^{(i)} = x_2$, then S sets $\hat{x}_1 = x_1$ with probability $\frac{1}{2}$ (and sets $\hat{x}_1 = \bar{x}_1$ otherwise).
   S then finishes the simulation as follows:
   (a) If $\hat{x}_1 \neq \hat{x}_2$, then S submits $(\hat{x}_1, \hat{x}_2)$ to the trusted party computing maj. Denote the output it receives from the trusted party by b_{out}. Then S sets $b_1^{(i-1)} = b_{\mathsf{out}}$, computes $b_{1|3}^{(i-1)} = b_1^{(i-1)} \oplus b_{1|1}^{(i-1)} \oplus b_{1|2}^{(i-1)}$, sends $b_{1|3}^{(i-1)}$ to P_2 (on behalf of P_3), outputs whatever A outputs, and halts.
   (b) If $\hat{x}_1 = \hat{x}_2$, then S sets $b_1^{(i-1)} = \hat{x}_1 = \hat{x}_2$, computes $b_{1|3}^{(i-1)} = b_1^{(i-1)} \oplus b_{1|1}^{(i-1)} \oplus b_{1|2}^{(i-1)}$, and sends $b_{1|3}^{(i-1)}$ to P_2 (on behalf of P_3). (We stress that this is done *before* sending anything to the trusted party computing maj.) If P_2 aborts, then S sends $(0, 1)$ to the trusted party computing maj. Otherwise, it sends $(\hat{x}_1, \hat{x}_2)$ to the trusted party computing maj. In both cases it outputs whatever A outputs, and then halts.
   If P_2 aborts in round $i < i^*$, then S acts analogously but swapping the roles of P_1 and P_2 as well as x_1 and x_2.
   If both parties abort in round $i < i^*$ (at the same time), then S sends $(0, 1)$ to the trusted party computing maj, outputs whatever A outputs, and halts.
8. In round i^*:
   (a) If $x_1 \neq x_2$, then S submits (x_1, x_2) to the trusted party. Let $b_{\mathsf{out}} = \mathsf{maj}(x_1, x_2, x_3)$ denote the output.
   (b) If $x_1 = x_2$, then S simply sets $b_{\mathsf{out}} = x_1 = x_2$ *without querying the trusted party* and continues. (Note that in this case, $b_{\mathsf{out}} = \mathsf{maj}(x_1, x_2, x_3)$ even though S did not query the trusted party.)
9. In rounds $i^*, \ldots, m - 1$, the simulator S sends b_{out} to A.
   If A aborts P_1 and P_2 simultaneously, then S submits $(1, 0)$ to the trusted party (if he hasn't already done so in step 8a), outputs whatever A outputs, and halts.
   If A aborts P_1 (only), then S sets $b_1^{(i-1)} = b_{\mathsf{out}}$, computes $b_{1|3}^{(i-1)} = b_1^{(i-1)} \oplus b_{1|1}^{(i-1)} \oplus b_{1|2}^{(i-1)}$, and sends $b_{1|3}^{(i-1)}$ to P_2 (on behalf of P_3). Then:
   Case 1: $x_1 \neq x_2$. Here S has already sent (x_1, x_2) to the trusted party. So S simply outputs whatever A outputs and ends the simulation.
   Case 2: $x_1 = x_2$. If P_2 does not abort, then S sends (x_1, x_2) to the trusted party. If P_2 aborts, then S sends $(0, 1)$ to the trusted party. In both cases S then outputs whatever A outputs and halts.

If $\mathcal{A}$ aborts P_2 (only), then $\mathcal{S}$ acts as above but swapping the roles of P_1, P_2 and x_1, x_2. If $\mathcal{A}$ does not abort anyone through round m, then $\mathcal{S}$ sends (x_1, x_2) to the trusted party (if he hasn't already done so), outputs what $\mathcal{A}$ outputs, and halts.

The above constitutes the full description of $\mathcal{S}$. Due to space limitations, the analysis of $\mathcal{S}$ is given in the full version of this paper [7].

4   Completely-Fair Computation of Boolean OR

The protocol in the previous section enables completely-fair computation of 3-party majority; unfortunately, we were not able to extend the approach to the case of $n > 3$ parties. In this section, we demonstrate feasibility of completely-fair computation of a non-trivial function for an arbitrary number of parties n, any $t < n$ of whom are corrupted. Specifically, we show how to compute boolean OR with complete fairness.

The basic idea behind our protocol is to have the parties repeatedly try to compute OR *on committed inputs* using a protocol that is secure-with-designated-abort where only the lowest-indexed party can force an abort. (See the full version [7] for further discussion.) The key observation is that in case of an abort, the dishonest players only "learn something" about the inputs of the honest players in case all the malicious parties use input 0. (If any of the malicious players holds input 1, then the output is always 1 regardless of the inputs of the honest parties.) So, if the lowest-indexed party is corrupt and aborts the computation of the committed OR, then the remaining parties simply recompute the committed OR using '0' as the effective input for any parties who have already been eliminated. The parties repeatedly proceed in this fashion, eliminating dishonest parties at each iteration. Eventually, when the lowest-indexed player is honest, the process terminates and all honest players receive (correct) output.

The actual protocol follows the above intuition, but is a bit more involved. A formal description of the protocol is given in Fig. 3, and the "committed OR" functionality is defined in Fig. 4.

Theorem 9. *Assume* Com *is a computationally-hiding, statistically-binding commitment scheme, and that* $\pi_{\mathcal{P}}$ *securely computes* CommittedOR$_{\mathcal{P}}$ *(with abort). Then the protocol of Fig. 3 computes* OR *with complete fairness.*

Proof. Let Π denote the protocol of Fig. 3. For simplicity we assume Com is perfectly binding, though statistical binding suffices. For any non-uniform, polynomial time adversary $\mathcal{A}$ in the hybrid world, we demonstrate a non-uniform polynomial-time adversary $\mathcal{S}$ corrupting the same parties as $\mathcal{A}$ and running in the ideal world with access to an ideal functionality computing OR (with complete fairness), such that

$$\left\{ \text{IDEAL}_{\text{OR},\mathcal{S}}(x_1, \ldots, x_n, k) \right\}_{x_i \in \{0,1\}, k \in \mathbb{N}}$$
$$\stackrel{c}{\equiv} \left\{ \text{HYBRID}_{\Pi,\mathcal{A}}^{\text{CommittedOR}_{\mathcal{P}}}(x_1, \ldots, x_n, k) \right\}_{x_i \in \{0,1\}, k \in \mathbb{N}}.$$

Protocol 2

Inputs: Each party P_i holds input $x_i \in \{0, 1\}$, and the security parameter is k.

Computation:

1. Let $\mathcal{P} = \{P_1, \ldots, P_n\}$ be the set of all players.
2. Each player P_i chooses random coins r_i and broadcasts $c_i = \mathsf{Com}(1^k, x_i, r_i)$, where Com denotes a computationally-hiding, statistically-binding commitment scheme. If any party P_i does not broadcast anything (or otherwise broadcasts an invalid value), then all honest players output 1. Otherwise, let $\boldsymbol{c} = (c_1, \ldots, c_n)$.
3. All players $P_i \in \mathcal{P}$ run a protocol $\pi_{\mathcal{P}}$ for computing $\mathsf{CommittedOR}_{\mathcal{P}}$, with party P_i using $(x_i, r_i, \boldsymbol{c}_{\mathcal{P}})$ as its input where $\boldsymbol{c}_{\mathcal{P}} \overset{\text{def}}{=} (c_i)_{i:P_i \in \mathcal{P}}$.
4. If players receive $\perp$ from the execution of $\mathsf{CommittedOR}_{\mathcal{P}}$, they set $\mathcal{P} = \mathcal{P} \setminus \{P^*\}$, where $P^* \in \mathcal{P}$ is the lowest-indexed player in $\mathcal{P}$, and return to Step 3.
5. If players receive a set $\mathcal{D} \subset \mathcal{P}$ from the execution of $\mathsf{CommittedOR}_{\mathcal{P}}$, they set $\mathcal{P} = \mathcal{P} \setminus \mathcal{D}$ and return to Step 3.
6. If players receive a binary output from the execution of $\mathsf{CommittedOR}_{\mathcal{P}}$, they output this value and end the protocol.

Fig. 3. A protocol computing OR for n players

CommittedOR$_{\mathcal{P}}$

Inputs: The functionality is run by parties in $\mathcal{P}$. Let the input of player $P_i \in \mathcal{P}$ be $(x_i, r_i, \boldsymbol{c}^i)$ where $\boldsymbol{c}^i = (c_j^i)_{j:P_j \in \mathcal{P}}$. The security parameter is k.

For each party $P_i \in \mathcal{P}$, determine its output as follows:

1. Say P_j *disagrees with* P_i if either (1) $\boldsymbol{c}^j \neq \boldsymbol{c}^i$ or (2) $\mathsf{Com}(1^k, x_j, r_j) \neq c_j^i$. (Note that disagreement is not a symmetric relation.)
2. Let $\mathcal{D}_i$ be the set of parties who disagree with P_i.
3. If there exist any parties that disagree with each other, return $\mathcal{D}_i$ as output to P_i. Otherwise, return $\bigvee_{j:P_j \in \mathcal{P}} x_j$ to all parties.

Fig. 4. Functionality $\mathsf{CommittedOR}_{\mathcal{P}}$, parameterized by a set $\mathcal{P}$

Applying the composition theorem of [9] then proves the theorem.

When no players are corrupt, security is trivial due to the assumed existence of a private broadcast channel. We now describe the execution of $\mathcal{S}$ assuming a set $\mathcal{C} \neq \emptyset$ of corrupted parties:

1. Let $\mathcal{H} = \{P_1, \ldots P_n\} \setminus \mathcal{C}$ denote the honest players. Initialize $\mathcal{I} = \mathcal{C}$. Looking ahead, $\mathcal{I}$ denotes the set of corrupted parties who have not yet been eliminated from the protocol.
2. $\mathcal{S}$ invokes $\mathcal{A}$ on the inputs $\{x_i\}_{i:P_i \in \mathcal{C}}$, the auxiliary input z, and the security parameter k.

3. For $P_i \in \mathcal{H}$, the simulator $\mathcal{S}$ gives to $\mathcal{A}$ a commitment $c_i = \mathsf{Com}(1^k, x_i, r_i)$ to $x_i = 0$ using randomness r_i. $\mathcal{S}$ then records the commitment c_i that is broadcast by $\mathcal{A}$ on behalf of each party $P_i \in \mathcal{C}$. If any corrupted player fails to broadcast a value c_i, then $\mathcal{S}$ submits 1's to the trusted party on behalf of all corrupted parties, outputs whatever $\mathcal{A}$ outputs, and halts.

4. If $\mathcal{I} = \emptyset$, $\mathcal{S}$ submits (on behalf of all the corrupted parties) 0's to the trusted party computing OR (unless it has already done so). It then outputs whatever $\mathcal{A}$ outputs, and halts. If $\mathcal{I} \neq \emptyset$, continue to the next step.

5. $\mathcal{S}$ sets $\mathcal{P} = \mathcal{H} \cup \mathcal{I}$ and obtains inputs $\{(r_i, x_i, c^i)\}_{i : P_i \in \mathcal{I}}$ for the computation of $\mathsf{CommittedOR}_{\mathcal{P}}$. For each $P_i \in \mathcal{P}$, the simulator $\mathcal{S}$ computes the list of players $\mathcal{D}_i$ that disagree with P_i (as in Fig. 4), using as the inputs of the honest parties the commitments defined in Step 3, and assuming that honest parties provide correct decommitments. Observe that if $P_i, P_j \in \mathcal{H}$ then $\mathcal{D}_i = \mathcal{D}_j \subseteq \mathcal{I}$. Let $\mathcal{D}_{\mathcal{H}} \subseteq \mathcal{I}$ be the set of parties that disagree with the honest parties.

   Let P^* be the lowest-indexed player in $\mathcal{P}$. If no parties disagree with each other, go to Step 6. Otherwise:
   (a) If $P^* \in \mathcal{I}$, then $\mathcal{A}$ is given $\{\mathcal{D}_i\}_{i : P_i \in \mathcal{I}}$. If P^* aborts, then $\mathcal{S}$ sets $\mathcal{I} = \mathcal{I} \setminus \{P^*\}$ and goes to Step 4. If P^* does not abort, then $\mathcal{S}$ sets $\mathcal{I} = \mathcal{I} \setminus \mathcal{D}_{\mathcal{H}}$ and goes to Step 4.
   (b) If $P^* \notin \mathcal{I}$, then $\mathcal{A}$ is given $\{\mathcal{D}_i\}_{i : P_i \in \mathcal{I}}$. Then $\mathcal{S}$ sets $\mathcal{I} = \mathcal{I} \setminus \mathcal{D}_{\mathcal{H}}$ and goes to Step 4.

6. $\mathcal{S}$ computes the value $b = \bigvee_{P_i \in \mathcal{I}} x_i$.
   (a) If $b = 0$, and $\mathcal{S}$ has not yet queried the trusted party computing OR, then $\mathcal{S}$ submits 0's (on behalf of all the corrupted parties) to the trusted party and stores the output of the trusted party as b_{out}. $\mathcal{S}$ gives b_{out} to $\mathcal{A}$ (either as just received from the trusted party, or as stored in a previous execution of this step).
   (b) If $b = 1$, then $\mathcal{S}$ gives the value 1 to $\mathcal{A}$ *without* querying the trusted party.
   $\mathcal{S}$ now continues as follows:
   (a) If $P^* \in \mathcal{I}$ and P^* aborts, then $\mathcal{S}$ sets $\mathcal{I} = \mathcal{I} \setminus \{P^*\}$ and goes to Step 4.
   (b) If $P^* \notin \mathcal{I}$, or if P^* does not abort, then $\mathcal{S}$ submits 1's to the trusted party if it has not yet submitted 0's. It outputs whatever $\mathcal{A}$ outputs, and halts.

We refer the reader to the full version of this paper [7] for an analysis of the above simulation.

References

1. Cleve, R.: Limits on the security of coin flips when half the processors are faulty. In: Proc. 18th Annual ACM Symposium on Theory of Computing (STOC), pp. 364–369 (1986)
2. Gordon, D., Hazay, C., Katz, J., Lindell, Y.: Complete fairness in secure two-party computation. In: Proc. 40th Annual ACM Symposium on Theory of Computing (STOC) (2008)

3. Chor, B., Ishai, Y.: On privacy and partition arguments. Inf. Comput. 167(1), 2–9 (2001)
4. Kilian, J., Kushilevitz, E., Micali, S., Ostrovsky, R.: Reducibility and completeness in private computations. SIAM J. Computing 29(4), 1189–1208 (2000)
5. Chor, B., Kushilevitz, E.: A zero-one law for boolean privacy. SIAM Journal of Discrete Math 4(1), 36–47 (1991)
6. Ishai, Y., Kushilevitz, E., Lindell, Y., Petrank, E.: On combining privacy with guaranteed output delivery in secure multiparty computation. In: Dwork, C. (ed.) CRYPTO 2006. LNCS, vol. 4117, pp. 483–500. Springer, Heidelberg (2006)
7. Gordon, S.D., Katz, J.: Complete fairness in multi-party computation without an honest majority. In: 6th Theory of Cryptography Conference, TCC (2009), http://eprint.iacr.org/2008/458
8. Goldreich, O.: Foundations of Cryptography, Basic Applications, vol. 2. Cambridge University Press, Cambridge (2004)
9. Canetti, R.: Security and composition of multiparty cryptographic protocols. Journal of Cryptology 13(1), 143–202 (2000)

A   Proofs

A.1   Proof of Security for Majority with a Single Corrupted Party

Claim 10. *For every non-uniform, polynomial-time adversary $\mathcal{A}$ corrupting P_1 and running Π in a hybrid model with access to ideal functionalities computing* ShareGen *(with abort) and* OR *(with complete fairness), there exists a non-uniform, poly-time adversary $\mathcal{S}$ corrupting P_1 and running in the ideal world with access to an ideal functionality computing* maj *(with complete fairness), such that*

$$\left\{ \mathrm{IDEAL}_{\mathsf{maj},\mathcal{S}}(x_1, x_2, x_3, k) \right\}_{x_i \in \{0,1\}, k \in \mathbb{N}}$$
$$\stackrel{\mathrm{s}}{\equiv} \left\{ \mathrm{HYBRID}_{\Pi,\mathcal{A}}^{\mathsf{ShareGen},\mathsf{OR}}(x_1, x_2, x_3, k) \right\}_{x_i \in \{0,1\}, k \in \mathbb{N}}.$$

Proof. Fix some polynomial-time adversary $\mathcal{A}$ corrupting P_1. We now describe a simulator $\mathcal{S}$ that also corrupts P_1 and runs $\mathcal{A}$ as a black box.

1. $\mathcal{S}$ invokes $\mathcal{A}$ on the input x_1, the auxiliary input z, and the security parameter k.
2. $\mathcal{S}$ receives input $x_1' \in \{0, 1\}$ on behalf of P_1 as input to ShareGen.
3. $\mathcal{S}$ computes $(sk, pk) \leftarrow \mathsf{Gen}(1^k)$, and gives to $\mathcal{A}$ the public key pk and values $b_{2|2}^{(0)}$, $b_{3|3}^{(0)}$, and $\left\{ b_{1|1}^{(i)}, b_{2|1}^{(i)}, b_{3|1}^{(i)} \right\}_{i=0}^{m}$ (along with their appropriate signatures) chosen uniformly at random.
4. If $\mathcal{A}$ aborts execution of ShareGen, then $\mathcal{S}$ sends 1 to the trusted party computing maj, outputs whatever $\mathcal{A}$ outputs, and halts. Otherwise, $\mathcal{S}$ picks a value i^* according to a geometric distribution with parameter $\alpha = \frac{1}{5}$.

   For simplicity in what follows, we ignore the presence of signatures and leave the following implicit from now on: (1) $\mathcal{S}$ always computes an appropriate signature when sending any value to $\mathcal{A}$; (2) $\mathcal{S}$ treats an incorrect signature as an abort; and (3) if $\mathcal{S}$ ever receives a valid signature on a previously unsigned message, then $\mathcal{S}$ outputs fail and halts.

5. S now simulates the rounds of the protocol one-by-one: for $i = 1$ to $m - 1$, the simulator chooses random $b_{2|2}^{(i)}$ and $b_{3|3}^{(i)}$ and sends these to A. During this step, an abort by A (on behalf of P_1) is treated as follows:
   (a) If P_1 aborts in round $i \leq i^*$, then S chooses a random value $\hat{x}_1$ and sends it to the trusted party computing maj.
   (b) If P_1 aborts in round $i > i^*$, then S submits x_1' to the trusted party computing maj.
   In either case, S then outputs whatever A outputs and halts.
6. If P_1 has not yet aborted, S then simulates the final round of the protocol. S sends x_1' to the trusted party, receives $b_{\mathsf{out}} = \mathsf{maj}(x_1', x_2, x_3)$, and chooses $b_{1|2}^{(m)}$ and $b_{1|3}^{(m)}$ at random subject to $b_{1|2}^{(m)} \oplus b_{1|3}^{(m)} \oplus b_{1|1}^{(m)} = b_{\mathsf{out}}$. S then gives these values to A, outputs whatever A outputs, and halts.

Due to the security of the underlying signature scheme, the probability that S outputs fail is negligible in k. Note that the view of P_1 is otherwise statistically close in both worlds. Indeed, until round m the view of P_1 is independent of the inputs of the other parties in both the real and ideal worlds. In round m itself, P_1 learns the (correct) output b_{out} in the ideal world and learns this value with all but negligible probability in the real world.

We therefore only have to argue that outputs of the two honest parties in the real and ideal worlds are statistically close. Clearly this is true if P_1 never aborts. As for the case when P_1 aborts at some point during the protocol, we divide our analysis into the following cases:

- If P_1 aborts the execution of ShareGen in step 4, then S submits '1' on behalf of P_1 to the trusted party computing maj. Thus, in the ideal world, the outputs of P_2 and P_3 will be $\mathsf{maj}(1, x_2, x_3)$. In the real world, if P_1 aborts computation of ShareGen, the honest parties output $\mathsf{OR}(x_2, x_3)$. Since $\mathsf{maj}(1, x_2, x_3) = \mathsf{OR}(x_2, x_3)$, their outputs are the same.
- If P_1 aborts in round i of the protocol (cf. step 5), then in both the real and ideal worlds the following holds:
  • If $i \leq i^*$, then P_2 and P_3 output $\mathsf{maj}(\hat{x}_1, x_2, x_3)$ where $\hat{x}_1$ is chosen uniformly at random.
  • If $i > i^*$, then P_2 and P_3 output $\mathsf{maj}(x_1', x_2, x_3)$
  Since i^* is identically distributed in both worlds, the outputs of P_2 and P_3 in this case are identically distributed as well.
- If P_1 aborts in round m (cf. step 6), then in the ideal world the honest parties will output $\mathsf{maj}(x_1', x_2, x_3)$. In the real world the honest parties output $\mathsf{maj}(x_1', x_2, x_3)$ as long as $i^* \leq m - 1$, which occurs with all but negligible probability.

This completes the proof.

Fairness with an Honest Minority and a Rational Majority*

Shien Jin Ong[1,**], David C. Parkes[2], Alon Rosen[3,***], and Salil Vadhan[4,†]

[1] Goldman, Sachs & Co., New York, NY
shienjin@alum.mit.edu
[2] Harvard School of Engineering and Applied Sciences, Cambridge, MA
parkes@eecs.harvard.edu
[3] Herzliya Interdisciplinary Center, Herzliya, Israel
alon.rosen@idc.ac.il
[4] Harvard School of Engineering and Applied Sciences and
Center for Research on Computation and Society, Cambridge, MA
salil@eecs.harvard.edu

Abstract. We provide a simple protocol for secret reconstruction in any threshold secret sharing scheme, and prove that it is *fair* when executed with many *rational* parties together with a small minority of *honest* parties. That is, all parties will learn the secret with high probability when the honest parties follow the protocol and the rational parties act in their own self-interest (as captured by a set-Nash analogue of trembling hand perfect equilibrium). The protocol only requires a *standard* (synchronous) broadcast channel, tolerates both early stopping and incorrectly computed messages, and only requires 2 rounds of communication.

Previous protocols for this problem in the cryptographic or economic models have either required an honest majority, used strong communication channels that enable simultaneous exchange of information, or settled for approximate notions of security/equilibria. They all also required a nonconstant number of rounds of communication.

Keywords: game theory, fairness, secret sharing.

1 Introduction

A major concern in the design of distributed protocols is the possibility that parties may deviate from the protocol. Historically, there have been two main paradigms for modeling this possibility. One is the cryptographic paradigm, where some parties are honest, meaning they will always follow the specified

---

* Earlier versions of this paper are [34,35].
** Work done while author was a graduate student at Harvard School of Engineering and Applied Sciences.
*** Part of this work done while the author was a postdoctoral fellow at Harvard University's Center for Research on Computation and Society. Work supported in part by ISF Grant 334/08.
† Supported by NSF grant CNS-0831289.

O. Reingold (Ed.): TCC 2009, LNCS 5444, pp. 36–53, 2009.

protocol, and others are malicious, meaning they can deviate arbitrarily from the protocol. The other is the economic paradigm, where all parties are considered to be rational, meaning that they will deviate from the protocol if and only if it is in their interest to do so.

Recently, some researchers have proposed studying mixtures of these traditional cryptographic and economic models, with various combinations of honest, malicious, and rational participants. One motivation for this that it may allow a more accurate modeling of the diversity of participants in real-life executions of protocols. Along these lines, the papers of Aiyer et al. [2], Lysyanskaya and Triandopoulos [30], and Abraham et al. [1] construct protocols that achieve the best of both worlds. Specifically, they take protocol properties that are known to be achievable in both the cryptographic model (with honest and malicious parties) and the economic model (with only rational parties), and show that protocols with the same properties can still be achieved in a more general model consisting of malicious and rational parties.

Our work is of the opposite flavor. We consider properties that are not achievable in either the cryptographic or economic models alone, and show that they can be achieved in a model consisting of both honest and rational parties. Specifically, we consider the task of secret reconstruction in *secret sharing*, and provide a protocol that is *fair*, meaning that all parties will receive the output, given many rational participants together with a small minority of honest participants. In standard communication models, fairness is impossible in a purely economic model (with only rational participants) [20,25] or in a purely cryptographic model (with a majority of malicious participants) [10]. Previous works in the individual models achieved fairness by assuming strong communication primitives that allow simultaneous exchange of information [20,19,1,25,28,29,21][1] or settled for approximate notions of security/equilibria [12,7,17,38,25], whereas we only use a standard (i.e. synchronous but not simultaneous) broadcast channel and achieve a standard notion of game-theoretic equilibrium (namely, a trembling hand perfect equilibrium).

Thus, our work illustrates the potential power of a small number of honest parties to maintain equilibria in protocols. These parties follow the specified strategy even when it is not in their interest to do so, whether out of altruism or laziness. While we study a very specific problem (secret sharing reconstruction, as opposed to general secure function evaluation), we hope that eventually the understanding developed in this clean setting will be leveraged to handle more complex settings (as has been the case in the past).

Below, we review the cryptographic and economic paradigms in more detail. We then introduce the secret-sharing problem we study and survey recent works on this problem in the purely economic model. We then describe our results and compare them to what was achieved before.

---

[1] Actually, the impossibility results of [20,25] also hold in the presence of a simultaneous broadcast channel and thus the works of [20,19,1,25] use additional relaxations, such as allowing the number of rounds and/or the sizes of the shares to be unbounded random variables.

1.1   The Cryptographic Paradigm

In the cryptographic paradigm, we allow for a subset of the parties to deviate from the protocol in an arbitrary, malicious manner (possibly restricted to computationally feasible strategies), and the actions of these parties are viewed as being controlled by a single adversary. Intuitively, this captures worst-case deviations from the protocol, so protocols protecting against such malicious and monolithic adversaries provide a very high level of security. Remarkably, this kind of security can be achieved for essentially every multiparty functionality, as shown by a series of beautiful results from the 1980's [45,18,8,5,39]. However, considering arbitrary (and coordinated) malicious behavior does have some important limitations. For example, it is necessary to either assume that a majority of the participants are honest (i.e. not controlled by the adversary) or allow for protocols that are unfair (i.e. the adversary can prevent some parties from getting the output). This follows from a classic result of Cleve [10], who first showed that there is no fair 2-party protocol for coin-tossing (even with computational security), and then deduced the general version by viewing a multiparty protocol an interaction between two super-parties, each of which controls half of the original parties. Lepinski et al. [28] bypass this impossibility result by assuming a strong communication primitive ("ideal envelopes") which allow simultaneous exchange of information, but it remains of interest to find ways of achieving fairness without changing the communication model.

1.2   The Economic Paradigm

In the economic paradigm, parties are modeled as rational agents with individual preferences, and will only deviate from the protocol if this is in their own self interest. This approach has become very popular in the CS literature in recent years, with many beautiful results. There are two aspects of this approach:

1. Design computationally efficient mechanisms (i.e. functionalities that can be implemented by a trusted mediator) that give parties an incentive to be truthful about their private inputs, while optimizing some *social choice function*, which measures the benefit to society and/or the mechanism designer [32,27,3].
2. Implement these mechanisms by distributed protocols, with computational efficiency emphasized in *distributed algorithmic mechanism design* [13,14,15] and extended to also emphasize additional equilibrium considerations in *distributed implementation* [42,36,37], so that parties are "faithful" and choose to perform message passing and computational tasks in *ex post* Nash equilibrium. More recent works achieve a strong form of distributed implementation, with provably no additional equilibria [29,21], but require strong communication primitives.

Note that distributed algorithmic mechanism design is different in spirit from the traditional problem considered in cryptographic protocols, in that parties have "true" private inputs (whereas in cryptography all inputs are considered equally valid) and there is freedom to change how these inputs are mapped to outcomes through choosing appropriate social choice functions to implement (whereas in cryptography, the functionality is pre-specified.) Nevertheless, recent works have explored whether we can use the economic model to obtain 'better' solutions to traditionally cryptographic problems, namely to compute some pre-specified functionalities. One potential benefit is that we may be able to incentivize parties to provide their "true" private inputs along the lines of Item 1 above; the papers [31,43] explore for what functionalities and kinds of utility functions this is possible in the presence of game-theoretic agents.

A second potential benefit is that rational deviations may be easier to handle than malicious deviations (thus possibly leading to protocols with better properties), while also preferable to assuming a mixture of players at the honest and malicious extremes. This has led to a line of work, started by Halpern and Teague [20] and followed by [19,1,25], studying the problems of secret sharing and multiparty computation in the purely economic model, with all rational participants. One can also require notions of equilibria that are robust against coalitions of rational players [1]. While this approach has proved to be quite fruitful, it too has limitations. Specifically, as pointed out in [19,25], it seems difficult to construct rational protocols that are fair in the standard communication model, because parties may have an incentive to stop participating once they receive their own output. The works [20,19,1,25], as well as [29,21] applied to appropriately designed mediated games, achieve fairness by using strong communication primitives (simultaneous broadcast, "ballot boxes") that allow simultaneous exchange of information.

As mentioned above, we achieve fairness in the standard communication model by considering a mix of many rational participants together with a *small* minority of honest participants. Note that Cleve's [10] proof that an honest majority is necessary in the cryptographic setting, by reduction to the two-party case, no longer applies. The reason is that we cannot view a subset of the rational parties as being controlled by a single super-party. Even when considering coalitions, it seems that each individual in that subset would only agree to a coordinated (joint) deviation if it is in its own interest to do so.

Our protocol is for the share reconstruction problem in secret sharing, which we now describe in more detail.

1.3   Secret Sharing

In a *t-out-of-n secret-sharing scheme* [41,6], a dealer takes a secret s and computes n (randomized) *shares* $s_1, \ldots, s_n$ of s, which are distributed among n parties. The required properties are that (a) any set of t parties can reconstruct the secret s from their shares, but (b) any set of fewer than t parties has no information about s (i.e. they would have been equally likely to receive the same shares for every possible value of s).

Secret sharing is a fundamental building block for cryptographic protocols [18,5,8,39]. Typically, these protocols are structured as follows. First, every party shares its private input among all the parties. Then the computation of the functionality is done on shares (to maintain privacy). And at the end, the parties reveal their shares of the output so that everyone can reconstruct it. Our focus is on this final reconstruction step. Typically, it is assumed that there are enough honest parties in the protocol to ensure that the secret can be reconstructed from the revealed shares, even if some parties refuse to reveal their shares or even reveal incorrect values. This turns out to be achievable if and only if more than a 2/3 fraction of the players are honest [9]. (In previous versions of the paper, we restricted attention to *fail-stop deviations* where a party may stop participating in the protocol early but otherwise follows the prescribed strategy, in which case only an honest majority is needed in the traditional cryptographic model.)

1.4   Rational Secret Sharing

It is natural to ask whether we can bypass this need for an honest majority by considering only *rational* deviations from the protocol. As noted above, the study of secret sharing with only rational participants was initiated by Halpern and Teague [20], and there have been several subsequent works [19,25,1]. In these works, it is assumed that participants prefer to learn the secret over not learning the secret, and secondarily, prefer that as few other agents as possible learn it. As pointed out in Gordon and Katz [19], any protocol where rational participants reveal their shares sequentially will not yield a Nash equilibrium. This is because it is rational for the t'th player to stop participating, as she can already compute the secret from the shares of the first $t - 1$ players and her own, and stopping may prevent the first $t - 1$ players from learning it.

One way to get around this difficulty is to assume a *simultaneous broadcast channel*, where all parties can broadcast values at the same time, without the option of waiting to see what values the other parties are broadcasting. All parties simultaneously revealing their shares is a Nash equilibrium. That is, assuming all of the other parties are simultaneously revealing their shares, no party can increase her utility by aborting (stopping early) instead of revealing. This basic protocol is instructive because it has several deficiences:

1. A simultaneous broadcast channel is a strong (and perhaps unrealistic) communication primitive, particularly in the context of trying to achieve fairness, where the typical difficulties are due to asymmetries in the times that parties get information. For example, fair coin-tossing is trivial with a simultaneous broadcast channel (everyone broadcasts a bit, and the output is the exclusive-or), in contrast to Cleve's impossibility result for synchronous broadcast channels [10].
2. Nash Equilibrium in this context is a very weak guarantee. As argued by Halpern and Teague [20], it seems likely that rational parties would actually abort. The reason is that aborting is never worse than revealing, and is sometimes better (if $t - 1$ other parties reveal, then the tth party will always learn the secret and can prevent the other parties from doing so by aborting.)

Halpern and Teague [20] and follow-up works [19,1,25] focus on the second issue. That is, they allow simultaneous broadcast, and explore whether stronger solution concepts than plain Nash equilibrium can be achieved. Halpern and Teague [20] propose looking for an equilibrium that survives "iterated deletion of weakly dominated strategies." They prove that no bounded-round protocol can achieve a fair outcome in equilibrium when adopting this solution concept. However, they and subsequent works by Gordon and Katz [19] and Abraham et al. [1] show that fair outcomes are possible even with this equilibrium refinement using a probabilistic protocol whose number of rounds has finite expectation. Moreover, Abraham et al. [1] show how to achieve an equilibrium that is resistant to deviations by coalitions of limited size. Kol and Naor [25] argue that "strict equilibria" is a preferable solution concept to the iterated deletion notion used by Halpern and Teague [20], and show how to achieve it with a protocol where the size of shares dealt is an unbounded random variable with finite expectation. (They also show that a strict equilibrium cannot be achieved if the shares are of bounded size.) In all of the above works, the protocols' prescribed instructions crucially depend on the utilities of the various players.

The works of Lepinski et al. [29] and Izmalkov et al. [21] also can be used to obtain fair protocols for secret sharing by making an even stronger physical assumption than a simultaneous broadcast channel, namely "ballot boxes." Specifically, they show how to compile any game with a trusted mediator into a fair ballot-box protocol with the same incentive structure. Since the share-reconstruction problem has a simple fair solution with a trusted mediator (the mediator takes all the inputs, and broadcasts the secret iff *all* players reveal their share), we can apply their compiler to obtain a fair ballot-box protocol. But our interest in this paper is on retaining standard communication models.

1.5   Our Results

In this paper, we address both issues above. Specifically, we assume that there is at least some *small* number k of honest participants, and along with many rational players. In this setting, we exhibit a simple protocol that only requires a standard communication model, namely *synchronous broadcast*, and in cases where the total number of players is sufficiently large, achieves fair outcomes with high probability with respect to a strong solution concept, namely (a set-Nash analogue of) *trembling hand perfect equilibrium*. We describe both aspects of our result in more detail below.

Synchronous Broadcast. With a *synchronous* (as opposed to *simultaneous*) broadcast channel, the protocol proceeds in rounds, and only one party can broadcast in each round.[2] When all parties are rational, the only previous positive results in this model are in works by Kol and Naor [25,24], who achieve a fair solution with an approximate notion of Nash equilibrium — no party can improve her utility by ε by deviating from the protocol. However, it is unclear whether such ε-Nash

---

[2] For round efficiency, sometimes people use a slightly more general channel where many parties can broadcast in a single round, but deviating parties are can perform 'rushing' — wait to see what others have broadcast before broadcasting their own values. We describe how to extend our results to this setting below.

equilibria are satisfactory solution concepts because they may be unstable. In particular, how can everyone be sure that some parties will not try to improve their utility by ε? Once this possibility is allowed, it may snowball into opportunities for even greater gains by deviation. Indeed, Kol and Naor argue in favor of *strict* Nash equilibria, where parties will obtain strictly less utility by deviating (and show how to achieve strict equilibria in the presence of a simultaneous broadcast channel).

In our work, we achieve an exact notion of equilibrium (i.e. $\varepsilon = 0$). However, we allow a negligible probability that the honest parties will fail to compute the secret correctly, so our notion of "fairness" is also approximate. Nevertheless, we feel that the kind of error we achieve is preferable to ε-Nash. Indeed, the equilibrium concept is supposed to ensure that parties have an incentive to behave in a particular manner; if it is too weak, then parties may ignore it entirely and whatever analysis we do may be rendered irrelevant. On the other hand, if we achieve a sufficiently strong notion of exact equilibrium, then we may be confident that players will behave as predicted, and we are unlikely to see any bad events that are shown to occur with small probability under equilibrium play.

Trembling-Hand Equilibrium. In order to establish the equilibrium properties of the protocol, we introduce a framework of "extensive form games with public actions and private outputs," and use the formalism of incomplete information games to model players' uncertainties about the inputs (i.e. shares) of other players as well as uncertainty about which players are honest and which are rational. (For simplicity, we assume that each player is honest independently with some probability p, but with small modifications, the result should extend to other distributions on the set of honest players.) The solution concept of *Bayesian Nash equilibrium* handles the uncertainty that a player has about the shares dealt to other players and requires that beliefs are updated according to Bayes rule "whenever possible," meaning that this occurs when the observed actions are consistent with the equilibrium. A standard refinement is that of Bayesian *subgame perfect* Nash equilibrium, which captures the idea that the strategy is rational to follow regardless of the previous history of messages; intuitively, this means that the equilibrium does not rely on irrational empty threats (where a player will punish another player for deviating even at his own expense). In fact, we achieve the additional refinement of *trembling hand perfect equilibrium* [40], which strengthens this notion by requiring that players update their beliefs in a consistent and meaningful manner even when out-of-equilibrium play occurs. It is one of the strongest solution concepts studied for extensive form games; related notions were advocated in this context by Peter Bro Miltersen (personal communication) and Jonathan Katz [23].

Our Protocol. The protocol that we instruct honest players to follow is simple to describe. The participants take turns broadcasting their shares in sequence. However, if any of the first $t-1$ parties deviates from the protocol by stopping and refusing to broadcast her share, then the protocol instructs all parties subsequent to the first $t-1$ to do the same. The intuition behind this protocol is that if there is likely to be at least one *honest* party after the first $t - 1$ parties, then each rational party in the first $t - 1$ parties will also have an incentive to reveal its share because by doing so, the honest party will also reveal her share and enable the rational parties to reconstruct the secret. Then we observe that as long as the set of honest parties is a random subset of $k = \omega(\log n)$ parties, and assuming that the total number, n, of players is sufficiently large, then there will be an honest party after party $t - 1$ with all but negligible probability, as long as $t \leq (1 - \Omega(1)) \cdot n$. Thus, assuming that parties have a nonnegligible preference to learn the secret, we obtain an *exact* equilibrium in which *everyone* learns the secret with all but negligible probability.

In order to deal with the possibility that some players may try to reveal incorrect shares, we use information-theoretic message authentication codes (MACs) to authenticate the shares, following Kol and Naor [25]. Intuitively, we can tolerate the (negligible) forgery probability of the MACs (without getting an ε-Nash equilibrium) because the first $t - 1$ players actually achieve *strictly* higher utility by revealing a valid share than by not doing so.

In addition, the incentives in our protocol hold regardless of what information the first $t - 1$ players have about each others' actions, and similarly for the last $n - t + 1$ players. Thus, our protocol can actually be implemented with only 2 rounds of communication (in contrast to all previous protocols, which required a super-constant number of rounds); we discuss how to formalize this below.

Modeling Contributions. While the intuition for our protocol is quite natural, modeling it game-theoretically turns out to be quite delicate. As discussed above, we introduce a Bayesian framework for capturing the uncertainty that players have about each others' secrets and which other players are honest vs. rational. Additional modeling contributions include:

<u>Set Nash.</u> We find it useful to avoid specifying the exact actions that rational players should take in situations where the choice is irrelevant to the overall strategic and fairness properties of our protocol. We do this by developing a variant of the notions of Set-Nash [26] and CURB (Closed Under Rational Behavior) Sets [4] for extensive-form games and trembling-hand perfect equilibrium. Roughly speaking, this notion allows us to specify the equilibrium actions only in cases that we care about, and argue that players have no incentive to deviate from the specified actions provided that all other players are playing according to the specified actions (even if they may act arbitrarily when the action is unspecified) and given the existence of a small number of honest players. Since the honest strategy is consistent with the specified equilibrium actions, this solution concept ensures that even repeated rational deviations from the honest strategy (which we envision to be initial "program" distributed to all players) by all but a small number of players will keep everyone consistent with the specified actions. When this occurs as predicted, we show that all honest players will learn the secret with all but negligible probability, and thus fairness is maintained.

<u>Modeling Rushing.</u> To save on rounds, the cryptography literature often allows protocols that specify messages for several players at once, but allows the

possibility that deviating players may wait to see other players' messages before computing their own (i.e. simultaneity is allowed but not enforced). Modelling such "rushing" game-theoretically was posed as a challenge in the survey talk of Katz [23]. As mentioned above, we argue that our protocol can be collapsed to two rounds of communication. To capture the possibility of rushing game-theoretically, we follow an idea of Kalai [22], and argue that the specified strategy remains an equilibrium for every ordering of players within each round. Thus, players have no incentive to wait for other players' messages; sending the same message will maximize their utility regardless of what other players send in the same round.

1.6   Future Directions and Independent Work

We view our work as but one more step in the line of work understanding the benefits of bringing together cryptography and algorithmic mechanism design. (See the survey [23].) While our main theorem is admittedly far from achieving an end goal that one would want to implement as is, we hope that our high-level message (regarding the benefit of a few honest players with many rational players) and our game-theoretic modelling (e.g. the Bayesian framework, the use of set-Nash, and the modelling of rushing) prove useful in subsequent work. Some specific ways in which our results could be improved are:

- Handling other distributions on (i.e. beliefs about) the set of honest players. Intuitively, this should be possible by having the dealer randomly permuting the order of the players and including the permutation in the shares (or publishing it).
- Achieving solution concepts that are robust even to coalitional deviations from the protocol. In an earlier version of our paper [34], we demonstrated coalition-proofness (against "stable" coalitions) in a model that is even more simplified than the fail-stop one. As we have mentioned, Abraham et al. [1] show how to handle arbitrary, not necessarily stable, collusions of a small number of players with a simultaneous broadcast channel.
- Generalizing from secret sharing to secure multiparty computation. Indeed, this is the main application for secret sharing and their reconstruction protocols.
- Getting stronger impossibility results for the entirely rational setting (prior impossibility results either require players to learn the secret with probability 1 [25], or suffered other restrictive constraints [35]) or, alternatively, finding a purely rational protocol.

O'Neill and Sangwan [33] extend the results from a preliminary version of our paper [35] in several ways, including achieving a strict trembling-hand perfect equilibrium for a restricted deviation model (which is still more general than the fail-stop deviation model we considered in [35]) and handling a small number of malicious players in this model. Fuchsbauer et al. [16] have recently shown how to obtain a computational analogue of trembling-hand ε-equilibrium on a standard communication channel when all players are rational.

2   Definitions

2.1   Games with Public Actions and Private Outputs

To cast protocol executions into a game-theoretic setting, we introduce the notion of *extensive games with public actions and private outputs*. The basis of this new notion is the more standard definition of *extensive form games with perfect information*. Extensive form games enable us to model the *sequential* aspect of protocols, where each player considers his plan of action only following some of the other players' messages (the *"actions"* of the game-theoretic model). The perfect information property captures the fact that each player, when making any decision in the public phase of the protocol, is perfectly informed of all the actions by other players that have previously occurred. Thus, extensive form games with perfect information are a good model for communication on a *synchronous broadcast channel*.

We build upon extensive form games with perfect information and augment them with an additional final *private* stage. This additional stage models the fact that at the end of the game, each player is allowed to take some arbitrary action as a function of the "terminal" history $h \in Z$ of messages so far. This action, along with the "non-terminal history" $h \in H \setminus Z$ of public actions that have taken place during the execution of the game (as well as the players' inputs) has a direct effect on players' payoffs.

Working in the framework of Bayesian games of incomplete information, players $i \in N$ are handed private inputs θ_i (a.k.a. "*types*") that belong to some pre-specified set Θ_i and specify a distribution μ according to which the inputs are chosen. Players' inputs can be thought of as the shares for the secret-sharing scheme, and are generated jointly with the secret. The secret is thought of as a "reference" value $\delta \in \Delta$ that is not given to the players at the beginning of the protocol (but may be determined by them through messages exchanged), and is used at the output stage along with private actions to determine player utilities.

A game $\Gamma = (N, H, P, A, L, \Delta, \Theta, \mu, u)$ proceeds as follows: the reference value and the types are selected according to a joint distribution μ. The type $\theta_i \in \Theta_i$ is handed to player $i \in N$ and the value $\delta \in \Delta$ remains secret and affects the players' utilities. This is followed by a sequence of actions that are visible by all players. After any history $h \in H$, player $i = P(h)$ whose turn to play is next chooses a public action $a \in A_i(\theta_i, h)$. This choice determines the next actions of the players, and so on until a terminal history $h \in Z$ is reached. At this point, all players $i \in N$ simultaneously pick an action, $b_i \in L_i(\theta_i, h)$, where $L_i(\theta_i, h) \subseteq \Delta$. The utility (or payoff) of player i for an execution of the game is then determined to be the value $u_i(\delta, \theta, h, b_1, \ldots, b_n)$.

2.2   Set Valued Strategies and Set Nash Equilibrium

The action chosen by a player for every history after which it is her turn to move, is determined by her *strategy* function. As is required in extensive-form games, the strategy is defined for all histories, even ones that would not be reached if the strategy is followed.

To enable a simple description of our protocol, and in order to minimize the difference between the description of the behavior of an honest player and that of a rational player, we allow each player to have a strategy that actually maps each information set (i.e. view of the player) into a *set* of possible actions. More precisely, a *set-valued strategy* for player $i \in N$ is a pair $S_i = (M_i, F_i)$, where:

- The *public set-valued strategy* M_i is a function that takes a pair $(\theta_i, h) \in \Theta_i \times (H \setminus Z)$ and defines a set of public messages, $M_i(\theta_i, h)$.
- The *private set-valued strategy* F_i is a function that takes a pair $(\theta_i, h) \in \Theta_i \times Z$ and defines a set of private outputs $F_i(\theta_i, h)$.

We write $s_i = (m_i, f_i) \in S_i$ to indicate that strategy s_i is consistent with S_i, i.e. with $m_i(\theta_i, h) \in M_i(\theta_i, h)$ for all $(\theta_i, h) \in \Theta_i \times (H \setminus Z)$ and $f_i(\theta_i, h) \in F_i(\theta_i, h)$ for all $(\theta_i, h) \in \Theta_i \times Z$ (where these inclusions should hold with probability 1 in case s_i is a mixed strategy).

We will allow the public and private strategy functions to be mixed, where the randomization of the strategy is interpreted to be done independently at each application of the function, if a player has multiple moves in the game. Strategies that consist of deterministic functions are called *pure*, whereas strategies whose functions have full support on the player's action set are said to be *fully mixed*. We achieve fairness (with high probability) in a pure strategy equilibrium but use fully mixed strategies in defining the concept of trembling hand equilibrium.

The *outcome o* of the game Γ under a strategy $s \in S$ is the random variable $(\delta, \theta, h, b_1, \ldots, b_n)$, where $(\delta, \theta) \in \Delta \times \Theta$ are sampled according to μ, $h \in Z$ is the terminal history that results when each player $i \in N$ is given her type $\theta_i \in \Theta_i$, publicly follows the actions chosen by m_i, and computes her final private output b_i using f_i. The value of player i's utility is totally determined by o. The initial distribution, μ, of the secret and the shares, along with the strategies $s_i = (m_i, f_i)$ induce a distribution on o, and thus on the utilities. Define $u_i(\mu, s)$ to be the expected value of the utility of player $i \in N$, when the types are selected according to the distribution μ and all players follow strategy s. We assume that rational players seek to maximize expected utility.

Definition 2.1 (Set Nash equilibrium). *A profile* $S = (S_1, \ldots, S_n)$ *of set-valued strategies is a* (Bayesian) Set Nash equilibrium *for a game* Γ *if for all* $i \in N$, *every (possibly mixed)* $s_{-i} \in S_{-i}$, *there exists a strategy* $s_i \in S_i$ *so that for all strategies* s_i', $u_i(\mu, (s_{-i}, s_i')) \leq u_i(\mu, s)$.

Our definition of set-Nash equilibrium is stronger than the set-Nash equilibrium definition introduced by Lavi and Nisan [26], who require only that for every pure $s_{-i} \in S_{-i}$ there exists some $s_i \in S_i$ for which $u_i(\mu, s) \geq u_i(\mu, (s_{-i}, s_i'))$ for all possible strategies s_i'. This earlier definition of set-Nash is insufficient to ensure that there is a Nash equilibrium consistent with the set-valued strategy profile S. The problem is that upon restricting the game to S the only Nash equilibrium may be a mixed equilibrium, yet there may be some strategy s_i' outside of S_i that is strictly better than all $s_i \in S_i$ given that players $\neq i$ play a mixed strategy consistent with S_{-i}. On the other hand, our definition of set-Nash is weaker than the CURB (Closed Under Rational Behavior) sets of Basu and Weibull [4]; see also a recent discussion in Benisch et al. [11]. A CURB set requires that for every mixed strategy s_{-i} consistent with S_{-i}, *all* best-responses for player i are consistent with set-valued strategy S_i whereas we require only that there is *some* best-response that is consistent with S_i.

2.3 Trembling-Hand Perfect Set Equilibrium

At the heart of a solution concept for extensive-form games with incomplete information is a requirement about the way in which the players update their *beliefs* about the values of other players' types. The beliefs are distributions from which players think that the types of other players were drawn. At the beginning of the game, the belief corresponds to the initial distribution μ conditioned on the player's knowledge of her own type. As the game progresses, players update their beliefs as a function of other players' actions.

A straightforward approach for a player to update her beliefs is to use Bayes rule to condition on her own view of the actions taken in the game. This is the basic approach taken in the game theory literature, and the one pursued in a previous version of this paper [35]. But such an approach suffers from the drawback that updating is not well-defined for views that occur with zero probability, i.e. following out of equilibrium play.

A stronger approach, also discussed in the game theory literature, is the one of *trembling-hand perfect equilibrium* [40]. The idea behind trembling-hand perfect equilibrium is that updating is not problematic if the strategies under consideration are fully mixed (since such strategies would never incur an updating that conditions on a zero probability event). It thus becomes natural to require that the equilibrium strategy is a best response in every subgame to some sequence of fully mixed strategies that converge to equilibrium, and this is indeed the definition of trembling hand equilibria.

The trembling-hand solution concept builds on the notion of a subgame $\Gamma(s, h)$, which is defined in the natural way to be the restriction of an extensive form game with public actions and private outputs Γ, at a history $h \in H$. This definition implicitly captures the way in which players update their beliefs as a result of past players' actions, assuming previous play according to strategy s.

Given a strategy profile s and a strategy s_i' for player i, we denote by $u_i|_h(\mu|_h, s_{-i}|_h, s_i'|_h)$ the expected value of player i's utility under strategy vector $(s_{-i}|_h, s_i'|_h)$ in the game $\Gamma(s, h)$. This is interpreted as considering player i's expected utility when all players except player i follow strategies s_{-i}, and assuming that until the history h has been reached player i has played according to strategy s_i, and from that point on according to strategy s_i'.

We define a set-Nash analogue of trembling-hand perfect equilibrium. To the best of our knowledge, such a combination has not been previously considered in the literature. Given a history $h \in H$ and a set valued strategy S_i, we define $S_i|_h$ in the natural way (i.e., if $S_i = (M_i, F_i)$ then $S_i|_h = (M_i|_h, F_i|_h)$ where $M_i|_h(\theta_i, h') = M_i(\theta_i, (h, h'))$ and $F_i|_h(\theta_i, h') = F_i(\theta_i, (h, h')))$.

Definition 2.2 (Trembling-hand perfect set equilibrium). *Let Γ be an extensive form game with public actions and private outputs. A profile of set-valued strategies $S = (S_1, \ldots, S_n)$ is said to be a* trembling-hand perfect set equilibrium *for Γ if for every $s \in S_1 \times \cdots \times S_n$ there exists a sequence of fully mixed strategy profiles $(s^k)_{k=0}^\infty$ converging to s so that for every history $h \in H$ and every $i \in N$, there exists a strategy $s_i' \in S_i|_h$ such that for all strategies s_i'' in the game $\Gamma(s^k, h)$ it holds that $u_i|_h(\mu|_h, s_{-i}^k|_h, s_i') \geq u_i|_h(\mu|_h, s_{-i}^k|_h, s_i'')$ in the game $\Gamma(s^k, h)$ for all k.*

3   Secret Sharing

A secret sharing scheme $(N, t, \Delta, \Theta, \mu, g)$ is implemented by letting a trusted *dealer* jointly pick the secret and shares according to the distribution μ, and then distributing share $\theta_i \in \Theta_i$ to player $i \in N$. The reconstruction functions are what enables any set S of at least t players to use their shares $(\theta_i)_{i \in S}$ in order to jointly reconstruct the secret (by using a function $g_S \in g$). The scheme should also guarantee secrecy against any subset S of less than t players.

To prevent players from revealing shares that are different than the ones they were dealt, we will want to work with a secret sharing scheme that is *authenticated*. One can use standard "information theoretic" techniques for authenticating shares in any (plain) secret sharing scheme (cf. [44,39,25]).

3.1   Reconstruction Protocols

Once shares are distributed among the players, it is required to specify a protocol according to which the players can jointly reconstruct the secret at a later stage. The reconstruction protocol prescribes a way in which the players compute their "messages", which are chosen from a given fixed "alphabet," and are then broadcast to all other players. The protocol also specifies an output function that is used by the players to compute their (private) output.

A reconstruction protocol $\Pi = (\Sigma, H, P, m^*, f^*)$ for a given secret sharing scheme is implemented under the assumption that the secret and shares $(\delta, \theta_1, \ldots, \theta_n)$ are chosen according to the distribution μ. Player i's type is θ_i. The protocol is interpreted as follows: player $i = P(h)$ chooses a message $m = m_i^*(\theta_i, h) \in \Sigma$; this choice determines the next player to move, and so on until a terminal history $h \in Z$ is reached. At this point all players can determine the value of their private output functions, $f_i^*(\theta_i, h)$.

3.2   Reconstruction Games

A secret sharing protocol induces a *reconstruction game* in a natural way. Loosely speaking, this is an interpretation of a reconstruction protocol as an extensive form game with public messages and private outputs, in which arbitrary deviations from the protocol's instructions are allowed. The interpretation of the protocol as a game is straightforward: protocol histories correspond to game histories, messages in the protocol correspond to actions, next message functions correspond to strategies, and the outputs correspond to output actions.

The reconstruction game allows player i the choice between continuing with the protocol's prescribed instructions (and in particular choosing an action according to m_i^*), and deviating from Π (by sending some other message from Σ).

We require that the utility functions are *linear* in the sense there are (real) parameters $\{a_{ij}\}$ such that the utility received by player i is equal to the sum of a_{ij} over all players j that correctly compute the secret at the end of the protocol. We define the players' *preference* for learning the secret to be $\rho = \min_i a_{ii}/(-\sum_{j \neq i} a_{ij})$. We require that $a_{ii} > 0$, $a_{ij} < 0$ for $i \neq j$, and $\rho > 1$. These assumptions correspond to the assumptions (also made in previous works) that players prefer to learn the secret over all else, and secondarily prefer that as few other players learn the secret as possible.

4 Our Protocol

4.1 Introducing an Honest Minority

Our goal is to show that every authenticated secret-sharing scheme has a reconstruction protocol so that any reconstruction game that corresponds to it has an equilibrium strategy in which all players learn the secret. To do this, we require that a small subset of *honest* players in the reconstruction game always follows the strategy prescribed by the reconstruction protocol (whether or not this is the best response to other players' actions). We model this scenario by assuming that the set of honest players is selected according to some distribution that specifies to each player whether she is to act honestly or rationally. The set of actions of an honest player coincides with the strategy prescribed by the reconstruction protocol. The set of actions of a rational player remains unchanged.

The private type of player $i \in N$ in a reconstruction game with honest players consists of a pair $(\theta_i, \omega_i) \in \Theta_i \times \Omega_i$ that is drawn along with other player's types and the reference value δ according to the distribution $\mu \times \zeta$. The value of $\omega_i \in \{\text{honest}, \text{rational}\}$ determines whether player i is bound to follow the honest strategy (as prescribed by Π), or will be allowed to deviate from it. We constraint the set of actions of each player in order to create a situation in which rational players are indeed free to deviate from the public strategy vector m^* (since they are allowed to choose any action in Σ), whereas the honest players are in fact restricted to the single action prescribed by m^*.

4.2 Main Result

We show that assuming the existence of a small number of honest players, there is a reconstruction protocol such that every corresponding reconstruction game has an equilibrium such that with high probability all players learn the secret, provided that the set of honest players is uniform among all sets of a sufficiently large size and every player has a nonnegligible preference for learning the secret. Specifically, our theorem is the following:

Theorem 4.1. *Every authenticated secret-sharing scheme* $(N, t, \Delta, \Theta, \mu, g)$, *with* $t < |N| = n$, *has a reconstruction protocol* Π *such that the following holds. Let* $\Gamma = (N, H, P, A, L, \Delta, \Theta \times \Omega, \mu \times \zeta_m, u)$ *be a reconstruction game that corresponds to* Π *with honest players and linear utility functions, where* ζ_m *is a distribution over tuples* $(\omega_1, \ldots, \omega_n) \in \Omega$ *for which* $\omega_i = \mathtt{honest}$ *with probability* m/n *independently for all* $i \in N$, *for some real number* $m \in [0, n]$. *Suppose further that the players' preference* ρ *for learning the secret satisfies:*

$$\rho > \frac{1 - 1/|\Delta|}{1 - 1/|\Delta| - p(n, m) - \gamma} \tag{1}$$

where $p(n, m) = (1 - m/n)^{n-t+1} \leq \exp(-m \cdot (n - t)/n)$ *and* γ *is the forgery probability of the authenticated secret-sharing scheme. Then* Γ *has a profile* $S = (S_1, \ldots, S_n)$ *of set-valued "rational" strategies such that:*

1. *The honest strategy profile* $s^* = (m^*, f^*)$ *is consistent with* S,
2. S *is a trembling-hand perfect set equilibrium in* Γ,
3. *For every strategy vector* $s \in S$, *the probability that all honest players compute the secret correctly in* Γ *is at least* $1 - (n - t + 1)\gamma - p(n, m)$, *when the players' types are chosen according to* μ *and they follow strategy vector* s.
4. *For every Nash equilibrium* $s \in S$, *the probability that all players compute the secret correctly in* Γ *is at least* $1 - (n - t + 1)\gamma - p(n, m)$, *when players' types are chosen according to* μ *and they follow strategy* s.
5. S *does not depend on the utility functions* u *in* Γ *(provided they satisfy (1)).*

To interpret this theorem, consider a setting in which we distribute to each player software that is programmed to play the honest strategy, which is in consistent with S by Item 1. Rational players may then decide to deviate from this strategy (i.e. reprogram their software) in order to improve their utility. The fact that S is a trembling-hand perfect set equilibrium (Item 2), however, guarantees that there is no incentive for the rational players to deviate from S, even if this process is iterated. As long as all players remain within S, Items 3 and 4 say that fairness is maintained (with high probability).

In case that $t \leq (1 - \Omega(1)) \cdot n$, observe that $p(n, m) = \exp(-\Omega(m))$ is negligible provided that $m = \omega(\log n)$, i.e. the expected number of honest players is superlogarithmic. If, in addition, the forgery probability γ is negligible then (1) simply says that a player's preference for learning the secret should not be negligible.

4.3   The Reconstruction Protocol

The protocol proceeds in two stages, where in the first stage a subset of $t - 1$ players is instructed to reveal their share to all other parties in some sequence, and in the second stage the remaining $n - t + 1$ players are instructed to reveal their share in some sequence, provided that none of the $t - 1$ players in the first stage has failed to reveal her share. The individual parties will reveal their share using a *synchronous broadcast* channel.

The stage in which a player is instructed to broadcast is fixed in some arbitrary manner. For concreteness, suppose that at stage 1 of the protocol, it is the turn of players $1, \ldots, t-1$ to broadcast, and that at stage 2 it is the turn of players $t, \ldots, n$. The protocol will instruct player i to either *reveal* her share $\theta_i \in \Theta_i$ or not to reveal anything (symbolized by a special action denoted $\perp \in \Sigma$). Specifically, we will require that player i reveals θ_i unless she is one of the stage 2 players and one of the first $t-1$ parties to speak has chosen not to reveal their share. In the latter case player i does not reveal her share either.

In addition to revealing her share, player i is required to send along the authentication information that was provided to her by the dealer. In case that either the authentication fails, or that the player has refused to broadcast her message,[3] player i will be considered as having failed the authentication and chosen the special $\perp$ action.

After the two stages are completed, each player locally uses a reconstruction function $g_S \in g$ in order to try and compute the secret given the shares that have been revealed during the protocol's execution. By the properties of secret sharing, it follows that a party will be able to compute the secret at the end of the protocol if a set $S \subseteq N$ of at least $t-1$ *other* parties have revealed their shares, and otherwise she has no information about the secret.

The protocol requires that the players in each stage reveal their shares in sequential order. However, the order in which the first $t-1$ players broadcast has no effect on the strategic properties of the protocol, and similarly for the last $n-t+1$ players. Thus, the protocol can effectively be implemented with two rounds of communication (see full version for details).

4.4 Rational Strategies for Corresponding Reconstruction Games

The rational set-valued strategy S instructs both honest and rational players to follow the strategy prescribed by Π, except that it does not specify how rational players should act in cases when the honest strategy may not be in their self interest. Specifically, we allow arbitrary action by a rational player $i \geq t$ when the first $t-1$ players have all revealed valid shares (whereas honest players must reveal in this case). The honest strategy (equivalent to the earlier 'protocol') is itself consistent with the rational set-valued strategy profile. In the full version of the paper, we show that S satisfies all the requirements of Theorem 4.1.

Acknowledgements

We thank Drew Fudenberg, Jonathan Katz, Silvio Micali, Peter Bro Miltersen, Moni Naor, Adam O'Neill and the anonymous reviewers for helpful discussions and comments.

---

[3] In an implementation, a player that fails to broadcast her value within some predetermined amount of time might be considered to have refused to broadcast.

References

1. Abraham, I., Dolev, D., Gonen, R., Halpern, J.Y.: Distributed computing meets game theory: robust mechanisms for rational secret sharing and multiparty computation. In: PODC 2006, pp. 53–62 (2006)
2. Aiyer, A.S., Alvisi, L., Clement, A., Dahlin, M., Martin, J.-P., Porth, C.: Bar fault tolerance for cooperative services. In: SOSP, pp. 45–58 (2005)
3. Babaioff, M., Lavi, R., Pavlov, E.: Mechanism design for single-value domains. In: Proc. Nat. Conf. on Artificial Intelligence, AAAI 2005 (2005)
4. Basu, K., Weibull, J.W.: Strategy subsets closed under rational behavior. Economics Letters 36, 141–146 (1991)
5. Ben-Or, M., Goldwasser, S., Wigderson, A.: Completeness theorems for non-cryptographic fault-tolerant distributed computation (extended abstract). In: STOC 1988, pp. 1–10 (1988)
6. Blakely, G.: Safeguarding cryptographic keys. In: AFIPS, vol. 48, p. 313 (1979)
7. Boneh, D., Naor, M.: Timed commitments. In: Bellare, M. (ed.) CRYPTO 2000. LNCS, vol. 1880, pp. 236–254. Springer, Heidelberg (2000)
8. Chaum, D., Crépeau, C., Damgård, I.: Multiparty unconditionally secure protocols (extended abstract). In: STOC 1988, pp. 11–19 (1988)
9. Chor, B., Goldwasser, S., Micali, S., Awerbuch, B.: Verifiable secret sharing and achieving simultaneity in the presence of faults (extended abstract). In: FOCS, pp. 383–395. IEEE, Los Alamitos (1985)
10. Cleve, R.: Limits on the security of coin flips when half the processors are faulty (extended abstract). In: STOC, pp. 364–369. ACM, New York (1986)
11. Davis, G.B., Sandholm, T.W.: Algorithms for Rationalizability and CURB Sets. In: AAAI 2006 (2006)
12. Even, S., Goldreich, O., Lempel, A.: A randomized protocol for signing contracts. Commun. ACM 28(6), 637–647 (1985)
13. Feigenbaum, J., Papadimitriou, C., Sami, R., Shenker, S.: A BGP-based mechanism for lowest-cost routing. In: PODC, pp. 173–182 (2002)
14. Feigenbaum, J., Papadimitriou, C.H., Shenker, S.: Sharing the cost of multicast transmissions. Journal of Computer and System Sciences 63, 21–41 (2001)
15. Feigenbaum, J., Shenker, S.: Distributed Algorithmic Mechanism Design: Recent Results and Future Directions. In: Proc. 6th Int'l Workshop on Discrete Algorithms and Methods for Mobile Computing and Communications, pp. 1–13 (2002)
16. Fuchsbauer, G., Katz, J., Levieil, E., Naccache, D.: Efficient rational secret sharing in the standard communication model. Cryptology ePrint Archive, Report 2008/488 (2008), http://eprint.iacr.org/
17. Garay, J.A., Jakobsson, M.: Timed release of standard digital signatures. In: Proc. Financial Cryptography 2002, pp. 168–182 (2002)
18. Goldreich, O., Micali, S., Wigderson, A.: How to play any mental game or a completeness theorem for protocols with honest majority. In: STOC, pp. 218–229. ACM, New York (1987)
19. Gordon, S.D., Katz, J.: Rational secret sharing, revisited. In: De Prisco, R., Yung, M. (eds.) SCN 2006. LNCS, vol. 4116, pp. 229–241. Springer, Heidelberg (2006)
20. Halpern, J.Y., Teague, V.: Rational secret sharing and multiparty computation: extended abstract. In: Babai, L. (ed.) STOC, pp. 623–632. ACM, New York (2004)
21. Izmalkov, S., Micali, S., Lepinski, M.: Rational secure computation and ideal mechanism design. In: FOCS, pp. 585–595. IEEE Computer Society, Los Alamitos (2005)
22. Kalai, E.: Large robust games. Econometrica 72(6), 1631–1665 (2004)

23. Katz, J.: Bridging game theory and cryptography: Recent results and future directions. In: Canetti, R. (ed.) TCC 2008. LNCS, vol. 4948, pp. 251–272. Springer, Heidelberg (2008)
24. Kol, G., Naor, M.: Cryptography and game theory: Designing protocols for exchanging information. In: Canetti, R. (ed.) TCC 2008. LNCS, vol. 4948, pp. 320–339. Springer, Heidelberg (2008)
25. Kol, G., Naor, M.: Games for exchanging information. In: STOC, pp. 423–432. ACM, New York (2008)
26. Lavi, R., Nisan, N.: Online ascending auctions for gradually expiring goods. In: SODA 2005 (2005)
27. Lehmann, D., O'Callaghan, L.I., Shoham, Y.: Truth revelation in approximately efficient combinatorial auctions. Journal of the ACM 49(5)
28. Lepinski, M., Micali, S., Peikert, C., Shelat, A.: Completely fair sfe and coalition-safe cheap talk. In: PODC 2004, pp. 1–10 (2004)
29. Lepinski, M., Micali, S., Shelat, A.: Collusion-free protocols. In: Gabow, H.N., Fagin, R. (eds.) STOC, pp. 543–552. ACM, New York (2005)
30. Lysyanskaya, A., Triandopoulos, N.: Rationality and adversarial behavior in multi-party computation. In: Dwork, C. (ed.) CRYPTO 2006. LNCS, vol. 4117, pp. 180–197. Springer, Heidelberg (2006)
31. McGrew, R., Porter, R., Shoham, Y.: Towards a general theory of non-cooperative computation. In: TARK, pp. 59–71 (2003)
32. Nisan, N., Ronen, A.: Algorithmic mechanism design. Games and Economic Behavior 35, 166–196 (2001)
33. O'Neill, A., Sangwan, A.: Honesty, rationality, and malice in secret sharing and MPC: Robust protocols for real-world populations (manuscript, 2008)
34. Ong, S.J., Parkes, D., Rosen, A., Vadhan, S.: Fairness with an honest minority and a rational majority (April 2007),
    http://eecs.harvard.edu/~salil/Fairness-abs.html
35. Ong, S.J., Parkes, D., Rosen, A., Vadhan, S.: Fairness with an honest minority and a rational majority. Cryptology ePrint Archive, Report 2008/097 (March 2008),
    http://eprint.iacr.org/
36. Parkes, D.C., Shneidman, J.: Distributed implementations of Vickrey-Clarke-Groves mechanisms. In: Proc. 3rd AAMAS, pp. 261–268 (2004)
37. Petcu, A., Faltings, B., Parkes, D.: M-dpop: Faithful distributed implementation of efficient social choice problems. In: AAMAS 2006, pp. 1397–1404 (May 2006)
38. Pinkas, B.: Fair secure two-party computation. In: Biham, E. (ed.) EUROCRYPT 2003. LNCS, vol. 2656, pp. 87–105. Springer, Heidelberg (2003)
39. Rabin, T., Ben-Or, M.: Verifiable secret sharing and multiparty protocols with honest majority (extended abstract). In: STOC, pp. 73–85. ACM, New York (1989)
40. Selten, R.: A reexamination of the perfectness concept for equilibrium points in extensive games. International Journal of Game Theory 4, 25–55 (1975)
41. Shamir, A.: How to share a secret. Commun. ACM 22(11), 612–613 (1979)
42. Shneidman, J., Parkes, D.C.: Specification faithfulness in networks with rational nodes. In: PODC 2004, St. John's, Canada (2004)
43. Shoham, Y., Tennenholtz, M.: Non-cooperative computation: Boolean functions with correctness and exclusivity. Theor. Comput. Sci. 343(1-2), 97–113 (2005)
44. Wegman, M.N., Carter, L.: New hash functions and their use in authentication and set equality. J. Comput. Syst. Sci. 22(3), 265–279 (1981)
45. Yao, A.C.-C.: How to generate and exchange secrets (extended abstract). In: FOCS, pp. 162–167. IEEE, Los Alamitos (1986)

Purely Rational Secret Sharing
(Extended Abstract)

Silvio Micali[1] and abhi shelat[2]

[1] MIT CSAIL
silvio@csail.mit.edu
[2] U. Virginia
abhi@virginia.edu

Abstract. Rational secret sharing is a problem at the intersection of cryptography and game theory. In essence, a dealer wishes to engineer a communication game that, when rationally played, guarantees that each of the players learns the dealer's secret. Yet, all solutions proposed so far did not rely solely on the players' rationality, but also on their *beliefs*, and were also *quite inefficient*.

After providing a more complete definition of the problem, we exhibit a very efficient and purely rational solution to it with a verifiable trusted channel.

1 Introduction

In [LMPS04], Lepinski, Micali, Peikert and shelat put forward the notion and the first implementation of *Fair Secure Function Evaluation* . This is a communication protocol extending the traditional notion of secure function evaluation [GMW87]. In essence, a Fair SFE is an SFE in which either (1) all players learn the result of evaluating a given function on their secret inputs (but no other information about their inputs) or (2) none of them learns anything. The first outcome is reached when all players want it, and the second one when at least one of the players wants it. The difficulty lies in the fact that such objectives must be reached no matter what the function may be and no matter how many the player are, provided that at least one of the players is *honest*, that is sticking to his communication instructions in all cases.

In [HT04], Halpern and Teague put forward the notion of *rational secret sharing* (RSS), aiming at distilling separately, and in purely game theoretic terms, the last stage of a Fair SFE (where the players attempt to reconstruct the specified output from their shares of it). We believe this to be a very valuable contribution, but we also believe that the notion of an RSS can be improved.

In this extended abstract we shall solely deal with the two-player version of the notion, arguably the best way to highlight the novel and most poignant aspects of the problem.[1]

---

[1] Our approach easily extends to n players, where the dealer wishes that n out of n of them learn the secret. The k-out-of-n definition of traditional secret sharing is very relevant for robustness, and protects against the potential loss of shares, but is quite distracting and orthogonal to the rationality problem at hand. Indeed, in a "k-out-of-n" rational secret sharing (assuming as usual that the fewer the players knowing the secret, the more value to them), k players will presumably prevent the others from learning their secret. Is this the natural wish of the dealer?

O. Reingold (Ed.): TCC 2009, LNCS 5444, pp. 54–71, 2009.

1.1   Rational Secret Sharing as a Special Form of Mechanism Design

The Intuitive Notion. At the verge of dyeing, a *dealer* possessing a secret string S wishes to ensure that two players will later on cooperate so as to *both* learn S. To this end, he provides each player i with his share of the secret, a string S_i. Each share is individually meaningless (i.e., its distribution is independent of S), while together the two shares reveal S. If the players were both honest, the dealer's goal could be trivially achieved. Unfortunately, honesty is not an available commodity: each player is assumed to be *rational* (i.e., always trying to maximize his own utility), and the utilities that the players attach to the possible ways of learning S are quite problematic. In particular, each player prefers most to be the only one learning S, prefers less learning S together with the other player, and even less not learning S at all. Accordingly, the dealer wishes to chose the shares such that,

> for a suitable communication channel, there exists a communication game that, when rationally played, yields the secret to both players.

It is thus worth to recall quickly the traditional notions of solving a game.

Game Solution Concepts. Given a game G, a solution concept essentially is a way of predicting how G will be played. From the cryptographic perspective of the authors, traditional solution concepts are only partially meaningful, as they are stated from the perspective of *individual* players, disregarding *collusion* altogether. Nonetheless their meaningfulness is intact for the problem at hand restricted to just two players. (This is by itself a good reason to focus on the two-player case.)

The strongest, traditional solution concept is that of solvability in *dominant strategies*. Here, each player has a strategy σ_i that is best of him, no matter what strategies the other players may use. In such a case predicting that each player i will play σ_i is indeed the strongest form of prediction of play. Note that, in choosing σ_i, each player i does not need to rely on the rationality on the other players, but just be rational himself! Unfortunately, not all games admit dominant-strategy solutions.

The "next best" solution concept is *dominant solvability,* which now very roughly explain. In a game G, a strategy a for player A is said to weakly dominate another strategy a' for A if (1) for all possible strategies b of player B, A's utility under a is greater than or equal to his utility under a'; and (2) for at least one strategy b' of B, A's utility under a is strictly greater than his utility under a'. This being the case, a rational A should remove strategy a' and all of his weakly dominated strategies from consideration. And a rational B should do the same on his side. Trusting that B has done so, A should then eliminate from his remaining strategies all those that now become weakly dominated (relative to the strategies left over to B). And so on, until neither player can eliminate any more strategies. At that point, if A is left with a single strategy a and player B with a single strategy b, then G is called *dominant solvable*, and (a, b) is a very strong prediction for the way in which G will be rationally played. Notice, however, that this second solution concept is weaker than the first one, since each player must rely non only on his rationality, but also on that of his opponent. Again, not all games admit such a strong solution concept.

The next solution concept we wish to roughly recall is that a Nash equilibrium. This is a pair of strategies (a, b), such that a is the best strategy for A if he believes that B will play b (and symmetrically for B). The good news is that each game admits such equilibria, but the bad news is this is a *very distant third* among these solution concepts. The meaningfulness of a Nash equilibrium in fact depends not only on the rationality of both players, but also on their *beliefs*. Typically a game has a plurality of equilibria, often having symmetric payoffs, making it very uncertain to predict which of them will be played. Furthermore, the game may easily not end up in equilibrium at all. If A believes that equilibrium (a, b) will be played, while B believes (a', b'), then the strategy profile ultimately played may be (a, b') which needs not to be an equilibrium at all!

Mechanism Design. Very roughly said the goal of mechanism design is to engineer a game so that, "when rationally played", a given property P is guaranteed to hold. The quality of such design therefore crucial depends on the solution concept adopted: it is exceptionally meaningful (recall that we are focusing on the two-player case!) when the game has dominant strategy solution, it is very meaningful when the game is "dominant solvable", and it has only very limited meaningfulness when the game is "Nash solvable." Such limited meaningfulness persists even if P is *guaranteed to hold at each of possible Nash equilibria of the game.* In a sense, if the game has k Nash equilibria, then —due mismatched beliefs— it roughly has k^2 (k^n if there are n players) possible ways not to end in any equilibrium.

Another quality measure in mechanism design is the amount of knowledge about the players (e.g., knowledge about their utilities) required to engineer the game. Indeed, since precise knowledge about the players may not be available or too expensive to gather, the lesser the knowledge required from the designer the better.

Rational Secret Sharing and Mechanism Design. We propose to view *rational secret sharing as a special mechanism-design problem.* That is, one should try to guarantee the property "all players learn the secret" by means of a pure communication game. In essence, the game should be such that (1) all player actions consist of exchanging messages over a special channel, (2) no trusted party is involved, and (3) no exogenous punishments, fines, etc. can be triggered by the final outcome: the players' utilities must solely depend on who has, or has not, learned the secret.

This point of view enables us to extend to RSS the same quality analysis applicable to mechanism design, providing a more meaningful comparison among various RSS protocols.

1.2 Prior Solutions

In their quoted paper, Halpern and Teague present a protocol for the 3-out-of-3 case (and then show how to modify it for the m-out-of-n case, where $3 \leq m < n$). Their protocol guarantees that all players learn the secret at a Nash equilibrium whose strategies survive the iterated elimination of weakly dominated strategies (IEOWDS for short). Rather than the *swap channel* of Lepinski, Micali, Peikert, and shelat, they rely on *simultaneous-broadcast channels*, and prove that no rational secret sharing protocol can be fixed-round with such channels. A main limitation of their protocol is that the trusted dealer continues to be an active participant. (In most settings, such a dealer could directly inform the players of what the secret is.)

Gordon and Katz [GK06] present a protocol for just two players that dismisses the need for the periodic involvement of the dealer. Their protocol too relies on simultaneous-broadcast channels, and guarantees that all players learn the secret at a Nash equilibrium whose strategies survive IEOWDS. Abraham, Dolev, Gonen, and Halpern [ADGH06] present a similar protocol, but focus on defining (and protecting against) coalitions of rational players.

Lysyanskaya and Triandopoulos [LT06], with the same channels and implementation type, consider a model with a mix of rational and malicious players.

Kol and Naor [KN08b] present a quite different protocol with simultaneous-broadcast channels, which guarantees that all players learn the secret at a "strict Nash equilibrium", a locally stronger version of a Nash equilibrium. (In essence, any player deviating from his own equilibrium strategy expects to receive a strictly smaller utility.)

A Separate Protocol. We wish to mention an interesting and recent protocol of Ong, Parkes, Rosen, and Vadhan [OPRV08]. Their protocol however works in a quite different model. On one side, it does not require any special channels (that is, it relies on ordinary broadcast channels rather then simultaneous-broadcast ones). On the other, it relies on the *honesty* of a few players. (As we focus solely on rational players, we shall not include this protocol in any future discussion or comparison.)

1.3   Weaknesses of Prior Solutions

Protocol Inefficiency and Excessive Designer Knowledge. The prior protocols share the following logical structure. The players interact in several rounds, using some special channels. The protocol has a special round R, unknown to the players because it is secretly selected by the dealer according to a given distribution. If no player "cheats" then all players learn the secret. A player can successfully cheat only if he correctly guesses R. If a player i erroneously guesses R, then no one learns the secret (which gives i utility u_i). But if i guesses R correctly (and acts appropriately), then he is the only one to learn the secret (which gives him utility U_i).

In essence, therefore, to hope that it is rational to stick to the protocol's prescribed strategies without cheating, letting p be the probability of successfully guessing R, p needs to be so small that $p \cdot u_i \leq (1-p)U_i$. This shows two separate weaknesses of these protocols. First, because properly engineering the game implies properly selecting p, the designer needs to know the u_i's and the U_i's quite accurately. (Thus, from a mechanism design perspective, this diminishes the quality of these approaches.) Second, because the expected number of rounds must be greater or equal to p, this implies that all prior protocols run in exponential time. In fact, independent of the distribution according to which R is selected, the expected number of rounds of the prior protocols must be exponential in k, assuming as it is natural that all players utilities are presented in binary, and that their length is k. This inefficiency alone calls for new protocols.

Limited Guarantee for the Desired Property. Prior solutions ensure that the property "all players learn the secret" holds at a given Nash equilibrium of the engineered communication game. Again, however, this assurance is far from guaranteeing our property for two separate reasons: *equilibrium selection* and *equilibrium absence*. Let us discuss the first reason first. Even if one were certain that the engineered game will end up in an equilibrium, he could not be certain of which equilibrium would be actually selected. And since, in the engineered games of the previous works, the "all-know-the-secret" property was guaranteed only at one of the possible Nash equilibria, the equilibrium ultimately selected could very well be one in which not all players learn the dealer's secret. Let us now discuss the second reason. The meaningfulness of any Nash equilibrium is inextricably linked to the *assumption* that the players' beliefs are *"consistent"*, which of course needs not be the case. Thus, *even if* all players learned the secret at each Nash equilibrium of the engineered games, there is no guarantee at all that the engineered game ends up in equilibrium. Again: assume that (a, b) and (a', b') are Nash equilibria, that A believes that B will play b, and that B believes that A will play a'. Then, A will rationally (based on her belief!) play strategy a, and B will play b'. And since (a, b') may not be an equilibrium, let alone an outcome in which all players learn the secret.

To be sure, the prior protocols were engineered so that all players learned the dealer's secret not just at a generic Nash equilibrium, but at one whose strategies survived IEOWDS. But as long as multiple Nash survive IEOWDS (which is the case in prior protocols), then equilibrium selection and equilibrium absence will continue to poison the landscape.

To be sure too, some of the prior RSS protocols guaranteed that all players learned the secret at an even stronger type of equilibrium, such as the strict Nash of [KN08b]. But these equilibria are in a sense only "locally stronger." That is, if the players believe that a strict Nash equilibrium (a, b) will be played out, they would have "even less incentives" of deviating from it. But ensuring that A does not deviate from a if she believes that B chooses strategy b is not too meaningful, unless one can also ensure that B actually chooses b. If the game is engineered so that the best we can say about it is that it has a strict Nash equilibrium (at which the desired property holds) alongside with other additional equilibria, then equilibrium selection and equilibrium absence will continue to stand in the way.

In sum, *all prior RSS protocols did not solely depend on the players' rationality, but also on their beliefs.* Thus they could not guarantee that all players, if rational, learned the secret.

1.4   Our Contributions

Our contributions can be summarized as follows.

- *Modeling.* We put forward a more complete modeling of the RSS problem.
  In particular: we highlight the inputs available to the designer of protocol; provide a more comprehensive set of utilities —including the possibility of learning the wrong secret—; highlight the necessity of modeling RSS as a potentially infinite communication game; provide a very general definition of a communication channel; highlight the necessity of worrying about other channels even in a communication game designed for a specific channel; provide the first rationalization of aborting in a communication game; and bring to the fore the necessity of including bargaining into the definition of RSS.

- *Purely Rational Implementation.* Our RSS protocol is an *implementation in surviving strategies*, as put forward by Chen and Micali [CM08]. In essence, such an implementation is "equilibrium-less." It guarantees that the desired property holds for any combination of strategies surviving IEOWDS. Implementation in surviving strategies thus implies that *the desired property is guaranteed based solely on the rationality of the players, and not on their beliefs.* In a sense, as long no player chooses a dumb strategy, the desired property is guaranteed to hold.

  Actually, our protocol satisfies a stronger notion of implementation: namely, the surviving strategy of each player is essentially *unique*.[2] That is, in our RSS protocol, after iteratively deleting all weakly dominated strategies, essentially a single strategy survives for each player, and playing these two strategies guarantees that both players learn the secret. That is, *our RSS protocol essentially is a dominant solvable game.*

  Note that IEOWDS often eliminates very few strategies (a fact that has been used to argue that Nash equilibria that survive IEOWDS is a solution concept not really better than an ordinary Nash). Thus it is even more remarkable that our protocol is such that, for any player, all but one strategy is "rationally credible."

  Note too that, in general, which strategy survives depends on the order in which weakly dominated strategies are eliminated. In our case, however, the (essentially) unique surviving strategy of a player is the same irrespective of any possible elimination order. In sum, our solution concept is indeed very strong.

- *Communication Channel and Security.* Our communication channel uses only ordinary envelopes (as a way of temporarily and perfectly hiding a secret value) and the dealer's public key.

  The security depends on the ability of envelopes to perfectly hide their content and unforgeable digital signatures.

- *Operational Efficiency.* Ours is the first polynomial-time RSS protocol, fully accounting for all inputs. In fact, each surviving strategy requires a total of $10k$ envelope operations, $4kL$ bit operations, plus the time of verifying two signatures relative to k-bit public keys. Here L is an upperbound to the length of the binary representation (of the absolute value) of any of the players' utilities, and k is a security parameter. The security parameter k controls the probability that something goes wrong. (The probability of something going wrong is guaranteed to be exponentially small in k.)

  The dealer is required to perform a total of $4kL$ bit operations, to generating matching public and secret keys of a digital signature with security parameter k, and to produce two signatures relative to the selected public key.

---

[2] The reason that we do not say unique is that, as we shall argue, a pure communication game G should be modeled as a possibly infinite sequence of the same sub-game g. Thus, a strategy of any player in G actually consists of a sequence of strategies, $\sigma_1, \sigma_2, \ldots$, where σ_j is the player's intended strategy for jth copy of g, if reached. By saying that each player has an essentially unique surviving strategy in G we mean that any of his surviving strategies is of the form $s, \sigma_2, \ldots$, where s is fixed; that is first sub-strategy is the same for any surviving strategy of the player. And when all players play their first such strategies, G terminates.

- *Round Efficiency.* A play of our surviving strategies involves only 6 rounds (1 for the players, and 5 for the channel).

2 Selected Modeling Issues

Dealer Secret, Player Outputs, Player Utilities, and Designer Knowledge. For concreteness, we model the secret as a uniformly selected string of n bits. (Our protocol of course works for all kinds of other distributions as well.)

We assume that, upon termination, each player outputs either an n-bit string (interpretable as the player's guess for the dealer's secret) or the special symbol "?" (interpretable as the player's having no information about the secret). The protocol terminates when a prespecified stage is publicly reached, or when either one of the players *aborts,* that is stops communicating and for ever takes no further action —after setting his own output.[3]

We define an outcome of an RSS protocol to consist of three possible outputs for each player: (1) the correct secret the dealer, (2) the symbol "?", and (3) an incorrect string. We assume that each player prefers his outputs in this order, and prefers the inverse order for the outputs of the other player. That is, for each player i, denoting by K_i (for "i knows the secret") his first output, by W_i (for "i wrongly learns the secret) his third output, and by u_i his utility function, we assume that the utilities of the first player over the possible 9 outcomes are as follows:

$$u_1(K_1, W_2) \geq u_1(K_1, ?) \geq u_1(K_1, K_2) \geq$$
$$u_1(?, W_2), \geq u_1(?, ?) = 0 \geq u_1(?, K_2) \geq$$
$$u_1(W_1, W_2), \geq u_1(W_1, ?) \geq u_1(W_1, K_2).$$

Player 2's utilities are symmetrically defined. (Setting the players' utilities to 0 when both of them have no information about the secret is somewhat arbitrary, but concretely useful to fix our thoughts.) All of the above inequalities can be strict. But for our analysis it suffices that $u_1(K_1, ?) > u_1(K_1, K_2) > u_1(?, ?) > u_1(W_1, W_2)$, and symmetrically for player 2. That is, each player prefers learning the secret alone to learning together with the other player, prefers the latter outcome to not learning the secret, and prefers the latter outcome to learning the wrong secret.[4] It is also useful to assume that a player's expected utility when randomly guessing the secret is negative. (Alternatively, we must ensure that the utility of random guessing the dealer's secret is less than that of learning the secret together with the other player. Else, a player would not have any incentive to participate in an RSS protocol.)

---

[3] That is, we explicitly assume that one players' aborting is detectable by the other player. (After all, stopping all communications should be "eventually detectable" in practical settings, and immediately detectable in synchronous ones.) Alternatively, each player may keep track of his current output at all times (rather than producing his output at termination). This way if a player aborts without the "courtesy" of informing the other player, the latter' output is properly set.

[4] Indeed, if the secret were the combination of a safe with money and a bomb inside, and the safe exploded when the wrong combination were entered, learning the wrong secret could have truly negative utility for a player!

This structure of the utility is assumed to be known to the designer. And so is an upperbound to the number of bits necessary to write down the largest of the 16 possibilities of the players. (In other words, it suffices for the designer to know the players' utilities within an exponential accuracy, rather than the linear accuracy of the prior works.)

Ensuring the Rationality of Abort. Our protocol, if a special point in which a player i has not yet learned the secret is reached, calls for him to abort. By so doing, of course, the player looses any hope of learning the secret. Thus, in order to guarantee that the suggested strategy survives IEOWDS, we need to ensure that, at that point of our protocol, the player no longer has any rational hope of learning the secret (whether alone or together with the other player). What should this mean? In particular, of course, it should mean that i's expected utility when continuing the current execution of the protocol is worse than that of aborting outputting "?". But it *should not* mean just that. The dealer who has provided the players with their shares is now dead, and can no longer control what the players do from his grave. RSS is a pure communication game, the players have all the information they need to continue any given execution of our protocol (if they so want) and no authority is there to stop them from (or fine them for) doing so. In addition, the players also have the ability of starting another execution from scratch. (For instance, they may use their same shares, but different coin tosses for their strategies, if probabilistic. Alternatively, if reusing the old shares is not "rationally advisable," they may first resort to a secure function evaluation to "compute new, equivalent and, independently selected shares from their old ones, and then execute our protocol again. The possible alternatives abound.) Better yet, perhaps, they also have the ability to start a totally different RSS protocol using the same communication channel. More generally yet (unless one were ready to make the outlandish assumption that no other channel exists), they have the ability to execute a totally different RSS protocol with a totally different channel! In sum,

> To rationalize player i's aborting in an RSS protocol, we should **prove** that any chance of i's learning the secret has vanished.

Realizing, formalizing, and delivering this property is a main contribution of our work.

Modeling Special Channels. All RSS protocols with rational players must use some special communication channel, such as a swap channel or a simultaneous-broadcast channel. Since we have just argued that a proper analysis of RSS should include the possibility of running a different protocol over a different channel, it becomes imperative to model any possible special channel of communication. We do so by letting special channels consist of "mildly trusted parties in abundant supply." Let us explain. If some party T could be totally and universally trusted, then many problems (including rational secret sharing) would be trivialized. For instance, the dealer might as well confide his secret to T and ask him to reveal it when all the designated players show up together. Thus "mild trust" became imperative. As for abundant supply we mean that there is not a unique mildly trusted party in the world. (If this were the case, one might ask T to interact only once with a given group of players for a given task, and simplify a lot of things too.) By contrast, to model the fact that a special communication channel (if it exists at all) is indeed a commodity purchasable at any store, we envisage that

there is a plurality of mildly trusted parties, not aware of —or not in contact with—
each other.

Accordingly, following [ILM08], we model a mildly trusted party in abundant supply
as a *verifiable trusted party* (VTP for short) with no memory. By verifiable we mean
that every one can see the actions a VTP takes and verify that they are the prescribed
ones. That is, a VTP is not trusted to keep, nor to correctly make any secret actions. A
VTP knows nothing and acts publicly, so that he is trusted only to the extent that he will
indeed publicly perform his prescribed public actions.

For example, a VTP can trivially implement a swap channel between two parties as
follows. First each of A and B seals his message for the other into an opaque envelope
and publicly gives it to the VTP. Then the VTP publicly hands A's envelope to B and
B's envelope to A.

As for another example, a VTP can implement a simultaneous-broadcast channel as
follows. First, A and B seal their respective messages for the other in two envelopes
and publicly hand them to T, then T publicly opens both of them.

In sum, VTPs can be viewed as a formalization of a legal system. One may not
want to trust his secrets to —say— a judge, but should at least trust a judge to carry
out under public scrutiny a specified sequence of totally public actions. Since typically
there are multiple judges to choose from, the analogy with the legal system makes it
clear that the players can always walk to an new judge to execute their protocol one
more time. The analogy also makes it clear that if one type of channel is available, then
indeed other types are likely to be available too. Whether or not, as *functions*, the "swap
channel" is reducible to the "simultaneous-broadcast-channel" (or viceversa), from the
VTP perspective, both exist. (Indeed any judge can, with envelopes, implement both
channels and a host of similar ones.) This highlights the point that when a player is
asked to abort, then it really must be the case that no hope to resurrect the secret exists
for him, no matter what other protocol and channel might be considered.

Adding Costs to the Model. Consider a cryptographic rational secret sharing protocol
in which the dealer also announces an encryption E of the secret S. Then, a player, in
addition to any other strategy, also has available a computational-attack one: namely,
abort and try to decrypt E. A computational-attack strategy is also possible in our protocol, but in a more complex way. Indeed, successfully forging a given value enables a
player to learn the secret alone, and force the other to learn a false secret. Thus we too
need to argue that computational-attack strategies are not rational. One way to do so
is to define a computationally bounded version of rational secret sharing. A preferable
way is to *attach cost to computation* so as make it preferable for a player to play honestly our protocol rather than try to attack the signature scheme and then, if the attack
is successful, getting an advantage in the protocol. Details will be provided in the final
version. (In any case, as argued by Halpern and Pass [HP08] considering computational
costs may be meaningful even for more traditional —i.e., non-cryptographic— game
theoretic settings.)

We also associate a small additive cost of γ to each use of the channel. (E.g., every
one has the right to access the legal system, but incurs a fixed cost in doing so.

We note that additive (or multiplicative) discounts of the players final utilities are quite standard in game theoretical models in which the players could go on interacting (possible even for ever), typically by executing a given sub-game.[5]

The Issue of Bargaining. Finally, let us bring to the reader attention a point totally neglected so far. Traditionally, to guarantee the dealer's wish that all players learn the secret (at least when everyone behaves rationally), the only restrictions envisaged for the utilities are *local to each player* (e.g., each player must prefer reconstructing the secret alone to reconstructing it together with the other player, etc.). That is, the utilities of an individual player must be "compatible with each other," but not with those of other players. We wish to point out, however, that it is necessary to consider *inter-player* restrictions on the players' utilities, or be ready live with the consequences relative to the dealer's wishes. Let us explain.

A dealer providing players with shares of his secret S automatically enables them to *bargain*. In a bargaining situation, one player may get a better deal than others *without any failure of rationality*. For instance, in an RSS context, Player 1 may simply insist that unless everyone plays a protocol in which he learns the secret *alone* 99 times out of 100, he is not going to cooperate. (In a sense, if to Bill Gates learning the secret together with you and me is worth $1K, but learning it alone is worth $1B, then he would be wasting time and opportunity costs in participating with you and me in a "fair" reconstruction of the secret. Therefore, he may successfully bargain for a higher probability of learning the secret.) Now, if the dealer indeed has come up with shares and channels enabling the players to rationally reconstruct the secret together using a given special communication channel, then we should also expect that —whether with the same or with a different channel— the players can use their same shares to skew the payoffs so as to suit their bargaining needs. Truly unbelievable assumptions must be made to prevent the shares to be used in this alternative manner (especially in light of the result of [ILM08], that essentially enables the players to do rationally almost anything, although not too conveniently). Thus, either one must make the additional assumption that the players utilities are such that their bargaining game has a unique solution (e.g., some form of symmetry), or the dealer must be ready to die in peace with the comfort that either all players (if rational) will learn the secret, or that he has put all of them on a *technically* equal bargaining position.

The reader is free to pick the assumption he prefers. But always guaranteeing that all players together learn the secret may not be possible. For the rest of this extended abstract let us assume that the utilities are such that there is a unique bargaining solution.

3   Our Enriched Solution

It is simpler to explain our protocol assuming first that also special, *dealer-sealed*, envelopes are available: anyone can verify that such an envelope has been sealed by the dealer, and thus that its content is what the dealer wanted it to be, because any attempt to break the dealer's seal is guaranteed to be detectable by anyone.

---

[5] For instance, if a given contract is executed after i days of negotiation it is worth less to the players than executing the same contract as $i - 1$ days of negotiation.

Notice that, if such special envelopes were available, then a trivial solution to the RSS problem exists. In essence, letting s be the secret, the Dealer creates two random strings s_A, s_B such that $s = s_A \oplus s_B$, and then provides player A (respectively B) with infinitely many pink (respectively, blue) dealer-sealed envelopes, each containing s_A (respectively s_B). Players A and B then interact as follows. First, each player, simultaneously with the other, gives the VTP one of his dealer-sealed envelopes. Then, if the VTP receives both a pink and a blue dealer-sealed envelope, he publicly opens both of them. Else (e.g., one of the envelopes is ordinary, or has a broken seal), he destroys all envelopes received. In either case, the players incur a positive cost for this interaction.

The above indeed is an RSS protocol working in dominant-strategies. The fact that s becomes public is not a problem: the dealer could just give both players the same string r and choose s_A and s_B such that their bit-by-bit exclusive-or is $s \oplus r$. The problem is that we see no way of keeping its analysis by simulating its dealer-sealed envelopes with ordinary ones and digital signatures. We thus now describe a more complex protocol for which we can "simulate" dealer-sealed envelops as follows. Rather than handing to a player infinitely many dealer-sealed envelopes with content c, the dealer gives him a single digital signature of c, which then the player can —copy and— put into an ordinary envelope and give to the VTP as many times as necessary. (In the final version we shall prove that this simulation keeps our analysis essentially intact.)

In order to guarantee implementation in surviving strategies, our protocol critically introduces an asymmetry in the way the players are treated.

3.1 Dealer's Instructions

On input an ℓ-bit secret s and a security parameter k', do:
1. Choose a random string $\sigma \in \{0,1\}^\ell$ and compute $s' \leftarrow s \oplus \sigma$.
2. Choose a value k such that for all i
   (a) $u_i(K_1, K_2) > (2^{-k/2}) u_i(K_1, ?) + (1 - 2^{-k/2}) u_i(?, ?)$
3. For $i = 1, 2, \ldots, k$, repeat the following
   (a) Randomly select a four-tuple (a_0, a_1, b_0, b_1) such that a_0, b_0 are a random $\oplus$-sharing of the secret s' and a_1, b_1 are random and independent values of the same length as s.
   (b) Pick two random bits $e_1, e_2 \leftarrow \{0,1\}$.
   (c) Player 1's share is (a_{e_1}, a_{1-e_1}) and Player 2's share is (b_{e_2}, b_{1-e_2}).
   (d) Player 1's check value is $C_{1,i} = (e_2, b_1)$ and player 2's check value is $C_{2,i} = (e_1, a_1)$.
   (e) Place value a_j into envelope $E_{1,i,j}$ and place value b_j into envelope $E_{2,i,j}$ for $j \in \{0,1\}$.
4. Let C be the $k(\ell + 1)$-bit number corresponding to the check values $C_{2,1}, \ldots, C_{2,k}$. Choose random values $\alpha, \beta \in \mathbb{Z}_k$ and compute the message authentication code $\gamma = \alpha \cdot C + \beta$.
5. Place into an envelope $E_{1,0}$ the values $(C_{1,1}, \ldots, C_{1,k}, \alpha, \beta)$ and into an envelope $E_{2,0}$ the values $(C_{2,1}, \ldots, C_{2,k}, \gamma)$. Seal the envelope $E_{1,0}$.
6. Place into an envelope $E_{p,\sigma}$ the value σ for $p \in \{0,1\}$.
7. Send the player 1 the envelopes $E_{1,0}, E_{1,\sigma}$ and $E_{1,i,j}$ for $i \in [1, k]$ and $j \in \{0,1\}$. Send to player 2 the envelopes $E_{1,0}, E_{1,\sigma}$ and $E_{1,i,j}$ for $i \in [1, k]$ and $j \in \{0,1\}$.

3.2 Reconstruction Instructions

Recall that a player's strategy consists of a Turing machine that on input a history h outputs either a special symbol $\bot$ to indicate abort, an output string s, or a sequence of $2k + 1$ strings to place into envelopes that are submitted to the VTP. We use the symbol ε to denote the initial history consisting of only the envelopes received from the dealer.

Player p instructions $T(h)$:

1. If $h = \varepsilon$, then submit envelopes $E_{p,0}$ and $E_{p,i,j}$ for $i \in [1, k]$ to the VTP. If the VTP destroy the envelopes, output $\bot$ and stop. Else, after the VTP completes all of its steps, reconstruct n candidates of s by xor'ing the non-check values that have been opened. Let s' be the majority candidate. If no majority exists, then output $\bot$. Otherwise, privately open envelope E_σ and output $s' \oplus \sigma$.
2. For all other histories, output $\bot$ (i.e. do not invoke the VTP).

VTP Instructions:

1. Publicly verify envelope $E_{1,0}$. If the envelope's seal does not verify, then destroy all envelopes. Otherwise, publicly open the envelope to reveal the values $(C_{1,1}, \ldots, C_{1,k})$ and α, β.
2. Publicly open envelope $E_{2,0}$ to reveal values $C = (C_{2,1}, \ldots, C_{2,k})$ and γ. If $\gamma \neq \alpha \cdot C + \beta$, then destroy all envelopes.
3. Open the check envelopes (left or right) of player two indicated by $C_{1,i}, \ldots, C_{1,k}$. If there exists an opened envelope $E_{2,i,j}$ that does not match its stated value in $C_{1,i}$, the check fails: destroy all envelopes.
4. Open the check envelopes (left or right) of player one indicated by $C_{2,i}, \ldots, C_{2,k}$. If there exists an opened envelope $E_{1,i,j}$ that does not match its stated value in $C_{2,i}$, the check fails: destroy all envelopes.
5. If all k checks succeed, open the remaining $2k$ envelopes (corresponding to shares of the secret s').

3.3 Analysis

Theorem 1. *The strategy profile (T, T) for players 1 and 2 constitute a profile that uniquely survives the iterated deletion of weakly dominated strategies in the given VTP model.*

The main idea of the proof. Unless the first envelope submitted by the first player is sealed correctly, the VTP destroys envelopes. Once the one-and-only sealed envelope $E_{1,0}$ is opened, the second player knows which of her share values are check values, and which are values that are used in the sharing of s'. If the VTP succeeds in the same use that $E_{1,0}$ is opened, then both players learn the secret. If it does not, then some check envelope has failed and therefore no share value has yet been opened. In subsequent uses of the VTP, the second player can then modify all of her share values by XORing a random string r to them. This action is undetectable by the first player. Moreover, this action is the weakly dominant response for player 2 since player 2 prefers to learn the secret alone. Therefore, the first player has no hope to recover the secret (since any future opened share values will be independent of the real secret s'. Thus, the first player will abort in every subsequent use of the VTP. As a result, it is best for the second player to submit the envelopes received from the dealer on the first use (since either her envelopes are never opened, or they are opened in the first and only rational opportunity there will be to recover the secret). In this case, the first player should follow T since each use of the VTP incurs a small cost. Then finally, the second player should also play T.

Definition 1. A revealing history h is a history in which the envelope $E_{1,0}$ has been opened and verified in some use of the VTP, but in every use of the VTP, all envelopes have been destroyed.

Let X_1 be the set of all player-one strategies, and X_2 be the set of all player-two strategies. Notice that for all $\sigma \in X_1$, $u_2(\sigma, T) \geq u_2(?, ?)$ and for all $\tau \in X_2$, $u_1(T, \tau) \geq u_1(?, ?)$. Therefore in the first step of removal, all guessing strategies that have expected utility less than $u_i(?, ?)$ can be removed.

For any player-two strategy τ, define $\Gamma(\tau)$ as the following strategy:

1. For the first use of the VTP, follow $\tau(\varepsilon)$. If the first use of the VTP results in all envelopes being opened, output the same as strategy τ.
2. If the first use of the VTP does not result in all envelopes being opened, for the subsequent uses of the VTP, follow strategy τ with the following exception: for any revealing history h, compute which of player 2's envelopes are non-check envelopes, choose a random value r and XOR r to each of these non-check values. Use these new non-check envelope values in place of the original non-check values received from the dealer to compute $\tau(h)$ for all subsequent histories h. If in this use or any subsequent use of the VTP, all envelopes are opened, compute the output O as per τ using the original non-check envelope values.

Claim. The player-two strategy $\Gamma(\tau)$ weakly dominates τ whenever $\tau \neq \Gamma(\tau)$.

For any player-one strategy σ, the player-two strategies τ and $\Gamma(\tau)$ are the same for the first use of the VTP, and thus result in similar utilities in any execution that succeeds.

For any revealing history, $\Gamma(\tau)$ never does worse than τ since $\Gamma(\tau)$ is both perfectly indistinguishable from τ to player one, and the share values produced by $\Gamma(\tau)$ do not have any information about the secret s'. Since $\Gamma(\tau) \neq \tau$, then there is some σ and some execution for which $\Gamma(\tau)$ will be strictly better than τ.

Set $X_2^1 = \{\Gamma(\tau)\}_{\tau \in X_2}$. For any player-one strategy σ, let $\Pi(\sigma)$ be the strategy that does the following: If the input history h is not revealing, then follow $\sigma(h)$. If input history h is revealing, then (a) never use the VTP in any subsequent round and (b) if $\sigma(h)$ outputs a string s, then output s and otherwise output $\perp$.

Claim. The player-one strategy $\Pi(\sigma)$ weakly dominates σ whenever (1) $\sigma(\varepsilon)$ submits the sealed envelope $E_{1,0}$, and (2) there exists $\tau' \in X_2^1$ such that (σ, τ') produces a revealing history h with positive probability and $\sigma(h)$ does not instruct to abort.

Consider any profile (σ, τ) were $\tau \in X_2^1$. The strategies σ and $\Pi(\sigma)$ are equivalent on the first use of the VTP and therefore result in the same history h. If h is successful, then both σ and $\Pi(\sigma)$ result in reconstructing the secret. Similarly, if h is not successful and also not revealing, then the two strategies are equivalent. If h is revealing, but $\sigma(h)$

produces an output, then both are equivalent. Finally, if h is a revealing history and $\sigma(h)$ uses the VTP again, then $\Pi(\sigma)$ is strictly better. This follows because τ survives the first step of removal, and therefore τ produces envelopes for the second (and future) uses of the VTP that are independent of the secret s'. This upper-bounds player 1's utility $u_1(\sigma, \tau)$ by $-\epsilon + u_1(?, \cdot)$. However, $u_1(\Pi(\sigma), \tau') = u_1(?, \cdot)$ which is strictly greater. (Similar analysis for the case when σ outputs s instead of $\bot$.)

The second condition of the claim ensures this situation occurs for some τ, and therefore therefore $\Pi(\sigma)$ weakly dominates σ.

Set X_1^1 to be the set of player-one stratgies in which after the sealed envelope is submitted, the VTP is never used again. Let $\Theta^i(\tau)$ be the player-two strategy that plays $T(\varepsilon)$ in the first i uses of the VTP, and follows τ for all subsequent uses.

Claim. If $\tau \neq \Theta^1(\tau)$, then $\Theta^1(\tau)$ weakly dominates τ.

Consider any player-one strategy $\sigma \in X_1^1$.

For those executions of σ in which player 1 submits an unsealed envelope in the first use of the VTP, all envelopes are immediately destroyed and therefore it holds that $u_2(\sigma, \Theta(\tau)) = u_2(\sigma, \tau)$ since both strategies are equivalent for all second and subsequent uses of the VTP.

We now consider those executions of σ in which the sealed $E_{1,0}$ is submitted. (This can only happen once.) Let $p_{\sigma,\tau}$ be the probability that under profile (σ, τ), the first use of the VTP results in destroyed envelopes. Observe that $p_{\sigma,\tau} \geq p_{\sigma,\Theta(\tau)}$ for all σ. Since $\sigma \in X_1^1$, the VTP is never used again by σ, and therefore $u_2(\sigma, \tau) = p_{\sigma,\tau} u_2(\cdot, K_2)$ which is less than or equal to $p_{\sigma,\Theta(\tau)} u_2(\cdot, K_2) = u_2(\sigma, \Theta(\tau))$. The condition that $\Theta(\tau) \neq \tau$ implies that the inequality is strict for some player one strategy σ which establishes the claim. Induction can be used to show that the claim holds for all i.

Set $X_2^2 = \{\Theta(\tau)\}_{\tau \in X_1^1}$.

Claim. The player-one strategy T weakly dominates every surviving strategy σ.

Observe that $u_1(T, \tau) = u_1(K_1, K_2)$ for any $\tau \in X_2^2$. Any other player one strategy has a positive probability of causing the VTP to destroy all envelopes, and therefore incurring a cost of $-\epsilon$.

A similar argument with Π can be applied to every player-two strategy. Thus, in any use of the VTP that reveals the dealer-received envelope E_2, the player-two strategy no longer uses the VTP. This implies that the player-two strategy T weakly dominates every surviving strategy.

Acknowledgements

Many thanks to Sergei Izmalkov for his characteristically generous and insightful help.

References

[ADGH06] Abraham, I., Dolev, D., Gonen, R., Halpern, J.: Distributed Computing Meets Game Theory: Robust Mechanisms for Rational Secret Sharing and Multiparty Computation. In: PODC 2006 (2006)

[BP98]   Ben-Porath, E.: Correlation without mediation: Expanding the set of equilibria out-comes by "cheap" pre-play procedures. J. of Economic Theory 80, 108–122 (1998)

[CM08]   Chen, J., Micali, S.: Resilient Mechanisms For Truly Combinatorial Auctions. MIT-CSAIL-TR-2008-067 (November 2008)

[GK06]   Gordon, S.D., Katz, J.: Rational Secret Sharing, Revisited. In: De Prisco, R., Yung, M. (eds.) SCN 2006. LNCS, vol. 4116, pp. 229–241. Springer, Heidelberg (2006)

[GMW87]   Goldreich, O., Micali, S., Wigderson, A.: How to play any mental game. In: STOC 1987 (1987)

[HT04]   Halpern, J., Teague, V.: Rational secret sharing and multiparty computation. In: STOC 2004 (2004)

[HP08]   Halpern, J., Pass, R.: Game Theory with Costly Computation (manuscript, 2008)

[IML05]   Izmalkov, S., Micali, S., Lepinski, M.: Rational Secure Computation and Ideal Mechanism Design. In: FOCS 2005 (2005)

[ILM08]   Izmalkov, S., Lepinski, M., Micali, S.: Verifiably secure devices. In: Canetti, R. (ed.) TCC 2008. LNCS, vol. 4948, pp. 273–301. Springer, Heidelberg (2008)

[KN08a]   Kol, G., Naor, M.: Cryptography and Game Theory: Designing Protocols for Ex-changing Information. In: Canetti, R. (ed.) TCC 2008. LNCS, vol. 4948, pp. 320–339. Springer, Heidelberg (2008)

[KN08b]   Kol, G., Naor, M.: Games for Exchanging Information. In: STOC 2008 (2008)

[LMPS04]   Lepinski, M., Micali, S., Peikert, C., Shelat, A.: Completely Fair SFE and Coalition-Safe Cheap Talk. In: PODC 2004 (2004)

[LT06]   Lysyanskaya, A., Triandopoulos, N.: Rationality and adversarial behavior in multi-party computation. In: Dwork, C. (ed.) CRYPTO 2006. LNCS, vol. 4117, pp. 180–197. Springer, Heidelberg (2006)

[OPRV08]   Ong, S.J., Parkes, D., Rosen, A., Vadhan, S.: Fairness with an Honest Minority and a Rational Majority. On Eprint, 2008/097 (2008)

[OR]   Osborne, M.J., Rubinstein, A.: A Course in Game Theory. MIT Press, Cambridge (1994)

A   The Ballot-Box Model

Ballot-box mechanisms are extensive-form, imperfect-information mechanisms with Nature. Accordingly, to specify them we must specify who acts when, the actions and the information available to the players, when the play terminates, and how the outcome is determined upon termination.

A ballot-box mechanism ultimately is a mathematical abstraction, but possesses a quite natural physical interpretation. The physical setting is that of a group of players, seated around a table, acting on a set of *ballots*. Within this physical setting, one has considerable latitude in choosing reasonable actions available to the players. In this paper, we make a specific choice, sufficient for our present goals.

A.1   Intuition

Ballots. Externally, all ballots of the same kind are identical. (Unlike [ILM08], we do not need super-envelopes here.) An envelope may contain a symbol from a finite alphabet. An envelope perfectly hides and guarantees the integrity of the symbol it contains until it is opened. Initially, all ballots are empty and in sufficient supply.

Ballot-Box Operations. We only need 3 classes of ballot-box operations. Each opera-tion except for the first type is referred to as a *public action*, because it is performed in plain view, so that all players know exactly which action has been performed. These classes are: (1) writing a symbol on a piece of paper and sealing it into a new, empty envelope; (2) publicly opening an envelope to reveal its content to all players; (3) publicly destroying a ballot; and (4) do nothing.

Public Information. Conceptually, the players observe which actions have been performed on which ballots. Formally, (1) we associate to each ballot a unique identifier, a positive integer that is common information to all players (these identifiers correspond to the order in which the ballots are placed on the table for the first time or returned to the table —e.g., after being ballot-boxed); and (2) we have each action generate, when executed, a public string of the form "A, j, k, l, ..."; where A is a string identifying the action and $j, k, l, ...$ are the identifiers of the ballots involved. The *public record* is the concatenation of the public strings generated by all actions executed thus far.

A.2 Formalization

Basic Notation. We denote by Σ the alphabet consisting of English letters, arabic numerals, and punctuation marks; by Σ^* the set of all finite strings over Σ; by $\mathbb{S}_k$ the group of permutations of k elements; by $x := y$ the operation that assigns value y to variable x; by $p := rand(\mathbb{S}_k)$ the operation that assigns to variable p a randomly selected permutation in $\mathbb{S}_k$; and by $\emptyset$ the empty set.

If S is a set, by S^0 we denote the empty set, and by S^k the Cartesian product of S with itself k times. If x is a sequence, by either x^i or x_i we denote x's ith element,[6] and by $\{x\}$ the set $\{z : x^i = z$ for some $i\}$. If x and y are sequences, respectively of length j and k, by $x \circ y$ we denote their concatenation (i.e., the sequence of $j + k$ elements whose ith element is x^i if $i \leq j$, and y^{i-j} otherwise). If x and y are strings (i.e., sequences with elements in Σ), we denote their concatenation by xy.

If A is a probabilistic algorithm, the distribution over A's outputs on input x is denoted by $A(x)$. A probabilistic function $f : X \to Y$ is *finite* if X and Y are both finite sets and, for every $x \in X$ and $y \in Y$, the probability that $f(x) = y$ has a finite binary representation.

Ballots and Actions. An *envelope* is a triple $(j, c, 0)$, where j is a positive integer, and c a symbol of Σ. A *ballot* is an envelope. If (j, c, L) is a ballot, we refer to j as its *identifier*, to c as its *content*, and to L as its *level*.

A set of ballots B is *well-defined* if distinct ballots have distinct identifiers. If B is a well-defined set of ballots, then I_B denotes the set of identifiers of B's ballots. For $j \in I_B$, B_j (or the expression *ballot j*) denotes the unique ballot of B whose identifier is j. For $J \subset I_B$, B_J denotes the set of ballots of B whose identifiers belong to J.

Relative to a well-defined set of ballots B: if j is an envelope in B, then $cont_B(j)$ denotes the content of j; if $x = j^1, \ldots, j^k$ is a sequence of envelope identifiers in I_B, then $cont_B(x)$ denotes the concatenation of the contents of these envelopes, that is, the string $cont_B(j^1) \cdots cont_B(j^k)$.

---

[6] For any given sequence, we shall solely use superscripts, or solely subscripts, to denote all of its elements.

A *global memory* consists of a pair (B, R), where

- B is a well defined set of ballots; and
- R is a sequence of strings in Σ^*, $R = R^1, R^2, \ldots$.

We refer to B as the *ballot set;* to R as the *public record;* and to each element of R as a *record*. The *empty global memory* is the global memory for which the ballot set and the public record are empty. We denote the set of all possible global memories by GM.

Ballot-box actions are functions from GM to GM. The subset of ballot-box actions available at a given global memory gm is denoted by $\mathcal{A}_{gm}$. The actions in $\mathcal{A}_{gm}$ are described below, grouped in 8 classes. For each $a \in \mathcal{A}_{gm}$ we provide a formal identifier; an informal reference (to facilitate the high-level description of our constructions); and a functional specification. If $gm = (B, R)$, we actually specify $a(gm)$ as a program acting on variables B and R. For convenience, we include in R the auxiliary variable ub, the *identifier upper-bound*: a value equal to 0 for an empty global memory, and always greater than or equal to any identifier in I_B.

1. (NEWEN, c) —where $c \in \Sigma$.
   "Make a new envelope with content c."
   $\text{ub} := \text{ub} + 1$; $B := B \cup \{(\text{ub}, c, 0)\}$; and $R := R \circ (\text{NEWEN}, c, \text{ub})$.
2. (OPENEN, j) —where j is an envelope identifier in I_B.
   "Publicly open envelope j to reveal content $cont_B(j)$."
   $B := B \setminus \{B_j\}$ and $R := R \circ (\text{OPENEN}, j, cont_B(j), \text{ub})$.
3. $(\text{DESTROY}, j)$ —where j is a ballot identifier in I_B.
   "Destroy ballot j"
   $B := B \setminus \{B_j\}$ and $R := R \circ (\text{DESTROY}, j, \text{ub})$.
4. (DONOTHING).
   "Do nothing"
   $B := B$ and $R := R \circ (\text{DONOTHING}, \text{ub})$.

Remarks

- All ballot-box actions are deterministic functions.
- The variable ub never decreases and coincides with the maximum of all identifiers "ever in existence." Notice that we never re-use the identifier of a ballot that has left, temporarily or for ever, the table. This ensures that different ballots get different identifiers.

Definition 2. *A global memory gm is feasible if there exists a sequence of global memories $gm^0, gm^1, \ldots, gm^k$, such that gm^0 is the empty global memory; $gm^k = gm$; and, for all $i \in [1, k]$, $gm^i = a^i(gm^{i-1})$ for some $a^i \in \mathcal{A}_{gm^{i-1}}$.*

If (B, R) is a feasible memory, we refer to R as a feasible public record.

Notice that if $gm = (B, R)$ is feasible, then $\mathcal{A}_{gm}$ is easily computable from R alone. Indeed, what ballots are in play, which ballots are envelopes and which are super-envelopes, *et cetera*, are all deducible from R. Therefore, different feasible global memories that have the same public record also have the same set of available actions. This motivates the following definition.

Definition 3. *If R is a feasible public record, by $\mathcal{A}_R$ we denote the set of available actions for any feasible global memory with public record R.*

B   The Notion of a Public Ballot-Box Mediator (VTP in Our Language)

Definition 4. *Let $\mathcal{P}$ be a sequence of K functions. We say that $\mathcal{P}$ is a public ballot-box mediator (of length K) if, for all $k \in [1, K]$ and public records R, $P^k(R)$ is a public ballot-box action in $\mathcal{A}_R$.*

An execution of $\mathcal{P}$ on an initial feasible global memory (B^0, R^0) is a sequence of global memories
$(B^0, R^0), \dots, (B^K, R^K)$ *such that* $(B^k, R^k) = a^k(B^{k-1}, R^{k-1})$ *for all* $k \in [1, K]$, *where* $a^k = P^k(R^{k-1})$.[7]

If e is an execution of $\mathcal{P}$, by $B^k(e)$ and $R^k(e)$ we denote, respectively, the ballot set, the public record, and the private history profile of e at round k. By $R_{\mathcal{P}}^k(e)$ we denote the last k records of $R^k(e)$ (i.e., "the records appended to R^0 by executing $\mathcal{P}$").

Remarks

- Note that the above definition captures our intuitive desideratum that no special trust is bestowed on a public mediator. Because he performs a sequence of public ballot-box actions, any one can verify that
    - (i) he performs the right sequence of actions;
    - (ii) he does not choose these actions; and
    - (iii) he does not learn any information that is not publicly available.
- Note too that if $\mathcal{P} = P^1, \dots, P^K$ and $\mathcal{Q} = Q^1, \dots, Q^L$ are public mediators, then their concatenation, that is, $P^1, \dots, P^K, Q^1, \dots, Q^L$ is a public mediator too.

---

[7] Note that the executions of $\mathcal{P}$ are, in general, random since $P^k(R)$ may return an action of Nature.

Some Recent Progress in Lattice-Based Cryptography

Chris Peikert

SRI International

Abstract. The past decade in computer science has witnessed tremendous progress in the understanding of *lattices*, which are a rich source of seemingly hard computational problems. One of their most promising applications is to the design of cryptographic schemes that enjoy exceptionally strong security guarantees and other desirable properties.

Most notably, these schemes can be proved secure assuming only the *worst-case* hardness of well-studied lattice problems. Additionally, and in contrast with number-theoretic problems typically used in cryptography, the underlying problems have so far resisted attacks by *subexponential-time* and *quantum* algorithms. Yet even with these security advantages, lattice-based schemes also tend to be remarkably simple, asymptotically efficient, and embarrassingly parallelizable.

This tutorial will survey the foundational results of the area, as well as some more recent developments. Our particular focus will be on the core hard cryptographic (average-case) problems, some recurring techniques and abstractions, and a few notable applications.

O. Reingold (Ed.): TCC 2009, LNCS 5444, p. 72, 2009.

Non-malleable Obfuscation

Ran Canetti[1,*] and Mayank Varia[2,**]

[1] School of Computer Science, Tel Aviv University
canetti@cs.tau.ac.il
[2] Massachusetts Institute of Technology
varia@csail.mit.edu

Abstract. Existing definitions of program obfuscation do not rule out malleability attacks, where an adversary that sees an obfuscated program is able to generate another (potentially obfuscated) program that is related to the original one in some way.

We formulate two natural flavors of non-malleability requirements for program obfuscation, and show that they are incomparable in general. We also construct non-malleable obfuscators of both flavors for some program families of interest. Some of our constructions are in the Random Oracle model, whereas another one is in the common reference string model. We also define the notion of verifiable obfuscation which is of independent interest.

1 Introduction

The problem of program obfuscation has recently received a lot of attention in cryptography. Informally, the goal of obfuscation is to transform a program in such a way that its code becomes unintelligible while its functionality remains the same. This intuitive idea was formalized in [1] using a simulation-based definition.

In [1] it is shown that there do not exist generic algorithms that obfuscate any program family. These results are extended in [2,3]. However, positive results have been shown for some program families of interest, such as the family of "point circuits," which accept a single input string (that is explicitly given in the circuit description) and reject all other inputs [4,5,6,7]. These constructions can be generalized to form obfuscators for two more families: "multi-point circuits," which accept a constant number of input strings, and "point circuits with multibit output," which store a hidden string that is revealed only for a single input value [8]. Finally, a different definition of obfuscation has been formulated [9,10] in which it is possible to obfuscate the family of re-encryption programs [10].

---

* Supported by NSF grant 0635297, US-Israel Binational Science Foundation grant 20006317, a European Union Marie Curie grant, and the Check Point Institute for Information Security.
** Supported by the Department of Defense through the NDSEG Program and by NSF grant 0635297.

However, the question of malleability attacks on obfuscated programs has not been addressed. In fact, many of the above constructions are malleable. We provide an overview of the definition of obfuscation and then describe its malleability concerns.

Virtual black-box obfuscation. At a high level, the concept of "obfuscating" a program is to produce a new program with the same functionality but with "garbled" code. Of course, it is impossible for the garbled code to hide all useful information, because at the very least one can run the program and observe its input-output behavior. In this way, the code of a program must be at least as useful as access to an oracle for the program. At a high level, obfuscation ensures the converse: that access to the code of an obfuscated program is *no more* useful than access to the oracle.

The formalization of this idea provided by [1], called the *virtual black-box* property, considers two different worlds. In the real world, an efficient adversary has access to the code of an obfuscated program, and attempts to learn a single-bit predicate about the underlying program. Now consider an imaginary world in which the code of an obfuscated program is not provided, but rather only oracle access to the program is provided. Obfuscation ensures that there exists an efficient algorithm known as a simulator that can learn the same predicate in the imaginary world that the adversary learns in the real world.

Malleability concerns. If an adversary has access to obfuscated code, the virtual black-box property guarantees that she cannot "understand" the underlying program. However, suppose the adversary instead uses the obfuscated code to create a new program in such a way that she controls the relationship between the input-output functionality of the two programs.

Intuitively, one might expect that virtual black-box obfuscation already prevents malleability attacks. The simulator only has oracle access to the obfuscated code, so any program that it makes can only depend on the input-output functionality of the obfuscated code at a polynomial number of locations. Therefore, obfuscation should guarantee that the adversary is also restricted to these trivial malleability attacks. However, the virtual black-box definition in [1] does not carry this guarantee. Upon close inspection, the problem is that the virtual black-box definition only considers adversaries and simulators that output a single bit, not adversaries and simulators that output programs.

A naïve solution to this problem is to extend the virtual black-box definition to hold even when the adversary and simulator output long strings. However, in this case obfuscation becomes unrealizable for any family of interest. Consider the adversary that outputs its input. Then, a corresponding simulator has oracle access to a program and needs to write the code for this program, which is usually impossible.

In this paper, we demonstrate two different methods to incorporate non-malleability guarantees into obfuscation. Both non-malleability definitions extend the virtual black-box definition by allowing the adversary and simulator to produce multiple bit strings, but only in a restricted manner. There are many subtleties involved in constructing a proper definition, such as deciding the appropriate restrictions to impose on the adversary and simulator, and creating relations to test the similarities between the adversary's input and output programs. We defer treatment of these important details to Section 3. Here, we motivate and describe the two definitions at a high level.

Functional non-malleability. Imagine that Alice, Bob, and Charles are three graduate students in an office that receives a new computer. The department's network administrator wishes to configure the computer to allow the grad students root access to the computer. The administrator receives the students' desired passwords, and she needs to write a login program that accepts these three passwords and rejects all other inputs. The administrator knows that she has to be careful in designing the login program because the students will be able to read the program's code. As a result, she forms the login program using an obfuscator for the family of three-point circuits, which ensures that a dictionary attack is the best that the graduate students can do to learn their officemates' passwords.

However, obfuscation does not alleviate all of the administrator's fears, because the students will have root access to the computer so they can alter the login program as well. The administrator would like to prevent tampering of the login program, but the virtual black-box definition does not provide this guarantee. For instance, suppose Alice wants to remove Bob's access to the computer. There exist obfuscators of the three-point circuit such that Alice can succeed in this attack with noticeable probability [8].

Intuitively, the goal of obfuscation is to turn a program into a "black box," so the only predicates that Alice can learn from the program are those she could learn from a black box. We extend this intuition to cover modifications as well. We say that an obfuscator is *functionally non-malleable* if the only programs that Alice can create given obfuscated code are the programs she could create given black-box access to the obfuscated code.

This definition provides a guarantee on the possible attacks Alice can apply. For instance, if Alice only has black-box access to the login program, then she can only remove Bob's access to the computer with negligible probability. In this sense, functional non-malleability provides stronger security for Bob because it protects all aspects of his access to the computer, whereas the virtual black-box property only protects his password.

We wish to define functional non-malleability using a simulation-based definition: for every adversary that receives obfuscated code and uses it to create a new program, there exists a simulator that only has oracle access to the obfuscated code and produces a program that is functionally equivalent to the adversary's program. However, this definition is too strong: given the trivial adversary that outputs its input, the simulator gets oracle access to a program and needs to output the code of this program, which is usually impossible. But it is unfair to demand that the simulator do this much work. After all, the adversary's input is a program but the simulator's input is just an oracle. At the very least, the adversary can output a program that uses its input program in a black-box manner, and the simulator should have the same ability. Therefore, we allow the program that the simulator outputs to have oracle access to the obfuscated code.

Verifiable non-malleability. Functional non-malleability is a nice property for obfuscators, but there are some scenarios where even this does not suffice. For example, suppose that Alice wishes to play an April fools' prank on her office-mates by altering the login program to accept their old passwords appended to the string "Alice is great." Alice only knows her own password, so she cannot run the obfuscator to produce this modified program. Nevertheless, she can write the following program: "on input a string s, check that s begins with 'Alice is great,' and if so, remove it from s and send the rest of the string to the administrator's login program." Functional non-malleability does not prevent this prank. In fact, it is impossible to prevent this prank because Alice only uses the obfuscated login program in a black-box manner.

Still, this attack is not "perfect": after Alice performs this attack, the new login program "looks" very different from a program that the network administrator would create. As a result, we may not be able to prevent Alice from performing her prank, but we may be able to detect Alice's modification afterward and restore the original program. Alice's job is now harder, since she has to modify obfuscated code in such a way that the change is undetectable. We say that an obfuscator is *verifiably non-malleable* if the only programs Alice can create that pass a verification procedure are the programs she could create given black-box access to the obfuscated code. This approach gives us hope to detect attacks that we cannot prevent, although it requires a stronger model in which a verification procedure routinely audits the program.

In the setting of our example, one simple way to achieve non-malleability is for the network administrator to digitally sign every program she makes, and for the verification procedure to check the validity of the signature attached to an obfuscated program before running it. By the existential unforgeability of the signature scheme, Alice cannot make any modifications, so the non-malleability goal is achieved.

However, this solution requires that the verification procedure can find and store the network administrator's verification key, which may not be practical. We want the non-malleability guarantee to be an intrinsic property of the ob-fuscation, without relying on an external public key infrastructure. As a result, in this paper we consider "public" verifiers that depend only on the obfuscation algorithm, and not on the party performing the obfuscation. Informally, a *veri-fier* algorithm V accepts programs if and only if they could have been produced by running the obfuscation algorithm. We stress that V does not receive any party-specific information (such as public keys), so it does not depend on the person that runs the obfuscator.

Verifiability has interesting applications in and of itself, as we describe in the Discussion section below, but in this paper we only use it to create a simulation-based definition of verifiable non-malleability. The definition guarantees that an adversary cannot maul obfuscated code into a new program that passes the verification test unless there exists a simulator that can perform the same attack given only oracle access to the obfuscated code. (Note that the adversary must create a new program and not simply output the obfuscated code it receives.) Because we hope to detect attacks that operate in a black-box manner (which we could not hope to prevent in the functional setting), we no longer give the simulator the extra help that we gave it in the definition of functional non-malleability. Instead, the simulator must output a fully-functional program that does not have an oracle.

Comparison. We show that both forms of non-malleability imply the virtual black-box property. Intuitively, this relationship holds because an adversary that outputs programs should easily be able to encode a single bit of information in the output. As a result, all known impossibility results regarding the virtual black-box property continue to hold for both types of non-malleability [1,2].

Additionally, we compare the two flavors of non-malleability. The goal of functional non-malleability is to *prevent* as many malleability attacks as possible, whereas the goal of verifiable non-malleability is to *detect* as many attacks as possible. Intuitively, these goals are incomparable: the verifiable definition is stronger because we can detect more attacks than we can prevent, but on the other hand it is weaker because the model requires its participants to understand and apply the verification algorithm. We justify this intuition by showing that in the random oracle model, the two definitions of non-malleability are indeed incomparable.

Constructions. In the random oracle model, we show that the obfuscator for point circuits in [6] satisfies both functional and verifiable non-malleability. Next, we study the family of multi-point circuits, which accept a constant number of inputs. One idea to obfuscate the program that accepts values $x_1, \ldots, x_m$ is as follows:

1. Use several instantiations of a single-point circuit obfuscator in order to create obfuscated programs $P_1, P_2, \ldots, P_m$, where each P_i accepts only the value x_i
2. Create the program P that contains P_1 through P_m as subroutines, and on input x, iteratively feeds x into the P_i and accepts if any one of these programs accept.

This methodology is known as *concatenation*, and it is shown in [8] that concatenation preserves obfuscation. That is, given any obfuscator for the family of single-point circuits, concatenation produces an obfuscator for the family of multi-point circuits. However, concatenation does not preserve non-malleability. The program P stores the subroutines P_1 through P_m in a readily identifiable way, so an adversary can modify one accepted point by changing one of the subroutines. This is true even if the obfuscator for the family of single-point circuits is non-malleable.

In the verifiable setting, we resolve the problem with concatenation by using a self-signing technique to ensure that the subroutines are not modified. The verification algorithm associated with this construction runs the verification algorithm for the signature scheme. This approach does not suffice in the functional setting, where the self-signing technique is useless because there is no guarantee that anybody checks the signature. Instead, we "glue" the accepted points together in such a way that any attempt to change the obfuscated code destroys information on all of the points simultaneously.

We also give a construction that does not use random oracles. Instead, it uses the common reference string (CRS) model, in which a sequence of bits is chosen uniformly at random and published in a public location that all participants can access. (Note that this is different from a public key infrastructure because the CRS is not tied to the specific identity of the party performing the obfuscation.) We construct a verifiably non-malleable obfuscation for the family of point circuits by providing any (potentially malleable) obfuscation along with a non-malleable NIZK proof of knowledge [11,12] that the obfuscator knows the point that is accepted.

Informally, a non-malleable NIZK proof of knowledge considers an adversary that can request multiple proofs for statements of its choice and then produces a new proof. The non-malleability guarantee requires that the adversary knows a witness to its constructed proof, so it cannot simply modify the old proofs to prove a new statement.

Intuitively, our construction is verifiably non-malleable because an adversary can only make a program that passes the verification test if she knows its functionality, so she cannot produce a program that is related to a given obfuscated point circuit or else she would learn information about the obfuscated circuit, which is impossible by the virtual black-box property. However, the actual proof turns out to be delicate. Using proof techniques from [4], we achieve a somewhat weaker variant of verifiable non-malleability. Specifically, we show that for a large class of relations, no adversary can perform a modification that satisfies the relation with noticeable probability.

Discussion on verifiable obfuscation. The concept of verifiable obfuscation is useful even in situations where malleability is not a concern. For example, suppose you create a new computer program that solves an important problem. You wish to profit from your research by selling this program to others, but you also want to protect the algorithm that you discovered. Therefore, you sell an obfuscated version of the program to your customers. This protects your intellectual property, but another problem has presented itself. Your customers do not wish to install the obfuscated program on their computers because they no longer have any guarantees about what this program does. For all they know, the program could contain a virus, and because the program is obfuscated there is no hope for a virus checker to detect the presence of a virus. Hence, you need a verification algorithm that proves to your customers that you are selling them a program from the proper family.

Future work. First, the constructions in this paper use the random oracle model or common reference string model. It remains an open question to construct a non-malleable obfuscator (of either flavor) without trusted setup.

Second, we provide a verifiably non-malleable obfuscation of single-point circuits in the CRS model for a large class of relations. Unfortunately, extending this construction to the multi-point setting is insufficient, as it only succeeds for a small class of relations. It remains open to find a better construction in the multi-point setting.

Organization. In Section 2, we provide an overview of virtual black-box obfuscation [1,2]. In Section 3, we provide rigorous definitions of the two notions of non-malleability. In Sections 4 and 5, we present non-malleable obfuscators of both flavors for the family of multi-point circuits.

2   Obfuscation

In [1], [2], and other works, an obfuscator is defined as a compiler that takes a circuit as input and returns another circuit. The output circuit should be equivalent in functionality to the input circuit, but the output circuit should be unintelligible in the sense that any information that can be obtained from the output circuit can also be obtained with oracle access to the circuit. In this paper we will be interested in obfuscation with dependent auxiliary information, as defined in [2].

Throughout this work, the adversaries and simulators are assumed to be non-uniform.

Definition 1 (Obfuscation). *Let $C = \{C_n\}_{n\in\mathbb{N}}$ be a family of polynomial-size circuits, where C_n denotes all circuits of input length n. A probabilistic polynomial time (PPT) algorithm $\mathcal{O}$ is an* obfuscator *for the family C operating over randomness $\mathcal{R} = \{\mathcal{R}_n\}$ if the following three conditions are met.*

- Approximate functionality: *There exists a negligible function ε such that for every n, every circuit $C \in C_n$, and for all $x \in \{0,1\}^n$,*

$$\Pr[r \leftarrow \mathcal{R}_n, C' \leftarrow \mathcal{O}(C, r) : C(x) = C'(x)] > 1 - \varepsilon(n) \ .$$

  If this probability is always 1, then we say that $\mathcal{O}$ has exact functionality.
- Polynomial slowdown: *There exists a polynomial p such that for every n, circuit $C \in C_n$, and $r \in \mathcal{R}_n$, the description length $|\mathcal{O}(C, r)| \le p(|C|)$.*
- Virtual black-box: *For every polynomial ρ and every PPT adversary A, there exists a PPT simulator S such that for all sufficiently large n, for all $C \in C_n$, and for all auxiliary information $z \in \{0,1\}^*$,*

$$\left|\Pr[A(\mathcal{O}(C), z) = 1] - \Pr[S^C(1^n, z) = 1]\right| < \frac{1}{\rho(n)} \ ,$$

  where the first probability is taken over the coin tosses of A and $\mathcal{O}$, and the second probability is taken over the coin tosses of S. Furthermore, we require that A and S operate in time polynomial in the length of their first input.

We define obfuscation without auxiliary information in the same manner, except that the auxiliary information is removed from the virtual black-box definition. Unless otherwise specified, in this paper we assume that obfuscations are secure with respect to dependent auxiliary information.

3 Defining Non-malleable Obfuscation

In this section, we rigorously define the two variants of non-malleable obfuscation.

3.1 Functionally Non-malleable Obfuscation

We obtain functionally non-malleable obfuscation by generalizing the virtual black-box definition to allow the adversary and simulator to output programs instead of bits. Intuitively, a functionally non-malleable obfuscation has the property that an adversary, given the obfuscated code to a program, can only make a related program if it could have already done so given only black-box access to the program.

This is problematic in general, because the simulator cannot emulate all programs that the adversary can produce [1,7]. For example, consider the adversary that outputs its input. Then, the simulator has oracle access to a circuit and has to produce a program that is functionally equivalent to its oracle. This is impossible unless the circuit is learnable with oracle queries, in which case the entire concept of obfuscation is uninteresting. To make the definition meaningful, we allow the program that the simulator produces to make oracle queries to the original circuit as well.

To capture the effectiveness of an adversary's modification, we introduce a polynomial-time computable relation E that receives the adversary's input program and output program. The adversary succeeds in the modification if E accepts it. The definition of non-malleability ensures that for every relation E, the simulator can perform a successful modification with the approximately the same probability as the adversary.

One technical concern about the relation E is the manner in which it receives the adversary's input and output programs. The goal of functional non-malleability is to compare the functionality of these programs, and not their underlying code, so E should operate in the same manner when given functionally equivalent inputs. Our definition resolves this issue by giving the relation a "canonical" member of the family that is equivalent to the adversary's output program. (See the Discussion section below for more detail on this issue.)

Additionally, in many situations, the adversary knows some a-priori useful information on the obfuscated program, so we allow dependent auxiliary information in the definition of non-malleability. For instance, in the motivating example from the Introduction in which Alice wishes to modify a login program, she possesses the knowledge of her own password.

Definition 2 (Functional equivalence). *We say that two circuits C_1 and C_2 are* functionally equivalent, *and write $C_1 \equiv C_2$, if for all inputs x it holds that $C_1(x) = C_2(x)$.*

Definition 3 (Functionally non-malleable obfuscation). *Let C and D be families of circuits, and let O be a PPT algorithm. We say that O is an* obfuscator *for C that is functionally non-malleable over D if the following three conditions hold:*

- Almost exact functionality: *There exists a negligible function ε such that for every n and every circuit $C \in \mathcal{C}_n$, $\Pr[r \leftarrow \mathcal{R}_n : \mathcal{O}(C, r) \equiv C] > 1 - \varepsilon(n)$.*
- Polynomial slowdown: *There exists a polynomial p such that for every n, circuit $C \in \mathcal{C}_n$, and $r \in \mathcal{R}_n$, the description length $|\mathcal{O}(C, r)| \leq p(|C|)$.*
- Functional non-malleability: *for every polynomial ρ and PPT adversary A, there exists a PPT simulator S such that for all sufficiently large n, for all circuits $C \in \mathcal{C}_n$, for all auxiliary information $z \in \{0, 1\}^*$, and for all polynomial time computable relations $E : \mathcal{C}_n \times \mathcal{D}_n \to \{0, 1\}$ (that may depend on the circuit C),*

$$| \Pr[P \leftarrow A(\mathcal{O}(C), z) : \exists D \in \mathcal{D}_n \ s.t. \ D \equiv P \ and \ E(C, D) = 1]$$

$$-\Pr[Q \leftarrow S^C(1^n, z) : \exists D \in \mathcal{D}_n \ s.t. \ D \equiv Q^C \ and \ E(C, D) = 1]| < \frac{1}{\rho(n)} \ ,$$

where the probabilities are over the coin tosses of A, $\mathcal{O}$, and S. We require that A and S run in time polynomial in the length of their first inputs.

If $\mathcal{D} = \mathcal{C}$, we say that $\mathcal{O}$ is a functionally non-malleable obfuscator for $\mathcal{C}$.

Discussion. We make several remarks about this definition.

Almost exact functionality. The functionality requirement here is stronger than the one used in Definition 1 above. Approximate functionality only guarantees that an obfuscated program $\mathcal{O}(C, r)$ is "close" in functionality to C. However, $\mathcal{O}(C, r)$ might never have the same functionality as C does (for any choice of r). By contrast, almost exact functionality requires that the two circuits have identical functionality for most choices of r.

We note that most of the constructions in this paper satisfy exact functionality.

Bivariate relation. In this definition, the bivariate relation E is allowed to depend on the choice of circuit $C \in \mathcal{C}_n$. Thus, restricting attention to univariate relations $E(D)$ results in an equivalent definition. We use a bivariate relation only to emphasize the fact that E depends on both C and D.

Possible definitions for E. As mentioned above, an important feature of the definition is that E only depends on the functionality of the adversary's output P and simulator's output Q^C, and not on the code of these circuits. We found three possible ways to enforce this condition on E.

First, we can constrain E to receive only oracle access to the program P or Q^C. As a result, it follows immediately that E only depends on the functionality of these programs, and not on their underlying code. Unfortunately, this definition is too weak, because there are many natural predicates that cannot be tested by relations of this type.

For instance, consider the family of point circuits, where I_w is the circuit that accepts only the string w. Suppose the adversary is given an obfuscation of the point circuit I_x and wishes to create a new point circuit I_y such that the first bit of x and y are equal. No polynomial-time relation E (even ones that know x, since E can depend on x) can test the adversary's probability of success given only oracle access to I_y. We believe that relations of this type are meaningful, and therefore we want a definition that can test for them.

Second, we can give the relation E full access to the code of P or Q^C, but restrict our attention to relations that have identical output when given two functionally equivalent programs. Specifically, we only consider relations E such that given any programs $C \in \mathcal{C}_n$, $D \in \mathcal{D}_n$, and P, P' such that $D \equiv P \equiv P'$, it follows that $E(C, P) = E(C, P')$. This definition does allow E access to the code of its input programs. The advantage of this definition is that E finally gets access to the code of the programs. The disadvantage is that the condition we impose on relations is very restrictive. As a result, it is still impossible to compute many relations, such as the one described in the previous paragraph.

Specifically, the virtual black-box definition guarantees that any relation $E(I_x, \mathcal{O}(I_y))$ cannot compute whether x and y have the same first bit with probability greater than $\frac{1}{2}$. Thus, the condition that we impose on E is that it has the same probability of success even when it is given I_x and I_y as inputs, in which case E has enough information to perform the computation with probability 1 but must fail half of the time anyway in order to be consistent with the condition.

Third, we can allow all polynomial-time relations E, but instead of providing the code of P or Q^C as input to E, we provide the code of a functionally equivalent member in $\mathcal{D}_n$. This is the option we use in Definition 3 above, because it clearly satisfies the requirement that E only depend on the functionality of the adversary and simulator's output, and it is a stronger definition that can test for many relations that the previous two definitions cannot. For these reasons, we choose to use relations of this type in the definition of functional non-malleability.

One technical point to keep in mind is that the relation E takes the description of circuits in $\mathcal{C}_n$ and $\mathcal{D}_n$ as input. As a result, this definition is dependent upon the representation of the circuits in these families, and not just the functionality of these circuits. Therefore, we should choose a representation of the circuit families that enables relations to extract important information easily from the description of a circuit.

As a result, we define the families of multi-point circuits as follows. Given $w_1, w_2, \ldots, w_m \in \{0,1\}^n$, let $I_{\{w_1,\ldots,w_m\}}$ be the circuit that stores $w_1, \ldots, w_m$ in some canonical, explicit manner, and on input x returns 1 if and only if $x = w_i$ for some i. In particular, relations can extract the strings $w_1, \ldots, w_m$ from the description of $I_{\{w_1,\ldots,w_m\}}$ in polynomial time. Note that the w_i need not be distinct, so the circuit $I_{\{w_1,\ldots,w_m\}}$ accepts between 1 and m points. For technical reasons, we may also want to consider the circuit $I_\emptyset$ that immediately rejects all inputs. Let

$$\mathcal{P}_n^m = \{I_{\{w_1,\ldots,w_m\}} : w_1, \ldots, w_m \in \{0,1\}^n\} \cup \{I_\emptyset\}$$

be the set of all m-point circuits on n bits, and let $\mathcal{P}^m = \{\mathcal{P}_n^m\}_{n \in \mathbb{N}}$ be the family of m-point circuits. Also, let $\mathcal{P}^{m+}$ be the subfamily that does not include $I_\emptyset$.

Output family $\mathcal{D}$. According to our definition, an adversary succeeds only if it outputs a circuit that is equivalent to a circuit in the family $\mathcal{D}$. The most natural family to choose is $\mathcal{D} = \mathcal{C}$, but we allow $\mathcal{D}$ to be different from $\mathcal{C}$ in order to consider a wider range of adversaries. For instance, perhaps $\mathcal{C}$ is the family of point circuits, but we are concerned with adversaries that produce two-point circuits as output as well. The definition of functional non-malleability allows us to form a larger family $\mathcal{D}$ to acknowledge this.

Of course, there is no reason to stop there: we may also be concerned with an adversary that produces a three-point circuit, or a four-point circuit, or any circuit for that matter. In fact, the presence of the circuit family $\mathcal{D}$ in the definition seems restrictive. It would be nice if our definition simultaneously covered all possible outputs of the adversary, and not just those in a specific family. In other words, we would like a functionally non-malleable obfuscator when $\mathcal{D}$ is the family of all circuits, but unfortunately this is impossible. Intuitively, the family of all circuits is so big that it allows A to output the obfuscated code that it receives as input, which the simulator cannot do. A formal proof can be found in the full version of this paper [13].

Comparison to virtual black-box obfuscation. Now that we have introduced a new definition of obfuscation, it is natural to compare it to the old one. We show that the functional non-malleability property implies the virtual black-box property (at least for reasonable choices of the circuit family $\mathcal{D}$). This justifies our terminology of using the word "obfuscation" in Definition 3.

Theorem 4. *Let $\mathcal{C}$ and $\mathcal{D}$ be circuit families, and let $\mathcal{O}$ be an obfuscator for $\mathcal{C}$ that is functionally non-malleable over $\mathcal{D}$. Furthermore, suppose that for sufficiently large n, there exist circuits $D_0, D_1 \in \mathcal{D}_n$ such that $D_0 \not\equiv D_1$. Then, $\mathcal{O}$ satisfies the virtual black-box property. As a result, $\mathcal{O}$ is an obfuscator for $\mathcal{C}$.*

This theorem also holds if neither the virtual black-box property nor functional non-malleability allows auxiliary information.

Intuitively, this theorem holds because an adversary that outputs programs can use this channel to transmit a single bit of information b by outputting the program D_b. See the full version of this paper [13] for a rigorous proof.

One consequence of this theorem is that all impossibility results pertaining to the virtual black-box property immediately carry over to the non-malleability setting [1,2].

3.2   Verifiably Non-malleable Obfuscation

In this section, we develop the notion of verifiable obfuscation and use it to define another definition of non-malleability.

Definition 5 (Verifier). *Given a pair of PPT algorithms $\mathcal{O}$ and V and a circuit family $\mathcal{C}$, we say that V is a* verifier *for $\mathcal{O}$ applied to $\mathcal{C}$ if there exists a negligible function ε such that for all n and for all $C \in \mathcal{C}_n$, $\Pr[V(\mathcal{O}(C)) = 1] > 1 - \varepsilon(n)$, where the probability is taken over the randomness of V and $\mathcal{O}$.*

If $\mathcal{O}$ is an obfuscator for the family of circuits $\mathcal{C}$, then we say that the pair $(\mathcal{O}, V)$ constitutes a verifiable obfuscator *for $\mathcal{C}$.*

We do not place any restrictions on V when its input is not the result of the obfuscator applied to a circuit in the family. In particular, given any obfuscator $\mathcal{O}$, the pair $(\mathcal{O}, \mathbb{1})$ is a verifiable obfuscator, where $\mathbb{1}$ is the algorithm that accepts all inputs. In many cases, however, we can create much better verification algorithms. For example, the (r, r^x) construction of [4] can simply be verified by checking whether r and r^x are elements in the desired group G of prime order, because there is a unique discrete log of r^x so the program does implement a point circuit as desired. This results in a *perfect verifier* that accepts its input program if and only if it has the form of a program produced by the obfuscator.

Now we create a definition of non-malleability for verifiable obfuscators. As before, we consider an adversary that takes an obfuscated circuit as input and outputs a program. In this model, the adversary succeeds only if her output program passes the verification test and is related to the input program. Our definition of non-malleability requires that a simulator succeeds with approximately the same probability, so it must also output a program that passes the verification test. In particular, we no longer give an oracle to the program constructed by the simulator. A formal definition follows.

Definition 6 (Verifiable non-malleability). *Let $\mathcal{C}$ and $\mathcal{D}$ be a families of circuits such that $\mathcal{C} \subseteq \mathcal{D}$, and let $\mathcal{O}$ and V be PPT algorithms. We say that $(\mathcal{O}, V)$ is an obfuscator for $\mathcal{C}$ that is verifiably non-malleable over $\mathcal{D}$ if the following four conditions hold:*

- *Verification: V is a verifier for $\mathcal{O}$ applied to $\mathcal{C}$. Additionally, for every n and every circuit P with n bits of input such that $V(P) = 1$, there exists $D \in \mathcal{D}_n$ such that $P \equiv D$.*
- *Almost exact functionality: There exists a negligible function ε such that for every n and every circuit $C \in \mathcal{C}_n$, $\Pr[r \leftarrow \mathcal{R}_n : \mathcal{O}(C, r) \equiv C] > 1 - \varepsilon(n)$.*
- *Polynomial slowdown: There exists a polynomial p such that for every n, circuit $C \in \mathcal{C}_n$, and $r \in \mathcal{R}_n$, the description length $|\mathcal{O}(C, r)| \leq p(|C|)$.*
- *Verifiable non-malleability: for every polynomial ρ and PPT adversary A, there exists a PPT simulator S such that for all sufficiently large n, for all circuits $C \in \mathcal{C}_n$, for all auxiliary information $z \in \{0, 1\}^*$, and for all polynomial time computable relations $E : \mathcal{C}_n \times \mathcal{D}_n \to \{0, 1\}$,*

$$| \Pr[P \leftarrow A(\mathcal{O}(C), z) : P \neq \mathcal{O}(C), V(P) = 1, \exists D \in \mathcal{D}_n \text{ s.t. } D \equiv P, E(C, D) = 1]$$

$$- \Pr[Q \leftarrow S^C(1^n, z) : V(Q) = 1, \exists D \in \mathcal{D}_n \text{ s.t. } D \equiv Q, E(C, D) = 1]| < \frac{1}{\rho(n)} \ ,$$

where A and S run in time polynomial in the length of their first inputs.

If $\mathcal{D} = \mathcal{C}$, we say that $(\mathcal{O}, V)$ is a verifiably non-malleable obfuscator *for $\mathcal{C}$.*

It is potentially reasonable to relax the definition by not requiring the simulator to pass the verification text. We choose not to do so because the current definition puts the adversary and simulator on more equal footing and because all of our constructions satisfy the stronger notion.

We also note that the definition requires that V only accept circuits that are equivalent to members of $\mathcal{D}$. The benefit of this restriction is that the adversary can efficiently test whether she outputs a circuit in $\mathcal{D}_n$, which is a requirement for her to succeed under this definition.

Additionally, the remarks pertaining to functional non-malleability also apply here:

1. Restricting to univariate relations $E(D)$ results in an equivalent definition.
2. It is usually impossible to achieve verifiable non-malleability if $\mathcal{D}$ is the family of all circuits.
3. Verifiable non-malleability also implies the virtual black-box property.

Theorem 7. *Let C and D be circuit families, and let $(\mathcal{O}, V)$ be an obfuscator for C that is verifiably non-malleable over $\mathcal{D}$. Then, $\mathcal{O}$ is an obfuscator for C. This theorem also holds if neither definition allows auxiliary information.*

3.3   Comparison

We conclude this section by showing that the two non-malleability definitions are incomparable. Intuitively, functional non-malleability *prevents* more attacks. This is reflected in the definition by the fact that an adversary attacking functional non-malleability does not have to pass a verification test, so the simulator must emulate more potential attacks. A concrete example of this separation is the obfuscator described in Algorithm 2 below, which does not prevent the attack described in the Introduction in which Alice removes Bob's password.

On the other hand, verifiable non-malleability *detects* more attacks. This is reflected in the definitions by the fact that the program Q constructed by the simulator is not given oracle access to C in the verifiable definition. A concrete example of this separation is the functionally non-malleable obfuscator for multi-point functions described in the full version of the paper [13], which is vulnerable to a slightly modified form of Alice's April fools' prank described in the Introduction.

4   Constructions of Functionally Non-malleable Obfuscators

In this section, we present a functionally non-malleable obfuscator for the family of multi-point circuits in the random oracle model.

We begin with the single-point case. Algorithm 1 describes an obfuscator $\mathcal{O}_{\mathcal{P}^1}$ for the family $\mathcal{P}^1$ [6]. Informally, $\mathcal{O}_{\mathcal{P}^1}$ obfuscates the point circuit I_w by recording $R(w)$. This provides an information-theoretic way to hide the point w while still making it easy to check whether the input to the obfuscated program is w. However, the obfuscator should not be deterministic [4], so $\mathcal{O}_{\mathcal{P}^1}$ uses some randomness as well.

Algorithm 1. Obfuscator $\mathcal{O}_{\mathcal{P}^1}$ for the family of point circuits $\mathcal{P}^1$

Input: a circuit of the form I_w or $I_\emptyset$
  1: **if** the input circuit is $I_\emptyset$ **then**
  2:    choose $r \overset{U}{\leftarrow} \{0,1\}^{3|w|}$ and $t \overset{U}{\leftarrow} \{0,1\}^{4|w|}$ (that is, uniformly at random)
  3: **else**
  4:    extract the point w and choose randomness $r \overset{U}{\leftarrow} \{0,1\}^{3|w|}$
  5:    set $t = R(w \circ r)$, where $\circ$ denotes the string concatenation operation
  6: **end if**
  7: **output** the circuit $\Phi_{r,t}$ that stores r and t in some clearly identifiable manner, and
     on input a string x, outputs 1 if $R(x \circ r) = t$ and 0 otherwise

---

Theorem 8. *In the random oracle model, the algorithm $\mathcal{O}_{\mathcal{P}^1}$ is a functionally non-malleable obfuscator for the family of point circuits $\mathcal{P}^1$.*

Due to space constraints, rigorous proofs of non-malleability for all of our constructions are deferred to the full version of this paper [13].

Constructing a functionally non-malleable obfuscator for the family of multi-point circuits $\mathcal{P}^m$ is significantly more difficult. Roughly speaking, the principal issue is that the obfuscated program must "bundle together" the m points in a way that would prevent the adversary from changing any point in the bundle without applying the exact same change to all points in the bundle. For instance, simply concatenating m obfuscations of a single-point circuit (even obfuscations that are individually non-malleable) does not suffice because an adversary will be able to change some of the points at will. Instead, the obfuscated program must be a "house of cards" in the sense that an adversary cannot change the code without destroying information about all of the accepted points simultaneously. Our construction is described in the full version of this paper [13].

5   Constructions of Verifiably Non-malleable Obfuscators

In this section, we present verifiably non-malleable obfuscators for the family of multi-point circuits in the random oracle and common reference string models.

5.1   Random Oracle Model

In the single-point case, the obfuscator $\mathcal{O}_{\mathcal{P}^1}$ from Algorithm 1 can also be used to create a verifiably non-malleable obfuscation. Let $V_{\mathcal{P}^1}$ be the verification algorithm that accepts if and only if its input is a program of the form $\Phi_{r,t}$. It is clear from Algorithm 1 that $V_{\mathcal{P}^1}$ always accepts proper obfuscations of point circuits.

Theorem 9. *In the random oracle model, $(\mathcal{O}_{\mathcal{P}^1}, V_{\mathcal{P}^1})$ is a verifiably non-malleable obfuscator for $\mathcal{P}^1$.*

In the multi-point setting, we can concatenate m copies of $\mathcal{O}_{\mathcal{P}^1}$ and "glue" them together using a self-signing technique in order to construct the obfuscator $\mathcal{O}_{\mathcal{P}^m}$

described in Algorithm 2. The associated verification algorithm $V_{\mathcal{P}^m}$ checks that its input program has the proper structure and validates the signature.

The self-signing technique ensures that an adversary will be detected if she tries to re-use any of the $\mathcal{O}_{\mathcal{P}^1}$ obfuscations given to her, because she will not be able to forge the required signature. For instance, using the example described in the Introduction, Alice cannot keep the pieces of the obfuscated circuit that accept Charles and herself but remove the part that accepts Bob. However, the scheme does not *prevent* Alice from implementing this attack, so the scheme is malleable in the functional sense.

The one-time signature scheme can be constructed from any one-way function [14] so in particular it can be constructed from the random oracle.

Algorithm 2. Obfuscator $\mathcal{O}_{\mathcal{P}^m}$ for the family of m-point circuits

---

Input: a circuit of the form $I_{\{w_1,\ldots,w_m\}}$ or $I_\emptyset$

1: let k be the number of distinct accepted points for the input circuit (possibly 0)
2: extract the k distinct points $w_1, \ldots, w_k$
3: choose randomness $r_1, \ldots, r_m \xleftarrow{U} \{0,1\}^{3mn}$
4: choose a signature-verification key pair (s, v) for a one-time signature scheme
5: **for** $i = 1$ to k **do**
6:     set $t_i = R(w_i \circ r_i \circ v)$
7: **end for**
8: **for** $i = k + 1$ to m **do**
9:     choose $t_i \xleftarrow{U} \{0,1\}^{n+3mn+|v|}$
10: **end for**
11: choose a random permutation π on m elements, and permute the r_i and t_i by π
12: compute the signature $\sigma = \mathrm{sign}_s(t_1, \ldots, t_m)$
13: **output** the circuit that stores the r_i, t_i, v, and σ in a clearly identifiable manner, and on input x does the following: "for i from 1 to m, accept if $R(x \circ r_i \circ v) = t_i$"

Theorem 10. *In the random oracle model, $(\mathcal{O}_{\mathcal{P}^m}, V_{\mathcal{P}^m})$ is a verifiably non-malleable obfuscation for $\mathcal{P}^m$. However, $\mathcal{O}_{\mathcal{P}^m}$ is malleable in the functional sense.*

The self-signing technique can also be applied to the functionally non-malleable obfuscator for the family of multi-point circuits described in the full version of the paper [13], producing an obfuscator that simultaneously satisfies both forms of non-malleability.

5.2   Common Reference String Model

In the common reference string (CRS) model, we provide verifiably non-malleable obfuscators for the family $\mathcal{P}^{m+}$ of multi-point circuits that does not include $I_\emptyset$. This is a slightly different family than we used in the random oracle constructions, because the constructions in this section have the property that it is easy to tell that obfuscated circuits accept at least one input, which was not the case in the random oracle constructions.

Also, in this section we can only prove a slightly weaker form of verifiable non-malleability. We first present an obfuscator in the single-point setting, where we believe the weaker non-malleability property is meaningful. Then, we generalize the obfuscator to operate in the multi-point setting, but we also show that the weaker form of non-malleability is insufficient in this setting.

Single-point circuits. Our construction uses two building blocks:

1. Let $\hat{\mathcal{O}}_{\mathcal{P}^{1+}}$ be any obfuscator for $\mathcal{P}^{1+}$ without auxiliary information, along with a perfect verifier $\hat{V}_{\mathcal{P}^{1+}}$ for $\hat{\mathcal{O}}_{\mathcal{P}^{1+}}$, such as the obfuscator of [4] described in Section 3.2.
2. Let Π be a non-malleable non-interactive zero-knowledge (NIZK) proof of knowledge system [11,12]. Informally, the proof system Π has the property that given an adversary that sees a proof π and then creates a proof π' with $\pi' \neq \pi$, there exists an extractor that extracts the witness to the proof of π'.

Using these building blocks, we form the obfuscator $\tilde{\mathcal{O}}_{\mathcal{P}^{1+}}(I_w)$ that outputs $\hat{\mathcal{O}}_{\mathcal{P}^{1+}}(I_w)$ along with a proof that it knows the point w. The verification algorithm $\tilde{V}_{\mathcal{P}^{1+}}$ associated to $\tilde{\mathcal{O}}_{\mathcal{P}^{1+}}$ runs $\hat{V}_{\mathcal{P}^{1+}}$ and the verification algorithm of the proof system Π to check the validity of the proof.

More formally, we define an NP relation $R_{\hat{\mathcal{O}}_{\mathcal{P}^{1+}}}$ based on $\hat{\mathcal{O}}_{\mathcal{P}^{1+}}$ as follows:

$$R_{\hat{\mathcal{O}}_{\mathcal{P}^{1+}}}(P, w) = 1 \text{ if and only if } \exists r \text{ s.t. } P = \hat{\mathcal{O}}_{\mathcal{P}^{1+}}(I_w, r) \ .$$

That is, the first input to the relation must be a valid output of the obfuscator $\hat{\mathcal{O}}_{\mathcal{P}^{1+}}$ (which can be efficiently verified by $\hat{V}_{\mathcal{P}^{1+}}$), and the second input must be the unique point that is accepted by this circuit. The obfuscator $\tilde{\mathcal{O}}_{\mathcal{P}^{1+}}$, described in Algorithm 3, uses the NIZK Π on this NP relation.

Algorithm 3. Obfuscator $\tilde{\mathcal{O}}_{\mathcal{P}^{1+}}$ for the family of point circuits in the CRS model

Input: a circuit of the form I_w and a common reference string Σ
1: produce the obfuscated circuit $\hat{I}_w = \hat{\mathcal{O}}_{\mathcal{P}^{1+}}(I_w)$
2: use Π to prove that the obfuscator knows a witness w to the statement that $R_{\hat{\mathcal{O}}_{\mathcal{P}^{1+}}}(\hat{I}_w, w) = 1$, and call the resulting proof π_w
3: **output** the circuit $\tilde{I}_w$ that is equal to $\hat{I}_w$ except that it also stores π_w in some clearly visible way

---

Intuitively, the non-malleability of the obfuscation follows from the non-malleability of the NIZK. Unfortunately, the proof turns out to be quite delicate, and we can only prove a weaker version of the verifiable non-malleability property.

Definition 11 (Weakly verifiable non-malleability in the CRS model).
Let $\mathcal{C}$ be a family of circuits and $(\mathcal{O}, V)$ be a pair of algorithms. We say that $(\mathcal{O}, V)$ is weakly verifiably non-malleable for relation E if for every polynomial

ρ and *PPT adversary A, there exists a PPT simulator S such that for all sufficiently large n and for all circuits $C \in \mathcal{C}_n$,*

$$|\Pr[P \leftarrow A(\mathcal{O}(C), \Sigma) : P \neq \mathcal{O}(C), V(P, \Sigma) = 1, \exists D \in \mathcal{C}_n \ s.t. \ D \equiv P, \ E(C, D) = 1]$$

$$-\Pr[(Q, \Sigma) \leftarrow S^C(1^n) : V(Q, \Sigma) = 1, \exists D \in \mathcal{C}_n \ s.t. \ D \equiv Q, \ E(C, D) = 1]| < \frac{1}{\rho(n)} \ ,$$

*where the first probability is taken over the coin tosses of A and $\mathcal{O}$, along with
the uniformly random choice of the common reference string Σ, and the second
probability is taken over the coin tosses of S.*

Note that this definition is weaker than the verifiable non-malleability property
in Definition 6 in two ways: the simulator S is allowed to depend on the relation
E, and there is no auxiliary information in this definition. We can prove that our
construction satisfies this weaker variant of non-malleability for many interesting
relations E.

Definition 12 (Invertible relation). *A bivariate relation E is invertible if
there exists a polynomial time algorithm $\bar{E}$ such that for every y, $\bar{E}(y)$ returns
a list of all x such that $E(x, y) = 1$.*

In particular, because E is a polynomial time algorithm, it can only output a
list that is polynomially long in length. Therefore, for every y, there must be
only polynomially many x such that $E(x, y) = 1$.

Theorem 13. *Let E be an invertible relation. In the common reference string
model, $(\tilde{\mathcal{O}}_{\mathcal{P}1+}, \tilde{V}_{\mathcal{P}1+})$ is a weakly verifiably non-malleable obfuscator for E.*

Multi-point circuits. The obfuscator $\tilde{\mathcal{O}}_{\mathcal{P}1+}$ can be generalized to the multi-point
setting, as follows. Let $\tilde{\mathcal{O}}_{\mathcal{P}m+}$ be the obfuscator that, when given $I_{\{w_1,...,w_m\}}$ as
input, outputs the concatenation of m single-point obfuscations $\hat{\mathcal{O}}_{\mathcal{P}1+}(I_{w_1})$, ...,
$\hat{\mathcal{O}}_{\mathcal{P}1+}(I_{w_m})$ followed by a non-malleable NIZK proof of knowledge that it knows
all of the accepted points $w_1, ..., w_m$. As before, let $\tilde{V}_{\mathcal{P}m+}$ be the verification
algorithm that checks the structure of the program and the validity of the proof.
It is shown in [8] that $\tilde{\mathcal{O}}_{\mathcal{P}m+}$ is an obfuscator, and we show that this obfuscator
is weakly verifiably non-malleable for invertible relations.

Theorem 14. *Let E be an invertible relation. In the common reference string
model, $(\tilde{\mathcal{O}}_{\mathcal{P}m+}, \tilde{V}_{\mathcal{P}m+})$ is a weakly verifiably non-malleable obfuscator for E.*

Unfortunately, in the multi-point setting, the set of invertible relations is too
small. For example, the simple relation $E(I_{\{w_1,...,w_m\}}, I_{\{w'_1,...,w'_m\}})$ that accepts
if any of the w_i equal any of the w'_j is not invertible. As a result, Theorem 14 is
a promising result but still unsatisfactory. Future research is needed to find an
obfuscator that is verifiably non-malleable for a wider class of relations.

Acknowledgment. We thank Ronny Dakdouk for his useful comments. In particular, an improvement in Theorem 7 and an error in an earlier proof of Theorem
9 were found by Ronny.

References

1. Barak, B., Goldreich, O., Impagliazzo, R., Rudich, S., Sahai, A., Vadhan, S., Yang, K.: On the (im)possibility of obfuscating programs. In: Kilian, J. (ed.) CRYPTO 2001. LNCS, vol. 2139, pp. 1–18. Springer, Heidelberg (2001)
2. Goldwasser, S., Kalai, Y.T.: On the impossibility of obfuscation with auxiliary input. In: FOCS, pp. 553–562. IEEE Computer Society, Los Alamitos (2005)
3. Goldwasser, S., Rothblum, G.N.: On best-possible obfuscation. In: Vadhan, S.P. (ed.) TCC 2007. LNCS, vol. 4392, pp. 194–213. Springer, Heidelberg (2007)
4. Canetti, R.: Towards realizing random oracles: Hash functions that hide all partial information. In: Kaliski Jr., B.S. (ed.) CRYPTO 1997. LNCS, vol. 1294, pp. 455–469. Springer, Heidelberg (1997)
5. Canetti, R., Micciancio, D., Reingold, O.: Perfectly one-way probabilistic hash functions. In: Proceedings of the 30th ACM Symposium on Theory of Computing, pp. 131–140 (1998)
6. Lynn, B., Prabhakaran, M., Sahai, A.: Positive results and techniques for obfuscation. In: Cachin, C., Camenisch, J.L. (eds.) EUROCRYPT 2004. LNCS, vol. 3027, pp. 20–39. Springer, Heidelberg (2004)
7. Wee, H.: On obfuscating point functions. In: Proceedings of the 37th ACM Symposium on Theory of Computing, pp. 523–532 (2005)
8. Canetti, R., Dakdouk, R.R.: Obfuscating point functions with multibit output. In: Smart, N.P. (ed.) EUROCRYPT 2008. LNCS, vol. 4965, pp. 489–508. Springer, Heidelberg (2008)
9. Hofheinz, D., Malone-Lee, J., Stam, M.: Obfuscation for cryptographic purposes. In: Vadhan, S.P. (ed.) TCC 2007. LNCS, vol. 4392, pp. 214–232. Springer, Heidelberg (2007)
10. Hohenberger, S., Rothblum, G.N., Shelat, A., Vaikuntanathan, V.: Securely obfuscating re-encryption. In: Vadhan, S.P. (ed.) TCC 2007. LNCS, vol. 4392, pp. 233–252. Springer, Heidelberg (2007)
11. Sahai, A.: Non-malleable non-interactive zero knowledge and adaptive chosen-ciphertext security. In: FOCS, pp. 543–553 (1999)
12. De Santis, A., Di Crescenzo, G., Ostrovsky, R., Persiano, G., Sahai, A.: Robust non-interactive zero knowledge. In: Kilian, J. (ed.) CRYPTO 2001. LNCS, vol. 2139, pp. 566–598. Springer, Heidelberg (2001)
13. Canetti, R., Varia, M.: Non-mallable obfuscation. Cryptology ePrint Archive, Report 2008/495 (2008), http://eprint.iacr.org/2008/495
14. Lamport, L.: Constructing digital signatures from a one-way function. Technical Report SRI-CSL-98, SRI International Computer Science Laboratory (1979)

Simulation-Based Concurrent Non-malleable Commitments and Decommitments

Rafail Ostrovsky[1], Giuseppe Persiano[2], and Ivan Visconti[2]

[1] Department of Computer Science and Department of Mathematics,
UCLA, Los Angeles, CA 90095, USA
rafail@cs.ucla.edu
[2] Dipartimento di Informatica ed Applicazioni, Università di Salerno,
84084 Fisciano (SA), Italy
{giuper,visconti}@dia.unisa.it

Abstract. In this paper we consider commitment schemes that are secure against *concurrent* man-in-the-middle (cMiM) attacks. Under such attacks, two possible notions of security for commitment schemes have been proposed in the literature: concurrent non-malleability with respect to *commitment* and concurrent non-malleability with respect to *decommitment* (i.e., opening).

After the original notion of non-malleability introduced by [Dolev, Dwork and Naor STOC 91] that is based on the independence of the committed messages, a new and stronger simulation-based notion of non-malleability has been proposed with respect to openings or with respect to commitment [1,2,3,4] by requiring that for any man-in-the-middle adversary there is a stand-alone adversary that succeeds with the same probability. When commitment schemes are used as sub-protocols (which is often the case) the simulation-based notion is much more powerful and simplifies the task of proving the security of the larger protocols.

The main result of this paper is a commitment scheme that is simulation-based concurrent non-malleable with respect to both commitment and *decommitment*. This property protects against cMiM attacks mounted during both commitments and decommitments which is a crucial security requirement in several applications, as in some digital auctions, in which players have to perform both commitments and decommitments. Our scheme uses a constant number of rounds of interaction in the plain model and is the first scheme that enjoys all these properties under the simulation-based definitions.

1 Introduction

Commitment schemes are fundamental two-party protocols and have been used in the design of more complex cryptographic protocols since the early 80's (e.g., for coin flipping [5] and for zero-knowledge for NP [6]).

The basic setting in which commitment schemes are defined only requires the hiding and binding properties. However several different scenarios need stronger notions of commitment schemes. In some application scenarios, one wants to be able to guarantee that an adversary $\mathcal{A}$, playing as a receiver in an execution in

O. Reingold (Ed.): TCC 2009, LNCS 5444, pp. 91–108, 2009.

which a honest committer commits to message m, is not able to commit to a related value $\tilde{m}$ to a honest receiver in another execution in which $\mathcal{A}$ plays as a committer. It is easy to observe that the hiding property does not guarantee this extra property. This type of adversary is called a *man-in-the-middle* adversary (as the adversary plays in between two honest players). Commitment schemes secure with respect to these attacks are called *non-malleable* commitments.

Two notions of non-malleable commitments have been considered in the literature. A commitment scheme that is *non-malleable with respect to commitment* (in short NMc), first defined by Dolev, Dwork, and Naor [7] guarantees that no polynomial-time man-in-the-middle adversary $\mathcal{A}$ can *commit* to a message $\tilde{m}$ that is related to the message m committed by the honest committer. Instead, a commitment scheme that is *non-malleable with respect to decommitment* (also known as non-malleable with respect to opening), (in short NMd), first defined by Di Crescenzo, Ishai and Ostrovsky [1] guarantees that after the commitment phase, no polynomial-time man-in-the-middle adversary $\mathcal{A}$, observing the decommitment to m of the honest committer, obtains an advantage to *decommit* its commitment to a message $\tilde{m}$ that is related to m.

The need for non-malleable cryptography has been first pointed out in the seminal paper by Dolev, Dwork and Naor [7] who also gave constructions for non-malleable encryption, non-malleable zero-knowledge proofs and non-malleable commitments. The constructions for non-malleable commitments of [7] required $O(\log k)$ rounds, where k is the security parameter. The non-malleability notion of [7] is based on the independence of the committed/decommitted messages played by the man-in-the-middle with respect to the ones played by the sender.

The first non-interactive non-malleable commitment scheme (in the common random string model) was shown by Di Crescenzo, Ishai and Ostrovsky [1] (with further efficiency improvement in [2]). They also introduced a new notion of non-malleability by requiring that for any man-in-the-middle adversary there exists a stand-alone simulator with essentially the same success probability. This new simulation-based notion is stronger than the one of Dolev, Dwork and Naor [7] and is much more useful when a commitment scheme is used as sub-protocol since the security of the larger protocol can be proved more easily by using the simulator associated with the commitment scheme.

The first constant-round non-malleable commitment scheme in the plain model (i.e., without setup assumptions such as a common reference string) has been given by Barak [8] under the assumption of the existence of trapdoor permutations and hash functions that are collision resistant against subexponential-time adversaries. Pass and Rosen [3] reduced the assumption to the existence of hash functions that are collision resistant against polynomial-time adversaries using simulation-based definition. Pass and Rosen [3] gave two different simulation-based schemes: one that is NMc and one that is NMd.

More recently, Pass and Rosen [4] have considered *concurrent* man-in-the-middle attacks (cMiM attacks) where the man-in-the-middle can be active in any polynomial number of executions as a receiver and as a committer. A commitment scheme that is secure against cMiM attacks is called *concurrent non-malleable.* As before, we can have two notions of concurrent non-malleable commitment schemes: concurrent NMc and NMd commitment schemes. Pass and Rosen in [4] showed that the NMc scheme of [3] is actually a simulation-based *concurrent* NMc. This implies that simulation-based security is guaranteed if the commitments are concurrently executed but decommitments are not. Their paper leaves as an open problem the construction of constant-round commitment schemes that are simulation-based concurrent NMd. The scheme of [4] enjoys a weaker notion of non-malleability with respect to decommitment that only focuses on the independence of the opened messages [7].

The security of the scheme of [4] relies on the assumption that commitments and decommitments do not overlap in time. We retain this assumption in our schemes. This assumption is motivated by the fact that several important applications have such a separation (e.g., electronic auctions where first all parties send their hidden bids, and only in a second phase they decommit their bids).

Our results. Our main result consists in the construction in the plain model of a constant-round commitment scheme that is simultaneously concurrent NMc and NMd under the simulation-based definition of [3,4]. This implies that security is preserved when polynomially many commitment phases are concurrently executed and when subsequently polynomial many decommitment phases are concurrently executed. This solves a problem left open by the results of [4] and allows one to securely run some commitment-based applications (e.g., digital auctions) by only requiring a constant number of rounds. We follow [4] in that concurrent non-malleability is guaranteed only if commitments and decommitments do not overlap in time (which is the case in several applications).

Our scheme builds and extends multiple techniques. In particular, our scheme uses the perfect NMZK argument of knowledge of [3,4,9] but in a critically different manner. Indeed, whereas in [3,4] the perfect NMZK argument of knowledge is simply combined with a (potentially malleable) commitment scheme and a signature scheme, to achieve security in a concurrent setting we also employ a technique by Feige [10] and a more sophisticated rewind technique. Furthermore, the simulator used by [3,4] works in a straight-line fashion including non-black-box techniques. Our result, instead, combines the straight-line simulation with a new rewinding simulation that still avoids the well known problems of using rewinds in concurrent settings [11]. Our approach also includes and extends some of the techniques developed for building concurrent NMZK in the bare public-key model [12,13]. Finally we stress that in [3] non-malleability with respect to commitment is considered only with respect to statistically binding commitments. Here, perhaps somewhat surprisingly, we show that it is possible to have non-malleable commitments with respect to commitments that are not statistically binding. This is crucially used in our main result since the constant-round NMc and NMd commitment scheme that we show is not statistically binding.

We also remark that the recent work of Barak et al. [14] obtains concurrent non-malleable zero-knowledge with a poly-logarithmic round complexity, and thus does not seem to help for achieving constant-round simulation-based concurrent non-malleable commitments.

2   Simulation-Based Non-malleable Commitments

Since all our results concern the simulation-based notion of non-malleability, we will concentrate on this notion only. Here we start by considering concurrent non-malleable commitment schemes; that is, commitments schemes that are secure under *Concurrent Man-in-the-Middle* attacks (cMiM attacks). Informally speaking, a non-malleable commitment scheme guarantees that the value committed to (or the value that is decommitted) by a polynomial-time adversary $\mathcal{A}$ is independent of the value simultaneously committed (or the value that is decommitted) to $\mathcal{A}$ by a honest committer. We assume that $\mathcal{A}$ has full power over the scheduling of the messages in the two sessions (the one in which $\mathcal{A}$ is a committer and the one in which $\mathcal{A}$ is a receiver). Following [1,2,3,4], we formalize this notion by comparing two executions: the *man-in-the-middle* execution (the MiM execution) and the *simulated* execution. We denote the security parameter by k and consider the concurrent case where the adversary $\mathcal{A}$ receives and send a polynomial number of commitments.

The Dolev-Dwork-Naor notion of non-malleability. Informally speaking, non-malleability with respect to commitment guarantees that the commitment computed by the MiM adversary corresponds to a message that is independent from the one committed to by the honest committer. In [7], dependency of the values m and $\tilde{m}$ has been formalized through a poly-time computable relation $\mathcal{R}$ for which $\mathcal{R}(m, \tilde{m}) = 1$. Specifically, Dolev, Dwork and Naor[7] defined non-malleability with respect to commitment by requiring that for any man-in-the-middle adversary $\mathcal{A}$ and any polynomial time computable relation $\mathcal{R}$, there exists a poly-time stand-alone adversary S whose success probability in committing to a value $\tilde{m}$ so that $\mathcal{R}(m, \tilde{m}) = 1$ is at least as good as $\mathcal{A}$'s success probability. Non-malleability with respect to decommitment [1] instead considers the ability of $\mathcal{A}$ to decommit to a value $\tilde{m}$ that is related to m. Notice that under the definition of [7], if $\mathcal{A}$ is no more likely to commit to a related value than S and the commitment is statistically binding, then $\mathcal{A}$ is also no more likely to decommit to a related value. This is true regardless of whether $\mathcal{A}$ is given the decommitment information or not. So under this definition, any (statistically binding) commitment that is NMc is also NMd.

A (stronger) simulation-based notion of non-malleable commitments. In this work we adopt the simulation-based definition [1,2,3,4], which requires that the value $\tilde{m}$ committed to by S in the stand-alone execution is computationally indistinguishable from the value committed to by $\mathcal{A}$ in the man-in-the-middle execution. To have a meaningful definition of non-malleability with respect to decommitment, since $\mathcal{A}$ obtains the message m committed by the honest sender before decommitting its commitment, S is assumed to obtain m before decommitting its commitment. It is not clear that with respect to the simulation-based definition, non-malleability with respect to commitment still implies non-malleability with respect to decommitment. The problem here is that (unlike in the [7] definition), one would like the success probability of S

(i.e., the probability that the stand-alone simulator playing with a honest receiver correctly completes the decommitment phase) to be only negligibly far from $\mathcal{A}$'s success probability. Indeed, in the $\mathcal{NM}c$ schemes in the plain model of [3,4], the simulator S generates a bogus commitment that is being fed to $\mathcal{A}$. However, after having committed to some value, S is stuck with the bogus value and it is not clear how to enable S to decommit it to $\mathcal{A}$ as m. (In the common reference string (CRS) model, the situation is easier, as CRS could be arranged so that S can use "equivocal" commitments that can be decommitted to any value, while $\mathcal{A}$ forced to use statistically-binding on CRS commitment [1,2]. Here,we concentrate on the plain model without the CRS, and hence this approach does not work).

From the above discussion we have that the constant-round commitment scheme $\mathcal{NM}c$ of [3] that is proved to be NMc, does not seem to be NMd (according to the simulation-based definition of [3]) or, at least, no evidence of this is provided by the proof of [3]. Specifically, the simulator that computes $c = \mathsf{SBCom}(0^k, s)$ in the commitment phase cannot open c as m since the decommitment phase simply consists in the decommitment phase of SBCom which is *statistically* binding. Therefore under the simulation-based notion of non-malleability, the proof that $\mathcal{NM}c$ is an NMc commitment scheme does not seem to extend to prove that $\mathcal{NM}c$ is also NMd. We stress that in [3,4], only the commitment phase is considered for proving NMc, and since the decommitment phase as discussed above is quite problematic, their security proof implicitly requires that the commitment and decommitment phases do not overlap in time.

NMc does not necessarily require statistical binding. When statistically binding commitments are considered, the commitment phase encodes the unique message to which the commitment can be later decommitted. Indeed, even in case the adversarial committer is unbounded there is no way for him to violate the binding property. Since NMc considers the message committed in the commitment phase, the statistical binding property guarantees that this non-malleability notion is well defined, and indeed in [3] the authors consider the notion of NMc only for statistically binding commitment schemes. Intuitively, NMc seems far more problematic in case the scheme is not statistically binding (but only computationally binding), since the commitment phase does not uniquely specify the message that is going to be decommitted. Therefore, the meaning of NMc for an unbounded adversarial committer is unclear. We observe though that NMc commitments are meant to be secure against polynomial-time MiM adversaries for which the computational binding property still holds. It is therefore potentially possible to have a commitment scheme that is not statistically binding (i.e., binding does necessarily hold in case the adversarial committer is unbounded) but however still is NMc as at the end of the commitment phase it is always possible to determine the message committed by the polynomial-time MiM and by the honest sender. Indeed, in this paper, we show commitment schemes that are *not* statistically binding but that are NMc commitment schemes and, at the same time, NMd. To define NMc we will use the concept of "message committed to by an adversary $\mathcal{A}$ during the commitment phase." By this we mean the following. We will consider commitment schemes in which, for all adversaries $\mathcal{A}$, and for each possible transcript trans of the interaction between adversary $\mathcal{A}$ and a honest receiver R such that R accepts the commitment, there exists (statistically) only one message m that is consistent with trans; that is, for which there exist random coin tosses that give trans. We stress that statistically hiding commitment schemes do not have the above property and thus our definition is not suitable for these commitment schemes.

For lack of space, in the full version of this paper [15] we review the two schemes of [3] for non-malleable commitments: $\mathcal{NM}c$ that is NMc and $\mathcal{NM}d$ that is NMd and we also show a commitment scheme that combines $\mathcal{NM}c$ and $\mathcal{NM}d$.

3   Simulation-Based cNM Commitments

Following [3,4], we now formalize the concept of a (simulation-based) *concurrent* non-malleable commitment scheme by comparing two executions: the *concurrent man-in-the-middle* execution (the cMiM execution) and the *simulated* execution. We denote the security parameter by k.

The cMiM execution. In the cMiM execution, the cMiM adversary $\mathcal{A}$ is simultaneously participating in $\mathsf{poly}(k)$ left and $\mathsf{poly}(k)$ right interactions.

Consider a cMiM execution in which the cMiM adversary $\mathcal{A}$ with auxiliary information z interacts in the i-th left interaction with a honest committer running on input a message m_i of length $\mathsf{poly}(k)$ and in the right interactions with honest receivers. We denote by $\mathsf{cmim}^{\mathcal{A}}_{\mathsf{Com}}(M, z)$, where $M = (m_1, \ldots, m_{\mathsf{poly}(k)})$, the random variable that associates to the cMiM execution a vector $\tilde{M}$ whose i-th component $\tilde{m}_i$ is defined as follows. If the commitment phase of the i-th right interaction ends successfully and its transcript is different from the commitment phase of all the left interactions, then $\tilde{m}_i$ is the message that $\mathcal{A}$ has *committed to* in the i-th right interaction. Otherwise, $\tilde{m}_i = \perp$.

Similarly, we denote by $\mathsf{cmim}^{\mathcal{A}}_{\mathsf{Dec}}(M, z)$ the vector $\tilde{M}$ whose i-th component $\tilde{m}_i$ is the message that $\mathcal{A}$ has *decommitted* in the right interaction. If the i-th right interaction is not successful or its transcript (including commitment and decommitment phase) is identical to the transcript of one of the left interactions then $\tilde{m}_i = \perp$.

The simulated execution. In the simulated execution we have one party S (called the *simulator*) that interacts with $\mathsf{poly}(k)$ honest receivers. S works in two phases: in the commitment phase S receives security parameter 1^k and auxiliary information z and interacts with the honest receivers. We denote by $\mathsf{csis}^{S}_{\mathsf{Com}}(1^k, z)$ the vector $\tilde{M}$ whose i-th component $\tilde{m}_i$ is the value committed to by S if the i-th commitment phase has been successfully completed. Otherwise $\tilde{m}_i$ is set equal to $\perp$.

Once the commitment phases have been completed, S receives input vector M and interacts with the honest receiver to complete the decommitment phase. We denote by $\mathsf{csis}^{S}_{\mathsf{Dec}}(M, z)$ the vector $\tilde{M}$ whose i-th component $\tilde{m}_i$ is the value decommitted by S in the i-th decommitment phase if it has been successfully completed. Otherwise $\tilde{m}_i$ is set equal to $\perp$.

We have the following definitions (see also [3,4]).

Definition 1. *A commitment scheme is simulation-based concurrent non-malleable with respect to commitment (a concurrent NMc commitment scheme) if, for every probabilistic polynomial-time cMiM adversary $\mathcal{A}$, there exists a probabilistic polynomial time simulator S such that following ensembles are computationally indistinguishable:*

$$\{\mathsf{cmim}_{\mathsf{Com}}^{\mathcal{A}}(M,z)\}_{M\in(\{0,1\}^{\mathsf{poly}(k)})^{\mathsf{poly}(k)},z\in\{0,1\}^\star} \text{ and } \{\mathsf{csis}_{\mathsf{Com}}^{S}(1^k,z)\}_{z\in\{0,1\}^\star}.$$

Definition 2. *A commitment scheme is simulation-based concurrent non-malleable with respect to decommitment (a concurrent NMd commitment scheme) if, for every probabilistic polynomial-time cMiM adversary $\mathcal{A}$, there exists a probabilistic polynomial time simulator S such that the following ensembles are computationally indistinguishable:*

$$\{\mathsf{cmim}_{\mathsf{Dec}}^{\mathcal{A}}(M,z)\}_{M\in(\{0,1\}^{\mathsf{poly}(k)})^{\mathsf{poly}(k)},z\in\{0,1\}^\star}$$

and

$$\{\mathsf{csis}_{\mathsf{Dec}}^{S}(M,z)\}_{M\in(\{0,1\}^{\mathsf{poly}(k)})^{\mathsf{poly}(k)},z\in\{0,1\}^\star}.$$

3.1 Commitment Scheme $c\mathcal{NM}cd$

In this section we present a constant-round commitment scheme $c\mathcal{NM}cd$ that enjoys both simulation-based concurrent NMc and simulation-based concurrent NMd. We will use a constant-round tag-based perfect NMZK argument of knowledge $\mathsf{nmZK} = \{\mathcal{P}_t, \mathcal{V}_t\}_t$ for all NP [3], a constant-round witness indistinguishable $(\mathsf{wi}\mathcal{P}, \mathsf{wi}\mathcal{V})$ proof of knowledge (WIPoK) for all NP [16,17], a non-interactive statistically binding commitment scheme Com and a secure signature scheme $SS = (\mathsf{SG}, \mathsf{Sig}, \mathsf{SVer})$. The most sophisticated tool that we use is obtained from a sequence of works by Pass and Rosen.

Theorem 1 ([3,4,9]). *Assume that there exists a family of claw-free permutations. Then for any NP language L there exists a constant-round tag-based one-left many-right perfect cNMZK arguments of knowledge $\mathsf{nmZK} = \{\langle \mathcal{P}_{\mathsf{tag}}, \mathcal{V}_{\mathsf{tag}} \rangle\}_{\mathsf{tag}}$ for all NP.*

According to the above definition, this theorem says that for any efficient one-left many-right concurrent man-in-the-middle adversary $\mathcal{A}$ that is restricted to one left session there exists an efficient simulator S that guarantees: 1) the view (including the left proof and all the right proofs) given in output by S is perfectly indistinguishable from the interaction of $\mathcal{A}$ with honest prover and honest verifiers; 2) the extraction succeeds for all accepting right proofs in which the one-left many-right concurrent man-in-the-middle adversary has used a tag not appearing in the left proof.

See the full version [15] of this paper for details about the other tools.

A description of commitment scheme $c\mathcal{NM}cd$ is found in Figure 1.

Security Parameter: 1^k.
Input to Committer: $m \in \{0,1\}^k$.
Commitment Phase:

> $C \to R$: pick $s \in \{0,1\}^k$, set $c = \mathsf{Com}(m,s)$ and send c to R.
> $C \to R$: set $(\mathsf{PK},\mathsf{SK}) \leftarrow \mathsf{SG}(1^k)$ and send PK to R.
> $C \leftrightarrow R$: C executes the code of $\mathcal{P}_{\mathsf{PK}}$ on input c to prove knowledge of $m,s \in \{0,1\}^k$ such that $c = \mathsf{Com}(m,s)$. R executes the code of $\mathcal{V}_{\mathsf{PK}}$ on input c. If $\mathcal{V}_{\mathsf{PK}}$ rejects then R aborts.
> $R \to C$: pick $m_0,s_0,m_1,s_1 \in \{0,1\}^k$, set $c_0 = \mathsf{Com}(m_0,s_0), c_1 = \mathsf{Com}(m_1,s_1)$ and send c_0 and c_1 to C.
> $R \leftrightarrow C$: R select a random bit b and executes the code of $\mathsf{wi}\mathcal{P}$ on input (c_0,c_1) to prove knowledge of $\hat{m},\hat{s} \in \{0,1\}^k$ such that $c_0 = \mathsf{Com}(\hat{m},\hat{s})$ or $c_1 = \mathsf{Com}(\hat{m},\hat{s})$ using (m_b,s_b) as witness. C executes the code of $\mathsf{wi}\mathcal{V}$ on input (c_0,c_1). If $\mathsf{wi}\mathcal{V}$ rejects then C aborts.
> $C \to R$: let trans_0 be the transcript so far with an extra bit 0 at the end. Set $\sigma_0 \leftarrow \mathsf{Sig}(\mathsf{trans}_0,\mathsf{SK})$ and send σ_0 to R.
> R: if $\mathsf{SVer}(\mathsf{trans}_0,\sigma_0,\mathsf{PK}) \neq 1$ abort.

Decommitment Phase:

> $C \to R$: send m.
> $C \leftrightarrow R$: C executes the code of $\mathcal{P}_{\mathsf{PK}}$ on input (c,c_0,c_1) to prove knowledge of $\hat{m},\hat{s} \in \{0,1\}^k$ such that $c = \mathsf{Com}(m,\hat{s})$ or $c_0 = \mathsf{Com}(\hat{m},\hat{s})$ or $c_1 = \mathsf{Com}(\hat{m},\hat{s})$, using (m,s) as witness. R executes the code of $\mathcal{V}_{\mathsf{PK}}$ on input (c,c_0,c_1). If $\mathcal{V}_{\mathsf{PK}}$ rejects then R aborts.
> $C \to R$: let trans_1 be the transcript so far with an extra bit 1 at the end. Set $\sigma_1 \leftarrow \mathsf{Sig}(\mathsf{trans}_1,\mathsf{SK})$ and send σ_1 to R.
> R: if $\mathsf{SVer}(\mathsf{trans}_1,\sigma_1,\mathsf{PK}) \neq 1$ abort.

Fig. 1. Our concurrent NMc and concurrent NMd commitment scheme $c\mathcal{N}\mathcal{M}cd$

How we achieve concurrent NMd. First of all, we notice that a straight-forward combination of the two commitment schemes of [3] produces a commitment scheme that we call $\mathcal{N}\mathcal{M}cd$ that achieves non-malleability with respect to both commitment and decommitment, when concurrency is not considered. In proving the NMd property one crucially relies on the existence of a simulator extractor for the NMZK argument nmZK. If one tries to argue that $\mathcal{N}\mathcal{M}cd$ is a concurrent NMd commitment scheme along the same lines, one would need a simulator that simulates concurrent executions; in other words, one would need a concurrent NMZK argument of knowledge. Unfortunately, the existence of a constant-round concurrent NMZK argument system in the plain model is still an open problem.

We use instead a more sophisticated protocol and prove its properties by blending the straight-line simulator of the concurrent NMc commitment scheme of [4] with a sophisticated rewind technique. In using rewinding we have to be careful as the nested sessions can potentially make the running time super polynomial[1]. Instead, we perform rewinds "in advance," to extract information from the adversary. The simulator is then able to simulate in a straight-line fashion the decommitment phase by using the information extracted by means of rewinds. Our security proof also employs the two-witness technique by [10] and the well known FLS-technique [18].

In somewhat more details, we extend the commitment phase of the concurrent non-malleable commitment scheme of [4] by requiring that the receiver gives a proof of knowledge of a secret. The decommitment phase consists in sending a message and in proving with a NMZK proof that either the message corresponds to the committed one or the sender knows the secret (this is the FLS-technique [18]). Our simulator will extract the secrets of all receivers in the commitment phase and will use them as fake witnesses in the decommitment phase. Note that one could think that the rewinding technique used by the simulator during the commitment phase could blow up its running time since the adversarial receiver could adaptively play different messages when the transcript changes. Fortunately we adopt a non-dangerous rewind technique that does not harm the running time of the simulator. Indeed, the simulator will first play the commitment phase running the honest sender algorithm. Then it will extract the secrets encoded by the receiver in that specific transcript by running an extractor sequentially for each commitment, one-by-one, starting each time from the same transcript. During this extraction procedure the simulator will not be interested in re-committing again or in simulating concurrent sessions, it will simply play again the honest sender procedure in all sessions with the only exception of the one in which it extracts the secret of the receiver. The extracted secret will only be kept in memory by the simulator and will not be used in the commitment phase. Instead, the decommitment phase will be crucially based on the knowledge of the secrets of the receiver, and will allow the simulator to play in straight-line, opening the committed messages as any messages.

We show that an adversary will not be able to use such a secret, since we prove that any successful adversary can be used to break a standard complexity-theoretic assumption by using the two-witness technique of [10] and the non-malleability of nmZK.

In the next section we prove the properties of commitment scheme $c\mathcal{NM}cd$. We will often use the simulation-extractability property of nmZK. Notice that this property is guaranteed only in case the tag used by the adversary is different from the one used by the other parties. Since in our scheme we use as tag the public key of a signature scheme, and since each phase is only correctly completed if there is a signature under that public key of the transcript of the phase, we assume that the simulation-extractability property always holds, since otherwise the security of the signature scheme is broken. We will detail this argument only when we prove the NMc property for the one-left many-right case (see the discussion below the description of Expt_2), in the other cases the argument is quite similar and is omitted.

---

[1] The study of this problem started with the notion of concurrent zero knowledge [11].

Binding. In the proof of concurrent NMd we show that any man-in-the-middle adversary that completes the commitment phase, can later open that commitment only in one way. This property is even stronger than binding (since the classical adversary for the binding property can not play as receiver) thus that proof properly contains the proof of the binding property.

Hiding. Assume by contradiction that there exists an adversarial receiver $\mathcal{A}$ that, after the commitment phase distinguishes a commitment to m_0 from a commitment to m_1 with non-negligible advantage. We show how to reduce $\mathcal{A}$ to an adversary $\mathcal{A}'$ that breaks the hiding property of Com. Indeed, $\mathcal{A}'$ on input a challenge com (i.e., a commitment of either m_0 or m_1), plays the honest committer algorithm with the following two exceptions: com is sent in the commitment phase and the simulator for nmZK$_{\mathsf{PK}}$ is used instead of the honest prover algorithm. Since the simulation for nmZK$_{\mathsf{PK}}$ is perfect, the only chance $\mathcal{A}$ has to guess concerns the value of com. Therefore, $\mathcal{A}'$ by simply giving in output the same bit given in output by $\mathcal{A}$ succeeds in guessing with non-negligible advantage the message committed in com.

Simulation-Based Concurrent NMc. We start by considering the simpler case in which the adversary $\mathcal{A}$ is active in one left commitment and in polynomially many right commitments (a one-left many-right adversary).

The one-left many-right case. For every one-left many-right MiM adversary $\mathcal{A}$, we consider simulator $S(z)$ that internally runs $\mathcal{A}(z)$ and provides $\mathcal{A}$ with a left commitment by executing the code of the honest committer to commit to 0^k (k is the security parameter). For the right commitments instead S relays messages between the polynomially many honest receivers and $\mathcal{A}$. We stress that for NMc we only have to consider the commitment phase.

We now prove that for all messages $m \in \{0,1\}^k$ and all z, we have that

$$\left| \mathrm{Prob}[\ D(m, \mathsf{cmim}_{\mathsf{Com}}^{\mathcal{A}}(m, z)) = 1\ ] - \mathrm{Prob}[\ D(m, \mathsf{csis}_{\mathsf{Com}}^{S}(1^k, z)) = 1\ ] \right|$$

is negligible in k for all distinguishers D. We consider hybrid experiments starting with $\mathsf{Expt}_0(v, z)$.

$\mathsf{Expt}_0(v, z)$ is the experiment in which $\mathcal{A}(z)$ interacts in the left commitment with a honest committer committing to v and with honest receivers in the right commitments. We denote by $\tilde{M}$ the vector whose i-th component $\tilde{m}_i$ is defined as follows. If the i-th right commitment is successfully completed by $\mathcal{A}$ and its transcript differs from the one of the left commitment then $\tilde{m}_i$ is the message $\mathcal{A}$ has committed to[2] in the i-th right commitment. Otherwise $\tilde{m}_i = \bot$. $\mathsf{Expt}_0(v, z)$ returns $D(v, \tilde{M})$. We set $p_0(v, z) = \mathrm{Prob}[\ \mathsf{Expt}_0(v, z) = 1\ ]$. Obviously, we have that for all z, k and $m \in \{0,1\}^k$, $p_0(m, z) = \mathrm{Prob}[\ D(m, \mathsf{cmim}_{\mathsf{Com}}^{\mathcal{A}}(m, z)) = 1\ ]$ and that $p_0(0^k, z) = \mathrm{Prob}[\ D(m, \mathsf{csis}_{\mathsf{Com}}^{S}(1^k, z)) = 1\ ]$.

---

[2] This is the message that is consistent with the transcript. Since we use a statistically binding commitment scheme there is a unique such message.

To define the next experiment, we observe that $\mathcal{A}$ naturally defines a one-left many-right MiM adversary $\mathcal{A}'$ for nmZK. Specifically, consider the following adversary $\mathcal{A}'$. $\mathcal{A}'(z)$ internally runs $\mathcal{A}(z)$. $\mathcal{A}'$ forwards externally all $\mathcal{A}$'s messages of all the executions of nmZK. For the execution of (wi$\mathcal{P}$, wi$\mathcal{V}$) of each right commitment (here $\mathcal{A}$ acts as a verifier), $\mathcal{A}'$ computes the commitment of two random messages and executes the code of wi$\mathcal{P}$. For the executions of (wi$\mathcal{P}$, wi$\mathcal{V}$) of the left commitments, $\mathcal{A}'$ executes the code of wi$\mathcal{V}$. Now let $\mathcal{S}'$ be the simulator-extractor of nmZK for adversary $\mathcal{A}'$.

Experiment $\mathsf{Expt}_1(v, z)$ differs from $\mathsf{Expt}_0(v, z)$ in that we have the simulator $\mathcal{S}'$ for adversary $\mathcal{A}'$ instead of $\mathcal{A}$ that is playing with the honest prover and honest verifiers for nmZK. More precisely, in the left commitment of $\mathsf{Expt}_1(v, z)$, we first compute $\mathsf{com} = \mathsf{Com}(v, s)$ and $(\mathsf{PK}, \mathsf{SK}) = \mathsf{SG}(1^k)$ and then run $\mathcal{S}'$ on input com, tag PK and z. All other steps (executions of (wi$\mathcal{P}$, wi$\mathcal{V}$) and signatures) are performed just like in $\mathsf{Expt}_0(v, z)$. Let View be the view output by $\mathcal{S}'$ and define vector $\tilde{M}$ as follows. If the i-th right commitment in View is successfully completed and its transcript differs from the one of the left commitment, then set $\tilde{m}_i$ equal to the message committed to (again, this message is unique since Com is statistically binding) by $\mathcal{A}$. Otherwise, set $\tilde{m}_i = \perp$. Finally, $\mathsf{Expt}_1(v, z)$ outputs $D(v, \tilde{M})$. We set $p_1(v, z) = \mathrm{Prob}[\ \mathsf{Expt}_1(v, z) = 1\ ]$. By the perfect NMZK property of nmZK, we have that $p_0(v, z) = p_1(v, z)$ for all v and z.

Experiment $\mathsf{Expt}_2(v, z)$ differs from $\mathsf{Expt}_1(v, z)$ in the way in which vector $\tilde{M}$ (and consequently the output) is computed. Specifically, in $\mathsf{Expt}_2(v, z)$ we set $\tilde{m}_i$ as the message that has been extracted by $\mathcal{S}'$ as part of the witness for the i-th right execution of nmZK. If no message is extracted then $\tilde{m}_i = \perp$. We set $p_2(v, z) = \mathrm{Prob}[\ \mathsf{Expt}_2(v, z) = 1\ ]$.

Denote by $\tilde{\mathsf{PK}}_i$ the signature public key used as a tag for the i-th right execution of nmZK in View and by PK the signature public key used as a tag for the left execution of nmZK in View. First of all observe that, for all i, if the transcript of the i-th right commitment of View differs from the one of the left commitment then, by the security of the signature scheme, the probability that $\tilde{\mathsf{PK}}_i = \mathsf{PK}$ is negligible. Therefore, for each i, only two cases have non-negligible probability. In the first case the transcript of the i-th right commitment is equal to the one of the left commitment (and thus $\tilde{\mathsf{PK}}_i = \mathsf{PK}$). Then we observe that in this case $\tilde{m}_i = \perp$ both in $\mathsf{Expt}_1(v, z)$ and in $\mathsf{Expt}_2(v, z)$. If instead the transcript of the i-th right commitment differs from the one of the left commitment and $\tilde{\mathsf{PK}}_i \neq \mathsf{PK}$ then, by the extraction properties of $\mathcal{S}'$, the value $\tilde{m}_i$ extracted by $\mathcal{S}'$ is not the value committed to by $\mathcal{A}$ in View with negligible probability. Therefore we conclude that $|p_2(v, z) - p_1(v, z)|$ is negligible for all v and z.

We now conclude the proof by showing that for all k and for all $v \in \{0, 1\}^k$, $|p_2(v, z) - p_2(0^k, z)|$ is negligible. Suppose that it is not and thus for infinitely many k there exists $v_k \in \{0, 1\}^k$ and z such that $|p_2(v_k, z) - p_2(0^k, z)| \geq 1/\mathsf{poly}(k)$. Then, we can construct the following adversary B that breaks the hiding of Com. B receives $\hat{c}$ that is a commitment to either 0^k or v_k and executes $\mathsf{Expt}_2(v_k, z)$ by setting in the left commitment phase $c = \hat{c}$. We notice that Expt_2 can be executed in polynomial time even though the message committed to by c in the left interaction is not known. From the output of the experiment B has a non-negligible advantage in guessing the committed bit.

We have shown that both $|p_0(v_k, z) - p_2(v^k, z)|$ and $|p_2(v_k, z) - p_2(0^k, z)|$ are negligible, Using again the same arguments, it follows that $|p_2(0^k, z) - p_0(0^k, z)|$ is negligible. Therefore, we have that $\Big|\text{Prob}[\ D(m, \text{cmim}^{\mathcal{A}}_{\text{Com}}(m, z)) = 1\ ] -$ $\text{Prob}[\ D(m, \text{csis}^S_{\text{Com}}(1^k, z)) = 1\ ]\Big| = |p_0(v^k, z) - p_0(0^k, z)|$ is negligible.

The many-left many-right case for concurrent NMc. We now consider the many-left many-right case. For concurrent MiM adversary $\mathcal{A}$, we consider simulator $S(z)$ that runs $\mathcal{A}(z)$ internally and executes the code of the honest committer on input 0^k for all left commitments. For the right interactions, S relays messages between the external receivers and $\mathcal{A}$. Notice that if we have only one left commitment S coincides with the simulator we used for proving non-malleability with respect to one-left many-right MiM.

Assume by contradiction that there exists a distinguisher D that distinguishes $\text{cmim}^{\mathcal{A}}_{\text{Com}}(M, z)$ and $\text{csis}^S_{\text{Com}}(1^k, z)$. Let $l = \text{poly}(k)$ be the number of left commitments and, for $i = 0, \ldots, l$, consider hybrid experiment $\text{Expt}^{\mathcal{A}}_i$ defined as follows. Let $M = (m_1, \ldots, m_l)$ be a vector of messages. In $\text{Expt}^{\mathcal{A}}_i(M, z)$, adversary $\mathcal{A}$ is run on input z and the j-th honest left committer commits to m_j if $j \leq i$ and to 0^k otherwise. $\text{Expt}^{\mathcal{A}}_i(M, z)$ outputs a vector whose i-th component consists of the messages committed to by $\mathcal{A}$ in the i-th right commitment if it has been successfully completed by $\mathcal{A}$ and if its transcript differs from the transcripts of all the left commitments. If this is not the case then the i-th component of the output of $\text{Expt}^{\mathcal{A}}_i(M, z)$ is set equal to $\bot$. Obviously, for all M and z, $\text{Expt}^{\mathcal{A}}_0(M, z)$ coincides with $\text{csis}^S_{\text{Com}}(1^k, z)$ and $\text{Expt}^{\mathcal{A}}_l(M, z)$ with $\text{cmim}^{\mathcal{A}}_{\text{Com}}(M, z)$. If there exists a probabilistic polynomial time distinguisher D that distinguishes between $\text{csis}^S_{\text{Com}}(1^k, z)$ and $\text{cmim}^{\mathcal{A}}_{\text{Com}}(M, z)$ then there must be $i \in \{0, \ldots l - 1\}$ such that D distinguishes the output of $\text{Expt}^{\mathcal{A}}_i(M, z)$ and the output of $\text{Expt}^{\mathcal{A}}_{i+1}(M, z)$. We stress that the only difference between experiment $\text{Expt}^{\mathcal{A}}_i$ and experiment $\text{Expt}^{\mathcal{A}}_{i+1}$ is that in the $(i+1)$-st left commitment of $\text{Expt}^{\mathcal{A}}_i$ we are committing to 0^k (just like the simulator) whereas in $\text{Expt}^{\mathcal{A}}_{i+1}$ we are committing to m_{i+1}. We can therefore construct a successful MiM adversary $\mathcal{A}'$ for the one-left many-right case. Adversary $\mathcal{A}'$ internally runs all left sessions with the only exception of the $(i+1)$-st session that is played either with a honest committer committing to m_{i+1} or with the simulator of the one-left many-right case. Therefore $\mathcal{A}'$ breaks the one-left many-right non-malleability which is a contradiction.

Simulation-Based Concurrent NMd. For every cMiM adversary $\mathcal{A}$, we describe a simulator S that interacts with polynomially many honest receivers and performs with each of them a commitment and a decommitment phase. To satisfy Definition 2, we will show that, for every vector M of messages, S decommits its commitments to a vector $\tilde{M}$ of messages that is indistinguishable from the messages decommitted by $\mathcal{A}$ when interacting on the left with honest committers committing to M.

The simulator. Since now we also have to care about decommitments, we extend the simulator in the following way. S first runs the left and the right commitment phases with $\mathcal{A}$ executing the code of the honest receiver in the right commitment phases and the code of the honest committer on input message 0^k in the left commitment phase. Notice that $\mathcal{A}$ is interacting solely with S and no honest receiver is involved. Then S runs the extractors for all the proofs (both in left and right commitment phases) provided by $\mathcal{A}$ in order to get the corresponding witnesses. More precisely, for each right commitment phase, S runs the extractor of nmZK and we denote by (m_i, s_i) the witness extracted in the i-th right commitment phase; for each left commitment phase, S runs the extractor of the WIPoK and we denote by $(m_{b_i,i}, s_{b_i,i})$, with $b_i \in \{0, 1\}$, the witness extracted in the i-th left commitment phase. Extractions are executed sequentially and thus the running time of S is polynomial.

Next, S plays the commitment phases with the honest receivers. S does so by executing the code of the honest committer and using, for the i-th commitment phase, message m_i as input.

After the commitment phases have been completed, S receives vector $M^\star = (m_1^\star, \ldots, m_l^\star)$ and has to perform the decommitment phases with $\mathcal{A}$. S does so by resuming the interactions with $\mathcal{A}$ in the following way. In the left decommitment phase corresponding to the i-th left commitment phase, S uses knowledge of $m_{b_i,i}$ to open the commitment (that was originally computed by S as a commitment to 0^k) to $m_i^\star$. In the right decommitment phases, S acts as a honest receiver. Then, for each i, if $\mathcal{A}$ has successfully completed the i-th right decommitment phase, then S completes the i-th decommitment phase with the honest receiver decommitting the commitment to m_i (notice that in the i-th commitment phase with honest receivers, S had committed to m_i). This ends the description of the simulator S.

The above simulator combines the techniques we propose in this paper to overcome the limitations of the [4] result. Our simulator not only guarantees concurrent NMc as we proved previously, but it will also guarantee concurrent NMd. Notice that the [4] simulator only works for concurrent NMc, while for NMd it immediately fails when a single decommitment phase is executed. We now turn to proving that the described simulator S satisfies Definition 2.

We now prove that the distribution of the messages decommitted by $\mathcal{A}$ when interacting with honest committers and honest receivers is indistinguishable from the distribution of the messages decommitted by $\mathcal{A}$ when interacting with S.

Indistinguishability of the simulation. We start with the one-left many-right case and then we will consider the many-left many-right case. We consider a sequence of experiments $\mathsf{Expt}_i^{\mathcal{A}}(m, z)$ and show that any distinguisher D between the experiments can be used to produce a contradiction. Therefore, the output of each experiment is the output of a distinguisher D (which existence is assumed by contradiction) on input a message m and a vector $\tilde{M}$ whose i-th component $\tilde{m}_i$ is defined as follows. If the decommitment phase of the i-th right interaction terminates successfully and its transcript is different from all the left interactions, then $\tilde{m}_i$ is the message that $\mathcal{A}$ has decommitted in the i-th right interaction. Otherwise, $\tilde{m}_i = \perp$. We also set $p_i^{\mathcal{A}}(m, z) = \mathrm{Prob}[\,\mathsf{Expt}_i^{\mathcal{A}}(m, z) = 1\,]$.

$\mathsf{Expt}_0^{\mathcal{A}}(m, z)$ is the experiment in which $\mathcal{A}$ plays with S that behaves as a honest receivers in the right interactions and as a honest committer on input m in the left interaction. We notice that, since S is acting as honest receiver and honest committer, $p_0^{\mathcal{A}}(m, z)$ is the probability that D outputs 1 on input distributed according to $\mathsf{cmim}_{\mathsf{Dec}}^{\mathcal{A}}(m, z)$.

Experiment $\mathsf{Expt}_1^{\mathcal{A}}(m, z)$ differs from Expt_0 only because in the left commitment phase, S runs the extractor of the WIPoK used by $\mathcal{A}$. Since there is no other deviation, we have that $p_1^{\mathcal{A}}(m, z) = p_0^{\mathcal{A}}(m, z)$.

Experiment $\mathsf{Expt}_2^{\mathcal{A}}(m, z)$ differs from Expt_1 in that in the left decommitment phase, S executes the code of the honest prover but uses a fake witness (that is the witness extracted in the left commitment phase from $\mathcal{A}$'s WIPoK). Next we prove that $|p_2^{\mathcal{A}}(m, z) - p_1^{\mathcal{A}}(m, z)|$ is negligible. Assume by contradiction that this difference is non-negligible; as the only difference between the two games consists in the witness used in the nmZK played in the decommitment phase, we show how to break the witness indistinguishability of nmZK. Specifically, we play the following game with an external prover P. We perform the commitment phase like in game $\mathsf{Expt}_1^{\mathcal{A}}(m, z)$. In particular, in the left commitment phase S has computed and sent to $\mathcal{A}$ commitment $c = \mathsf{Com}(m, s)$ and $\mathcal{A}$ has produced commitments c_0 and c_1 and proved knowledge of the message committed to by one of the two. We denote by (m_b, s_b) the witness extracted by S from $\mathcal{A}$'s WIPoK. The decommitment phase proceeds as in game Expt_1 with the exception of the execution of nmZK in the left decommitment phase which is performed by the external prover P. P is fed with the real witness (m, s) and the fake witness (m_b, s_b) and performs the code of the honest prover using one of the two. Notice that the decommitment phase is straight-line. We observe that if P uses the fake witness then we are actually playing game $\mathsf{Expt}_2^{\mathcal{A}}(m, z)$ whereas if P uses the real witness we are playing $\mathsf{Expt}_1^{\mathcal{A}}(m, z)$. Therefore if D distinguishes these two games, we break the witness indistinguishability of nmZK. We stress that in this reduction we have not used the extractor of the nmZK of the decommitment phase, therefore we can relay messages with P without rewinding it.

Next we consider $\mathsf{Expt}_3^{\mathcal{A}}(m, z)$ in which S uses the simulator of nmZK in the left commitment phase. Since the simulation is perfect we have that $p_3^{\mathcal{A}}(m, z) = p_2^{\mathcal{A}}(m, z)$.

Next we consider $\mathsf{Expt}_4^{\mathcal{A}}(m, z)$ in which S commits to 0^k in the left commitment phase. Any distinguisher between $\mathsf{Expt}_4^{\mathcal{A}}(m, z)$ and $\mathsf{Expt}_3^{\mathcal{A}}(m, z)$ can be easily reduced to a distinguisher between a commitment of 0^k and a commitment of m using Com, by simply playing this commitment as c, completing the experiment and then giving in output the same output of the distinguisher. Therefore by the computational hiding of Com we have that $|p_4^{\mathcal{A}}(m, z) - p_3^{\mathcal{A}}(m, z)|$ is negligible.

Next we consider $\mathsf{Expt}_5^{\mathcal{A}}(m, z)$ in which S runs the honest prover of nmZK in the left commitment phase. Since the simulation is perfect we have that $p_5^{\mathcal{A}}(m, z) = p_4^{\mathcal{A}}(m, z)$.

This sequence of experiments shows that the distribution of the messages decommitted by $\mathcal{A}$ during the man-in-the-middle game when the honest sender commits and decommits to m and $\mathcal{A}$ commits and decommits with the honest receiver R (i.e., $\mathsf{Expt}_0^{\mathcal{A}}(m, z)$), is indistinguishable from the distribution of the messages that $\mathcal{A}$ decommits in the simulated game where S plays both as sender committing to 0^k and as receiver (i.e., $\mathsf{Expt}_5^{\mathcal{A}}(m, z)$).

Epilogue. We now show that S is actually a stand-alone adversary, i.e., it can commit and open to a honest receiver R the same messages that $\mathcal{A}$ can open and decommit during a man-in-the-middle game.

Following the description of S, we know that S commits to R the messages that it extracts from $\mathcal{A}$ at the end of the commitment phase of the simulated game. The proof of non-malleability with respect to commitment given previously, says that the messages committed by S to R have the same distribution of the ones committed by $\mathcal{A}$ in the real game. Then the description of S says that S decommits to R the commitments that correspond to the ones that $\mathcal{A}$ decided to decommit to S in the decommitment phase of the simulated game. Since the indistinguishability of the simulation proved so far says that $\mathcal{A}$ decommits to S the same messages that $\mathcal{A}$ decommits in the real game, we have that S decommits to R the same messages decommitted by $\mathcal{A}$ in the real game, unless $\mathcal{A}$ in the real game decommits messages different with respect to the committed ones (indeed, S never decommits to R a message that is different from the committed one).

Therefore we now show that in the real game $\mathcal{A}$ can not open to different messages, this will imply that S decommits to R messages with the same distribution of the ones decommitted by $\mathcal{A}$.

In the real game $\mathcal{A}$ cannot open in a different way. Assume by contradiction that, with some non-negligible probability, in the real game (i.e., when $\mathcal{A}$ plays with a honest prover committing to m and with honest receivers) there exists i such that the decommitted message m_i' is different from the committed message m_i[3]. We denote by c_0 and c_1 the two commitments computed by R in the i-th commitment phase of $\mathcal{A}$ and by $b \in \{0, 1\}$ the bit such that the receiver R used knowledge of the message committed to by c_b to perform the WIPoK of the i-th commitment phase. Given that $\mathcal{A}$ successfully completes the i-th decommitment phase then, we can consider the following experiment. Adversary $\mathcal{A}$ plays with a real sender and a receiver-extractor. The real sender commits to m, while the receiver-extractor runs the honest receiver algorithm for all right commitments and runs the extractor of nmZK of the i-th decommitment phase. The receiver-extractor with overwhelming probability outputs a pair $(\hat{m}, \hat{s})$ such that either $c_b = \mathsf{Com}(\hat{m}, \hat{s})$ or $c_{1-b} = \mathsf{Com}(\hat{m}, \hat{s})$ (i.e., since $\mathcal{A}$ decommitted to a different message, the witness must be a fake one).

Suppose that with some non-negligible probability it happens that $c_{1-b} = \mathsf{Com}(\hat{m}, \hat{s})$. Then we break the hiding property of Com. Consider the following

---

[3] The committed message is the one uniquely specified by the statistically binding commitment scheme used as sub-protocol.

adversary $\mathcal{B}$ that receives a commitment $\hat{c}$ and would like to compute the message committed to by $\hat{c}$ with some non-negligible probability. $\mathcal{B}$ interacts with $\mathcal{A}$ and plays all commitment phases as the honest senders and receivers, with the only exception of the i-th commitment phase played as receiver. Here $\mathcal{B}$ picks a random $b \in \{0, 1\}$, a random $m_b \in \{0, 1\}^k$ and random $s_b \in \{0, 1\}^k$ and computes commitment $c_b = \mathsf{Com}(m_b, s_b)$ and sets $c_{1-b} = \hat{c}$. Then $\mathcal{B}$ continues the commitment phase by running the code of the honest prover wiP of the WIPoK using (m_b, s_b) as witness. By our hypothesis, with some non-negligible probability, the extractor gives the message committed to by $\hat{c}$, this gives to $\mathcal{B}$ a non-negligible advantage for breaking the hiding property of Com.

Suppose instead that, except with negligible probability, it happens that $c_b = \mathsf{Com}(\hat{m}, \hat{s})$. We show that the witness indistinguishability of the WIPoK is violated. More specifically, we consider a WI adversary $\mathcal{B}$ that executes internally all the previous interactions with the only exception that the WIPoK of the i-th right commitment phase is played by relaying messages with an external prover (that uses a witness for $c_{b^\star}$ for some $b^\star \in \{0, 1\}$). $\mathcal{B}$ then plays internally the decommitment phases with the exception of the i-th decommitment phase for which the extractor is used. By looking at the extracted witness, $\mathcal{B}$ will guess the witness used by the external prover.

Summing up. We have therefore shown that $\mathcal{A}$ decommits successfully only the committed messages. Moreover, we have shown that in the simulated game $\mathcal{A}$'s choices for which commitment have to be decommitted are indistinguishable from its choices in the simulated game. These two properties guarantee that S decommits to R messages indistinguishable from the ones decommitted by $\mathcal{A}$ in the real game.

This terminates the proof for the one-left many-right case.

The many-left many-right case for concurrent NMd. Let $l = \mathsf{poly}(k)$ be the size of the vector of messages M, we consider the hybrid games $\{\mathsf{Expt}_i^{\mathcal{A}}\}_{0 \le i \le l}$, where $\mathsf{Expt}_i^{\mathcal{A}}$ for $i = 0, \ldots l$ is defined as follows. In the game $\mathsf{Expt}_i^{\mathcal{A}}$ the committer commits to m_j as the j-th commitments if $j \le i$, and to 0^k if $j > i$. Moreover in $\mathsf{Expt}_i^{\mathcal{A}}$ the i-th commitment is decommitted using a legal witness if $j \le i$ and a fake witness if $j > i$. Obviously $\mathsf{Expt}_0^{\mathcal{A}}$ corresponds to the game played by the simulator (including both the commitment and decommitment phases) while $\mathsf{Expt}_l^{\mathcal{A}}$ corresponds to game played by the honest committer (again, including both the commitment and decommitment phases). For all M and z we denote by $\{\mathsf{csis}_{\mathsf{Dec}}^{\mathsf{Expt}_i^{\mathcal{A}}}(M, z)\}$ the random variable that associates to each successfully completed decommitment phase of $\mathsf{Expt}_i^{\mathcal{A}}$ the messages decommitted by $\mathcal{A}$. Instead $\{\mathsf{csis}_{\mathsf{Dec}}^{\mathsf{Expt}_i^{\mathcal{A}}}(M, z)\}$ associates the value $\perp$ to interactions that have not been completed by $\mathcal{A}$.

Assume by contradiction that the scheme is not concurrent non-malleable with respect to decommitment. It follows that there must be an index $i \in \{0, \ldots l - 1\}$ such that D distinguishes with non-negligible probability between $\{\mathsf{csis}_{\mathsf{Dec}}^{\mathsf{Expt}_i^{\mathcal{A}}}(M, z)\}$ and $\{\mathsf{csis}^{\mathsf{Expt}_{i+1}^{\mathcal{A}}}\mathsf{Dec}(M, z)\}$. The only difference between game

$\mathsf{Expt}_i^{\mathcal{A}}$ and game $\mathsf{Expt}_{i+1}^{\mathcal{A}}$ for $i \in \{0, \ldots l-1\}$ is that the $i+1$ commitment is computed for message 0^k in $\mathsf{Expt}_i^{\mathcal{A}}$ while it is computed for message m_i in $\mathsf{Expt}_{i+1}^{\mathcal{A}}$. Moreover the corresponding decommitment uses a fake witness in $\mathsf{Expt}_i^{\mathcal{A}}$ and a legal witness in $\mathsf{Expt}_{i+1}^{\mathcal{A}}$.

We can therefore construct a successful MiM adversary $\mathcal{A}'$ for the one-left many-right case. Adversary $\mathcal{A}'$ internally runs all left sessions with the only exception of the $(i + 1)$-st commitment and the corresponding decommitment that is played either with a honest committer committing to m_{i+1} or with the simulator of the one-left many-right case. Therefore $\mathcal{A}'$ breaks the one-left many-right non-malleability which is a contradiction.

From the previous discussion and by observing that existence of a family of claw-free permutations is sufficient for the tools we use, we have the following theorem and corollary.

Theorem 2. *Under the assumption of existence of a tag-based one-left many-right perfect cNMZK arguments of knowledge for all* NP, *of a secure signature scheme and of a statistically-binding non-interactive commitment scheme, commitment scheme $\mathcal{NM}cd$ is both simulation-based concurrent NMc and simulation-based concurrent NMd.*

Corollary 1. *Under the existence of a family of claw-free permutations there exists a constant-round commitment scheme that is both simulation-based concurrent NMc and simulation-based concurrent NMd.*

Acknowledgments

We thank the anonymous reviewers for their suggestions. The work of the first author has been supported in part by IBM Faculty Award, Xerox Innovation Group Award, NSF grants 0430254, 0716835, 0716389, 0830803 and U.C. MICRO grant. The work of the authors has been supported in part by the European Commission through the EU IST program under Contract IST-2002-507932 ECRYPT, and the one of the last two authors through the the EU ICT program under Contract ICT-2007-216646 ECRYPT II and through the FP6 program under contract FP6-1596 AEOLUS.

References

1. Di Crescenzo, G., Ishai, Y., Ostrovsky, R.: Non-interactive and non-malleable commitment. In: 30th Annual ACM Symposium on Theory of Computing, Dallas, Texas, USA, pp. 141–150. ACM Press, New York (1998)
2. Di Crescenzo, G., Katz, J., Ostrovsky, R., Smith, A.: Efficient and non-interactive non-malleable commitment. In: Pfitzmann, B. (ed.) EUROCRYPT 2001. LNCS, vol. 2045, pp. 40–59. Springer, Heidelberg (2001)
3. Pass, R., Rosen, A.: New and Improved Constructions of Non-Malleable Cryptographic Protocols. In: 37th Annual ACM Symposium on Theory of Computing, pp. 533–542. ACM Press, New York (2005)

4. Pass, R., Rosen, A.: Concurrent non-malleable commitments. In: 46th Annual Symposium on Foundations of Computer Science, pp. 563–572. IEEE Computer Society Press, Los Alamitos (2005)
5. Blum, M.: Coin flipping by telephone. In: Proc. IEEE Spring COMPCOM, pp. 133–137 (1982)
6. Goldreich, O., Micali, S., Wigderson, A.: Proofs that yield nothing but their validity and a methodology of cryptographic protocol design. In: 27th Annual Symposium on Foundations of Computer Science, Toronto, Ontario, Canada, pp. 174–187. IEEE Computer Society Press, Los Alamitos (1986)
7. Dolev, D., Dwork, C., Naor, M.: Non-malleable cryptography. In: 23rd Annual ACM Symposium on Theory of Computing, New Orleans, Louisiana, USA, pp. 542–552. ACM Press, New York (1991)
8. Barak, B.: Constant-round coin-tossing with a man in the middle or realizing the shared random string model. In: 43rd Annual Symposium on Foundations of Computer Science, Vancouver, British Columbia, Canada, pp. 345–355. IEEE Computer Society Press, Los Alamitos (2002)
9. Pass, R., Rosen, A.: Concurrent nonmalleable commitments. SIAM Journal on Computing 37, 1891–1925 (2008)
10. Feige, U.: Alternative Models for Zero Knowledge Interactive Proofs. Weizmann Institute of Science (1990)
11. Dwork, C., Naor, M., Sahai, A.: Concurrent zero-knowledge. In: 30th Annual ACM Symposium on Theory of Computing, Dallas, Texas, USA, pp. 409–418. ACM Press, New York (1998)
12. Ostrovsky, R., Persiano, G., Visconti, I.: Constant-round concurrent non-malleable zero knowledge in the bare public-key model. In: Aceto, L., Damgård, I., Goldberg, L.A., Halldórsson, M.M., Ingólfsdóttir, A., Walukiewicz, I. (eds.) ICALP 2008, Part II. LNCS, vol. 5126, pp. 548–559. Springer, Heidelberg (2008)
13. Ostrovsky, R., Persiano, G., Visconti, I.: Concurrent non-malleable witness indistinguishability and its applications. Technical Report ECCC Report TR06-095, ECCC (2006)
14. Barak, B., Prabhakaran, M., Sahai, A.: Concurrent non-malleable zero knowledge. In: 47th Annual Symposium on Foundations of Computer Science. IEEE Computer Society Press, Los Alamitos (2006)
15. Ostrovsky, R., Persiano, G., Visconti, I.: Constant-round concurrent non-malleable commitments and decommitments. Technical Report 2008/235, Cryptology ePrint Archive (2008)
16. Blum, M.: How to Prove a Theorem So No One Else Can Claim It. In: Proceedings of the International Congress of Mathematicians, pp. 1444–1451 (1986)
17. Feige, U., Shamir, A.: Witness indistinguishable and witness hiding protocols. In: 22nd Annual ACM Symposium on Theory of Computing, Baltimore, Maryland, USA, pp. 416–426. ACM Press, New York (1990)
18. Feige, U., Lapidot, D., Shamir, A.: Multiple NonInteractive Zero Knowledge Proofs under General Assumptions. SIAM Journal on Computing 29, 1–28 (1999)

Proofs of Retrievability via Hardness Amplification

Yevgeniy Dodis[1], Salil Vadhan[2], and Daniel Wichs[1]

[1] Department of Computer Science, New York University
{dodis,wichs}@cs.nyu.edu
[2] Harvard School of Engineering & Applied Sciences and Center for Research on Computation
and Society, Cambridge, MA
salil@eecs.harvard.edu

Abstract. *Proofs of Retrievability (PoR)*, introduced by Juels and Kaliski [JK07], allow the client to store a file F on an untrusted server, and later run an efficient audit protocol in which the server proves that it (still) possesses the client's data. Constructions of PoR schemes attempt to minimize the client and server storage, the communication complexity of an audit, and even the number of file-blocks accessed by the server during the audit. In this work, we identify several different variants of the problem (such as bounded-use vs. unbounded-use, knowledge-soundness vs. information-soundness), and giving nearly optimal PoR schemes for each of these variants. Our constructions either improve (and generalize) the prior PoR constructions, or give the first known PoR schemes with the required properties. In particular, we

- Formally prove the security of an (optimized) variant of the bounded-use scheme of Juels and Kaliski [JK07], without making any simplifying assumptions on the behavior of the adversary.
- Build the first unbounded-use PoR scheme where the communication complexity is linear in the security parameter and which does not rely on Random Oracles, resolving an open question of Shacham and Waters [SW08].
- Build the first bounded-use scheme with *information-theoretic* security.

The main insight of our work comes from a simple connection between PoR schemes and the notion of *hardness amplification*, extensively studied in complexity theory. In particular, our improvements come from first abstracting a purely information-theoretic notion of *PoR codes*, and then building nearly optimal PoR codes using state-of-the-art tools from coding and complexity theory.

1 Introduction

Many organizations and even average computer users generate huge quantities of electronic data. Although advances in hard-disk capacity have mostly kept up, allowing most users to store their data locally, there are many reasons not to do so. Users worried about reliability want to have replicated copies of their files stored remotely in case their local storage fails. Remotely stored data can be made accessible from many locations making it more convenient for many users. Some companies provide useful functionality on remotely stored data using the "software as a service" model. For example, many web-based e-mail services provide tools for searching and managing remotely stored e-mails, making it beneficial for users to store these remotely. Lastly, some organizations

O. Reingold (Ed.): TCC 2009, LNCS 5444, pp. 109–127, 2009.

create large data sets that must be archived for many years, but are rarely accessed and so there is little reason to store such data locally.

PROOFS OF RETRIEVABILITY. One problem with remote storage is that of accountability. If remotely stored data is rarely accessed, how can users be sure that it is being stored honestly? For example, if a remote storage provider experiences hardware failure and loses some data, it might reason that there is no need to notify its clients, since there is a good chance that the data will never be accessed and, hence, the client would never find out! Alternatively, a malicious storage provider might even choose to delete rarely accessed files to save money. To assuage such concerns, we would like a simple auditing procedure for clients to verify that their data is stored correctly.

Such audits, called *Proofs of Retrievability (PoR)*, were first formalized by Juels and Kaliski in [JK07]. In a PoR protocol a client stores a file F on a server an keeps only a very short private verification string locally. Later, the client can run an audit protocol in which it acts as a verifier while the server proves that it possesses the client's data. The security of a PoR protocol is formalized by the existence of an extractor that *retrieves* the original file F from *any* adversarial server that can pass an audit with some reasonable probability. One simple PoR protocol would be for the client to sign the file F and store only the verification key locally. Then, to run an audit, the server would send the file along with the signature. Of course, for practical use, we are interested in schemes with significantly better efficiency. In particular, we want to minimize the communication between the client and the server, and even wish that the amount of data read by the server to run an audit should be much smaller than (essentially independent of) the size of the original file.

In the rest of the introduction, we introduce a general "PoR framework" and show how prior PoR constructions fit into it. We then describe our contributions by showing how to optimize the components of this framework. We also explain a connection between PoR and "hardness amplification", which will allow us to get qualitatively stronger results for some of our schemes.

1.1   The PoR Framework

Our framework consists of two parts: first, we define a purely information-theoretic primitive which we call a *PoR code* and, second, we give several options for converting PoR codes into "full" PoR schemes.

POR CODES. A PoR code consists of three procedures Init, Read and Resp. The function Init specifies the *initial encoding* of the original *client file* F into the *server file* $F' = \text{Init}(F)$ which is stored on the server. The functions Read, Resp are used to specify a *challenge-response* audit protocol. The client sends a *random challenge* e which consists of two parts $e = (e_1, e_2)$. The first part of the challenge identifies a set of t indices $(i_1, \ldots, i_t) = \text{Read}(e_1)$, which correspond to t locations in F' that the server should read to compute its response. We refer to t as the *locality parameter* and attempt to minimize it. The server reads the sub-string $x = F'[i_1]||\ldots||F'[i_t]$ of the server file F' and computes a *response* $\mu = \text{Resp}(x, e_2)$, which it sends to the client. The PoR code specifies a natural but incomplete PoR protocol, as depicted in Figure 1.

1.	The client starts out with the *client file* F and computes a *server file* $F' = \mathsf{Init}(F)$, to store on the server.
2.	To run an audit, the client picks a random challenge $e = (e_1, e_2)$ and sends it to the server.
3.	The server reads t locations $(i_1, \ldots, i_t) = \mathsf{Read}(e_1)$ of F', resulting in a sub-string x of length t and sends a response $\mu = \mathsf{Resp}(x, e_2)$ to the client.

Fig. 1. An Incomplete PoR Protocol based on a PoR Code ($\mathsf{Init}, \mathsf{Read}, \mathsf{Resp}$)

EXTRACTION PROPERTY. For the security of a PoR code, we want to ensure that any (even *computationally unbounded*) adversary $\mathcal{A}$, which provides the correct value μ with some "reasonable" probability ε, must indeed "know" the file F. Towards that goal we require the existence of a *decoder* $\mathcal{D}$ which decodes the file F given oracle access to some such adversary $\mathcal{A}$. We distinguish between two types of "ε-adversaries": an ε-*erasure* adversary answers correctly with probability at least ε and *does not answer* the rest of the time, while an ε-*error* adversary answers correctly with probability at least ε and can *answer incorrectly* the rest of the time. As the names suggest, there is a clear relation between our problem and the erasure/error decoding of *error-correcting codes (ECC)*. In other words, we can think of the list of all correct responses μ as the challenge e varies as comprising a (possibly exponential) encoding of F and an ε-(erasure/error) adversary as defining a corrupted codeword. The extraction property requires that we construct PoR codes which are (erasure/error)-decodable from an ε fraction of correct responses. Since we want to allow $\varepsilon \ll \frac{1}{2}$, we will need to rely on the notion of *list-decoding* (i.e. the decoder $\mathcal{D}$ outputs a small list of L candidates, one of which is the actual file F) for the case of *errors*. Notice that the functions $\mathsf{Init}, \mathsf{Read}, \mathsf{Resp}$ give our PoR codes a special structure, which is not usually present in general ECCs: for any client file F, the server file $F' = \mathsf{Init}(F)$ allows the server to compute a response μ for a challenge e (i.e. any arbitrary position in the full codeword) efficiently by accessing only t "blocks" of F' for some small t. The server file F' should not be much larger than the original file F, while the full codeword, which consists of all responses μ, could be exponentially long and is never computed or stored in full.

BASIC PoR CODE CONSTRUCTION. We now describe a basic PoR code construction, which is the basis of most prior work. The function Init is simply an encoding of F under some appropriate ECC, so that F can be recovered from any δ fraction of the blocks of the server file F'. The challenge is a random t-tuple of locations in F', and the response is the value of F' at those locations. We can think of this in our framework as e_1 explicitly listing t locations, e_2 being empty, and the $\mathsf{Read}, \mathsf{Resp}$ functions being identity. On an intuitive level, in order for the server to "forget" *any* part of F, it must "forget" *at least* $(1 - \delta)$-fraction of the blocks of F', in which case it should not be able to respond correctly with probability better than $\varepsilon = \delta^t$ (which can be made exponentially small by choosing a constant $\delta < 1$ and setting t to be proportional to the security parameter). However, this is only intuition and our actual proof needs to work the other way — given an adversarial server that responds correctly with probability $\varepsilon > \delta^t$, we need to decode the original file.

FULL PoR SCHEMES. The protocol shown in Figure 1 is incomplete, since the server can give any answer in step 3 and we did not specify how the client decides if to accept or reject the audit! We need to ensure that any adversarial server that passes a single audit with probability ε *is* an ε-(erasure/error) adversary. To that end we can have several possible techniques for converting a PoR code into a full PoR scheme.

1. INFORMATION-THEORETIC BOUNDED-USE PoR. The simplest technique is to have the client precompute several random challenge-response pairs $\{(e^{(i)}, \mu^{(i)})\}$ and store them locally before giving F' to the server. Later, in the i-th audit, the client sends the challenge $e^{(i)}$ and directly verifies the server's response by comparing it against the correct stored value $\mu^{(i)}$. To argue the security of this construction, we notice that a prover who can pass an audit with probability ε must be an ε-error adversary. The advantage of this solution comes from the fact that it does not need to rely on any computational assumptions, and, hence, we get *information-theoretic security*. The downside comes from the fact that a fresh challenge-response pair is needed for each audit. Thus, we will only get a *bounded-use information-theoretic PoR*, where the client's storage will be proportional to the maximum number of audits ℓ.

2. COMPUTATIONAL BOUNDED-USE PoR. It is simple to reduce the client's storage in the above protocol, and make it independent of the number of audits that the client runs, by settling for *computational security*. Firstly, the client picks the challenges $e^{(i)}$ pseudorandomly so that it only needs to remember a short key k_1 for a *pseudorandom function (PRF)* f and can efficiently recreate the challenge $e^{(i)} = f_{k_1}(i)$ later on at the time of an audit. Secondly, the client does not store the responses $\mu^{(i)}$ at all, but instead computes a *tag* $\sigma_i = f_{k_2}((i, \mu^{(i)}))$ for each response, and stores the tags σ_i on the server. During the ith audit protocol, the server computes a a response and sends it along with the ith tag σ_i, so that the client can verify that the response is correct. Note that the client only stores two short keys k_1, k_2 and the rest of the storage is relegated to the server. If the file F is much larger than the maximum number of audits ℓ, the extra server storage will also be relatively insignificant, resulting in a very efficient *bounded-use computational PoR*. The tags σ_i hide future challenge values while ensuring that the server's response is correct, and thus the security analysis is similar to that of the information-theoretic scheme.

3. COMPUTATIONAL UNBOUNDED-USE PoR. To get an *unbounded* (computational) PoR scheme, where the client can run an unlimited number of audits, we need a slightly more complicated technique. The basic idea is for the client to provide the server with some *authenticator-tags* for the blocks of F' in addition to the actual server file F', and keep only some small *verification key* locally. The authenticator-tags must be designed specifically to fit the PoR codes in such a way that the server can use them to authenticate the response μ for an *arbitrary* challenge e and convince the client that μ is correct. For example, in the basic PoR code construction, where the response $\mu = F'[i_1]\|\ldots\|F'[i_t]$ consists of a subset of blocks, the authenticators-tags can simply be the tags of the individual blocks of F' under some *message authentication code (MAC)* so that the server sends the response μ together with the tags of *each of* the blocks $F'[i_1], \ldots, F'[i_t]$. For more complicated PoR codes, we will require smarter authenticator-tag constructions which allow the server to authenticate a short response μ by aggregating the authenticator-tags of the individual blocks in some clever way.

BOUNDED OR UNBOUNDED? Let us compare this last solution with the bounded-use approaches from before. On the positive side, unbounded-use schemes do not force us to choose the number of audits ahead of time. On the negative side, the main problem with the authenticator-based solution is that, in all known schemes, *the authenticators require a significant amount of extra storage on the server*, resulting in a large server storage overhead. Intuitively, each block of the server file F' usually needs to be separately authenticated and therefore the solution usually *doubles* server storage (or increases communication complexity by settling for large blocks). Bounded-use schemes can often be significantly more efficient in terms of server storage, especially when the number of audits to be run is much smaller than the file-size. For example, for a 1 GB file, current unbounded-use schemes only become more efficient than bounded-use ones if the client wants to run more than 33 million audits (at one audit per hour, this won't occur for 3,800 years)![1] Therefore, there is often a good reason to study bounded-use schemes in practice. Moreover, bounded-use schemes also allow us to get *information-theoretic* security (at a cost in client storage).

In summary, there are tradeoffs in parameters and security between bounded and unbounded use schemes, and which one is better depends on the particular application. However, there is also another fundamental difference between these schemes. The use of authenticators in unbounded schemes allows us to restrict our attention to ε-*erasure* adversaries, since the decoder can always detect if the adversary answers incorrectly for any challenge. As we shall see, this will translate to a relatively simple form of (unique-)decoding with a straightforward proof. In bounded-use schemes, the decoder cannot verify if arbitrary responses are correct and therefore we need to analyze the (list-)decoding of PoR codes with respect to an ε-*error* adversary, which will make our analysis more difficult. Although we only require list-decoding for the *PoR code*, we would like the extractor for the *full PoR scheme* to output a single candidate which matches the client's file F with overwhelming probability. To do so, we will also make the client store a short, private, "almost-universal" hash h of the file F and the extractor outputs the only possibility in the list which matches the hash.

1.2 Prior Work

Naor and Rothblum [NR05] studied a primitive called *sublinear authenticators* which is closely related to PoR. Although the motivation for sublinear authenticators is somewhat different than PoR, we can also think of sublinear authenticators as PoR schemes that provide security against an *restricted class of adversarial servers*. In particular, for sublinear authenticators, we assume that the adversarial server runs the *honest code* of an audit protocol, but does so using some possibly *modified* server file $\widetilde{F}' \neq F'$. In the PoR setting, the adversarial server may use an *arbitrary strategy* to run an audit protocol and its responses may not be consistent with *any* particular server file. Hence, security for sublinear-authenticators is strictly weaker and does *not necessarily* imply security for PoR. The main result of Naor and Rothblum is a lower bound for *information theoretic unbounded-use* sublinear authenticators, which translated to a lower-bound for *information theoretic unbounded-use* PoR schemes, essentially showing that

---

[1] We assume an additional server storage of (at least) 1 GB for the unbounded-use schemes versus 256 bits per audit for bounded-use schemes, on top of the file F'.

schemes of this type *cannot* be efficient. In addition, Naor and Rothblum proposed two constructions for sublinear authenticators: a bounded-use information theoretic scheme (corresponding to construction 1 in our framework) and an unbounded-use computational scheme (corresponding to construction 3). Both schemes use the basic PoR code that we described. In [NR05], these schemes were shown to be secure as sublinear authenticators *but not* as PoR schemes.

Juels and Kaliski [JK07] were the first to define PoR schemes formally and gave two PoR constructions. First, they show that the unbounded-use computational scheme of [NR05] is also secure as a PoR scheme. Second, Juels and Kaliski propose a computational bounded-use scheme as their main construction. This scheme is essentially equivalent to construction 2 within our framework and also uses the basic PoR code. However, to prove the security of the scheme, [JK07] resorts to a simplifying assumption on the behavior of the adversary called *block-independence*, more or less assuming that the PoR attacker follows the restrictive syntax of the sublinear-authenticator attacker mentioned above. Put differently, they only show the security of their scheme as a sublinear authenticator, and make the "simplifying assumption" that this is enough for full PoR security. In a follow-up work, Bowers et al. [BJO08] use a slightly different simplifying assumption, requiring that the adversary cannot update its state during a multi-round protocol. Neither work gives any security guarantees against fully Byzantine adversarial servers.

Shacham and Waters [SW08] notice that, in the *unbounded-use* scheme of [NR05, JK07] (corresponding to construction 3, with basic PoR code), the communication between server and client is unnecessarily large; in particular, it is $O(\lambda^2)$ where λ is the security parameter. This is because a server response consists of $t = O(\lambda)$ file blocks, each of which is of size $O(\lambda)$ (being the tag for the block under the message authentication code). In our language, the problem is that the underlying PoR code has a large *output alphabet*. Shacham and Waters showed that this is not necessary, by constructing a new scheme which implicitly uses an *improved* PoR code with a smaller output alphabet. This scheme improves the server's response size but, unfortunately, at the cost of increasing the client's challenge size to $O(\lambda^2)$ — a problem which was remedied in [SW08] through the use of Random Oracles. The main contribution of Shacham and Waters is the construction of *homomorphic linear authenticators*, following a similar (but informal and less efficient) approach of Ateniese et al. [ABC$^+$07]. Such authenticators allow the server to aggregate the tags of individual file blocks and authenticate a response under the improved PoR code (actually, any linear functions of the blocks) using a single short tag. Shacham and Waters (again following [ABC$^+$07]) also design a PoR scheme with *public verifiability* (i.e. the client need not store any private data), using a clever construction of publicly verifiable homomorphic authenticators.

1.3   Our Results

We make two observations about the prior work. Firstly, although all constructions implicitly use some form of PoR codes, such codes have not been defined or studied explicitly. In particular, [NR05, JK07] use the basic PoR code, while the recent work of [SW08] has a clever but ad-hoc optimization which improves some parameters (the response size) at the cost of others (the challenge size). Secondly, we notice that *none* of the prior *bounded-use* schemes are known to be to secure against fully Byzantine adversarial server strategies, even though such schemes are often more efficient and practical than unbounded-use constructions in many scenarios.

In this work, we undertake a thorough study of PoR codes and give a construction of PoR codes based on *hitting samplers* and *error-correcting codes*. Since all prior PoR code constructions (which implicitly appeared in prior work) employ sub-optimal variants of these primitives, we can improve the efficiency of all known constructions. In particular, we show how to construct a variant of the computational unbounded-use scheme of [SW08], where the challenge size *and* response size are short, without the need of the random-oracle model. We also show that our optimizations improve the prior bounded-use schemes and allow for a *flexible tradeoff* between the locality t and the server storage overhead, without increasing the communication complexity.

As we have already mentioned, proving the security of bounded-use schemes (information theoretic and computational) relies on the security of PoR codes with respect to ε-*error* adversaries, which was never shown fully in prior work. It turns out that most of the difficulty lies in decoding the basic PoR code. Interestingly, this problem is intimately connected to *hardness amplification* and, more specifically, to *direct product theorems* [Yao82, Lev87, Imp95, GNW95, IW97, Tre03, IJK06, IJKW08]. Informally, direct product theorems state that if a given task T is hard to accomplish (for a certain class of attackers) with probability more than δ, then t independently sampled tasks $T_1, \ldots, T_t$ are hard to *simultaneously* accomplish (for a slightly weaker class of attackers) with probability significantly greater than the "expected" $\varepsilon = \delta^t$. To prove such a statement, one starts with the attacker A who can solve $T_1, \ldots, T_t$ with probability greater than ε, and builds a more complicated attacker B who can solve a single task T with probability more than δ. The connection between hardness amplification and error-decoding for the basic PoR code is as follows: one simply defines T to be the task of predicting a random location of F'. Then, the attacker A above becomes an adversarial server who answers ε-fraction of the t-tuples in F', and the constructed attacker B becomes an "extractor" who recovers a δ-fraction of F' (which should suffice in recovering all of F). Using this connection, we will be able construct an efficient extractor for bounded-use schemes, and, thus, provide the *first formal proofs of security* for such schemes, including the first information-theoretic bounded-use scheme.

In Table 1, we compare the efficiency of our schemes with prior construction. We assume that the client file size is k, the security parameter is λ, and choose to parameterize the schemes by a value γ in the range $1 < \gamma \leq 2$, which allows us to formulate a flexible tradeoff between parameters. In all of the schemes the locality $t = \mathcal{O}(\lambda/(\gamma - 1))$ and, as we will see, all bounded-use schemes achieve a server storage of roughly γk, while the use of authenticators in the unbounded-use schemes roughly doubles the server storage.[2] We see that γ highlights a (necessary) tradeoff between server storage overhead and the locality t. For example, setting $\gamma = 2$ we double the server storage and get locality $t = O(\lambda)$, while setting $\gamma = 1.01$ we can achieve bounded-use schemes were the server only stores 1% of additional information, at the expense of $100\times$ increase in

---

[2] In prior unbounded-use schemes, one might reduce the overhead by making the block size larger. However, this increases communication (and most other parameters, too). Thus, to keep our comparison fair, we assume fixed block size and do not reflect this tradeoff in Table 1.

locality (but no other parameter degradation). We look at both information-theoretic (I.T.) and computational (Comp) security. We also consider both efficient and inefficient "extractors". Intuitively, an efficient extractor provides "knowledge soundness" (Know), guaranteeing that the adversary stores the file in some reasonable format and can efficiently recover it. An inefficient extractor only provides "information soundness" (Inf), guaranteeing that the server still has all of the "information" in the file, but may store it in some inefficient format. In the table we consider all possibilities with (I.T. / Know) being the strongest security guarantee.

Due to the space constraints, all the proofs are deferred to the full version [DVW09].

Table 1. PoR Schemes: Prior Results and Our Improvements. k is the file size, λ is the security parameter, and $1 < \gamma \leq 2$ is a flexible parameter. All schemes have locality $\mathcal{O}(\lambda/(\gamma - 1))$.

Scheme	Bounded?	Security	Server Storage	Client Storage	Challenge	Response
[JK07] †	ℓ-time	Comp/Know	$\gamma k + \mathcal{O}(\ell\lambda)$	$\mathcal{O}(\lambda)$	$\mathcal{O}\left(\frac{\lambda}{\gamma-1}\log k\right)$	$\mathcal{O}\left(\frac{\lambda^2}{\gamma-1}\right)$
[JK07]	No	Comp/Know	$2\gamma k$	$\mathcal{O}(\lambda)$	$\mathcal{O}\left(\frac{\lambda}{\gamma-1}\log k\right)$	$\mathcal{O}\left(\frac{\lambda^2}{\gamma-1}\right)$
[SW08] ‡	No	Comp/Know	$2\gamma k$	$\mathcal{O}(\lambda)$	$\mathcal{O}(\lambda)$	$\mathcal{O}(\lambda)$
Our	ℓ-time	I.T./Know	γk	$\mathcal{O}\left(\ell\frac{\lambda}{\gamma-1}\log k\right)$	$\mathcal{O}\left(\frac{\lambda}{\gamma-1}\log k\right)$	$\mathcal{O}(\lambda)$
Our	ℓ-time	I.T./Inf	γk	$\mathcal{O}(\ell\lambda)$	$\mathcal{O}(\lambda)$	$\mathcal{O}(\lambda)$
Our	ℓ-time	Comp/Know	$\gamma k + \mathcal{O}(\ell\lambda)$	$\mathcal{O}(\lambda)$	$\mathcal{O}\left(\frac{\lambda}{\gamma-1}\log k\right)$	$\mathcal{O}(\lambda)$
Our	ℓ-time	Comp/Inf	$\gamma k + \mathcal{O}(\ell\lambda)$	$\mathcal{O}(\lambda)$	$\mathcal{O}(\lambda)$	$\mathcal{O}(\lambda)$
Our	No	Comp/Know	$2\gamma k$	$\mathcal{O}(\lambda)$	$\mathcal{O}(\lambda)$	$\mathcal{O}(\lambda)$

†= Not proven secure as PoR scheme ‡= Random Oracle Model

2 Preliminaries

We assume the basic familiarity with (linear) error-correcting codes. In particular, the standard notation $[n, k, d]_q$ denotes an error-correcting codes over a q-ary alphabet with minimal (Hamming) distance d, block length n and message k. We also assume familiarity with Reed-Solomon codes, which are $[n, k, n - k + 1]_q$-codes for a prime power $q \geq n$. For an integer N, we let $[N]$ denote the set $\{1, \ldots, N\}$. Given a string $F \in \Sigma^N$ we let $F[i] \in \Sigma$ denote the ith symbol of F for $i \in [N]$.

A *hitting sampler*, or just *hitter* for short, provides a randomness-efficient way of sampling t elements. We review the definition and parameters achieved by known efficient constructions (see the survey [Gol97] for more details).

Definition 1. *Let* $\mathsf{Hit} : [M] \to [n]^t$ *be a function and interpret the output* $\mathsf{Hit}(e)$ *as a sample of t elements in $[n]$. We say that* $\mathsf{Hit}(e)$ *hits* $W \subseteq [n]$ *if it includes at least one member of W. A function* Hit *is a* (δ, ρ)-*hitter if for every subset* $W \subseteq [n]$ *of size* $|W| \geq (1 - \delta)n$, $\Pr_{e \leftarrow [M]}[\mathsf{Hit}(e)$ *hits* $W] \geq (1 - \rho)$.

A simple hitter construction involves choosing t uniformly random and independent elements of $[n]$. This results in a (δ, ρ)-hitter with *sample complexity* $t = \mathcal{O}(\log(1/\rho)/ (1 - \delta))$ and *randomness complexity* $\log(M) = t \log(n)$. It is known how to reduce the randomness complexity significantly. Indeed, the survey of Goldreich [Gol97] shows how to achieve the following parameters using a construction based on expander graphs.

Theorem 1. *There exists an efficient hitter family such that, for any integer n and any δ, ρ, we get sample complexity $t = \mathcal{O}(\log(1/\rho)/(1 - \delta))$ and randomness complexity $\log(M) = \log(n) + 3\log(1/\rho)$.*

3   PoR Codes

A PoR code consists of a triple of functions (Init, Read, Resp) with domains and ranges:

$$\text{Init} : (\Sigma_c)^k \to (\Sigma_c)^n \quad , \quad \text{Read} : [M] \to [n]^t \quad , \quad \text{Resp} : (\Sigma_c)^t \times [n'] \to \Sigma_r$$

for some alphabets Σ_c, Σ_r of sizes q_c, q_r respectively. The function Init is an *initial encoding* which converts a *client file* F into a *server file* $F' = \text{Init}(F)$. We let $\widetilde{k} = \lfloor k \log(q_c) \rfloor$ denote the initial file size in bits. The function Read, Resp are used by the server to run an audit. The client picks a challenge $e = (e_1, e_2) \in [N]$ where $N = Mn'$ and we identify $[N]$ with $[M] \times [n']$. The function $\text{Read}(e_1)$ determines a tuple of t positions $(i_1, \ldots, i_t)$ in F' which the server must read to produce the response $\mu = \text{Resp}(F'[(i_1, \ldots, i_t)], e_2)$. The functions Init, Read, Resp are actually used by the client/server and hence we require that they are all polynomial time computable.

For our understanding of PoR codes, it makes sense to think of functions Read, Resp as defining a *challenge-response* encoding of the server file F'. Firstly, we can think of the Read function as defining a simpler *direct-product encoding* DPE of the server file F' into the codeword $C' \in ((\Sigma_c)^t)^M$ defined by:

$$C' = \text{DPE}(F') = (F'[\text{Read}(1)], F'[\text{Read}(2)], \ldots, F'[\text{Read}(M)])$$

so that each position of C' consists of a concatenation of t positions in F'. The function Read, Resp together define the function $\text{Answer} : (\Sigma_c)^n \times [N] \to \Sigma_r$ as $\text{Answer}(F', e) = \text{Resp}(F'[\text{Read}(e_1)], e_2)$ where $e = (e_1, e_2)$. The *challenge-response encoding* CRE of the server file F' results in a codeword $C \in (\Sigma_r)^N$ defined by:

$$C = \text{CRE}(F') = (\text{Answer}(F', 1), \ldots, \text{Answer}(F', N)) .$$

Note that neither the direct-product encoding C' nor the challenge-response encoding C are ever stored explicitly and hence the functions DPE, CRE need not be efficient. In our construction, the values N, M, n' will be exponential. Of course, as is usually the case for error-correcting codes, we want to have a family of PoR codes which allows for many flexible choices of the parameters. In particular, we would like to have a family of codes parameterized by the bit size $\widetilde{k} = k \log(q_c)$ of the initial client file and the security parameter λ.

Definition 2. *For any PoR code, an ε-oracle $\mathcal{O}_F$ is an oracle parameterized by some $F \in (\Sigma_c)^k$ such that, letting $F' = \text{Init}(F)$, $C = \text{CRE}(F')$, we have $\Pr_{e \in_R [N]}[\mathcal{O}_F(e) = C[e]] \geq \varepsilon$. We say that $\mathcal{O}_F$ is an* erasure *oracle if, $\Pr[\mathcal{O}_F(e) \notin \{C[e], \bot\}] = 0$. Otherwise we say that $\mathcal{O}_F$ is an* error *oracle.*

Definition 3. *We say that* (Init, Read, Resp) *is a $(\alpha, \beta, \gamma, t)_{q_c}$-PoR code if the* alphabet *size is q_c, challenge size is $\alpha = \log_2(N)$, the* response *size is $\beta = \log_2(|\Sigma_r|)$, the* storage overhead *is $\gamma = n/k$ and the* locality *is t. For a PoR code family, all of these values are functions of the parameters $\widetilde{k}, \lambda$. We say that a PoR code is*

- ε_0-erasure decodable *if there is a oracle-access decoder $\mathcal{D}^{(\cdot)}$ such that, for any ε-erasure oracle $\mathcal{O}_F$, with $\varepsilon \geq \varepsilon_0$, the decoder $\mathcal{D}^{\mathcal{O}_F}(\widetilde{k}, \lambda, \varepsilon)$ outputs F with probability at least $1 - 2^{-\lambda}$.*
- $(\varepsilon_0, L(\cdot))$-error decodable *if there is a oracle-access decoder $\mathcal{D}^{(\cdot)}$ such that, for any ε-error oracle $\mathcal{O}_F$ with $\varepsilon \geq \varepsilon_0$, the decoder $\mathcal{D}^{\mathcal{O}_F}(\widetilde{k}, \lambda, \varepsilon)$ outputs a list of size at most $L(\varepsilon)$ containing the element F, with probability at least $1 - 2^{-\lambda}$.*

For both erasures and errors, we say that the scheme is efficiently decodable *if $\mathcal{D}$ runs in time* poly$(\widetilde{k}, \lambda, 1/\varepsilon)$.

3.1 Constructions of PoR Codes

In all of our constructions, the initial encoding Init is an $[n, k, d]_{q_c}$ error-correcting code with a "good" distance d over the appropriate alphabet Σ_c. The initial encoding defines the server storage overhead $\gamma = n/k$. For the functions Read, Resp we first present a *basic construction* followed by two orthogonal improvements.

BASIC CONSTRUCTION. In the basic construction, the challenge is simply a random t-tuple of positions in F', and the response is the value of F' at those positions. More concretely, the number of challenges is $N = M = n^t$ and the function Read(e_1) simply identifies the value $e_1 \in [N]$ with the tuples $(i_1, \ldots, i_t) \in [n]^t$. The function Resp does not get any portion e_2 of the challenge and is the identity function on the first argument: Resp$(x, 1) = x$. Thus, the challenge-response encoding CRE is equivalent here to the direct product encoding DPE. This construction yields an $(\alpha, \beta, \gamma, t)_{q_c}$-PoR code where the challenge size is $\alpha = t \log(n)$, the response size is $\beta = \log(q_r) = t \log(q_c)$. This basic PoR code is implicitly used in the schemes of [NR05, JK07].

IMPROVEMENT 1: FLEXIBLE RESPONSE SIZE. One problem with the basic construction that the response size $\beta = t \log(q_c)$ increases proportionally to the locality t. Indeed, in order to achieve good (list-)decoding, we will see that there is an advantage to having a large alphabet Σ_c: i.e., $\log(q_c) = \Omega(\lambda)$. On the other hand, the locality t must also be (at least) proportional to λ, making $\beta = \Omega(\lambda^2)$. In fact, we already mentioned that there is a necessary tradeoff between the locality t and the server storage overhead γ, making t even larger if one is concerned with minimizing the server storage. Thus, we would like to avoid the dependence of the response size β on t.

We improve our basic construction so that, instead of responding with all of the read blocks $x = F'[i_1] || \ldots || F'[i_t]$, the server responds with a randomly chosen position in an encoding of x under some ECC, which we call a *secondary encoding*. More precisely, we instantiate a secondary encoding Sec which is an $[n', k', d']_{q_r}$ ECC over the alphabet Σ_r of size $|\Sigma_r| = q_r$. We assume that it is easy to compute any one position in the codeword without computing the entire codeword. In particular, we define Resp : $(\Sigma_r)^{k'} \times [n'] \rightarrow \Sigma_r$ so that Resp$(x, e_2) = \text{Sec}(x)[e_2]$ computes the value in position e_2 of the secondary encoding of x. We require that Resp is efficiently computable (but allow Sec to be inefficient, and n' to be exponential). Also, we assume that $(q_r)^{k'} \geq q_c{}^t$ so that we can easily interpret elements in $(\Sigma_c)^t$ as elements in $(\Sigma_r)^{k'}$. The read function remains unchanged from the basic scheme. This yields a PoR code where $\beta = \log(q_r)$ is flexible and does not need to depend on t. For example, we can set $\Sigma_r = \Sigma_c$ and $k' = t$ so that $(\Sigma_r)^{k'} = (\Sigma_c)^t$ and $\beta = \log(q_c)$. As we will see, such setting will not degrade the (list-)decoding capabilities too much, as long as we set the value of n' and the alphabet size $\log q_c$ to be (appropriately) exponential in the security parameter λ. With such a setting, even a negligible fraction of the responses in a secondary encoding allow us to (list-)decode the original t-tuple x, more or less bringing us back to the basic construction, while reducing the response size β to be $\mathcal{O}(\lambda)$.

We also note that a variant of this improvement was implicitly used by [SW08], which employed the Hadamard code (over a large alphabet) as the secondary encoding.

IMPROVEMENT 2: REDUCING THE CHALLENGE SIZE. We would also like to get rid of the dependence between t and the challenge size α (currently, $\alpha = t \log(n)$). We do so by using derandomization techniques to choose the indices $(i_1, \ldots, i_t)$. Instead of just choosing these indices uniformly at random, we just use a function Read which is a (δ, ρ)-*hitter* (see Definition 1). Intuitively, hitters are useful in our context since, if an adversarial server "forgets" many blocks of F', then the indices chosen by $\mathsf{Read}(e_1)$ are likely to "hit" at least one such block. We can think of the basic PoR code construction as simply employing a "naive" hitter which is not randomness-efficient.

THE GENERAL CONSTRUCTION AND INSTANTIATIONS. To recap, our construction is parameterized by:

- An initial encoding Init : $(\Sigma_c)^k \to (\Sigma_c)^n$ which is a $[n, k, d]_{q_c}$-ECC.
- A (δ, ρ)-hitter Read : $[M] \to [n]^t$.
- A secondary encoding Sec : $(\Sigma_r)^{k'} \to (\Sigma_r)^{n'}$ which is a $[n', k', d']_{q_r}$-ECC and $q_r^{k'} \geq q_c^{t}$.

Most of our analysis will only use generic properties of the above primitives. However, when discussing parameters, we will rely on the following two concrete instantiations of PoR codes. For simplicity, we hide the concrete constants in the "Big-$\mathcal{O}$" notation. Both instantiations are parameterized by the security parameter λ, the file bit-size $\widetilde{k}$, and the server storage overhead γ, and will ensure locality $t = \mathcal{O}(\lambda/(\gamma - 1))$. In fact, they will use the identical Reed-Solomon Codes for their initial and secondary encodings, setting $q_c = 2^{\mathcal{O}(\lambda)}$, $k = \widetilde{k}/\log(q_c)$ and $n = \gamma k$ for the initial Reed-Solomon code, and $q_r = q_c$, $k' = t$ and $n' = 2^{\mathcal{O}(\lambda)}$ for the secondary Reed-Solomon code (recall, this defines $d = n - k + 1$ and $d' = n' - k' + 1$). In fact, the only difference will be in the choice of the (δ, ρ)-hitter, where $\delta \approx \frac{1}{\gamma}$ is roughly the fraction of the initial encoding sufficient to recover the file and $\rho = 2^{-\Omega(\lambda)}$. The first instantiation will use a randomness-efficient hitter construction from Theorem 1, while the second instantiation will use the "naive" hitter, where the t samples are chosen uniformly at random. As we can see, both hitters will indeed achieve sample complexity $t = \mathcal{O}(\log(1/\rho)/(1 - \delta)) = \mathcal{O}(\lambda/(\gamma - 1))$. However, the first "clever" hitter will achieve randomness complexity $\log M = \log n + 3 \log(1/\rho) = \mathcal{O}(\lambda)$, while the second "naive" hitter will achieve $\log M = t \log n = \mathcal{O}(\lambda \log(\widetilde{k})/(\gamma - 1))$. We summarize the (easily verified) resulting efficiency parameters below, and then state our main technical theorem.

Parameters: $\lambda, \widetilde{k}$ and γ, where $1 < \gamma \leq 2$ (and $\gamma \widetilde{k} + 1 \leq 2^{\lambda/2}$).
First Instantiation: Our first instantiation is an $(\alpha, \beta, \gamma, t)_{q_c}$-PoR code family with:

$$t = \mathcal{O}\left(\frac{\lambda}{\gamma - 1}\right), \quad \log(q_c) = \mathcal{O}(\lambda), \quad \alpha = \mathcal{O}(\lambda), \quad \beta = \mathcal{O}(\lambda) \tag{1}$$

Second Instantiation: Our second instantiation is a $(\alpha, \beta, \gamma, t)_{q_c}$-PoR code family with:

$$t = \mathcal{O}\left(\frac{\lambda}{\gamma - 1}\right), \quad \log(q_c) = \mathcal{O}(\lambda), \quad \alpha = \mathcal{O}\left(\frac{\lambda \log(\widetilde{k})}{\gamma - 1}\right), \quad \beta = \mathcal{O}(\lambda) \quad (2)$$

Theorem 2. *For appropriately selected constants, our PoR code family instantiations have the following security properties.*

1. *The <u>first</u> instantiation is <u>efficiently</u> $(2^{-\lambda})$-<u>erasure</u> decodable.*
2. *The <u>first</u> instantiation is <u>inefficiently</u> $(2^{-\lambda}, \mathcal{O}(\lambda/\varepsilon^3))$-<u>error</u> decodable.*
3. *The <u>second</u> instantiation is <u>efficiently</u> $(2^{-\lambda}, \mathcal{O}(\lambda/\varepsilon^3))$-<u>error</u> decodable.*

Thus, the first construction achieves (essentially) optimal parameters on all fronts, but in the case of errors is only known to be inefficiently decodable, while the second construction is only marginally suboptimal in the challenge size α, but achieves efficient error decodability instead. The proof of this theorem is given in the full version [DVW09]. In the following subsections, we only briefly sketch the high-level outline for the proof of the two efficient decoding variants. The latter variant for the case of errors will use the state-of-the-art direct product theorem of [IJK06, IJKW08] to remove the "simplifying assumption" on the behavior of the adversary made by [JK07].

3.2 Efficient Erasure Decodability

We now show that our first construction is efficiently erasure decodable. To do so, we assume that both the primary and secondary encodings are efficiently erasure-decodable up to the maximum radii d and d', respectively. This is true of the Reed-Solomon code employed by our concrete instantiation. First, we show how to (efficiently) convert an ε-erasure oracle $\mathcal{O}_F$ for the full PoR codeword $C = \mathsf{CRE}(\mathsf{Init}(F))$, into an ε'-erasure oracle $\widetilde{O}_F$ for the direct-product encoding $C' = \mathsf{DPE}(\mathsf{Init}(F))$.

Lemma 1. *Let $c_0 = n' - d' + 1$ and let $\mathcal{O}_F$ be an ε-erasure oracle with $\varepsilon \geq 4(c_0/n')$ for the full PoR codeword $C = \mathsf{CRE}(\mathsf{Init}(F))$. Then there is an (efficient) machine $\mathcal{D}_2^{\mathcal{O}_F}(\lambda, \varepsilon)$ which is an ε'-erasure oracle for the codeword $C' = \mathsf{DPE}(\mathsf{Init}(F))$ where $\varepsilon' \geq \varepsilon/4$. Moreover, on a query $e_1 \in [M]$, the machine $\mathcal{D}_2$ runs in time $\mathsf{poly}(\lambda, 1/\varepsilon)$.*

Now we show how to (efficiently) recover $n - d + 1$ values in the server file $F' = \mathsf{Init}(F)$ given access to the ε'-oracle $\widetilde{O}_F$. Using erasure-decoding of the initial code, we can then efficiently recover F.

Lemma 2. *Let $\widetilde{O}_F$ be an ε'-erasure oracle for the codeword $C' = \mathsf{DPE}(\mathsf{Init}(F)) = (F[\mathsf{Read}(1)], \ldots, F[\mathsf{Read}(M)])$ and assume that the function Read is (δ, ρ)-hitter where $\delta = \frac{d+1}{n}$ and $\varepsilon' \geq 2\rho$. Then there is an (efficient) algorithm $\mathcal{D}_1^{\widetilde{O}_F}(\widetilde{k}, \lambda, \varepsilon')$ such that $\mathcal{D}_1^{\mathcal{O}_F} = F$ with probability $1 - 2^{-\lambda}$ and $\mathcal{D}_1$ runs in time $\mathsf{poly}(\widetilde{k}, \lambda, 1/\varepsilon')$.*

Putting Lemma 1 and Lemma 2 together we easily get Part 1 of Theorem 2.

3.3 Efficient Error Decodability

We now show that our PoR code is also efficiently error decodable. Unfortunately, this forces us to sacrifice some of our generality. We cannot use a general hitter construction and must instead rely on the "naive" hitter, which samples the t positions uniformly at random. In addition, we need an efficient list-decodability for the secondary encodings and efficient error-correction for the initial encoding. All these properties are met by our second instantiation. Again, we first show how to convert an ε-error oracle that answers with values in the full encoding $C = \mathsf{CRE}(F')$ (which includes the secondary encoding) into an ε'-error oracle that answers with values in the code $C' = \mathsf{DPE}(F')$.

Lemma 3. *Let $\mathcal{O}_F$ be an ε-error oracle where $\varepsilon \geq 8k'/(n')^{1/4}$ for the full PoR codeword $C = \mathsf{CRE}(\mathsf{Init}(F))$. Then there is an (efficient) machine $\mathcal{D}_2^{\mathcal{O}_F}(\varepsilon, \lambda)$ which accepts queries $e_1 \in [M]$ and is an ε'-error oracle for the codeword $C' = \mathsf{DPE}(F')$, where $\varepsilon' = \varepsilon^3/256$. Moreover, $\mathcal{D}_2$ runs in time $\mathsf{poly}(\widetilde{k}, \lambda, 1/\varepsilon)$.*

Now we need to efficiently list-decode the direct-product code $C' = \mathsf{DPE}(F')$. Furthermore, we need to do so by reading only a small number of position in C', and certainly far fewer than the entire codeword. This is a highly non-trivial problem which is intimately related to *direct product theorems* in hardness amplification. We now identify codewords C', F' with functions $C' : [n]^t \rightarrow (\Sigma_c)^t$ and $F' : [n] \rightarrow (\Sigma_c)$ which map a position in the codeword to the value of the codeword at that position. Direct product theorems [Yao82, Lev87, Imp95, GNW95, IW97, Tre03, IJK06, IJKW08] essentially show that, if there exists an efficient algorithm which ε-computes the direct product function C' then there also exists an efficient algorithm which δ-computes the original function F'. Unfortunately, most direct product theorems (in particular, [Yao82, Lev87, Imp95, GNW95, IW97, Tre03]) are in the context of circuits and the reductions are *not* fully constructive. Instead, the reductions show the *existence* of some *advice* which, if hardwired into a circuit, would allow it to δ-compute F'. They do not provide a way of efficiently finding such advice (except is special restrictive cases) and, hence, these results are not appropriate for our setting where we need to actually *run the reduction* as an extractor. Fortunately, direct-product theorems for uniform adversaries (where all advice is efficiently self-generated) recently appeared in [IJK06, IJKW08]. Let us restate their main result of Impagliazzo et al. [IJKW08] in our language.

Theorem 3. *(Theorem 1.3 of [IJKW08]) Given an ε'-error oracle $\widetilde{O}_F$ for the direct-product function C', there exists an efficient algorithm $\mathcal{D}_1^{\widetilde{O}_F}$ which outputs a list of L candidate oracle-access functions $g_1^{\widetilde{O}_F}, \ldots, g_L^{\widetilde{O}_F}$ such that, with probability $(1 - 2^{-\lambda})$, there exists an $i \in \{1, \ldots, L\}$ for which the function g_i is a δ-error oracle for the function F'. In particular this means that $\left| \{ j \mid g_i^{\widetilde{O}_F}(j) = F'[j] \} \right| \geq \delta$. Moreover the functions $g_1, \ldots, g_L$ are efficient, $L = \mathcal{O}\left(\frac{\lambda}{\varepsilon}\right)$ and $\delta = 1 - \mathcal{O}(\log\left(\frac{1}{\varepsilon}\right)/t)$.*

Combining Lemma 3 and Theorem 3, we can then show that our second instantiation of a PoR code family is efficiently list decodable, proving Part 3 of Theorem 2. Note that it is an interesting open problem if we can modify the result of [IJKW08] to work with some hitter having parameters similar to Theorem 1. This would also lead to some nice derandomization results for hardness-amplification and seems like a difficult problem. Indeed, the proof of Theorem 3 relies on certain efficient sampling properties of the naive hitter construction that the construction from Theorem 1 does not have.

4 PoR Schemes from PoR Codes

A PoR scheme consists of a generation algorithm Gen and an audit protocol defined by two ITMs $\mathcal{P}, \mathcal{V}$ for the prover (server) and verifier (client) respectively. All of the algorithms are parameterized by a security parameter λ which we will omit from the description in the sequel. The Gen algorithm is a randomized algorithm which takes as input a file $F \in \{0,1\}^*$ and produces $(\widetilde{F}, \mathsf{ver}) \leftarrow \mathsf{Gen}(F)$. In the audit protocol, the prover $\mathcal{P}$ is given $\widetilde{F}$ and verifier $\mathcal{V}$ is given ver. At the conclusion of the protocol, the verifier outputs a verdict $v \in \{\mathsf{accept}, \mathsf{reject}\}$. In general we allow the verifier to be stateful and update the value of ver during the protocol and thus, for example, keep track of how many proofs have been run. For unbounded-use schemes, we will give constructions where the verifier is stateless. Our definition essentially follows that of [JK07] but is slightly more general.

COMPLETENESS. We require that in any interaction $\left\{ \mathcal{P}(\widetilde{F}) \rightleftharpoons \mathcal{V}(\mathsf{ver}) \right\}$ between honest prover and and honest verifier, the verifier outputs a verdict $v = \mathsf{accept}$.

SOUNDNESS. We define the soundness game $\mathbf{Sound}_{\mathcal{A}}^{\mathcal{E}}(k, \ell)$ between an adversary $\mathcal{A}$ and a challenger. In the soundness game, the adversary gets to create an adversarial prover and the challenger runs the extractor $\mathcal{E}$ on it.

1. The adversary $\mathcal{A}$ chooses a file $F \in \{0,1\}^k$.
2. The challenger produces $(\widetilde{F}, \mathsf{ver}) \leftarrow \mathsf{Gen}(F)$ and gives $\widetilde{F}$ to $\mathcal{A}$. In addition, the challenger initializes a verifier $\mathcal{V}(\mathsf{ver})$.
3. We first have a *test stage*, where the adversary $\mathcal{A}$ gets protocol access to $\mathcal{V}(\mathsf{ver})$ and can run at most $\ell - 1$ proofs with it. For each proof, the adversary gets the output $v \in \{\mathsf{accept}, \mathsf{reject}\}$ of the verifier $\mathcal{V}$.
4. At the end of the test stage, the adversary produces code for an (probabilistic) ITM prover $\widetilde{\mathcal{P}}$ and gives this code to the challenger.
5. Let $\varepsilon = \Pr\left[\left\{ \widetilde{P} \rightleftharpoons \mathcal{V}(\mathsf{ver}) \right\} = \mathsf{accept} \right]$ be the success probability of the adversarial prover $\widetilde{P}$, and $\bar{F} = \mathcal{E}^{\widetilde{P}}(\mathsf{ver}, k, \varepsilon)$ be the output of the extractor.

For complexity classes $\mathcal{C}_1, \mathcal{C}_2$, we say that an *unbounded-use* PoR scheme is *sound* if there exists an extractor $\mathcal{E} \in \mathcal{C}_1$ and two negligible functions $\varepsilon_0(\cdot), \varepsilon_1(\cdot)$ such that, for any adversary $\mathcal{A} \in \mathcal{C}_2$ and any polynomials p_1, p_2

$$\Pr\left[\varepsilon > \varepsilon_0(\lambda) \ \wedge \ \bar{F} \neq F \mid \varepsilon, \bar{F} \leftarrow \mathbf{Sound}_{\mathcal{A}}^{\mathcal{E}}(p_1(\lambda), p_2(\lambda)) \right] \leq \varepsilon_1(\lambda).$$

We say that an ℓ-*time* PoR scheme sound if the above holds with $p_1(\lambda)$ replaced by ℓ.

The definition guarantees that the adversary cannot "lose" the file and still succeed in an audit. We give four interesting variants of this definition based on the the complexity classes $\mathcal{C}_1, \mathcal{C}_2$ of the extractor and adversary.

- If the definition holds for the class $\mathcal{C}^{\mathsf{all}}$ of all ITM adversaries, we say that the scheme has *information-theoretic security*. Otherwise, if the definition holds for the class $\mathcal{C}^{\mathsf{poly}}$ of all ITMs running in time $\mathsf{poly}(\lambda)$, we say the scheme has *computational security*.
- If the run time of $\mathcal{E}$ is $\mathsf{poly}(1/\varepsilon, k, \lambda)$ then we say that our scheme has *knowledge-soundness*. Otherwise, if there is no bound on the running time of $\mathcal{E}$, we say that the scheme has *information-soundness*.

Remark 1. As in Proofs of Knowledge, the extractor is *not* part of the protocol but rather serves as a thought-experiment that helps us define security. The adversary, after running some arbitrary audits with the verifier, should not be able to cleverly "lose" parts of the file F (represented by construction of the prover $\widetilde{\mathcal{P}}$ on which $\mathcal{E}$ fails) and yet still succeed in the subsequent audit with reasonable probability $\varepsilon \geq \varepsilon_0$. Of course, the adversarial server might correctly run all audits, but still refuse to give the full file back to the client. This attack, unfortunately, cannot be prevented if the audits are significantly shorter than the size of the file. Instead, we are satisfied if the server is guaranteed to *know* the file at the conclusion of an audit; whether or not the server will actually *give* that file back to the client is an orthogonal concern.

Remark 2. The four variants of soundness are all meaningful. For example, information soundness is already a strong notion which ensures that the adversarial server cannot save on space (by deleting a portion of the file F) and still pass an audit. Knowledge soundness is a stronger notion, which also guarantees that the server stores the file F in some efficiently recoverable representation. The strongest notion — information-theoretic security with knowledge soundness — means that any adversary (regardless of computational power) must store the file in some efficiently recoverable representation.

FROM POR CODES TO POR SCHEMES. In the following subsections, we will briefly sketch how to build a secure PoR scheme (for any of the four variants) from an appropriate $(\alpha, \beta, \gamma, t)_{q_c}$-PoR code (Init, Read, Resp). Intuitively, the key step of the extractor $\mathcal{E}$ will be to simply run the corresponding decoder $\mathcal{D}$, giving $\mathcal{D}$ oracle access to the adversarial prover $\tilde{P}$. Then, if $\mathcal{D}$ is efficient, we will get knowledge-soundness; if not, we will only settle for information-soundness. As for the attacker's efficiency, recall that PoR codes are information-theoretically secure. Thus, as long as we do not introduce any additional computationally-secure primitives into the final PoR scheme, the resulting security will be information-theoretic. Also, for *bounded-use* schemes we will be using *error-decodable* PoR codes, while for the unbounded-use schemes we can use *erasure-decodable* schemes. In our presentation below, we will be only concentrating on the main ideas, primarily focusing on justifying the parameters claimed in Section 1.3.

4.1 Bounded-Use Information-Theoretic Schemes

First, we present a very simple and efficient construction of an ℓ-time, information-theoretically secure PoR scheme from an *error*-decodable PoR code. We also use a family of *almost-universal* hash functions $\mathcal{H} = \{h\}$. Recall, a function family $\mathcal{H}$ is ψ-universal, if for any inputs $x \neq y$, $\Pr_{h \leftarrow \mathcal{H}}(h(x) = h(y)) \leq \psi$. It is well known that one can construct such families on $\tilde{k}$-bit messages (say, based on polynomial evaluation at a random point) having the description of h and the output length of h be at most $\log(\tilde{k}/\psi)$ bits each. We will set the value ψ later. Bellow we give a detailed description, which corresponds to construction 1 from the introduction.

Gen: —   Let $F' = \mathsf{Init}(F)$.

    —   Choose ℓ uniformly random challenges $e^{(1)}, \ldots, e^{(\ell)}$ with $e^{(i)} \in [N]$ and compute the responses $\mu^{(1)}, \ldots, \mu^{(\ell)}$ where $\mu^{(i)} = \mathsf{Answer}(F', e^{(i)}) = \mathsf{Resp}(F'[\mathsf{Read}(e_1^{(i)})], e_2^{(i)})$.

    —   Chooses a uniformly random hash function $h \leftarrow \mathcal{H}$ and compute $\omega := h(F)$.

    —   Set $\widetilde{F} = F'$, counter $i = 1$ and ver $:= \left(\langle (e^{(1)}, \mu^{(1)}), \ldots, (e^{(\ell)}, \mu^{(\ell)}) \rangle, h, \omega, i \right)$

$\mathcal{P}, \mathcal{V}$: To run an audit $i \in \{1, \ldots, \ell\}$, the verifier $\mathcal{V}$ sends $e^{(i)}$ to $\mathcal{P}$. Upon the receipt of a challenge e, the prover $\mathcal{P}$ computes $\mu = \mathsf{Answer}(F', e)$ and sends μ to $\mathcal{V}$. When the verifier $\mathcal{V}$ receives a value μ', it checks if $\mu' = \mu^{(i)}$ and outputs accept if yes and reject otherwise (updating $i := i + 1$ in both cases).

Notice that the function h and the value $\omega = h(F)$ are not even used in the audit! Of course, they are used by the extractor $\mathcal{E}$ instead, to check which of the L possibilities returned by the decoder $\mathcal{D}$ is the actual file F.

Theorem 4. *Let* $(\mathsf{Init}, \mathsf{Read}, \mathsf{Resp})$ *be an* $(\alpha, \beta, \gamma, t)_{q_c}$-*PoR code family which is efficiently (resp. inefficiently)* $(\varepsilon_0 \varepsilon_1, L)$-*error decodable, where* ε_1 *is any negligible function. Let* $\mathcal{H}$ *be a* ψ-*universal hash family. Then the above scheme is a information-theoretically secure* ℓ-*time PoR protocol with knowledge (resp. information) soundness error at most* $\max(\varepsilon_0, L\psi + 2^{-\lambda})$. *In addition, the scheme has locality* t, *server storage overhead* γ, *communication complexity* $(\alpha + \beta)$ *and client storage overhead* $\ell(\alpha + \beta) + \mathcal{O}(\log(k \log(q_c)/\psi))$.

PARAMETERS. We can now instantiate this scheme with particular error-decodable PoR codes constructed in Theorem 2 (parts 2 and 3). Notice, for both variants, we have $L = \mathcal{O}(\lambda/\varepsilon^3)$ and can achieve $\varepsilon_0 = 2^{-\lambda}$ (by changing constants, if needed). Thus, we can set $\psi = 2^{-2\lambda}\varepsilon_0^3/\lambda$, so that $\psi L \le 2^{-2\lambda}$ and the description of h and $h(F)$ are only $\mathcal{O}(\log(k \log(q_c)/\psi)) = \mathcal{O}(\lambda)$ bits long. Then, we get the final PoR soundness $\max(\varepsilon_0, L\psi + 2^{-\lambda}) \approx 2^{-\lambda}$. In fact, comparing the parameters for the efficient and inefficient variant (see Equations (1) and (2)), the only noticeable difference is the client challenge size α, equal to $\mathcal{O}(\lambda)$ for the inefficient case, and $\mathcal{O}(\lambda \log(\widetilde{k})/(\gamma - 1))$ for the efficient case. Overall, we get an ℓ-time information-theoretically secure PoR with knowledge/information soundness, matching the parameters claimed in Section 1.3.

4.2   Reducing Client Storage: Bounded-Use Computational Schemes

Using computational assumptions, we now show that it is possible to "transfer" the client's storage of the ℓ challenge/response pairs to the server. Overall, the client's storage becomes $\mathcal{O}(\lambda)$, and the server storage becomes $O(\ell\lambda)$. Below we give a detailed description of construction 2 as outlined in the introduction. Let $\mathcal{F}_1, \mathcal{F}_2$ be two PRF families, where the output size of $\mathcal{F}_1$ is equal to the client's challenge size α, and the output size size of $\mathcal{F}_2$ is $\mathcal{O}(\lambda)$.

Gen: —   Let $F' = \mathsf{Init}(F)$.
- Choose a random function $f_{k_1} \in \mathcal{F}_1$ and compute ℓ challenges $e^{(1)} = f_{k_1}(1), \ldots, e^{(\ell)} = f_{k_1}(\ell)$. Compute the responses $\mu^{(1)}, \ldots, \mu^{(\ell)}$, where $\mu^{(i)} = \mathsf{Answer}(F', e^{(i)})$.
- Choose a random function $f_{k_2} \in \mathcal{F}_2$ and compute
$$\sigma_1 := f_{k_2}\left(1, \mu^{(1)}\right), \ldots, \sigma_\ell := f_{k_2}\left(\ell, \mu^\ell\right). \text{ Set } \widetilde{F} := (F', \sigma_1, \ldots, \sigma_\ell).$$
- Choose a uniformly random hash function $h \leftarrow \mathcal{H}$ and set $\omega = h(F')$.
- Initialize count $i = 1$ and set $\mathsf{ver} := (k_1, k_2, h, \omega, i)$.

$\mathcal{P}, \mathcal{V}$: To run an audit $i \in \{1, \ldots, \ell\}$, the verifier $\mathcal{V}$ computes $e^{(i)} = f_{k_1}(i)$ and sends $(e^{(i)}, i)$ to $\mathcal{P}$. Upon the receipt of a challenge $e = (e_1, e_2)$ and an index i, the prover $\mathcal{P}$ computes $\mu = \mathsf{Answer}(F', e^{(i)})$ and sends (μ, σ_i) to $\mathcal{V}$. When the verifier $\mathcal{V}$ receives a value μ', σ', it verifier $\sigma' = f_{k_2}(i, \mu')$ and rejects if this check fails. Otherwise, it accepts (updating $i := i + 1$ in either case).

COMMENTS AND PARAMETERS. The analysis of this scheme is very similar to the bounded-use information-theoretic scheme. In particular, consider an adversarial prover $\widetilde{P}$ which answers with probability ε. Then this must be an ε'-error adversary where $\varepsilon - \varepsilon'$ is some negligible distinguishing advantage for the PRF families $\mathcal{F}_1, \mathcal{F}_2$. We omit this analysis. It is also easy to see that the computational PoR protocols with information/knowledge soundness, resulting by using our particular error-decodable PoR codes constructed in Theorem 2 (parts 2 and 3), will match the parameters claimed in Section 1.3. In particular, although the client's storage is now reduced to $\mathcal{O}(\lambda)$ even for the case of knowledge soundness, the client's challenge cannot be "pseudorandomly compressed" below $\mathcal{O}(\lambda \log(\widetilde{k})/(1 - \gamma))$, since it needs to look random to the server.

4.3   An Unbounded-Use Computational Scheme

We now show how to construct an unbounded use scheme in our framework using the techniques of [SW08]. The construction is based on the concept of a *homomorphic linear authenticator scheme* — a notion we abstract away from the works of [ABC+07, SW08]. On a high level, this is a scheme in which a verifier computes a *vector-tag* $\overline{\sigma} = (\sigma_1, \ldots, \sigma_n)$ for a vector-message $\overline{x} = (x_1, \ldots, x_n)$ consisting of n field values, using some secret key K. A prover, who is given the vector-message $\overline{x}$, the corresponding vector-tag $\overline{\sigma}$, and a vector-challenge $\overline{a} = (a_1, \ldots, a_n)$, but *not the secret key* K, can then efficiently compute an *authenticator* σ for the field element $\mu = \sum_{i=1}^{n} a_i x_i$. The verifier, when given μ', σ' from the prover, can then run a *verification* procedure (using K) and, if it accepts, be convinced that $\mu' = \mu$.

More precisely, a homomorphic authenticator scheme consists of three algorithms (LinTag, LinAuth, LinVer), a key domain $\mathcal{K}$ and a field $\mathbb{F}$. For a key $K \in \mathcal{K}$, and a vector-message $\overline{x} = (x_1, \ldots, x_n) \in \mathbb{F}^n$ the algorithm $\mathsf{LinTag}_K(\overline{x})$ produces a vector-tag $\overline{\sigma} = (\sigma_1, \ldots, \sigma_n)$. For a vector-challenge of n coefficients $\overline{a} = (a_1, \ldots, a_n)$, the *un-keyed* function LinAuth computes an authenticator $\sigma = \mathsf{LinAuth}(\overline{x}, \overline{a}, \overline{\sigma})$. Moreover, the LinAuth algorithm is "local"; i.e., it only reads values x_i, σ_i for which $a_i \neq 0$. Lastly, the verification algorithm computes $b = \mathsf{LinVer}_K(\overline{a}, \mu', \sigma')$, where $b \in \{0, 1\}$, decides if the algorithm *accepts* or *rejects*.

For completeness, we require that for any $K \in \mathcal{K}, \overline{x} \in \mathbb{F}^n, \overline{a} \in \mathbb{F}^n$ letting $\overline{\sigma} \leftarrow \mathsf{LinTag}_K(\overline{x})$, $\sigma \leftarrow \mathsf{LinAuth}(\overline{x}, \overline{a}, \overline{\sigma})$ and $\mu = \sum_{i=1}^{n} x_i a_i$ then $\mathsf{LinVer}_K(\overline{a}, \mu, \sigma) = 1$. For security, given an adversary $\mathcal{A}$, we define the *unforgeability* game as follows:

1.   The adversary $\mathcal{A}$ chooses a vector-message $\overline{x} \in \mathbb{F}^n$.
2.   The challenger chooses a uniformly random key $K \leftarrow \mathcal{K}$ and computes $\overline{\sigma} \leftarrow \mathsf{LinTag}_K(\overline{x})$. The adversary is given $\overline{\sigma}$.
3.   The adversary $\mathcal{A}$ produces a vector-challenge $\overline{a} \in \mathbb{F}^n$ and a tuple (μ', σ').
4.   If $\mathsf{LinVer}_K(\overline{a}, \mu', \sigma') = 1$ and $\mu' \neq \sum_{i=1}^{n} a_i x_i$ then the adversary wins.

We require that for every efficient adversary $\mathcal{A}$, the probability that $\mathcal{A}$ succeeds in the unforgeability game is negligible in the security parameter λ. We refer to [SW08] for a particular, very elegant and efficient construction of such homomorphic linear authenticators. We note that the definition can also be extended to the public-key setting and an efficient construction for such setting was also given in [SW08].

We can employ linear-homomorphic authenticators along with any PoR code in which the response function $\mathsf{Resp} : (\Sigma_c)^t \to (\Sigma_c)$ is *linear* (as is the case in our constructions) to construct a full PoR scheme as follows:

Gen: Let $F' = \mathsf{Init}(F)$. Choose a key K for the linear authenticator scheme and compute $\overline{\sigma} = \mathsf{LinTag}_K(F'[1], \ldots, F'[n])$. Set $\widetilde{F} := (F', \sigma_1, \ldots, \sigma_n)$ and ver $:= K$.

$\mathcal{P}, \mathcal{V}$:   – The verifier $\mathcal{V}$ chooses a uniformly random value $e \in [N]$ and sends e to $\mathcal{P}$.

    – The prover $\mathcal{P}$, upon receiving $e = (e_1, e_2)$, computes $(i_1, \ldots, i_t) = \mathsf{Read}(e_1)$ and $\overline{x} = (x_1, \ldots, x_t) = F'[(i_1, \ldots, i_t)] \in (\Sigma_c)^t$. Since the function $\mathsf{Resp}(\overline{x}, e_2)$ linear, we can write $\mu = \mathsf{Resp}(\overline{x}, e_2) = \sum_{i=1}^{t} a_i x_i$ for some coefficients $\overline{a} = (a_1, \ldots, a_t)$. Let $\overline{\sigma} = (\sigma_{i_1}, \ldots, \sigma_{i_t})$.

    The prover computes $\sigma = \mathsf{LinAuth}(\overline{x}, \overline{a}, \overline{\sigma})$ and sends (μ, σ) to $\mathcal{V}$.

  – Upon receipt of a value (μ', σ') the client accepts iff $\mathsf{LinVer}_K(\mu', \sigma') = 1$.

In terms of parameters, we see that the client storage is $\mathcal{O}(\lambda)$, client communication is still α, and the server communication is only increased by the the length of the authenticator, which is $\mathcal{O}(\beta)$ (i.e., the size of a field element) for the homomorphic scheme in [SW08]. However, the main price one pays is in the server's storage, to store the tag-vector $\overline{\sigma} = (\sigma_1, \ldots, \sigma_n)$. For the scheme of [SW08], the length of $\overline{\sigma}$ is equal to the length of F, meaning that the server storage is doubled (see also Footnote 2). In terms of security, we get computational *unbounded-use* PoR scheme with *knowledge-soundness*, as long as our PoR code is *efficiently ε_0-erasure decodable*, where ε_0 is negligible in λ.

Finally, to obtain our actual parameters claimed in Section 1.3, we use the same linear-authenticator construction as [SW08], but plug in our improved *erasure-decodable* PoR Code from Equation (1) and Theorem 2 (part 1). In particular, by using the (linear) Reed-Solomon code in place of the Hadamard code for the secondary encoding, and a randomness-efficient hitter as our Read function, our construction is an unbounded-use PoR scheme with communication complexity $\mathcal{O}(\lambda)$ in the *standard model*, and without the use of Random Oracles.

Acknowledgments. Yevgeniy Dodis was supported in part by NSF Grants CNS-0831299, CNS-0716690, CCF-0515121, CCF-0133806, and completed part of this work while visiting Center for Research on Computation and Society at Harvard University. Salil Vadhan was supported in part by NSF Grant CNS-0831289.

References

[ABC+07] Ateniese, G., Burns, R.C., Curtmola, R., Herring, J., Kissner, L., Peterson, Z.N.J., Song, D.X.: Provable data possession at untrusted stores. In: Ning, et al. (eds.) [NdVS07], pp. 598–609 (2007)

[BJO08] Bowers, K.D., Juels, A., Oprea, A.: Proofs of retrievability: Theory and implementation. Cryptology ePrint Archive, Report 2008/175 (2008), http://eprint.iacr.org/

[DVW09] Dodis, Y., Vadhan, S., Wichs, D.: Proofs of retrievability via hardness amplification (full version). Cryptology ePrint Archive (2009), http://eprint.iacr.org/

[GNW95] Goldreich, O., Nisan, N., Wigderson, A.: On yao's xor-lemma. Electronic Colloquium on Computational Complexity (ECCC) 2(50) (1995)

[Gol97]   Goldreich, O.: A sample of samplers - a computational perspective on sampling (survey). Electronic Colloquium on Computational Complexity (ECCC) 4(20) (1997)

[IJK06]   Impagliazzo, R., Jaiswal, R., Kabanets, V.: Approximately list-decoding direct product codes and uniform hardness amplification. In: FOCS, pp. 187–196. IEEE Computer Society, Los Alamitos (2006)

[IJKW08]   Impagliazzo, R., Jaiswal, R., Kabanets, V., Wigderson, A.: Uniform direct product theorems: simplified, optimized, and derandomized. In: Ladner, R.E., Dwork, C. (eds.) STOC, pp. 579–588. ACM, New York (2008)

[Imp95]   Impagliazzo, R.: Hard-core distributions for somewhat hard problems. In: FOCS, pp. 538–545 (1995)

[IW97]   Impagliazzo, R., Wigderson, A.: P = BPP if EXP requires exponential circuits: Derandomizing the xor lemma. In: STOC, pp. 220–229 (1997)

[JK07]   Juels, A., Kaliski, B.S.: Pors: proofs of retrievability for large files. In: Ning, et al. (eds.) [NdVS07], pp. 584–597 (2007)

[Lev87]   Levin, L.A.: One-way functions and pseudorandom generators. Combinatorica 7(4), 357–363 (1987)

[NdVS07]   Ning, P., De Capitani di Vimercati, S., Syverson, P.F.: Proceedings of the 2007 ACM Conference on Computer and Communications Security, CCS 2007, Alexandria, Virginia, USA, October 28-31, 2007. ACM, New York (2007)

[NR05]   Naor, M., Rothblum, G.N.: The complexity of online memory checking. In: FOCS, pp. 573–584 (2005)

[SW08]   Shacham, H., Waters, B.: Compact proofs of retrievability. In: Pieprzyk, J. (ed.) ASIACRYPT 2008. LNCS, vol. 5350, pp. 90–107. Springer, Heidelberg (2008)

[Tre03]   Trevisan, L.: List-decoding using the xor lemma. In: FOCS, pp. 126–135. IEEE Computer Society, Los Alamitos (2003)

[Yao82]   Chi-Chih Yao, A.: Theory and applications of trapdoor functions (extended abstract). In: FOCS, pp. 80–91. IEEE, Los Alamitos (1982)

Security Amplification for *Interactive* Cryptographic Primitives

Yevgeniy Dodis[1], Russell Impagliazzo[2], Ragesh Jaiswal[3], and Valentine Kabanets[4]

[1] New York University
dodis@cs.nyu.edu
[2] University of California at San Diego and IAS
russell@cs.ucsd.edu
[3] Columbia University
rjaiswal@cs.columbia.edu
[4] Simon Fraser University
kabanets@cs.sfu.ca

Abstract. Security amplification is an important problem in Cryptography: starting with a "weakly secure" variant of some cryptographic primitive, the goal is to build a "strongly secure" variant of the same primitive. This question has been successfully studied for a variety of important cryptographic primitives, such as one-way functions, collision-resistant hash functions, encryption schemes and weakly verifiable puzzles. However, all these tasks were non-interactive. In this work we study security amplification of *interactive* cryptographic primitives, such as message authentication codes (MACs), digital signatures (SIGs) and pseudorandom functions (PRFs). In particular, we prove direct product theorems for MACs/SIGs and an XOR lemma for PRFs, therefore obtaining nearly optimal security amplification for these primitives.

Our main technical result is a new Chernoff-type theorem for what we call *Dynamic Weakly Verifiable Puzzles*, which is a generalization of ordinary Weakly Verifiable Puzzles which we introduce in this paper.

1 Introduction

Security amplification is a fundamental cryptographic problem: given a construction C of some primitive P which is only "weakly secure", can one build a "strongly secure" construction C' from C? The first result in this domain is a classical conversion from weak one-way functions to strong one-way function by Yao [Yao82] (see also [Gol01]): if a function f is only mildly hard to invert on a random input x, then, for appropriately chosen n, the function $F(x_1, \ldots, x_n) = (f(x_1), \ldots, f(x_n))$ is very hard to invert. The above result is an example of what is called the *direct product theorem*, which, when true, roughly asserts that simultaneously solving many independent repetitions of a mildly hard task is a much harder "combined task". Since the result of Yao, such direct product theorems have have been successfully used to argue security amplification of several other important cryptographic primitives, such as collision-resistant hash functions [CRS+07], encryption schemes [DNR04] and weakly verifiable puzzles [CHS05, IJK08].

O. Reingold (Ed.): TCC 2009, LNCS 5444, pp. 128–145, 2009.

However, all the examples above are non-interactive: namely, after receiving its challenge, the attacker needs to break the corresponding primitives without any further help or interaction. This restriction turns out to be important, as security amplification, and, in particular, direct product theorems become much more subtle for interactive primitives. For example, Bellare, Impagliazzo and Naor [BIN97] demonstrated that that parallel repetition does *not*, in general, reduce the soundness error of multiround (computationally sound) protocols, and this result was further strengthened by Pietrzak and Wikstrom [PW07]. On the positive side, parallel repetition is known to work for the special case of three-round protocols [BIN97] and constant-round public-coin protocols [PV07]. However, considerably less work has been done in the security amplification of more "basic" cryptographic primitives requiring interaction, such as block ciphers, message authentications codes (MACs), digital signatures (SIGs) and pseudorandom functions (PRFs). For example, Luby and Rackoff [LR86] (see also [NR99, Mye99]) showed how to improve the security of a constant number of pseudorandom permutation generators by composition, while Myers [Mye03] showed that a (non-standard) variant of the XOR lemma [Yao82, Lev87, Imp95, GNW95] holds for PRFs. In particular, the known results for the interactive case are either weaker or more specialized than those for the non-interactive case. The difficulty is that, for instance in the case of MACs, the attacker has oracle access to the corresponding "signing" and "verification" oracles, and the existing techniques do not appear to handle such cases.

In this work we study the question of security amplification of MACs, SIGs and PRFs, showing how to convert a corresponding weak primitive into a strong primitive. In brief, we prove a direct product theorem for MACs/SIGs (and even a Chernoff-type theorem to handle MACs/SIGs with *imperfect completeness*), and a (regular) XOR lemma for PRFs. Before describing these results in more details, however, it is useful to introduce our main technical tool for all these cases — a Chernoff-type theorem for what we call *Dynamic Weakly Verifiable Puzzles* (DWVPs) — which is of independent interest.

Dynamic Weakly Verifiable Puzzles. Recall, (non-dynamic) weakly verifiable puzzles (WVPs) were introduced by Canetti, Halevi and Steiner [CHS05] to capture the class of puzzles whose solutions can only be verified efficiently by the party generating the instance of the puzzle. This notion includes, as special cases, most previously mentioned non-interactive primitives, such as one-way functions, collision-resistant hash functions, one-way encryption schemes, CAPTCHAs, etc. To handle also interactive primitives, such as MACs and SIGs (and also be useful later for PRFs), in Section 3 we generalize this notion to that of *dynamic* WVPs (DWVPs) as follows. Just like in WVPs, one samples a pair (x, α) from some distribution $\mathcal{D}$, where α is the secret advice used to verify proposed solutions r to the puzzle x. Unlike WVPs, however, each x actually defines a *set* of related puzzles, indexed by some value $q \in Q$, as opposed to a single puzzle (which corresponds to $|Q| = 1$). An efficient verification algorithm R for the DWVP uses α and the puzzle index q to test if a given solution r is correct. An attacker B has oracle access to this verification procedure. Additionally, the attacker has oracle access to the *hint oracle*: given an index q, the hint oracle returns some hint value $H(\alpha, q)$, presumably "helping" the attacker to solve the puzzle q. The attacker wins the DWVP game if it ever causes the verification oracle to succeed on a query $q \in Q$

not previously queried to the hint oracle. As we see, this abstraction clearly includes MACs and SIGs as special cases. It also generalizes ordinary WVPs, corresponding to $|Q| = 1$. We say that the DWVP is δ-hard, if no (appropriately bounded) attacker can win the above game with probability more than $(1 - \delta)$.

Our main technical result is the following (informally stated) Chernoff-type theorem for DWVPs. Given n independently chosen δ-hard DWVPs on some index set Q, the chance of solving more than $(n - (1 - \gamma)\delta n)$ DWVPs — on the *same* value $q \in Q$ and using less than h "hint" queries $q' \neq q$ — is proportional to $h \cdot e^{-\Omega(\gamma^2 \delta n)}$; the exact statement is given in Theorem 3. Notice, the value $0 < \gamma \leq 1$ measures the "slackness parameter". In particular, $\gamma = 1$ corresponds to the direct product theorem where the attacker must solve all n puzzles (on the same q). However, setting $\gamma < 1$ allows to handle the setting where even the "legitimate" users, — who have an advantage over the attacker, like knowing α or being humans, — can also fail to solve the puzzle with some probability slightly less than $(1 - \gamma)\delta$.

This result generalizes the corresponding Chernoff-type theorem of Impagliazzo, Jaiswal and Kabanets [IJK08] for standard, *non-dynamic*, WVPs. However, the new theorem involves a considerably more complicated proof. The extra difficulties are explained in Section 3.1. In essence, in order to amplify security, the attacker B for the single DWVP must typically execute the assumed attacker A for the "threshold" variant several times, before obtaining sufficient "confidence" in the quality of the solutions output by A. In each of these "auxiliary" runs, however, there is a chance that A will ask a hint query for the index q which is equal to the one that A is going to solve in the "actual" run leading to the forgery, making B's forgery value q "old". Thus, a new delicate argument has to be made to argue security in this scenario. At a high level, the argument is somewhat similar to Coron's improved analysis [Cor00] of the full domain hash signature scheme, although the details differ. See Theorem 3 for the details.

Applications to MACs, SIGs and PRFs. Our main technical result above almost immediately implies security amplification for MACs and SIGs, even with imperfect completeness. For completeness, we briefly state the (asymptotic) result for the MAC case. (The case of SIGs and the exact security version for both cases are immediate.) We assume that the reader is familiar with the basic syntax and the standard Chosen Message Attack (CMA) scenario for a MAC, which is given by a tagging algorithm Tag and the verification algorithm Ver. We denote the secret key by s, and allow the tagging algorithm to be probabilistic (but not stateful). Given the security parameter k, we say that the MAC has *completeness error* $\beta = \beta(k)$ and *unforgeability* $\delta = \delta(k)$, where $\beta < \delta$, if for any message m, $\Pr(\text{Ver}(s, m, \text{Tag}(s, m)) = 1) \geq 1 - \beta$, and that no probabilistic polynomial-time attacker B can forge a valid tag for a "fresh" message m with probability greater than $(1 - \delta)$ during the CMA attack.

The MAC Π is said to be *weak* if $\delta(k) - \beta(k) \geq 1/\text{poly}(k)$, for some polynomial poly, and is said to be *strong* if, for sufficiently large k, $\beta(k) \leq \text{negl}(k)$ and $\delta(k) \geq 1 - \text{negl}(k)$, where $\text{negl}(k)$ is some negligible function of k. Given an integer n and a number $\gamma > 0$, we can define the "threshold direct product" MAC Π^n in the natural way: the key of Π^n consists of n independent keys for the basic MAC, the tag of m contains the concatenation of all n individual tags of m, and the verification accepts an

n-tuples of individual tags if at least $(n - (1 - \gamma)\delta n)$ individual tags are correct. Then, a straightforward application of Theorem 3 gives:

Theorem 1. *Assume Π is a weak MAC. Then one can choose $n = \text{poly}(k)$ and $\gamma > 0$ so that Π^n has completeness error $2^{-\Omega(k)}$ and unforgeability $(1 - 2^{-\Omega(k)})$. In particular, Π^n is a strong MAC.*

We then use our direct product result for MACs to argue the XOR lemma for the security amplification of PRFs. Namely, in Section 4.2 we show that the XOR of several independent weak PRFs results in a strong PRF (see Section 4.2 for definitions). It is interesting to compare this result with a related XOR lemma for PRFs by Myers [Mye03]. Meyers observed that the natural XOR lemma above cannot hold for δ-pseudorandom PRFs, where $\delta \geq \frac{1}{2}$. In particular, a PRF one of whose output bits is a constant for some input can potentially reach security (almost) $1/2$, but can never be amplified by a simple XOR. Because of this counter-example, Meyers had a more complicated XOR lemma for PRFs, where a separate pad was selected for each δ-pseudorandom PRF, and showed that this variant worked for any $\delta < 1$. In this work, we show that Meyers' counter-example is the worst: the simple XOR lemma holds for δ-pseudorandom PRFs, for any $\delta < \frac{1}{2}$.

The PRF result generally follows the usual connection between the direct product theorems and the XOR lemmas first observed by [GNW95], but with a subtlety. First, it is easy to see that it suffices to consider *Boolean* PRFs. For those, we notice that a δ-pseudorandom PRF is also a $(1-2\delta)$-unforgeable (Boolean) MAC (this is where $\delta < \frac{1}{2}$ comes in). Then, we apply the direct product theorem to obtain a strong (non-Boolean) MAC. At this stage, one typically applies the Goldreich-Levin [GL89] theorem to argue that the XOR of a random subset of (strong) MACs is a PRF. Unfortunately, as observed by Naor and Reingold [NR98], the standard GL theorem *does not work in general* for converting unpredictability into pseudorandomness, at least when the subset is *public* (which will ultimately happen in our case). However, [NR98] showed that the conversion *does* work when r is kept *secret*. Luckily, by symmetry, it is easy to argue that for "direct product MACs", keeping r secret or public does not make much difference. Indeed, by slightly adjusting the analysis of [NR98] to our setting, we will directly obtain the desired XOR lemma for PRFs.

Finally, in Section 4.1 we observe a simple result regarding the security amplification of pseudorandom generators (PRGs). This result does not use any new techniques (such as our Chernoff-type theorem). However, we state it for completeness, since it naturally leads to the (more complicated) case of PRFs in Section 4.2 and, as far as we know, it has never explicitly appeared in the literature before.

2   Preliminaries

For a natural number k, we will denote by $[k]$ the set $\{1, \ldots, k\}$.

Lemma 1 (Hoeffding bound). *Let $X_1, \ldots, X_t$ be independent identically distributed random variables taking values in the interval $[0, 1]$, with expectation μ. Let $\chi = (1/t)\sum_{i=1}^{t} X_i$. For any $0 < \nu \leq 1$, we have $\mathbf{Pr}[\chi < (1 - \nu)\mu] < e^{-\nu^2 \mu t/2}$.*

Theorem 2 ([GL89]). *There is a probabilistic algorithm Dec with the following property. Let* $a \in \{0,1\}^k$ *be any string, and let* $O : \{0,1\}^k \to \{0,1\}$ *be any predicate such that* $|\mathbf{Pr}_{z \in \{0,1\}^k}[O(z) = \langle a, z \rangle] - 1/2| \geq \nu$ *for some* $\nu > 0$. *Then, given* ν *and oracle access to the predicate* O, *the algorithm Dec runs in time* $\mathrm{poly}(k, 1/\nu)$, *and outputs a list of size* $O(1/\nu^2)$ *such that, with probability at least* $3/4$, *the string* a *is on the list.*

2.1   Samplers

We will consider bipartite graphs $G = G(L \cup R, E)$ defined on a bipartition $L \cup R$ of vertices; we think of L as left vertices, and R as right vertices of the graph G. We allow graphs with multiple edges. For a vertex v of G, we denote by $N_G(v)$ the multiset of its neighbors in G; if the graph G is clear from the context, we will drop the subscript and simply write $N(v)$. Also, for a vertex x of G, we denote by E_x the set of all edges in G that are incident to x. We say that G is *bi-regular* if the degrees of vertices in L are the same, and the degrees of vertices in R are the same.

Let $G = G(L \cup R, E)$ be any bi-regular bipartite graph. For a function $\lambda : [0,1] \times [0,1] \to [0,1]$, we say that G is a λ-*sampler* if, for every function $F : L \to [0,1]$ with the average value $\mathbf{Exp}_{x \in L}[F(x)] \geq \mu$ and any $0 < \nu < 1$, there are at most $\lambda(\mu, \nu) \cdot |R|$ vertices $r \in R$ where $\mathbf{Exp}_{y \in N(r)}[F(y)] \leq (1 - \nu)\mu$.

We will use the following properties of samplers (proved in [IJKW08, IJK08]). The first property says that for any two large vertex subsets W and F of a sampler, the fraction of edges between W and F is close to the product of the densities of W and F.

Lemma 2 ([IJKW08]). *Suppose* $G = G(L \cup R, E)$ *is a* λ-*sampler. Let* $W \subseteq R$ *be any set of measure at least* τ, *and let* $V \subseteq L$ *be any set of measure at least* β. *Then, for all* $0 < \nu < 1$ *and* $\lambda_0 = \lambda(\beta, \nu)$, *we have* $\mathbf{Pr}_{x \in L, y \in N(x)}[x \in V \ \& \ y \in W] \geq \beta(1 - \nu)(\tau - \lambda_0)$, *where the probability is for the random experiment of first picking a random node* $x \in L$ *uniformly at random, and then picking a uniformly random neighbor* y *of* x *in the graph* G.

The second property deals with edge-colored samplers. It basically says that removing some subset of right vertices of a sampler yields a graph which (although not necessarily bi-regular) still has the following property: Picking a random left node and then picking its random neighbor induces roughly the same distribution on the edges as picking a random right node and then its random neighbor.

Lemma 3 ([IJKW08]). *Suppose* $G = G(L \cup R, E)$ *is a* λ-*sampler, with the right degree* D. *Let* $W \subseteq R$ *be any subset of density at least* τ, *and let* $G' = G(L \cup W, E')$ *be the induced subgraph of* G *(obtained after removing all vertices in* $R \setminus W$*), with the edge set* E'. *Let* $Col : E' \to \{red, green\}$ *be any coloring of the edges of* G' *such that at most* $\eta D|W|$ *edges are colored red, for some* $0 \leq \eta \leq 1$. *Then, for all* $0 < \nu, \beta < 1$ *and* $\lambda_0 = \lambda(\beta, \nu)$, *we have*

$$\mathbf{Pr}_{x \in L, y \in N_{G'}(x)}[Col(\{x, y\}) = red] \leq \max\{\eta/((1 - \nu)(1 - \lambda_0/\tau)), \beta\},$$

where the probability is for the random experiment of first picking a uniformly random node $x \in L$, *and then picking a uniformly random neighbor* y *of* x *in the graph* G'.

3   Dynamic Weakly Verifiable Puzzles

We consider the following generalization of weakly verifiable puzzles (WVP) [CHS05], which we call *dynamic weakly verifiable puzzles (DWVP)*.

Definition 1 (Dynamic Weakly Verifiable Puzzle). *A DWVP Π is defined by a distribution $\mathcal{D}$ on pairs of strings (x, α); here α is the advice used to generate and evaluate responses to the puzzle x. Unlike the case of WVP, here the string x defines a set of puzzles, (x, q) for $q \in Q$ (for some set Q of indices). There is a probabilistic polynomial-time computable relation R that specifies which answers are solutions for which of these puzzles: $R(\alpha, q, r)$ is true iff response r is correct answer to puzzle q in the collection determined by α. Finally, there is also a probabilistic polynomial-time computable hint function $H(\alpha, q)$.*

A solver can make a number of queries: query $hint(q)$ asks for $H(\alpha, q)$, the hint for puzzle number q; a verification query $V(q, r)$ asks whether $R(\alpha, q, r)$. The solver succeeds if it makes an accepting verification query for a q where it has not previously made a hint query on q.

Clearly, WVP is a special case of DWVP when $|Q| = 1$. A MAC is also a special case of DWVP where α is a secret key, x is the empty string, queries q are messages, a hint is to give the MAC of a message, and correctness is for a (valid message, MAC) pair. Signatures are also a special case with α being a secret key, x a public key, and the rest similar to the case of MACs.

We give hardness amplification for such weakly verifiable puzzle collections, using direct products. First, let us define an n-wise direct product for DWVPs.

Definition 2 (n-wise direct-product of DWVPs). *Given a DWVP Π with $\mathcal{D}$, R, Q, and H, its n-wise direct product is a DWVP Π^n with the product distribution $\mathcal{D}^n$ producing n-tuples $(\alpha_1, x_1), \ldots, (\alpha_n, x_n)$. For a given n-tuple $\bar{\alpha} = (\alpha_1, \ldots, \alpha_n)$ and a query $q \in Q$, the new hint function is $H^n(\bar{\alpha}, q) = (H(\alpha_1, q), \ldots, H(\alpha_n, q))$. For parameters $0 \le \gamma, \delta \le 1$, we say that the new relation $R^k((\alpha_1, \ldots, \alpha_n), q, (r_1, \ldots, r_n))$ evaluates to true if there is a subset $S \subseteq [n]$ of size at least $n - (1 - \gamma)\delta n$ such that $\wedge_{i \in S} R(\alpha_i, q, r_i)$.*

A solver of the n-wise DWVP Π^n may ask hint queries $hint^n(q)$, getting $H^n(\bar{\alpha}, q)$ as the answer. A verification query $V^n(q, \bar{r})$ asks if $R^n(\bar{\alpha}, q, \bar{r})$, for an n-tuple $\bar{r} = (r_1, \ldots, r_n)$. We say that the solver succeeds if it makes an accepting verification query for a q where it has not previously made a hint query on q.[1]

Theorem 3 (Security amplification for DWVP (uniform version)). *Suppose a probabilistic t-time algorithm A succeeds in solving the n-wise direct-product of some DWVP Π^n with probability at least ϵ, where $\epsilon \ge (800/\gamma\delta) \cdot (h + v) \cdot e^{-\gamma^2 \delta n / 40}$, and h is the number of hint queries [2], and v the number of verification queries made by A.*

---

[1] We don't allow the solver to make hint queries $(q_1, \ldots, q_n)$, with different q_i's, as this would make the new k-wise DWVP completely insecure. Indeed, the solver could ask cyclic shifts of the query $(q_1, \ldots, q_n)$, and thus learn the answers for q_1 in all n positions, without actually making the hint query $(q_1, \ldots, q_1)$.

[2] Note that when $h = 0$, we're in the case of WVPs.

Then there is a uniform probabilistic algorithm B that succeeds in solving the original DWVP Π with probability at least $1 - \delta$, while making $O((h(h + v)/\epsilon) \cdot \log(1/\gamma\delta))$ hint queries, only one verification query, and having a running time

$$O\left(((h + v)^4/\epsilon^4) \cdot t + (t + \omega h) \cdot (h + v)/\epsilon \cdot \log(1/\gamma\delta)\right).$$

Here ω denotes the maximum time to generate a hint for a given query. The success probability of B is over the random input puzzle of Π and internal randomness of B.

Note that B in the above theorem is a uniform algorithm. We get a reduction in running time of an algorithm for attacking Π if we allow it to be non-uniform. The algorithm B above (as we will see later in the proof) samples a suitable hash function from a family of pairwise independent hash functions and then uses the selected function in the remaining construction. In the non-uniform version of the above theorem, we can skip this step and assume that the suitable hash function is given to it as advice. Following is the non-uniform version of the above theorem.

Theorem 4 (Security amplification for DWVP (non uniform version)). *Suppose a probabilistic t-time algorithm A succeeds in solving the n-wise direct-product of some DWVP Π^n with probability at least ϵ, where $\epsilon \geq (800/\gamma\delta) \cdot (h+v) \cdot e^{-\gamma^2\delta n/40}$, h is the number of hint queries, and v the number of verification queries made by A. Then there is a probabilistic algorithm B that succeeds in solving the original DWVP Π with probability at least $1 - \delta$, while making $O((h \cdot (h+v)/\epsilon)) \cdot \log(1/\gamma\delta))$ hint queries, only one verification query, and having the running time $O((t + \omega h) \cdot ((h + v)/\epsilon) \cdot \log(1/\gamma\delta))$, where ω denotes the maximum time to generate a hint for a given query. The success probability of B is over the random input puzzle of Π and internal randomness of B.*

3.1   Intuition

We want to solve a single instance of DWVP Π, using an algorithm A for the n-wise direct-product Π^n, and having access to the hint-oracle and the verification-oracle for Π. The idea is to "embed" our unknown puzzle into an n-tuple of puzzles, by generating the $n - 1$ puzzles at random by ourselves. Then we simulate algorithm A on this n-tuple of puzzles. During this simulation, we can answer the hint queries made by A by computing the hint function on our own puzzles and by making the appropriate hint query to the hint-oracle for Π. We will answer all verification queries of A by 0 (meaning "failure"). At the end, we see if A made a verification query which was "sufficiently" correct in the positions corresponding to our own puzzles; if so, we make a probabilistic decision to output this query (for the position of our unknown input puzzle).

To decide whether to believe or not to believe the verification query made by A, we count the number of correct answers it gave for the puzzles we ourselves generated (and hence can verify), and then believe with probability inverse-exponentially related to the number of incorrect answers we see (i.e., the more incorrect answers we see, the less likely we are to believe that A's verification query is correct for the unknown puzzle); since we allow up to $(1-\gamma)\delta n$ incorrect answers, we will discount these many incorrect answers, when making our probabilistic decision.

Such a "soft" decision algorithm for testing if an n-tuple is good has been proposed in [IW97], and later used in [BIN97, IJK08]. Using the machinery of [IJK08], we may assume, for the sake of intuition, that we can decide if a given verification query $(q, \bar{r})$ made by A is correct (i.e., is correct for at least $n - (1 - \gamma)\delta n$ of r_i's in the n-tuple $\bar{r}$).

Since A is assumed to succeed on at least ϵ fraction of n-tuples of puzzles, we get from A a correct verification query with probability at least ϵ (for a random unknown puzzle, and random $n - 1$ self-generated puzzles). Hence, we will produce a correct solution to the input puzzle of Π with probability at least ϵ.

To improve this probability, we would like to repeatedly sample $n - 1$ random puzzles, simulate A on the obtained n-tuple of puzzles (including the input puzzle in a random position), and check if A produces a correct verification query. If we repeat for $O(\log 1/\delta)/\epsilon)$ iterations, we should increase our success probability for solving Π to $1 - \delta$.

However, there is a problem with such repeated simulations of A on different n-tuples of puzzles: in some of its later runs, A may make a successful verification query for the same q for which it made a hint query in an earlier run. Thus, we need to make sure that a successful verification query should not be one of the hint queries asked by A in one of its previous runs. We achieve this by randomly partitioning the query space Q into the "attack" queries P, and "hint" queries. Here P is a random variable such that any query has probability $\frac{1}{2(h+v)}$ of falling inside P. We will define the set P by picking a random hash function $hash$ from Q to $\{0, 1, \ldots, 2(h + v) - 1\}$, and setting $P = P_{hash}$ to be the preimages of 0 of $hash$.

We say that the *first success query* for A is the first query where a successful verification query without a previous hint query is made. A *canonical success* for attacker A with respect to P is an attack so that the first successful verification query is in P and all earlier queries (hint or verification) are not in P.

We will show that the expected fraction of canonical successes for P_{hash} is at least $\frac{\epsilon}{4(h+v)}$. We will also give an efficient algorithm (the **Pick-hash** procedure below) that finds a hash function $hash$ so that the fraction of canonical successes for P_{hash} is close to the expected fraction. Then we test random candidates for being canonical successes with respect to P_{hash}.

Due to this extra complication (having to separate hint and verification queries), we lose on our success probability by a factor of $8(h + v)$ compared to the case of WVPs analyzed in [IJK08]. The formal proof of the theorem is given in the following subsection.

3.2   Proof of Theorems 3 and 4

Proof (proof of Theorem 3). For any mapping $hash : Q \to \{0, \ldots, 2(h + v) - 1\}$, let P_{hash} denote the preimages of 0. Also, as defined in the previous section, a canonical success for an attacker A with respect to $P \subseteq Q$ is an attack so that the first successful verification query is in P and all earlier queries (hint or verification) queries are not in P. The proof of the main theorem follows from the following two lemmas.

Lemma 4. *Let A be an algorithm which succeeds in solving the n-wise direct-product of some DWVP Π^n with probability at least ϵ while making h hint queries, v verification* *queries and have a running time t. Then there is a probabilistic algorithm which runs in time $O(((h + v)^4/\epsilon^4) \cdot t)$ and with high probability outputs a function $hash : Q \to \{0, ..., 2(h + v) - 1\}$ such that the canonical success probability of A with respect to the set P_{hash} is at least $\frac{\epsilon}{8(h+v)}$.*

Lemma 5. *Let $hash : Q \to \{0, ..., 2(h + v) - 1\}$ be a function. Let A be an algorithm such that the canonical success probability of A over an n-wise DWVP Π^n with respect to P_{hash} is at least $\epsilon' = (100/\gamma\delta) \cdot e^{-\gamma^2\delta n/40}$. Furthermore, let A makes h hint queries and v verification queries and have a running time t. Then there is a probabilistic algorithm B that succeeds in solving the original DWVP Π with probability at least $1 - \delta$, while making $O((h(h + v)/\epsilon) \cdot \log(1/\gamma\delta))$ hint queries, only one verification query, and having the running time $O((t + \omega h) \cdot (h/\epsilon) \cdot \log(1/\gamma\delta))$, where ω denotes the maximum time to generate a hint for a given query.*

Proof (proof of Theorem 4). The proof follows from Lemmas 4 and 5.

In the remaining subsection, we give the proof of Lemmas 4 and 5. The proof of Lemma 5 is very similar to the analysis of WVPs in [IJK08].

Proof (proof of Lemma 4). Let $\mathcal{H}$ be a pairwise independent family of hash functions mapping Q into $\{0, ..., (2(h + v) - 1)\}$. First note that for a randomly chosen function $hash \xleftarrow{\$} \mathcal{H}$, P_{hash} is a random variable denoting the partition of Q into two parts which satisfies the following properties:

$$\forall q_1, q_2 \in Q, \; \mathbf{Pr}[q_1 \in P_{hash} \mid q_2 \in P_{hash}] = \mathbf{Pr}[q_1 \in P_{hash}] = \frac{1}{2(h + v)} \quad (1)$$

For any fixed choice of $\bar{\alpha} = (\alpha_1, ..., \alpha_n)$, let $q_1^{\bar{\alpha}}, ..., q_h^{\bar{\alpha}}$ denote the hint queries of A and $(q_{h+1}^{\bar{\alpha}}, \bar{r}_{h+1}^{\bar{\alpha}}), ..., (q_{h+v}^{\bar{\alpha}}, \bar{r}_{h+v}^{\bar{\alpha}})$. Also, let $(q_j^{\bar{\alpha}}, \bar{r}_j^{\bar{\alpha}})$ denote the first successful verification query, in case A succeeds, and let it denote any arbitrary verification query in the case A fails. Furthermore, let $E_{\bar{\alpha}}$ denote the event that $q_1^{\bar{\alpha}}, ..., q_h^{\bar{\alpha}}, q_{h+1}^{\bar{\alpha}}, ..., q_{j-1}^{\bar{\alpha}} \notin P_{hash}$ and $q_j^{\bar{\alpha}} \in P_{hash}$. We bound the probability of the event $E_{\bar{\alpha}}$ as follows.

Claim. For each fixed $\bar{\alpha}$, we have $\mathbf{Pr}_{P_{hash}}[E_{\bar{\alpha}}] \geq \frac{1}{4(h+v)}$.

Proof. We have

$$\mathbf{Pr}_{P_{hash}}[E_{\bar{\alpha}}] = \mathbf{Pr}[q_j^{\bar{\alpha}} \in P_{hash} \; \& \; \forall i < j, q_i^{\bar{\alpha}} \notin P_{hash}]$$
$$= \mathbf{Pr}[q_j^{\bar{\alpha}} \in P_{hash}] \cdot \mathbf{Pr}[\forall i < j, q_i^{\bar{\alpha}} \notin P_{hash} \mid q_j^{\bar{\alpha}} \in P_{hash}].$$

By (1), we get that the latter expression is equal to $\frac{1}{2(h+v)} \cdot (1 - \mathbf{Pr}[\exists i < j, q_i^{\bar{\alpha}} \in P_{hash} \mid q_j^{\bar{\alpha}} \in P_{hash}])$, which, by the union bound, is at least

$$\frac{1}{2(h + v)} \cdot \left(1 - \sum_{i<j} \mathbf{Pr}[q_i^{\bar{\alpha}} \in P_{hash} \mid q_j^{\bar{\alpha}} \in P_{hash}]\right).$$

Finally, using the pairwise independence property (1), we conclude that

$$\mathbf{Pr}[E_{\bar{\alpha}}] \geq \frac{1}{2(h+v)} \cdot \left(1 - \sum_{i<j} \mathbf{Pr}[q_i^{\bar{\alpha}} \in P_{hash}]\right) \geq \frac{1}{4(h+v)},$$

as required.

Let T denote the "good" set corresponding to A's attack, that is, $T = \{\bar{\alpha} : A's \; attack \; succeeds\}$. We have $Pr[\bar{\alpha} \in T] \geq \epsilon$. Consider the following random variable: $G_{P_{hash}} = \{\bar{\alpha} \mid \bar{\alpha} \in T \; and \; E_{\bar{\alpha}}\}$. So, $G_{P_{hash}}$ contains those $\bar{\alpha}$'s for which A has canonical success.

Since $\forall \bar{\alpha} \in G, \mathbf{Pr}_{P_{hash}}[E_{\bar{\alpha}}] \geq \frac{1}{4(h+v)}$, using linearity of expectation we get $\mathbf{Exp}_{P_{hash}}[\mathbf{Pr}_{\bar{\alpha}}[\bar{\alpha} \in G_{P_{hash}}]] \geq \frac{\epsilon}{4(h+v)}$. Hence, by averaging, we get that with probability at least $\frac{\epsilon}{8(h+v)}$ over the randomness of P_{hash}, there is at least $\frac{\epsilon}{8(h+v)}$ chance that a randomly chosen $\bar{\alpha} \in G_{P_{hash}}$. Let us call such P_{hash}'s "good". The subroutine **Pick-hash** (see figure 2) uses sampling and runs in time $O(((h+v)^4/\epsilon^4) \cdot t)$ to return a mapping $hash$ such that P_{hash} is good.

Pick-hash

00.	Let $\mathcal{H}$ be a pairwise independent family of hash functions which maps Q into $\{0, 1, ..., (2h-1)\}$.
01.	Repeat lines $(2-15)$ for at most $64(h+v)^2/\epsilon^2$ times:
02.	$\quad hash \stackrel{\$}{\leftarrow} \mathcal{H}$
03.	$\quad$ Let P_{hash} denote the subset of all queries q such that $hash(q) = 0$
04.	$\quad count \leftarrow 0$
05.	$\quad$ Repeat for at most $64(h+v)^2/\epsilon^2$ times:
06.	$\quad\quad$ Pick $\bar{\alpha} = (\alpha_1, ..., \alpha_n)$ randomly
07.	$\quad\quad$ Execute A
08.	$\quad\quad\quad$ When A asks a hint query q
09.	$\quad\quad\quad\quad$ **If** $(q \in P_{hash})$, then abort A and continue with step 5
10.	$\quad\quad\quad\quad$ Let $(r_1, ..., r_n)$ be hints to query q for puzzle sets $x_1, ..., x_n$
11.	$\quad\quad\quad\quad$ return $(r_1, ..., r_n)$ to A
12.	$\quad\quad\quad$ When A asks a verification query $(q, \bar{r})$
13.	$\quad\quad\quad\quad$ **If** $(R^n(\bar{\alpha}, q, \bar{r}) = 1)$ and $q \in P_{hash}$ then
14.	$\quad\quad\quad\quad\quad$ increase $count$ by 1 and continue at step 5
15.	$\quad\quad$ **If** $(count \geq 4(h+v)/\epsilon)$ then return $hash$

Fig. 1. Algorithm for picking a *good* hash function

Proof (Proof of Lemma 5)

Due to space restriction, we only give an intuitive sketch of the proof in this paper. The detailed proof of this lemma can be found in the full version of the paper. Figure (2) gives the formal description of the algorithm B which uses the algorithm A.

For any fixed $\bar{\alpha}$, let $q_1^{\bar{\alpha}}, ..., q_h^{\bar{\alpha}}$ denote the hint queries made by A and $(q_{h+1}^{\bar{\alpha}}, r_{h+1}^{\bar{\alpha}}), ..., (q_{h+v}^{\bar{\alpha}}, r_{h+v}^{\bar{\alpha}})$ denote the verification queries. Let $j \in [h+v]$ be the first query such that $q_v^{\bar{\alpha}} \in P_{hash}$. For all simulations of A, B correctly answers every hint query of A by itself

B(x)

00.	Let $\rho = (1 - \gamma/10)$ and $\Theta = (1 - \gamma)\delta n$
01.	Let $hash$ denote the function as in Lemma 4. For Theorem 3, $hash$ is chosen using the subroutine **Pick-hash**. For Theorem 4, we can assume that $hash$ is given as advice.
02.	Let P_{hash} denote the subset of all queries q such that $hash(q) = 0$
03.	Repeat lines $(4 - 23)$ for at most $timeout = O(((h + v)/\epsilon) \cdot \log(1/\gamma\delta))$ steps:
04.	Pick $i \xleftarrow{\$} [1..n]$
05.	Pick $(n - 1)$ α's randomly. Let $(\alpha_1, ..., \alpha_{i-1}, \alpha_{i+1}, ..., \alpha_n)$ denote these α's and let $(x_1, ..., x_{i-1}, x_{i+1}, ..., x_n)$ denote the puzzle sets corresponding to these α's.
06.	$S_v \leftarrow \emptyset$
07.	Execute $A(x_1, ..., x_{i-1}, x, x_{i+1}, ..., x_n)$
08.	When A asks its hint query q
09.	If $q \in P_{hash}$ then return $\perp$ to A and halt the current simulation of A
10.	B makes a hint query q to get the answer r
11.	Let $(r_1, ..., r_{i-1}, r_{i+1}, ..., r_{n-1})$ be the hints for query q for puzzle sets $(x_1, ..., x_{i-1}, x_{i+1}, ..., x_n)$
12.	$\bar{r} \leftarrow (r_1, .., r_{i-1}, r, r_{i+1}.., r_n)$
13.	return $\bar{r}$ to A
14.	When A asks a verification query $(q, \bar{r})$
15.	**If** $q \notin P_{hash}$ then return 0 to A
16.	**else**
17.	Parse $\bar{r}$ as $(r_1, \ldots, r_n)$
18.	$m \leftarrow |\{j : R(\alpha_j, q, r_j) = 1, j \neq i\}|$
19.	**If** $(m \geq n - \Theta)$ then
20.	with probability 1, B makes a verification query (q, r_i) and halts
21.	**else**
22.	with probability $\rho^{m-\Theta}$, B makes a verification query (q, r_i) and halts
23.	Halt the current simulation of A and continue at line (03)
25.	return $(\perp, \perp)$

Fig. 2. Algorithm for solving Π

making a hint query with respect to the DWVP Π which it is trying to solve (lines 10–13). Note that for a single simulation of A (lines 6–25), the simulation is aborted if $j \leq h$ (line 9). Otherwise, B makes a "soft" decision using $(q_j^{\bar{\alpha}}, \bar{r}_j^{\bar{\alpha}})$ to produce its verification query (lines 19–24). Lemma 4 tells that there are at least ϵ' fraction of $\bar{\alpha}$'s such that $(q_j^{\bar{\alpha}}, \bar{r}_j^{\bar{\alpha}})$ is a correct verification query for A(these are $\bar{\alpha}$'s on which A has canonical success). So, intuitively there is a fair chance that B produces a correct verification query.

Let $Good$ denote the set of $\bar{\alpha}$'s on which A succeed canonically. From the assumption of the Lemma we know that $Good$ contains at least ϵ' fraction of $\bar{\alpha}$'s. The remaining task is to argue that this is sufficient to show that B succeeds with high probability. The rest of the analysis, apart from minor details, is similar to the proof of the Direct Product theorem for WVPs from [IJK08], essentially arguing that the lines 17–22 of the algorithm B act as a decision procedure for the set $Good$. Next we give some details of these arguments.

Consider the following bipartite graph $G = G(L \cup R, E)$: the set of left vertices L is the set of α's; the right vertices R are all n-tuples $\bar{\alpha} = (\alpha_1, ..., \alpha_n)$; for every

$y = (u_1, \ldots, u_n) \in R$, there are n edges $(y, u_1), \ldots, (y, u_n) \in E$. Using Lemma 1, we see that this graph is a λ-sampler for $\lambda(\mu, \nu) = e^{-\nu^2 \mu k / 2}$.

For an unknown secret key α, let $\bar{\alpha} = (\alpha_1, \ldots, \alpha_{i-1}, \alpha, \alpha_i, \ldots, \alpha_{n-1})$ be the n-tuple of secret keys that corresponds to the n-tuple of puzzles $(x_1, \ldots, x_n)$ that B will feed to A in line 7. Let $q_1^{\bar{\alpha}}, \ldots, q_h^{\bar{\alpha}}$ be the A's hint queries and $(q_{h+1}^{\bar{\alpha}}, \bar{r}_{h+1}^{\bar{\alpha}}), \ldots,$ $(q_{h+v}^{\bar{\alpha}}, \bar{r}_{h+v}^{\bar{\alpha}})$ be the verification queries. Let $j \in [h + v]$ be the first index such that $q_j^{\bar{\alpha}} \in P_{hash}$. Let $(q^{\bar{\alpha}}, \bar{r}^{\bar{\alpha}})$ denote $(q_j^{\bar{\alpha}}, \bar{r}_j^{\bar{\alpha}})$ in case $j > h$ and $(\perp, \perp)$ otherwise.

In the case $(q^{\bar{\alpha}}, \bar{r}^{\bar{\alpha}}) \neq (\perp, \perp)$, B makes a probabilistic decision about using $(q^{\bar{\alpha}}, \bar{r}^{\bar{\alpha}})$ to produce its verification queries. It does that by verifying the answers to the query at all positions other than position i where the unknown α has been planted. Let $(q^{\bar{\alpha}}, r^{\bar{\alpha}})$ denote the verification query made by the algorithm B in this simulation of A. If no verification query is made or if $(q^{\bar{\alpha}}, \bar{r}^{\bar{\alpha}}) = (\perp, \perp)$, then $(q^{\bar{\alpha}}, r^{\bar{\alpha}}) = (\perp, \perp)$. Let (q^B, r^B) denote the single verification query made by B.

First we bound the probability of timeout of B or in other words the probability that $(q^{\bar{\alpha}}, r^{\bar{\alpha}}) = (\perp, \perp)$ in all iterations of B. If $\bar{\alpha} \in Good$, then lines 7–23 will return a verification query with probability 1. Hence, the probability of timeout is at most the probability that B never samples a neighbor $\bar{\alpha} \in Good$ of α in the graph G.

Consider the set H of all those left vertices α of G such that α has less than $\epsilon'/4$ fraction of its neighbors falling into the set $Good$. These are precisely those α's for which B is likely to time out. The next lemma shows that the set H is small.

Lemma 6. *The set H has density at most $\gamma\delta/5$.*

Proof. Suppose that the density of H is greater than $\beta = \gamma\delta/5$. Let $H' \subseteq H$ be any subset of H of density exactly β. By our assumption, we have that $\mathbf{Pr}_{\alpha \in L, w \in N(\alpha)}[\alpha \in H'$ & $w \in Good] < \beta\epsilon'/4$. On the other hand, by Lemma 2 we get that the same probability is at least $\beta(\epsilon' - \lambda_0)/3$ for $\lambda_0 = \lambda(\beta, 2/3)$. This is a contradiction if $\lambda_0 \leq \epsilon'/4$.

Lemma 7. *For every $\alpha \notin H$, we have $\mathbf{Pr}[B$ timeouts $] \leq \gamma\delta/20$, where the probability is over the internal randomness of B.*

Proof. By the definition of H, we get that the probability of timeout on any given $\alpha \notin H$ is at most $(1 - \epsilon'/4)^{4 \ln(20/\gamma\delta)/\epsilon'} \leq \gamma\delta/20$.

Next, we need to show that the probability of $R(\alpha, q^B, r^B) = 0$ conditioned on the event that $(q^B, r^B) \neq (\perp, \perp)$, is small. Note that this conditional probability remains the same across all the simulations of A in lines 4–23. Consider any fixed simulation of A (lines 8–23) such that $(q^{\bar{\alpha}}, \bar{r}^{\bar{\alpha}}) \neq (\perp, \perp)$. Let err be the number of incorrect answers in $\bar{r}^{\bar{\alpha}}$ for the query $q^{\bar{\alpha}}$. Then if $err \leq (1 - \gamma)\delta n$, then lines $7 - 25$ of B produces a verification query with probability 1. Otherwise a verification query is produced with probability that decreases exponentially (by a factor of ρ) as err increases. For this intuitive sketch of the proof, let us make a simplifying assumption there is an oracle $\mathcal{O}$ which tells whether $\bar{\alpha} \in Good$ [3] (in some sense lines 17–22 is an approximation of such an oracle). Given such an oracle, consider the the algorithm $B^{\mathcal{O}}$, which is same as B except we replace lines 17–23 with the following line:

---

[3] Note that B does not have access to α and hence does not know $\bar{\alpha}$

> **If** $\mathcal{O}$ tells that the hidden $\bar{\alpha} \in Good$, then B makes verification query (q, r_i)

Intuitively, lines 17–22 in B is an approximation of the line above and hence $\mathbf{Pr}[R(\alpha, q^B, r^B) = 0 | (q^B, r^B) \neq (\bot, \bot)]$ should be close to $\mathbf{Pr}[R(\alpha, q^{B^{\mathcal{O}}}, r^{B^{\mathcal{O}}}) = 0 | (q^{B^{\mathcal{O}}}, r^{B^{\mathcal{O}}}) \neq (\bot, \bot)]$. On the other hand, analyzing the conditional probability $\mathbf{Pr}[R(\alpha, q^{B^{\mathcal{O}}}, r^{B^{\mathcal{O}}}) = 0 | (q^{B^{\mathcal{O}}}, r^{B^{\mathcal{O}}}) \neq (\bot, \bot)]$ is simple and is done by analyzing the following graph: Let G' be the induced subgraph of G obtained after removing all vertices in $R \backslash Good$. For each edge $((\alpha_1, \ldots, \alpha_n), \alpha_l)$ of the graph G', we color this edge green if the lth answer in $\bar{r}^{\bar{\alpha}}$ is correct for the query $q^{\bar{\alpha}}$, and we color it red otherwise. Consider the following random experiment $\mathcal{E}$ defined on the graph G':

"Pick a random $\alpha \in L$, and its random incident edge $e = (\alpha, \bar{\alpha})$ in G', for $\bar{\alpha}$ containing α in position $l \in [n]$. If $\bar{\alpha} \in Good$, then output e with probability 1 else output $\bot$."

For each α, we have

$$\mathbf{Pr}[R(\alpha, q^{B^{\mathcal{O}}}, r^{B^{\mathcal{O}}}) = 0 \mid (q^{B^{\mathcal{O}}}, r^{B^{\mathcal{O}}}) \neq (\bot, \bot)] =$$
$$\mathbf{Pr}[\mathcal{E} \text{ outputs red edge incident to } \alpha \mid \mathcal{E} \text{ outputs some edge incident to } \alpha], \quad (2)$$

where the first probability is over internal randomness of $B^{\mathcal{O}}$, and the second probability is over the random choices of $\mathcal{E}$ for the fixed α (i.e., over the random choice of an edge e incident to α, and the random choice whether e is output).

From Lemma 3 we get that:

$$\mathbf{Pr}[\mathcal{E} \text{ outputs red edge incident to } \alpha \mid \mathcal{E} \text{ outputs some edge incident to } \alpha] \leq$$
$$\max \left(\frac{\eta}{(1 - \nu)(1 - \lambda_0/\tau)}, \beta \right)$$

which is at most $\delta - \gamma\delta/2$ for $\eta = (1 - \gamma)\delta$, $\beta = \delta/2$, $\nu = \gamma/4$, and $\frac{\lambda_0}{\tau} = \frac{\lambda(\delta/2, \gamma/4)}{\epsilon'} \leq \gamma/4$.

Summing up, from Lemma 6 we get that the fraction of α's for which B might time out is small. From Lemma 7 we get that for the remaining α's, it does not time out with high probability. Furthermore, from the above argument, the conditional probability of failing to produce a correct verification query is small. Hence, the probability that B fails is small.

4   XOR Lemmas for PRGs and PRFs

In this section, we show how to amplify security of pseudorandom (function) generators, using Direct Products (old and new) and the Goldreich-Levin decoding algorithm from Theorem 2.

4.1 Amplifying PRGs

We start with PRGs. Let $G : \{0,1\}^k \rightarrow \{0,1\}^{\ell(k)}$ be a polynomial-time computable generator, stretching n-bit seeds to $\ell(k)$-bit strings, for $\ell(k) > k$, such that G is $\delta(k)$-*pseudorandom*. That is, for any probabilistic polynomial-time algorithm A, and all sufficiently large k, we have $|\mathbf{Pr}_s[A(G(s)) = 1] - \mathbf{Pr}_x[A(x) = 1]| \leq \delta(k)$, where s is chosen uniformly at random from $\{0,1\}^k$, and x from $\{0,1\}^{\ell(k)}$.

We say that a PRG G is *weak* if it is δ-pseudorandom for a constant $\delta < 1/2$. We say that a PRG G is *strong* if it is $\delta(n)$-pseudorandom for $\delta(n) < 1/k^c$ for any constant $c > 0$ (i.e., negligible).

For the rest of this subsection, let $n > \omega(\log k)$ and let $n' = 2n$. We show that any weak PRG G_{weak} of stretch $\ell(k) > kn$ can be transformed into a strong PRG G_{strong} as follows: The seed to G_{strong} is a n-tuple of seeds to G_{weak}, and the output of $G_{strong}(s_1, \ldots, s_n)$ is the bit-wise XOR of the n strings $G_{weak}(s_1), \ldots, G_{weak}(s_n)$.

Theorem 5 (Security amplification for PRGs). *If G_{weak} is a weak PRG with stretch $\ell(k) > kn$, then the generator G_{strong} defined above is a strong PRG, mapping nk-bit seeds into $\ell(k)$-bit strings.*

Proof. Since the proof uses standard techniques, we will only sketch it here. Let G_{weak} be δ-pseudorandom for $\delta < 1/2$. The proof is by a sequence of the following steps.

1. Use Yao's "pseudorandom implies unpredictable" reduction to argue that, for a random seed s, each output bit $G_{weak}(s)_i$ (for $i \in [\ell(k)]$) is computable from the previous bits $G_{weak}(s)_{1..i-1}$ with probability at most $1/2 + \delta$, which is some constant $\alpha < 1$ since $\delta < 1/2$ (this is where we need that $\delta < 1/2$).
2. Use a Direct-Product lemma (say the one from [GNW95], or the one from the present paper, Theorem 3) to argue that, for each $i \in [\ell(k)]$, computing the direct-product $(G_{weak}(s_1)_i, ..., G_{weak}(s_{n'})_i)$ from $(G_{weak}(s_1)_{1..i-1}, ..., G_{weak}(s_{n'})_{1..i-1})$ for *independent* random seeds $s_1, \ldots, s_{n'}$ can't be done better than with probability $\epsilon \leq e^{-\Omega(n)}$, which is negligible.
3. Use the Goldreich-Levin decoding algorithm from Theorem 2 to argue that, for each $i \in [\ell(k)]$, computing the XOR $G_{weak}(s_1)_i \oplus \cdots \oplus G_{weak}(s_n)_i$ (i.e., $G_{strong}(s_1, \ldots, s_n)_i$) from the given bit-wise XOR of $G_{weak}(s_1)_{1..i-1}, ...,$ $G_{weak}(s_n)_{1..i-1}$ (i.e., from $G_{strong}(s_1, \ldots, s_n)_{1..i-1}$), for independent random seeds $s_1, \ldots, s_n$, can't be done better than with probability $1/2 + \text{poly}(\epsilon n)$, which is negligibly better that random guessing.
4. Finally, using Yao's "unpredictable implies pseudorandom" reduction, conclude that G_{strong} is $(\ell(k) \cdot \text{poly}(\epsilon n))$-pseudorandom, which means that G_{strong} is $\delta'(k)$-pseudorandom for negligible $\delta'(k)$, as required.

4.2 Amplifying PRFs

Here we would like to show similar security amplification for pseudorandom function generators (PRFs).

First we recall the definition of a PRF. Let $\{f_s\}_{s \in \{0,1\}^*}$ be a function family, where, for each $s \in \{0,1\}^*$, we have $f_s : \{0,1\}^{d(|s|)} \rightarrow \{0,1\}^{r(|s|)}$. This function family is called *polynomial-time computable* if there is polynomial-time algorithm that on inputs s and $x \in \{0,1\}^{d(|s|)}$ computes $f_s(x)$. It is called $\delta(k)$-*pseudorandom function* family if, for every probabilistic polynomial-time oracle machine M, and all sufficiently large k, we have

$$|\mathbf{Pr}_s[M^{f_s}(1^k) = 1] - \mathbf{Pr}_{h_k}[M^{h_k}(1^k) = 1]| \leq \delta(k),$$

where s is chosen uniformly at random from $\{0,1\}^k$, and h_k is a uniformly random function from $\{0,1\}^{d(k)}$ to $\{0,1\}^{r(k)}$. Finally, we say that a PRF is *weak* if it is δ-pseudorandom for some constant $\delta < 1/2$, and we say a PRF is *strong* if it is $\delta(k)$-pseudorandom for some $\delta(k) < 1/k^c$ for any constant $c > 0$.

Let $\{f_s\}_s$ be a weak PRF. By analogy with the case of PRGs considered above, a natural idea for defining a strong PRF from $\{f_s\}_s$ is as follows: For some parameter n, take n independent seeds $\bar{s} = (s_1, \ldots, s_n)$, and define $g_{\bar{s}}(x)$ to be the bit-wise XOR of the strings $f_{s_1}(x), \ldots, f_{s_n}(x)$.

We will argue that the defined function family $\{g_{\bar{s}}\}_{\bar{s}}$ is a strong PRF. Rather than proving this directly, we find it more convenient to prove this first for the case of weak PRF $\{f_s\}_s$ of *Boolean* functions f_s, and use a simple reduction to get the result for general weak PRFs.

For the rest of this subsection, let $n > \omega(\log k)$ and let $n' = 2n$.

Theorem 6 (XOR Lemma for Boolean PRFs). *Let $\{f_s\}_s$ be a δ-pseudorandom Boolean function family for some constant $\delta < 1/2$. Let $\bar{s} = (s_1, \ldots, s_n)$ be a n-tuple of k-bit strings. Then, for some constant c_0 dependent on δ, the following function family $\{g_{\bar{s}}\}_{\bar{s}}$ is ϵ-pseudorandom for $\epsilon \leq \mathrm{poly}(k) \cdot e^{-(\delta n)/c_0}$:*

$$g_{\bar{s}}(x) = f_{s_1}(x) \oplus \cdots \oplus f_{s_n}(x).$$

Proof. The idea is to view $\{f_s\}$ also as a MAC, which is a special case of a DWVP and hence we have a direct-product result (our Theorem 3). We will argue that if $g_{\bar{s}}$ is not a strong PRF, then one can break with non-negligible probability the direct product of MACs $(f_{s_1}, \ldots, f_{s_{n'}})$ for independent random seeds $s_1, \ldots, s_{n'}$, and hence (by Theorem 3), one can break a single MAC f_s with probability close to 1. The latter algorithm breaking f_s as a MAC will also be useful for breaking f_s as a PRF, with the distinguishing probability $\delta' > \delta$, which will contradict the assumed δ-pseudorandomness of the PRF $\{f_s\}_s$.

In more detail, suppose that A is a polynomial-time adversary that distinguishes $g_{\bar{s}}$ from a random function, with a distinguishing probability $\epsilon > \mathrm{poly}(k) \cdot e^{-\Omega(\delta n)}$. Using a standard hybrid argument, we may assume that the first query m of A is decisive. That is, answering this query with $g_{\bar{s}}(m)$ and all subsequent queries m_i with $g_{\bar{s}}(m_i)$ makes A accept with higher probability than answering this query randomly and all subsequent queries m_i with $g_{\bar{s}}(m_i)$. Let $\delta_1(k) \geq \epsilon/\mathrm{poly}(k)$ be the difference between the two probabilities.

Since $g_{\bar{s}}$ is a Boolean function, we can use Yao's "distinguisher-to-predictor" reduction [Yao82] to predict $g_{\bar{s}}(m)$ with probability $1/2 + \delta_1(n)$ over random n-tuples $\bar{s}$, and for the same fixed input m (since m is *independent* from the choice of $\bar{s}$).

By a standard argument, we get an algorithm A' for computing the following inner product

$$\langle f_{s_1}(m) \ldots f_{s_{n'}}(m), z \rangle, \tag{3}$$

for random $s_1, \ldots, s_{n'}$ and a random $z \in \{0,1\}^{n'}$, whose success probability is at least $1/2 + \delta_2(k) \geq 1/2 + \Omega(\delta_1(k)/\sqrt{n'})$; the idea is that a random $n' = 2n$-bit string z is balanced with probability $\Omega(1/\sqrt{n'})$, in which case we run the predictor for n-XOR, and otherwise (for non-balanced z) we flip a fair random coin. Next, by averaging, we get that, for at least $\delta_2(k)/2$ fraction of n'-tuples $s_1, \ldots, s_{n'}$, our algorithm A' correctly computes the inner product in (3) for at least $1/2 + \delta_2(k)/2$ fraction of random z's.

Applying the Goldreich-Levin algorithm from Theorem 2 to our algorithm A', we get an algorithm A'' that, for each of at least $\delta_2(k)/2$ fraction of n'-tuples $s_1, \ldots, s_{n'}$, computes $(f_{s_1}(m), \ldots, f_{s_{n'}}(m))$ with probability at least $\text{poly}(\delta_2(k))$. Hence, this algorithm A'' computes $(f_{s_1}(m), \ldots, f_{s_{n'}}(m))$ for a non-negligible fraction of n'-tuples $s_1, \ldots, s_{n'}$.

Next, we view A'' as an algorithm breaking the n'-wise direct-product of the MAC f_s, with non-negligible probability. Using Theorem 3, we get from A'' an algorithm B that breaks the single instance of the MAC f_s with probability at least $1 - \delta'$ for $\delta' \leq O((\log(\text{poly}(k)/\delta_2(k)))/n)$, which can be made less than $1/2 - \delta$ for $n > \omega(\log k)$ and sufficiently large constant c_0 in the bound on ϵ in the statement of the theorem (this is where we need the assumption that $\delta < 1/2$).

Note the algorithm B has $1 - \delta' > 1/2 + \delta$ probability over a secret key s to compute a correct message-tag pair (msg, tag) such that $f_s(msg) = tag$. Also note that the algorithm B makes some signing queries $f_s(q_i) =?$ for $q_i \neq msg$, but no verification queries (other than its final output pair (msg, tag)). We can use this algorithm B to distinguish $\{f_s\}_s$ from random in the obvious way: simulate B to get (msg, tag) (using the oracle function to answer the signing queries of B); query the oracle function on msg; if the answer is equal to tag, then accept, else reject.

Clearly, the described algorithm accepts with probability $1/2$ on a random oracle, and with probability greater than $1/2 + \delta$ on a pseudorandom function f_s. This contradicts the assumption that $\{f_s\}_s$ is δ-pseudorandom.

As a corollary, we get the following.

Theorem 7 (Security amplification for PRFs). *Let $\{f_s\}_s$ be a weak PRF. For a parameter $n > \omega(\log k)$, take n independent seeds $\bar{s} = (s_1, \ldots, s_n)$, and define $g_{\bar{s}}(x)$ to be the bit-wise XOR of the strings $f_{s_1}(x), \ldots, f_{s_n}(x)$. The obtained function family $\{g_{\bar{s}}\}_{\bar{s}}$ is a strong PRF.*

Proof. Note that given a non-Boolean weak PRF $\{f_s\}_s$, we can define a Boolean function family $\{f'_s\}_s$ where $f'_s(x, i) = f_s(x)_i$, i.e., f'_s treats its input as an input x to f_s and an index $i \in [r(|s|)]$, and outputs the ith bit of $f_s(x)$. Clearly, if $\{f_s\}_s$ is δ-pseudorandom, then so is $\{f'_s\}_s$.

Then we amplify the security of $\{f'_s\}_s$, using our XOR Theorem for PRFs (Theorem 6). We obtain a strong PRF $\{g'_{\bar{s}}\}_{\bar{s}}$, where $\bar{s} = (s_1, \ldots, s_n)$ and $g'_{\bar{s}}(x, i) = f'_{s_1}(x, i) \oplus \cdots \oplus f'_{s_n}(x, i)$.

Finally, we observe that our function $g_{\bar{s}}(x)$ is the concatenation of the values $g'_{\bar{s}}(x, i)$ for all $1 \leq i \leq r(|k|)$. This function family $\{g_{\bar{s}}\}_{\bar{s}}$ is still a strong PRF, since we can simulate each oracle access to $g_{\bar{s}}$ with $d(|s|)$ oracle calls to $g'_{\bar{s}}$.

5 Conclusions

We have established security amplification theorems for several interactive cryptographic primitives, including message authentication codes, digital signature and pseudorandom functions. The security amplifications for MACs and SIGs follow the direct product approach and work even for the weak variants of these primitives with imperfect completeness. For δ-pseudorandom PRFs, we have shown that the standard XOR lemma works for any $\delta < \frac{1}{2}$, which is optimal, complementing the non-standard XOR lemma of [Mye03], which works even for $\frac{1}{2} \le \delta < 1$.

Of independent interest, we abstracted away the notion of dynamic weakly verifiable puzzles (DWVPs), which generalize a variety of known primitives, including ordinary WVPs, MACs and SIGs. We have also shown a very strong Chernoff-type security amplification theorem for DWVPs, and used it to establish our security amplification results for MACs, SIGs and PRFs.

Acknowledgments. Yevgeniy Dodis was supported in part by NSF Grants 0831299, 0716690, 0515121, 0133806. Part of this work was done while the author was visiting the Center for Research on Computation and Society at Harvard University. Russell Impagliazzo was supported in part NSF Grants 0716790, 0835373, 0832797, and by the Ellentuck Foundation. Ragesh Jaiswal was supported in part by NSF Grant 0716790, and completed part of this work while being at the University of California at San Diego.

References

[BIN97]   Bellare, M., Impagliazzo, R., Naor, M.: Does parallel repetition lower the error in computationally sound protocols? In: Proceedings of the Thirty-Eighth Annual IEEE Symposium on Foundations of Computer Science, pp. 374–383 (1997)

[CHS05]   Canetti, R., Halevi, S., Steiner, M.: Hardness amplification of weakly verifiable puzzles. In: Kilian, J. (ed.) TCC 2005. LNCS, vol. 3378, pp. 17–33. Springer, Heidelberg (2005)

[Cor00]   Coron, J.S.: On the exact security of full domain hash. In: Bellare, M. (ed.) CRYPTO 2000. LNCS, vol. 1880, pp. 229–235. Springer, Heidelberg (2000)

[CRS+07]   Canetti, R., Rivest, R., Sudan, M., Trevisan, L., Vadhan, S., Wee, H.: Amplifying collision resistance: A complexity-theoretic treatment. In: Menezes, A. (ed.) CRYPTO 2007. LNCS, vol. 4622, pp. 264–283. Springer, Heidelberg (2007)

[DNR04]   Dwork, C., Naor, M., Reingold, O.: Immunizing encryption schemes from decryption errors. In: Cachin, C., Camenisch, J.L. (eds.) EUROCRYPT 2004. LNCS, vol. 3027, pp. 342–360. Springer, Heidelberg (2004)

[GL89]   Goldreich, O., Levin, L.A.: A hard-core predicate for all one-way functions. In: Proceedings of the Twenty-First Annual ACM Symposium on Theory of Computing, pp. 25–32 (1989)

[GNW95]   Goldreich, O., Nisan, N., Wigderson, A.: On Yao's XOR-Lemma. Electronic Colloquium on Computational Complexity, TR95-050 (1995)

[Gol01]   Goldreich, O.: Foundations of Cryptography: Basic Tools. Cambridge University Press, New York (2001)

[IJK08]   Impagliazzo, R., Jaiswal, R., Kabanets, V.: Chernoff-type direct product theorems. Journal of Cryptology (published online September 2008); preliminary version in CRYPTO 2007

[IJKW08]   Impagliazzo, R., Jaiswal, R., Kabanets, V., Wigderson, A.: Uniform direct-product theorems: Simplified, optimized, and derandomized. In: Proceedings of the Fortieth Annual ACM Symposium on Theory of Computing, pp. 579–588 (2008)

[Imp95]   Impagliazzo, R.: Hard-core distributions for somewhat hard problems. In: Proceedings of the Thirty-Sixth Annual IEEE Symposium on Foundations of Computer Science, pp. 538–545 (1995)

[IW97]   Impagliazzo, R., Wigderson, A.: P=BPP if E requires exponential circuits: Derandomizing the XOR Lemma. In: Proceedings of the Twenty-Ninth Annual ACM Symposium on Theory of Computing, pp. 220–229 (1997)

[Lev87]   Levin, L.A.: One-way functions and pseudorandom generators. Combinatorica 7(4), 357–363 (1987)

[LR86]   Luby, M., Rackoff, C.: Pseudorandom permutation generators and cryptographic composition. In: Proceedings of the Eighteenth Annual ACM Symposium on Theory of Computing, pp. 356–363 (1986)

[Mye03]   Myers, S.: Efficient Amplification of the Security of Weak Pseudo-Random Function Generators. J. Cryptology 16(1), 1–24 (2003)

[Mye99]   Myers, S.: On the development of block-ciphers and pseudorandom function generators using the composition and XOR operators. Master's thesis, University of Toronto (1999)

[NR98]   Naor, M., Reingold, O.: From unpredictability to indistinguishability: A simple construction of pseudo-random functions from MACs. In: Krawczyk, H. (ed.) CRYPTO 1998. LNCS, vol. 1462, pp. 267–282. Springer, Heidelberg (1998)

[NR99]   Naor, M., Reingold, O.: On the construction of pseudorandom permutations: Luby-Rackoff revisited. Journal of Cryptology, 29–66 (1999)

[PV07]   Pass, R., Venkitasubramaniam, M.: An efficient parallel repetition theorem for Arthur-Merlin games. In: Proceedings of the Thirty-Ninth Annual ACM Symposium on Theory of Computing, pp. 420–429 (2007)

[PW07]   Pietrzak, K., Wikstrom, D.: Parallel repetition of computationally sound protocols revisited. In: Vadhan, S.P. (ed.) TCC 2007. LNCS, vol. 4392, pp. 86–102. Springer, Heidelberg (2007)

[Yao82]   Yao, A.C.: Theory and applications of trapdoor functions. In: Proceedings of the Twenty-Third Annual IEEE Symposium on Foundations of Computer Science, pp. 80–91 (1982)

Composability and On-Line Deniability of Authentication

Yevgeniy Dodis[1,*], Jonathan Katz[2,**], Adam Smith[3,***],
and Shabsi Walfish[4,†]

[1] Dept. of Computer Science, New York University
[2] Dept. of Computer Science, University of Maryland
[3] Dept. of Computer Science and Engineering, Pennsylvania State University
[4] Google, Inc.

Abstract. Protocols for *deniable authentication* achieve seemingly para-doxical guarantees: upon completion of the protocol the receiver is con-vinced that the sender authenticated the message, but neither party can convince anyone else that the other party took part in the protocol. We introduce and study *on-line deniability*, where deniability should hold even when one of the parties colludes with a third party during execu-tion of the protocol. This turns out to generalize several realistic scenarios that are outside the scope of previous models.

We show that a protocol achieves our definition of on-line deniability if and only if it realizes the message authentication functionality in the *gen-eralized universal composability* framework; any protocol satisfying our definition thus automatically inherits strong composability guarantees. Unfortunately, we show that our definition is impossible to realize in the PKI model if adaptive corruptions are allowed (even if secure erasure is assumed). On the other hand, we show feasibility with respect to static corruptions (giving the first separation in terms of *feasibility* between the static and adaptive setting), and show how to realize a relaxation termed *deniability with incriminating abort* under adaptive corruptions.

1   Introduction

Message authentication allows a sender S to authenticate a message m to a receiver R. If S has a public key, message authentication is usually handled using digital signatures. A well-known drawback of digital signatures, however, is that they leave a trace of the communication and, in particular, allow R (or, in fact, any eavesdropper) to prove to a third party that S authenticated the message in question. In some scenarios such non-repudiation is essential, but in many other cases *deniability* is desired.

---

[*] Supported by NSF grants CNS-0831299, CNS-0716690, CCF-0515121, and CCF-0133806. A portion of this work was done while visiting CRCS at Harvard Uni-versity.
[**] Supported by NSF CNS-0447075 and NSF CNS-0627306.
[***] Supported by NSF TF-0747294 and NSF CAREER award 0729171.
[†] A portion of this work was done while at New York University.

O. Reingold (Ed.): TCC 2009, LNCS 5444, pp. 146–162, 2009.

Deniable authentication, introduced in [16, 18] and studied extensively since then, achieves the seemingly paradoxical guarantees that (1) the receiver is convinced that the message originated from the sender, yet (2) the receiver, even if malicious, cannot *prove* to anyone else that the sender authenticated the given message. Furthermore, (3) the receiver cannot be incriminated as having been involved, even by a malicious sender (this is meaningful when the receiver has a public key, as will be the case in our work; see further below).

Deniability is a fundamental concept in cryptography. Non-repudiation is sometimes crucial for the free exchange of ideas: without the assurance of remaining "off the record", individuals may be discouraged from discussing subversive (or embarrassing) topics. Deniability is also intimately tied to the *simulation paradigm* that is central to our understanding of cryptographic protocols.

Indeed, deniability is typically formalized via the simulation paradigm introduced in the context of zero-knowledge (ZK) proofs [20]. Zero-knowledge proofs, however, do not automatically provide deniability. Pass [27] points out that *non-interactive* ZK proofs are not deniable, nor are many existing ZK proofs in the *random oracle model*. Furthermore, ZK proofs for which simulation requires rewinding may not suffice to achieve *on-line deniability* which protects each party even when the other party colludes with an on-line entity that cannot be "rewound" (see below for an example).[1] Looking ahead, we note that on-line deniability is only potentially feasible if receivers hold public keys, and we assume this to be the case in our work. Once receivers have public keys, however, protocols can realize the stronger semantics by which a sender can authenticate a message *for a specific receiver R* but not for anyone else.

One might question whether on-line deniability is too strong. To see why it might be essential, consider a setting where Bob talks to Alice while relaying all messages to/from an external party (such as a law-enforcement agent). Ideally, a deniable authentication protocol would not permit the agent to distinguish the case when Bob is having a real conversation with Alice from the case when Bob is fabricating the entire interaction. Previous, off-line models of deniability provide no guarantees in this setting. Alternatively, imagine a publicly readable/writeable bulletin board (e.g., a wiki) where all entries are time-stamped and assigned unpredictable identifiers. A corrupt receiver running a protocol with an honest sender can post all the messages it receives to the bulletin board, and then generate its responses based on the identifiers assigned to the resulting posts. Again, off-line deniability would not suffice since the bulletin board cannot be rewound; in this case, the contents of the bulletin board would prove that the interaction occurred. More generally, one can imagine a "chosen-protocol attack" by which someone designs and deploys a public service specifically targeted to destroying the deniability of a particular authentication protocol.

With the above motivation in mind, we introduce a strong notion of deniability that, in particular, implies on-line deniability. We then show that a protocol

---

[1] In contrast, previous notions of deniable authentication only guarantee *off-line deniability* which protects against a malicious party who records the transcript and shows it to a third party after the fact.

satisfies our definition if and only if it securely realizes the message authentication functionality $\mathcal{F}_{auth}$ in the recently introduced *generalized UC* (GUC) framework [7]. (This is an extension of Canetti's UC framework [5] that models globally-available, "external" functionalities like a common reference string, PKI, etc.) Protocols proven secure with respect to our definition thus inherit all the strong composability properties of the (G)UC framework. We stress that protocols realizing $\mathcal{F}_{auth}$ in the UC framework do *not* necessarily provide deniability; in particular, digital signatures — which are clearly not deniable — realize $\mathcal{F}_{auth}$ in the UC framework [6]. Similarly, protocols realizing $\mathcal{F}_{auth}$ in the UC framework may be problematic when composed with other protocols that are allowed to depend on parties' public keys (see Section 2.3). In both cases, the reason is that the UC framework treats public keys as *local* to a particular session. When this condition is enforced, the expected security properties hold; when public keys are truly public, the expected security properties may not hold.

1.1   Our Results

We propose a definition of deniable authentication which, in comparison to prior work, guarantees stronger security properties such as on-line deniability and security under concurrent executions with arbitrary other protocols. Unfortunately, we show that our notion of deniable authentication is *impossible* to achieve in the PKI model if *adaptive* corruptions are allowed. This holds even if secure erasure is assumed; if we are unwilling to allow erasure then we can rule out even the weaker notion of forward security (where, informally, honest parties' secret keys and state might be compromised after completion of the protocol). In the full version, we show reductions from deniable authentication to deniable key exchange and vice versa; thus, our impossibility results imply that deniable key exchange is impossible (with regard to adaptive corruptions) as well.

Our impossibility result is very different from prior impossibility results in the UC setting [5,9,11,7]. Previous impossibility results assume secure channels as a primitive, and show that additional setup assumptions (such as a PKI) are necessary to realize other, more advanced functionalities. Here, we show that the basic functionality of authenticated channels cannot be realized even given the setup assumption of a PKI.

Faced with this strong negative result, we ask whether relaxed definitions of deniable authentication can be achieved. In this direction, we show several positive results based on standard assumptions and without random oracles. First, we observe that our definition can be satisfied with respect to *static* adversaries. This appears to give the first separation between the static and adaptive settings with regard to *feasibility*.[2] Second, we observe that our definition can be achieved, with respect to adaptive corruptions, in the *symmetric-key* setting where all pairs of parties share a key.

---

[2] Nielsen [26] shows a separation between the static and adaptive settings with regard to *round complexity*, but not feasibility.

The symmetric-key setting is less appealing than the public-key setting. To partially bridge the two we suggest that symmetric keys for deniable authentication can be established using a weak form of deniable key exchange termed *key exchange with incriminating abort* (KEIA). Intuitively, KEIA guarantees deniability as long as the protocol terminates successfully; once a shared key is established, deniability is guaranteed even if corruptions occur at any later time. If a malicious party aborts the protocol, however, this party may obtain some incriminating evidence against the other party; all this proves, however, is that the two parties attempted to establish a key. In light of our impossibility result, realizing KEIA (and hence a weak form of deniable authentication) seems to be a reasonable compromise.

As our third and most technically interesting feasibility result, we show how to realize KEIA in the PKI model (without erasure) with respect to *semi-adaptive* adversaries who corrupt parties either before or after (but not during) an execution of the protocol.

Due to space constraints, this extended abstract discusses the proposed modeling of online deniability and states our main results. Proofs, additional results, and further discussions are deferred to the full version.

1.2 Previous Work in Relation to Our Own

Deniable authentication was formally introduced by Dwork, Naor, and Sahai [18] (though it was also mentioned in [16]) and it, along with several extensions and generalizations, has received significant attention since then [4, 31, 3, 18, 19, 17, 24, 27, 28, 14, 7, 25]. This prior work all assumes that only the sender has a public key; thus, this work implicitly assumes that "guaranteed delivery" of messages to a specific, intended recipient is possible, and/or that the sender is willing to authenticate a given message for anyone. In such cases on-line deniability does not make sense. Since we are specifically interested in on-line deniability, we consider the setting where the receiver also has a public key. (This setting was also considered in concurrent work done independently of our own [23, 32].) Once the receiver holds a public key, the sender can meaningfully authenticate a message *for a particular receiver* without being willing to authenticate the message for all other parties.

As previously mentioned, our definition implies very strong notions of deniability. In particular, our protocols remain secure under concurrent composition, something that was an explicit goal of prior work [18, 19, 24, 28]. To the best of our knowledge all prior constructions achieving concurrent security use timing assumptions [18] which, though reasonable, seem preferable to avoid.[3] Our protocols also remain secure when run concurrently with arbitrary *other* protocols, something not addressed by previous work.

Designated-verifier proofs [22] and two-party ring signatures [30, 2] also provide authentication without non-repudiation. These primitives do not provide

---

[3] It is not clear whether plugging a generic concurrent ZK proof [29] into any existing deniable authentication protocol would yield a concurrently secure protocol. In any case, this approach would yield protocols with very high round complexity [10].

deniable authentication, however, since they incriminate the sender S and receiver R *jointly*; that is, they *do* leave evidence that either S or R was involved in authenticating some message. Deniable authentication, in contrast, does not implicate either party. Similarly, although there has been extensive work constructing and analyzing various deniable key-exchange protocols such as as SIGMA, SKEME, and HMQV (see [13,14]), none of these protocols meets our definition of deniability. For example, SIGMA leaves a trace that the sender and receiver communicated, even if it does not reveal exactly what message was authenticated. (HMQV might satisfy our definition with respect to static adversaries, though we have not verified the details. The HMQV protocol is not, however, forward-secure unless erasure by honest parties is allowed).

2 Defining Deniable Authentication

In this section we define our notion of deniable authentication. We begin by giving a self-contained definition whose primary aim is to model *on-line* deniability. Our definition is based on an interactive distinguisher, much like the "environment" in the UC framework. Indeed, we observe that a protocol satisfies our definition if and only if it securely realizes the message authentication functionality $\mathcal{F}_{auth}$ in the GUC framework. This means that any protocol satisfying our definition automatically inherits the strong composability guarantees of the UC framework, and also provides some justification of the claim of Canetti et al. [7] that the GUC-framework models deniability.

2.1   The Basic Definition

We start by introducing the relevant parties. We have a *sender S* who is presumably sending a message m to a *receiver R*, a *judge $\mathcal{J}$* who will eventually rule whether or not the transmission was attempted, an *informant $\mathcal{I}$* who witnesses the message transmission and is trying to convince the judge, and a *misinformant $\mathcal{M}$* who did not witness any message transmission but still wants to convince the judge that one occurred. Jumping ahead, a protocol is secure if the judge is unable to distinguish whether it is talking to a true informant $\mathcal{I}$ (interacting with S and R while they are running the protocol), or a misinformant $\mathcal{M}$.

We assume the sender and receiver are part of a network environment that includes some trusted parties (e.g., trusted setup like the PKI), some means of communication between the sender and receiver (e.g., a direct unauthenticated channel), and a direct, private channel between the judge and the informant (or misinformant, depending on the setting). Intuitively, this on-line channel, coupled with the fact that $\mathcal{J}$ cannot be "rewound", is what guarantees on-line deniability. Additionally, we assume that the judge does not have direct access to the players (in particular, the judge does not know whether S really intends to send a message, or whether R really received one); instead, the judge must obtain information about the parties through the (mis)informant. However, the judge $\mathcal{J}$ *does* have direct access to any global setup (for example, in the case of a PKI it can reliably obtain the public keys of S and R), and so the misinformant cannot necessarily lie arbitrarily without being caught. Both the informant $\mathcal{I}$ and the misinformant $\mathcal{M}$ can adaptively corrupt either the sender S or the receiver R at any time, and thereby learn the entire state of the corrupted party (if this party has a public key, this state includes the corresponding secret key). Additionally, once either S or R is corrupt the judge learns about the corruption, and the (mis)informant can totally control the actions of this party going forward. We assume the (mis)informant cannot corrupt the global setup; for example, in the case of a PKI, the (mis)informant does not know the secret keys of any uncorrupted party (but does know all the public keys). Finally, the (mis)informant has partial control over the network: it can totally control all unauthenticated links, and can block messages from authenticated links.

Roughly, a protocol π achieves *on-line deniable authentication* if for any efficient informant $\mathcal{I}$, there exists an efficient misinformant $\mathcal{M}$ such that that no efficient judge $\mathcal{J}$ can distinguish the following two experiments with non-negligible probability.

1. **Informant experiment.** S and R run π in the presence of the informant $\mathcal{I}$ (who in turn interacts with the judge $\mathcal{J}$ in an on-line manner). R informs $\mathcal{J}$ upon accepting any message m' as having been authenticated by S (the message m' need not be the same as the input to S if, say, S is corrupt and ignores its input).

2. **Misinformant experiment.** S and R do nothing, and $\mathcal{J}$ only interacts with the misinformant $\mathcal{M}$. (Here, $\mathcal{M}$ is allowed to falsely inform $\mathcal{J}$ that R accepted some arbitrary message m' as having been authenticated by S.)

(See the full version for a precise definition.) As in the UC model, we can take the informant $\mathcal{I}$ to be a "dummy" attacker who follows the instructions of the judge and truthfully reports everything it sees.

2.2 Deniable Authentication in the GUC Framework

The ideal message authentication functionality $\mathcal{F}_{auth}$ (essentially from [5]) is given in Figure 1. (In all our ideal functionalities, delivery of messages to parties is scheduled by the adversary.) $\mathcal{F}_{auth}$ is "deniable" because, although the adversary learns that a message transmission took place, the adversary is not provided with any "evidence" of this fact that would convince a third party. Since $\mathcal{F}_{auth}$ is deniable, we expect that a protocol π realizing $\mathcal{F}_{auth}$ (with respect to a sufficiently strong notion of "realizing") would be deniable as well. Such a claim is not, however, immediate; in particular, recall that protocols realizing $\mathcal{F}_{auth}$ in the UC framework are *not* necessarily deniable.

Thus, we turn instead to a recently-proposed extension to the UC framework called *generalized universal composability* (GUC) [7] which enables direct modeling of *global setup*. Canetti et al. claim [7], informally, that modeling global setup in this way provides a means of capturing additional security concerns, including deniability, within a UC-style security framework. We validate their claim (at least in our context) via the following result:

Proposition 1. *A protocol π achieves on-line deniable authentication if and only if it realizes $\mathcal{F}_{auth}$ in the GUC framework.*

Functionality $\mathcal{F}_{auth}$

1. Upon receiving input (send, sid, m) from S, do: If sid = (S, R, sid') for some R, then give the message (sent, sid, m) to the adversary who then schedules delivery to R. Else ignore the input.
2. Upon receiving (corruptsend, sid, m') from the adversary, if S is corrupt and no message (sent, sid, m) was yet output, then give the message (sent, sid, m') to the adversary who then schedules delivery to R.

Fig. 1. The message authentication functionality of [5]

In the full version of this paper we define notions of on-line deniability for identification and key exchange, and show that protocols achieve these definitions if and only if they realize appropriate functionalities in the GUC framework.

2.3   PKI Setup and Comparison with Prior Models

We model a PKI as a shared functionality $\mathcal{F}_{krk}$ that enforces the following: Honest parties register with a central authority who generates the public and secret keys for them. Corrupt parties register an arbitrary, but consistent, pair of public/secret keys with the authority. Note that the central registration authority knows the secret keys of all parties (including the corrupt parties), and therefore the model is referred to as "Key Registration with Knowledge" [1]. Clearly this model is more involved than a "bare" PKI. The fact that we work in this model only strengthens our impossibility results. Moreover, the bare PKI model does not suffice for our feasibility results (cf. Proposition 2).

An additional requirement we impose is that honest parties protect their secret keys by using them only in some specified authentication protocol Φ. (Corrupt parties are allowed to use their keys in an arbitrary manner.) There are several ways to model this requirement. For concreteness, we parameterize the key-registration functionality with a description of Φ, and allow honest parties to run Φ via calls to the key-registration functionality. (See Figure 2.) In a real execution of the protocol, of course, honest parties will actually hold their secret key and it is up to them to restrict its use.

Comparison to prior models of a PKI. Our PKI functionality is defined similarly to that of [1]. However, unlike in [1], we restrict honest parties to only use their secret keys with the protocol Φ, and a single instance of $\mathcal{F}_{krk}$ persists across multiple sessions. In the terminology of [7], $\mathcal{F}_{krk}$ is a *shared functionality* as opposed to a *local* one. The shared nature of the functionality implies, in particular, that the environment has direct access to the public keys of all the parties (as well as the secret keys of corrupted parties). Local setup, in contrast, is not adequate for capturing deniability. As argued in [7], local setup is also not satisfactory with regard to composition. Indeed, local modeling of the PKI seemingly leaves two options: either a fresh instance of the PKI is required for every execution of the protocol (which is impractical), or one must use the joint UC (JUC) theorem [12]. Unfortunately, the latter option only guarantees security under composition with a restricted class of protocols. Specifically, security is

---

Shared Functionality $\mathcal{F}_{krk}^{\Phi}$

Parameterized by a security parameter λ, a protocol (or, more generally, a list of protocols) Φ, and a (deterministic) key generation function Gen, shared functionality $\mathcal{F}_{krk}$ proceeds as follows when running with parties $P_1, \ldots, P_n$:

Registration: When receiving a message (register) from an honest party P_i that has not previously registered, sample $r \leftarrow \{0,1\}^{\lambda}$ then compute $(PK_i, SK_i) \leftarrow \mathsf{Gen}^{\lambda}(r)$ and record the tuple (P_i, PK_i, SK_i).

Corrupt Registration: When receiving a message (register, r) from a corrupt party P_i that has not previously registered, compute $(PK_i, SK_i) \leftarrow \mathsf{Gen}^{\lambda}(r)$ and record the tuple (P_i, PK_i, SK_i).

Public Key Retrieval: When receiving a message (retrieve, P_i) from any party P_j (where $i = j$ is allowed), if there is a previously recorded tuple of the form (P_i, PK_i, SK_i), then return (P_i, PK_i) to P_j. Otherwise return $(P_i, \perp)$ to P_j.

Secret Key Retrieval: When receiving a message (retrievesecret, P_i) from a party P_i that is either *corrupt* or honestly running the protocol code for Φ, if there is a previously recorded tuple of the form (P_i, PK_i, SK_i) then return (P_i, PK_i, SK_i) to P_i. In all other cases, return $(P_i, \perp)$.

Fig. 2. The Φ-Key Registration with Knowledge shared functionality

not guaranteed under composition with protocols that may depend on honest parties' public keys. (We provide an example in the full version.)

2.4 Flavors of Protocols/Attackers

An *adaptive* attacker can corrupt parties before, during, and after execution of a protocol. A *static* attacker can only corrupt parties before the beginning of the protocol. A *semi-adaptive* attacker can corrupt parties before and after (but not during) a protocol execution. Finally, a *forward-secure attacker* can only corrupt the parties after the protocol. We will distinguish between the setting where (honest) parties are assumed to be able to securely erase information, and where they cannot. Erasures do not affect the model for static attackers, but are meaningful for semi-adaptive, forward-secure, and fully adaptive attackers.

3 Impossibility Result

In this section we prove our main result: adaptively secure deniable authentication is impossible in the PKI model, even if erasures are allowed, and even if each secret key is used only once. If secure erasure is not assumed, we can even rule out forward security for deniable authentication.

Before stating the precise impossibility results, it is instructive to consider some naïve strategies for ruling out adaptively secure on-line deniable authentication. At first, it appears that since the behavior of the sender S can be simulated in a straight-line manner *without* its secret key SK_S, there is an attacker who can impersonate the sender to the recipient R (by running the simulator).

One of the reasons this does not work is that R might use its own secret key SK_R for verification. In particular, a simulated transcript might be easily distinguishable from a real transcript to R (since R can employ knowledge of SK_R to distinguish the transcript) but be indistinguishable from a real transcript to the adversary. One "fix" to this problem is for the adversary to (adaptively) corrupt R and then check the simulated transcript from R's viewpoint. Unfortunately, if R is corrupted too early (say, at the very beginning), it could be the case that knowledge of R's secret key is subsequently employed by the simulator in order to simulate the proper transcript (without talking to S or obtaining SK_S). Notice that such a simulation does not contradict soundness since, in the real world, R would know that he is not simulating the conversation with S. On the other hand, if R is corrupted too late (say, at the very end), the initial flows from the "then-honest" party R were also chosen by the simulator, so there is no guarantee that they correspond to the behavior of a real R interacting with the sender's simulator.

In fact, a proof of the following theorem is more complicated and requires a sequence of "hybrid arguments" where corruption of one of the parties is delayed by one round each time. See the following section.

Theorem 1. *There does not exist a protocol Π realizing the deniable authentication functionality $\mathcal{F}_{auth}$ in the $\mathcal{F}_{krk}^{\Pi}$-hybrid model with respect to adaptive corruptions. Moreover, impossibility holds even under the following additional assumptions/constraints:*

- *Secure data erasures are allowed.*
- *Each honest party P uses its secret key sk_P for only a single execution of the protocol.*
- *The attacker $\mathcal{A}$ either impersonates a sender to a single honest receiver, or impersonates a receiver to a single honest sender. (In particular, $\mathcal{A}$ does not run a concurrent attack or a "man-in-the-middle attack" against an honest sender and receiver.)*

We also show how to extend the above impossibility result to rule out forward security if erasures are not allowed.

Theorem 2. *If data erasures are not allowed, it is impossible to realize $\mathcal{F}_{auth}$ in the $\mathcal{F}_{krk}$-hybrid model with respect to forward security. Moreover, impossibility holds even under the constraints of Theorem 1.*

The results above can also be extended to rule out the possibility of realizing deniable key exchange or identification.

Finally, we note that the "bare PKI" model (in which parties are allowed to post public keys without necessarily knowing a corresponding secret key) is not sufficient to realize deniable authentication at all, even with respect to security. This seems to imply that key registration with knowledge is an unavoidable requirement for deniability.

Proposition 2. *It is impossible to realize the identification, authentication, or key exchange functionalities in the bare public key model, even with respect to static corruption.*

3.1   Proof Sketch for Impossibility (Theorem 1)

At a high level, the proof is an inductive argument showing that each round of the protocol either incriminates one of the parties, or can be simulated entirely (from either side) without knowledge of *any* secret keys. Of course, if either party can simulate the *entire* protocol without knowledge of any secret keys, it cannot be sound (*i.e.*, an attacker without S's key can authenticate an arbitrary message to R). Thus, we show that either the protocol is not deniable, or it is not sound, contradicting our security requirements. The difficult part of the proof is the inductive step, which requires a delicate series of hybrid arguments. In particular, one must be careful about the order of corruptions in the various hybrids.

More formally, let Π be any protocol for deniable identification using $r = r(n)$ rounds, and assume toward a contradiction that Π is adaptively secure. Without loss of generality, we assume that the receiver goes first, and that the final message of the protocol is sent by the sender. In particular, we let $\alpha_1, \alpha_2, \ldots, \alpha_r$ denote the messages sent by the receiver R and $\beta_1, \ldots, \beta_r$ denote the response messages sent by the sender S. For convenience, we let α_{r+1} denote the binary decision bit of the receiver indicating whether or not R accepted. Throughout the protocol, we denote the current state of the sender and the receiver by ω_S and ω_R, respectively. This evolving state will include all the information currently stored by the given party, except for its secret key. Because we allow erasures, the current state does not include any information previously erased by this party.

We already stated that we only consider two kinds of attackers: sender impersonator $\mathcal{A}_S$ and receiver impersonator $\mathcal{A}_R$. The sender impersonator $\mathcal{A}_S$ will talk with an honest receiver R, while the receiver impersonator $\mathcal{A}_R$ will talk to an honest sender S. By assumption, there exists efficient simulators Sim_R and Sim_S for $\mathcal{A}_S$ and $\mathcal{A}_R$, respectively: the job of Sim_R is to simulate the behavior of R when talking to $\mathcal{A}_S$, while the job of Sim_S is to simulate the behavior of S when talking to $\mathcal{A}_R$. Moreover, the GUC security of Π implies that Sim_S and Sim_R have to work given only oracle access to R and S, respectively.[4] In particular, this means that in each round $1 \leq i \leq r$,

- As long as neither S nor R is corrupted, Sim_S (resp., Sim_R) will receive some arbitrary message α_i (resp., β_i) and must generate a "good-looking" response β_i (resp., α_{i+1}). Moreover, it must do so *without knowledge of the secret keys SK_S and SK_R, or any future messages $\alpha_{i+1}, \ldots$ (resp., $\beta_{i+1}, \ldots$).*
- If S (resp., R) is corrupted, Sim_S (resp., Sim_R) will be given the secret SK_S (resp., SK_R), and must then generate a "consistent-looking" internal state ω_S (resp., ω_R) for the corresponding party at round i. The pair (SK_S, ω_S) (resp., (SK_R, ω_R)) will then be given to the attacker and the environment.

From this description, we make our first key observation: as long as S and R are not corrupted, it is within the power of our attackers $\mathcal{A}_S$ and $\mathcal{A}_R$ to internally

---

[4] This is because, without loss of generality, $\mathcal{A}_S$ and $\mathcal{A}_R$ are simply the dummy parties forwarding the messages of the environment, and the simulator has to work for any environment. In fact, this property follows whenever there is an external "judge" with whom the adversary may interact when gathering evidence of protocol interactions.

run the simulators Sim_S and Sim_R, respectively. In particular, we can make meaningful experiments where $\mathcal{A}_S$ runs Sim_S against an honest receiver R, or $\mathcal{A}_R$ runs Sim_R against an honest sender S. Of course, *a priori* it is unclear what happens during these experiments, since Sim_S was only designed to work against attackers $\mathcal{A}_R$ who *do not know* SK_R (as opposed to R itself, who certainly knows it), and similarly for Sim_R. The bulk of the proof consists of showing that such "unintended" usages of Sim_S and Sim_R nevertheless result in the "expected" behavior. We give a sequence of hybrid experiments which show that, without knowing the secret key of the sender, the simulator Sim_S can still successfully imitate the sender to an honest receiver, contradicting the soundness of identification. The details are given in the full version.

4   Circumventing the Impossibility Result

In this section, we discuss several positive results that circumvent the impossibility result of the previous section. We exhibit:

- A 1-message deniable authentication protocol tolerating *adaptive* corruptions, assuming a symmetric key infrastructure (i.e., a symmetric key shared between the sender and receiver);
- A 1-message deniable authentication protocol tolerating *static* corruptions in the PKI model;
- A 4-message protocol achieving a relaxed notion of deniable authentication, dubbed *incriminating abort*, and tolerating *semi-adaptive* corruptions in the PKI model. The protocol we give also satisfies the non-relaxed definition with respect to a static adversary.

Key exchange and deniable authentication can be reduced to each other, and so the results above also imply the feasibility of corresponding notions of deniable key exchange.

The first two results above are quite simple, and mainly serve to illustrate the gap between the simpler settings (symmetric keys and static corruptions) and the more realistic setting of public keys and adaptive corruptions. The third feasibility result is significantly more involved. We feel it represents an interesting and reasonable compromise between realistic modeling and feasibility.

Deniability with symmetric keys. Suppose for a moment that players have access to a *symmetric* key infrastructure; i.e., every pair of participants shares a uniformly random long-term key that is unknown to other participants. Then S can authenticate a message m to R by appending a MAC (message authentication code) tag computed on the input (sid, S, R, m), where sid is a fresh random nonce. This is deniable roughly because the simulator for the protocol can make up a key for every pair of communicating players, and generate tags using the made-up key. In case of an adaptive corruption, the simulator can include the made-up key in the corrupted player's simulated memory contents.

This can be formalized in terms of the ideal functionalities $\mathcal{F}_{auth}$ and $\mathcal{F}_{ke}$. The SKI corresponds to granting every player of players one-time use of $\mathcal{F}_{ke}$, with key re-use modeled via the UC with joint state theorem [12]. The use of a MAC shows that $\mathcal{F}_{auth}$ can be reduced to a one-time $\mathcal{F}_{ke}$. In fact, the converse is also true: one can realize $\mathcal{F}_{ke}$ by encrypting a key using a protocol that is secure against *adaptive* but *passive* adversaries (known as *non-committing encryption* [8]). The flows of this protocol can be authenticated using $\mathcal{F}_{auth}$ to make it resistant to active attacks.

Lemma 1 ($\mathcal{F}_{auth} \iff \mathcal{F}_{ke}$). *If one way functions exist, then there exists a protocol that UC-realizes the natural multi-session extension of $\mathcal{F}_{auth}$ in the $\mathcal{F}_{ke}$ hybrid model, requiring only a single call to $\mathcal{F}_{ke}$. Conversely, if non-committing encryption exists, then there exists a protocol that UC-realizes $\mathcal{F}_{ke}$ in the $\mathcal{F}_{auth}$-hybrid model. These reduction hold even against adaptive adversaries.*

Static security with a PKI. We now turn to the public-key model. For certain types of public keys, players can use a PKI to generate a symmetric key k *non-interactively*. The idea of the protocol is then to use k to compute tags on messages as above. The authenticity of the message is derived from the authentication inherent in the PKI. The non-interactive key generation is not adaptively secure, and so the resulting protocols are only secure against static adversaries.

For example, suppose we operate in a cyclic group G generated by a generator g where the Decisional Diffie-Hellman (DDH) assumption holds. If each party P_i has a secret key $x_i \in \mathcal{Z}_q$ and a public key $y_i = g^{x_i}$, then P_i and P_j non-interactively share a key $k = g^{x_i x_j} = y_i^{x_j} = y_j^{x_i}$. Under the DDH asumption, k looks like a random group element to an attacker who only knows the public keys y_i and y_j. Either one of P_i or P_j can then use k as a MAC key to authenticate messages to the other.

This type of key exchange is abstracted as *non-interactive authenticated key exchange*. We model the PKI via the *registered keys with knowledge* functionality $\mathcal{F}_{krk}^{\Phi}$ (Figure 2) described in Section 2. (Key knowledge is necessary even for static security—see Proposition 2).

Theorem 3. *Assuming the existence of non-interactive authenticated key exchange, there exists an efficient protocol Φ such that $\mathcal{F}_{auth}$ can be UC-realized in the $\mathcal{F}_{krk}^{\Phi}$-hybrid model with respect to static adversaries.*

In contrast, the impossibility result of Section 3 rules out adaptively secure GUC realizations. To the best of our knowledge, this is the first example of a task that cannot be realized with respect to adaptive adversaries, but *can* be achieved with respect to static adversaries.

4.1   Deniability with Incriminating Abort

Given the impossibility results of Section 3 and the possibility of PKI-based deniable authentication for static adversaries, it is natural to ask just how strong a notion of deniability can be achieved in the public key setting. We show here that one can guarantee deniability as long as *(a)* the protocol does not abort, and *(b)* there is one round during which neither the sender S nor the receiver

R is adaptively corrupted. If the protocol aborts, the adversary can learn unsimulatable information depending on the secret keys of S and R—potentially enough to prove that one of them was trying to talk to the other. We call this notion *deniability with incriminating abort*.

We refer to an adversary that makes no corruptions during some phase of the protocol run as *semi-adaptive*. In particular, such an adversary will not corrupt any players during the protocol's single vulnerable round. However, the restriction to semi-adaptive security is also necessary to make the notion of abort meaningful: a fully adaptive adversary could always ensure that a protocol does not abort by corrupting a party immediately before it complains. Semi-adaptive security implies *forward security*; that is, a conversation that completes successfully is later deniable even if parties are forced to reveal the contents of their memories.

We phrase our results in terms of key exchange. This implies the corresponding feasibility results for authentication. However, forward security is especially meaningful for key exchange, because the protocol need only be run once, at setup time, for every pair of participants. If the key exchange protocol succeeds (with no adaptive corruption occurring during the protocol execution), then we can still use adaptively secure protocols realized in the $\mathcal{F}_{ke}$-hybrid model, and they will retain their adaptive security. In other words, the new protocol *almost* represents a deniable realization of $\mathcal{F}_{ke}$: if we could somehow guarantee that the protocol never aborts, then it would GUC-realize $\mathcal{F}_{ke}$.

Modeling incriminating abort. We model the PKI via a shared, "registered keys with knowledge" functionality $\mathcal{F}_{krk}^{\Phi}$ (Figure 2), as in the protocols for static adversaries. The key exchange with incriminating abort functionality $\mathcal{F}_{keia}$ is similar to $\mathcal{F}_{ke}$ except that the ideal-model adversary may explicitly request the protocol to abort, and in such a case the functionality will provide evidence that one of S and R was trying to talk to the other. It would be intuitively appealing to leak a single bit to the environment stating that a conversation occurred. We do not know of a way to ensure such limited leakage, and besides this gives up too much information: as we will see, our protocol only compromises deniability if the protocol aborts *and* one of S or R is corrupted at a later time. Instead, we parametrize $\mathcal{F}_{keia}$ with an "incrimination procedure" IncProc. In the case of an abort, $\mathcal{F}_{keia}^{\mathsf{IncProc}}$ allows the adversary to interact with $\mathsf{IncProc}(SK_S, ...)$, which essentially represents the potentially non-simulatable flows of the protocol. If the protocol doesn't abort, then a fresh symmetric key is distributed to S and R and nothing is leaked to the adversary. The functionality $\mathcal{F}_{keia}^{\mathsf{IncProc}}$ is described in Figure 3.

The incrimination procedure may at first be hard to interpret, and so we highlight some properties of protocols that realize $\mathcal{F}_{keia}^{\mathsf{IncProc}}$. First, if no abort occurs then the ideal protocol is forward secure, since the symmetric key is random and unconnected to other quantities in the protocol. Hence, if a protocol π GUC-realizes $\mathcal{F}_{keia}^{\mathsf{IncProc}}$ for any procedure IncProc then π is forward-secure. Second, if an abort does occur and the adversary learns information, this information depends only on the sender's secret key and public information. Because secret

---

Functionality $\mathcal{F}_{\text{keia}}^{\text{IncProc}}$

$\mathcal{F}_{\text{keia}}$, which is parameterized by an "incrimination procedure" IncProc, and a security parameter λ proceeds as follows, when running in the $\mathcal{F}_{krk}$-hybrid model with parties S and R (who have already registered secret keys SK_S and SK_R, respectively) and adversary $\mathcal{S}$:

1. Upon receiving a message of the form $(S, \text{keyexchange}, \text{sid}, S, R, SK_S)$ from party S, if there are no previous records, then record the value $(\text{keyexchange}, \text{sid}, S, R, SK_S)$, mark S "active", and a send public delayed output $(\text{keyexchange}, \text{sid}, S, R)$ to R. (Otherwise, ignore the message.)

2. Upon receiving a message of the form $(R, \text{keyexchange}, \text{sid}, S, R, SK_R)$ from party R, if R is not yet "active", mark R as "active" and send a public delayed output $(\text{active}, \text{sid}, S, R)$ to S. (Otherwise, ignore the message.)

3. Upon receiving a message of the form $(\text{setkey}, \text{sid}, S, R, k')$ from $\mathcal{S}$, if R is corrupt and S is "active", then output $(\text{setkey}, \text{sid}, S, R, k')$ to S and R, and halt. If R is "active" but not corrupt, then sample a fresh key $k \leftarrow \{0,1\}^\lambda$ and send the message $(\text{setkey}, \text{sid}, S, R, k)$ to R. Furthermore, if S is "active", then send the delayed message $(\text{setkey}, \text{sid}, S, R, k)$ to S as well. In all cases, this completes the protocol, and the functionality halts.

4. Upon receiving a message of the form $(\text{abort}, \text{sid}, S, R)$ from $\mathcal{S}$, if S is "active", send $(\text{abort}, \text{sid}, S, R)$ as a delayed message to S and mark S "aborted". If R is "active", send $(\text{abort}, \text{sid}, S, R)$ as a delayed message to R (*i.e.*, $\mathcal{S}$ need not notify either party that the protocol was aborted, and may still cause R to output a key using a setkey message, but cannot cause S to output a key once an abort has occurred).

5. Upon receiving a message of the form $(\text{incriminate}, \text{sid}, S)$ from $\mathcal{S}$, if this is the first time receiving such a message and S is currently "aborted" and honest, then run the procedure IncProc$(\text{sid}, S, R, PK_S, PK_R, SK_S)$.

Fig. 3. The ideal functionality for Key Exchange with Incriminating Abort, parameterized by an incrimination procedure IncProc which runs only if the key exchange is aborted by the adversary

keys are useless in the ideal model, *this has no impact on the security of other protocols*. In particular, the incrimination information cannot be used to fake authenticated messages in other conversations, or to convince the environment that S talked to anyone other than R. This last observation implies that not all incrimination oracles can be realized by real protocols: for example, if IncProc gives away the sender's secret key, then a real adversary would subsequently be able to fake arbitrary messages from the sender to other parties, contradicting the indistinguishability from the ideal model. We prove below that there exists a procedure IncProc for which $\mathcal{F}_{\text{keia}}^{\text{IncProc}}$ can, in fact, be realized.

Construction. At the core of our constructions for realizing $\mathcal{F}_{\text{keia}}$ is a chosen-ciphertext (CCA) secure variant of *Dual Receiver Encryption* (DRE) [15]. DRE allows anyone to encrypt a message to two parties with a single ciphertext, with the guarantee that attempts to decrypt a ciphertext by either of the two

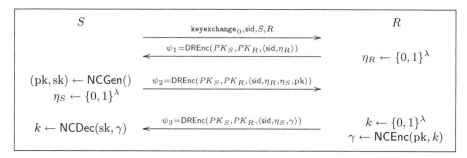

Fig. 4. A graphical illustration of Protocol Φ_{dre} for realizing $\mathcal{F}_{\mathsf{keia}}$. S and R check consistency of each flow immediately upon receiving it; if the flow is not consistent, the protocol is aborted. Notation: (Gen, DREnc, DRDec) is a Dual Receiver Encryption scheme and (NCGen, NCEnc, NCDec, NCSim, NCEqv) is a Non-Committing Encryption scheme (see full version).

recipients will produce the *same result*. In our protocol, this means that certain actions can be simulated using either S or R's secret key. We formally define CCA-secure DRE in the full version, and describe a DRE scheme that is similar to the plaintext-aware encryption of [21]. Our protocol also uses a 2-round non-committing encryption (NCE) scheme [8].

Our 4-message protocol for realizing $\mathcal{F}_{\mathsf{keia}}^{\mathsf{IncProc}}$ is summarized in Figure 4. Intuitively, the incrimination procedure IncProc will expect the adversary to supply a ciphertext that matches the form of the second flow (ψ_1) in the protocol below, and will then use S's secret key to decrypt ψ_1 and compute a corresponding third flow (ψ_2). The incrimination procedure hands ψ_2 to the adversary, along with the random coins used by a non-committing encryption scheme.

Notably, although we only realize $\mathcal{F}_{\mathsf{keia}}^{\mathsf{IncProc}}$ against a semi-adaptive adversary, *the same protocol* is also a *statically secure* realization of $\mathcal{F}_{ke}$. Therefore, we have achieved a strictly stronger notion of security than that achieved by the one-message protocol using NI-AKE and MACs, or HMQV. Honest parties are always guaranteed complete deniability when the protocol succeeds, and even if the protocol aborts, deniability is maintained until some future corruption of a party occurs. It is an open question whether this notion of deniability can be further improved upon.

Theorem 4. *Assuming the existence of dual-receiver and non-committing encryption, the protocol Φ_{dre} in Figure 4*

1. *realizes $\mathcal{F}_{\mathsf{keia}}^{\mathsf{IncProc}_{dre}}$ in the $\mathcal{F}_{krk}^{\Phi_{dre}}$-hybrid model with semi-adaptive security, for a suitable procedure $\mathsf{IncProc}_{dre}$ (defined in the full version); and*
2. *realizes $\mathcal{F}_{ke}$ in the $\mathcal{F}_{krk}^{\Phi_{dre}}$-hybrid model with static security.*

Moreover, the output of $\mathsf{IncProc}_{dre}$ can be simulated using the secret key of R instead of S.

As mentioned above, realizing $\mathcal{F}_{\mathsf{keia}}^{\mathsf{IncProc}}$ is meaningful for any procedure IncProc. However, the particular incrimination procedure of IncProc has additional properties. First, it can be faked without knowing either secret key, and the fake distribution is indistinguishable from the real one to a distinguisher who knows neither key. Second, it can be simulated exactly using either of the secret keys. These properties together mean that Φ_{dre} is statically secure (as stated in the theorem) and that even in the case of an abort with a subsequent corruption, it is only possible to incriminate one of the pair $\{S, R\}$, and not S specifically.

Acknowledgments. We would like to thank Ran Canetti for many extremely illuminating discussions about composable authentication. We would also like to thank Yehuda Lindell, Rafael Pass and Amit Sahai for useful comments.

References

1. Barak, B., Canetti, R., Nielsen, J., Pass, R.: Universally composable protocols with relaxed set-up assumptions. In: FOCS, pp. 186–195. IEEE Computer Society, Los Alamitos (2004)
2. Bender, A., Katz, J., Morselli, R.: Ring signatures: Stronger definitions, and constructions without random oracles. In: Halevi, S., Rabin, T. (eds.) TCC 2006. LNCS, vol. 3876, pp. 60–79. Springer, Heidelberg (2006)
3. Borisov, N., Goldberg, I., Brewer, E.: Off-the-record communication, or, why not to use PGP. In: WPES, pp. 77–84. ACM, New York (2004)
4. Boyd, C., Mao, W., Paterson, K.G.: Deniable authenticated key establishment for internet protocols. In: Christianson, B., Crispo, B., Malcolm, J.A., Roe, M. (eds.) Security Protocols 2003. LNCS, vol. 3364, pp. 255–271. Springer, Heidelberg (2005)
5. Canetti, R.: Universally composable security: A new paradigm for cryptographic protocols. In: FOCS, pp. 136–145. IEEE Computer Society, Los Alamitos (2001)
6. Canetti, R.: Universally composable signatures, certification, and authentication. In: Computer Security Foundations Workshop, pp. 219–235. IEEE Computer Society Press, Los Alamitos (2004)
7. Canetti, R., Dodis, Y., Pass, R., Walfish, S.: Universally composable security with global setup. In: Vadhan, S.P. (ed.) TCC 2007. LNCS, vol. 4392, pp. 61–85. Springer, Heidelberg (2007)
8. Canetti, R., Feige, U., Goldreich, O., Naor, M.: Adaptively secure multi-party computation. In: STOC, pp. 639–648. ACM, New York (1996)
9. Canetti, R., Fischlin, M.: Universally composable commitments. In: Kilian, J. (ed.) CRYPTO 2001. LNCS, vol. 2139, pp. 19–40. Springer, Heidelberg (2001)
10. Canetti, R., Kilian, J., Petrank, E., Rosen, A.: Black-box concurrent zero-knowledge requires (almost) logarithmically many rounds. SIAM J. Computing 32(1), 1–47 (2002)
11. Canetti, R., Kushilevitz, E., Lindell, Y.: On the limitations of universally composable two-party computation without set-up assumptions. J. Cryptology 19(2), 135–167 (2006)
12. Canetti, R., Rabin, T.: Universal composition with joint state. In: Boneh, D. (ed.) CRYPTO 2003. LNCS, vol. 2729, pp. 265–281. Springer, Heidelberg (2003)
13. Di Raimondo, M., Gennaro, R., Krawczyk, H.: Secure off-the-record messaging. In: WPES, pp. 81–89. ACM, New York (2005)
14. Di Raimondo, M., Gennaro, R., Krawczyk, H.: Deniable authentication and key exchange. In: Juels, A., Wright, R., De Capitani di Vimercati, S. (eds.) ACM Conf. Computer and Communications Security, pp. 400–409. ACM, New York (2006)

162   Y. Dodis et al.

15. Diament, T., Lee, H.K., Keromytis, A.D., Yung, M.: The dual receiver cryptosystem and its applications. In: Atluri, V., Pfitzmann, B., McDaniel, P.D. (eds.) ACM Conf. Computer and Communications Security, pp. 330–343. ACM, New York (2004)

16. Dolev, D., Dwork, C., Naor, M.: Nonmalleable cryptography. SIAM J. Computing 30(2), 391–437 (2000); Preliminary version in STOC 1991

17. Dwork, C., Naor, M.: Zaps and their applications. SIAM J. Computing 36(6), 1513–1543 (2007); Preliminary version in FOCS 2000

18. Dwork, C., Naor, M., Sahai, A.: Concurrent zero-knowledge. J. ACM 51(6), 851–898 (2004); Preliminary version in STOC 1998

19. Dwork, C., Sahai, A.: Concurrent zero-knowledge: Reducing the need for timing constraints. In: Krawczyk, H. (ed.) CRYPTO 1998. LNCS, vol. 1462, pp. 442–457. Springer, Heidelberg (1998); Full version available from the second author's webpage

20. Goldwasser, S., Micali, S., Rackoff, C.: The knowledge complexity of interactive proof systems. SIAM J. Computing 18(1), 186–208 (1989)

21. Herzog, J., Liskov, M., Micali, S.: Plaintext awareness via key registration. In: Boneh, D. (ed.) CRYPTO 2003. LNCS, vol. 2729, pp. 548–564. Springer, Heidelberg (2003)

22. Jakobsson, M., Sako, K., Impagliazzo, R.: Designated verifier proofs and their applications. In: Maurer, U.M. (ed.) EUROCRYPT 1996. LNCS, vol. 1070, pp. 143–154. Springer, Heidelberg (1996)

23. Jiang, S.: Deniable authentication on the internet. Cryptology ePrint Archive, Report 2007/082 (2007), http://eprint.iacr.org/

24. Katz, J.: Efficient and non-malleable proofs of plaintext knowledge and applications. In: Biham, E. (ed.) EUROCRYPT 2003. LNCS, vol. 2656, pp. 211–228. Springer, Heidelberg (2003)

25. Lim, M.-H., Lee, S., Park, Y., Moon, S.: Secure deniable authenticated key establishment for internet protocols. Cryptology ePrint Archive, Report 2007/163 (2007), http://eprint.iacr.org/

26. Nielsen, J.B.: Separating random oracle proofs from complexity theoretic proofs: The non-committing encryption case. In: Yung, M. (ed.) CRYPTO 2002. LNCS, vol. 2442, pp. 111–126. Springer, Heidelberg (2002)

27. Pass, R.: On deniability in the common reference string and random oracle model. In: Boneh, D. (ed.) CRYPTO 2003. LNCS, vol. 2729, pp. 316–337. Springer, Heidelberg (2003)

28. Di Raimondo, M., Gennaro, R.: New approaches for deniable authentication. In: Atluri, V., Meadows, C., Juels, A. (eds.) ACM Conf. Computer and Communications Security, pp. 112–121. ACM, New York (2005)

29. Richardson, R., Kilian, J.: On the concurrent composition of zero-knowledge proofs. In: Stern, J. (ed.) EUROCRYPT 1999. LNCS, vol. 1592, pp. 415–431. Springer, Heidelberg (1999)

30. Rivest, R., Shamir, A., Tauman, Y.: How to leak a secret. In: Boyd, C. (ed.) ASIACRYPT 2001. LNCS, vol. 2248, pp. 552–565. Springer, Heidelberg (2001)

31. Susilo, W., Mu, Y.: Non-interactive deniable ring authentication. In: Lim, J.-I., Lee, D.-H. (eds.) ICISC 2003. LNCS, vol. 2971, pp. 386–401. Springer, Heidelberg (2004)

32. Yao, A.C.-C., Yao, F., Zhao, Y., Zhu, B.: Deniable internet key-exchange. Cryptology ePrint Archive, Report 2007/191 (2007), http://eprint.iacr.org/

Authenticated Adversarial Routing[*]

Yair Amir[1], Paul Bunn[2], and Rafail Ostrovsky[3]

[1] Johns Hopkins University Department of Computer Science,
Baltimore, MD 21218, USA
yairamir@cs.jhu.edu
[2] UCLA Department of Mathematics,
Los Angeles, CA 90095, USA
paulbunn@math.ucla.edu
[3] UCLA Department of Computer Science and Department of Mathematics
Los Angeles, CA 90095, USA
rafail@cs.ucla.edu

Abstract. The aim of this paper is to demonstrate the feasibility of authenticated throughput-efficient routing in an unreliable and dynamically changing synchronous network in which the majority of malicious insiders try to destroy and alter messages or disrupt communication in any way. More specifically, in this paper we seek to answer the following question: Given a network in which the majority of nodes are controlled by a node-controlling adversary and whose topology is changing every round, is it possible to develop a protocol with polynomially-bounded memory per processor that guarantees throughput-efficient and correct end-to-end communication? We answer the question affirmatively for extremely general corruption patterns: we only request that the topology of the network and the corruption pattern of the adversary leaves at least one path each round connecting the sender and receiver through honest nodes (though this path may change at every round). Out construction works in the public-key setting and enjoys bounded memory per processor (that is polynomial in the network size and does not depend on the amount of traffic). Our protocol achieves *optimal transfer rate* with negligible decoding error. We stress that our protocol assumes no knowledge of which nodes are corrupted nor which path is reliable at any round, and is also fully distributed with nodes making decisions locally, so that they need not know the topology of the network at any time.

The optimality that we prove for our protocol is very strong. Given any routing protocol, we evaluate its efficiency (rate of message delivery) in the "worst case," that is with respect to the *worst* possible graph and against the *worst* possible (polynomially bounded) adversarial strategy (subject to the above mentioned connectivity constraints). Using this metric, we show that there does not exist *any* protocol that can be asymptotically superior (in terms of throughput) to ours in this setting.

We remark that the aim of our paper is to demonstrate via explicit example the feasibility of throughput-efficient authenticated adversarial

---

[*] Full version of the paper is available on-line [5].

O. Reingold (Ed.): TCC 2009, LNCS 5444, pp. 163–182, 2009.

routing. However, we stress that out protocol is *not* intended to provide a practical solution, as due to its complexity, no attempt thus far has been made to reduce constants and memory requirements.

Our result is related to recent work of Barak, Goldberg and Xiao in 2008 [9] who studied fault localization in networks assuming a private-key trusted setup setting. Our work, in contrast, assumes a public-key PKI setup and aims at not only fault localization, but also transmission optimality. Among other things, our work answers one of the open questions posed in the Barak et. al. paper regarding fault localization on multiple paths. The use of a public-key setting to achieve strong error-correction results in networks was inspired by the work of Micali, Peikert, Sudan and Wilson [14] who showed that classical error-correction against a polynomially-bounded adversary can be achieved with surprisingly high precision. Our work is also related to an interactive coding theorem of Rajagopalan and Schulman [15] who showed that in noisy-edge static-topology networks a constant overhead in communication can also be achieved (provided none of the processors are malicious), thus establishing an optimal-rate routing theorem for static-topology networks.

Finally, our work is closely related and builds upon to the problem of End-To-End Communication in distributed networks, studied by Afek and Gafni [1], Awerbuch, Mansour, and Shavit [8], and Afek, Awerbuch, Gafni, Mansour, Rosen, and Shavit [2], though none of these papers consider or ensure correctness in the setting of a node-controlling adversary that may corrupt the majority of the network.

Keywords: Network Routing; End-to-End Communication; Fault Localization; Error-Correction; Multi-Party Computation; Communication Complexity.

1   Introduction

Our goal is to design a routing protocol for an unreliable and dynamically changing synchronous network that is resilient against malicious insiders who may try to destroy and alter messages or disrupt communication in any way. We model the network as a communication graph $G = (V, E)$ where each vertex (node) is a processor and each edge is a communication link. We do not assume the topology of this graph is fixed or known by the processors. Rather, we assume a complete graph on n vertices, where some of the edges are "up" and some are "down", and the status of each edge can change dynamically at any time.

We concentrate on the most basic task, namely how two processors in the network can exchange information. Thus, we assume that there are two designated vertices, called the sender S and the receiver R, who wish to communicate with each other. The sender has an infinite read-once input tape of *packets* and the receiver has an infinite write-once *output tape* which is initially empty. We assume that packets are of some bounded size, and that any edge in the system that is "up" during some round can transmit only one packet (or control variables, also of bounded size) per round.

We will evaluate our protocol using the following three considerations:

1. **Correctness.** A protocol is *correct* if the sequence of packets output by the receiver is a prefix of packets appearing on the sender's input tape, without duplication or omission.
2. **Throughput.** This measures the number of packets on the output tape as a function of the number of rounds that have passed.
3. **Processor Memory.** This measures the memory required of each node by the protocol, independent of the number of packets to be transferred.

All three considerations will be measured in the worst-case scenario as standards that are guaranteed to exist regardless of adversarial interference. One can also evaluate a protocol based on its dependence on global information to make decisions. The protocol that we present in this paper will not assume the internal nodes have a global view of the network. Such protocols are termed "local control," in that each node can make all routing decisions based only on the local conditions of its adjacent edges and neighbors.

Our protocol is designed to be resilient against a malicious, polynomially-bounded adversary who may attempt to impact the *correctness, throughput,* and *memory* of our protocol by disrupting links between the nodes or by taking direct control over the nodes and forcing them to deviate from our protocol in any manner the adversary wishes. In order to relate our work to previous results and to clarify the two main forms of adversarial interference, we describe two separate (yet coordinated with each other) adversaries:[1]

Edge-Scheduling Adversary. This adversary controls the *links* between nodes every round. More precisely, for each round, this adversary decides which edges in the network are up and which are down. We will say that the edge-scheduling adversary is *conforming* if for every round there is at least one path from the sender to the receiver (although the path may change each round).[2] The adversary can make any arbitrary poly-time computation to maximize interference in routing, so long as it remains *conforming*.

Node-Controlling Adversary. This adversary controls the *nodes* of the network that it has corrupted. More precisely, each round this adversary decides which nodes to corrupt. Once corrupted, a node is forever under complete adversarial control and can behave in an arbitrary malicious manner. We say that the node-controlling adversary is *conforming* if every round there is a connection between the sender and receiver consisting of edges that are "up" for the round (as specified by the edge-scheduling adversary) and that passes

---

[1] The separation into two separate adversaries is artificial: our protocol is secure whether edge-scheduling and corruption of nodes are performed by two separate adversaries that have different capabilities yet can coordinate their actions with each other, or this can be viewed as a single coordinated adversary.

[2] A more general definition of an edge-scheduling adversary would be to allow completely arbitrary edge failures, with the exception that in the limit there is no permanent cut between the sender and receiver. However, this definition (while more general) greatly complicates the exposition, including the definition of throughput rate, and we do not treat it here.

through *uncorrupted* nodes. We emphasize that this path can change each round, and there is no other restriction on which nodes the node-controlling adversary may corrupt (allowing even a vast majority of corrupt nodes).

There is another reason to view these adversaries as distinct: we deal with the challenges they pose to correctness, throughput, and memory in different ways. Namely, aside from the conforming condition, the edge-scheduling adversary cannot be controlled or eliminated. Edges themselves are not inherently "good" or "bad," so identifying an edge that has failed does not allow us to forever refuse the protocol to utilize this edge, as it may come back up at any time (and indeed it could form a crucial link on the path connecting the sender and receiver that the conforming assumption guarantees). In sum, we cannot hope to control or alter the behavior of the edge-scheduling adversary, but must come up with a protocol that works well regardless of the behavior of the ever-present (conforming) edge-scheduling adversary.

By contrast, our protocol will limit the amount of influence the node-controlling adversary has on correctness, throughput, and memory. Specifically, we will show that if a node deviates from the protocol in a sufficiently destructive manner (in a well-defined sense), then our protocol will be able to identify it as corrupted in a timely fashion. Once a corrupt node has been identified, it will be *eliminated* from the network. Namely, our protocol will call for honest nodes to refuse all communication with nodes that have been eliminated.[3] Thus, there is an inherent difference in how the two adversaries are handled: We can restrict the influence of the node-controlling adversary by eliminating the nodes it has corrupted, while the edge-scheduling adversary must be dealt with in a more ever-lasting manner.

1.1   Previous Work

To motivate the importance of the problem we consider in this paper, and to emphasize the significance of our result, it will be useful to highlight recent works in related areas. To date, routing protocols that consider adversarial networks have been of two main flavors: *End-to-End Communication* protocols that consider dynamic topologies (a notion captured by our "edge-scheduling adversary"), and *Fault Detection and Localization* protocols, which handle devious behavior of nodes (as modeled by our "node-controlling adversary").

END-TO-END COMMUNICATION: One of the most relevant research directions to our paper is the notion of End-to-End Communication in distributed networks, considered by Afek and Gafni [1], Awerbuch, Mansour and Shavit [8], Afek, Awebuch, Gafni, Mansour, Rosen, and Shavit [2], and Kushilevitz, Ostrovsky and Rosen [13]. Indeed, our starting point is the *Slide* protocol (also known in practical works as "gravitational flow" routing) developed in these works. It was designed to perform end-to-end communication with bounded memory in a model where (using our terminology) an edge-scheduling adversary controls the edges (subject to the constraint there is no permanent cut between the sender

---

[3] The *conforming* assumption guarantees that the sender and receiver are incorruptible, and in our protocol they will identify and eliminate corrupt nodes.

and receiver). The Slide protocol has proven to be incredibly useful in a variety of settings, including multi-commodity flow (Awerbuch and Leigthon [7]) and in developing routing protocols that compete well (in terms of packet loss) against an online bursty adversary ([4]). However, prior to our work there was no version of the Slide protocol that could handle malicious behavior of the nodes.

FAULT DETECTION AND LOCALIZATION PROTOCOLS: At the other end, there have been a number of works that explore the possibility of a node-controlling adversary that can corrupt nodes. In particular, there is a recent line of work that considers a network consisting of a *single path* from the sender to the receiver, culminating in the recent work of Barak, Goldberg and Xiao [9] (for further background on fault localization see references therein). In this model, the adversary can corrupt any node on the path (except the sender and receiver) in a dynamic and malicious manner. Since corrupting any node on the path will sever the honest connection between S and R, the goal of a protocol in this model is *not* to guarantee that all messages sent to R are received. Instead, the goal is to *detect* faults when they occur and to *localize* the fault to a single edge.

There have been many results that provide Fault Detection (FD) and Fault Localization (FL) in this model. In Barak et. al. [9], they formalize the definitions in this model and the notion of a secure FD/FL protocol, as well as providing lower bounds in terms of communication complexity to guarantee accurate fault detection/location in the presence of a node-controlling adversary. While the Barak et. al. paper has a similar flavor to our paper, we emphasize that their protocol does not seek to guarantee successful or efficient routing between the sender and receiver. Instead, their proof of security guarantees that if a packet is deleted, malicious nodes cannot collude to convince S that no fault occurred, nor can they persuade S into believing that the fault occurred on an honest edge. Localizing the fault in their paper relies on cryptographic tools, and in particular the assumption that one-way functions exist. Although utilizing these tools (such as MACs or Signature Schemes) increases communication cost, it is shown by Goldberg, Xiao, Barak, and Redford [12] that the existence of a protocol that is able to securely detect faults (in the presence of a node-controlling adversary) implies the existence of one-way functions, and it is shown in Barak et. al. [9] that *any* protocol that is able to securely localize faults necessarily requires the intermediate nodes to have a trusted setup. The proofs of these results do not rely on the fact that there is a single path between S and R, and we can therefore extend them to the more general network encountered in our model to justify our use of cryptographic tools and a trusted setup assumption (i.e. PKI) to identify malicious behavior.

Another paper that addresses routing in the Byzantine setting is the work of Awerbuch, Holmes, Nina-Rotary and Rubens [6], though this paper does not have a fully formal treatment of security, and indeed a counter-example that challenges its security is discussed in the appendix of [9].

ERROR-CORRECTION IN THE ACTIVE SETTING: Due to space considerations, we will not be able to give a comprehensive account of all the work in this area.

Instead we highlight some of the most relevant works and point out how they differ from our setting and results. For a lengthy treatment of error-correcting codes against polynomially bounded adversaries, we refer to the work of Micali at. al [14] and references therein. It is important to note that this work deals with a graph with a single "noisy" edge, as modelled by an adversary who can partially control and modify information that crosses the edge. In particular, it does not address throughput efficiency or memory considerations in a full communication network, nor does it account for malicious behavior at the vertices. Also of relevance is the work on Rajagopalan and Schulman on error-correcting network coding [15], where they show how to correct noisy edges during distributed computation. Their work does not consider actively malicious nodes, and thus is different from our setting. It should also be noted that their work utilizes Schulman's tree-codes [18] that allow length-flexible online error-correction. The important difference between our work and that of Schulman is that in our network setting, the amount of malicious activity of corrupt nodes is not restricted.

1.2   Our Results

To date, there has not been a protocol that has considered simultaneously a network susceptible to faults occurring due to edge-failures *and* faults occurring due to malicious activity of corrupt nodes. The end-to-end communication works are not secure when the nodes are susceptible to corruption, and the fault detection and localization works focus on a *single path* for some duration of time, and do not consider a fully distributed routing protocol that utilizes the entire network and attempts to maximize throughput efficiency while guaranteeing correctness. Indeed, our work answers one of the open questions posed in the Barak et. al. paper regarding fault localization on multiple paths. In this paper we bridge the gap between these two research areas and obtain the first routing protocol simultaneously secure against both an edge-scheduling adversary and a node-controlling adversary, even if these two adversaries attack the network using an arbitrary coordinated poly-time strategy. Furthermore, our protocol achieves comparable efficiency standards in terms of throughput and processor memory as state-of-the-art protocols that are not secure against a node-controlling adversary, and it does so using *local-control*. An informal statement of our result can be found below. We emphasize that the linear transmission rate that we achieve (assuming at least n^2 messages are sent) is asymptotically optimal, as *any* protocol operating in a network with a single path connecting sender and receiver can do no better than one packet per round.

A ROUTING THEOREM FOR ADVERSARIAL NETWORKS (Informal): If one-way functions exist, then for any n-node graph and k sufficiently large, there exists a trusted-setup *linear throughput* transmission protocol that can send n^2 messages in $O(n^2)$ rounds with $O(n^4(k + \log n))$ memory per processor that is resilient against **any** poly-time conforming Edge-Scheduling Adversary and **any** conforming poly-time Node-Controlling Adversary, with negligible (in k) probability of failure or decoding error.

| | Secure Against: | | Processor | Throughput Rate |
	Edge-Sched?	Node-Contr?	Memory	x rounds $\rightarrow$ $f(x)$ packets
Slide Protocol of [2]	YES	NO	$O(n^2 \log n)$	$f(x) = O(x - n^2)$
Slide Protocol of [13]	YES	NO	$O(n \log n)$	$f(x) = O(x/n - n^2)$
(folklore) (Flooding + Signatures)	YES	YES	$O(1)$	$f(x) = O(x/n - n^2)$
(folklore) (Signatures + Sequence No.'s)	YES	YES	$unbounded$	$f(x) = O(x - n^2)$
Our Protocol	YES	YES	$O(n^4(k+\log n))$	$f(x) = O(x - n^2)$

Fig. 1. Comparison of Our Protocol to Related Existing Protocols and Folklore

2 Challenges and Naïve Solutions

Before proceeding, it will be useful to consider a couple of naïve solutions that achieve the goal of *correctness* (but perform poorly in terms of *throughput*), and help to illustrate some of the technical challenges that our theorem resolves. Consider the approach of having the sender continuously *flood* a single signed packet into the network for n rounds. Since the *conforming* assumption guarantees that the network provides a path between the sender and receiver through honest nodes at every round, this packet will reach the receiver within n rounds, regardless of adversarial interference. After n rounds, the sender can begin flooding the network with the next packet, and so forth. Notice that this solution will require each processor to store and continuously broadcast a single packet at any time, and hence this solution achieves excellent efficiency in terms of *processor memory*. However, notice that the *throughput* rate is sub-linear, namely after x rounds, only $O(x/n)$ packets have been outputted by the receiver.

One idea to try to improve the throughput rate might be to have the sender streamline the process, sending packets with ever-increasing sequence numbers without waiting for n rounds to pass (or signed acknowledgments from the receiver) before sending the next packet. In particular, across each of his edges the sender will send every packet once, waiting only for the neighboring node's confirmation of receipt before sending the next packet across that edge. The protocol calls for the internal nodes to act similarly. Analysis of this approach shows that not only has the attempt to improve throughput failed (it is still $O(x/n)$ in the worst-case scenario), but additionally this modification requires arbitrarily large (polynomial in n and k) processor memory, since achieving correctness in the dynamic topology of the graph will force the nodes to remember all of the packets they see until they have broadcasted them across all adjacent edges or seen confirmation of their receipt from the receiver.

2.1 Challenges in Dealing with Node-Controlling Adversaries

In this section, we discuss some potential strategies that the node-controlling and edge-scheduling adversaries may incorporate to disrupt network communication. Although our theorem will work in the presence of *arbitrary* malicious activity of the adversarial controlled nodes (except with negligible probability), it will be instructive to list a few obvious forms of devious behavior that our protocol must protect against. It is important to stress that this list is *not* intended to be exhaustive. Indeed, we do not claim to know all the specific ways an arbitrary polynomially bounded adversary may force nodes to deviate from a given protocol, and we rigorously prove that our protocol is secure against all possible deviations.

Packet Deletion/Modification. Instead of forwarding a packet, a corrupt node "drops it to the floor" (i.e. deletes it or effectively deletes it by forever storing it in memory), or modifies the packet before passing it on. Another manifestation of this is if the sender requests fault localization information of the internal nodes, such as providing documentation of their interactions with neighbors. A corrupt node can then block or modify information that passes through it in attempt to hide malicious activity or implicate an honest node.

Introduction of Junk/Duplicate Packets. The adversary can attempt to disrupt communication flow and "jam" the network by having corrupted nodes introduce junk packets or re-broadcast old packets. Notice that junk packets can be handled by using cryptographic signatures to prevent introduction of "new" packets, but this does not control the re-transmission of old, correctly signed packets.

Disobedience of Transfer Rules. If the protocol specifies how nodes should make decisions on where to send packets, etc., then corrupt nodes can disregard these rules, including lying to adjacent nodes about their current state.

Coordination of Edge-Failures. The edge-scheduling adversary can attempt to disrupt communication flow by scheduling edge-failures in any manner that is consistent with the *conforming* criterion. Coordinating edge-failures can be used to impede correctness, memory, and throughput in various ways: e.g. packets may become lost across a failed edge, stuck at a suddenly isolated node, or arrive at the receiver out of order. A separate issue arises concerning fault localization: When the sender requests documentation from the internal nodes, the edge-scheduling adversary can slow progress of this information, as well as attempt to protect corrupt nodes by allowing them to "play-dead" (setting all of its adjacent edges to be *down*), so that incriminating evidence cannot reach the sender.

2.2   Highlights of Our Solution

Our starting point is the Slide protocol [2], which has enjoyed practical success in networks with dynamic topologies, but is not secure against nodes that are allowed to behave maliciously. Due to space constraints, we will only highlight the main ideas of the protocol here; the interested reader can find a full exposition in [5]. We begin by viewing the edges in the graph as consisting of two directed edges, and associate to each end of a directed edge a *stack* data-structure able to hold $2n$ packets and to be maintained by the node at that end. The protocol specifies the following simple, local condition for transferring a packet across a directed edge: if there are more packets in the stack at the originating end than the terminating end, transfer a packet across the edge. Similarly, within a node's local stacks, packets are shuffled to average out the stack heights along each of its edges. Intuitively, packet movement is analogous to the flow of water: high stacks create a pressure that force packets to "flow" to neighboring lower stacks. At the source, the sender maintains the pressure by filling his outgoing stacks (as long as there is room) while the receiver relieves pressure by consuming packets and keeping his stacks empty. Loosely speaking, packets traveling to nodes "near" the sender will therefore require a very large potential, packets traveling to nodes near the receiver will require a small potential, and packet transfers near intermediate nodes will require packages to have a moderate potential. Assuming these potential requirements exist, packets will pass from the sender with a high potential, and then "flow" downwards across nodes requiring less potential, all the way to the receiver.

Because the Slide protocol provides a fully distributed protocol that works well against an edge-scheduling adversary, our starting point was to try to extend the protocol by using digital signatures[4] to provide resilience against Byzantine attacks and arbitrary malicious behavior of corrupt nodes. This proved to be a highly nontrivial task that required us to develop a lot of additional machinery, both in terms of additional protocol ideas and novel techniques for proving correctness. We give a detailed explanation of our techniques in Section 3, but due to space considerations we have omitted the formal pseudo-code and rigorous proofs of security (these can be found in the full version, see [5]). Below we give a sample of some of the key ideas we used in ensuring our additional machinery would be provably secure against a node-controlling adversary, and yet not significantly affect throughput or memory, compared to the original Slide protocol:

ADDRESSING THE "COORDINATION OF EDGE-SCHEDULING" ISSUES. In the absence of a node-control- ling adversary, previous versions of the Slide protocol (e.g. [2]) are secure and efficient against an edge-scheduling adversary, and it will be useful to discuss how some of the challenges posed by a network with a dynamic topology are handled. First, note that the total capacity of the stack data-structure is bounded by $4n^3$. That is, each of the n nodes can hold at most $2n$ packets in each of their $2n$ stacks (along each directed edge) at any time.

- To handle the loss of packets due to an edge going down while transmitting a packet, a node is required to maintain a copy of each packet it transmits along an edge until it receives confirmation from the neighbor of successful receipt.

---

[4] In this paper we use public-key operations to sign individual packets with control information. Clearly, this is too expensive to do per-packet in practice. There are methods of amortizing the cost of signatures by signing "batches" of packets; using private-key initialization [9,12], or using a combination of private-key and public key operations, such as "on-line/off-line" signatures [10,17]. For the sake of clarity and since the primary focus of our paper is theoretical feasibility, we restrict our attention to the straight-forward public-key setting without considering these additional cost-saving techniques.

- To handle packets becoming stuck in some internal node's stack due to edge failures, *error-correction* is utilized to allow the receiver to decode a full message without needing every packet. In particular, if an error-correcting code allowing a fraction of λ faults is utilized, then since the capacity of the network is $4n^3$ packets, if the sender is able to pump $4n^3/\lambda$ codeword packets into the network and there is no malicious deletion or modification of packets, then the receiver will necessarily have received enough packets to decode the message.
- The Slide protocol has a natural bound in terms of memory per processor of $O(n^2 \log n)$ bits, where the bottleneck is the possibility of a node holding up to $2n^2$ packets in its stacks, where each packet requires $O(\log n)$ bits to describe its position in the code.

Of course, these techniques are only valid if nodes are acting honestly, which leads us to our first extension idea.

HANDLING PACKET MODIFICATION AND INTRODUCTION OF JUNK PACKETS. Before inserting any packets into the network, the sender will authenticate each packet using his digital signature, and intermediate nodes and the receiver never accept or forward messages not appropriately signed. This simultaneously prevents honest nodes becoming bogged down with junk packets, as well as ensuring that if the receiver has obtained enough authenticated packets to decode, a node-controlling adversary cannot impede the successful decoding of the message as the integrity of the codeword packets is guaranteed by the inforgibility of the sender's signature.

FAULT DETECTION. In the absence of a node-controlling adversary, our protocol looks almost identical to the Slide protocol of [2], with the addition of signatures that accompany all interactions between two nodes. First, the sender attempts to pump the $4n^3/\lambda$ codeword packets of the first message into the network, with packet movement exactly as in the original Slide protocol. We consider all possible outcomes:

1. The sender is able to insert all codeword packets and the receiver is able to decode. In this case, the message was transmitted successfully, and our protocol moves to transfer the next message.
2. The sender is able to insert all codeword packets, but the receiver has not received enough to decode. In this case, the receiver floods the network with a single-bit message indicating *packet deletion* has occurred.
3. The sender is able to insert all codeword packets, but the receiver cannot decode because he has received duplicated packets. Although the sender's authenticating signature guarantees the receiver will not receive junk or modified packets, a corrupt node can duplicate valid packets. Therefore, the receiver may receive enough packets to decode, but cannot because he has received duplicates. In this case, the receiver floods the network with a single message indicating the label of a duplicated packet.
4. After some amount of time, the sender still has not inserted all codeword packets. In this case, the duplication of old packets is so severe that the network

has become jammed, and the sender is prevented from inserting packets even along the honest path that the conforming assumption guarantees. If the sender believes the jamming cannot be accounted for by edge-failures alone, he will halt transmission and move to localizing a corrupt node.[5] One contribution this paper makes is to prove a lower bound on the insertion rate of the sender for the Slide protocol *in the absence of the node-controlling adversary.* This bound not only alerts the sender when the jamming he is experiencing exceeds what can be expected in the absence of corrupt nodes, but it also provides a mechanism for localizing the offending node(s).

The above four cases exhaust all possibilities. Furthermore, if a transmission is not successful, the sender is not only able to *detect* the fact that malicious activity has occurred, but he is also able to distinguish the *form* (i.e. Case 2-4) of the malicious activity. Meanwhile, for the top case, our protocol enjoys (within a constant factor) an equivalent throughput rate as the original Slide protocol.

Fault Localization. Once a fault has been detected, it remains to describe how to *localize* the problem to the offending node. To this end, we use digital signatures to achieve a new mechanism we call "Routing with Responsibility." By forcing nodes to sign key parts of every communication with their neighbors during the transfer of packets, they can later be held accountable for their actions. In particular, once the sender has identified the reason for failure (Cases 2-4 above), he will request all internal nodes to return *status reports*, which are signatures on the relevant parts of the communication with their neighbors. We then prove in each case that with the complete status report from every node, the sender can identify and eliminate a corrupt node. Of course, malicious nodes may choose not to send self-incriminating information. We handle this separately as explained below.

Processor Memory. The signatures on the communication a node has with its neighbors for the purpose of fault localization is a burden on the memory required of each processor that is not encountered in the original Slide protocol. One major challenge was to reduce the amount of signed information each node must maintain as much as possible, while still guaranteeing that each node has maintained "enough" information to identify a corrupt node in the case of arbitrary malicious activity leading to a failure of type 2-4 above. The content of Theorem 3.2 in Section 3 demonstrates that the extra memory required of our protocol is a factor of n^2 higher than that of the original Slide protocol.

---

[5] We emphasize here the importance that the sender is able to distinguish the case that the jamming is a result of the edge-scheduling adversary's controlling of edges verses the case that a corrupt node is duplicating packets. After all, in the case of the former, there is no reward for "localizing" the fault to an edge that has failed, as *all* edges are controlled by the edge-scheduling adversary, and therefore no edge is inherently better than another. But in the case a node is duplicating packets, if the sender can identify the node, it can eliminate it and effectively reduce the node-controlling adversary's ability to disrupt communication in the future.

INCOMPLETE INFORMATION. As already mentioned, we will show that as long as the sender has the complete status reports from every node, he will be able to identify a corrupt node, regardless of the reason for failure 2-4 above. However, this relies on the sender obtaining all of the relevant information; the absence of even a single node's information can prevent the localization of a fault. We address this challenge in the following ways:

1. We minimize the amount of information the sender requires of each node. This way, a node need not be connected to the sender for very many rounds in order for the sender to receive its information. Specifically, regardless of the reason for failure 2-4 above, a status report consists of only n pieces of information from each node, i.e. one packet for each of its edges.

2. If the sender does not have the n pieces of information from a node, it cannot afford to wait indefinitely. After all, the edge-scheduling adversary may keep the node disconnected indefinitely, or a corrupt node may simply refuse to respond. For this purpose, we create a *blacklist* for non-responding nodes, which will disallow them from transferring codeword packets in the future. This way, anytime the receiver fails to decode a codeword as in Cases 2-4 above, the sender can request the information he needs, blacklist nodes not responding within some short amount of time, and then re-attempt to transmit the codeword using only non-blacklisted nodes. Nodes should not transfer codeword packets to blacklisted nodes, but they do still communicate with them to transfer the information the sender has requested. If a new transmission again fails, the sender will only need to request information from nodes that were participating, i.e. he will *not* need to collect new information from blacklisted nodes (although the nodes will remain blacklisted until the sender gets the original information he requested of them). Nodes will be removed from the blacklist and re-allowed to route codeword packets as soon as the sender receives their information.

THE BLACKLIST. Blacklisting nodes is a delicate matter; we want to place malicious nodes "playing-dead" on this list, while at the same time we don't want honest nodes that are temporarily disconnected from being on this list for too long. We prove in the full version (see [5]) that the occasional honest node that gets put on the blacklist won't significantly hinder packet transmission. Intuitively, this is true because any honest node that is an important link between the sender and receiver will not remain on the blacklist for very long, as his connection to the sender guarantees the sender will receive all requested information from the node in a timely manner.

Ultimately, the blacklist allows us to control the amount of malicious activity to which a single corrupt node can contribute. Indeed, we show that each failed message transmission (Cases 2-4 above) can be localized (eventually) to (at least) one corrupt node. More precisely, the blacklist allows us to argue that malicious activity can cause at most n failed transmissions before a corrupt node can necessarily be identified and eliminated. Since there are at most n corrupt nodes, this bounds the number of failed transmissions at n^2. The result of this is that other than at most n^2 failed message transmissions, our protocol enjoys the same throughput efficiency of the old Slide protocol. The formal statement of this and a sketch of the proof are the contents of Theorem 3.3 in Section 3.

3 Routing against a Node-Controlling + Edge-Scheduling Adversary

3.1 Definitions

In this section, we briefly describe our protocol. Due to space constraints, a detailed presentation, including formal pseudo-code and rigorous proofs, has been omitted (these can be found in the full version [5]). As mentioned in the Introduction, our model considers *end-to-end communication* in a network consisting of n nodes in the presence of *conforming* edge-scheduling and node-controlling adversaries. We assume a synchronous network with discrete stages, where a *stage* is defined to be the unit of time in which every edge can transfer a single packet of P bits.[6] A *round* will consist of two consecutive stages during which packets are transferred between adjacent nodes (the *Routing Phase*), followed by the *Re-Shuffle Phase* in which nodes perform (instantaneous) local maintenance of their buffers. A *transmission* (usually denoted by T) will consist of $O(n^3)$ rounds during which the sender inserts packets corresponding to a single codeword. At the end of each transmission, the receiver will broadcast an *end of transmission* message, indicating whether it could successfully decode the codeword. In the case that the receiver cannot decode, we will say that the transmission *failed*, and otherwise the transmission was *successful*.

In the case a transmission fails, the sender will determine the reason for failure (Cases 2-4 from Section 2.2, and also F2-F4 below), and request nodes to return *status reports* that correspond to a particular piece of signed communication between each node and its neighbors. We will refer to status report packets as *parcels* to clarify discussion in distinguishing them from the codeword *packets*.

The first step in providing a guarantee of efficiency (in terms of *throughput*) is to prove that every failed transmission falls under one of the following cases (the number of packets per codeword, D, will be defined in the next section):

F2. The receiver could not decode, and the sender has inserted D packets

F3. The receiver could not decode, the sender has inserted D packets, and the receiver has *not* received any duplicated packets corresponding to the current codeword

F4. The receiver could not decode and cases F2 and F3 do not happen

We describe in Section 3.3 how we identify a corrupt node in each case. The primary tool that will be used to handle case F2 will be the notion of *potential*, defined now.

---

[6] We assume $P > O(k + \log n)$, where k is the security parameter used for the signature scheme and n is the number of nodes. In particular, this will allow packets to carry two signatures (requires $2k$ bits) and a codeword index (requires $\log n$ bits) in addition to the codeword information.

Definition 3.1. *The **height** H_B of any internal buffer B is the number of pack-ets currently stored in the buffer. The **potential** Φ_B of the buffer is the arithmetic sum up to H_B, i.e. $\Phi_B = \sum_{i=1}^{H} i = \frac{H(H+1)}{2}$.*

3.2   Description of the Node-Controlling+Edge-Scheduling Protocol

Setup. The sender has a sequence of messages $\{m_1, m_2, \ldots\}$ of uniform size $M = \frac{6\sigma(P-2k)n^3}{\lambda}$ that he will expand into *codewords* $\{b_1, b_2, \ldots\}$ of size $C = \frac{M}{\sigma}$ (σ is the *information rate* and λ the *error-rate* of any error-correcting code). The codewords are divided into packets of size $P - 2k$ (P is the number of bits that can be transferred by an edge in a single stage, k is the security parameter), which will allow packets to have enough room to hold two signatures of size k. Since the number of packets per codeword is $D := \frac{C}{P-2k} = \frac{6(P-2k)n^3}{(P-2k)\lambda} = \frac{6n^3}{\lambda}$, if R receives $(1 - \lambda)D$ distinct packets corresponding to the same codeword, he will be able to decode.

  Each internal node has the following buffers:

1. *Incoming and Outgoing Buffers.* For each incoming/outgoing edge, a node will have a buffer that has the capacity to hold $2n$ packets at any given time. The receiver has one large storage buffer, and the sender has a "Copy of Current Packets" buffer to be used any time a transmission fails and needs to be repeated.

2. *Signature Buffers.* Each node has a signature buffer along each edge to keep track of incoming (resp. outgoing) information exchanged with its neighbor along that edge. The signature buffers will hold information corresponding to changes in: 1) The net number of packets passed across each adjacent edge; 2) The cumulative change in potential due to packet transfers across each adjacent edge; and 3) For each packet p, the net number of times p has passed across each adjacent edge. Each of the three items above, together with the current round index and transmission index, will be signed by the adjacent node before they are stored.

3. *Broadcast Buffer.* This is where nodes will temporarily store their neigh-bor's (and their own) state information that the sender will need to identify malicious activity. A node's broadcast buffer can hold the *start* and *end of transmission* parcels (see below), blacklist information, and up to n parcels of status report information for each node in the network.

4. *Data Buffer.* This keeps track of eliminated and blacklisted nodes. The sender's data buffer will also be able to store information for up to n failed transmissions, including why they failed, blacklisted nodes, and up to n sta-tus report parcels per node per failed transmission.

Also as part of the Setup, all nodes learn the relevant parameters (P, n, λ, and σ), each node receives a private key from a trusted third party for signing, and each node receives public information that allows them to verify the signature of every other node in the network.

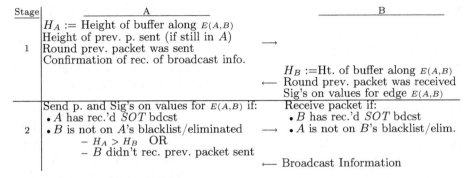

Fig. 2. Description of Communication Exchange Along Directed Edge $E(A, B)$ During the Routing Phase of Some Round

Routing Phase. This consists of two consecutive stages during which nodes transfer codeword packets and broadcast parcels that comprise status reports and auxiliary information. The manner in which packets and parcels are transferred across a directed edge[7] $E(A, B)$ is succinctly described in the figure below. We state once and for all that if a node ever receives inaccurate or mis-signed information, it will act as if no information was received at all (e.g. as if the edge had failed for that stage).

At the end of every transmission, the receiver will broadcast a parcel indicating if it was able to decode the previous codeword, as well as containing the label of a codeword packet he received twice (if one exists). From this, the sender will create the *start of transmission* (SOT) broadcast, which includes information concerning up to n failed transmissions, including why the transmission failed and which nodes are blacklisted (or eliminated) for those transmissions. We stress that *no node is allowed to transfer any codeword packets until it has received the complete SOT broadcast.*

Re-Shuffle Rules. At the end of each round, nodes will shuffle the packets they are holding to balance the distribution of packets in their incoming and outgoing buffers. After re-shuffling, all buffers will have the same number of packets, where preference will be given to outgoing buffers if perfect balancing is not possible. During the Re-Shuffle Phase, the sender will fill each of his outgoing buffers (in an arbitrary order) with packets corresponding to the current codeword. Meanwhile, the receiver will empty all of its incoming buffers into its storage buffer. If at any time R has received enough packets to decode a codeword b_i, then R outputs message m_i and empties his storage buffer.

3.3 Analysis of Our Node-Controlling + Edge-Scheduling Protocol

We state our results concerning the correctness, throughput, and memory of our adversarial routing protocol.

---

[7] For clarity, even though we are considering "directed edge" $E(A, B)$, we indicate communication that travels from B to A. In reality, this communication will pass across $E(B, A)$.

Theorem 3.2. *The memory required of each node is at most $O(n^4(k + \log n))$.*

Proof (Sketch). Looking at the information each node is required to store in their buffers (see *Setup* of Section 3.2), the dominant expense comes from maintaining the signature buffers. The theorem follows as there are $O(n)$ such buffers, and each has the capacity to hold $D=O(n^3)$ packets of $P=O(k + \log n)$ bits.

Theorem 3.3. *Except for the at most n^2 transmissions that may fail due to malicious activity, our Routing Protocol enjoys linear throughput. More precisely, after x transmissions, the receiver has correctly outputted at least $x - n^2$ messages. If the number of transmissions x is quadratic in n or greater, than the failed transmissions due to adversarial behavior become asymptotically negligible. Since a transmission lasts $O(n^3)$ rounds and messages contain $O(n^3)$ bits, information is transferred through the network at a linear rate.*

We begin with a sequence of lemmas:

Lemma 1. *Every failed transmission falls under Case F2, F3, or F4; the sender (with the aide of the end of transmission parcel) can determine at the end of each transmission which case occurred.*

Proof. That Cases F2-F4 cover all possibilities is clear. The sender will know Case F2 has occurred since the sender keeps track of how many packets he has inserted in each transmission. The sender will know Case F4 has occurred if the receiver returns the label of a packet received twice (in the *end of transmission* parcel). Otherwise, a failed transmission is Case F3.

Lemma 2. *If a transmission fails and Case F4 occurred, then if the sender has collected the complete status report from every **participating** node, then the sender can identify a corrupt node.*

Proof (Sketch). Case F4 roughly corresponds to a mixed adversarial strategy of packet deletion and packet duplication: a corrupt node has been replacing current codeword packets with duplicated packets. When a transmission T fails due to Case F4, the sender has the label of a packet p that has been received at least twice by the receiver, and a node's *status report* contains its signed communication with neighbors regarding the number of times p transferred between them. The idea is to use the status reports to find a node who *output* p more times than it *input* p. In the full version, we argue that if the sender has the complete status reports from all nodes who participated in this transmission, then he will be able to find such a node $N \in G$, and this node is necessarily corrupt.

Lemma 3. *If a transmission fails and Case F3 occurred, then if the sender has collected the complete status report from every **participating** node, then the sender can identify a corrupt node.*

Proof (Sketch). Case F3 roughly corresponds to an adversarial strategy of packet deletion. When a transmission fails due to Case F3, a node's *status report* contains its signed communication with neighbors regarding the net number of packets transferred between them. The idea is to use the status reports to find a node who *input* more packets than it *output*. In the full version, we argue that if the sender has the complete status reports from all nodes who participated in this transmission, then he will be able to find such a node $N \in G$, and this node is necessarily corrupt.

Lemma 4. *If a transmission fails and Case F2 occurred, then if the sender has collected the complete status report from every **participating** node, then the sender can identify a corrupt node.*

Proof (Sketch). Case F2 roughly corresponds to an adversarial strategy of packet duplication. When a transmission fails due to F2, a node's *status report* contains its signed communication with neighbors regarding the net change in potential due to the packet transfers between them.

Notice that a single packet in some internal buffer at *height* H should (if all nodes are honest) contribute this amount H to the buffer's *potential*. Since packets in the sender's buffers do not count towards potential, when a packet is *inserted* by the sender, the total potential in the network will *increase* by the height the packet assumes in the incoming buffer that receives this packet (which is at most $2n$). Since the sender inserted less than D packets in Case F2, (in the absence of malicious activity) the total potential in the network can have increased by *at most* $2nD$. Meanwhile, we argue in the full version [5] that in each of the $4D - D$ rounds in which the sender could not insert a packet, the packet movement along the active honest path for the round will necessarily cause a *decrease* of at least n in the total potential in the network. Since the maximum amount of potential added to the network (due to insertions by the sender and in the absence of malicious activity) is $2nD$, while the minimum *decrease* in potential is $3nD$, there would be a *negative* amount of potential in the network. By definition of potential, this is impossible, and thus there must be a corrupt node who is contributing to illegal increases in potential (e.g. by duplicating packets). We show in the full version [5] how the status reports (which contain information on potential changes across each edge) can be used by the sender to identify and eliminate a corrupt node.

Lemma 5. *There can be at most n failed transmissions before the sender necessarily has the complete status report from every node that participated in one of those n transmissions.*

Proof (Sketch). A node will only be allowed to *participate* in a transmission if it is in "good standing" with the sender; i.e. the sender is not missing any status report parcel from the node. Therefore, for every failed transmission for which the sender does not have the complete status report from all *participating* nodes, there will be a distinct node $N \in G$ whose status report the sender does not have. Since there are n nodes, there are at most n such transmissions.

Proof of Theorem 3.3 (Sketch). We provide here only a very brief sketch of the proof, leaving the details to the full version [5]. We proceed by making a sequence of Lemmas. Theorem 3.3 now follows from Lemmas 1-5 as follows. There are at most n^2 failed transmissions (Cases F2-F4) since Lemma F5 states that after

n failed transmissions, the sender will have the complete status report from every participating node for one of these transmissions, and then Lemmas 1-4 state that the sender can identify (and eliminate) a corrupt node. After a node has been eliminated, the network is reduced to $n - 1$ nodes, and the argument can be repeated recursively. Since there are at most n corruptible nodes, there are at most n^2 failed transmissions. Meanwhile, all successful transmissions enjoy linear throughput, as each transmission lasts $4D=O(n^3)$ rounds and successfully decoded codewords contain $M=O(n^3)$ bits.

4   Conclusion and Open Problems

In this paper, we have described a protocol that is secure simultaneously against conforming node-controlling and edge-scheduling adversaries. Our results are of a theoretical nature, with rigorous proofs of correctness and guarantees of performance. Surprisingly, our protocol shows that the additional protection against the node-controlling adversary, on top of protection against the edge-scheduling adversary, can be achieved without any additional asymptotic cost in terms of throughput.

While our results do provide a significant step in the search for protocols that work in a dynamic setting (edge-failures controlled by the edge-scheduling adversary) where some of the nodes are susceptible to corruption (by a node-controlling adversary), there remain important open questions. The original Slide protocol[8] requires each internal node to have buffers of size $O(n^2 \log n)$, while ours requires $O(n^4 \log n)$, though this can be slightly improved with additional assumptions.[9] In practice, the extra factor of n^2 may make our protocol infeasible for implementation, even for overlay networks. While the need for signatures inherently force an increase in memory per node in our protocol verses the original Slide protocol, this is not what contributes to the extra $O(n^2)$ factor. Rather, the only reason we need the extra memory is to handle the third kind of malicious behavior, which roughly corresponds to the mixed adversarial strategy of a corrupt node replacing a valid packet with an old packet that the node has duplicated. Recall that in order to detect this, for *every* packet a node sees and for every neighbor, a node must keep a (signed) record of how many times this packet has traversed the adjacent edge (the $O(n^3)$ packets per codeword and $O(n)$ neighbors per node yield the $O(n^4)$ bound on memory). Therefore, one open problem is finding a less memory-intensive way to handle this type of adversarial behavior.

Our model also makes additional assumptions that would be interesting to relax. In particular, it remains an open problem to find a protocol that provides

---

[8] In [13], it was shown how to modify the Slide protocol so that it only requires $O(n \log n)$ memory per internal node. We did not explore in this paper if and/or how their techniques could be applied to our protocol to similarly reduce it by a factor of n.

[9] If we are given an a-priori bound that a path-length of any conforming path is at most L, the $O(n^4 \log n)$ can be somewhat reduced to $O(Ln^3 \log n)$.

efficient routing against a node-controlling and edge-scheduling adversary in a network that is fully *asynchronous* (without the use of timing assumptions, which can be used to replace full synchrony in our solution) and/or does not restrict the adversaries to be *conforming*. As mentioned in the Introduction, if the adversary is not conforming, then he can simply permanently disconnect the sender and receiver, disallowing any possible progress. Therefore, results in this direction would have to first define some notion of *connectedness* between sender and receiver, and then state throughput efficiency results in terms of this definition.

Acknowledgments

We thank the anonymous reviewers for their suggestions. Part of the work of the authors was done while visiting IPAM and supported in part by NSF grant 0430254. The third author was also supported in part by IBM Faculty Award, Xerox Innovation Group Award, NSF grants 0430254, 0716835, 0716389, 0830803 and U.C. MICRO grant. The first author was also supported in part by NSF grants 0430271 and 0716620.

References

1. Afek, Y., Gafni, E.: End-to-End Communication in Unreliable Networks. In: PODC (1988)
2. Afek, Y., Awebuch, B., Gafni, E., Mansour, Y., Rosen, A., Shavit, N.: Slide– The Key to Poly. End-to-End Communication. J. of Algorithms 22, 158–186 (1997)
3. Afek, Y., Gafni, E., Rosén, A.: The Slide Mechanism With Applications In Dynamic Networks. In: Proc. of the 11th ACM Symp. on PoDC, pp. 35–46 (1992)
4. Aiello, W., Kushilevitz, E., Ostrovsky, R., Rosén, A.: Adaptive Packet Routing For Bursty Adversarial Traffic. J. Comput. Syst. Sci. 60(3), 482–509 (2000)
5. Amir, Y., Bunn, P., Ostrovsky, R.: Authenticated Adversarial Routing, Full Version. Cornell Univ. Library arXiv, Article No. 0808.0156 (2008), http://arxiv.org/abs/0808.0156
6. Awerbuch, B., Holmer, D., Nina-Rotaru, C., Rubens, H.: A Secure Routing Protocol Resilient to Byzantine Failures. In: WiSE, pp. 21–30. ACM, New York (2002)
7. Awerbuch, B., Leighton, T.: Improved Approximation Algorithms for the Multi-Commodity Flow Problem and Local Competitive Routing in Dynamic Networks. In: STOC (1994)
8. Awerbuch, B., Mansour, Y., Shavit, N.: End-to-End Communication With Polynomial Overhead. In: Proc. of the 30th IEEE Symp. on Foundations of Computer Science, FOCS (1989)
9. Barak, B., Goldberg, S., Xiao, D.: Protocols and Lower Bounds for Failure Localization in the Internet. In: Smart, N.P. (ed.) EUROCRYPT 2008. LNCS, vol. 4965, pp. 341–360. Springer, Heidelberg (2008)
10. Even, S., Goldreich, O., Micali, S.: On-Line/Off-Line Digital Signatures. J. Cryptology 9(1), 35–67 (1996)
11. Goldreich, O.: The Foundations of Cryptography, Basic Applications. Cambridge University Press, Cambridge (2004)
12. Goldberg, S., Xiao, D., Tromer, E., Barak, B., Rexford, J.: Path-Quality Monitoring in the Presence of Adversaries. ACM SIGMETRICS 36, 193–204 (2008)

13. Kushilevitz, E., Ostrovsky, R., Rosén, A.: Log-Space Polynomial End-to-End Communication. SIAM Journal of Computing 27(6), 1531–1549 (1998)
14. Micali, S., Peikert, C., Sudan, M., Wilson, D.A.: Optimal error correction against computationally bounded noise. In: Kilian, J. (ed.) TCC 2005. LNCS, vol. 3378, pp. 1–16. Springer, Heidelberg (2005)
15. Rajagopalan, S., Schulman, L.: A Coding Theorem for Distributed Computation. In: Proc. 26th STOC, pp. 790–799 (1994)
16. Shannon, C.E.: Communication in the presence of noise. Proc. Institute of Radio Engineers 37(1), 10–21 (1949)
17. Shamir, A., Tauman, Y.: Improved Online/Offline Signature Schemes. In: Kilian, J. (ed.) CRYPTO 2001. LNCS, vol. 2139, pp. 355–367. Springer, Heidelberg (2001)
18. Schulman, L.: Coding for interactive communication. Special issue on Codes and Comp. of IEEE Transactions on Info. Theory 42(6), Part I: 1745–1756 (1996)

Adaptive Zero-Knowledge Proofs and Adaptively Secure Oblivious Transfer

Yehuda Lindell and Hila Zarosim

Department of Computer Science
Bar-Ilan University, ISRAEL
{zarosih,lindell}@cs.biu.ac.il

Abstract. In the setting of secure computation, a set of parties wish to securely compute some function of their inputs, in the presence of an adversary. The adversary in question may be *static* (meaning that it controls a predetermined subset of the parties) or *adaptive* (meaning that it can choose to corrupt parties during the protocol execution and based on what it sees). In this paper, we study two fundamental questions relating to the basic zero-knowledge and oblivious transfer protocol problems:

- *Adaptive zero-knowledge proofs:* We ask whether it is possible to construct adaptive zero-knowledge *proofs* (with unconditional soundness). Beaver (STOC 1996) showed that known zero-knowledge proofs are not adaptively secure, and in addition showed how to construct zero-knowledge arguments (with computational soundness).
- *Adaptively secure oblivious transfer:* All known protocols for adaptively secure oblivious transfer rely on seemingly stronger hardness assumptions than for the case of static adversaries. We ask whether this is inherent, and in particular, whether it is possible to construct adaptively secure oblivious transfer from enhanced trapdoor permutations alone.

We provide surprising answers to the above questions, showing that achieving adaptive security is sometimes harder than achieving static security, and sometimes not. First, we show that assuming the existence of one-way functions only, there exist adaptive zero-knowledge proofs for all languages in $\mathcal{NP}$. In order to prove this, we overcome the problem that all adaptive zero-knowledge protocols known until now used equivocal commitments (which would enable an all-powerful prover to cheat). Second, we prove a black-box separation between adaptively secure oblivious transfer and enhanced trapdoor permutations. As a corollary, we derive a black-box separation between adaptively and statically securely oblivious transfer. This is the first black-box separation to relate to adaptive security and thus the first evidence that it is indeed harder to achieve security in the presence of adaptive adversaries than in the presence of static adversaries.

1 Introduction

In the setting of secure two-party and multiparty computation, parties with private inputs wish to securely compute some joint function of their inputs, where

O. Reingold (Ed.): TCC 2009, LNCS 5444, pp. 183–201, 2009.

"security" must hold in the presence of adversarial behavior by some of the parties. An important parameter in any definition of security relates to the adversary's power. Is the adversary computationally bounded or all powerful? Is the adversary semi-honest (meaning that it follows all protocol instructions but tries to learn more than it's supposed to by analyzing the messages it receives) or is it malicious (meaning that it can arbitrarily deviate from the protocol specification)? Finally, are the adversarial corruptions *static* (meaning that the set of corrupted parties is fixed) or *adaptive* (meaning that the adversary can corrupt parties throughout the computation and the question of who to corrupt and when may depend on the adversary's view in the protocol execution). It is desirable to achieve security in the presence of adaptive adversaries where possible, since it models the real-world phenomenon of "hackers" actively breaking into computers, possibly while they are executing secure protocols. However, it seems to be technically harder to achieve security in the presence of adaptive adversaries. Among other things, it requires the ability to construct a simulator who can first generate a transcript blindly (without knowing any party's input) and then later, upon receiving inputs, "explain" the transcript as an execution of honest parties with those inputs.

In this paper, we ask two basic questions related to the feasibility of achieving security in the presence of adaptive adversaries. Our questions were borne out of the following two observations:

1. *Adaptive zero-knowledge proofs:* It has been shown that the zero-knowledge proof system of [15] (and all others known) is not secure in the presence of adaptive adversaries, or else the polynomial hierarchy collapses [1]. Due to this result, all known zero-knowledge protocols for $\mathcal{NP}$ in the adaptive setting are *arguments*, meaning that soundness only holds in the presence of a polynomial-time prover (adaptive zero-knowledge arguments were presented by [1] and later in the context of universal composability; e.g., see [7,8]). However, the question of whether or not adaptive zero-knowledge *proofs* exist for all $\mathcal{NP}$ has not been addressed.

2. *Adaptively secure oblivious transfer:* One of the goals of the theory of cryptography is to understand what assumptions are necessary and sufficient for carrying out cryptographic tasks; see for example [16]. Despite this, no such study has been carried out regarding adaptively secure protocols. In particular, we do not know what assumptions are necessary for achieving adaptively secure oblivious transfer (since oblivious transfer is complete for secure computation, this question has important ramifications to adaptively secure computation in general). Currently, what is known is that although statically secure oblivious transfer can be constructed from enhanced trapdoor permutations [10,14], all constructions for adaptively secure oblivious transfer use additional assumptions like the ability to sample a permutation without knowing its trapdoor [2,8].

Our results – adaptive zero-knowledge proofs. All known zero-knowledge protocols for $\mathcal{NP}$ essentially follow the same paradigm: the prover sends the verifier commitments that are based on the statement being proved (and its witness), and the verifier then asks the prover to open part or all of the commitments. Based on the prover's answer, the verifier is either convinced that the statement is true or detects the prover cheating. It therefore follows that soundness only holds if the commitment scheme used is *binding*, and this is a problem in the setting of adaptive security. Consider an adversary that corrupts the verifier at the beginning of the execution and the prover at the end. In this case, the zero-knowledge simulator must generate a transcript without knowing the NP-witness. However, at the end, after the prover is corrupted (and the simulator then receives a witness), it must be able to show that the commitments were generated using that witness. Until now, this has been solved by using *equivocal* commitments that can be opened to any value desired (in order for soundness to hold, the ability to equivocate is given to the simulator and not the real prover). However, this means that the protocol has only computational soundness, because an all-powerful prover is able to equivocate like the simulator. Indeed, the above observation led us to initially conjecture that adaptive zero-knowledge proofs exist only for $\mathcal{SZK}$. However, our conjecture was wrong, and in this paper we prove the following theorem:

Theorem 1. *Assuming the existence of one-way functions that are hard to invert for non-uniform adversaries, there exist adaptive zero-knowledge proofs for all $\mathcal{NP}$.*

We prove Theorem 1 by constructing a new type of *instance-dependent* commitment scheme. Instance-dependent commitment schemes are commitments whose properties depend on whether the instance (or statement) in question is in the language or not [3,18]. Typically, they are defined for a language L as follows. Let x be a statement. If $x \in L$ then the commitment associated with x is computationally hiding and if $x \notin L$ then the commitment associated with x is perfectly binding. This has proven very useful in the context of zero-knowledge where hiding alone is needed for the case of $x \in L$, and binding alone is needed in the case of $x \notin L$; see for example [21,26,22]. We construct an instance-dependent commitment scheme with the additional property that if $x \in L$ then the commitment is *equivocal* and the simulator can open it to any value it wishes. To be more exact, we need the commitment itself to be *adaptively secure*, meaning that it must be possible to generate a commitment value c and then later find "random coins" r for any bit b so that c is a commitment string generated by an honest committer with input b and random coins r.[1] In contrast to the above, if $x \notin L$ then the commitment is still perfectly binding. Given such a commitment (which is actually very similar to the commitment schemes presented in [11]and [8]) we

---

[1] We stress that this is a strictly stronger requirement than equivocality. In most equivocal commitments, the committer reveals only some of its coins upon decommitting. This does not suffice for achieving adaptive commitments.

are able to construct the first computational zero-knowledge *proof* for all $\mathcal{NP}$ that is secure also in the case of adaptive corruptions.[2]

Our results – adaptively secure oblivious transfer. As we have mentioned, all known protocols for adaptively secure oblivious transfer require assumptions of the flavor that it is possible to sample a permutation without its trapdoor. In contrast, standard trapdoor permutations do not have this property. We remark that enhanced trapdoor permutations do have the property that it is possible to sample an element in the domain of the permutation without knowing its preimage. This begs the question as to whether such "oblivious sampling" of the permutation's domain suffices for achieving adaptively secure oblivious transfer, or is something stronger needed (like oblivious sampling of permutations themselves). We remark that oblivious sampling is used in this context by having the simulator sample unobliviously and then "lie" in its final transcript by claiming to have sampled in the regular way. However, this strategy is problematic when the oblivious sampling is carried out on elements in the domain because if the trapdoor is known then it may be possible to see if the preimage of the sampled value appears implicitly in the protocol transcript. (For example, in the protocol of [10], the preimages fully define the sender's input and so if the trapdoor is known, the values can be checked.) Of course, such arguments do not constitute any form of evidence. In order to demonstrate hardness, we use the methodology of black-box separations, introduced by [17] and later used in [25,19,12,24] amongst others. We prove the following informally stated theorem:

Theorem 2. *There exists an oracle relative to which enhanced trapdoor permutations exist but adaptively secure oblivious transfer does not exist.*

Recalling that statically secure oblivious transfer can be constructed from any enhanced trapdoor permutation in a black-box way [10,14], we obtain the following corollary:

Corollary 3. *There exists an oracle relative to which statically secure oblivious transfer exists but adaptively secure oblivious transfer does not exist.*

This is the first evidence that it is strictly harder to achieve security in the presence of adaptive adversaries than to achieve security in the presence of static adversaries. We prove Theorem 2 by showing that if it is possible to achieve adaptively secure oblivious transfer using only enhanced trapdoor permutations, then it is possible to achieve statically secure oblivious transfer using only symmetric encryption (this is very inexact but sufficient for intuition). We then show that statically secure oblivious transfer does not exist relative to most symmetric

---

[2] In [22], adaptively secure commitment schemes are constructed for the languages of Graph Isomorphism and Quadratic Residuosity (although they are not presented in this way nor for this purpose). The constructions in [22] are incomparable to ours. On the one hand, they require no hardness assumptions whereas we use one-way functions. On the other hand, our construction is for all languages in $\mathcal{NP}$ whereas they are restricted to the above two specific languages (which are also in $\mathcal{SZK}$).

encryption oracles. We prove this using the recent result of [9] that shows the equivalence of the random oracle and ideal cipher models, to replace a symmetric encryption oracle by a "plain" random oracle (using six rounds of the Luby-Rackoff construction [20]). This enables us to extend the black-box separation of [17] to show that key agreement does not exist relative to most symmetric encryption oracles. We conclude the proof by recalling that key agreement can be constructed from oblivious transfer [12]. Thus, adaptively secure oblivious transfer cannot be constructed in a black-box way from enhanced trapdoor permutations. We remark that all of our results for oblivious transfer are proven for semi-honest adversaries (and thus hold also for malicious adversaries).

Our proof makes no use of the fact that the functionality being computed is oblivious transfer and holds for any functionality. We conclude that either a given function can be securely computed statically assuming only the existence of one-way functions (or to be more exact, only given a "symmetric" random oracle), or enhanced trapdoor permutations do not suffice for computing it adaptively.

Organization. Due to lack of space in this extended abstract, we present only proof sketches of our results; the full proofs can be found in the full version. Likewise, we do not present the definitions of secure computation and refer to [14] for the definition of security in the presence of semi-honest static adversaries (needed in Section 3.2), and to [6] for the definition of security in the presence of adaptive adversaries. Very briefly, the formulation of security for adaptive adversaries in [6] includes an environment $\mathcal{Z}$ that communicates with the adversary before and after the execution. An important property of the definition is that of *post-execution corruption*, meaning that the environment can ask the adversary to corrupt parties after the protocol transcript has already been fixed. This property is crucial for proving sequential composition; see [6].

2   Adaptive Zero-Knowledge Proofs

In this section we show how to construct adaptive zero-knowledge proof for the language of Hamiltonicity (HC). Our construction is based on Blum's zero-knowledge proof for Hamiltonicity [5]. In this protocol, the prover first commits to a random permutation of G, and the verifier then chooses randomly whether to check that the committed graph is indeed a permutation of G or that the committed graph contains a Hamiltonian cycle. Soundness holds because a non-Hamiltonian graph cannot simultaneously be a permutation of G and contain a Hamiltonian cycle. The simulator for this proof system does not know the witness and so cannot decommit to a Hamiltonian cycle after committing to a permutation of G. Therefore, it works by randomly choosing whether to send commitments to a permutation of G or to a graph containing only a random cycle of length n. Note that in the latter case, the commitments generated by the simulator are to different values than those generated by the real prover. This is not a problem when considering static corruptions because the hiding property of the commitments means that this cannot be distinguished. However, in the setting of adaptive corruptions, the prover can be corrupted after the simulation ends. In this case, the simulator must be able to provide random coins that demonstrate that the commitments sent initially are those that an honest prover would have sent. However, when the simulator commits to a graph containing only a Hamiltonian cycle, it cannot do this (because an honest prover never sends such a commitment). Thus, the commitment scheme used must be such that the simulator – given an appropriate trapdoor – can "explain" commitments to 1 as commitments to 0 and vice versa (actually it suffices that commitments to 0 be explainable as commitments to 1). However, if we use this type of commitment scheme, then we can no longer achieve statistical soundness (since an all-powerful cheating prover can find the trapdoor and use the adaptive property of the commitment scheme to fool the verifier).

We overcome this problem and construct adaptive zero-knowledge *proofs* for all $\mathcal{NP}$ by constructing an *adaptive* instance-dependent commitment scheme. Informally speaking, an adaptive instance-dependent commitment scheme (AIDCS) for an NP-language L is comprised of 3 algorithms: **(1)** An ordinary commitment algorithm Com that on an instance x, a bit b, and random coins r returns a commitment c to b, denoted $\mathsf{Com}(x, b; r)$; **(2)** A "fake" commitment algorithm Com' that on an instance x and random coins r' returns a commitment $c' = \mathsf{Com}'(x; r')$; **(3)** An "adaptive-opening" algorithm Adapt that given $x \in L$ and a witness $w \in R_x$, can present every output c' of Com' as a valid commitment to any bit b. That is, given c', the random coins r' used by Com' to generate c' and any bit b, algorithm Adapt outputs "random coins" r such that $c' = \mathsf{Com}(x, b; r)$. Note the difference between Com and Com': While Com is an ordinary committing algorithm (creating a commitment value for a given bit), when $x \in L$, algorithm Com' creates commitment values that are not associated to any specific bit. However, given a witness attesting to the fact that $x \in L$, these commitments can later be claimed to be commitments to 0 or to 1 by using algorithm Adapt. We stress that without such a witness, a commitment generated by Com' cannot necessarily be decommitted to any bit.

The security requirements of (Com, Com', Adapt) are as follows. For every $x \in L$, the commitment scheme must be hiding, meaning that commitments to 0, to 1 and fake commitments are all indistinguishable (i.e., $\{\mathsf{Com}(x, 0)\}_{x \in L} \overset{\mathrm{c}}{\equiv} \{\mathsf{Com}(x, 1)\}_{x \in L} \overset{\mathrm{c}}{\equiv} \{\mathsf{Com}'(x)\}_{x \in L}$). Furthermore, the output of Adapt must be indistinguishable from the output of an "honest committer" using algorithm Com. More specifically, for every c' in the range of Com' and every bit b, when given a witness $w \in R_x$ and coins r' such that $c' = \mathsf{Com}'(x; r')$, algorithm Adapt outputs a string r that is computationally indistinguishable from a uniformly distributed string and for which $c' = \mathsf{Com}(x, b; r)$. In addition to the above hiding properties we require that for every $x \notin L$, the commitment scheme Com is *perfectly* binding (i.e., $\{\mathsf{Com}(x, 0)\}_{x \notin L} \cap \{\mathsf{Com}(x, 1)\}_{x \notin L} = \phi$). A formal definition appears in the full version of this paper.

Constructing adaptive instance-dependent schemes. Our construction is almost identical to the trapdoor commitment scheme of [11] (as adapted by [8]), with one small but crucial difference. We begin by describing the adaptation by [8] of the trapdoor commitment of [11]. Let C be a perfectly binding commitment scheme with pseudorandom range and let G be a graph (in [11] G is a Hamiltonian graph generated by the receiver whereas in [8] it is a Hamiltonian graph that is placed in the common reference string). Then, in order to commit to 0, the committer chooses a random permutation π of the vertices of G and commits to the adjacency matrix of $\pi(G)$ using C. To decommit, it opens all entries and sends π. To commit to 1, the committer chooses a random n-cycle and for all entries in the adjacency matrix corresponding to the edges of the n-cycle, it uses C to commit to 1. In contrast, all other entries are set to a random string (recall that the commitment scheme has a pseudorandom range and thus this is indistinguishable from a commitment to 0). To decommit, it opens only the entries corresponding to the edges of the n-cycle. As stated, this scheme is computationally hiding due to the underlying commitment scheme C. In addition, it is computational binding as long as the sender does not know any Hamiltonian cycle in G. We stress that the scheme is not perfectly binding because an all-powerful corrupted committer can find the Hamiltonian cycle in G and send commitments that it can later open to both 0 and 1.

Our key observation is that in the setting of zero-knowledge we can use the graph G that is the statement being proven as the graph in the above commitment scheme. This implies that if $G \in HC$, then the commitment scheme is computationally hiding and if $G \notin HC$ then it is perfectly binding, as required. (As an added bonus, the graph need not be generated by the protocol.) Regarding adaptivity, when $G \in HC$ a commitment to 0 can be opened as a 0 or 1 given a cycle in G. This is due to the fact that when a cycle is known in G, it is possible to decommit to the cycle only (and claim that the rest of the commitments are just random coins), or to decommit to the entire graph.

In summary, we construct the following tuple of probabilistic polynomial-time algorithms: Com works as described above. Algorithm Com′ simply generates a commitment to 0 (that is, $\text{Com}'(G; r) = \text{Com}(G, 0; r)$). Given a witness $w \in R_G$ (a Hamiltonian cycle in G) and a commitment c in the range of Com′, if Adapt has to explain c as a commitment to 0, then it simply outputs the random coins used by Com′. In contrast, in order to explain c as a commitment to 1, Adapt outputs the randomness used by C for the edges in the Hamiltonian cycle in $\pi(G)$ (recall Adapt receives a Hamiltonian cycle w in G as input) and simply claims that all the other commitments are merely random strings (recall that C has pseudorandom range and therefore the output of Adapt is computationally indistinguishable from the uniform distribution).

Adaptive zero-knowledge proof for Hamiltonicity. Our adaptive ZK proof system is exactly that of [5], with the ordinary commitment scheme replaced by an adaptive instance-dependent commitment scheme. The fact that this scheme has unconditional soundness follows from the fact that when $x \notin L$, the commitment is perfectly binding. The simulation works like the standard zero-knowledge simulator for Hamiltonicity except that Com′ is used to commit to all edges outside of the n-cycle (in the case that the simulator sends a graph containing only a cycle). This enables the simulator to use Adapt later in case the prover is corrupted, and show that the commitment was "really" to a permutation of G.

3   Adaptive Oblivious Transfer

We prove our black-box separation of adaptively secure oblivious transfer from enhanced trapdoor permutations in the following steps. First, in Section 3.1 we define Γ and Δ oracles, where a Γ-oracle essentially represents an enhanced trapdoor permutation and a Δ-oracle is essentially a type of symmetric encryption scheme. Then, in Section 3.2 we show that if there exists a protocol for securely computing any functionality in the presence of *adaptive* adversaries relative to Γ-oracles, then there exists a protocol for securely computing the same functionality in the presence of *static* adversaries relative to Δ-oracles. The next step of the proof is to then show that for measure 1 of random Δ-oracles no statically secure OT_1^2 exists. This is done by using the original black-box separation of key agreement from one-way functions [17], and the fact that key agreement can be obtained from statically secure oblivious transfer; see Section 3.3. We conclude that for measure 1 of random Γ-oracles no adaptively secure OT_1^2 exists (see Section 3.4).

3.1   Oracle Definitions

We begin by defining (asymmetric) Γ and (symmetric) Δ oracles which are used in our proof.

Γ-oracles. Informally speaking, a Γ oracle is supposed to model an enhanced trapdoor permutation. Thus, it has an oracle for specifying a function and its trapdoor, and an oracle for computing the function (given the function identifier) and inverting it (given the trapdoor). The functions themselves are all over $\{0,1\}^n$ and thus it is trivial to sample an element without knowing its inverse (as is required for enhanced trapdoor permutations). Formally, we define a Γ-oracle to be an oracle containing the following functions:

- $G_\Gamma(\cdot) = (G_\Gamma^1, G_\Gamma^2)$ is a pair of injective functions such that on an input $r \in \{0,1\}^n$, $G_\Gamma(r) = (G_\Gamma^1(r), G_\Gamma^2(r)) = (fid, tid) \in \{0,1\}^{2n} \times \{0,1\}^{2n}$. Note that a party can query only G_Γ and cannot query one of its components separately.
- A function $F(\cdot, \cdot)$, such that for every $fid \in Range(G_\Gamma^1)$, $F(fid, \cdot)$ is a permutation over $\{0,1\}^n$ and for every $fid \notin Range(G_\Gamma^1)$ and every $x \in \{0,1\}^n$, $F(fid, x) = \perp$.
- A function F^{-1} satisfying $F^{-1}(tid, F(fid, x)) = x$ for every $x \in \{0,1\}^n$ and every $(fid, tid) \in Range(G_\Gamma)$. If tid is not in the range of $G_\Gamma^2(\cdot)$, then F^{-1} returns $\perp$. Note that since G_Γ^1 and G_Γ^2 are injective functions, pairs of the form (fid, tid) and (fid', tid), where $fid \neq fid'$ do not exist and F^{-1} is well defined.

Uniform distribution over oracles – notation: We denote by $\mathcal{U}_\Gamma$ the uniform distribution over Γ-oracles. Namely, an oracle $O_\Gamma = (G_\Gamma, F, F^{-1})$ is distributed according to $\mathcal{U}_\Gamma$ if G_Γ^1 and G_Γ^2 are two uniformly distributed injective functions from $\{0,1\}^n$ to $\{0,1\}^{2n}$ and for every $fid \in Range(G_\Gamma^1)$, $F(fid, \cdot)$ is a uniformly distributed permutation over $\{0,1\}^n$. We write "O_Γ is a random Γ-oracle" as shorthand for "O_Γ is distributed according to $\mathcal{U}_\Gamma$".

Δ-oracles. Informally, a Δ oracle is a symmetric oracle, meaning that anyone with the ability to compute the function also has the ability to invert it. Specifically, we define a function P and its inverse that is analogous to F and F^{-1} in a Γ oracle. For reasons that will become apparent later, we also define a function Q and its inverse (this has no analogue in a Γ oracle). Formally, we define a "Δ-oracle" to be an oracle containing the following functions:

- G_Δ is an injective function from $\{0,1\}^n$ to $\{0,1\}^{2n}$.
- A function $P(\cdot, \cdot)$ such that for every $fid \in Range(G_\Delta)$, $P(fid, \cdot)$ is a permutation over $\{0,1\}^n$. For $fid \notin Range(G_\Delta)$ and every $x \in \{0,1\}^n$, $P(fid, x) = \bot$.
- P^{-1} is the inversion algorithm of P. Namely for every $fid \in Range(G_\Delta)$ and $x \in \{0,1\}^n$, $P^{-1}(fid, P(fid, x)) = x$. For $fid \notin Range(G_\Delta)$ and every $x \in \{0,1\}^n$, $P^{-1}(fid, x) = \bot$.
- Q is an injective function from the range of G_Δ to $\{0,1\}^{2n}$. Namely, for every $fid \in Range(G_\Delta)$, $Q(fid) \in \{0,1\}^{2n}$, for every $fid \neq fid' \in Range(G_\Delta)$, $Q(fid) \neq Q(fid')$ and for every $fid \notin Range(G_\Delta)$, $Q(fid) = \bot$.
- Q^{-1} is the inversion algorithm of Q. Namely, for every $fid \in Range(G_\Delta)$, $Q^{-1}(Q(fid)) = fid$. for every $y \notin Range(Q)$, $Q^{-1}(y) = \bot$.

We denote by $\mathcal{U}_\Delta$ the uniform distribution over Δ-oracles. Namely, the oracle $O_\Delta = (G_\Delta, P, P^{-1}, Q, Q^{-1})$ is distributed according to $\mathcal{U}_\Delta$, if G_Δ is a uniformly distributed injective function from $\{0,1\}^n$ to $\{0,1\}^{2n}$, for every $fid \in Range(G_\Delta)$, $P(fid, \cdot)$ is a uniformly distributed permutation over $\{0,1\}^n$ and Q is a uniformly distributed injective function from the range of G_Δ to $\{0,1\}^{2n}$.

Note the difference between Γ-oracles and Δ-oracles. Γ-oracle have an asymmetric nature: F and its inversion oracle F^{-1} use different keys. On the contrary, Δ-oracles have a symmetric nature: identical keys are used by P and its inversion oracle P^{-1}. (For this reason, we used a "symmetric" character Δ for Δ-oracles and an "asymmetric" character Γ for Γ-oracles.)

Γ-oracles versus Δ-oracles. We now show a bijection ϕ that maps every Γ-oracle to a corresponding Δ-oracle. Let $O_\Gamma = (G_\Gamma, F, F^{-1})$ be a Γ-oracle. $\phi(O_\Gamma)$ is the tuple of functions $(G_\Delta, P, P^{-1}, Q, Q^{-1})$ satisfying:

- For every $r \in \{0,1\}^n$, it holds that $G_\Delta(r) = G_\Gamma^1(r)$.
- For every $r \in \{0,1\}^n$, $Q(G_\Delta(r)) = G_\Gamma^2(r)$, and for every $fid \notin Range(G_\Delta)$, $Q(fid) = \bot$.
- For every $fid \in \{0,1\}^{2n}$ and $x \in \{0,1\}^n$, it holds that $P(fid, x) = F(fid, x)$.
- P^{-1} and Q^{-1} are the inversion algorithms of P and Q.

Claim 1. *ϕ is a bijection from the set of Γ-oracles to the set of Δ-oracles.*

The above claim is proven in the full version of this paper and immediately implies the following:

Corollary 2. *The random variables* $\mathcal{U}_\Delta$ *and* $\phi(\mathcal{U}_\Gamma)$ *are identically distributed.*

Enhanced trapdoor permutations relative to Γ-oracles. It is not difficult to show there exist enhanced trapdoor permutations, as defined in [14], relative to random Γ-oracles. Indeed, it can be shown that there exist enhanced trapdoor permutations relative to measure 1 of the Γ-oracles. This is shown in the full version. We remark also that semi-honest oblivious transfer with static corruptions can be constructed from any enhanced trapdoor permutation [10] and thus exists relative to measure 1 of the Γ-oracles.

3.2   Static OT_1^2 Relative to Δ-Oracles from Adaptive OT_1^2

In this section we prove that if there exists an adaptively secure OT_1^2 relative to random Γ-oracles, then there exists a statically secure OT_1^2 relative to random Δ-oracles. We actually prove a more general theorem that if there exists a protocol for securely computing a functionality f in the presence of adaptive adversaries relative to a random Γ-oracle, then there exists a protocol for securely computing f in the presence of static adversaries relative to a random Δ-oracle. We restrict our proof to two-party protocols only, but stress that the claim can be proved similarly for multiparty protocols as well.

Let $\Pi_1 = \langle Alice_1, Bob_1 \rangle$ be a protocol for securely computing a functionality f in the presence of *adaptive* adversaries relative a Γ-oracle. We use Π_1 to construct a new protocol $\Pi_2 = \langle Alice_2, Bob_2 \rangle$ for securely computing f in the presence of *static* adversaries relative to a Δ-oracle.

Recall that the parties $Alice_2$ and Bob_2 have access to a Δ-oracle, while in the original protocol, $Alice_1$ and Bob_1 have access to a Γ-oracle. There is a fundamental difference between these two cases because a Γ-oracle is inherently *asymmetric* (it is possible to send a party fid while keeping tid secret, thereby enabling them to compute the permutation but not invert it), while a Δ-oracle is inherently *symmetric* (the same fid is used to compute and invert the permutation). The idea behind our proof is to eliminate the asymmetric nature of the Γ-oracle by using the fact that in the adaptive setting (e.g., in the post-execution corruption phase), the distinguisher can ultimately corrupt all parties. If it does so, it obtains the entire view of all parties and in particular the view of any party who samples a permutation using G_Γ. The critical observation is that the probability of a party finding an fid in the range of G_Γ without explicitly querying it is negligible. However, if it does make such a query, then its view contains *both* fid and tid and this will be obtained by the distinguisher upon corrupting the parties. Thus, the distinguisher is able to compute and invert the permutation, just like in the case of a Δ-oracle. The fact that the adaptive simulator must simulate well even when the distinguisher works in this way (learning all fid, tid pairs) is the basis for constructing a simulator for the static case when using a Δ-oracle.

We begin by defining $\Pi_2 = \langle Alice_2, Bob_2 \rangle$ which is constructed from Π_1 by replacing the Γ-oracle with a Δ-oracle:

Protocol Π_2: *On input x_A, $Alice_2$ invokes $Alice_1$ on x_A. On input x_B, Bob_2 invokes Bob_1 on x_B. The execution is described below for a party $\mathcal{P}_2$ emulating $\mathcal{P}_1$, and is the same for both $Alice_2$ and Bob_2. In each round:*

- *When $\mathcal{P}_2$ gets the message sent by the other party in the previous round, it hands it to $\mathcal{P}_1$.*
- *If $\mathcal{P}_1$ makes a query r to the oracle G_Γ, $\mathcal{P}_2$ first queries $G_\Delta(r)$ and gets an output fid. Then, $\mathcal{P}_2$ queries $Q(fid)$ and gets an output tid. $\mathcal{P}_2$ hands the pair (fid, tid) to $\mathcal{P}_1$.*
- *If $\mathcal{P}_1$ makes a query (fid, x) to F, $\mathcal{P}_2$ queries $P(fid, x)$, receives an output y and hands y to $\mathcal{P}_1$ (note that y may equal $\bot$).*
- *If $\mathcal{P}_1$ makes a query (tid, y) to F^{-1}, $\mathcal{P}_2$ first queries its oracle Q^{-1} on tid and receives an output fid. If the outputs is $\bot$, it hands $\bot$ to $\mathcal{P}_1$. Otherwise, $\mathcal{P}_2$ queries $P^{-1}(fid, y)$, obtains an output x and hands x to $\mathcal{P}_1$.*
- *If $\mathcal{P}_1$ writes a string m on its outgoing communication tape, $\mathcal{P}_2$ sends m to the other party.*
- *At the end of the simulation, $\mathcal{P}_2$ outputs the output of $\mathcal{P}_1$.*

We now prove that Π_2 securely computes the functionality f in the presence of semi-honest static adversaries.

Theorem 3. *If Π_1 securely computes the functionality f in the presence of adaptive adversaries relative to a random Γ-oracle O_Γ, then Π_2 securely computes f in the presence of static semi-honest adversaries relative to the Δ-oracle $\phi(O_\Gamma)$.*

Proof Sketch: The intuition has already been described above and we therefore proceed directly to the proof. Let O_Γ be an oracle that is distributed according to $\mathcal{U}_\Gamma$. We show that if Π_1 is a secure adaptive protocol for computing f relative to O_Γ, then Π_2 is a secure static semi-honest protocol for computing f relative to $O_\Delta = \phi(O_\Gamma)$. It is easy to see that Π_2 computes f relative to O_Δ because an execution of Π_2 is, in fact, an execution of Π_1 with a simulated Γ-oracle which is exactly $\phi^{-1}(O_\Delta) = O_\Gamma$.

Next, we show that Π_2 is a statically secure protocol relative to O_Δ. We use the ideal-process simulator $\mathcal{SIM}$ of Π_1 for the adaptive setting to construct two probabilistic polynomial-time simulators $\mathcal{S}_{Alice_2}$ and $\mathcal{S}_{Bob_2}$ for Π_2 in the static setting. Due to space restrictions, we present below only $\mathcal{S}_{Bob_2}$ (the simulator $\mathcal{S}_{Alice_2}$ is almost identical). Let $\mathcal{A}$ and $\mathcal{Z}$ be the following adversary strategy and environment: $\mathcal{Z}$ starts with an input $z \in \{0, 1\}$. At the onset of the run of Π_1, $\mathcal{A}$ corrupts Bob_1 and at the end of the computation outputs the entire view of Bob_1. In the postexecution phase, if $z = 0$, no corruptions are made and if $z = 1$, $\mathcal{Z}$ creates a "corrupt $Alice_1$" message, hands it to $\mathcal{A}$ who corrupts Alice. Eventually $\mathcal{Z}$ outputs the entire view of the corrupted parties (that is: if $z = 0$, the view of Bob alone and if $z = 1$, the view of both parties). No auxiliary information is sent by $\mathcal{Z}$ to $\mathcal{A}$. Let $\mathcal{SIM}$ be the ideal-process adversary guaranteed to exist for $\mathcal{A}$ and $\mathcal{Z}$ by the security of Π_1. We now use $\mathcal{A}$, $\mathcal{Z}$ and $\mathcal{SIM}$ to define $\mathcal{S}_{Bob_2}$ (the static simulator for the case that Bob is corrupted). $\mathcal{S}_{Bob_2}$ receives the input

x_B and output y_B of Bob as defined by the functionality f and emulates the run of $\mathcal{SIM}$ in the adaptive ideal model with environment $\mathcal{Z}$ with input $z = 0$. Note that $\mathcal{SIM}$ must corrupt only Bob, because in the real world only Bob is corrupted when $z = 0$. We also can assume, w.l.o.g. that $\mathcal{SIM}$ corrupts Bob in the first corruption phase.

$\mathcal{S}_{Bob_2}$ receives input (x_B, y_B) and works as follows, simulating a Γ-oracle for $\mathcal{SIM}$ using its Δ-oracle:

- If $\mathcal{SIM}$ makes a query r to the oracle G_Γ, $\mathcal{S}_{Bob_2}$ queries its oracle $G_\Delta(r)$ and receives an output fid. It then queries Q on fid, gets an output tid and hands the pair (fid, tid) to $\mathcal{SIM}$.
- If $\mathcal{SIM}$ makes a query (fid, x) to F, $\mathcal{S}_{Bob_2}$ queries it oracle $P(fid, x)$, gets an output y and hands it to $\mathcal{SIM}$.
- If $\mathcal{SIM}$ makes a query (tid, y) to F^{-1}, $\mathcal{S}_{Bob_2}$ first queries its oracle Q^{-1} on tid, gets an output fid. If the outputs is $\perp$, it hands $\perp$ to $\mathcal{SIM}$. Otherwise, $\mathcal{S}_{Bob_2}$ queries $P^{-1}(fid, y)$, gets an output x and hands x to $\mathcal{SIM}$.
- When $\mathcal{SIM}$ decides to corrupt Bob_1, $\mathcal{S}_{Bob_2}$ plays the role of $\mathcal{Z}$ by sending x_B to $\mathcal{SIM}$.
- In the computation phase, $\mathcal{S}_{Bob_2}$ plays the role of the trusted party and sends y_B to $\mathcal{SIM}$ (recall that $\mathcal{S}_{Bob_2}$ gets y_B as input).
- At the end of the simulation, $\mathcal{S}_{Bob_2}$ outputs the output of $\mathcal{SIM}$.

Informally speaking, we show that a distinguisher D_2 for Π_2 and $\mathcal{S}_{Bob_2}$ (relative to O_Δ) implies the existence of a distinguisher D_1 for Π_1 and $\mathcal{SIM}$ (relative to O_Γ). The idea is to have D_1 simulate the run of D_2 on the view of Bob. However, D_2 has oracle access to a Δ-oracle O_Δ, while D_1 has oracle access to a Γ-oracle O_Γ. This might be problematic for example if D_2 wishes to compute $P^{-1}(fid, y)$ but D_1 doesn't know the corresponding tid (recall that D_1 can only invert y in the Γ-oracle world if it holds the trapdoor tid whereas D_2 can invert y given fid only). Despite the above, we use the fact that the range of G_Γ is a negligible fraction of $\{0,1\}^{2n} \times \{0,1\}^{2n}$, and therefore any fid used in the protocol (except with negligible probability) must have been generated via a query to G_Γ, as described in the intuition above. More specifically, we show that if there exists a distinguisher D_2 that distinguishes the output of $\mathcal{S}_{Bob_2}$ from the output of a corrupted Bob_2 in a real execution of Π_2, then there exists a distinguisher D_1 that distinguishes the result of an ideal execution with $\mathcal{SIM}$ from a real execution of Π_1 with adversary $\mathcal{A}$ and environment $\mathcal{Z}$ *with input* $z = 1$, meaning that Alice is also corrupted at the end. (Note that we set $z = 0$ in order to define $\mathcal{S}_{Bob_2}$, but now set $z = 1$ to construct the distinguisher. Since $\mathcal{SIM}$ has to work for all inputs z to $\mathcal{Z}$, this suffices.) Since both Alice and Bob are corrupted in this execution, D_1 obtains all of the (fid, tid) pairs generated by queries to G_Γ and so it can invert always, enabling it to run D_2 and use its Γ-oracle to answer all of D_2's Δ queries.

Formally, the distinguisher D_1 begins by initializing a table T_Q that will hold all pairs (fid, tid) generated by queries to the oracle. D_1 invokes $\langle Alice_1^{O_\Gamma}, Bob_1^{O_\Gamma} \rangle$ on the appropriate input and random tapes (recall that they are a part of D_1's input) and for every access of one of the parties to G_Γ, namely a query $G_\Gamma(r) = (fid, tid)$, D_1 records the entry (fid, tid) in T_Q. D_1 starts simulating D_2 on the view of Bob and proceeds as follows:

- If D_2 makes a query $G_\Delta(r)$, D_1 makes a query $G_\Gamma(r)$, gets a pair (fid, tid), records the entry (fid, tid) in T_Q and hands fid to D_2.
- If D_2 tries to compute $Q(fid)$, D_1 looks for an entry (fid, tid) in T_Q. If such an entry exists, it hands tid to D_2 and continues. Else, it hands $\perp$ to D_2.
- If D_2 tries to compute $Q^{-1}(tid)$, D_1 looks for (fid, tid) in T_Q. If such an entry exists, it hands fid to D_2 and continues. Otherwise, it hands $\perp$ to D_2.
- If D_2 tries to compute $P(fid, x)$, D_1 queries its oracle $F(fid, x)$ and returns its answer.
- If D_2 tries to compute $P^{-1}(fid, y)$, D_1 checks whether an entry (fid, tid) exists in T_Q. If not, it returns $\perp$. If yes, it queries $F^{-1}(tid, y)$ and returns its answer.

There are two cases: If the simulated D_2 does not make a query on an fid (or its corresponding tid) in the range of G_Δ that does not appear in T_Q, then the run of D_1 with O_Γ is identical to a run of D_2 with O_Δ and therefore D_1 outputs the same as D_2. On the other hand, if the simulated D_2 does make such a query, then the output of D_1 might be different than that of D_1 (since, for such queries D_1 replies by $\perp$, while the real O_Δ's reply is different). However, D_2 can find such an fid (or tid) with only negligible probability and therefore if D_2 is a distinguisher for Π_2, D_1 is a distinguisher for Π_1. ∎

Remark 4. *Theorem 3 is true only for* random *Γ-oracles. Specifically, if O_Γ is not a random Γ-oracle, then the claim that finding an fid in the range of G_Γ without making a query to it can happen only with negligible probability does not necessarily hold, and therefore the theorem is not necessarily true for an arbitrary Γ-oracle.*

Needless to say, Theorem 3 holds for oblivious transfer as a special case.

3.3 No Static OT_1^2 Relative to Δ Oracles

For the next step of our proof, we show that static OT_1^2 does not exist relative to most Δ oracles. In order to do this, we show that key agreement does not exist relative to most Δ oracles, and then derive the result from the fact that secure OT_1^2 implies key agreement. In order to show that key agreement does not exist relative to most Δ oracles, we show that a Δ-oracle can be replaced with a "plain random oracle", with at most a negligible difference. Thus, the results of [17] for key agreement relative to a plain random oracle hold also relative to a Δ oracle. We begin by formally defining a random oracle type, denoted ρ, and show its relationship to Δ-oracles.

ρ-oracles. We define a ρ-oracle to be an oracle with the following functions:

- G_ρ is an injective function from $\{0, 1\}^n$ to $\{0, 1\}^{2n}$.

- G_{TEST} is a function that returns a string in $\{0,1\}^n$ on inputs in the range of $G_\rho(\cdot)$. For any other input, it returns $\perp$. Note that G_{TEST} is in fact a tool for examining whether a string of size $2n$ is in the range of G_ρ or not.[3]
- F_P is a function that on a triple $(I, k, x) \in \{0, \ldots, 5\} \times \{0,1\}^{2n} \times \{0,1\}^{\frac{n}{2}}$ returns a string $y \in \{0,1\}^{\frac{n}{2}}$. Note that for a given I and $k \in \{0,1\}^{2n}$, $F_P(I, k, \cdot)$ is a function from $\{0,1\}^{\frac{n}{2}}$ to $\{0,1\}^{\frac{n}{2}}$.
- F_Q is a function that on a pair $(I, x) \in \{0, \ldots, 5\} \times \{0,1\}^n$ returns a string $y \in \{0,1\}^n$. Thus, for a given I, $F_Q(I, \cdot)$ is a function from $\{0,1\}^n$ to $\{0,1\}^n$.

Note that the output of G_ρ is an fid – or symmetric key k – of length $2n$ which defines 6 random functions $F_P(0, k, \cdot), \ldots, F_P(5, k, \cdot)$ which are then used to simulate the P permutation of a Δ-oracle, using Luby-Rackoff. Likewise, the index I in F_Q is used for deriving 6 different function for Luby-Rackoff (there is no "secret key" k for F_Q because it is used for simulating the Q permutation in a Δ oracle which is not keyed).

We denote by $\mathcal{U}_\rho$ the uniform distribution on ρ-oracles. Namely, we say that a ρ-oracle $O_\rho = (G_\rho, G_{TEST}, F_P, F_Q)$ is distributed according to $\mathcal{U}_\rho$ if G_ρ is a uniformly distributed injective function from $\{0,1\}^n$ to $\{0,1\}^{2n}$, G_{TEST} is a uniformly distributed function from the range of G_ρ to $\{0,1\}^n$ (and for inputs not in the range of G_ρ, it returns $\perp$), F_P is a uniformly distributed function from $\{0, \ldots, 5\} \times \{0,1\}^{2n} \times \{0,1\}^{\frac{n}{2}}$ to $\{0,1\}^{\frac{n}{2}}$ and F_Q is a uniformly distributed function from $(I, x) \in \{0, \ldots, 5\} \times \{0,1\}^n$ to $\{0,1\}^n$. We sometimes use the phrase "O_ρ is a random ρ-oracle" as an abbreviation for "O_ρ is distributed according to $\mathcal{U}_\rho$".

Δ-oracles versus ρ-oracles. We now use the Luby-Rackoff construction [20] to replace a random Δ-oracle with a random ρ-oracle. We stress that unlike Corollary 2, the distributions are only computationally indistinguishable.

Definition 5 (Feistel Permutation). *Let* $f : \{0,1\}^l \to \{0,1\}^l$ *be a function and let* $x_1, x_2 \in \{0,1\}^l$. *DES_f is the permutation defined by* $DES_f(x_1, x_2) \stackrel{\text{def}}{=} (x_2, x_1 \oplus f(x_2))$. *$DES_{f_1,\ldots,f_k}$ is the permutation defined by* $DES_{f_1,\ldots,f_k}(x_1, x_2) \stackrel{\text{def}}{=} DES_{f_2,\ldots,x_k}(DES_{f_1}(x_1, x_2))$.

Note that inverting a Feistel permutation is no harder than computing it, as $DES_f^{-1}(y_1, y_2) = (y_2 \oplus f(y_1), y_1)$. Intuitively, a Feistel permutation upon a random ρ-oracle can be used in order to obtain an oracle that behaves like a Δ-oracle. Formally, for a given ρ-oracle $O_\rho = (G_\rho, G_{TEST}, F_P, F_Q)$, an $fid \in \{0,1\}^{2n}$ and $x_1, x_2 \in \{0,1\}^{\frac{n}{2}}$, we define six functions: $f_0 = F_P(0, fid, \cdot)$, $f_1 = F_P(1, fid, \cdot)$, $f_2 = F_P(2, fid, \cdot)$, $f_3 = F_P(3, fid, \cdot)$, $f_4 = F_P(4, fid, \cdot)$ and $f_5 = F_P(5, fid, \cdot)$. Then, the permutation $PDES$ relative to a given oracle O_ρ, that simulates the P permutation in the Δ-oracle, is defined by

$$PDES_{O_\rho, fid}(x_1, x_2) \stackrel{\text{def}}{=} DES_{f_0,\ldots,f_5}(x_1, x_2)$$

---

[3] It was shown in [12] that the black-box separation of [17] holds when G_{TEST} is added to the oracle defined in [17].

Note that $PDES_{O_\rho, fid}$ is a permutation over $\{0,1\}^n$ (similar to $P(fid, \cdot)$ in a Δ-oracle). Let $PDES^{-1}_{O_\rho, fid}$ be the inverse permutation. Similarly, for a given ρ-oracle $O_\rho = (G_\rho, G_{TEST}, F_P, F_Q)$ and for $x_1, x_2 \in \{0,1\}^n$ we define $g_0 = F_Q(0, \cdot), \ldots, g_5 = F_Q(5, \cdot)$. (Recall that Q oracle queries in a Δ-oracle are not keyed and thus when simulated using F_Q in a ρ-oracle, no key is used.) We define:

$$QDES_{O_\rho}(x_1, x_2) \overset{\text{def}}{=} DES_{g_0, \ldots, g_5}(x_1, x_2)$$

As above, $QDES_{O_\rho}$ is a permutation over $\{0,1\}^{2n}$ (similar to Q in a Δ-oracle). Let $QDES^{-1}_{O_\rho}$ be the inverse permutation.

We define a mapping ψ from ρ to Δ oracles. Let $O_\rho = (G_\rho, G_{TEST}, F_P, F_Q)$ be a ρ-oracle. Then $\psi(O_\rho) = (G_\Delta, P, P^{-1}, Q, Q^{-1})$ is the following Δ-oracle:

- For every $r \in \{0,1\}^n$, $G_\Delta(r) = G_\rho(r)$
- For every $fid \in Range(G_\Delta)$ and all $x \in \{0,1\}^n$, $P(fid, x) = PDES_{O_\rho, fid}(x)$
- For every $fid \notin Range(G_\Delta)$ and for every $x \in \{0,1\}^n$, $P(fid, x) = \perp$
- For every $fid \in Range(G_\Delta)$, $Q(fid) = QDES_{O_\rho}(fid)$
- For every $fid \notin Range(G_\Delta)$, $Q(fid) = \perp$
- P^{-1} and Q^{-1} are the inverse functions of P and Q

We denote by $\psi(\mathcal{U}_\rho)$ the distribution where a random ρ-oracle is chosen and then ψ is applied to it. The following claim states that access to a random Δ-oracle O_Δ is essentially the same as access to a Δ-oracle $\psi(O_\rho)$, when O_ρ is random.

Theorem 6 ([9]). *There exists a simulator S and a negligible function μ, such that for every machine D with unbounded running time which makes a polynomial number of queries,*

$$\left| \Pr\left[D^{\mathcal{U}_\rho, \psi(\mathcal{U}_\rho)}(1^n) = 1 \right] - \Pr\left[D^{S^{\mathcal{U}_\Delta}, \mathcal{U}_\Delta}(1^n) = 1 \right] \right| < \mu(n)$$

We remark that [9] refer to a plain random oracle and a plain random permutation, without the additional fid generating and other functions. However, $G_\rho = G_\Delta$ by definition, and so clearly G_ρ can be simulated given G_Δ. Likewise, G_{TEST} can be simulated using P (because the latter returns $\perp$ if the fid is not in the range). We use Theorem 6 in order to prove the following theorem:

Theorem 7. *If $\mathcal{P} = \mathcal{NP}$, then relative to measure 1 of Δ-oracles, there does not exist any statically secure protocol for computing the OT^2_1 functionality.*

In order to prove Theorem 7, we recall the original black-box separation of key agreement from a random oracle, as proven in [17].

Theorem 8 ([17]). *If $\mathcal{P} = \mathcal{NP}$, then given any key-agreement protocol relative to a random ρ-oracle[4], for every polynomial $poly(\cdot)$, there exists a polynomial time Eve such that Eve finds all intersection queries with probability $1 - \frac{1}{poly(n)}$.*

---

[4] [17] refer to a single random permutation oracle; however, the same proof can be extended to ρ-oracles.

We first show that a similar argument holds relative to Δ-oracles (that is, every key agreement protocol relative to a random Δ-oracle can be broken with probability $1 - \frac{1}{poly(n)}$). Then, using the same methods as in [17], we show that relative to measure 1 of Δ-oracles, any key-agreement can be broken in polynomial time. As described in [12], it is possible to construct a secure key agreement from any static oblivious transfer protocol and it is easy to verify that this construction relativizes. Therefore, we conclude that relative to measure 1 of Δ-oracles, there does not exist any statically secure protocol for computing the OT_1^2 functionality. We begin by proving the following claim:

Proposition 9. *If $\mathcal{P} = \mathcal{NP}$, then given any key-agreement protocol relative to a random Δ-oracle, for every polynomial $poly(\cdot)$, there exists a polynomial time Eve such that Eve finds all intersection queries with probability $1 - \frac{1}{2poly(n)}$.*

Proof Sketch: Let $\langle \mathcal{A}_1, \mathcal{B}_1 \rangle$ be a key-agreement protocol relative to random Δ-oracles. We use $\langle \mathcal{A}_1, \mathcal{B}_1 \rangle$ to construct a key-agreement protocol $\langle \mathcal{A}_2, \mathcal{B}_2 \rangle$ relative to random ρ-oracles. Recall that $\mathcal{A}_2$ and $\mathcal{B}_2$ have oracle access to a ρ-oracle while $\mathcal{A}_1$ and $\mathcal{B}_1$ have oracle access to a Δ-oracle. The idea is to use the ρ-oracle in order to simulate a Δ-oracle while replacing queries to P, P^{-1}, Q and Q^{-1} by appropriate Feistel permutations obtained from F_P and F_Q.

Let $\langle \mathcal{A}_2, \mathcal{B}_2 \rangle$ be the following protocol:

Protocol 1. *On input 1^n, $\mathcal{A}_2$ invokes $\mathcal{A}_1$ on 1^n and $\mathcal{B}_2$ invokes $\mathcal{B}_1$ on 1^n. The execution is described below for a party $\mathcal{P}_2$ emulating $\mathcal{P}_1$, and is the same for both $\mathcal{A}_2$ and $\mathcal{B}_2$. In each round:*

- *When $\mathcal{P}_2$ gets the message sent by the other party in the previous round, it sends it to $\mathcal{P}_1$.*
- *If $\mathcal{P}_1$ makes a query r to oracle G_Δ, $\mathcal{P}_1$ queries it oracle $G_\rho(r)$, and hands the output to $\mathcal{P}_1$.*
- *If $\mathcal{P}_1$ makes a query $P(fid, x)$, $\mathcal{P}_1$ queries its oracle G_{TEST} on fid (recall that $G_{TEST}(fid)$ returns $\perp$ if and only if fid is not in the range of G_ρ). If the oracle returns $\perp$, $\mathcal{P}_1$ returns $\perp$ as well. Otherwise, uses its oracle F_P to compute $y = PDES_{O_\rho, fid}(x)$ and hands y to $\mathcal{P}_1$.*
- *If $\mathcal{P}_1$ makes a query $P^{-1}(fid, y)$, $\mathcal{P}_1$ queries G_{TEST} on fid. If it returns $\perp$, $\mathcal{P}_1$ returns $\perp$ as well. Otherwise, $\mathcal{P}_1$ uses its oracle F_P to compute $x = PDES_{O_\rho, fid}^{-1}(y)$ and hands x to $\mathcal{P}_1$.*
- *If $\mathcal{P}_1$ makes a query $Q(fid)$, $\mathcal{P}_1$ queries G_{TEST} on fid. If it returns $\perp$, $\mathcal{P}_1$ returns $\perp$ as well. Otherwise, it uses its oracle F_Q to compute $tid = QDES_{O_\rho}(fid)$ and hands tid to $\mathcal{P}_1$.*
- *If $\mathcal{P}_1$ makes a query $Q^{-1}(tid)$, $\mathcal{P}_1$ uses its oracle F_Q to compute $fid = QDES^{-1}(tid)$ and queries $G_{TEST}(fid)$. If it returns $\perp$, $\mathcal{P}_1$ returns $\perp$ as well. Otherwise, $\mathcal{P}_1$ hands fid to $\mathcal{P}_1$.*
- *If $\mathcal{P}_1$ writes a string m on its outgoing communication tape, $\mathcal{P}_1$ sends m to the other party.*
- *At the end of the protocol, $\mathcal{P}_1$ outputs the output of $\mathcal{P}_1$.*

Now, assume $\mathcal{P} = \mathcal{NP}$. Let $poly(\cdot)$ be some polynomial and let Eve_2 be as in Theorem 8. We use Eve_2 to construct an adversary Eve_1 for $\langle \mathcal{A}_1, \mathcal{B}_1 \rangle$. Eve_1 simply invokes Eve_2 and simulates the ρ-oracle using the simulator $\mathcal{S}$ guaranteed to exist by Theorem 6. Note that if Eve_1 outputs a list of intersection queries with probability less than $1 - \frac{1}{2poly(n)}$, then it is possible to distinguish oracles $\mathcal{U}_\rho, \psi(\mathcal{U}_\rho)$ from $\mathcal{S}^{\mathcal{U}_\Delta}, \mathcal{U}_\Delta$ with non-negligible probability. Specifically, given a pair of oracles $(\mathcal{O}_1, \mathcal{O}_2)$ that are distributed according to $\mathcal{U}_\rho, \psi(\mathcal{U}_\rho)$ or $\mathcal{S}^{\mathcal{U}_\Delta}, \mathcal{U}_\Delta$, distinguisher D first invokes a run of $\langle \mathcal{A}_1^{\mathcal{O}_2}, \mathcal{B}_1^{\mathcal{O}_2} \rangle$ and then invokes $Eve_2^{\mathcal{O}_1}$ on the transcript. D outputs 1 if and only if Eve_2 outputs all intersection queries. Now, if $(\mathcal{O}_1, \mathcal{O}_2)$ are distributed according to $\mathcal{U}_\rho, \psi(\mathcal{U}_\rho)$ then Eve_2 outputs all intersection queries with probability at least $1 - \frac{1}{poly(n)}$, and if $(\mathcal{O}_1, \mathcal{O}_2)$ are distributed according to $\mathcal{S}^{\mathcal{U}_\Delta}, U_\Delta$ then Eve_2 outputs all intersection queries with probability less than $1 - \frac{1}{2poly(n)}$. Thus D distinguishes with non-negligible probability. ∎

Remark 10. *Theorem 6 holds even when $\mathcal{P} = \mathcal{NP}$ since the running time of D is unbounded.*

The following corollary can be proved using the same methods as in [17] (the only difference between it and what was proven in [17] is the type of oracle used):

Corollary 11. *If $\mathcal{P} = \mathcal{NP}$, then for measure 1 of Δ-oracles, any key-agreement protocol can be broken in polynomial time.*

Recalling that the existence of a secure OT_1^2 relative to an oracle $\mathcal{O}$ implies the existence of a secure key agreement relative to $\mathcal{O}$, we obtain:

Corollary 12. *If $\mathcal{P} = \mathcal{NP}$, then for measure 1 of Δ-oracles, there does not exist any statically secure protocol for computing the OT_1^2 functionality.*

3.4   Concluding the Proof

Theorem 3 states that if there exists an adaptively secure protocol for OT_1^2 relative to a given Γ oracle $\mathcal{O}$, then there exists a statically secure protocol for OT_1^2 relative to the oracle $\phi(\mathcal{O})$. Now, by Theorem 7, for measure 1 of Δ oracles, there exists no statically secure OT_1^2. Using the fact that ϕ is a bijection (Claim 1), we conclude that for measure 1 of Γ oracles, there exists no adaptively secure OT_1^2. That is, we have the following:

Theorem 13. *If $\mathcal{P} = \mathcal{NP}$, then for measure 1 of Γ-oracles, there does not exist any adaptively secure protocol for computing the OT_1^2 functionality.*

Similarly to [17], we derive an oracle separation of enhanced trapdoor permutations form adaptively secure OT_1^2 (even for semi-honest adversaries):

Corollary 14. *There exists an oracle relative to which enhanced trapdoor permutations exist, but not adaptively secure OT_1^2.*

Proof: Let $\mathcal{O}$ be a $\mathcal{PSPACE}$-complete oracle combined with a random Γ-oracle. Enhanced trapdoor permutations exist relative to $\mathcal{O}$ whereas adaptively secure OT_1^2 does not, as we have shown. ∎

Acknowledgements. We thank Omer Reingold for helpful discussions.

References

1. Beaver, D.: Adaptive Zero Knowledge and Computational Equivocation. In: 28th STOC, pp. 629–638 (1996)
2. Beaver, D.: Adaptively Secure Oblivious Transfer. In: Ohta, K., Pei, D. (eds.) ASIACRYPT 1998. LNCS, vol. 1514, pp. 300–314. Springer, Heidelberg (1998)
3. Bellare, M., Micali, S., Ostrovsky, R.: Perfect Zero-Knowledge in Constant Rounds. In: 22nd STOC, pp. 482–493 (1990)
4. Blum, M.: Coin Flipping by Phone. In: IEEE Spring COMPCOM, pp. 133–137 (1982)
5. Blum, M.: How to Prove a Theorem So No One Else Can Claim It. In: Proceedings of the International Congress of Mathematicians, USA, pp. 1444–1451
6. Canetti, R.: Security and Composition of Multiparty Cryptographic Protocols. Journal of Cryptology 13(1), 143–202 (2000)
7. Canetti, R., Fischlin, M.: Universally Composable Commitments. In: Kilian, J. (ed.) CRYPTO 2001. LNCS, vol. 2139, pp. 19–40. Springer, Heidelberg (2001)
8. Canetti, R., Lindell, Y., Ostrovsky, R., Sahai, A.: Universally Composable Two-Party and Multi-Party Computation. In: 34th STOC, pp. 494–503 (2002), http://eprint.iacr.org/2002/140
9. Coron, J.S., Patarin, J., Seurin, Y.: The Random Oracle Model and the Ideal Cipher Model are Equivalent. In: Wagner, D. (ed.) CRYPTO 2008. LNCS, vol. 5157, pp. 1–20. Springer, Heidelberg (2008)
10. Even, S., Goldreich, O., Lempel, A.: A Randomized Protocol for Signing Contracts. Communications of the ACM 28(6), 637–647 (1985)
11. Feige, U., Shamir, A.: Zero Knowledge Proofs of Knowledge in Two Rounds. In: Brassard, G. (ed.) CRYPTO 1989. LNCS, vol. 435, pp. 526–544. Springer, Heidelberg (1990)
12. Gertner, Y., Kannan, S., Malkin, T., Reingold, O., Viswanathan, M.: The Relationship Between Public Key Encryption and Oblivious Transfer. In: 41st FOCS, pp. 325–335 (2000)
13. Gertner, Y., Malkin, T., Reingold, O.: On the Impossibility of Basing Trapdoor Functions on Trapdoor Predicates. In: The 42nd FOCS, pp. 126–135 (2001)
14. Goldreich, O.: Foundations of Cryptography: Basic Applications, vol. 2. Cambridge University Press, Cambridge (2004)
15. Goldreich, O., Micali, S., Wigderson, A.: Proofs that Yield Nothing but their Validity or All Languages in NP Have Zero-Knowledge Proof Systems. Journal of the ACM 38(1), 691–729 (1991)
16. Impagliazzo, R., Luby, M.: One-way Functions are Essential for Complexity Based Cryptography. In: The 30th FOCS, pp. 230–235 (1989)
17. Impagliazzo, R., Rudich, S.: Limits on the Provable Consequences of One-way Permutations. In: 21st STOC, pp. 44–61 (1989)
18. Itoh, T., Ohta, Y., Shizuya, H.: A Language-Dependent Cryptographic Primitive. Journal of Cryptology 10(1), 37–49 (1997)
19. Kim, J.H., Simon, D.R., Tetali, P.: Limits on the Efficiency of One-Way Permutation-Based Hash Functions. In: The 40th FOCS, pp. 535–542 (1999)
20. Luby, M., Rackoff, C.: How to Construct Pseudorandom Permutations from Pseudorandom Functions. SIAM Journal on Computing 17(2), 373–386 (1988)

21. Micciancio, D., Vadhan, S.: Statistical Zero-Knowledge Proofs with Efficient Provers: Lattice Problems and More. In: Boneh, D. (ed.) CRYPTO 2003. LNCS, vol. 2729, pp. 282–298. Springer, Heidelberg (2003)

22. Micciancio, D., Ong, S.J., Sahai, A., Vadhan, S.: Concurrent Zero Knowledge without Complexity Assumptions. In: Halevi, S., Rabin, T. (eds.) TCC 2006. LNCS, vol. 3876, pp. 1–20. Springer, Heidelberg (2006)

23. Naor, M.: Bit Commitment Using Pseudorandomness. Journal of Cryptology 4(2), 151–158 (1991)

24. Reingold, O., Trevisan, L., Vadhan, S.P.: Notions of Reducibility between Cryptographic Primitives. In: Naor, M. (ed.) TCC 2004. LNCS, vol. 2951, pp. 1–20. Springer, Heidelberg (2004)

25. Simon, D.R.: Finding Collisions on a One-Way Street: Can Secure Hash Functions Be Based on General Assumptions? In: Nyberg, K. (ed.) EUROCRYPT 1998. LNCS, vol. 1403, pp. 334–345. Springer, Heidelberg (1998)

26. Vadhan, S.P.: An Unconditional Study of Computational Zero Knowledge. In: The 45th FOCS, pp. 176–185 (2004)

On the (Im)Possibility of Key Dependent Encryption*

Iftach Haitner[1],[**] and Thomas Holenstein[2],[* * *]

[1] Microsoft Research, New England Campus
iftach@microsoft.com
[2] Department of Computer Science, Princeton University
tholenst@princeton.edu

Abstract. We study the possibility of constructing encryption schemes secure under messages that are chosen depending on the key k of the encryption scheme itself. We give the following separation results that hold both in the private and in the public key settings:

- Let $\mathcal{H}$ be the family of $\mathrm{poly}(n)$-wise independent hash-functions. There exists no fully-black-box reduction from an encryption scheme secure against key-dependent messages to one-way permutations (and also to families of trapdoor permutations) if the adversary can obtain encryptions of $h(k)$ for $h \in \mathcal{H}$.
- There exists no reduction from an encryption scheme secure against key-dependent messages to, essentially, *any* cryptographic assumption, if the adversary can obtain an encryption of $g(k)$ for an *arbitrary* g, as long as the reduction's proof of security treats both the adversary and the function g as black boxes.

Keywords: Key-dependent input, Black-box separations, One-way functions.

1 Introduction

A cryptographic primitive is *key-dependent input secure*, or KDI-secure for short, if it remains secure also in case where the input depends on the secret key. In the case of encryption schemes, KDI-security means that the adversary can obtain, in addition to the usual queries, encryptions of $h(k)$, where k is the key of the scheme and h is chosen by the adversary from some (hopefully large) family of functions $\mathcal{H}$.

On a first look it might seem that by using the right design it is possible to prevent any "KDI-attacks" on the encryption scheme, and thus achieving such strong security would be only of pure theoretical interest. It turns out, however,

---

* All omitted proofs can be found in [HH08].
** This work was performed while at Weizmann Institute of Science and at Microsoft Research, Silicon Valley Campus.
* * * This work was performed while at Microsoft Research, Silicon Valley Campus.

O. Reingold (Ed.): TCC 2009, LNCS 5444, pp. 202–219, 2009.

that such attacks might "naturally" arise when considering complex systems. For instance, in the BitLocker disk encryption utility (used in Windows Vista), the disk encryption key can end up being stored on the disk and thus encrypted along with the disk contents. For more details on the importance of KDI-security, see [BHHO08] and references within.

In this work we study the possibility of obtaining such an encryption scheme both from one-way functions and from other hardness assumptions. In particular, we exclude different types of reductions from a KDI-secure encryption scheme to different hardness assumption. Intuitively, a *black-box reduction* of a primitive P to a primitive Q is a construction of P out of Q that ignores the internal structure of the implementation of Q and just uses it as a "subroutine" (i.e., as a black box). The reduction is *fully-black-box* (following [RTV04]) if the proof of security (showing that an adversary that breaks the implementation of P implies an adversary that breaks the implementation of Q) is also black-box (i.e., the internal structure of the adversary that breaks the implementation of P is ignored as well).

Our first result shows that there is no fully-black-box reduction from KDI-secure encryption schemes to one-way permutations, even if the KDI-security is only against the relatively small class of poly(n)-wise independent hash-functions. When considering reduction from a KDI-secure encryption scheme, it is natural to ask whether the proof of security accesses the challenge function h in a black-box manner as well. Our second result, however, shows that under this restriction essentially no hardness assumption implies a KDI-secure encryption scheme.

1.1 Related Work

KDI security. The development of encryption secure against key-dependent inputs started by the works of Abadi and Rogaway [AR02]. They studied formal security proofs for cryptographic protocols (as described by [DY83]), and showed that these imply security by a reduction, as long as no *key-cycles* exist in the protocol, i.e., there is a partial order $\preceq$ on the keys exists such that a message depending on k_1 would only be encrypted with k_2 if $k_1 \preceq k_2$. Since this is a restriction (even though it may be a very natural one), the community became aware that it would be desirable to create encryption schemes that provide security even in the existence of such key cycles. Consequently, Black, Rogaway, and Shrimpton [BRS02] define the (possibly stronger) notion of KDI-security for symmetric encryption schemes, and show how to obtain this notion in the random-oracle model. In such a scheme, an adversary can obtain encryptions of $h(k)$ under the key k, where h is given as a circuit to an encryption oracle. Such a scheme implies the security of the scheme under cycles as well. Independently of [BRS02], a notion of circular security has been defined by Camenisch and Lysyanskaya [CL01], considering asymmetric encryption schemes as well.

Recently, Halevi and Krawczyk [HK07] generalized the notion of KDI-security to other cryptographic primitives, such as pseudorandom functions. They also coined the name KDI for this sort of security (previously, it was named *key-dependent message security*). Their results in this setting are mainly for the construction of pseudorandom functions. In addition, [HK07] shows that a *deterministic* encryption scheme cannot be KDI-secure. Independently and concurrently of [HK07], Hofheinz and Unruh [HU08] provided private-key encryption schemes secure under a limited class of KDI-attacks. The main limitation of their work is that the scheme only remains secure as long as $h(k)$ is significantly shorter than the key; also, after every application of the encryption scheme the key is updated. This makes the construction insufficient for the initial motivation of allowing key cycles in cryptographic protocols. Very recently, Boneh et al. [BHHO08] presented a public-key encryption scheme that is KDI-secure (assuming that the DDH assumption holds) against the family of affine transformations over the messages' domain. Their system remains secure also when key-cycles are allowed.

Black-box impossibility results. Impagliazzo and Rudich [IR89] showed that there is no black-box reduction of key-agrement to one-way permutations and additional work in this line followed (cf., [GKM⁺00, Rud88, Sim98]). Kim, Simon and Tetali [KST99] initiated a new line of impossibility results, by providing a lower bound on the *efficiency* any black-box reduction of universal one-way hash functions to one-way permutations, substantial additional work in this line followed (cf., [GGKT05, HHRS07, HK05, Wee07]). Dodis et al. [DOP05] (and also [Hof08]) give a black-box separation of a similar flavor to the one given in Theorem 2, in the sense that it excludes a large family of hardness assumptions.

1.2   Contributions of This Paper

In this paper we give two impossibility results for security proofs of constructions of KDI-secure private-key encryption schemes. However, since every public-key encryption scheme can be also viewed as a private-key scheme (i.e., both parties use the same private/public key), our impossibility results immediately extend to the public-key case. Our first result is a black-box separation of KDI encryption scheme from one-way permutations and from (even enhanced) family of trapdoor permutations.

Theorem 1. *Let* (Enc, Dec) *be an encryption scheme that is fully-black-box constructed from one-way permutations. Then there exists an efficient family* $\mathcal{H}$ *of* poly(n)*-wise independent hash functions such that the following holds: there exists no black-box reduction from breaking the KDI-security of* (Enc, Dec) *against* $\mathcal{H}$ *to inverting one-way permutations. Furthermore, the above holds also with respect to (enhanced) families of trapdoor permutations.*

For our second result, we assume that the challenge function itself under which the scheme should be KDI-secure is treated by the proof of security as a black box. Moreover, the proof of security does not forward an access to the challenge function to a "third party". We call such a reduction **strongly-black-box**.

Theorem 2 (informal). *There exists no reduction with strongly-black-box proof from the KDI-security of an encryption scheme to the security of "any cryptographic assumption".*

We stress that the construction of the encryption scheme considered in Theorem 2 can be arbitrary. The formal statement of Theorem 2 is given in Section 4.

1.3   Interpretation of Our Results

So what should we think on the possibility of building KDI-secure encryption scheme given the above negative results? Let us start with Theorem 1 and let's first consider the fully-black-box restrictions. We remark that while quite a few black-box impossibility results of these types are known (see Section 1.1), there is not even a single known example where we have an impossibility result of the type given in Theorem 1, and yet a "non-black-box" reduction was found.[1] We also remark that the reductions given in [BHHO08, HK07, HU08] are fully-black-box. The second issue is that we only rule out security against poly-wise independent hash function, where the value for "poly" is determined as a function of the encryption scheme. It seems, however, that in most settings one cannot limit the power of the queries used in the KDI attack (but merely assume that these functions should be efficiently computable). Typically, when designing an encryption scheme, the exact configuration of each of the systems in which the scheme is going to be used is unknown. These configurations, however, determine the challenge functions "used" in the KDI attacks.

In Theorem 2 we consider arbitrary constructions, but require only black-box access to the challenge functions. This additional restriction actually reflects three separate restrictions. The first is that the proof has only input/output access to the challenge function, the second is that the challenge function cannot be assumed to be efficient, and the third is that the reduction "knows" all the queries made to the challenge functions (we force the last restriction by disallowing the reduction to give a third party an handle to the challenge function). While the first two restrictions seem to be a real limitation on the generality of our second result, the third restriction is harmless in most settings. In particular, this is the case where the hardness assumption does not accept (even implicitly) handler to functions. This list includes all the "non-interactive hardness assumptions" such as one-way functions, factoring, DDH etc.[2]

1.4   Our Technique

In the proof of both our results, we are using the same oracle, Breaker, that helps us to break the KDI-security of every encryption scheme. Let (Enc, Dec) be some fixed encryption scheme. On input (h, c), where h is some *length doubling*

---

[1] The superiority of non-black-box techniques was demonstrated by Barak [Bar01] in the settings of zero-knowledge arguments for NP. In these settings, however, the black-box access is to the, possibly cheating, verifier and not to any underlying primitive.

[2] The only exception we could think of for a reduction that benefits from passing the handler to the challenge function to a third party, is a reduction from one KDI-secure encryption scheme B to another KDI-secure encryption scheme A. In such a reduction, the security proof of scheme B typically forwards the challenge function to the security proof of scheme A.

function and c is a ciphertext, Breaker considers all possible keys, and returns the first key k for which $\mathrm{Dec}(k, c) = h(k)$, or $\bot$ if no such key exists. It is not hard to see that (Enc, Dec) is not KDI-secure, with respect to h, in the presence of Breaker. Therefore, our impossibly results follow if Breaker does not help to violate the underlying hardness assumption. For this, we need to assume that Breaker is only called with functions h that are chosen uniformly from the respective set of challenge functions. We ensure this by restricting the functions h for which Breaker performs the above computation to one that is randomly chosen (and then give the adversary access to it). Under this restriction, we manage to prove that Breaker does not help in breaking the assumption. Proving this is our main technical contribution (note that Breaker cannot be implemented efficiently) and we prove it differently in each of our separation results.

One-way permutations. Let π be a random permutation and let $(\mathrm{Dec}^\pi, \mathrm{Enc}^\pi)$ be a candidate of a KDI-secure encryption scheme given π as the one-way permutation. We find a polynomial $p(n)$ (which depends on Dec and Enc) and use a family of length-doubling $p(n)$-wise independent hash-functions as the challenge functions. Imagine now that a call of A to Breaker helps A to invert π. Then, the behavior of Breaker must be very different for a large number of potential preimages of y, as otherwise the call gave roughly no information about the preimage of y. We show, however, that for all but a negligible fraction of the functions h, the behavior of Breaker will be the same for most possible preimages of y, no matter how the ciphertext is chosen.

Arbitrary assumptions. In this we use the family of *all* length-doubling functions as the challenge functions. The idea is that for a random h in the family, all calls to Breaker (done outside of the KDI game) are very likely to be answered with $\bot$. The reason is that for a fixed $k \in \{0, 1\}^t$, the probability that $\mathrm{Dec}(k, c) = h(k)$ is roughly 2^{-2t}. This somewhat naive intuition is actually false, as it can fail in the following way: A picks a key k' itself, queries $h(k')$, encrypt this with k' itself, and gives the resulting ciphertext to Breaker. We prove, however, that this is essentially the only way in which the above intuition fails. Thus, instead of calling Breaker, A can as well check the keys on which h was previously queried, which can be done efficiently. We conclude that if there is a reduction with strongly-black-box proof of security from a KDI encryption scheme to a given hardness assumption, then the hardness assumption is false.

2   Preliminaries

2.1   Notation

We denote the concatenation of two strings x and y by $x \circ y$. If X is a random variable taking values in a finite set $\mathcal{U}$, then we write $x \leftarrow \mathcal{U}$ to indicate that x is selected according to the uniform distribution over $\mathcal{U}$. We often use probabilities where we choose an oracle π from some uncountable set of oracles at random. It is possible to defined these using Lebesgue measure and an appropriate mapping of oracles to $[0, 1)$.

2.2   Many-Wise Independence

We use standard facts on s-independence, as well as the following upper bound on the probability that many s-wise independent events occur, where each event has low probability.[3]. The proof is omitted in this version.

Lemma 1. *For $s, V \in \mathbb{N}$ let $B_1, \ldots, B_V$ be s-wise independent Bernoulli random variables with $\Pr[B_i = 1] \leq \frac{1}{V}$. If $\alpha > s$, then $\Pr\left[\sum_{i=1}^{V} B_j \geq \alpha\right] < \frac{\log(V)}{\alpha^{s-1}}$.*

2.3   Encryption Schemes and KDI Security

We define a (private-key) encryption scheme as a pair of an encryption and a decryption algorithm (Enc, Dec). On security parameter n, the encryption algorithm Enc gets as input a key of length $t(n)$ and a message of length m, and outputs a ciphertext of length $\ell(n, m)$. The decryption algorithm Dec, gets a key and a ciphertext and outputs the message.[4] Informally, an encryption scheme (Enc, Dec) is KDI-secure against a family of functions $\mathcal{H} \subseteq \{h : \{0,1\}^t \to \{0,1\}^*\}$, if no efficient adversary can distinguish between an oracle that correctly returns an encryption of $h(k)$, given input h, and one that returns an encryption of the all zero string of the same length. Note that if $\mathcal{H}$ contains functions that map to constants, plain-text queries can be obtained as well.

Definition 1 (KDI-security). *Given an encryption scheme (Enc, Dec) and a key of length t, let $Q^{\mathrm{Enc},k}$ [resp. $\widetilde{Q}^{\mathrm{Enc},k}$] be an algorithm that gets as input a function $h : \{0,1\}^t \mapsto \{0,1\}^{m(t)}$, and returns $\mathrm{Enc}(k, h(k))$ [resp. $\mathrm{Enc}(k, 0^{m(t)})$] (if the schemes is randomized, it returns a random encryption). We say that (Enc, Dec) is KDI-secure for a class of functions $\mathcal{H}$, if*

$$\left| \Pr_{k \leftarrow \{0,1\}^{t(n)}} [A^{Q^{\mathrm{Enc},k}}(1^n) = 1] - \Pr_{k \leftarrow \{0,1\}^{t(n)}} [A^{\widetilde{Q}^{\mathrm{Enc},k}}(1^n) = 1] \right|$$

is negligible for every efficient algorithm A that only queries functions in $\mathcal{H}$.

2.4   Cryptographic Games

For reductions that treat the family $\mathcal{H}$ of query-functions as black-box, we are able to prove a very strong impossibility result. In this case, we show that essentially no cryptographic assumption is sufficient to guarantee the KDI-security of the scheme. In order to do this, we first define the set of cryptographic assumptions we consider.[5] For the sake of readability, however, we will not try to be as

---

[3] The usual bounds seem not strong enough in our setting, as they focus on the range where the probability of a single event is constant. Here, the probability of a single event decreases as the number of events increases, and we use a different bound.

[4] In some definitions, a private-key encryption scheme also includes a key-generation algorithm "Gen". We omit this since we are not concerned by polynomial factors, and in this case one can simply take the random-coins used by Gen as the private key.

[5] We remark that definitions of similar spirit to the one below were previously used in [DOP05, Hof08].

general as possible here. Yet, as far as we can see our definition still captures all natural hardness assumptions.

Definition 2 (cryptographic games). *A* cryptographic game *is a (possibly inefficient) random system Γ that on security parameter n interacts with an attacker A and may output a special symbol* win. *In case $\Gamma(1^n)$ outputs this symbol in an interaction with $A(1^n)$, we say that $A(1^n) \leftrightarrow \Gamma(1^n)$* wins. *The game is* secure *if $\Pr[A(1^n) \leftrightarrow \Gamma(1^n)$* wins*] is negligible for all PPT A, where the probability is over the randomness of A and Γ.*

Examples: One might define the security of a one-way function f by the following game. On security parameter n, the system Γ selects a random $x \in \{0,1\}^n$ and sends $y = f(x)$ to the adversary. Γ outputs win if A outputs $x' \in f^{-1}(y)$.

To define the DDH hardness assumption one needs a bit more work.[6] On security parameter n, the system Γ expects first a sequence of at least n ones, we denote the actual number received by α. The system Γ then sends A a description of an appropriately chosen group $\langle g \rangle$ of order $\Omega(2^n)$ and the generator g, as well as α randomly chosen triples $(g^{x_i}, g^{y_i}, g^{z_i})$, where $z_i = x_i y_i$ or a uniform random element, each with probability $\frac{1}{2}$. The attacker A wins, if the number of instances where he incorrectly predicts whether z_i was chosen independently of x_i and y_i, is at most $\frac{\alpha}{2} - \alpha^{2/3}$. Using [IJK07, Theorem 1], one can now show that winning the above is equivalent to the DDH assumption, we omit the details in this version.

2.5 Black-Box Reductions

A reduction from a primitive P to a primitive Q consists of showing that if there exists an implementation C of Q, then there exists an implementation M_C of P. This is equivalent to showing that for every adversary that breaks M_C, there exists an adversary that breaks C. Such a reduction is semi-black-box if it ignores the internal structure of Q's implementation, and it is fully-black-box (using the terminology of [RTV04]) if it also has black-box proof of security. That is, the adversary for breaking Q ignores the internal structure of both Q's implementation and of the (alleged) adversary breaking P. The following definition expands the above general discussion for the case of a fully-black-box reduction of a KDI-secure encryption scheme from a one-way permutation.

Definition 3 (fully-black-box reduction). *A fully-black-box reduction of a KDI-secure encryption scheme from a one-way permutation consists of polynomial-time oracle-aided algorithms $(\mathrm{Enc}^{(\cdot)}, \mathrm{Dec}^{(\cdot)})$ and a polynomial-time oracle-aided adversary $A_{\mathrm{OWP}}^{(\cdot)}$, such that the following hold:*

- *If f is a permutation, then $(\mathrm{Enc}^f, \mathrm{Dec}^f)$ is an encryption scheme.*
- *For any (possibly unbounded) A_{KDI} that breaks the KDI-security of the encryption scheme, $A_{\mathrm{OWP}}^{f,A_{\mathrm{KDI}}}$ inverts the permutation with non-negligible probability.*

---

[6] The same argument can be applied for many other assumptions, but we refrain from formalizing this in order no to get bogged down in unrelated details.

When considering reductions from a KDI-secure cryptosystem, it is natural to consider whether the proof of security accesses the challenge functions also as a black box. We say that a proof of KDI-security of a cryptosystem is **strongly-black-box**, if it treats the challenge function also as a black-box.

Definition 4 (strongly-black-box reduction). *A reduction from a KDI-secure encryption scheme to a cryptographic game Γ with strongly-black-box proof of security, consists of polynomial-time oracle-aided algorithms* (Enc, Dec) *and a polynomial-time oracle-aided adversary* $A_\Gamma^{(\cdot)}$ *such that the following holds:*

- (Enc, Dec) *is an encryption scheme.*
- *For any adversary* A_{KDI}^Q *that breaks the KDI-security of* (Enc, Dec), *the oracle-aided adversary* $A_\Gamma^{(A_{KDI})}$ *violates the security of* Γ. *Additionally,* A_Γ *treats the challenge functions provided by* A_{KDI} *as a black box.*

The requirement that A_Γ *treats the challenge function as black-box, means that* A_Γ *can only obtain evaluations of it at arbitrary chosen points and the reduction must work for* every *challenge function (not just efficiently computable ones). In addition,* A_Γ *does not provide* Γ *with a description of the function.[7]*

2.6   Extending KDI-Secure Encryption Schemes

We would like to make sure our impossibility results hold even for encryption schemes that encrypt messages of length one bit. For technical reasons, however, we will actually need to encrypt messages of length $2t$, where t is the key length. We therefore give a straightforward, but slightly tedious transformation that allows us to do that. (In fact, the following transformation does slightly more, in order to make the technical part in Sections 3 and 4 a bit easier.) We omit the proof of it in this version.

Proposition 1. *Let* (Enc, Dec) *be an encryption scheme for single bit messages. Assume* (Enc, Dec) *is KDI-secure for a given set* $\mathcal{H} \subseteq \{\{0,1\}^t \to \{0,1\}\}$, *then there exists an encryption scheme* (Enc$_1$, Dec$_1$) *with the following properties:*

(a) The key length t_1 of (Enc$_1$, Dec$_1$) *equals the security parameter. (b)* (Enc$_1$, Dec$_1$) *is defined for messages of arbitrary length. (c)* (Enc$_1$, Dec$_1$) *is KDI-secure for* $\mathcal{H}_1 := \{h : \{0,1\}^t \to \{0,1\}^* : \forall i, \forall \tau \in \{0,1\}^{t_1-t} : h_{|i}(x,\tau) \in \mathcal{H}\}$, *where $h_{|i}$ is the function that outputs the i'th bit of the output of h.[8] (d)* (Enc$_1$, Dec$_1$) *has perfect correctness. (e)* (Enc$_1$, Dec$_1$) *has deterministic decryption. (f) If* (Enc, Dec) *has a strongly-black-box [resp. black-box] proof of security to a cryptographic game* Γ, *then* (Enc$_1$, Dec$_1$) *has a strongly-black-box [resp. black-box] proof of security to* Γ.

---

[7] Alternatively, given A_Γ's (partial) view, it is possible to (efficiently) list all the queries done to the challenge function during the execution.

[8] Namely, $\mathcal{H}_1$ is the set of functions with the property that *every output bit* is described by a function in $\mathcal{H}$, after some appropriate padding of the input.

3   From One-Way Permutations

In this section we prove Theorem 1, but we only give the proof for the case of one-way permutations. The proof for (enhanced) family of trapdoor permutations follows immediately using standard techniques (cf., [GT00, HHRS07]). Let $(\mathrm{Enc}^{(\cdot)}, \mathrm{Dec}^{(\cdot)})$ be an encryption scheme with oracle access to a one-way permutation. By Proposition 1, we can assume that the encryption scheme is always correct, has a deterministic decryption algorithm, defined on messages of any polynomial length and has a security parameter t equal to it's key length. We let $\ell(t)$ be the length of an encryption of a message of length $2t$. In order to prove Theorem 1, we use the following inefficient algorithm $\mathrm{Breaker}^{f,h}$.

Algorithm 3 $\mathrm{Breaker}^{f,h}$.

Oracles: *A function* $f : \{0,1\}^t \times \{0,1\}^{\ell(t)} \mapsto \{0,1\}^{2t}$ *(defined for every* $t \in \mathbb{N}$*)*
and an infinite sequence of functions $h = \{h_t : \{0,1\}^t \mapsto \{0,1\}^{2t}\}_{t \in \mathbb{N}}$.
Input: *A pair* $(t, c) \in \mathbb{N} \times \{0,1\}^*$.
Operation: *Return the smallest* $k \in \{0,1\}^t$ *such that* $f(k, c) = h_t(k)$, *or* $\perp$ *if*
no such k *exists.*

Let $\Pi = \{\Pi_t\}_{t \in \mathbb{N}}$, where Π_t is the set of all possible permutations over $\{0,1\}^t$, and let $\mathcal{H} = \{\mathcal{H}_t\}_{t \in \mathbb{N}}$, where h_t is a family of $(\ell(t) + t)$-wise independent hash functions from $\{0,1\}^t$ to $\{0,1\}^{2t}$ with polynomial description size. We denote by $\pi = \{\pi_t\}_{t \in \mathbb{N}} \leftarrow \Pi$ [resp., $h = \{h_t\}_{t \in \mathbb{N}} \leftarrow \mathcal{H}$] the sequence of functions induced by selecting, for every $t \in \mathbb{N}$, π_t uniformly at random from Π_t [resp., h_t uniformly at random from $\mathcal{H}_t$]. In this section, we consider an instantiation of Breaker with $f = \mathrm{Dec}^\pi$, where π is chosen at random from Π, and h chosen at random from $\mathcal{H}$. In Section 3.1, we show how to use $\mathrm{Breaker}^{\mathrm{Dec}^\pi, h}$ for violating the KDI-security of $(\mathrm{Enc}^\pi, \mathrm{Dec}^\pi)$, where in Section 3.2 we show that $\mathrm{Breaker}^{\mathrm{Dec}^\pi, h}$ does not help inverting a random π. We prove Theorem 1 in Section 3.3.

3.1   Breaker Violates the KDI-Security of the Scheme

The following adversary uses $\mathrm{Breaker}^{\mathrm{Dec}^\pi, h}$ for breaking the KDI-security of $(\mathrm{Enc}^\pi, \mathrm{Dec}^\pi)$.

Algorithm 4 *Algorithm* $\mathsf{A}_{\mathrm{KDI}}^{\mathrm{Breaker}^{\mathrm{Dec}^\pi, h}, h}$.

Oracles: *An infinite sequence of functions* $h = \{h_t : \{0,1\}^t \to \{0,1\}^{2t}\}_{t \in \mathbb{N}}$ *and*
$\mathrm{Breaker}^{\mathrm{Dec}^\pi, h}$.
Input: *Security parameter* t.
Operation:
    Step 1: *Call* $Q(h_t)$ *[or* $\widetilde{Q}(h_t)$*] to obtain an encryption* c *of* $h_t(k)$ *[or* 0^{2t}*].*
    Step 2: *Call* $\mathrm{Breaker}^{\mathrm{Dec}^\pi, h}(t, c)$ *to obtain a candidate key* k' *or* $\perp$.
    Step 3: *Output* 1 *iff* Breaker *did not return* $\perp$.

Lemma 2. *For every value of* $\pi \in \Pi$, *algorithm* $\mathsf{A}_{\mathrm{KDI}}^{\mathrm{Breaker}^{\mathrm{Dec}^\pi, h}, h}$ *breaks the KDI-security of the* $(\mathrm{Enc}^\pi, \mathrm{Dec}^\pi)$ *with probability one over a random choice of* $h \in \mathcal{H}$.

Proof. Algorithm $\mathsf{A}_{\mathrm{KDI}}$ only gives the wrong answer if the oracle is $\widetilde{Q}$ and Breaker does not return $\bot$. Assume now that the oracle is $\widetilde{Q}$. Then, for any fixed k' and k we have $\Pr_{h_t}[h_t(k') = \mathrm{Dec}(k', \mathrm{Enc}(k, 0^{2t}))] = \frac{1}{2^{2t}}$, and using the union bound we have that the for a fixed k the probability that Breaker does not return $\bot$ is at most 2^{-t}.[9] Using an averaging argument, the probability that h_t is such that something else but $\bot$ is returned with probability higher than $2^{-t/2}$ is at most $2^{-t/2}$.

Since the $h_t \in h$'s are chosen independently from each other, the probability that there exists $t_0 \in \mathbb{N}$ for which $\mathsf{A}_{\mathrm{KDI}}$ breaks the scheme for no $t > t_0$ is zero. We conclude that with probability one over the random choice of $h \in \mathcal{H}$, it holds that $\mathsf{A}_{\mathrm{KDI}}$ breaks the KDI-security of $(\mathrm{Enc}, \mathrm{Dec})$ infinitely often.

3.2   Breaker Does Not Invert Random Permutations

We prove the following upper bound on the probability that an algorithm with access to Breaker inverts a random permutation. In the following let $\mu_{\mathsf{A}}(n)$ be an upper bound on number of π queries and the length of the maximal π query that A does on input $y \in \{0,1\}^n$ (either directly or through the calls to $\mathrm{Breaker}^{\mathrm{Dec}^\pi, h}$), and let $\mu_{\mathrm{Dec}}(t)$ the same bound with respect to the π queries that Dec does on input $(k, c) \in \{0,1\}^t \times \{0,1\}^{\ell(t)}$. We assume without loss of generality that both upper bounds are monotonically increasing, that $\mu_{\mathrm{Dec}}(t) \geq t + \ell(t)$ and that $\mu_{\mathsf{A}}(n) \geq n$. We also assume that $\mu_{\mathrm{Dec}}(t) < 2^t$.

Lemma 3. *Let A be an adversary that gets h as an auxiliary input[10] and has oracle access to π and $\mathrm{Breaker}^{\mathrm{Dec}^\pi, h}$. Then for every $y \in \{0,1\}^n$ it holds that*

$$\Pr_{\pi \leftarrow \Pi, h \leftarrow \mathcal{H}}[\mathsf{A}_h^{(\pi, \mathrm{Breaker})}(y) = \pi^{-1}(y)] < 3\mu_{\mathsf{A}}(n)\left(2^{-\mu_{\mathrm{Dec}}^{-1}(n)} + \mu_{\mathrm{Dec}}(\mu_{\mathsf{A}}(n))^2 2^{-n}\right),$$

where $\mu_{\mathrm{Dec}}^{-1}(n) := \min\{t \in \mathbb{N} : \mu_{\mathrm{Dec}}(t) \geq n\}$.

Applying the Borel-Cantelli lemma on the above we get the following corollary.

Corollary 1. *Assume that A and Dec are polynomially bounded, then there exists a negligible function ε such that with probability one over the choice of π and h, $\Pr_{y \leftarrow \{0,1\}^n}[\mathsf{A}_h^{(\pi, \mathrm{Breaker}^{\mathrm{Dec}^\pi, h})}(y) = \pi^{-1}(y)] < \varepsilon(n)$ for large enough n.*

In Appendix A, we give a proof of a non-uniform version of Lemma 3 (the adversary can use an arbitrary additional non-uniform advice) using the technique introduced by Gennaro and Trevisan [GT00]. Here, we use a different technique that is similar to the one used by Simon [Sim98]. The main idea is to study what happens if π is modified slightly by mapping a second, randomly chosen element to y (the element that A tries to invert). We show that such a change will likely go unnoticed by $\mathsf{A}(y)$, and it will not find the new preimage. After the change,

---

[9] For this lemma, we are only using the "one-wise" independence of h.

[10] We handle the fact that h is an infinite object, by only providing A the (description of the) first $q(n)$ functions in the sequence, where $q(n)$ is an upper bound on the running-time of $\mathsf{A}(y \in \{0,1\}^n)$.

however, both preimages of y are equally likely to be the original one, so $A(y)$ could not have found the original one either.[11]

For a given function $g : \{0,1\}^n \mapsto \{0,1\}^n$ and two strings $x^*, y \in \{0,1\}^n$, we define the function $g_{|x^* \to y}$ as $g_{|x^* \to y}(x) := \begin{cases} y & \text{if } x = x^*, \\ g(x) & \text{otherwise.} \end{cases}$ We assume that all calls to $\mathrm{Dec}^{\pi_{|x^* \to y}}$ are well defined. In particular, if Dec queries $\pi_{|x^* \to y}$ both at position $\pi^{-1}(y)$ and at $x^* \neq \pi^{-1}(y)$ (and thus might act arbitrarily as it "notices" that $\pi_{|x^* \to y}$ is not a permutation), we assume it stops and outputs 0. We now wish to consider the elements $\{x^* \in \{0,1\}^n\}$ for which $\mathrm{Breaker}^{\mathrm{Dec}^\pi,h}(t,c) \neq \mathrm{Breaker}^{\pi_{|x^* \to y},h}(t,c)$. The set $\mathrm{Diff}^\pi(t,c,h,y)$ is a (possibly proper) superset of this set.

Definition 5 (Diff). *For an oracle function* Dec, *an infinite sequence of functions* $h \in \mathcal{H}$, $t \in \mathbb{N}$, $c \in \{0,1\}^*$ *and* $y \in \{0,1\}^n$, *we let* $\mathrm{Diff}^\pi(t,c,h,y) := \{x^* \in \{0,1\}^n \mid \exists k \in \{0,1\}^t : (\mathrm{Dec}^\pi(k,c) \neq h_t(k) = \mathrm{Dec}^{\pi_{|x^* \to y}}(k,c)) \vee (\mathrm{Dec}^\pi(k,c) = h_t(k) \neq \mathrm{Dec}^{\pi_{|x^* \to y}}(k,c))\}$.

For $x^* \notin \mathrm{Diff}^\pi(t,c,h,y)$, it holds that $\mathrm{Breaker}^{\mathrm{Dec}^\pi,h}(t,c) = \mathrm{Breaker}^{\pi_{|x^* \to y},h}(t,c)$. To see this, let $k_0 \neq \perp$ be the lexicographic smaller output of the two calls. Clearly, k_0 must be the output of both calls to Breaker. The next claim states that if h is uniformly chosen from $\mathcal{H}$, then $\mathrm{Diff}^\pi(t,c,h,y)$ is very likely to be small for all possible c.

Claim. Let A be an adversary with oracle access to π and $\mathrm{Breaker}^{\mathrm{Dec}^\pi,h}$, which gets h as an auxiliary input. Then, for every π and $y \in \{0,1\}^n$:

$$\Pr_{h \leftarrow \mathcal{H}} \left[A_h^{(\pi, \mathrm{Breaker}^{\mathrm{Dec}^\pi,h})}(y) \text{ queries } \mathrm{Breaker}^{\mathrm{Dec}^\pi,h}(t,c) \text{ with} \right.$$

$$\left. |\mathrm{Diff}^\pi(t,c,h,y)| \geq \mu_{\mathrm{Dec}}(\mu_A(n))^2 \right] < \mu_A(n) 2^{-\mu_{\mathrm{Dec}}^{-1}(n)}$$

Proof. For $t \in \mathbb{N}$, $c \in \{0,1\}^*$ and $k \in \{0,1\}^t$, let $\mathcal{D}_{c,k}$ be the set of all possible images of $\mathrm{Dec}^{\pi_{|x^* \to y}}(k,c)$, enumerating over all $x^* \in \{0,1\}^n$ (i.e., the set $\mathcal{D}_{c,k} := \{\mathrm{Dec}^{\pi_{|x^* \to y}}(k,c) : x^* \in \{0,1\}^n\}$).[12] We first note that $|\mathcal{D}_{c,k}| \leq \mu_{\mathrm{Dec}}(t) + 1 \leq 2^t$ – in an execution of $\mathrm{Dec}^\pi(k,c)$ at most $\mu_{\mathrm{Dec}}(t)$ elements $x_1, \ldots, x_{\mu_{\mathrm{Dec}}(t)}$ are queried, and only if $x^* \in \{x_1, \ldots, x_{\mu_{\mathrm{Dec}}(t)}\}$ the image of Dec can be changed. Let $\mathcal{H}_t$ be the t'th entry in $\mathcal{H}$. Applying Lemma 1 with $V = 2^t$, $s = t + \ell(t)$, $\alpha = \mu_{\mathrm{Dec}}(t)$ and letting $B_k = 1$ iff $h_t(k) \in \mathcal{D}_{c,k}$, we have that $\Pr_{h_t \leftarrow \mathcal{H}_t} \left[|\{k \in \{0,1\}^t : h_t(k) \in \mathcal{D}_{c,k}\}| \geq \mu_{\mathrm{Dec}}(t) \right] < \frac{t}{\mu_{\mathrm{Dec}}(t)^{t+\ell(t)-1}} \leq 2^{-t-\ell(t)}$.

We next show that $|\mathrm{Diff}^\pi(t,c,h,y)| \leq \mu_{\mathrm{Dec}}(t) \cdot |\{k \in \{0,1\}^t : h_t(k) \in \mathcal{D}_{c,k}\}|$. We prove this by presenting an injective function ϕ from $\mathrm{Diff}^\pi(t,c,h,y)$ to $\{1, \ldots, \mu_{\mathrm{Dec}}(t)\} \times \{k : h_t(k) \in \mathcal{D}_{c,k}\}$. If $x^* \in \mathrm{Diff}^\pi(t,c,h,y)$, then there exists

---

[11] The main difference between our approach and the one in [Sim98], is that we do not insist on keeping π a permutation. It turns out that this slackness makes our proof significantly simpler.

[12] Note that the original image $\mathrm{Dec}^\pi(k,c)$ is in $\mathcal{D}_{c,k}$.

k_{x^*} such that $\mathrm{Dec}^\pi(k_{x^*}, c) \neq h_t(k_{x^*}) = \mathrm{Dec}^{\pi|_{x^*\to y}}(k_{x^*}, c)$ or $\mathrm{Dec}^\pi(k_{x^*}, c) = h_t(k_{x^*}) \neq \mathrm{Dec}^{\pi|_{x^*\to y}}(k_{x^*}, c)$ and therefore $k_{x^*} \in \{k : h_t(k) \in \mathcal{D}_{c,k}\}$. Furthermore, $\mathrm{Dec}^\pi(k_{x^*}, c)$ must query π on x^*. Let i_{x^*} denote the index (i.e., position) of the query $\pi(x^*)$ among the π queries that $\mathrm{Dec}^\pi(k_{x^*}, c)$ does, and let ϕ be the function that maps x^* to (i_{x^*}, k_{x^*}). Since a pair (i, k) specifies a single x (the one queried at i'th position in $\mathrm{Dec}^\pi(k, c)$), it follows that ϕ is indeed injective.

We note that for $t > \mu_{\mathrm{Dec}}^{-1}(n)$, it always holds that $x^* \notin \mathrm{Diff}^\pi(t, c, h, y)$ (Dec cannot invoke π on such a long input). Combining the above, we have that with probability at least $1 - \mu_{\mathsf{A}}(n)2^{-\mu_{\mathrm{Dec}}^{-1}(n)}$ over the choice of h, for each of the at most $\mu_{\mathsf{A}}(n)$ queries $\mathrm{Breaker}(t, c)$ that $\mathsf{A}(y)$ does, it holds that $|\mathrm{Diff}(t, c, h, y)| \leq \mu_{\mathrm{Dec}}(t)^2 \leq \mu_{\mathrm{Dec}}(\mu_{\mathsf{A}}(n))^2$.

For a given value of π and $x_0, x_1 \in \{0,1\}^n$, let $\pi|_{x_0 \leftrightarrow x_1} := \pi|_{x_0 \to \pi(x_1), x_1 \to \pi(x_0)}$. We next show that with high probability, $\mathsf{A}(y)$ behaves exactly the same given the oracle π or $\pi|_{x^* \leftrightarrow y}$.

Definition 6 (trace). *For a given oracle function Dec, the trace, $\mathrm{tr}(\pi, h, y, r_{\mathsf{A}})$, of an adversary A is the sequence of all queries $\mathsf{A}(y)$ makes to $\mathrm{Breaker}^{\mathrm{Dec}^\pi, h}$ and π (and their responses), when it uses r_{A} as its random-coins and gets h as an auxiliary input.*

Claim. Let $y \in \{0,1\}^n$ and let A be an adversary with oracle access to π and $\mathrm{Breaker}^{\mathrm{Dec}^\pi, h}$, and assume that A queries π on its output before returning it. Then, $\Pr_{\pi, h, r_{\mathsf{A}}, x^* \leftarrow \{0,1\}^n}[\mathrm{tr}(\pi, h, y, r_{\mathsf{A}}) \neq \mathrm{tr}(\pi|_{x^* \leftrightarrow \pi^{-1}(y)}, h, y, r_{\mathsf{A}})] < p(n)$, where $p(n) := 2\mu_{\mathsf{A}}(n)\big(2^{-\mu_{\mathrm{Dec}}^{-1}(n)} + \mu_{\mathrm{Dec}}(\mu_{\mathsf{A}}(n))^2 2^{-n}\big)$.

Proof. Let $\mathcal{X}_\pi$ be all the queries to π in $\mathrm{tr}(\pi, h, y, r_{\mathsf{A}})$, clearly $|\mathcal{X}_\pi| \leq \mu_{\mathsf{A}}(n)$. Let further $\mathcal{X}_{\mathrm{Diff}}$ be the union of the sets $\mathrm{Diff}^\pi(t, c, h, y)$ for all calls (t, c) made by A to Breaker. If $x^* \notin \mathcal{X}_\pi \cup \mathcal{X}_{\mathrm{Diff}}$, we have that $\mathrm{tr}(\pi, h, y, r_{\mathsf{A}}) = \mathrm{tr}(\pi|_{x^* \to y}, h, y, r_{\mathsf{A}})$ (cf., the remark after Definition 5). Claim 3.2 yields that with probability at least $1 - \mu_{\mathsf{A}}(n)2^{-\mu_{\mathrm{Dec}}^{-1}(n)}$ over the choice of h and r_{A}, it holds that $|\mathcal{X}_\pi \cup \mathcal{X}_{\mathrm{Diff}}| \leq \mu_{\mathsf{A}}(n)\mu_{\mathrm{Dec}}(\mu_{\mathsf{A}}(n))^2$. The union bound yields that $\Pr_{h, r_{\mathsf{A}}, x^* \leftarrow \{0,1\}^n}[\mathrm{tr}(\pi, h, y, r_{\mathsf{A}}) \neq \mathrm{tr}(\pi|_{x^* \to y}, h, y, r_{\mathsf{A}})] < \mu_{\mathsf{A}}(n)\big(2^{-\mu_{\mathrm{Dec}}^{-1}(n)} + \mu_{\mathrm{Dec}}(\mu_{\mathsf{A}}(n))^2 2^{-n}\big)$.

Finally, if $\mathrm{tr}(\pi, h, y, r_{\mathsf{A}})$ and $\mathrm{tr}(\pi|_{x^* \leftrightarrow \pi^{-1}(y)}, h, y, r_{\mathsf{A}})$ are different, then one of $\mathrm{tr}(\pi, h, y, r_{\mathsf{A}}) \neq \mathrm{tr}(\pi|_{x^* \to y}, h, y, r_{\mathsf{A}})$ or $\mathrm{tr}(\pi|_{x^* \to y}, h, y, r_{\mathsf{A}}) \neq \mathrm{tr}(\pi|_{x^* \leftrightarrow \pi^{-1}(y)}, h, y, r_{\mathsf{A}})$ must hold. Since $\pi|_{x^* \leftrightarrow \pi^{-1}(y)}$ is also a permutation, and x^* is a uniformly chosen element given $\pi|_{x^* \leftrightarrow \pi^{-1}(y)}$ and y, the inequality obtained before states that both these events have probability at most $p(n)/2$.

Proof. (of Lemma 3) Assuming without lost of generality that A queries it's output, the probability that $\mathsf{A}^{(\pi, \mathrm{Breaker}^{\mathrm{Dec}^\pi, h})}(y)_h = \pi^{-1}(y)$ is at most the probability that the traces $\mathrm{tr}(\pi, h, y, r_{\mathsf{A}})$ and $\mathrm{tr}(\pi|_{x^* \leftrightarrow \pi^{-1}(y)}, h, y, r_{\mathsf{A}})$ are different plus 2^{-n} (to handle the case $x^* = \pi^{-1}(y)$). By Claim 3.2, the latter probability is at most $2^{-n} + p(n)$.

3.3   Putting It Together

Proof. (of Theorem 1) Assume that there exists a black-box proof of security from breaking the KDI-security of $(\mathrm{Enc}^\pi, \mathrm{Dec}^\pi)$ using a poly-wise independent hash function to breaking the hardness of π, and let $M^{(\cdot)}$ be the algorithm for inverting π as guaranteed by this proof of security. Lemma 2 yields that $\mathsf{A}_{\mathrm{KDI}}^{\mathrm{Breaker}^{\mathrm{Dec}^\pi,h},h}$ breaks the KDI-security of $(\mathrm{Enc}^\pi, \mathrm{Dec}^\pi)$ with probability one over the choice of h. Thus, $M^{\mathsf{A}_{\mathrm{KDI}}^{\mathrm{Breaker}^{\mathrm{Dec}^\pi,h},h}}$ needs to break the one-way property of π with probability one over the choice of h as well. However, since $\mathsf{A}^{\pi,\mathsf{B}}$ can be efficiently emulated by an algorithm $\widetilde{\mathsf{A}}$ with oracle access to $\mathrm{Breaker}^{\mathrm{Dec}^\pi,h}$ and π, and given h as an auxiliary input, Corollary 1 yields that with probability one over the choice of π and h, algorithm $\mathsf{A}^{\pi,\mathsf{B}}$ does not break the one-wayness of π, and a contradiction is derived.

4   From Arbitrary Assumptions

In this section, we rule out the existence of reductions with strongly-black-box proof of security from the KDI-security of an encryption scheme, to a very large class of hardness assumptions. That is, we prove the following theorem.

Theorem 5 (formal restatement of Theorem 2). *There exists no reduction with strongly-black-box proof of security from the KDI-security of an encryption scheme to* any *cryptographic game.*

Let $(\mathrm{Enc}, \mathrm{Dec})$ be an encryption scheme. As in Section 3, we use Proposition 1 and assume without loss of generality that the encryption scheme is always correct, has deterministic decryption algorithm, is defined on messages of any length, and has security parameter t equal to the key length. We let $\ell(t)$ be the length of an encryption of a message of length $2t$.

Consider an instantiation of Breaker (Algorithm 3) with $f = \mathrm{Dec}$ and $h \in \mathcal{H} = \{\mathcal{H}_t\}_{t\in\mathbb{N}}$, where $\mathcal{H}_t$ is the set of *all* possible function from $\{0,1\}^t$ to $\{0,1\}^{2t}$. As in Section 3, we have that there exists an efficient algorithm, with oracle access to $\mathrm{Breaker}^{\mathrm{Dec},h}$ and h, that breaks the KDI-security of $(\mathrm{Enc}, \mathrm{Dec})$ with probability one over the choice of $h \in \mathcal{H}$. The following Lemma states that in many settings, having oracle access to $\mathrm{Breaker}^{\mathrm{Dec},h}$ does not yield any significant additional power.

Lemma 4. *Let* $\mathsf{A}^{\mathrm{Breaker}^{\mathrm{Dec},h},h}$ *be an algorithm with oracle access to* $\mathrm{Breaker}^{\mathrm{Dec},h}$ *and* h, *and let* $t_{\mathsf{A}}(n)$, *for security parameter* n, *be a polynomial-time computable upper bound on the running-time of* $\mathsf{A}^{\mathrm{Breaker}^{\mathrm{Dec},h},h}$.

Then for every polynomial computable function $\delta : \mathbb{N} \mapsto [0,1]$, *there exists an algorithm* $\widetilde{\mathsf{A}}_\delta^h$, *which has oracle access only to* h, *runs in time* $\mathrm{poly}(1/\delta(n), t_{\mathsf{A}}(n), n)$ *and uses random-coins of the same length as* $\mathsf{A}^{\mathrm{Breaker}^{\mathrm{Dec},h},h}$ *such that the following holds. If* $\mathsf{A}^{\mathrm{Breaker}^{\mathrm{Dec},h},h}$ *and* $\widetilde{\mathsf{A}}_\delta^h$ *are using the same random-coins, then* $\mathsf{A}^{\mathrm{Breaker}^{\mathrm{Dec},h},h}(1^n) = \widetilde{\mathsf{A}}_\delta^h(1^n)$ *with probability* $1 - \delta(n)$ *over a random choice of* h.

Proof. Algorithm $\widetilde{A}$ emulates A, while remembering all query and answer pairs to h. When A queries $\text{Breaker}^{\text{Dec},h}(t,c)$, algorithm $\widetilde{A}$ distinguishes two cases:

Case 1: $t < \log(t_A(n)) + \log(1/\delta(n))$. $\widetilde{A}$ fully emulates $\text{Breaker}^{\text{Dec},h}$. Namely, $\widetilde{A}$ evaluates $h_t(k)$ for all $k \in \{0,1\}^t$ and returns the first one for which $\text{Dec}(k,c) = h_t(k)$. It returns $\perp$ if no such k exists.

Case 2: $t \geq \log(t_A(n)) + \log(1/\delta(n))$. $\widetilde{A}$ checks all the previous queries to h of length t in lexicographic order. If for one of those queries it holds that $\text{Dec}(k,c) = h_t(k)$, it returns k, otherwise it returns $\perp$.

The bound on the running-time of $\widetilde{A}$ is clear, in the following we show $\widetilde{A}$ emulates A well. We first note that in Case 1, $\widetilde{A}$ always returns the same answer that $\text{Breaker}^{\text{Dec},h}$ would. To handle Case 2, let $k \in \{0,1\}^t$ and assume that the query $h_t(k)$ was not perviously asked by A. Since h is length doubling, the probability over the choice of h that $\text{Dec}(k,c) = h_t(k)$ is 2^{-2t}. Using a union bound we have that the probability, over the choice of h, that $\widetilde{A}$ returns a value different from what $\text{Breaker}^{\text{Dec},h}$ would (i.e., $\widetilde{A}$ returns $\perp$ where $\text{Breaker}^{\text{Dec},h}$ finds a consistent key) is at most $2^{-\log(t_A(n))-\log(1/\delta(n))} = \delta(n)/t_A(n)$. Since there are at most $t_A(n)$ calls to $\text{Breaker}^{\text{Dec},h}$, the probability that in any of those $\widetilde{A}$ returns a wrong value is at most $\delta(n)$, which proves the lemma.

Proof. (of Theorem 5) The proof follows the lines of the one of Theorem 1, but we need to work a little harder for proving that having access to h does not give the adversary additional power.[13]

Assume that there exists a strongly-black-box proof of security from (Enc, Dec) to a cryptographic game Γ and let $M^{(\cdot)}$ be the algorithm for breaking Γ as guaranteed by this proof of security. It easily follows from the proof of Lemma 2 that also in the setting of this section there is an efficient algorithm $A_{\text{KDI}}^{\text{Breaker}^{\text{Dec},h},h}$, with oracle access to $\text{Breaker}^{\text{Dec},h}$ and h breaking the KDI-security of (Enc, Dec) with probability one over the choice of h. Thus, $M^{A_{\text{KDI}}^{\text{Breaker}^{\text{Dec},h},h}}$ breaks Γ with probability 1 over the choice of h. Namely,

$$\Pr_h\left[\Pr_{r_A,r_\Gamma}[M^{A_{\text{KDI}}^{\text{Breaker}^{\text{Dec},h},h}} \leftrightarrow \Gamma(1^n) \text{ wins}] > \tfrac{1}{p_h(n)} \text{ for infinitely many } n\right] = 1, \quad (1)$$

where r_A and r_Γ denote the random-coins of A and Γ, respectively, and p_h is some polynomial that may depend on h. In the following we first remove the dependence of the polynomial p_h from h. For this let $\varepsilon(n) := \Pr_{h,r_A,r_\Gamma}[M^{A_{\text{KDI}}^{\text{Breaker}^{\text{Dec},h},h}}$ $\leftrightarrow \Gamma(1^n) \text{ wins}] = E_h\left[\Pr_{r_A,r_\Gamma}[M^{A_{\text{KDI}}^{\text{Breaker}^{\text{Dec},h},h}} \leftrightarrow \Gamma(1^n) \text{ wins}]\right]$. We show that ε is non negligible. Using Markov's inequality we get for every $n \in \mathbb{N}$ that $\Pr_h[\Pr_{r_A,r_\Gamma}[M^{A_{\text{KDI}}^{\text{Breaker}^{\text{Dec},h},h}} \leftrightarrow \Gamma(1^n) \text{ wins}] < n^2\varepsilon(n)] > 1 - \tfrac{1}{n^2}$, and therefore[14] $\Pr_h[\Pr_{r_A,r_\Gamma}[M^{A_{\text{KDI}}^{\text{Breaker}^{\text{Dec},h},h}} \leftrightarrow \Gamma(1^n) \text{ wins}] < n^2\varepsilon(n) \text{ for all } n > 2]$

---

[13] One gets this property "for free", when the underlying hardness assumption is inverting a random permutation.

[14] The σ-additivity of the measure implies that the event in the next probability is measurable.

$\geq 1 - \sum_{n=2}^{\infty} \frac{1}{n^2} > \frac{1}{3}$ Combining this with Equation (1) we get that $\Pr_h \left[\frac{1}{p_h(n)} < \right.$
$\Pr_{r_A, r_\Gamma} [M^{A_{\mathrm{KDI}}^{\mathrm{Breaker}^{\mathrm{Dec},h,h}}} \leftrightarrow \Gamma(1^n) \text{ wins}] < n^2 \varepsilon(n) \text{ for infinitely many } n \left.\right] > \frac{1}{3}$,
which implies that there is a polynomial $p(n)$ such that $\varepsilon(n) > \frac{1}{p(n)}$ infinitely
often.

In order to finish the proof, we will now to apply Lemma 4 on $M^{A_{\mathrm{KDI}}^{\mathrm{Breaker}^{\mathrm{Dec},h,h}}}$.
Recall that Lemma 4 was proved only in the stand alone settings, where in partic-
ular no interaction with a random system is considered. Since the proof of security of (Enc, Dec) is strongly-black-box, we have that Γ does not access, through
interaction with $M^{A_{\mathrm{KDI}}^{\mathrm{Breaker}^{\mathrm{Dec},h,h}}}$, the function h. Therefore, Γ's answers are determined by the output behavior of $M^{A_{\mathrm{KDI}}^{\mathrm{Breaker}^{\mathrm{Dec},h,h}}}$ and the proof of Lemma 4
goes through also in this setting. Hence, Lemma 4 yields, letting $\delta(n) = \frac{1}{2p(n)}$,
the existence of an efficient algorithm $\widetilde{M}^h$ with oracle access *only* to h, such that
$\Pr_{h, r_A, r_\Gamma} [\widetilde{M}^h \leftrightarrow \Gamma(1^n) \text{ wins}] > \frac{1}{2p(n)}$ for infinitely many n's.

Our final step is to emulate $\widetilde{M}^h$, where rather than accessing h we randomly
chooses the answer of each time one is requested (and cache it). The latter
emulation breaks the cryptographic assumption with probability at least $\frac{1}{2p(n)}$
for infinitely many n's and since it is also efficient, it implies that Γ is not secure.

5   Applying Our Technique to Other Primitives

It seems tempting to try and use the above Breaker also to show the impossibility of constructing other KDI-secure primitives. Consider for instance pseudorandom functions or permutations that are supposed to be secure even if
the adversary can obtain its value on a function of its secret key. Halevi and
Krawczyk [HK07] show that a deterministic construction cannot exist, but give
a construction in case the permutation has an additional public parameter (i.e.,
salt) chosen after the challenge function is fixed. Their construction, however,
compresses (e.g., maps n bits to $n/2$).

It is indeed possible to generalize our techniques to this case, as long for as the
pseudorandom functions are injective for *every key*. In this case, Breaker finds
a key k such that $f^{k,r}(h(k)) = c$, where f is the pseudorandom function and r
is the random salt. The reason this method fails if the construction compresses
(as the one given by Halevi and Krawczyk [HK07]), is that Breaker as defined
above does not seem to give useful information about the key anymore, since it
is unlikely that the correct key is the lexicographically smallest.

It seems that we also cannot utilize our Breaker for the general case of length
increasing (non-injective) pseudorandom functions (or equivalently, for the case
that we are allowed to make several KDI queries). Consider the question whether
a given pseudorandom function is constant on a negligible fraction of the keys
(e.g., on a single key k it holds that $f^{k,r}(\cdot) := 0^\ell$). Deciding whether a given
function has this property or not might be infeasible. Yet, using for instance
the Breaker of Section 4, we can easily find the right answer: ask the Breaker
on $(h, 0^\ell)$, where h is a random hash function, and answer "Yes" is the Breaker

finds some consistent key. Thus, in this setting our Breaker gives us an extra power that we cannot emulate.

Acknowledgments

We are very grateful to Oded Goldreich, Jonathan Hoch, Gil Segev, Omer Reingold and Udi Wieder for useful discussions. We thank the anonymous referees for many useful comments.

References

[AR02]   Abadi, M., Rogaway, P.: Reconciling two views of cryptography (the computational soundness of formal encryption). JoC 15(2), 103–127 (2002)

[Bar01]   Barak, B.: How to go beyond the black-box simulation barrier. In: 42nd FOCS, pp. 106–115. IEEE Computer Society, Los Alamitos (2001)

[BHHO08]   Boneh, D., Halevi, S., Hamburg, M., Ostrovsky, R.: Circular-secure encryption from decision diffie-hellman. In: Wagner, D. (ed.) CRYPTO 2008. LNCS, vol. 5157, pp. 108–125. Springer, Heidelberg (2008)

[BRS02]   Black, J., Rogaway, P., Shrimpton, T.: Encryption-scheme security in the presence of key-dependent messages. In: Nyberg, K., Heys, H.M. (eds.) SAC 2002. LNCS, vol. 2595, pp. 62–75. Springer, Heidelberg (2003)

[CL01]   Camenisch, J.L., Lysyanskaya, A.: An efficient system for non-transferable anonymous credentials with optional anonymity revocation. In: Pfitzmann, B. (ed.) EUROCRYPT 2001. LNCS, vol. 2045, p. 93. Springer, Heidelberg (2001)

[CW79]   Carter, J.L., Wegman, M.N.: Universal classes of hash functions. JCSS 18(2), 143–154 (1979)

[DOP05]   Dodis, Y., Oliveira, R., Pietrzak, K.: On the generic insecurity of the full domain hash. In: Shoup, V. (ed.) CRYPTO 2005. LNCS, vol. 3621, pp. 449–466. Springer, Heidelberg (2005)

[DY83]   Dolev, D., Yao, A.C.: On the security of public key protocols. IEEE Transactions on Information Theory 29(2), 198–208 (1983)

[GGKT05]   Gennaro, R., Gertner, Y., Katz, J., Trevisan, L.: Bounds on the efficiency of generic cryptographic constructions. S. J. on Comp. 35(1), 217–246 (2005)

[GKM+00]   Gertner, Y., Kannan, S., Malkin, T., Reingold, O., Viswanathan, M.: The relationship between public key encryption and oblivious transfer. In: FOCS 2000 (2000)

[GT00]   Gennaro, R., Trevisan, L.: Lower bounds on the efficiency of generic cryptographic constructions. In: FOCS 2000 (2000)

[HHRS07]   Haitner, I., Hoch, J.J., Reingold, O., Segev, G.: Finding collisions in interactive protocols – A tight lower bound on the round complexity of statistically-hiding commitments. In: FOCS 2007 (2007)

[HH08]   Haitner, I., Holenstein, T.: On the (Im) Possibility of Key Dependent Encryption (full version), http://eprint.iacr.org/2008/164

[HK05]   Horvitz, O., Katz, J.: Bounds on the efficiency of "black-box" commitment schemes. In: Caires, L., Italiano, G.F., Monteiro, L., Palamidessi, C., Yung, M. (eds.) ICALP 2005. LNCS, vol. 3580, pp. 128–139. Springer, Heidelberg (2005)

[HK07]   Halevi, S., Krawczyk, H.: Security under key-dependent inputs. In: 14th ACM CCS (2007)

[Hof08]   Hofheinz, D.: Possibility and impossibility results for selective decommitments. Technical Report 2008/168, eprint.iacr.org (April 2008)

[HU08]   Hofheinz, D., Unruh, D.: Towards key-dependent message security in the standard model. In: Smart, N.P. (ed.) EUROCRYPT 2008. LNCS, vol. 4965, pp. 108–126. Springer, Heidelberg (2008)

[IJK07]   Impagliazzo, R., Jaiswal, R., Kabanets, V.: Chernoff-type direct product theorems. In: Menezes, A. (ed.) CRYPTO 2007. LNCS, vol. 4622, pp. 500–516. Springer, Heidelberg (2007)

[IR89]   Impagliazzo, R., Rudich, S.: Limits on the provable consequences of one-way permutations. In: STOC 1989 (1989)

[KST99]   Kim, J.H., Simon, D.R., Tetali, P.: Limits on the efficiency of one-way permutation-based hash functions. In: FOCS 1999 (1999)

[RTV04]   Reingold, O., Trevisan, L., Vadhan, S.P.: Notions of reducibility between cryptographic primitives. In: Naor, M. (ed.) TCC 2004. LNCS, vol. 2951, pp. 1–20. Springer, Heidelberg (2004)

[Rud88]   Rudich, S.: Limits on the Provable Consequences of One-Way Functions. PhD thesis, U.C. Berkeley (1988)

[Sim98]   Simon, D.R.: Findings collisions on a one-way street: Can secure hash functions be based on general assumptions? In: Nyberg, K. (ed.) EUROCRYPT 1998. LNCS, vol. 1403, pp. 334–345. Springer, Heidelberg (1998)

[Wee07]   Wee, H.M.: One-way permutations, interactive hashing and statistically hiding commitments. In: Vadhan, S.P. (ed.) TCC 2007. LNCS, vol. 4392, pp. 419–433. Springer, Heidelberg (2007)

A   Gennaro-Trevisan Style Proof of Lemma 3

In this section we prove an alternative non-uniform version of Lemma 3.

Lemma 5 (non-uniform version of Lemma 3). *Let* A *be a non-uniform adversary that gets* h *as an auxiliary input and has oracle access to* π *and* $\mathrm{Breaker}^{\mathrm{Dec}^{\pi},h}$. *Assume that* A *and* Dec *satisfy the bounds* $\mu_{\mathsf{A}}(n) = \mu_{\mathrm{Dec}}(n) = n^C$ *for some* $C \in \mathbb{N}$. *For* $\varepsilon(n) = 2^{-n^{1/(2C)}}$ *we have that* $\Pr_{h \leftarrow \mathcal{H}, \pi \leftarrow \Pi, y \leftarrow \{0,1\}^n}[\mathsf{A}^{(\pi, \mathrm{Breaker}^{\mathrm{Dec}^{\pi},h})}(y, h) = \pi^{-1}(y)] < 3\varepsilon$.

Our main tool is the following lemma.

Lemma 6. *Let* A *be a non-uniform adversary that gets* h *as an auxiliary input, and has oracle access to* π *and* $\mathrm{Breaker}$, *and assume* $\Pr_y\big[\mathsf{A}^{(\pi, \mathrm{Breaker}^{\mathrm{Dec}^{\pi},h})}(y, h) = \pi^{-1}(y) \wedge \overline{\mathrm{Bad}(y)}\big] > \varepsilon(n)$, *where* $\mathrm{Bad}(y)$ *is the event that* $\mathsf{A}(y, h)$ *makes a query* $\mathrm{Breaker}^{\mathrm{Dec}^{\pi},h}(t, c)$ *for which* $|\mathrm{Diff}^{\pi}(t, c, h, y)| \geq \mu_{\mathrm{Dec}}(\mu_{\mathsf{A}}(n))^2$. *Then,* π *can be described using* $\log((2^n - s(n))!) + 2s(n)\log\Big(\frac{e2^n}{s(n)}\Big) + \mu_{\mathsf{A}}(n)^2\mu_{\mathrm{Dec}}(\mu_{\mathsf{A}}(n))$ *bits, where* $s(n) = \varepsilon(n)2^n / \big(2\mu_{\mathsf{A}}(n)(\mu_{\mathrm{Dec}}(\mu_{\mathsf{A}}(n)))^2\big)$.

We omit the proof of Lemma 5 in this version, it is obtained from Lemma 6.

Proof. (of Lemma 6) We assume w.l.o.g. $\varepsilon(n) > 2\mu_{\mathsf{A}}(n)2^{-n}$, since otherwise the statement is trivial. Our description of π consists of the following parts: the description of a set $S \subseteq \{0,1\}^n$, the description of the image of S under π (which roughly corresponds to the y's on which A succeeds in inverting), the description of the permutation that π implies if restricted on $\{0,1\}^n \setminus S$ (i.e., the elements not in S) and finally the description of $\{h_m \in h \mid m \leq \mu_{\mathsf{A}}(n)\}$.

The description of S and the image of S both require $\log\left(\binom{2^n}{|S|}\right) \leq |S|\log(e\frac{2^n}{|S|})$ bits. The description of the permutation requires at most $\log((2^n - |S|)!)$ bits. To store the functions h_m takes $\mu_{\mathsf{A}}(n)^2\mu_{\mathrm{Dec}}(\mu_{\mathsf{A}}(n))$ bits, for some appropriate family $\mathcal{H}$. Thus, in total $\log((2^n - |S|)!) + 2|S|\log(e\frac{2^n}{|S|}) + \mu_{\mathsf{A}}(n)^2\mu_{\mathrm{Dec}}(\mu_{\mathsf{A}}(n))$ bits are sufficient. In the following we succeed in making S as big as $|S| = \varepsilon(n)2^n/(2\mu_{\mathsf{A}}(n)(\mu_{\mathrm{Dec}}(\mu_{\mathsf{A}}(n)))^2)$, which implies our description size of π.

Defining the set S. We use the following, inefficient, algorithm to create S: we start by letting $I = \{y \in \{0,1\}^n : \mathsf{A}(y) = \pi^{-1}(y) \wedge \overline{\mathrm{Bad}(y)}\}$ and iteratively do the following. First, remove the lexicographic smallest element y from I and add $\pi^{-1}(y)$ to S. Next, emulate $\mathsf{A}^{(\pi,\mathrm{Breaker}^{\mathrm{Dec}^\pi,h})}(y,h)$ and remove all queries A makes whose answers are in I from I (without putting them into S). In addition, for each query to $\mathrm{Breaker}^{\mathrm{Dec}^\pi,h}(t,c)$ done by $\mathsf{A}^{(\pi,\mathrm{Breaker}^{\mathrm{Dec}^\pi,h})}(y,h)$, remove the images $\pi(x)$ from I for all $x^* \in \mathrm{Diff}^\pi(t,c,h,y)$ (note that y itself is not removed from I, as we already removed it). Once the emulation is over, repeat with the next element. Since for every emulation we remove at most $\mu_{\mathsf{A}}(n) + \mu_{\mathsf{A}}(n) \cdot \mu_{\mathrm{Dec}}(\mu_{\mathsf{A}}(n))^2$ elements from I before moving another element to S, we have that $|S| \geq \varepsilon(n)2^n/2\mu_{\mathsf{A}}(n)(\mu_{\mathrm{Dec}}(\mu_{\mathsf{A}}(n)))^2$.

The reconstruction of π. We now show that we can reconstruct π from the given information. For this, we first reconstruct the oracle outside of S from the given information. Then pick the lexicographic smallest element $y \in S$ whose preimage is not yet known, and emulate $\mathsf{A}^{\pi,\mathrm{Breaker}^{\mathrm{Dec}^\pi,h}}(y,h)$. We first consider the queries $\pi(x)$ done by $\mathsf{A}(y,h)$. The definition of S yields that we either know the answer for this query, or we are guaranteed that $\pi(x) = y$ (and we can stop the emulation). So it is left to consider the queries $\mathrm{Breaker}^{\mathrm{Dec}^\pi,h}(t,c)$. We note the that if $k \in \{0,1\}^t$ is the value we should return as the answer of $\mathrm{Breaker}^{\mathrm{Dec}^\pi,h}(t,c)$, then the answers to all π-queries made by $\mathrm{Breaker}^{\mathrm{Dec}^\pi,h}(t,c)$ when it calls $\mathrm{Dec}^\pi(k,c)$ (see Algorithm 3) should be known, where the only exception is a query on $\pi^{-1}(y)$ if such occurs. Therefore, we try all candidates x^* (i.e., the elements whose image we don't know at this point) for $\pi^{-1}(y)$, and emulate $\mathrm{Dec}^{\pi|x^* \rightarrow y}(k,c)$. The latter emulation succeeds unless a query is made whose answer we don't know. In this case, we know by the above observation that the current pair (k,x^*) is not the one we are looking for, and we can safely move to the next candidate for x^*. Finally, note that if a successful emulation of $\mathrm{Dec}^{\pi|x^* \rightarrow y}(k,c) = h(k)$ done by $\mathrm{Breaker}^{\mathrm{Dec}^\pi,h}(t,c)$ satisfies $\mathrm{Dec}^{\pi|x^* \rightarrow y}(k,c) = h(k)$, then k must be the correct answer to $\mathrm{Breaker}^{\mathrm{Dec}^\pi,h}(t,c)$. The reason for that x^* would be in $\mathrm{Diff}^\pi(t,c,h,y)$, and therefore cannot be in S. All in all, we can emulate $\mathsf{A}(y,h)$'s run correctly and obtain the correct $\pi^{-1}(y)$ as the output of A.

On the (Im)Possibility of Arthur-Merlin Witness Hiding Protocols

Iftach Haitner[1,*], Alon Rosen[2,**], and Ronen Shaltiel[3,***]

[1] Microsoft Research, New England Campus
iftach@microsoft.com
[2] Herzliya Interdisciplinary Center, Herzliya, Israel
alon.rosen@idc.ac.il
[3] University of Haifa
ronen@cs.haifa.ac.il

Abstract. The concept of *witness-hiding* suggested by Feige and Shamir is a natural relaxation of zero-knowledge. In this paper we identify languages and distributions for which many known constant-round public-coin protocols with negligible soundness cannot be shown to be witness-hiding using black-box techniques. One particular consequence of our results is that parallel repetition of either 3-Colorability or Hamiltonicity cannot be shown to be witness hiding with respect to some probability distribution over the inputs assuming that:

1. the distribution assigns positive probability only to instances with *exactly one* witness.
2. Polynomial size circuits cannot find a witness with noticeable probability on a random input chosen according to the distribution.
3. The proof of security relies on a black-box reduction that is *independent* of the choice of the commitment scheme used in the protocol.

These impossibility results conceptually match results of Feige and Shamir that use such black-box reductions to show that parallel repetition of 3-Colorability or Hamiltonicity *is* witness-hiding for distributions with "two independent witnesses".

We also consider black-box reductions for parallel repetition of 3-Colorability or Hamiltonicity that *depend* on a specific implementation of the commitment scheme. While we cannot rule out such reductions completely, we show that "natural reductions" cannot bypass the limitations above.

Our proofs use techniques developed by Goldreich and Krawczyk for the case of zero knowledge. The setup of witness-hiding, however, presents new technical and conceptual difficulties that do not arise in the zero-knowledge setting. The high level idea is that if a black-box reduction establishes the witness-hiding property for a protocol, and the protocol also happens to be a proof of knowledge, then this latter property can be actually used "against the reduction" to find witnesses unconditionally.

Keywords: Zero-Knowledge, Witness-Hiding, Arthur Merlin protocols, Black-box reductions.

_____

[*] Part of this work performed while at the University of Haifa.

[**] Research supported by BSF grant 2006317.

[***] Research supported by BSF grant 2004329 and ISF grant 686/07.

O. Reingold (Ed.): TCC 2009, LNCS 5444, pp. 220–237, 2009.

1   Introduction

In a proof the *prover* tries to convince the *verifier* that a certain statement is true. The basic requirements are *completeness* and *soundness*. The former means that the prover is always able to convince the verifier in the validity of a true statement, while the latter means that the prover is not able to convince the verifier in the validity of a false statement.

In cryptography, the statement typically belongs to NP, and the proof is required to maintain the prover's "privacy". As a consequence, the proof is interactive and randomized, and the verifier only gets statistical confidence in the validity of the statement.

The privacy requirement usually refers to some information about the NP-witness used by the prover. Given the difficulty in capturing what exactly is the information that needs to be hidden, the tendency is to be conservative. This gives rise to the notion of *zero-knowledge* proofs: protocols that do not reveal *anything* beyond the validity of the statement being proved [15].

1.1   Zero Knowledge

Zero-knowledge (ZK) proofs are usually constructed using smaller "atomic" ZK protocols as a building block. The typical atomic protocol is "public coin", requires 3 rounds of interaction, and may convince the verifier in the validity of a false statement with constant probability.[1] Well known examples are the protocols for Quadratic Residuosity [15], 3-Colorability [13], and Hamiltonicity [5].

In order to gain higher statistical confidence in the validity of the statement, the verifier requests to repeat the execution of the atomic ZK protocol multiple times independently. To retain the zero-knowledge property of the atomic protocol, the verifier requests that the repetitions be conducted *sequentially* [14]. This results in high round complexity, and is highly undesirable.

An alternative way to increase the verifier's confidence (preferable in terms of round complexity) is to repeat the sub-protocol in *parallel*. Demonstrating that parallel repetition of the atomic protocols is zero-knowledge, however, appears to be a challenging task. Indeed, as shown by Goldreich and Krawczyk only trivial languages have a black-box zero-knowledge constant-round public-coin proof with negligible soundness error [12].[2]

The Goldreich-Krawczyk impossibility result can be bypassed by considering private-coin protocols [12], or by employing non black-box simulation techniques [3]. Nevertheless, it is still interesting to ask whether black-box techniques can be used to establish security of public-coin protocols. First of all, public-coin protocols tend to be simpler than private-coin ones, and are easier

---

[1] A protocol is public-coin if the verifier's messages consist of his own random coins. Constant-round, public-coin proofs are sometimes referred to as *Arthur-Merlin proofs* (AM) in the literature (cf., [2])

[2] Black-box zero-knowledge essentially means that when establishing the zero-knowledge property of the protocol, the protocol designer is restricted to only observing the "input-output" behavior of a given malicious verifier.

to work with when used as sub-protocols. Secondly, current non black-box techniques are only known to achieve soundness against computationally bounded provers, whereas the "atomic" ZK protocols retain their soundness even in face of a computationally unbounded prover. Thirdly, many of the known public-coin protocols require only 3 rounds of communication, whereas the best private-coin and non-black protocols require 5 and 7 rounds respectively. Finally, when available, black-box techniques are preferable over their "non-black-box" counterparts, mainly because they offer a better tradeoff of security vs. efficiency.

1.2   Witness Indistinguishability

Despite the failure in establishing the zero-knowledge property of constant-round public-coin protocols with negligible soundness, and in particular of the parallel repetition of "atomic" ZK, there is still no evidence that these specific protocols are insecure. This suggests an alternative approach: identify a weaker (yet meaningful) security property and prove that it is satisfied by the protocols.

Feige and Shamir define a protocol to be *witness-indistinguishable* (WI) if the verifier cannot identify the witness that was actually used by the prover in the interaction [9]. Witness-indistinguishability is implied by zero knowledge. Unlike zero-knowledge, however, witness indistinguishability is preserved under parallel repetition. As a consequence, repeating atomic ZK protocols in parallel results in a 3-round WI public-coin protocol with negligible soundness error.

Witness-indistinguishability has turned out to be highly applicable as a building block for higher level protocols. It is sometimes unclear, however, what exactly is hidden by such protocols. On the one hand, WI gives no security guarantee in case that the statement has only one witness associated with it. On the other hand, if a statement has at least two independent witnesses, then any WI protocol for that statement does not reveal the witness being used by the prover in the interaction [9].[3]

1.3   Witness Hiding

Motivated by the above observation, Feige and Shamir put forward the notion of *witness-hiding* [9]. Loosely speaking, a protocol is said to be witness-hiding (WH)

---

[3] The precise formulation of this implication is somewhat technical to state. See [11], Sec 4.6.3.2 for more details. Let us briefly review the argument in a special case. Let f be a one way function and consider the distribution X defined by $x = f(s)$ for a uniformly chosen s. Now consider the distribution X' that consists of two independent copies of X and let $L' = \{x_1, x_2 : \exists i, s_i \text{ s.t. } f(s_i) = x_i\}$. Note that $L' \in NP$ and L' has at least two witnesses on any input in the support of X'. [9] show that a WI proof for L' is hiding witnesses for the distribution X'. Loosely speaking, this follows because if a verifier V^* finds witnesses on X' then V^* can be used to invert f as follows: Given x chosen from X, one can sample another x' from X together with a witness s'. We then set (x_1, x_2) to be a random ordering of (x, x') and prove that the pair $(x_1, x_2) \in L$ to the verifier using the witness s'. We have assumed that the verifier produces a preimage for one of the two x's with noticeable probability. Since the proof is WI the verifier "does not know" which of the two x's was used, V^* can be used to invert the one-way function with noticeable probability.

if at the end of the protocol the verifier cannot compute any new witness that he did not know before the protocol began. This is a natural security requirement, and can replace zero-knowledge in many cryptographic protocols.

Clearly, any ZK protocol is also WH, for any distribution of instances and for any efficiently computable function of the witness. The converse, however, is not true in general.[4] A well known question in this context is whether the parallel repetition of any of the "classical" 3-round protocols hides "interesting" functions of the witness for "interesting" distributions on the instances.

The results of [9] exhibit languages and distributions for which WI protocols are also WH, and thus provide an example where the answer is affirmative. In particular, this is an example where black-box techniques give 3-round public-coin WH protocols whereas, by the results of [12], black-box techniques cannot give such efficient ZK protocols. We remark that WI is not sufficiently strong to *always* imply WH. For example, WI is meaningless for languages that have only one witness, whereas WH is not.

1.4   Our Contributions

The main question we investigate is for what choices of languages and distributions there exist constant-round public-coin WH protocols with negligible soundness. We identify settings in which black-box techniques *cannot* establish the WH hiding property of such protocols. The precise model and results are described below. Before going into these details let us describe some consequences: Suppose that L is a non-trivial language in NP (meaning that $L \notin$ BPP) and suppose that any input $x \in L$ has exactly one witness. Let X be an arbitrary distribution over inputs in L that is hard in the sense that no polynomial size circuit can produce a witness when given x sampled according to X with noticeable probability. We show that:

- It is impossible to show that parallel repetition of 3-colorability or Hamiltonicity is WH for X using certain black-box techniques. This result stands in contrast to the case where L has two or more witnesses, as in this case there exist distributions X for which there are black-box WH proofs (of the type we ruled out above) for the these protocols [9].
- It is impossible to show that generic constructions of Zaps given in [7] are WH for X using black-box techniques.

A consequence of our results is that there exist pairs of indistinguishable distributions for which parallel repetition of 3-colorability (or Hamiltonicity) cannot be shown to have strong witness indistinguishability using certain black-box techniques. The precise details appear in the full version.

Another consequence concerns a recent paper by Pass [17] that, following earlier work [10,6,1], investigates the possibility of constructing a one way function whose inversion task is at least as hard as deciding some NP complete language (via Turing reductions). Pass shows a relationship between the existence of a Turing reduction as above and the existence of constant-round, negligible soundness

---

[4] E.g., 2-round WI protocols (ZAPs) [7] cannot be ZK, but are WH in some special cases. E.g., the case that the statement has "two independent witnesses".

public-coin interactive proofs for SAT that hide the bits of the satisfying assignment. The main point in our context is that Pass's relationship requires that the WH property is established using *black-box techniques*. Our results provide limitations on this approach. The precise details appear in the full version.

1.5 The Notion of Black-Box Witness Hiding

We now explain what we mean by "black-box techniques". For a precise description of the model the reader is referred to Sections 2,3. Our definitions of "black-box WH" follow the framework of "black-box ZK" as defined in [12].

Loosely speaking, the definition of black-box WH requires that the WH property of the protocol is established using a reduction R (the reduction can be thought of as the "black-box simulator" in black-box ZK) that satisfies the following property: When R is given oracle access to a cheating verifier V^* that learns the witness following an interaction with the prover P, and an input x sampled from the distribution X, then either R is able to learn the witness with noticeable probability, or R is able to violate the security assumption on which the protocol is based. We consider several flavors of reductions.

Fully vs. weakly black-box reductions We distinguish between two kinds reductions depending on whether the reductions relies on a generic security assumption, such as "there exist one-way functions" or "there exist bit-commitment schemes", or on a specific security assumption, such as "factoring is a one-way function". To illustrate this distinction consider parallel repetition of the classical protocols for 3-Colorability [15] or Hamiltonicity [5]. These protocols require a bit-commitment scheme and can be seen as generic constructions of proof systems that can use *any* bit-commitment scheme. When considering black-box reductions for showing that these protocols are WH, we distinguish between two cases: A ***fully-black-box reduction*** is oblivious to the choice of commitment scheme and should work for any choice of commitment scheme. This is modeled by giving the reduction oracle access to the commitment scheme and requiring that the reduction works for any implementation of commitment schemes. A ***weakly-black-box reduction*** may be tailored for a specific implementation of the commitment scheme that relies on the hardness of a specific function (e.g., we can consider the protocol for 3-Colorability when implemented using Blum's commitment instantiated with Discrete Log). Naturally, it is more difficult to rule out weakly-black-box reductions than fully black box ones.[5]

Oblivious versus Tailored reductions. Recall that the witness-hiding property is defined with respect to some distribution X on inputs in L. A reduction R may

---

[5] Since in all the reductions considered in this paper access the cheating verifier (e.g., the adversary) as a "black box", the above definition of fully-black-box reduction coincides with the standard use of this notion (cf., [18]). In our definition of weakly-black-box reduction, however, the reduction treats the adversary as a black-box and treats the "hardness assumption" arbitrarily. This is with contrast to the standard definition of weakly-black-box reduction, where the reduction treats the adversary arbitrarily and treats the hardness assumption as a black box.

be **tailored** for a specific distribution X. Alternatively, it may be **oblivious** and work for any distribution X over inputs in L.

To illustrate this point note that in the case of ZK, a black-box simulator is an oblivious reduction (as the simulator is able to simulate a transcript of $(P, V^*)(x)$ on every input $x \in L$ and thus on every distribution over such inputs). In the case of witness-hiding the reduction of [9] is an example of a black-box reduction that is tailored to the specific distribution. Specifically, this reduction *critically relies* on the ability to query V^* on inputs x' that are different than the input x given to the reduction. It is easy to show that oblivious reductions (and in particular black-box simulators) do not benefit form such behavior, as V^* may be chosen as a function of x and refuse to answer in interactions on inputs $x' \neq x$. In contrast, in the setup of witness-hiding the verifier V^* must agree to participate in the protocol on a noticeable fraction of inputs x in the support of X in order to break the witness-hiding property with noticeable probability.

The fact that tailored reductions can benefit from querying V^* on many different inputs is a new consideration that does not come up in the setting of ZK. The main technical difficulty dealt with in this paper is the development of techniques that handle such reductions.

Embedding reductions. For tailored reductions we say that R is **non-embedding** if for every pair of different inputs, if R queries V^* on both inputs then all queries for one input are made before the first query to the other input. A reduction is *embedding* if it does not obey the previous requirement.

Reductions in the literature. The reductions that establish the ZK property of 3-Colorability and Hamiltonicity are fully-black-box oblivious reductions. In fact, as explained earlier all black-box simulators that establish ZK are *oblivious* reductions. There are examples in the literature where the WH property is established using *tailored* reductions (e.g. the reduction of [9] that we sketched earlier that is a tailored and fully-black-box). However, to the best of our knowledge all the reductions in the literature are *non embedding*.

1.6   Statement of Our Results

We now state our results more precisely and explain which kind of black-box reductions we can rule out. Recall that WH is defined with respect to a language $L \in$ NP and a distribution X over instances in L that is hard in the following sense. No polynomial-size circuit can produce a witness when given x sampled according to X with noticeable probability. We are assuming that $L \notin$ BPP and that every $x \in L$ has exactly one witness (0therwise WH may follow from WI).[6]

---

[6] We can relax this assumption and consider protocols where the goal is to hide some specific function g of the witness. We refer to g as a "feature" of the witness. We say that g is uniquely determined if for every input $x \in L$ and every two witnesses w_1, w_2 for x, $g(w_1) = g(w_2)$. (A special case is the function $g(w) = w$ that is uniquely determined in the case where every $x \in L$ has one witness). Our lower bounds apply to any reduction that establishes witness-hiding of some uniquely determined feature of the input, and our results in Section 3 are stated using this terminology.

Our results apply for any constant-round public-coin protocol that has negligible soundness. We now describe our precise results for various kinds of reductions:

Oblivious reductions: We show that oblivious reductions cannot be used to establish black-box WH even if they are weakly-black-box. This is a simple extension of the lower bounds of [12] on black-box ZK. The precise formulation of this result appears in Theorem 3.1.

Tailored non-embedding reductions: We show that tailored reductions that are non-embedding cannot be used to establish WH of protocols that are proofs of knowledge. This result also applies for weakly-black-box reductions. The precise formulation of this result appears in Theorem 3.2. For this result we need to develop new techniques that can handle tailored reductions. Recall that parallel repetition of 3-Colorability or Hamiltonicity *are* proofs of knowledge and therefore we obtain results on these specific protocols. As we can handle weakly-black-box reductions, these results apply to *any* implementation of these protocols using any choice of commitment schemes.

Embedding fully-black-box reductions: While we do not know how to handle embedding reductions in general, we can handle embedding reductions that are fully-black-box for protocols with an additional property, which we refer to as TKE for "Transcript Knowledge Extractor". Loosely speaking, such protocols have the property that for any prover P^* that convinces V on some input x with probability that is larger than the soundness of the protocol, V can learn a witness for x at the end of the interaction assuming it can break the security assumption on which the protocol is based. We elaborate on this property below (a precise definition appears in Section 3.4).

What we show is that such protocols cannot have a fully-black-box reduction even if the reduction is tailored and embedding. The precise formulation of this result appears in Theorem 3.3. Many generic constructions in the literature of interactive proofs for NP-complete languages have the TKE property. In particular, the protocols for parallel repetition of 3-colorability or Hamiltonicity are fully-black-box (in the sense that they can use any bit-commitment scheme) and have the TKE property (in the sense that a verifier that can break the commitment scheme can learn the witness). It follows that these protocols cannot have fully-black-box reductions even if the reductions are tailored and embedding.

Another interesting case is that of Zaps [7]. These are 2-round WI proofs. Generic constructions of Zaps are not known to be proofs of knowledge (and therefore the lower bound in the previous item does not apply). Nevertheless, we observe that the generic constructions of Zaps in [7] have the TKE property. It follows that these protocols cannot have fully-black-box reductions even if the reductions are tailored and embedding.

1.7   Transcript Knowledge Extractors

We now discuss the TKE assumption mentioned earlier. Loosely speaking a reduction for an interactive proof from an hardness assumption (e.g., the existence of bit commitment schemes) has a ***Transcript Knowledge Extractor*** (TKE) if the following holds: There is a polynomial-time machine E that has access to an oracle that breaks commitments, and E is able to extract witnesses from "most" accepting transcripts between any prover P^* and the verifier V. The precise definition is given in Definition 3.8. In many cases (e.g., 3-colorability and Hamiltonicity) the soundness analysis of the protocol implicitly presents a transcript knowledge extractor. More details are given in Section 3.4.

What about non-black-box techniques? Our results only apply when the proof establishing witness-hiding is done by a black-box reduction. As we explained earlier, following the breakthrough paper of Barak [3] there are examples of protocols where the reduction proving zero-knowledge is non-black-box and relies on the code of V^*. We remark that [4] shows that parallel repetition of Hamiltonicity *is* ZK assuming that CIRCUIT-SAT has "small" circuits. Thus, we cannot expect unconditional results that rule out non black-box reductions establishing WH of this protocol. A natural question (that is not addressed in this paper) is to try and prove impossibility results for non-black-box reductions under hardness assumptions.

Organization of this paper. Due to space limitations this extended abstract does not contain all our results and there are no proofs. The reader is referred to the full version for more details. We give formal definitions for WH in Section 2 and our results are stated in Section 3.

2 Definitions of Witness Hiding

2.1 Preliminaries on Interactive Proofs

We use standard definitions of interactive machines and protocols. The reader is referred to [11] for an extensive treatment that also introduces this notation.

In this paper we are only interested in interactive proofs for languages L in NP. Such languages are defined using a *witness relation* R_L (that is L contains all inputs x such that there exist $w \in R_L(x)$). When we consider $L \in$ NP we always assume that it comes with some specific witness relation R_L and that on input $x \in L$ the prover in the interactive proof is provided with some witness $w \in R_L(x)$. We want that completeness, soundness and privacy requirements are maintained for every choice of this witness. We use the following definition.

Definition 2.1 (interactive proofs for NP languages). *Let L be a language in NP. A witness choice is a function that maps every input x in L to a random variable W that is distributed over $R_L(x)$. A pair (P, V) of interactive machines is an* interactive proof *for L with completeness $c(n)$ and soundness $s(n)$ if V is probabilistic-polynomial-time and the following two conditions hold:*

- *Completeness: For every $x \in L$ and witness choice W, $\Pr[(P(W(x)), V)(x) = 1] \geq c(|x|)$.*
- *Soundness: For every $x \notin L$ and machine P^*, $\Pr[(P^*, V)(x) = 1] \leq s(|x|)$.*

If the completeness and soundness parameters are omitted then we mean perfect completeness (that is $c(n) = 1$) and negligible soundness (that is $s(n) = neg(n)$).

An interactive proof is public-coin if every message of V consists of independent random coins. The number of rounds in the protocol is the overall number of messages.

2.2   The Concept of Witness Hiding

Witness-hiding interactive proofs (defined by [9]) have the following property: if a verifier can find a witness to the NP statement that is being proven following an interaction with a prover, then he could have done so without such an interaction. This notion is defined with respect to a distribution ensemble over inputs in L.

Definition 2.2. *Let L be a language in* NP. *A distribution ensemble $X = \{X_n\}$ is over* positive instances with respect to *L if for every n, X_n assigns positive probability only to instances in $L \cap \{0,1\}^n$.*

Definition 2.3 (witness-hiding). *Let $L \in$ NP and let R_L be its witness rela-tion. Let $X = \{X_n\}$ be a distribution ensemble over positive instances. An interac-tive proof (P, V) for L is* witness-hiding *with respect to X if the following condition holds: If for every sufficiently large n and every polynomial size circuit C,*

$$\Pr_{X \leftarrow X_n}[C(X) \in R_L(X)] = neg(n)$$

then for every polynomial-time V^*, every witness choice W, sufficiently large n and every auxiliary input z_n,

$$\Pr_{X \leftarrow X_n}[(P(W(X)), V^*(z_n))(X) \in R_L(X)] = neg(n)$$

Note that there is an inherent difference between the definition of witness-hiding proofs and zero-knowledge proofs in the sense that the definition is with respect to an ensemble X, whereas in zero-knowledge proofs (or witness-indistinguishable proofs) information should not leak on *any* input x.

2.3   Hiding Features of the Witness

Definition 2.3 is only concerned with whether V^* can recover an *entire* wit-ness. A stronger privacy requirement is that V^* does not learn some efficiently computable feature of a witness. Our results are stated using this more general notion. We use the following definition.

Definition 2.4 (feature function). *Let L be a language in* NP *and let $m(n)$ denote the length of witnesses $w \in R_L(x)$ for inputs $x \in L$ of length n. Let $\ell(n)$ be an integer function. A feature function g is a polynomial-time computable function $g : \{0,1\}^{m(n)} \to \{0,1\}^{\ell(n)}$. We say that g is* uniquely determined *on an input $x \in L$, if $w_1, w_2 \in R_L(x)$ implies $g(w_1) = g(w_2)$. In that case, we sometimes abuse the notation and write $g(x)$ rather than $g(w)$. We say that g is* uniquely determined *on a distribution X that is distributed over $L \cap \{0,1\}^n$, if it is uniquely determined on every input in the support of X.*

Using this terminology, we can define the following notion of witness-hiding proofs that hides a uniquely determined feature g of the witness. (We restrict our attention to uniquely determined features as otherwise the feature of a witness $W(X)$ *depends* on the witness choice W). Loosely speaking, the definition below says that if the verifier V^* can distinguish the feature $g(X)$ from uniform following an interaction with the prover then he can do that prior to the interaction.

Definition 2.5 (witness-hiding for a uniquely determined feature g).
Let $L \in$ NP and let R_L be its witness relation. Let $X = \{X_n\}$ be a distribution ensemble over positive instances of L. An interactive proof (P, V) for L is witness-hiding *a feature g that is uniquely determined with respect to X, if the following condition holds: If for every sufficiently large n and every polynomial size circuit C,*

$$| \Pr_{X \leftarrow X_n} [C(X) = g(X)] - 2^{-\ell(n)}| = neg(n)$$

then for every polynomial-time V^*, every witness choice W, sufficiently large n and every auxiliary input z_n,

$$| \Pr_{X \leftarrow X_n} [(P(W(X)), V^*(z_n))(X) = g(X)] - 2^{\ell(n)}| = neg(n)$$

Remark 2.1 (The case of one witness). In the case that a language $L \in$ NP is defined using a witness relation R_L where every $x \in L$ has exactly one witness then any feature g is uniquely determined. This in particular applies to the feature $g(w) = w$. With this choice definitions 2.5 and 2.3 coincide. Our lower bounds apply to every uniquely determined feature of witnesses and in particular apply to the standard notion of witness-hiding in the case that there is only one witness.

3   Black-Box Witness-Hiding and Our Results

We study reductions that establish the conditions of Definitions 2.3 and 2.5. Consider an interactive proof (P, V). A black-box reduction R that establishes the witness-hiding property of (P, V) is a polynomial-time machine that receives oracle access to a "cheating verifier" V^* (that is not necessarily efficient). It is assumed that V^* is able to break the witness-hiding property of the proof system and learn the feature $g(X)$ following an interaction with P. As our goal is to prove lower bounds on reductions we make it easier for the reduction and assume that V^* learns $g(X)$ with probability one (this only makes our results stronger). The reduction R is required to perform one of the following two tasks when given oracle access to such a V^*:

- Learn the feature $g(X)$ with noticeable advantage when given X as input. (This shows that V^* could have learned $g(X)$ without interacting with the prover).
- Break the security assumption on which the protocol is based. (This gives a contradiction in case V^* is efficient).

We distinguish between two kinds of constructions of interactive protocols depending on whether the protocol relies on a generic security assumption (e.g., "there exist one-way functions" or "there exist bit-commitment schemes") or on a specific security assumption (e.g., "factoring is a one-way function"). This distinction is described in the next sections.

3.1   Weakly Black-Box Reductions

In this paper we consider the following notion of weakly-black-box reductions. The "protocol designer" chooses a specific "hardness assumption" (which we model, without lost of generality, as a one-way function) and designs specific machines P, V to be used by the prover and verifier. The designer also chooses a language $L \in \text{NP}$, an ensemble X over positive instances and a feature g that is uniquely determined for X. His goal is to show that (P, V) is witness-hiding for these specific choices and this allows the reduction R to depend in an arbitrary (non black-box) way on all the previous choices. A precise definition follows:

Definition 3.1 (weakly black-box reduction establishing witness hiding). *Let f be a length preserving function. Let L be a language in NP. Let (P, V) be a proof system for L. Let $X = \{X_n\}$ be a distribution ensemble over positive instances of L, and g be a feature that is uniquely determined for X. We say that R is a* weakly-black-box *WH reduction from f if R is a polynomial-time oracle machine and there exist polynomials $p(n)$ and $k(n)$ such that for every input length n, and every deterministic (not necessarily efficient) algorithm V^*: If there is a witness choice $W(x)$ such that $\Pr_{X \leftarrow X_n}[(P(W(X)), V^*)(X) = g(X)] = 1$ then*

- *either R^{V^*} inverts f on random inputs of length $k(n)$ with probability $1/p(n)$,*
- *or $\Pr_{X \leftarrow X_n}[R^{V^*}(X) = g(X)] \geq 2^{-\ell(n)} + 1/p(n)$.*

Remark 3.1 (Relationship to black-box simulation). It is natural to compare our definitions to that of "black-box simulation" introduced in [12]. The notion of black-box simulation corresponds to a specific protocol (P, V) and requires that there is one reduction R (called black-box simulator) so that for every *efficient* V^*, $R^{V^*}(x)$ simulates a transcript that is indistinguishable from $(P, V^*)(x)$. It turns out that all black-box simulators in the literature satisfy a stronger requirement: For every V^* (not necessarily efficient) either $R^{V^*}(x)$ simulates a transcript or it is able to invert some one-way function f. Note that every such reduction is a weakly-black-box WH reduction from f.

3.2   Our Results on Weakly Black-Box Reductions

We consider several notions of weakly-black-box reductions. A reduction R is *oblivious* if it does not depend on the choice of the distribution ensemble X and one reduction applies to any distribution ensemble. We remark that all proofs of "black-box ZK" [12] in the literature are oblivious reductions.

Definition 3.2 (oblivious reductions). *Let L, P, V, g, f be as in Definition 3.1. Let R be a polynomial-time oracle procedure. We say that R is an* oblivious reduction *if for every distribution ensemble X over positive instances, R is a weakly-black-box* WH *reduction from f with respect to X.*

We show that assuming NP $\neq$ BPP, oblivious reductions cannot show witness-hiding for constant-round public-coin protocols with negligible soundness, with respect to NP complete languages L, where every input $x \in L$ has exactly one witness. This result (stated below) is an easy extension of the negative results of [12] for black-box ZK.

Theorem 3.1. *Let L be a language in* NP *and let R_L be its witness relation. Let (P, V) be a constant-round public-coin interactive proof for L with negligible soundness. Assume that the feature $g(w) = w$ is uniquely determined on every input x (that is that every $x \in L$ has exactly one witness). Let R be an oblivious weakly-black-box* WH *reduction from some one-way function f. Then $L \in$ BPP.*

We now consider reductions that can be tailored to a specific distribution ensemble X. Such a weakly-black-box reduction R receives an input x and oracle access to a "cheating verifier" V^* that breaks the witness-hiding property. When R queries V^* it supplies some partial history of the protocol (P, V) and V^* replies with his next message in the protocol. A part of the partial history is the input x' to the protocol (P, V). Note that R may query V^* on partial histories that contain inputs x' that are different from the input x given to R. A reduction R is non-embedding if it finishes all queries to V^* on one input before it queries V^* on some other input.

Definition 3.3 (non-embedding reductions). *Let L, P, V, g, f, X be as in Definition 3.1 and let R be a weakly-black-box* WH *reduction from f. We say that R is a* non-embedding *reduction if for every input x and every oracle V^* and every two inputs $x_1 \neq x_2$, if $R^{V^*}(x)$ makes a query containing x_1 before making a query containing x_2 then all queries that contain x_1 are made before the first query that contains x_2.*

We show that if a non-embedding reduction is used to show witness-hiding for a constant-round public-coin protocol with negligible soundness with respect to some distribution X and uniquely determined feature g, and if furthermore the protocol is also a proof of knowledge, then it is possible to efficiently predict $g(x)$ with noticeable advantage when given x sampled from X. This means that it is impossible to use such reductions to hide features that are hard to predict.

Theorem 3.2. *Let $L \in$ NP, let R_L be its witness relation. Let X be a distribution ensemble over positive instances of L and let g be a feature that is uniquely determined with respect to X. Let (P, V) be a constant-round public-coin interactive proof for L with negligible soundness and assume that (P, V) is a proof of knowledge with negligible knowledge error (see Definition 3.7). Let R be a non-embedding weakly-black-box* WH *reduction from a one-way function f. Then there is a polynomial-time machine M and a polynomial p such that for every sufficiently large n, $\Pr_{X \leftarrow X_n}[M(X) = g(X)] \geq 2^{-\ell(n)} + 1/p(n)$.*

In particular, if every input $x \in L$ has one witness then the feature $g(w) = w$ is uniquely determined. The theorem says that if (P, V) is a proof of knowledge then the existence of a non-embedding reduction R gives that one can efficiently find witnesses when given x sampled from X with noticeable probability (and thus X is not a "hard distribution").

Corollaries on specific protocols. Consider parallel repetition of 3-Colorability [13] and Hamiltonicity [5] using any choice of commitment scheme (that may be based on an arbitrary one-way function). These protocols are constant-round public-coin interactive proofs with negligible soundness for complete languages in NP. Furthermore, both these protocols are proofs of knowledge with negligible knowledge error. Thus, Theorems 3.1 and 3.2 apply and give limitations on reductions that establish the WH property of these protocols.

3.3   Fully-Black-Box Reductions

In a fully-black-box construction the protocol designer is given a cryptographic primitive as a black-box. (In this paper we consider the primitives: one-way function, one-way permutation and information theoretically binding bit commitments). In this setup the protocol designer receives a black-box that implements the basic primitive. He designs oracle machines $P^{(\cdot)}, V^{(\cdot)}$ to be used by the prover and verifier. We start by formally defining this setup.

Definition 3.4 (black-box interactive proofs). *Let L be language in* NP. *Let $\mathcal{F}$ be a set of functions from strings to strings. A pair $(P^{(\cdot)}, V^{(\cdot)})$ of oracle machines is a $\mathcal{F}$-black-box interactive proof for L if V is probabilistic polynomial time and for every $f \in \mathcal{F}$, the pair (P^f, V^f) satisfy the completeness and soundness properties in Definition 2.1.*

We now consider several families of possible oracles that model one-way functions, one-way permutations and bit-commitment schemes. The same framework, however, can be used to describe most cryptographic primitives.

Definition 3.5 (oracles for primitives). *Let O_{OWF} denote the set of all length preserving functions. Let O_{OWP} be the subset of all functions in O_{OWF} that are permutations on every input length. Given $f \in O_{\mathrm{OWF}}$, an algorithm T η-breaks f on security parameter k if $\Pr_{X \leftarrow U_k}[T(f(X)) \in f^{-1}(f(X))] \geq \eta$.*
Let O_{BC} denote the set of all functions f that given a bit b and a string $r \in \{0,1\}^k$ produce a string $c \in \{0,1\}^k$. We furthermore require that f is binding, namely that for every k and $r_1, r_2 \in \{0,1\}^k$, $f(0, r_1) \neq f(1, r_2)$. Given $f \in O_{\mathrm{BC}}$, an algorithm T η-breaks f on security parameter k if $\Pr_{B \leftarrow U_1, R \leftarrow U_k}[T(f(B, R) = B] \geq 1/2 + \eta/2$.[7]

---

[7] The choice of dividing η by 2 is so that the success probability of T is one when $\eta = 1$. This way, for both $O_{\mathrm{OWF}}, O_{\mathrm{BC}}$ an algorithm T that 1-breaks f succeeds with probability one.

Remark 3.2 (interactive commitment schemes). The family O_{BC} defined above corresponds to perfectly binding non-interactive commitment schemes. In such a scheme the sender commits to a bit b by sending $f(b, r)$ for a randomly chosen r. The sender can later reveal the bit b by sending r and our definition requires that the commitment is binding.

One can consider more relaxed notion of commitment schemes in which the commitment phase is an interactive protocol between the sender and receiver. In such a scheme the binding property can be statistical rather than perfect (namely, binding only holds with high probability over the receiver's coins). We have chosen the more simple version of commitment schemes in order to simplify the presentation. All our results, however, apply also for the more general notion of interactive statistically binding commitments (and this holds by exactly the same proofs).

Remark 3.3 (3-Colorability and Hamiltonicity). Using this framework the classical protocols for 3-Colorability and Hamiltonicity can be viewed as O_{BC}-blackbox interactive proofs. (This also applies if we modify O_{BC} to capture interactive commitments as explained in Remark 3.2).

We can now give the definition of a fully-black-box reduction. We consider two flavors depending on whether the black-box interactive proof starts from one-way functions or bit-commitment (that is whether f is assumed to come from O_{OWF} or O_{BC}). The definition below is identical to definition 3.1 with the following modifications: all parties (including the verifier V^* and the reduction R) get oracle access to f and the reduction should work for every choice of f in the family of relevant oracles.

Definition 3.6 (fully-black-box reduction establishing witness hiding).
Let L be a language in NP. Let $(P^{(\cdot)}, V^{(\cdot)})$ be a O_{OWF}-black-box interactive proof for L (resp., O_{BC}-black-box interactive proof for L). Let $X = \{X_n\}$ be a distribution ensemble over positive instances of L, and g be a feature that is uniquely determined for X. We say that R is a fully-black-box WH reduction from OWF *(resp.,* fully-black-box WH reduction from BC) *if R is a polynomial-time oracle machine and there exist polynomials $p(n)$ and $k(n)$ such that for every $f \in O_{OWF}$ (resp., every $f \in O_{BC}$) and every input length n, and every deterministic (not necessarily efficient) algorithm V^*: If there is a witness choice $W(x)$ such that $\Pr_{X \leftarrow X_n}[(P^f(W(X)), V^{*f})(X) = g(X)] = 1$ then*

- *either $R^{V^{*f}, f}$ $1/p(n)$-breaks f on security parameter $k(n)$,*
- *or $\Pr_{X \leftarrow X_n}[R^{V^{*f}, f}(X) = g(X)] \geq 2^{-\ell(n)} + 1/p(n)$.*

We note that any fully-black-box reduction R gives a weakly-black-box reduction for any specific choice of f.

3.4 Transcript Knowledge Extractors

We introduce a non-standard notion of proofs of knowledge (which is incomparable to the standard one) and show that black-box interactive proofs from commitment schemes that are constant-round public-coin protocols with negligible soundness, and in addition have "transcript knowledge extractors" cannot have fully-black-box reductions establishing WH. Before we define this new notion, let us first recall the definition of "standard" knowledge extractors.

Definition 3.7 (knowledge extractor [11]). *Let* (P, V) *be an interactive proof system for* $L \in \mathrm{NP}$ *and let* R_L *be its witness relation. A probabilistic machine* E *is a* knowledge extractor *for* (P, V) *and* R_L *with error* $\eta \colon \mathbb{N} \mapsto \mathbb{R}$, *if there exists a polynomial* q_E *such that for every input* $x \in L_n$ *and every deterministic algorithm* P^*, $E^{P^*}(x)$ *runs in expected number of step bounded by* $\frac{q_E(n)}{\delta(x) - \eta(|x|)}$ *and outputs* $w \in R_L(x)$, *where* $\delta(x) = \Pr[(P^*, V)(x) = 1]$.

The new notion applies to black-box interactive proofs (See definition 3.4) and allow the extractor to access an oracle that breaks the security assumption on which the protocol is based. The extractor gets as input a transcript on which V accepts and is required to extract a witness from the transcript (we stress that the extractor does not get oracle access to the prover P^*). A precise definition follows. The definition has two flavors depending on whether the black-box interactive proofs is from one-way functions or bit-commitment.

Definition 3.8 (transcript knowledge extractor (TKE)). *Let* L *be a language in* NP *and let* $(P^{(\cdot)}, V^{(\cdot)})$ *be a* O_{OWF}-*black-box interactive proof for* L *(resp., a* O_{BC}-*black-box interactive proof for* L*). A polynomial-time oracle machine* E *is a* transcript knowledge extractor *with error* $\eta(n)$ *if for every* $f \in O_{\mathrm{OWF}}$ *(resp., every* $f \in O_{\mathrm{BC}}$*) and every algorithm* T *that* 1-*breaks* f *on every security parameter* k *it holds that: For every input* $x \in L$ *and for every deterministic algorithm* P^*, *let* $\tau(x)$ *be the random variable that is the transcript of* $(P^{*f}, V^f)(x)$ *then:*

$$\Pr[\tau(x) \text{ is accepting and } E^{f,T}(\tau(x)) \notin R_L(x)] \leq \eta(|x|)$$

We allow E to access both f and an oracle T that completely breaks f. While transcript knowledge extractors require an oracle that breaks the security assumption, they have the advantage that they do not require oracle access to the prover P^*. This in particular means that they do not rely on rewinding P^* and that the extraction process is efficient even if P^* is inefficient.

3.5   Our Results on Fully-Black-Box Reductions

We now state our main result on fully-black-box reductions. We consider black-box interactive proofs that use one-way functions or commitment schemes. This result applies to any reduction (even one that is embedding) whenever the black-box interactive proof has a transcript knowledge extractor.

Theorem 3.3. *Let* $L \in \mathrm{NP}$ *and let* R_L *be its witness relation. Let* X *be a distribution ensemble over positive instances of* L *and let* g *be a feature that is uniquely determined with respect to* X. *Let* $(P^{(\cdot)}, V^{(\cdot)})$ *be a constant-round public-coin* O_{OWF}-*black-box interactive proof for* L *(resp.,* O_{BC}-*black-box interactive proof for* L*). Assume that the proof system has negligible soundness and a* TKE *with* *negligible error. Let R be a fully-black-box* WH *reduction from* OWF *(resp.,* BC*). Then, there is a polynomial-time machine M and a polynomial p such that for every sufficiently large n,* $\Pr_{X \leftarrow X_n}[M(X) = g(X)] \geq 2^{-\ell(n)} + 1/p(n)$.

The theorem above is very similar to Theorem 3.2 with the exception that it handles general fully-black-box reductions (rather than non-embedding weakly-black-box ones) and requires transcript knowledge extractors (rather than standard knowledge extractors). In the next section we observe that many protocols in the literature have transcript knowledge extractors. In particular, when considering a language L in which every $x \in L$ has exactly one witness, the feature $g(w) = w$ is uniquely determined and the Theorem asserts that one cannot use a fully-black-box reduction to establish WH for distributions X for which finding a witness is hard.

3.6   Prevalence of Transcript Knowledge Extractors

On an intuitive level one can expect that any interactive proof where the privacy of the prover is based on a hardness assumption (e.g., the existence of bit-commitment schemes) has a transcript knowledge extractor as otherwise the hardness assumption is "not really needed" and the security of the protocol follows unconditionally. We do not make such a formal statement and do not know whether a statement of this flavor is true. In the discussion below we observe that many specific interactive proofs in the literature have transcript knowledge extractors. The impossibility results of Theorem 3.3 apply to all these protocols.

3-Colorability. Consider the ZK proof of [13] for 3-colorability. This is a O_{BC}-black-box interactive proof that is a 3-round protocol with perfect completeness and soundness $1 - 1/m$ (where m is the number of edges in the input graph). It is known that this protocol is zero-knowledge. The soundness analysis of this protocol shows that if in the first message of the protocol the prover does not send a commitment to a witness (a legal coloring) then with probability $1/m$ (where m is the number of edges in the input graph) the verifier rejects. It follows that this protocol has a transcript knowledge extractor with $\eta = 1 - 1/m$ as E can open the commitment using the fact it has oracle access to an algorithm T that breaks the commitment. Recall that we are interested in investigating the security of the parallel repetition of this atomic protocol when repeated t times. It is easy to see that after repetition there is a transcript knowledge extractor with error $\eta = (1 - 1/m)^t$. (This follows as if the extractor cannot find a witness in any of the commitments sent in the first round then the probability that the verifier accepts is the expression above).

Graph Hamiltonicity. Consider the ZK proof of [5] for Graph Hamiltonicity. This is a O_{BC}-black-box interactive proof that is a 3-round protocol with perfect completeness and soundness $1/2$. The soundness analysis of this protocol shows that if B does not commit to a graph that is isomorphic the input graph G in its first message then with probability $1/2$ he is caught in the third message. On the other hand if B commits to a graph that is isomorphic to the correct graph then with probability half he reveals a cycle in the graph in the third message and the knowledge extractor can "break" the commitment and find a cycle in the original graph when given the transcript. These properties give a knowledge extractor with $\eta = 1/2$ and parallel repetition reduces η at an exponential rate.

Zaps. These are 2-message WI protocols [7], and are not known to be a (standard) proof of knowledge. Zaps can be either constructed based on non-interactive zero-knowledge (NIZK) proofs, or based on a verifiable pseudo-random generator (VPRG). The "generic" versions of both of these primitives are constructed using trapdoor permutations, where the role of trapdoor permutations in all known constructions is to implement the *hidden bits* (or *hidden random string*) model [8,16,7]. A close examination of these constructions reveals that if one is able to invert the underlying trapdoor permutation then the bits (random string) becomes completely revealed. As observed in [19], such information can be used to extract the witness for the statement. With appropriately chosen parameters (i.e., if the soundness error is small enough), this can be done with all but negligible probability. The same applies for VPRG based constructions. Thus, many of the "generic" zap constructions have transcript knowledge extractors.

Acknowledgements. We thank Oded Goldreich and Rafael Pass for helpful discussions.

References

1. Akavia, A., Goldreich, O., Goldwasser, S., Moshkovitz, D.: On basing one-way functions on np-hardness. In: Proceedings of the 38th Annual ACM Symposium on Theory of Computing (STOC), pp. 701–710 (2006)
2. Babai, L., Moran, S.: Arthur-merlin games: A randomized proof system, and a hierarchy of complexity classes. J. Comput. Syst. Sci. 36(2), 254–276 (1988)
3. Barak, B.: How to go beyond the black-box simulation barrier. In: Proceedings of the 42nd Annual Symposium on Foundations of Computer Science (FOCS), pp. 106–115 (2001)
4. Barak, B., Lindell, Y., Vadhan, S.: Lower bounds for non-black-box zero knowledge. Journal of Computer and System Sciences 72(2), 321–391 (2006)
5. Blum, M.: How to prove a theorem so no one else can claim it. In: Proceedings of the International Congress of Mathematicians, pp. 1444–1451 (1987)
6. Bogdanov, A., Trevisan, L.: On worst-case to average-case reductions for np problems. SIAM Journal on Computing 36(4), 1119–1159 (2006)
7. Dwork, C., Naor, M.: Zaps and their applications. SIAM Journal on Computing 36(6), 1513–1543 (2007)
8. Feige, U., Lapidot, D., Shamir, A.: Multiple noninteractive zero knowledge proofs under general assumptions. SIAM Journal on Computing 29(1), 1–28 (1999)
9. Feige, U., Shamir, A.: Witness indistinguishable and witness hiding protocols. In: Proceedings of the 22nd Annual ACM Symposium on Theory of Computing (STOC), pp. 416–426. ACM, New York (1990)
10. Feigenbaum, J., Fortnow, L.: Random-self-reducibility of complete sets. SIAM Journal on Computing 22(5), 994–1005 (1993)

11. Goldreich, O.: Foundations of Cryptography: Basic Tools. Cambridge University Press, Cambridge (2001)
12. Goldreich, O., Krawczyk, H.: On the composition of zero-knowledge proof systems. SIAM J. Comput. 25(1), 169–192 (1996); Preliminary version in ICALP 1990
13. Goldreich, O., Micali, S., Wigderson, A.: Proofs that yield nothing but their validity and a methodology of cryptographic protocol design (extended abstract). In: FOCS, pp. 174–187. IEEE, Los Alamitos (1986)
14. Goldreich, O., Oren, Y.: Definitions and properties of zero-knowledge proof systems. Journal of Cryptology 7(1), 1–32 (1994)
15. Goldwasser, S., Micali, S., Rackoff, C.: The knowledge complexity of interactive proof systems. SIAM J. Comput. 18(1), 186–208 (1989); Preliminary version in STOC 1985
16. Kilian, J., Petrank, E.: An efficient noninteractive zero-knowledge proof system for np with general assumptions. J. Cryptology 11(1), 1–27 (1998)
17. Pass, R.: Parallel repetition of zero-knowledge proofs and the possibility of basing cryptography on np-hardness. In: IEEE Conference on Computational Complexity, pp. 96–110 (2006)
18. Reingold, O., Trevisan, L., Vadhan, S.P.: Notions of reducibility between cryptographic primitives. In: Naor, M. (ed.) TCC 2004. LNCS, vol. 2951, pp. 1–20. Springer, Heidelberg (2004)
19. Santis, A.D., Persiano, G.: Zero-knowledge proofs of knowledge without interaction. In: Proceedings of the 33rd Annual Symposium on Foundations of Computer Science (FOCS), pp. 427–436 (1992)

Secure Computability of Functions in the IT Setting with Dishonest Majority and Applications to Long-Term Security

Robin Künzler[1], Jörn Müller-Quade[2,*], and Dominik Raub[1,**]

[1] ETH Zurich, Department of Computer Science, CH-8092 Zurich, Switzerland
robink@student.ethz.ch, raubd@inf.ethz.ch
[2] IKS/EISS, Fakultät für Informatik, Universität Karlsruhe (TH), Germany
muellerq@ira.uka.de

Abstract. While general secure function evaluation (SFE) with information-theoretical (IT) security is infeasible in presence of a corrupted majority in the standard model, there are SFE protocols (Goldreich et al. [STOC'87]) that are computationally secure (without fairness) in presence of an actively corrupted majority of the participants. Now, computational assumptions can usually be well justified at the time of protocol execution. The concern is rather a potential violation of the privacy of sensitive data by an attacker whose power increases over time. Therefore, we ask which functions can be computed with long-term security, where we admit computational assumptions for the duration of a computation, but require IT security (privacy) once the computation is concluded.

Towards a combinatorial characterization of this class of functions, we also characterize the classes of functions that can be computed IT securely in the authenticated channels model in presence of passive, semi-honest, active, and quantum adversaries.

Keywords: long-term security, information-theoretic security, corrupted majority, secure function evaluation.

1   Introduction

In cryptography one distinguishes *computational (CO) security* which could in principle be broken by a very powerful adversary and *information theoretical (IT) security* which withstands even an unlimited attacker. However, general IT secure protocols fail in presence of an adversary that may corrupt a majority of the participants. On the other hand, an unlimited attacker is not a realistic threat and the problem with CO assumptions is not so much that these could be unjustified right now, but that concrete CO assumptions could *eventually* be broken by an attacker whose power increases over time. With such a more realistic threat model in mind an interesting question arises: **Which cryptographic tasks can be realized with long-term (LT) security?** I.e., which tasks can be realized in presence of an attacker (potentially corrupting a majority of protocol

---

* Thanks for financial support from the European Commission (SECOQC).
** Supported by the Swiss National Science Foundation (SNF), project no. 200020-113700/1.

O. Reingold (Ed.): TCC 2009, LNCS 5444, pp. 238–255, 2009.

participants) who is CO limited during the protocol execution, but becomes unlimited afterwards?

In this work we study multi-party secure function evaluation (SFE). The main result is a classification of the functions which can be computed with LT security over a network of authenticated channels. Furthermore we give a classification of all the 2-party functions which can securely be computed in presence of an adversary who is unlimited from the start. This class is strictly contained in the class of functions which can be computed with LT security and the notion of LT security hence lies strictly between CO security and IT security.

Quantum cryptography can achieve tasks, like IT secure key distribution, which cannot be achieved classically. For the task of secure function evaluation it is not known if quantum cryptography can achieve anything beyond the classically possible[1]. However, in this work we show that the class of 2-party functions which can be realized with quantum cryptography is strictly contained in the class of 2-party functions realizable with LT security. From this inclusion novel impossibility results for quantum cryptography arise that are no direct consequences of the results by Mayers [25] or Kitaev [1].

All results in this paper are constructive (whenever it is claimed that a class of functions is securely computable a protocol is given) and proven in a stand-alone simulatability based security model with a synchronous communication network (see e.g. [15]).

1.1 Contributions

To combinatorially characterize the class of functions which are computable with LT security we first characterize the class of *passively computable functions* $\mathfrak{F}_{pas}^{aut}$ which can securely be computed by parties connected by authenticated channels in presence of a CO unlimited passive[2] adversary who must behave according to the protocol. Next we characterize the class of *semi-honestly computable functions* $\mathfrak{F}_{sh}^{aut}$ which are securely computable in the same setting as above but in presence of a stronger, *semi-honest*[3] adversary, that has to stick to the protocol, but may replace his inputs or lie about his local output.

To prove a separation between the notion of LT security and IT security we characterize the class $\mathfrak{F}_{2act}$ of all 2-party functions which are securely computable in presence of an unlimited active adversary. We furthermore provide a necessary condition (which we conjecture to be also sufficient) for membership in the class of *actively computable functions* $\mathfrak{F}_{act}^{aut}$ that are securely computable in presence of an active adversary in the authenticated channels model with broadcast (BC). Next we consider the class of 2-party functions $\mathfrak{F}_{2qu}$ that can securely be computed where the parties may use quantum cryptographic protocols and the attacker is an unlimited active quantum adversary. We show that the class $\mathfrak{F}_{2qu}$ is strictly contained in the class $\mathfrak{F}_{2sh}$ of semi-honestly computable 2-party functions which gives rise to novel impossibility results beyond those of Mayers [25] or Kitaev [1].

---

[1] However, quantum bit commitment is impossible [25] and hence no function implying bit commitment is computable.

[2] In the literature our notion of passive is also occasionally referred to as semi-honest.

[3] In the literature our notion of semi-honest is also sometimes referred to as *weakly* semi-honest or *weakly* passive.

To obtain the desired result on LT security we prove that the class of semi-honestly computable functions $\mathfrak{F}_{sh}^{aut}$ equals the class $\mathfrak{F}_{lts}^{bc}$ of functions which can LT securely be computed given an authenticated BC channel. Furthermore, we show that the class of LT securely computable functions remains unchanged if we replace the authenticated BC channel by a network of authenticated channels or by a realistic communication infrastructure consisting of a network of insecure channels with a given public-key infrastructure (PKI). Hence our classification applies to a very practical internet-like setting.

Unlike our IT secure protocols the LT secure protocols given in this work do not achieve robustness or fairness. We show that this is optimal in the sense that generally functions implementable with LT security cannot be implemented with fairness. However, we present protocols which guarantee that only a specific designated party can abort the computation after learning the output. I.e. the fairness property can only be violated by this designated party. Interestingly these protocols make use of CO secure oblivious transfer (OT) protocols even though OT itself cannot be achieved with LT security.

Summarizing our results and importing the treatment of complete two party functions from [21] (i.e. functions which are cryptographically as powerful as oblivious transfer) we arrive at a complete classification of two party functions. Interestingly, there is a class of functions which cannot securely be computed but still are not complete. This shows that for non-boolean functions there is no zero-one law for privacy [9].

1.2   Related Work

Secure computability of functions was first discussed by [9]. They characterize the symmetric boolean functions (all parties receive the same output $y \in \{0, 1\}$) that can be computed with IT security in presence of passive adversaries in the private channels model. In this scenario functions are either computable or complete (zero-one law for privacy).

Kushilevitz [23,24] and Beaver [2] presented the first results for non-boolean functions describing the symmetric 2-party functions which can be computed with perfect security in presence of an unbounded passive adversary. Our protocols and proof techniques draw heavily upon [24]. Also, in the 2-party setting, [26] sketches a generalization of [24] to the asymmetric, IT case, connections to LT security and discusses quantum aspects, though without proper formalization or proofs. Our work goes beyond the results of [9,2,24,26] in that we consider IT secure computability of asymmetric, non-boolean functions, in presence of passive, semi-honest, active, and quantum adversaries, for the most part in the multi-party setting.

Gordon et al. [17] characterize the boolean functions computable with CO fairness in the 2-party setting in presence of active adversaries. Our protocols for active adversaries are robust (and hence fair) and being applicable to asymmetric, non-boolean functions, pertain to a larger class of functions than those of [17], but in the IT scenario instead of the CO setting.

Other works that deal with the computability of 2-party functions in the perfect or IT setting are [19,20,3,21]. However, these papers focus mostly on reducibility and completeness, while we are more interested in computability in the authenticated channels model and implications for LT security. Computability of a few interesting special functions in presence of dishonest majorities is discussed in [6].

Our impossibility result in the quantum case makes use of a result of Kitaev showing the impossibility of quantum coin flipping which is published in [1].

Everlasting security from temporary assumptions has been investigated in cryptographic research for some time. It was shown that a bound on the memory available to the adversary allows key exchange and OT protocols [8,7] which remain secure even if the memory bound holds only during the execution of the protocol. This idea has been pursued further to achieve everlasting security from a network of distributed servers providing randomness [28]. In [13] it was shown that using a CO secure key exchange in the bounded storage model need not yield everlasting security. For some time general quantum cryptographic protocols were sought which obtain everlasting security from a temporary assumption. Such protocols are now generally accepted to be impossible [5]. Additional assumptions, like a temporary bound on the quantum memory can again provide everlasting security for secure computations [11].

In this paper we investigate the power of temporary CO assumptions in the standard model. This is along the lines of [27]. However, in [27] strong composability requirements are imposed under which little is possible without additional setup assumptions, like the temporary availability of secure hardware.

2 Security Definitions and Notation

In *secure function evaluation* (SFE) the goal is to compute a function $f : \mathcal{X}_1 \times \ldots \times \mathcal{X}_n \to \mathcal{Y}_1 \times \ldots \times \mathcal{Y}_n$ securely among n parties $\mathfrak{P} = \{P_1, \ldots, P_n\}$.[4] Each party $P_i \in \mathfrak{P}$ ($i \in [n] := \{1, \ldots, n\}$) holds an input $x_i \in \mathcal{X}_i$ from a finite set $\mathcal{X}_i$ and is supposed to receive output $f_i(x_1, \ldots, x_n) := y_i \in \mathcal{Y}_i$, where $(y_1, \ldots, y_n) = f(x_1, \ldots, x_n)$. We extend this notation to sets $M = \{P_{m_1}, \ldots, P_{m_{|M|}}\} \subseteq \mathfrak{P}$ and write $f_M(x_1, \ldots, x_n) := y_M := (y_{m_1}, \ldots, y_{m_{|M|}})$ and $x_M := (x_{m_1}, \ldots, x_{m_{|M|}})$. We call the set of all n-party functions $\mathfrak{F}_n$ and the set of multi-party functions $\mathfrak{F} := \bigcup_{n \geq 1} \mathfrak{F}_n$.

In order to compute the function f the parties may execute a protocol π, utilizing a set of resources[5] (communication primitives) R. We designate by $\mathfrak{H} \subset \mathfrak{P}$ the set of honest parties, that execute their protocol machine π_i as specified by protocol π, and by $\mathfrak{E} := \mathfrak{P} \setminus \mathfrak{H}$ the set of corrupted parties that may deviate from the protocol. We generally make the worst case assumption that corrupted parties are controlled by a central adversary E. The adversary (if present, i.e. if at least one party is corrupted) acts for the corrupted parties, sees messages sent over authenticated channels, and can manipulate messages sent over insecure channels. If no party is corrupted, we assume that no adversary is present. External adversaries that can listen on authenticated channels and manipulate insecure channels even when no party is corrupted are easily modelled by adding an additional party, that has constant function output and whose input is ignored.

We define *security* using a simulation based stand-alone[6] model (see e.g. [15]) with synchronous message passing. The security of a protocol (the real model) is defined with respect to an ideal model, where f is evaluated by a *trusted third party* or *ideal*

---

[4] In the 2-party setting we will occasionally use A and B instead of P_1 and P_2.

[5] In this work these are most often a complete network of authenticated channels or an authenticated broadcast (BC) channel.

[6] As opposed to a universally composable model.

Table 1. Basic Security Paradigms

Paradigm	Short	$\mathcal{D} =$	$\mathcal{E} =$	$\mathcal{S} =$	$\epsilon(\kappa)$	Notation
Perfect security	PF	Algo	Algo	Algo	$\epsilon(\kappa) = 0$	$\pi \succcurlyeq^{\mathsf{PF}} I$
Information-theoretical security	IT	Algo	Algo	Algo	$\epsilon(\kappa) < \mathsf{negl}$	$\pi \succcurlyeq^{\mathsf{IT}} I$
PF security with efficient simulator	PFE	Algo	Algo	Poly	$\epsilon(\kappa) = 0$	$\pi \succcurlyeq^{\mathsf{PFE}} I$
IT security with efficient simulator	ITE	Algo	Algo	Poly	$\epsilon(\kappa) < \mathsf{negl}$	$\pi \succcurlyeq^{\mathsf{ITE}} I$
Computational security	CO	Poly	Poly	Poly	$\epsilon(\kappa) < \mathsf{negl}$	$\pi \succcurlyeq^{\mathsf{CO}} I$
Long-term security	LT	Algo	Poly	Poly	$\epsilon(\kappa) < \mathsf{negl}$	$\pi \succcurlyeq^{\mathsf{LT}} I$

functionality I. A protocol π achieves security according to the simulation paradigm if whatever an adversary E controlling a subset $\mathfrak{E} \subseteq \mathfrak{P}$ of parties can do in the real model, a simulator (or ideal adversary) S (connected to the interfaces of the corrupted parties to the ideal functionality I) could replicate in the ideal model.

This is formalized by means of a distinguisher D which provides inputs x_i ($P_i \in \mathfrak{H}$) for the honest parties and x_{E} for the adversary E. In the ideal setting the x_i ($P_i \in \mathfrak{H}$) are input to I, while x_{E} is passed to the simulator S (which in turn computes inputs $x'_{\mathfrak{E}}$ to I for the corrupted parties). In the real setting the protocol machines π_i are run on input x_i ($P_i \in \mathfrak{H}$) with the adversary E on input x_{E} and with resources R. Finally the outputs $y_{\mathfrak{H}}$ of the honest parties and y_{E} of the adversary or simulator are passed to D which then has to guess if it is connected to the real system $\mathsf{E} \circ \pi_{\mathfrak{H}} \circ \mathsf{R}$ or the ideal system $\mathsf{S}(\mathsf{E}) \circ I$. To facilitate a unified treatment of different corruption models, we will wlog assume that $x_{\mathsf{E}} = (x_{\mathfrak{E}}, x'_{\mathsf{E}})$ and $y_{\mathsf{E}} = (y_{\mathfrak{E}}, y'_{\mathsf{E}})$ where $x_{\mathfrak{E}} \in \mathcal{X}_{\mathfrak{E}}$ and $y_{\mathfrak{E}} \in \mathcal{Y}_{\mathfrak{E}}$ are function inputs and outputs respectively, x'_{E} is an auxiliary input and y'_{E} is the protocol transcript observed by the adversary.[7] If now for any adversaries E from a class $\mathcal{E}$ controlling a set $\mathfrak{E}$ of parties there is a simulator S from a class $\mathcal{S}$ such that the advantage of any distinguisher D from a class $\mathcal{D}$ in distinguishing the real system $\mathsf{E} \circ \pi_{\mathfrak{H}} \circ \mathsf{R}$ and the ideal system $\mathsf{S}(\mathsf{E}) \circ I$ is bounded by an advantage function $\epsilon(\kappa)$, then we say that protocol π securely implements the ideal functionality I. The type of security is dependent on the choice of $\mathcal{D}$, $\mathcal{E}$, $\mathcal{S}$, and $\epsilon(\kappa)$, and on the ideal functionality I. Denoting the class of efficient[8] algorithms by Poly, the class of arbitrary unbounded algorithms by Algo, and negligibility in the security parameter κ as $\epsilon(\kappa) < \mathsf{negl}$ we arrive at the security paradigms listed in Table 1.

We refine these further by defining adversarial models, i.e. restrictions that we can impose on the adversaries and simulators for any of the above paradigms. We discuss active (act) adversaries, where adversaries and simulators in the classes $\mathcal{E}_{\mathsf{act}}$, $\mathcal{S}_{\mathsf{act}}$ are not restricted further; semi-honest (sh) adversaries[3], where adversaries in the class $\mathcal{E}_{\mathsf{sh}}$ are restricted to generate messages according to the prescribed protocol π with the inputs $x_{\mathfrak{E}}$ provided by the distinguisher D, and simulators in the class $\mathcal{S}_{\mathsf{sh}}$ are not restricted further; and passive (pas) adversaries[2], where adversaries are in the class $\mathcal{E}_{\mathsf{pas}} = \mathcal{E}_{\mathsf{sh}}$ and simulators in the class $\mathcal{S}_{\mathsf{pas}}$ are restricted to forward the inputs $x_{\mathfrak{E}}$ provided by the distinguisher D to the ideal functionality I.

---

[7] Arbitrary inputs can be passed to the adversary via x'_{E} and whatever the adversary might compute from its observations can also be computed from the protocol transcript y'_{E} directly.

[8] By efficient we mean polynomially bounded in the security parameter κ.

We briefly motivate our definition of sh adversaries. When CO tools are applied to force active adversaries to behave passively, they can, in contrast to the pas setting, still substitute inputs. The sh setting is intended to model this scenario. However, for simplicity, the definition above only allows for simulators (and not adversaries) to substitute inputs, as this is actually equivalent under the distinguisher classes $\mathcal{D}$ we consider: For any $D \in \mathcal{D}$ we can find a distinguisher $D' = D \circ \sigma \in \mathcal{D}$ that incorporates the input substitution of the adversary $E = E' \circ \sigma$. So we can find a passive adversary $E' \in \mathcal{E}_{sh} = \mathcal{E}_{pas}$ and a distinguisher D' that yield the same advantage as E and D.

Security paradigms and adversarial models as defined above are combined by intersecting their defining sets, i.e. IT security against sh adversaries is described by $\mathcal{D}_{sh}^{IT} = \mathcal{D}^{IT} \cap \mathcal{D}_{sh}$, $\mathcal{S}_{sh}^{IT} = \mathcal{S}^{IT} \cap \mathcal{S}_{sh}$, $\mathcal{E}_{sh}^{IT} = \mathcal{E}^{IT} \cap \mathcal{E}_{sh}$, $\epsilon(\kappa) < $ negl and denoted $\pi \succcurlyeq_{sh}^{IT} I$. By definition we have the following implications among security paradigms and adversarial models respectively:

$$PFE \Longrightarrow ITE \Longrightarrow LT \Longrightarrow CO \qquad act \Longrightarrow sh \Longleftarrow pas$$

$$\Big\Downarrow \qquad\qquad \Big\Downarrow$$

$$PF \Longrightarrow IT$$

We can now formalize the computation of a function f with a specific set of security properties under each of the definitions above by providing an appropriate ideal functionality. Let $f \in \mathfrak{F}_n$ be a function[9] and let $\mathfrak{C} \subset \mathfrak{P}$ be a set of corrupted players.

Demanding *privacy, correctness and agreement on abort* only for the computation of f is captured by the ideal functionality I_f^{ab}, which operates as follows: I_f^{ab} accepts an input x_i from each party P_i. If a party P_i provides no input, a default input x_i^{def} is used. I_f^{ab} then computes the outputs $(y_1, \ldots, y_n) = f(x_1, \ldots, x_n)$ and outputs $y_{\mathfrak{C}}$ to the adversary (simulator). If $|\mathfrak{C}| > 0$, the adversary may decide whether the other parties also receive the output (output flag $o = 1$) or not (output flag $o = 0$). Finally, I_f^{ab} sends either the outputs y_i or the empty value $\bot$ to the honest parties, depending on the output flag received from the adversary.[10]

The ideal functionality I_f^{fair} specifying *privacy, correctness and fairness* (which implies agreement on abort) works like I_f^{ab} but takes an output flag *before* making output to the adversary. Then for output flag 1 the functionality I_f^{fair} sends the result y to *all* parties and for output flag 0 it sends $\bot$ to *all* parties.

Computing function f with *full security* (including robustness), which implies all the security notions mentioned above, is specified by means of the ideal functionality I_f. The functionality I_f operates like I_f^{fair} but takes no output flag and instead directly delivers the output y to *all* parties.

---

[9] In this work we take the function f to be independent of the security parameter κ. As such the efficiency of protocols is always discussed for a fixed function in terms of the security parameter κ. This is the most relevant case for applications, however, our proofs still hold for a family of functions f_κ, where the input domain grows at most polynomially fast in the security parameter κ.

[10] We could relax the definition further by allowing the adversary to send one output flag for each party, dropping agreement on abort. However, all our protocols will achieve agreement on abort.

Computing function f with a *designated aborter* (DA) is a slightly weaker notion of security than fairness in that only the *designated party* P_1 can abort the protocol after receiving output. The corresponding ideal functionality I_f^{des} operates as follows: I_f^{des} accepts an input x_i from each party P_i. If a party P_i provides no input, a default input x_i^{def} is used. I_f^{des} then computes the outputs $(y_1, \ldots, y_n) = f(x_1, \ldots, x_n)$. If $P_1 \in \mathfrak{C}$ the functionality I_f^{des} outputs $y_{\mathfrak{C}}$ to the adversary (simulator). If $|\mathfrak{C}| > 0$, the adversary may decide whether the other parties also receive the output (output flag $o = 1$) or not (output flag $o = 0$). Finally, I_f^{des} either delivers the remaining outputs y_i or the empty value $\bot$, depending on the output flag received from the adversary.

In the 2-party setting (but not for $n > 2$ parties) given I_f^{fair} we can implement I_f by having P_i output $f_i(x_i, x_{2-i}^{\mathrm{def}})$ when it receives $\bot$. Conversely, given I_f, we can directly use it as implementation of I_f^{fair}. Thus robustness and fairness amount to the same:

Lemma 1. *In the 2-party setting, I_f^{fair} and I_f are efficiently and PFE securely locally mutually reducible, even in presence of active adversaries.*

Finally we show that computability by public discussion (authenticated BC only as resources R) and in the authenticated channels model (complete network of authenticated channels as resources R) lead to identical results for semi-honest or passive adversaries. In the authenticated channels model we can securely (against sh and pas adversaries) implement BC by simply sending messages to all other parties. Conversely in the authenticated BC model, authenticated channels can be implemented by broadcasting messages and instructing parties other than the intended recipient to ignore the messages. By the same argument computability by public discussion and in the authenticated channels model *with BC* lead to identical results for active adversaries also.

Lemma 2. *In presence of semi-honest or passive adversaries, a function $f \in \mathfrak{F}$ is securely computable in the authenticated channels model if and only if it is computable by public discussion (authenticated BC only). In presence of passive, semi-honest, or active adversaries, a function $f \in \mathfrak{F}$ is securely computable in the authenticated channels model with BC if and only if it is computable by public discussion.*

3 The Class $\mathfrak{F}_{\mathrm{pas}}^{\mathrm{aut}}$ of Passively Computable Functions

We subsequently characterize the class $\mathfrak{F}_{\mathrm{pas}}^{\mathrm{aut}}$ of functions $f \in \mathfrak{F}$ that are computable IT securely in the authenticated channels model in presence of a passive adversary.

Definition 1 ($\mathfrak{F}_{\mathrm{pas}}^{\mathrm{aut}}$: Passively Computable Functions). *The class of passively computable functions $\mathfrak{F}_{\mathrm{pas}}^{\mathrm{aut}}$ consists of the functions $f \in \mathfrak{F}$ for which an efficient protocol $\pi \in \mathsf{Poly}$ exists that implements I_f with IT security in presence of a passive adversary in the authenticated channels model.*

Note that by Lem. 2 we have $\mathfrak{F}_{\mathrm{pas}}^{\mathrm{aut}} = \mathfrak{F}_{\mathrm{pas}}^{\mathrm{bc}}$, where $\mathfrak{F}_{\mathrm{pas}}^{\mathrm{bc}}$ denotes the functions computable by public discussion in the setting above. Hence we may, for the sake of simplicity, assume an authenticated BC channel instead of authenticated channels as the sole underlying resource in the following discussion.

An important subset $\mathfrak{F}_{\mathrm{pas}}^{\mathrm{aut}}$ is the set $\mathfrak{F}_{\mathrm{loc}}$ of locally computable n-party functions.

Definition 2 ($\mathfrak{F}_{\text{loc}}$: Locally Computable Functions). *A function $f \in \mathfrak{F}$ is called locally computable ($f \in \mathfrak{F}_{\text{loc}}$) if each party P_i can compute its function value $y_i = f_i(x_1, \ldots, x_n)$ locally, without interacting with a resource or another party.*

Obviously, for f to be locally computable, f_i cannot depend on the inputs of parties other then P_i:

Lemma 3 (Characterization of $\mathfrak{F}_{\text{loc}}$). *A function $f \in \mathfrak{F}$ is locally computable ($f \in \mathfrak{F}_{\text{loc}}$) iff for every $i \in [n]$, $x_i \in \mathcal{X}_i$ the restriction $f_i \mid_{\mathcal{X}_1 \times \ldots \times \mathcal{X}_{i-1} \times \{x_i\} \times \mathcal{X}_{i+1} \times \ldots \times \mathcal{X}_n}$ of f is constant.*

Towards a characterization of $\mathfrak{F}_{\text{pas}}^{\text{aut}}$, we give a combinatorial definition of a set $\mathfrak{F}_{\text{pas}}'$ of functions that we call *passively decomposable*. Passive decomposability captures the fact that a party can send a message about its input such that no adversary can learn anything that is not implied by its own input and function output.

Definition 3 ($\mathfrak{F}_{\text{pas}}'$: Passively Decomposable Functions). *A function $f \in \mathfrak{F}_n$ is called passively decomposable, denoted $f \in \mathfrak{F}_{\text{pas}}'$, if for any restriction $f \mid_{\tilde{\mathcal{X}}_1 \times \ldots \times \tilde{\mathcal{X}}_n}$ of f to subsets $\tilde{\mathcal{X}}_j \subseteq \mathcal{X}_j$ ($j \in [n]$) we have:*

1. *$f \mid_{\tilde{\mathcal{X}}_1 \times \ldots \times \tilde{\mathcal{X}}_n}$ is locally computable ($f \mid_{\tilde{\mathcal{X}}_1 \times \ldots \times \tilde{\mathcal{X}}_n} \in \mathfrak{F}_{\text{loc}}$) or*
2. *there is an $i \in [n]$ and a partition (K-Cut) of $\tilde{\mathcal{X}}_i$ into non-empty sets $\mathcal{X}_i' \dot\cup \mathcal{X}_i'' = \tilde{\mathcal{X}}_i$ such that for all $P_e \in \mathfrak{P} \setminus \{P_i\}$ and all $x_e \in \tilde{\mathcal{X}}_e$ ($\mathfrak{E} := \{P_e\}$, $\mathfrak{H}' := \mathfrak{H} \setminus \{P_i\}$):*
   $$f_e(x_e, \tilde{\mathcal{X}}_{\mathfrak{H}'}, \mathcal{X}_i') \cap f_e(x_e, \tilde{\mathcal{X}}_{\mathfrak{H}'}, \mathcal{X}_i'') = \emptyset.$$

The above definition only discusses adversary sets $\mathfrak{E}$ of cardinality $|\mathfrak{E}| = 1$. As we show next this is actually equivalent to quantifying over all sets $\mathfrak{E} \subseteq \mathfrak{P}$.

Lemma 4 (An Equivalent Characterization of $\mathfrak{F}_{\text{pas}}'$). *A function $f \in \mathfrak{F}_n$ is passively decomposable if and only if for any restriction $f \mid_{\tilde{\mathcal{X}}_1 \times \ldots \times \tilde{\mathcal{X}}_n}$ of f to subsets $\tilde{\mathcal{X}}_j \subseteq \mathcal{X}_j$ ($j \in [n]$) we have:*

1. *$f \mid_{\tilde{\mathcal{X}}_1 \times \ldots \times \tilde{\mathcal{X}}_n}$ is locally computable ($f \mid_{\tilde{\mathcal{X}}_1 \times \ldots \times \tilde{\mathcal{X}}_n} \in \mathfrak{F}_{\text{loc}}$) or*
2. *there is an $i \in [n]$ and a partition (K-Cut) of $\tilde{\mathcal{X}}_i$ into non-empty sets $\mathcal{X}_i' \dot\cup \mathcal{X}_i'' = \tilde{\mathcal{X}}_i$ such that for all $\emptyset \neq \mathfrak{E} \subseteq \mathfrak{P} \setminus \{P_i\}$ and all $x_{\mathfrak{E}} \in \tilde{\mathcal{X}}_{\mathfrak{E}}$ ($\mathfrak{H}' := \mathfrak{H} \setminus \{P_i\}$):*
   $$f_{\mathfrak{E}}(x_{\mathfrak{E}}, \tilde{\mathcal{X}}_{\mathfrak{H}'}, \mathcal{X}_i') \cap f_{\mathfrak{E}}(x_{\mathfrak{E}}, \tilde{\mathcal{X}}_{\mathfrak{H}'}, \mathcal{X}_i'') = \emptyset.$$

The proof of Lemma 4 is by induction over the size of the adversary set $\mathfrak{E}$ and can be found in the full version [22].

We now show that passive decomposability as defined above indeed characterizes the passively computable n-party functions:

Theorem 1. *A function $f \in \mathfrak{F}$ is passively computable if and only if it is passively decomposable. In short $\mathfrak{F}_{\text{pas}}^{\text{aut}} = \mathfrak{F}_{\text{pas}}'$. Furthermore, any function $f \in \mathfrak{F}_{\text{pas}}^{\text{aut}}$ can efficiently (in the security parameter κ) be computed with PFE security.*

The full proof of Thm. 1 can be found in [22]. $\mathfrak{F}_{\text{pas}}^{\text{aut}} \subseteq \mathfrak{F}_{\text{pas}}'$ is shown by demonstrating that in absence of a K-cut no protocol participant can send a message that bears any information about his input without losing security. The proof of $\mathfrak{F}_{\text{pas}}^{\text{aut}} \supseteq \mathfrak{F}_{\text{pas}}'$ is constructive in the sense that it inductively describes an efficient passively PFE secure protocol π_f to compute a function $f \in \mathfrak{F}_{\text{pas}}'$. The protocol π_f generalizes the approach of [24] to asymmetric n-party functions:

Wlog assume that there is a partition of $\mathcal{X}_i = \mathcal{X}_i^{(1)} \dot{\cup} \mathcal{X}_i^{(2)}$ as described in Definition 3. The protocol π_f then proceeds as follows: The party P_i determines the message $m_1 \in \{0, 1\}$ such that for the input $x_i \in \mathcal{X}_i$ of P_i we have $x_i \in \mathcal{X}_i^{(m_1)}$ and broadcasts m_1. The parties $\mathfrak{P}$ then restrict the function f to $f|_{\mathcal{X}_1 \times \mathcal{X}_2 \times \ldots \times \mathcal{X}_i^{(m_1)} \times \ldots \times \mathcal{X}_n}$ and proceed with a partition for the restricted function in the same fashion. The process is iterated until the parties arrive at a locally computable restriction of f, at which point they can determine the output locally.

We conjecture that the above protocol achieves the optimal round complexity if it is refined to use the finest possible decomposition (according to [24]) of the input domains in every round.

4   The Class $\mathfrak{F}_{\text{sh}}^{\text{aut}}$ of Semi-honestly Computable Functions

Next we characterize the class $\mathfrak{F}_{\text{sh}}^{\text{aut}}$ of n-party functions that are IT securely computable in the authenticated channels model in presence of a semi-honest adversary. Here, in order to obtain extra information, the corrupted parties are allowed to exchange their inputs for different ones, but must still behave according to the prescribed protocol. The results in this chapter will later help us to characterize LT secure functions in a very practical setting.

Definition 4 ($\mathfrak{F}_{\text{sh}}^{\text{aut}}$: Semi-Honestly Computable Functions). *The class of* semi-hon-estly computable *functions* $\mathfrak{F}_{\text{sh}}^{\text{aut}}$ *consists of the functions* $f \in \mathfrak{F}$ *for which an efficient protocol* $\pi \in$ Poly *exists that implements* I_f *with IT security in presence of a semi-honest adversary in the authenticated channels model.*

Note that by Lem. 2 we have $\mathfrak{F}_{\text{sh}}^{\text{aut}} = \mathfrak{F}_{\text{sh}}^{\text{bc}}$, where $\mathfrak{F}_{\text{sh}}^{\text{bc}}$ denotes the functions computable by public discussion in the setting above. Hence we may, for the sake of simplicity, assume an authenticated BC channel instead of authenticated channels as the sole underlying resource in the following discussion.

We intend to characterize the class $\mathfrak{F}_{\text{sh}}^{\text{aut}}$ combinatorially. To this end we introduce the concept of redundancy-freeness for n-party functions, generalizing the 2-party definitions of [21]. For a party P_i, two of its possible inputs x_i and x_i' to f may be completely indistinguishable to the other parties (by their output from f), while the input x_i may yield a more informative output from f for P_i than x_i'. We then say the input x_i yielding more information dominates the input x_i' giving less information. As semi-honest (and active) adversaries can select their inputs, generally with the goal to obtain as much information as possible, the dominated input x_i' giving less information is not useful to a corrupted P_i. Along the same lines an ideal adversary (simulator) can always use the dominating input x_i instead x_i' of for simulation. As such the input x_i' is redundant, irrelevant in terms of security, and we can eliminate it from the function f under consideration. This procedure yields a *redundancy-free version* $\hat{f}$ of f, with new, smaller, dominating input sets.

Definition 5 (Domination and Redundancy-Freeness). *Given an n-party function $f \in \mathfrak{F}_n$ we say $x_i \in \mathcal{X}_i$ dominates $x'_i \in \mathcal{X}_i$ iff for all $x_{\mathfrak{P}'} \in \mathcal{X}_{\mathfrak{P}'}$: $f_{\mathfrak{P}'}(x_i, x_{\mathfrak{P}'}) = f_{\mathfrak{P}'}(x'_i, x_{\mathfrak{P}'})$ and for all $x_{\mathfrak{P}'}, x'_{\mathfrak{P}'} \in \mathcal{X}_{\mathfrak{P}'}$ (where $\mathfrak{P}' := \mathfrak{P} \setminus \{P_i\}$): $f_i(x'_i, x_{\mathfrak{P}'}) \neq f_i(x'_i, x'_{\mathfrak{P}'}) \implies f_i(x_i, x_{\mathfrak{P}'}) \neq f_i(x_i, x'_{\mathfrak{P}'})$.*

We define sets of dominating inputs $\widetilde{\mathcal{X}}_j := \{\mathcal{X} \subseteq \mathcal{X}_j \mid \forall x' \in \mathcal{X}_j \exists x \in \mathcal{X} : x \text{ dominates } x'\}$ ($j \in [n]$). Take the dominating set $\hat{\mathcal{X}}_j$ as (some) element of minimal cardinality in $\widetilde{\mathcal{X}}_j$. We then call $\hat{f} := f|_{\hat{\mathcal{X}}_1 \times \dots \times \hat{\mathcal{X}}_n}$ the redundancy-free version of f. Furthermore, for $x_j \in \mathcal{X}_j$ let $\hat{x}_j \in \hat{\mathcal{X}}_j$ be the (unique) element that dominates x_j.

The redundancy-free version $\hat{f}$ of f is uniquely defined up to a renaming of input and output values (also see Sec. 8 or [22]). Domination is a reflexive and transitive relation. Furthermore it is antisymmetric up to renaming of input and output symbols. Hence two different dominating sets $\hat{\mathcal{X}}_i$ and $\hat{\mathcal{X}}'_i$ are sets of maximal elements under the domination relation and equal up to renaming of input and output values.

Since corrupted parties can cooperate to choose their inputs to obtain as much information as possible, it is important to note that the above Def. 5 generalizes to the combined input of the corrupted parties $\mathfrak{E}$ as stated in Lem. 5 below. So if each corrupted party P_{e_j} chooses an input x_{e_j} dominating input x'_{e_j}, then the combined adversarial input $x_{\mathfrak{E}}$ actually dominates $x'_{\mathfrak{E}}$.

Lemma 5. *Let $x_{\mathfrak{E}} = (x_{e_1}, \dots, x_{e_{|\mathfrak{E}|}})$, $x'_{\mathfrak{E}} = (x'_{e_1}, \dots, x'_{e_{|\mathfrak{E}|}})$ such that each x_{e_j} dominates x'_{e_j} ($j \in [|\mathfrak{E}|]$). Then we have for all $x_{\mathfrak{H}} \in \mathcal{X}_{\mathfrak{H}}$: $f_{\mathfrak{H}}(x_{\mathfrak{E}}, x_{\mathfrak{H}}) = f_{\mathfrak{H}}(x'_{\mathfrak{E}}, x_{\mathfrak{H}})$ and for all $x_{\mathfrak{H}}, x'_{\mathfrak{H}} \in \mathcal{X}_{\mathfrak{H}}$: $f_{\mathfrak{E}}(x'_{\mathfrak{E}}, x_{\mathfrak{H}}) \neq f_{\mathfrak{E}}(x'_{\mathfrak{E}}, x'_{\mathfrak{H}}) \implies f_{\mathfrak{E}}(x_{\mathfrak{E}}, x_{\mathfrak{H}}) \neq f_{\mathfrak{E}}(x_{\mathfrak{E}}, x'_{\mathfrak{H}})$. Again we say that $x_{\mathfrak{E}}$ dominates $x'_{\mathfrak{E}}$.*

The proof of Lem. 5 is by induction on $|\mathfrak{E}|$ and can be found in [22].

The following lemma states that the functions f and $\hat{f}$ are locally[11] and efficiently mutually reducible. This means that it does not matter in terms of security which of the two functions is used and redundant inputs can safely be eliminated.

Lemma 6. *The functions f and $\hat{f}$ are efficiently and PFE securely locally mutually reducible, even in presence of active adversaries.*

The proof of Lem. 6 is fairly straightforward, by showing how to implement $I_{\hat{f}}$ when I_f is given and vice versa. It can be found in [22]. One essentially replaces inputs x_i with dominating inputs $\hat{x}_i$.

As PFE security in presence of active adversaries implies IT security in presence of semi-honest adversaries, we can derive the following simple corollary:

Corollary 1. *For any function $f \in \mathfrak{F}$ we have: $f \in \mathfrak{F}_{\mathsf{sh}}^{\mathsf{aut}} \iff \hat{f} \in \mathfrak{F}_{\mathsf{sh}}^{\mathsf{aut}}$.*

An n-party function f is then sh computable if and only if its redundancy-free version $\hat{f}$ is pas computable.

Theorem 2. *For a function $f \in \mathfrak{F}$ we have: $f \in \mathfrak{F}_{\mathsf{sh}}^{\mathsf{aut}} \iff \hat{f} \in \mathfrak{F}_{\mathsf{pas}}^{\mathsf{aut}}$.*

---

[11] without using any communication resources

$f^{(1)}$	0	1
0	0/0	0/0
1	0/0	1/0

$f^{(2)}$	0	1
0	0	1
1	0	2
2	3	2

$f^{(3)}$	0	1	2
0	0/0	1/1	1/0
1	0/0	2/2	2/0
2	3/3	2/2	2/0

$f^{(4)}$	0	1	2	3
0	1/1	1/1	2/2	2/0
1	4/4	5/5	2/2	2/0
2	4/4	3/3	3/3	3/0

$f^{(5)}$	0	1
0	0	0
1	0	1

$f^{(6)}$	0	1	2
0	1	1	2
1	4	5	2
2	4	3	3

$f^{(7)}$	4	2	0
3	4	3	3
1	4	2	1
0	4	2	0

$f^{(8)}$	0	1
0	0	1
1	0	2
2	3	2
3	3	1

$f^{(9)}$	z_1	z_2	z_3
x_1	5/d	5/e	6/e
x_2	8/a	5/b	9/c
x_3	8/a	9/b	8/c

Fig. 1. Examples. Inputs for A are shown to the right, inputs for B on top. For asymmetric functions, outputs are denoted y_A/y_B; for symmetric functions only the common output of both parties is listed.

The full proof of Thm. 2 can be found in [22], we only give a sketch here. By Cor. 1 we know that $f \in \mathfrak{F}_{sh}^{aut} \iff \hat{f} \in \mathfrak{F}_{sh}^{aut}$. Therefore it suffices to show for redundancy-free functions f where $f = \hat{f}$ that we have $f \in \mathfrak{F}_{sh}^{aut} \iff f \in \mathfrak{F}_{pas}^{aut}$. The implication $f \in \mathfrak{F}_{pas}^{aut} \implies f \in \mathfrak{F}_{sh}^{aut}$ is then clear by definition. The implication $f \in \mathfrak{F}_{sh}^{aut} \implies f \in \mathfrak{F}_{pas}^{aut}$ is shown along the lines of the proof of Thm. 1, demonstrating that $f \in \mathfrak{F}_{sh}^{aut} \implies f \in \mathfrak{F}_{pas}'$. The proof exploits the redundancy-freeness of f due to which a (working) simulator in the sh setting cannot actually substitute inputs.

The functions $f^{(5)}$ and $f^{(6)}$ in Fig. 1 are examples of *not* sh computable functions taken from [24]. The function $f^{(6)}$ is of particular interest as it is of strictly less cryptographic strength than oblivious transfer. Function $f^{(9)}$ is sh computable: After eliminating the redundant input x_3, the function is pas computable (as indicated by the horizontal and vertical lines).

5   The Class $\mathfrak{F}_{act}^{aut}$ of Actively Computable Functions

We give a sufficient criterion for a function f to be in the class $\mathfrak{F}_{act}^{bc}$ of functions which can securely be computed by public discussion in presence of an unlimited active adversary. We conjecture that this criterion is also necessary and prove this fact for the 2-party case. As such we only obtain a full characterization of the class $\mathfrak{F}_{2act}$ of actively computable 2-party functions, but this suffices to see that $\mathfrak{F}_{2act}$ is strictly contained in $\mathfrak{F}_{2sh}$ and hence the notion of LT security lies strictly between IT security and CO security.

Definition 6 ($\mathfrak{F}_{act}^{aut}$: Actively Computable Functions). *The class of actively computable functions $\mathfrak{F}_{act}^{aut}$ consists of the functions $f \in \mathfrak{F}$ for which an efficient protocol $\pi \in$ Poly exists that implements I_f with IT security in presence of an active adversary in the authenticated channels model with broadcast.*

Note that by Lem. 2 we have $\mathfrak{F}_{act}^{aut} = \mathfrak{F}_{act}^{bc}$, where $\mathfrak{F}_{act}^{bc}$ denotes the functions computable by public discussion in the setting above. Hence we may in the following assume an authenticated BC channel as the sole underlying resource.

Interestingly there are some useful functions in the class $\mathfrak{F}_{act}^{aut}$, e.g. $f^{(7)}$ in Fig. 1 which is a formalization of a Dutch flower auction, where the price is lowered in every round until a party decides to buy.

We next give a combinatorial characterization of actively computable functions, which essentially states that a party P_i must be able to send a message about its input such that the corrupted parties $\mathfrak{E}$ reacting to this new information by changing their input from $x'_{\mathfrak{E}}$ to $x''_{\mathfrak{E}}$ could have achieved the same effect on the output by selecting a third input $x_{\mathfrak{E}}$ a priori:

Definition 7 ($\mathfrak{F}'_{act}$: Actively Decomposable Functions). *A function $f \in \mathfrak{F}$ is called actively decomposable, denoted $f \in \mathfrak{F}'_{act}$, if and only if $\hat{f} \in \widehat{\mathfrak{F}_{act}}$. We have $f \in \widehat{\mathfrak{F}_{act}}$ if one of the following holds:*

1. *f is locally computable ($f \in \mathfrak{F}_{loc}$);*
2. *there is an $i \in [n]$ and a partition (T-Cut) of $\mathcal{X}_i$ into non-empty sets $\mathcal{X}'_i \,\dot{\cup}\, \mathcal{X}''_i = \mathcal{X}_i$ such that*
   (i) $f|_{\mathcal{X}_1 \times ... \times \mathcal{X}'_i \times ... \times \mathcal{X}_n}, f|_{\mathcal{X}_1 \times ... \times \mathcal{X}''_i \times ... \times \mathcal{X}_n} \in \widehat{\mathfrak{F}_{act}}$ and
   (ii) for all $\mathfrak{E} \subseteq \mathfrak{P} \setminus \{P_i\}$ and $\mathfrak{H}' := \mathfrak{H} \setminus \{P_i\}$ we have

$$\forall \bar{x}_{\mathfrak{E}} \in \mathcal{X}_{\mathfrak{E}} : f_{\mathfrak{E}}(\bar{x}_{\mathfrak{E}}, \mathcal{X}_{\mathfrak{H}'}, \mathcal{X}'_i) \cap f_{\mathfrak{E}}(\bar{x}_{\mathfrak{E}}, \mathcal{X}_{\mathfrak{H}'}, \mathcal{X}''_i) = \emptyset \quad \text{(K-cut)} \quad \text{and}$$

$$\forall x'_{\mathfrak{E}}, x''_{\mathfrak{E}} \in \mathcal{X}_{\mathfrak{E}} \, \exists x_{\mathfrak{E}} \in \mathcal{X}_{\mathfrak{E}} \, \forall x_{\mathfrak{H}'} \in \mathcal{X}_{\mathfrak{H}'}$$
$$\forall x'_i \in \mathcal{X}'_i : f_{\mathfrak{H}}(x'_{\mathfrak{E}}, x_{\mathfrak{H}'}, x'_i) = f_{\mathfrak{H}}(x_{\mathfrak{E}}, x_{\mathfrak{H}'}, x'_i) \quad \wedge$$
$$\forall x''_i \in \mathcal{X}''_i : f_{\mathfrak{H}}(x''_{\mathfrak{E}}, x_{\mathfrak{H}'}, x''_i) = f_{\mathfrak{H}}(x_{\mathfrak{E}}, x_{\mathfrak{H}'}, x''_i)$$

Active decomposability indeed characterizes the actively computable functions:

Theorem 3. *A function $f \in \mathfrak{F}$ is actively computable if it is actively decomposable. In short $\mathfrak{F}_{act}^{aut} \supseteq \mathfrak{F}'_{act}$. In the 2-party case[12] we even have $\mathfrak{F}_{2act} \subseteq \mathfrak{F}'_{2act}$, i.e. $\mathfrak{F}_{2act} = \mathfrak{F}'_{2act}$. Furthermore, any function $f \in \mathfrak{F}'_{act}$ can be computed efficiently with PFE security.*

Furthermore, we conjecture:

Conjecture 1. $\mathfrak{F}_{act}^{aut} = \mathfrak{F}'_{act}$.

The full proof of Thm. 3 can be found in [22]. The implication $f \in \mathfrak{F}'_{act} \implies f \in \mathfrak{F}_{act}^{aut}$ is proven by showing the protocol for the semi-honest scenario secure against active adversaries, when applied to the T-cuts of a function $f \in \mathfrak{F}'_{act}$ instead of the K-cuts of a function in $\mathfrak{F}_{sh}^{aut}$. To obtain $f \in \mathfrak{F}_{2act} \implies f \in \mathfrak{F}'_{2act}$ we observe that for $f \notin \mathfrak{F}'_{2act}$ the adversary can in any protocol induce an output distribution that is impossible to achieve in the ideal setting. The adversary does this by extracting information on the inputs of other participants from the protocol messages and adjusting his input according to that information.

The functions $f^{(7)}$ and $f^{(8)}$ in Fig. 1 are examples of actively computable functions. Especially compare $f^{(8)}$ with $f^{(2)} \in \mathfrak{F}_{2sh}$ which is *not* actively computable. The lines in the tables for $f^{(7)}$ and $f^{(8)}$ represent messages which are to be sent in the protocol.

---

[12] For a function class $\mathfrak{F}_{name}^{chan}$ we denote the 2-party subclass $\mathfrak{F}_{name}^{chan} \cap \mathfrak{F}_2$ by $\mathfrak{F}_{2name}$. We drop the communication model specification chan as it is irrelevant for the 2-party setting.

6   Quantum Protocols

In this section we will relate the class $\mathfrak{F}_{2sh}$ of sh computable 2-party functions with the class of 2-party functions computable with quantum cryptography in presence of an active adversary. A similar result has been obtained by Louis Salvail, but is not published yet. Naturally, we have to adapt our model of security to the quantum case. All machines except for the distinguisher D will be quantum machines able to exchange quantum messages. Furthermore, all inputs and outputs must be classical and the distinguisher must try to distinguish the real and the ideal model based on this classical information.

Let $\mathfrak{F}_{2qu}$ denote the set of functions $f \in \mathfrak{F}_2$ which can, with the help of a quantum channel, securely and efficiently be computed in presence of an unbounded active adversary. Then the following result holds.

Theorem 4. *The class $\mathfrak{F}_{2qu}$ of quantum computable functions is strictly contained in the class of* sh *computable functions $\mathfrak{F}_{2sh}$.*

A proof of this theorem is sketched in [22]. The strict inclusion $\mathfrak{F}_{2qu} \subsetneq \mathfrak{F}_{2sh}$ gives rise to new impossibility results. For instance, the function $f^{(6)} \notin \mathfrak{F}_{2sh}$ in Fig. 1 cannot be computed by means of quantum cryptography. An interesting still open question is the power of temporary CO assumptions together with a quantum channel. It is known that this does not suffice to securely implement any function which could in turn be used to implement an IT secure bit commitment. However, a secure implementation of the function $f^{(6)}$ in Fig. 1 is not precluded by this impossibility result.

7   Long-Term Security

Subsequently we characterize the n-party functions that can be computed LT securely (without fairness) in presence of active adversaries. LT security means we are willing to make CO assumptions, but only for the duration of the protocol interaction. Once the protocol has terminated we demand IT security. We look at different classes of LT securely computable functions, defined by different channel models. The most practical model, corresponding to the class $\mathfrak{F}_{lts}^{ins, \, pki}$, is an internet-like setting, where insecure channels and a PKI are available to the parties. Furthermore we also discuss the classes $\mathfrak{F}_{lts}^{aut}$ where authenticated channels and $\mathfrak{F}_{lts}^{bc}$ where an authenticated BC channel are given. We find that all these classes $\mathfrak{F}_{lts}^{ins, \, pki} = \mathfrak{F}_{lts}^{bc} = \mathfrak{F}_{lts}^{aut}$ are equal to the class $\mathfrak{F}_{sh}^{aut}$ of sh computable functions.

Definition 8 ($\mathfrak{F}_{lts}^{bc}, \mathfrak{F}_{lts}^{ins, \, pki}, \mathfrak{F}_{lts}^{aut}$**: LT Computable Functions).** *The classes of LT computable functions (i) $\mathfrak{F}_{lts}^{bc}$, (ii) $\mathfrak{F}_{lts}^{ins, \, pki}$, (iii) $\mathfrak{F}_{lts}^{aut}$ consists of the functions $f \in \mathfrak{F}$ for which an efficient protocol $\pi \in$ Poly exists that implements I_f^{ab} with LT security in presence of an active adversary from (i) an authenticated broadcast channel; (ii) a complete network of insecure channels and a PKI; (iii) a complete network of authenticated channels; respectively.*

We now show that the classes defined in the previous section are all equivalent to $\mathfrak{F}_{sh}^{aut}$. First, we observe that once we allow CO assumptions during the protocol execution, we can force semi-honest behavior (i.e. that the adversary behaves according to the protocol) using an unconditionally hiding commitment scheme [18] and zero-knowledge arguments of knowledge:

Theorem 5. *If one-way functions (OWF) exist, we have* $\mathfrak{F}_{sh}^{aut} = \mathfrak{F}_{lts}^{bc}$.

A full proof of Thm. 5 can be found in [22]. We show that the semi-honest to active protocol compiler of [16] can be applied to a semi-honestly secure protocol in such a way that it becomes CO secure against active adversaries, while maintaining IT security against semi-honest adversaries. Furthermore we claim:

Theorem 6. *We have* $\mathfrak{F}_{lts}^{ins,\,pki} = \mathfrak{F}_{lts}^{bc} = \mathfrak{F}_{lts}^{aut} = \mathfrak{F}_{sh}^{aut}$.

We prove this by showing $\mathfrak{F}_{lts}^{bc} \subseteq \mathfrak{F}_{lts}^{ins,\,pki}$, $\mathfrak{F}_{lts}^{ins,\,pki} \subseteq \mathfrak{F}_{lts}^{aut}$, $\mathfrak{F}_{lts}^{aut} \subseteq \mathfrak{F}_{lts}^{bc}$. First, $\mathfrak{F}_{lts}^{bc} \subseteq \mathfrak{F}_{lts}^{ins,\,pki}$ holds as we can use the Dolev-Strong-Protocol [12] to obtain authenticated BC in the PKI setting. $\mathfrak{F}_{lts}^{ins,\,pki} \subseteq \mathfrak{F}_{lts}^{aut}$ holds as using detectable precomputation [14] we can establish a PKI in the authenticated channels model.[13] $\mathfrak{F}_{lts}^{aut} \subseteq \mathfrak{F}_{lts}^{bc}$ follows from Lem. 2.

Thm. 6 is optimal in the sense that we cannot hope to implement all functions $f \in \mathfrak{F}_{lts}^{ins,\,pki}$ with robustness or even fairness. Of course we have (by definition) robust LT (even IT) secure protocols for the functions $f \in \mathfrak{F}_{act}^{aut}$. But e.g. the symmetric XOR function $f_{XOR}(x_1, x_2) := (x_1 \,\text{XOR}\, x_2, x_1 \,\text{XOR}\, x_2)$ is by the combinatorial characterizations of the previous sections $f_{XOR} \in \mathfrak{F}_{2sh} \setminus \mathfrak{F}_{act} \subset \mathfrak{F}_{lts}^{bc} \setminus \mathfrak{F}_{act}^{aut}$. Now a fair implementation of f_{XOR} would clearly imply a fair cointoss, which by [10] cannot be implemented in the model under consideration. As such the security without fairness as guaranteed by Thm. 6 is indeed the best we can hope for.

7.1 Long Term Security with Designated Aborter

As mentioned above we cannot generally guarantee robustness or even fairness for a LT secure protocol π_f computing $f \in \mathfrak{F}_{2lts}$. However, under stronger CO assumptions, we can guarantee that only a specific designated party can abort the protocol after obtaining output and before the honest parties can generate output. This may be of practical relevance where a specific party is not trusted, but can be relied upon not to abort the protocol. For instance a party may have a vested interest in the successful termination of the protocol regardless of the outcome. One may think of an auctioneer that gets paid only if the auction terminates successfully. Or a party may act in an official capacity and cannot abort the protocol for legal reasons.

We will show that stronger guarantees of this type are obtainable if the underlying CO assumption allows for an oblivious transfer (OT) protocol which is LT secure against one of the participants. Enhanced trapdoor one-way permutations are an example of such an assumption [15]. It is generally believed that OT is *not* implied by OWFs, meaning that LT security with designated aborter appears to require strictly stronger assumptions than plain LT security.

---

[13] Note that robustness is not required here: The establishment of the PKI may fail, but then the protocol simply aborts.

Lemma 7. *Any* sh *computable function* $f \in \mathfrak{F}_{\text{sh}}^{\text{aut}} = \mathfrak{F}_{\text{lts}}^{\text{ins, pki}}$ *can be computed using a protocol* π *which is LTS-DA, i.e. implements* I_f^{des} *with CO security and simultaneously* I_f^{ab} *with LT security in the insecure channels model with PKI iff CO oblivious transfer LT-secure against one party (CO-OT+) exists.*

A proof of this lemma is sketched in [22]. Essentially we apply the protocol compiler of [16] to the distributed circuit of the sh secure protocol for f in such a fashion that gates owned by a specific party P_i are computed with CO primitives that IT protect P_i. Reconstruction is in the end done toward the designated party P_1, which then ensures that the remaining parties can reconstruct. As a result the protocol is CO correct, and IT no one learns more than in the sh secure protocol for f.

8   Classification of 2-Party Functions

Combining the results of this work and of [21], we can derive a complete combinatorial classification of the 2-party functions $\mathfrak{F}_2$ by completeness and computability.

We first define an equivalence relation *renaming* on $\mathfrak{F}_2$ by $f^{(1)} \equiv f^{(2)}$ iff $f^{(2)}$ is obtained from $f^{(1)}$ by locally renaming input and output values. A formal definition can be found in [21] or [22]. It is easy to see that renamings are locally mutually reducible under all security paradigms considered in this work. In particular $f^{(1)} \equiv f^{(2)}$ implies $I_{f^{(1)}} \succeq_{\text{act}}^{\text{PFE}} I_{f^{(2)}} \succeq_{\text{act}}^{\text{PFE}} I_{f^{(1)}}$ and $I_{f^{(1)}} \succeq_{\text{pas}}^{\text{PFE}} I_{f^{(2)}} \succeq_{\text{pas}}^{\text{PFE}} I_{f^{(1)}}$.

Next we define an equivalence relation *matching* on the set of classes $\mathfrak{F}_2/\equiv$ (and thereby on $\mathfrak{F}_2$) by isolating inputs that lead to identical behavior and regarding functions as matching if, after eliminating such trivially redundant inputs, they are renamings:

Definition 9. *Given a 2-party function* $f \in \mathfrak{F}_2$ *we say* x_A *matches* x'_A *for inputs* $x_A, x'_A \in \mathcal{X}_A$, *iff* x_A *dominates* x'_A *and* x'_A *dominates* x_A. *The matching relation is an equivalence relation on* $\mathcal{X}_A$. *By* $\bar{\mathcal{X}}_A$ *we designate a set of representatives.* $\bar{\mathcal{X}}_B$ *is defined analogously.*

We then call $\bar{f} := f|_{\bar{\mathcal{X}}_A \times \bar{\mathcal{X}}_B}$ *the* weakly redundancy-free version *of* f *and for* $f^{(1)}, f^{(2)} \in \mathfrak{F}_2$ *we write* $f^{(1)} \cong f^{(2)}$ *if* $\bar{f}^{(1)} \equiv \bar{f}^{(2)}$ *Furthermore for* $x_A \in \mathcal{X}_A$ *and* $x_B \in \mathcal{X}_B$ *let* $\bar{x}_A \in \bar{\mathcal{X}}_A$ *and* $\bar{x}_B \in \bar{\mathcal{X}}_B$ *be the (unique) elements that match* x_A *respectively* x_B.

Like the redundancy-free version $\hat{f}$ of f, the weakly redundancy-free version $\bar{f}$ of f is well defined up to renaming. Before we can state the actual classification, we have to reiterate another result of [21]:

Theorem 7 (Complete Functions [21]). *The classes* $\mathfrak{C}_{2\text{act}}$, $\mathfrak{C}_{2\text{sh}}$ *and* $\mathfrak{C}_{2\text{pas}}$ *of actively, semi-honestly, and passively complete 2-party functions are the classes of functions* $f \in \mathfrak{F}_2$ *to which all other 2-party functions can be securely reduced in presence of an active, semi-honest or passive adversary respectively. The classes* $\mathfrak{C}_{2\text{act}} = \mathfrak{C}_{2\text{sh}}$ *consist of exactly the functions* $f \in \mathfrak{F}_2$ *where* $\hat{f} \in \mathfrak{C}_{2\text{pas}}$. *The class* $\mathfrak{C}_{2\text{pas}}$ *consists of exactly the functions* $f \in \mathfrak{F}_2$ *where* $\exists\, a_1, a_2 \in \mathcal{X}_A$, $b_1, b_2 \in \mathcal{X}_B$:

$$\exists\, a_1, a_2 \in \mathcal{X}_A, b_1, b_2 \in \mathcal{X}_B : \ f_A(a_1, b_1) = f_A(a_1, b_2) \wedge f_B(a_1, b_1) = f_B(a_2, b_1)$$
$$\wedge \ (\ f_A(a_2, b_1) \neq f_A(a_2, b_2) \vee f_B(a_1, b_2) \neq f_B(a_2, b_2)\ ).$$

We refer to this combinatorial structure as minimal OT.

Note that $f \in \mathfrak{C}_{2pas}$ iff $f \in \mathfrak{C}_{2act}$ or $\hat{f} \not\equiv \bar{f}$. This is clear from Kraschewski's result as stated above and from the observation that $\hat{f} \not\equiv \bar{f}$ implies a minimal OT. We then arrive at the following

Theorem 8 (Classification). *The class of 2-party functions is a disjoint union of three sets* $\mathfrak{F}_2 = \mathfrak{C}_{2act} \cup \mathfrak{F}_{2act} \cup \mathfrak{F}_{2act}^{nct}$ *or* $\mathfrak{F}_2 = \mathfrak{C}_{2sh} \cup \mathfrak{F}_{2sh} \cup \mathfrak{F}_{2sh}^{nct}$ *or* $\mathfrak{F}_2 = \mathfrak{C}_{2pas} \cup \mathfrak{F}_{2pas} \cup \mathfrak{F}_{2pas}^{nct}$
where nct *stand for "neither complete nor computable". Now*

$$\emptyset \neq \mathfrak{F}_{2act}, \mathfrak{F}_{2pas} \subsetneq \mathfrak{F}_{2act} \cup \mathfrak{F}_{2pas} \subsetneq \mathfrak{F}_{2sh}$$

$$\emptyset \neq \mathfrak{F}_{2pas}^{nct} \subseteq \mathfrak{F}_{2sh}^{nct} \subsetneq \mathfrak{F}_{2act}^{nct}$$

$$\emptyset \neq \mathfrak{C}_{2act} = \mathfrak{C}_{2sh} \subsetneq \mathfrak{C}_{2pas}$$

The above results are directly derived from the combinatorial descriptions of the function classes that can be found in the preceding sections and, as far as complete functions are concerned, in [21]. Additional details and examples can be found in [22].

9 Conclusions

We defined the notion of long-term (LT) security, where we assume that the adversary is CO bounded *during* the execution of the protocol *only*. That is, we rely on CO assumptions, but only for the duration of the protocol execution; thereafter, a failure of the CO assumptions must not compromise security. We then gave a combinatorial description of the class $\mathfrak{F}_{lts}^{ins,\,pki}$ of functions that can be computed LT securely in an internet-like setting, where a complete network of insecure channels and a PKI are available. Towards this goal, we characterized the classes $\mathfrak{F}_{pas}^{aut}$, $\mathfrak{F}_{sh}^{aut}$ and $\mathfrak{F}_{act}^{aut}$ of functions that can be computed with information theoretic (IT) security in the authenticated channels model (with broadcast) in presence of passive, semi-honest and active adversaries. Our results are constructive in that, for every function proven computable in a given setting, one can deduce a secure protocol.

More precisely, we showed that semi-honest computability and LT secure computability amount to the same, i.e. $\mathfrak{F}_{sh}^{aut} = \mathfrak{F}_{lts}^{bc} = \mathfrak{F}_{lts}^{aut} = \mathfrak{F}_{lts}^{ins,\,pki}$, where the classes $\mathfrak{F}_{lts}^{aut}$ and $\mathfrak{F}_{lts}^{bc}$ are defined analogously to $\mathfrak{F}_{lts}^{ins,\,pki}$, but rely on a network of authenticated channels or authenticated broadcast respectively as communication resources. We then characterized the class $\mathfrak{F}_{2act}$ of actively computable 2-party functions in order to offset IT secure computability against LT secure computability. Indeed, we found $\mathfrak{F}_{act}^{aut} \subsetneq \mathfrak{F}_{lts}^{ins,\,pki}$, meaning that in presence of corrupted majorities strictly more functions are computable with LT security than with IT security. We furthermore gave a necessary condition (that we conjecture also to be sufficient) for an n-party function to be in $\mathfrak{F}_{act}^{aut}$. As the functions in $\mathfrak{F}_{act}^{aut}$ are robustly (and therefore fairly) computable, these results can be interpreted along the lines Gordon et al. [17], who discuss the fair computability of binary 2-party functions in the CO setting. Our results apply to the IT scenario instead of the CO setting, there however, our results are much more general in that they pertain to arbitrary n-party functions. We showed that for the functions $\mathfrak{F}_{lts}^{ins,\,pki}$ fairness is generally not achievable. However, for the functions $\mathfrak{F}_{lts}^{ins,\,pki}$ we can guarantee LT security with designated aborter, where only a specific designated party can prematurely abort the protocol after having learned the output. Astonishingly, CO secure oblivious transfer (OT) is used in our construction, even though OT itself cannot be realized with full LT security.

We remark, that from a practical point of view, LT security is a useful notion if we deal with sensitive data that has to remain private beyond a limited time frame in a setting where a majority of the parties may be corrupted. In such a setting general IT secure SFE protocols like [4] fail, as they do not tolerate corrupted majorities. CO protocols can tolerate corrupted majorities (if fairness is not required) but, as time passes, progress in hardware or algorithms may invalidate our CO assumptions and jeopardize the privacy of our computation. As the problem with CO assumptions is not so much that these could be unjustified right now, but rather their possible future invalidation, LT security is a viable alternative to IT security in this case. And indeed we could show that $\mathfrak{F}_{act}^{aut} \subsetneq \mathfrak{F}_{lts}^{ins,\ pki}$, i.e. there are functions that cannot be computed with IT security in presence of dishonest majorities, but can be computed with LT security.

Furthermore, we found that quantum cryptography is not helpful in our context, i.e. the class $\mathfrak{F}_{2qu}$ of 2-party functions which can be implemented with quantum cryptography is strictly contained in $\mathfrak{F}_{2sh}$. This inclusion implies novel impossibility results beyond those of Mayers [25] or Kitaev [1]. However, quantum cryptography can solve classically impossible problems in other models of security, like achieving a certain robustness to abort in a model with guaranteed message delivery or implementing deniable key exchange.

Finally, collecting results from the literature, especially [24,21], and adding the results of this work, we obtain a complete taxonomy of 2-party functions by computability and completeness in the IT setting.

Acknowledgments

The authors wish to thank Daniel Kraschewski for helpful comments and discussions, and Ueli Maurer for encouragement and insightful comments on security models.

References

1. Ambainis, A., Buhrman, H., Dodis, Y., Röhrig, H.: Multiparty quantum coin flipping. In: IEEE Conference on Computational Complexity, pp. 250–259. IEEE, Los Alamitos (2004)
2. Beaver, D.: Perfect privacy for two-party protocols. In: Proceedings of the DIMACS Workshop on Distributed Computing and Cryptography (1989)
3. Beimel, A., Malkin, T., Micali, S.: The all-or-nothing nature of two-party secure computation. In: Wiener, M. (ed.) CRYPTO 1999. LNCS, vol. 1666, pp. 80–97. Springer, Heidelberg (1999)
4. Ben-Or, M., Goldwasser, S., Wigderson, A.: Completeness theorems for non-cryptographic fault-tolerant distributed computation. In: STOC 1988, pp. 1–10 (1988)
5. Brassard, G., Crépeau, C., Mayers, D., Salvail, L.: Defeating classical bit commitments with a quantum computer. Los Alamos preprint archive quant-ph/9806031 (May 1999)
6. Broadbent, A., Tapp, A.: Information-theoretic security without an honest majority. In: Kurosawa, K. (ed.) ASIACRYPT 2007. LNCS, vol. 4833, pp. 410–426. Springer, Heidelberg (2007)

7. Cachin, C., Crépeau, C., Marcil, J.: Oblivious transfer with a memory-bounded receiver. In: STOC 2002, pp. 493–502. ACM Press, New York (2002)
8. Cachin, C., Maurer, U.: Unconditional security against memory-bounded adversaries. In: Kaliski Jr., B.S. (ed.) CRYPTO 1997. LNCS, vol. 1294, pp. 292–306. Springer, Heidelberg (1997)
9. Chor, B., Kushilevitz, E.: A zero-one law for boolean privacy. In: STOC 1989 (1989)
10. Cleve, R.: Limits on the security of coin flips when half the processors are faulty. In: STOC 1986, pp. 364–369. ACM Press, New York (1986)
11. Damgård, I., Fehr, S., Salvail, L., Schaffner, C.: Cryptography in the bounded quantum-storage model. In: FOCS 2005, pp. 449–458. IEEE, Los Alamitos (2005)
12. Dolev, D., Strong, R.: Authenticated algorithms for byzantine agreement. SICOMP: SIAM Journal on Computing, 12 (1983)
13. Dziembowski, S., Maurer, U.: On generating the initial key in the bounded-storage model. In: Cachin, C., Camenisch, J.L. (eds.) EUROCRYPT 2004. LNCS, vol. 3027, pp. 126–137. Springer, Heidelberg (2004)
14. Fitzi, M., Hirt, M., Holenstein, T., Wullschleger, J.: Two-threshold broadcast and detectable multi-party computation. In: Biham, E. (ed.) EUROCRYPT 2003. LNCS, vol. 2656, pp. 51–67. Springer, Heidelberg (2003)
15. Goldreich, O.: Foundations of Cryptography: Basic Applications, vol. 2. Cambridge University Press, Cambridge (2004)
16. Goldreich, O., Micali, S., Wigderson, A.: How to play any mental game — a completeness theorem for protocols with honest majority. In: STOC 1987, pp. 218–229 (1987)
17. Gordon, S.D., Hazay, C., Katz, J., Lindell, Y.: Complete fairness in secure two-party computation. In: STOC 2008, pp. 413–422. ACM, New York (2008)
18. Haitner, I., Reingold, O.: Statistically-hiding commitment from any one-way function. In: STOC 2007, pp. 1–10. ACM, New York (2007)
19. Kilian, J.: A general completeness theorem for two-party games. In: STOC 1991, pp. 553–560. ACM Press, New York (1991)
20. Kilian, J.: More general completeness theorems for secure two-party computation. In: STOC 2000, pp. 316–324. ACM Press, New York (2000)
21. Kraschewski, D., Müller-Quade, J.: Completeness theorems with constructive proofs for symmetric, asymmetric and general 2-party-functions (unpublished manuscript, 2008)
22. Künzler, R., Müller-Quade, J., Raub, D.: Secure computability of functions in the IT setting with dishonest majority and applications to long-term security. Cryptology ePrint Archive, Report 2008/264 (2008), http://eprint.iacr.org/2008/264
23. Kushilevitz, E.: Privacy and communication complexity. In: FOCS 1989, pp. 416–421. IEEE, Los Alamitos (1989)
24. Kushilevitz, E.: Privacy and communication complexity. SIAM Journal on Discrete Mathematics 5(2), 273–284 (1992)
25. Mayers, D.: Unconditionally secure bit commitment is impossible. Phys. Rev. Letters 78, 3414–3417 (1997)
26. Müller-Quade, J.: Temporary assumptions—quantum and classical. In: The 2005 IEEE Information Theory Workshop on Theory and Practice in Information-Theoretic Security, pp. 31–33 (2005)
27. Müller-Quade, J., Unruh, D.: Long-term security and universal composability. In: Vadhan, S.P. (ed.) TCC 2007. LNCS, vol. 4392, pp. 41–60. Springer, Heidelberg (2007)
28. Rabin, M.: Hyper-encryption by virtual satellite. Science Center Research Lecture Series (2003)

Complexity of Multi-party Computation Problems: The Case of 2-Party Symmetric Secure Function Evaluation⋆

Hemanta K. Maji, Manoj Prabhakaran, and Mike Rosulek

Department of Computer Science, University of Illinois, Urbana-Champaign
{hmaji2,mmp,rosulek}@uiuc.edu

Abstract. In symmetric secure function evaluation (SSFE), Alice has an input x, Bob has an input y, and both parties wish to securely compute $f(x, y)$. We show several new results classifying the feasibility of securely implementing these functions in several security settings. Namely, we give new alternate characterizations of the functions that have (statistically) secure protocols against passive and active (standalone), computationally unbounded adversaries. We also show a strict, infinite hierarchy of complexity for SSFE functions with respect to universally composable security against unbounded adversaries. That is, there exists a sequence of functions $f_1, f_2, \ldots$ such that there exists a UC-secure protocol for f_i in the f_j-hybrid world if and only if $i \leq j$.

The main new technical tool that unifies our unrealizability results is a powerful protocol simulation theorem, which may be of independent interest. Essentially, in any adversarial setting (UC, standalone, or passive), f is securely realizable if and only if a very simple (deterministic) "canonical" protocol for f achieves the desired security. Thus, to show that f is unrealizable, one need simply demonstrate a single attack on a single simple protocol.

1 Introduction

In the classical setting of secure two-party computation, Alice and Bob have private inputs x and y respectively, and they want to to jointly compute a common value $f(x, y)$ in a secure way. Starting from Yao's millionaire's problem [21], such *symmetric secure function evaluation* (SSFE) problems have remained the most widely studied multi-party computation problems, in many security models. SSFE problems are fully specified by their associated function tables (i.e., a matrix M with $M_{x,y} = f(x, y)$); studying this matrix can tell us everything about the corresponding SSFE problem. Despite this apparent simplicity, and several works carefully exploring SSFE problems, the landscape of such problems has remained far from complete.

⋆ Partially supported by NSF grants CNS 07-47027 and CNS 07-16626.

O. Reingold (Ed.): TCC 2009, LNCS 5444, pp. 256–273, 2009.

On "cryptographic complexity." One expects different cryptographic tasks to have different levels of cryptographic sophistication. For instance public-key encryption is more complex than symmetric-key encryption. One indication of this is that in a computationally unbounded setting one-time pads provide (a limited version of) symmetric-key encryption, but public-key encryption is simply impossible. Impagliazzo and Rudich [8] provide a separation between the complexity of these two primitives by demonstrating another primitive (namely random oracles) which is sufficient to realize (full-fledged) symmetric-key encryption, but not enough for public-key encryption (in, say, a computationally unbounded setting). Our goal in this work is to understand such complexity separations among 2-party SSFE functionalities (and more generally, among multi-party computation functionalities).

The natural tool for comparing qualitative complexity of two tasks is a *reduction*. In the context of cryptographic feasibility, the most natural reduction is a black-box security reduction — we say that an SSFE functionality f reduces to another functionality g if it is possible to securely realize f using calls to a trusted party that securely implements g. As in computational complexity, the most fine-grained distinctions in complexity are made by considering the most restricted kinds of reductions. In fact, fine-grained complexity distinctions often disappear when using more generous reductions. In this work, we consider a very strong formulation of black-box security reduction: universally composable security in computationally unbounded environments against active (and adaptive) adversaries.

Our goal is to identify broad complexity classes of SSFE functionalities under this notion of reduction. This involves identifying various functionalities that reduce to each other, and — this is typically the more difficult part — establishing separations (non-reducibility) between functionalities. A complexity class can be *understood* by providing alternate (say combinatorial) characterizations of the functionalities in the class. Another approach to understanding a class is to identify a *complete* functionality, which provides a concrete embodiment of the complexity of all functionalities in that class. Conversely, the inherent cryptographic qualities of a given functionality can be understood by studying its "degree," namely, the class of all functionalities that can be reduced to it. We pursue all these approaches in this paper.

Finally, a systematic study of multi-party computation functionalities, under a stringent notion of reduction, unifies several prior advances in different security models. In particular, the two main classes that we identify and combinatorially characterize in this paper, which are downward-closed under this reduction, correspond to realizability in weaker security models — standalone security and passive security (a.k.a, semi-honest or honest-but-curious security). We emphasize that in plotting these classes in our complexity map, we do not change our notion of reduction.

Our results only start to unveil the rich landscape of cryptographic complexity of multi-party computation functionalities. We believe it will be of great theoretical — and potentially practical — value to further uncover this picture.

1.1   Previous Work

Cryptographic complexity of multi-party computation functionalities (though not necessarily under that name) has been widely studied in various security models. We restrict our focus mostly to work on 2-party SSFE functions in computationally unbounded settings. Complexity questions studied were limited to realizability (least complex), completeness (most complex) and whether there exist functions of intermediate complexities.

Realizability. The oldest and most widely studied model for SSFE is security against passive (honest-but-curious) adversaries. Chor and Kushilevitz [6] characterized SSFE functions with *boolean output*. Beaver [2] and Kushilevitz [17] independently extended this to general SSFE function, but restricted to the case of *perfect* security (as opposed to statistical security). These characterizations were given in the standalone security model, but do extend to the the the universal composition (UC) framework [4] that we use.

However, in the case of security against active (a.k.a malicious) adversaries, demanding composability does affect realizability. The following hold for both computationally bounded and unbounded settings. In the UC-setting, Canetti et al. [5] characterized securely realizable SSFE functions as those in which the function is insensitive to one party's input. Lindell [18] showed that UC security is equivalent to concurrent self-composable security, for a class of functionalities that includes SSFE. But Backes et al. [1] gave an example of a function that is realizable in the standalone setting, but not in the UC-setting. The problem of identifying all such functions remained open.

Completeness. The question of completeness for SSFE was essentially settled by Kilian [12], who showed that a function is complete if and only if it contains a generalized "OR-minor." This relies on the completeness of the SFE functionality of oblivious transfer, a result originally proven in [10], and proven in the UC setting in [9].[1] The reduction in [12] was reconsidered in the UC setting by [15].

Intermediate Complexities. In some security settings, there are only two distinct levels of complexity: the realizable tasks and the complete tasks, with nothing in between. Indeed, such a dichotomy holds in the case of *asymmetric* SFE (in which only one party receives any output), both for passive [3] and active security [13],[2] and also in the case of passive security for *boolean output* SSFE [14]. In [20] it is conjectured that such a dichotomy holds for general functionalities *in a computationally bounded setting*. However, there is no such simple dichotomy in the setting of SSFE. Indeed the characterizations of complete and realizable

---

[1] The protocol in [10] is not UC-secure, but an extension presented in [11] is likely to be.

[2] [3] also considers a notion of active security for computationally bounded setting, and extends their dichotomy using a stronger notion of realizability and a weaker, non-black-box notion of completeness; this result draws the line between realizability and completeness differently from [13]. The dichotomy does not extend to the UC-setting.

SSFE functions [12,5] leaves much gap between them. Further, [2,17,14] give an example SSFE function which is neither complete nor even passively realizable.

1.2   Our Results

A visual overview of our results is given in Figure 1.

First, we show that SSFE functions with *perfect* passive-secure protocols (as characterized by Beaver and Kushilevitz) are exactly those with *statistically secure*, passive-secure protocols (Theorem 3). They are also exactly the functions that are UC-realizable in the $\mathcal{F}_{com}$-hybrid world — i.e., realizable against active adversaries in the UC framework, using calls to an ideal commitment functionality (Theorem 4). Thus, $\mathcal{F}_{com}$ exactly captures the difficulty of passively realizing SSFE functions (it cannot be said to be complete, since it is not SSFE itself). We also show that the perfectly secure deterministic protocols used by Kushilevitz achieve optimal round complexity, even among randomized, statistically secure protocols (Corollary 2).

Next, we give an explicit and simple combinatorial characterization of the standalone-realizable SSFE functions, as the uniquely decomposable functions which are "maximal" (Theorem 5). We call such functions *saturated* functions. We also show that every SSFE function which is standalone-realizable but not UC-realizable also has no protocol secure under concurrent self-composition (Theorem 6), strengthening a negative result from [1], and yielding a much simpler proof of Lindell's characterization [18] for the special case of SSFE.

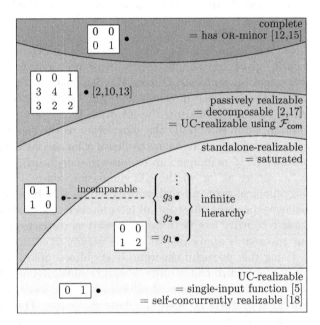

Fig. 1. Cryptographic complexity of 2-party SSFE

Finally, we focus our investigation on the vast area between the two complexity extremes of completeness and UC-realizability. We leverage ideas from passive security and standalone security to obtain a new technique for obtaining impossibility results in the UC framework. Namely, we describe a purely combinatorial criterion of two functions f, g (which has to do with the round complexity of realizing f and g) which implies that there is no UC-secure protocol for f that uses calls to an ideal functionality for g. (Theorem 7).

We apply this new separation technique to obtain several new results. We first demonstrate an infinite hierarchy of complexity — that is, SSFE functions $g_1, g_2, \ldots$ such that there is a secure protocol for g_i using calls to an ideal g_j if and only if $i \leq j$ (Corollary 8). We also show that there is no complete function (complete under our strong notion of reduction) for the class of standalone-realizable or passively realizable SSFE functions (Corollary 9). Finally, we show that there exist SSFE functions with *incomparable* complexities (Corollary 10), answering an open problem from [20].

We note that our characterizations of passive (Theorem 3) and standalone security (Theorem 5) were also independently discovered by Künzler, Müller-Quade, and Raub [16]. They also extend these results to a multi-party setting, and beyond the symmetric-output case.

About our techniques. The characterization of Beaver and Kushilevitz for passive security gives very simple deterministic protocols for SSFE functions, which we call *canonical protocols*. Our results demonstrate the special privilege of canonical protocols in the universe of SSFE. Our main technical tool is Theorem 1, which strongly formalizes the intuition that *every* protocol for f must disclose information in essentially the same order as the canonical protocol for f. Stated more formally, for any secure protocol π for an SSFE function f, if the canonical protocol for f is unique (up to some isomorphism), then the canonical protocol is "as secure as" π, in the UC simulation sense. That is, for any adversary in the canonical-protocol-real-world, there is an adversary in the π-real-world that achieves the same effect for all environments. Note that standalone security and passive security can both be expressed as restrictions of the UC definitions, so our theorem is applicable to a wide variety of security settings.

Using this powerful theorem, it is quite simple to demonstrate impossibility results for secure realizability. Roughly speaking, an SSFE function f satisfying the above condition is realizable in a given security model if and only if its canonical protocol achieves the desired security. Thus, to show f is unrealizable, we simply describe a feasible attack against the (simple) canonical protocol.

We crucially use Theorem 1 as the unifying component when proving impossibility of secure realization in the standalone (Theorem 5), concurrent self-composition (Theorem 6), and even UC hybrid world settings (Theorem 7).

2   Preliminaries

In this section we present some standard notions, and also introduce some concepts and definitions used in our constructions and proofs.

MPC Problems and Secure Realization. Following the Universal Composition [4] framework, a multi-party computation problem is defined by the code of a trusted (probabilistic, interactive) entity called a *functionality*. A protocol π is said to securely realize an MPC functionality $\mathcal{F}$, if for all (static) active adversaries, there exists a simulator such that for all environments, $|\Pr[\textsc{exec}_{\mathcal{F}} = 1] - \Pr[\textsc{exec}_{\pi} = 1]| \leq \epsilon(k)$, for some negligible function ϵ. Here k is the *security parameter* that is

---

- Wait for (and record) input x from Alice and input y from Bob.
- When both x and y have been received, if either party is corrupt, then send (OUTPUT, $f(x,y)$) to the adversary.
- On input DELIVER from the adversary, or if neither party is corrupt, send $f(x,y)$ to both Alice and Bob.

Fig. 2. Functionality $\mathcal{F}_f$: Symmetric Secure Function Evaluation of f with abort

an input to the protocol and simulator. EXEC$_{\mathcal{F}}$ denotes the environment's output distribution in the "ideal" execution: involving $\mathcal{F}$, the environment, and the simulator; EXEC$_\pi$ denotes the environment's output in the "real" execution: involving an instance of the protocol, the environment, and the adversary. Throughout this paper, we consider a UC model in which all entities are computationally unbounded.

We also consider *hybrid worlds*, in which protocols may also have access to a particular ideal functionality. In a $\mathcal{G}$-hybrid world, parties running the protocol π can invoke up any number of independent, asynchronous instances of $\mathcal{G}$. Regular protocols could be considered to use the (authenticated) communication functionality. If a protocol π securely realizes a functionality $\mathcal{F}$ in the $\mathcal{G}$-hybrid world, we shall write $\mathcal{F} \sqsubseteq \mathcal{G}$. The $\sqsubseteq$ relation is transitive and reflexive, and it provides our basis for comparing the complexity of various functionalities.

We also consider two common restrictions of the security model, which will prove useful in our results. If we restrict the security definition to environments which do not interact with the adversary during the protocol execution (i.e., simulators are allowed to rewind the adversary), we get a weaker notion of security, called *standalone security*. If we restrict to adversaries and simulators which receive an input from the environment and run the protocol honestly, we get a different relaxation of security called *passive security*. Note that in this definition, simulators must behave honestly in the ideal world, which means they simply pass the input directly to the functionality.

Symmetric SFE Functionalities. In this paper we focus exclusively on classifying 2-party *symmetric secure function evaluation* (SSFE) functionalities. An SSFE functionality $\mathcal{F}_f$ is parameterized by a function $f : X \times Y \rightarrow Z$, where X, Y, Z are finite sets.[3] The functionality $\mathcal{F}_f$ simply receives the inputs from the two parties, computes f on the inputs and sends the result to both the parties. However, as is standard, we also allow the adversary to first receive the output and block delivery if desired (see Figure 2). For convenience, we shall often use f and $\mathcal{F}_f$ interchangeably.

---

[3] We restrict our attention to *finite* functions. In particular, the size of the domain of a function does not change with the security parameter. This is in line with previous works. Further, in the computationally unbounded setting, it is reasonable to have the ideal world not involve the security parameter. Nevertheless, all of our results can be extended to the case where f's input domain is polynomially bounded in the security parameter. We can show that this restriction is necessary for most of our results.

2.1 Structure of Functions and Protocols

We say that two functions are isomorphic if each one can be computed using a single call to the other with no other communication (with only local processing of the inputs and output). That is, Alice and Bob can independently map their inputs for f function to inputs for g, carryout the computation for g, and then locally (using their private inputs) map the output of g to the output of f (and vice-versa).

Definition 1 (Function Isomorphism, $\cong$). *Let $f : X \times Y \to D$ and $g : X' \times Y' \to D'$ be two functions. We say $f \leq g$, if there exist functions $I_A : X \to X'$, $I_B : Y \to Y'$, $M_A : X \times D' \to D$ and $M_B : Y \times D' \to D$, such that for any $x \in X$ and $y \in Y$ $f(x,y) = M_A(x, g(I_A(x), I_B(y))) = M_B(y, g(I_A(x), I_B(y)))$. If $f \leq g$ and $g \leq f$ then $f \cong g$ (f is isomorphic to g).*

For example, the XOR function $\begin{bmatrix} 0 & 1 \\ 1 & 0 \end{bmatrix}$ is isomorphic to the function $\begin{bmatrix} 1 & 2 \\ 3 & 4 \end{bmatrix}$.

Definition 2 (Decomposable [2,17]). *A function $f : X \times Y \to D$ is row decomposable if there exists a partition $X = X_1 \cup \cdots \cup X_t$ ($X_i \neq \emptyset$), $t \geq 2$, such that the following hold for all $i \leq t$:*

- *for all $y \in Y$, $x \in X_i$, $x' \in (X \setminus X_i)$, we have $f(x,y) \neq f(x'y)$; and*
- *$f|_{X_i \times Y}$ is either a constant function or column decomposable, where $f|_{X_i \times Y}$ denotes the restriction of f to the domain $X_i \times Y$.*

We define being column decomposable symmetrically with respect to X and Y. We say that f is simply decomposable if it is either constant, row decomposable, or column decomposable.

For instance, $\begin{bmatrix} 0 & 0 \\ 1 & 2 \end{bmatrix}$ and $\begin{bmatrix} 0 & 1 \\ 1 & 0 \end{bmatrix}$ are decomposable, but $\begin{bmatrix} 0 & 0 \\ 0 & 1 \end{bmatrix}$ and $\begin{bmatrix} 0 & 0 & 1 \\ 3 & 4 & 1 \\ 3 & 2 & 2 \end{bmatrix}$ are not.

Note that our definition differs slightly from [2,17], since we insist that row and column decomposition steps strictly alternate. We say that a function f is *uniquely decomposable* if all of its decompositions are equivalent up to re-indexing $X_1, \ldots, X_t$ ($Y_1, \ldots, Y_t$) at each step. Thus $\begin{bmatrix} 0 & 0 \\ 1 & 2 \end{bmatrix}$ is uniquely decomposable, but $\begin{bmatrix} 0 & 1 \\ 1 & 0 \end{bmatrix}$ is not.

Canonical protocols [2,17]. If f is decomposable, then a *canonical protocol* for f is a deterministic protocol defined inductively as follows:

- If f is a constant function, both parties output the value, without interaction.
- If $f : X \times Y \to Z$ is row decomposable as $X = X_1 \cup \cdots \cup X_t$, then Alice announces the unique i such that her input $x \in X_i$. Then both parties run a canonical protocol for $f|_{X_i \times Y}$.
- If $f : X \times Y \to Z$ is column decomposable as $Y = Y_1 \cup \cdots \cup Y_t$, then Bob announces the unique i such that his input $y \in Y_i$. Then both parties run a canonical protocol for $f|_{X \times Y_i}$.

It is a simple exercise to see that a canonical protocol is a perfectly secure protocol for f against passive adversaries (cf. [2,17]).

Normal form for protocols. For simplicity in our proofs, we will often assume that a protocol is given in the following normal form:

1. At the end of the interaction, both parties include their outputs in the transcript as their last message before terminating, so that each party's output is a function of the transcript alone (i.e., not a function of their input and random tape, etc.). Since both parties should receive the same output, this is without loss of generality, even for standalone or UC security.

2. If $u_1, u_2, \ldots$ are the messages exchanged in a run of the protocol, then $(u_1, u_2, \ldots)$ can be uniquely and unambiguously obtained from the string $u_1 u_2 \cdots$.

3. The honest protocol does not require the parties to maintain a persistent state besides their private input (in particular, no random tape). Instead, the protocol is simply a mapping $P : \{0,1\}^* \times \{0,1\}^* \times \{0,1\}^* \to [0,1]$, indicating that if t is the transcript so far, and a party's input is x, then its next message is u with probability $P(t, x, u)$. In other words, randomness can be sampled as needed and immediately discarded. This requirement is without loss of generality for computationally unbounded parties.[4]

Deviation revealing. In [20] it is shown that for a class of functionalities called "deviation revealing", if a protocol π is a UC-secure realization of that functionality, then the same protocol is also secure against passive adversaries. Note that this property is not true in general for all SFE functionalities. For example, the SFE where Alice gets no output but Bob gets the boolean-OR of both parties' inputs is not passively realizable. However, the protocol where Alice simply sends her input to Bob is UC-secure, since a malicious Bob can always learn Alice's input in the ideal world by choosing 0 as its input to the functionality. It turns out SSFE functions are deviation revealing. We include an adapted version of the argument in [20]:

Lemma 1 ([20]). *Let π be a UC-secure (perhaps in a hybrid world) or standalone-secure protocol for an SSFE f. Then π itself is a passive-secure protocol for f as well (in the same hybrid world as π).*

Proof. We show that, without loss of generality, the simulator for π maps passive real-world adversaries to passive ideal-world adversaries. A passive adversary $\mathcal{A}$ for π is one which receives an input x from the environment, runs π honestly, and then reports its view to the environment. Note that the relevant kinds of environments comprise a special class of standalone environments, so that even if π is only standalone secure, its security still holds with respect to the environments we consider for passive security.

Suppose $\mathcal{S}$ is the simulator for $\mathcal{A}$. In the ideal world, both parties produce output with overwhelming probability, and so $\mathcal{S}$ must also allow the other party

---

[4] Note that because of this, security against adaptive corruption and static corruption are the same for this setting.

to generate output in the ideal world with overwhelming probability. Thus with overwhelming probability, $\mathcal{S}$ must receive x from the environment, send some x' to the ideal functionality f, receive the output $f(x', y)$ and deliver the output. Without loss of generality, we may assume $\mathcal{S}$ does so with probability 1.

Suppose x' is the input sent by $\mathcal{S}$ to f. If $f(x, y) \neq f(x', y)$ for some input y, then consider an environment that uses y for the other party's input. In this environment, the other party will report $f(x, y)$ in the real world, but $f(x', y)$ in the ideal world, so the simulation is unsound. Thus with overwhelming probability, $\mathcal{S}$ sends an input x' such that $f(x, \cdot) \equiv f(x', \cdot)$. We may modify $\mathcal{S}$ by adding a simple wrapper which ensures that x (the input originally obtained from the environment) is always sent to f. With overwhelming probability, the reply from f is unaffected by this change. Conditioned on these overwhelming probability events, the output of the wrapped $\mathcal{S}$ is identical to that of the original $\mathcal{S}$. However, the wrapped $\mathcal{S}$ is a *passive* ideal-world adversary: it receives x from the environment, sends x to f, and delivers the output.   □

3   Simulation of Canonical Protocol in a General Protocol

In this section, we develop our main new technical tool, the protocol simulation theorem. Throughout the section we fix an SSFE f with domain $X \times Y$, and fix a secure protocol π for f.

Definition 3. *We say that* x, x', y, y' *forms a* $\boxplus$-minor *(resp.* $\boxminus$-minor*) in* f *if:*

$$f(x, y) = f(x, y')$$
$$\neq \qquad \neq$$
$$f(x', y) \neq f(x', y')$$

$$\left(resp.\ if\ \begin{array}{ccc} f(x, y) & \neq & f(x, y') \\ = & & \neq \\ f(x', y) & \neq & f(x', y') \end{array} \right)$$

In the canonical protocol for a function that is entirely a $\boxplus$-minor, Alice must completely reveal her input before Bob reveals (anything about) his input. We show that, in general, this intuition carries through for *any* protocol, with respect to *any* embedded $\boxplus$ or $\boxminus$-minor. That is, there must be a point at which Alice has completely made the distinction between two of her inputs, but Bob has not made any distinction between two of his inputs.

Definition 4. *Let* $\Pr[u|x, y]$ *denote the probability that* π *generates a transcript that has* u *as a prefix, when executed honestly with* x *and* y *as inputs.*

Let F *be a set of strings that is prefix-free.*[5] *Define* $\Pr[F|x, y] = \sum_{u \in F} \Pr[u|x, y]$. *We call* F *a* frontier *if* F *is maximal – that is, if* $\Pr[F|x, y] = 1$ *for all* x, y. *We denote as* $\mathcal{D}_F^{x, y}$ *the probability distribution over* F *where* $u \in F$ *is chosen with probability* $\Pr[u|x, y]$.

Lemma 2 ($\boxplus$ Frontiers). *For all* $x \neq x' \in X$, *there is a frontier* F *and negligible function* ν *such that, for all* $y, y' \in Y$:

---

[5] That is, no string in F is a proper prefix of another string in F.

- if $f(x,y) \neq f(x',y)$, then $SD\left(\mathcal{D}_F^{x,y}, \mathcal{D}_F^{x',y}\right) \geq 1 - \nu(k)$, and

- if x, x', y, y' form a $\boxplus$-minor, then $SD\left(\mathcal{D}_F^{x,y}, \mathcal{D}_F^{x,y'}\right), SD\left(\mathcal{D}_F^{x',y}, \mathcal{D}_F^{x',y'}\right) \leq \nu(k)$.

We defer the technical proof of Lemma 2 to the full version [19].

Our main protocol simulation theorem extends this intuition to show that the information disclosed during a protocol must come in the same order as in the canonical protocol, provided that the canonical protocol is unique. This restriction on the canonical protocol is necessary, since different (non-isomorphic) canonical protocols for the same f can admit completely different kinds of attacks (e.g., for the XOR function, depending on which party speaks first).

Theorem 1 (Protocol Simulation). *If f has a unique decomposition, then for any protocol π for f, the canonical protocol for f is "as secure as" π. That is, for every adversary attacking the canonical protocol, there is an adversary attacking π which achieves the same effect in every environment.*

Proof (Sketch). The proof (presented in [19]) involves a careful inductive generalization of Lemma 2. Consider a step in the decomposition of $X \times Y$, say $X = X_1 \cup \cdots \cup X_k$. Roughly, if the function is uniquely decomposable, then for each $i \neq j$, there is a witnessing $\boxplus$-minor x, x', y, y' with $x \in X_i$, $x' \in X_j$. Thus we may apply Lemma 2 to obtain a frontier with respect to these inputs. We can combine these individual frontiers to obtain a frontier representing the entire decomposition step $X = X_1 \cup \cdots \cup X_k$. We show inductively that the transcript distribution at this combined frontier statistically reveals (in this case) which X_i contains Alice's input, while at the same is nearly independent of any further distinctions among either party's inputs within $X \times Y$.

Given such frontiers, the required simulation is fairly natural. The simulator's job is to simulate the canonical protocol to an adversary $\mathcal{A}$, while interacting with the honest party in the protocol π. The simulator $\mathcal{S}$ simply keeps track of the current step in the decomposition of f, and runs π honestly with any representative input. Each time $\mathcal{S}$ reaches a frontier for a decomposition step by the honest party, the transcript distribution at the frontier statistically determines with great certainty which part of the decomposition the honest party's input belongs to. Thus $\mathcal{S}$ can simulate to $\mathcal{A}$ what the honest party would send next in the canonical protocol. Then the simulator receives the adversary's next move in the canonical protocol. $\mathcal{S}$ changes its own input for π to any input consistent with the canonical protocol transcript so far, if necessary. Since the π-transcript so far is nearly independent of such distinctions among the adversary's input, it is indeed possible to swap inputs to π at this point and maintain a sound simulation. It is also for this reason that we consider protocols to be in a normal form, so that the protocol's next-message function depends only on the (currently considered) input and the transcript so far — i.e., not on any other persistent state. $\square$

Let $R(\pi, x, y)$ denote the random variable indicating the number of rounds taken by π on inputs x, y, and let $R(f, x, y)$ denote the same quantity with respect to the canonical protocol for f (which is deterministic).

Corollary 2. *If f is uniquely decomposable, then its canonical protocol achieves the optimal round complexity. That is, for every secure protocol π for f, we have $\mathbb{E}[R(\pi, x, y)] \geq R(f, x, y) - \nu$, where ν is a negligible function in the security parameter of π.*

Proof. The proof of Theorem 1 constructs for π a frontier for each step in the decomposition of f (corresponding to each step in the canonical protocol). By the required properties of the frontiers, the transcript for π must visit all the relevant frontiers in order, one *strictly* after the next, with overwhelming probability.   □

4   Characterizing Passive Security

In this section, we apply Lemma 2 to extend the characterization of Beaver [2] and Kushilevitz [17] to the case of statistical security. We also show a new characterization of passive security in terms of the ideal commitment functionality.

Theorem 3. *f is decomposable if and only if f has a (statistically secure) passive-secure protocol.*

Proof (Sketch). ($\Rightarrow$) Trivial by the (perfect) security of canonical protocols.

($\Leftarrow$) Suppose f is not decomposable. Without loss of generality, we may assume that f is not even row- or column-decomposable at the top level. Then for each way that Alice might start revealing information about her input, there is a ⊞-minor that witnesses the fact that this information cannot be revealed first — Bob must reveal a particular distinction of his inputs before it is safe for Alice to reveal information this way. However, the converse statement holds as well, and so neither party can be the first to safely reveal any information about their input.

More formally, we suppose a secure protocol exists for f. We consider all ⊞-minors, and take the upper envelope of their associated frontiers from Lemma 2. This new frontier corresponds to the first point at which Alice reveals significant information about her input. Similarly, we construct a frontier corresponding to the first point at which Bob reveals significant information. The properties of Lemma 2 imply that with overwhelming probability, the protocol must visit the first frontier after the second one, *and* visit the second frontier after the first one (i.e., neither party can be the first to significantly reveal information). This is a contradiction, since the two frontiers cannot coincide — they consist of points where different parties have just spoken. Thus no secure protocol is possible.   □

Theorem 4. *f has a passive-secure protocol if and only if f has a UC-secure protocol in the $\mathcal{F}_{\text{com}}$-hybrid world, where $\mathcal{F}_{\text{com}}$ denotes the ideal commitment functionality defined in Figure 3.*

> – On input (COMMIT, x, P_2) from party P_1, send (COMMITTED, P_1) to party P_2, and remember x.
> – On input REVEAL from party P_1, if x has already been recorded, send x to party P_2.

Fig. 3. Commitment functionality $\mathcal{F}_{\mathsf{com}}$

Proof (Sketch). ($\Leftarrow$) By Lemma 1, any UC-secure protocol π for a symmetric SFE f is also passively secure. There is a trivial passive-secure protocol for $\mathcal{F}_{\mathsf{com}}$ (the committing party sends "COMMITTED" in the commit phase and sends the correct value in the reveal phase). We can compose π with the passive-secure $\mathcal{F}_{\mathsf{com}}$ protocol to obtain a passive-secure protocol for f in the plain model.

($\Rightarrow$) We will give a general-purpose "compiler" from passive-secure protocols to the UC-secure $\mathcal{F}_{\mathsf{com}}$-hybrid protocols. Suppose π is a passive-secure protocol for f, in normal form. In fact, we need to consider only the canonical protocol for f. Below we consider an arbitrary deterministic protocol π. (In the full version [19] we extend this compiler to randomized protocols as well.)

Suppose Alice's input is $x \in \{1, \ldots, n\}$, and let $\chi \in \{0,1\}^n$ be the associated characteristic vector, where $\chi_i = 1$ if and only if $i = x$. Alice commits to both $\chi_{\sigma(1)}, \ldots, \chi_{\sigma(n)}$ and $\sigma(1), \ldots, \sigma(n)$ for many random permutations σ. For each pair of such commitments, Bob will (randomly) ask Alice to open either all of $\chi_{\sigma(1)}, \ldots, \chi_{\sigma(n)}$ (and verify that χ has exactly one 1) or open all $\sigma(i)$ (and verify that σ is a permutation). Bob also commits to his own input in a similar way, with Alice giving challenges. Then both parties simulate the π protocol one step at a time. When it is Alice's turn in π, she computes the appropriate next message $b \in \{0,1\}$ and sends it. For a deterministic protocol, the next message function is a function of the transcript so far and the input. Given the partial transcript so far t, both parties can compute the set $X_b = \{x' \mid \pi(t, x') = b\}$; i.e., the set of inputs for which the protocol instructs Alice to send b at this point. Then Alice can open enough of her commitments to prove that $\chi_i = 0$ for all $i \in X_{1-b}$ to prove that the message b was consistent with her committed input and the honest protocol. $\square$

Note that much like the well-known GMW compiler [7], we convert an *arbitrary* passive-secure protocol to a protocol secure in a stronger setting. In particular, we do not use the fact that the protocol realizes a functionality with symmetric, deterministic outputs. Thus the compiler may be of more general interest. Unlike the GMW compiler, our compiler operates in an unbounded setting, given ideal commitments. The GMW compiler relies on the existence of commitment protocols and zero-knowledge proofs, while in an unbounded setting, commitment protocols are impossible and zero-knowledge proofs are trivial.

5   Characterizing Standalone Security

From Lemma 1, we know that standalone-realizability for SSFE is a special case of passive-realizability, and hence by Theorem 3, any standalone realizable SSFE must be decomposable. In this section we identify the additional properties that exactly characterize standalone realizability.

Decomposition strategies. Fix some decomposition of an SSFE function $f :
X \times Y \to D$. We define an *Alice-strategy* as a function that maps every row decomposition step $X_0 = X_1 \cup \cdots \cup X_t$ to one of the X_i's. Similarly we define a *Bob-strategy* for the set of column decomposition steps. If A, B are Alice and Bob-strategies for f, respectively, then we define $f^*(A, B)$ to be the subset $X' \times Y' \subseteq X \times Y$ obtained by "traversing" the decomposition of f according to the choices of A and B.

The definition of f^* is easy to motivate: it denotes the outcome in a canonical protocol for f, as a function of the strategies of (possibly corrupt) Alice and Bob.

Definition 5 (Saturation). *Let f be a uniquely decomposable function. We say that f is* saturated *if $f \cong f^*$.*

To understand this condition further, we provide an alternate description for it. For every $x \in X$ we define an Alice-strategy A_x such that at any row decomposition step $X_0 = X_1 \cup \cdots \cup X_t$ where $x \in X_0$, it chooses X_i such that $x \in X_i$. (For X_0 such that $x \notin X_0$, the choice is arbitrary, say X_1.) Similarly for $y \in Y$ we define a Bob-strategy B_y. Note that in the canonical protocol, on inputs x and y, Alice and Bob traverse the decomposition of f according to the strategy (A_x, B_y), to compute the set $f^*(A_x, B_y)$ (where f is constant). If f is saturated, then all Alice strategies should correspond to some x that Alice can use as an input to f. That is, for all Alice-strategies A, there exists a $x \in X$ such that for all $y \in Y$, we have $f^*(A, B_y) = f^*(A_x, B_y)$; similarly each Bob strategy B is equivalent to some B_y.

As an example, $\begin{array}{|cc|} \hline 0 & 1 \\ 2 & 3 \\ \hline \end{array}$ is not uniquely decomposable. $\begin{array}{|ccc|} \hline 0 & 1 & 1 \\ 2 & 3 & 2 \\ \hline \end{array}$ is uniquely decomposable, but not saturated. Finally, $\begin{array}{|cccc|} \hline 0 & 1 & 1 & 0 \\ 2 & 3 & 2 & 3 \\ \hline \end{array}$ is saturated.

Note that there is exactly one saturated function (up to isomorphism) for each decomposition structure.

Theorem 5. *f is standalone-realizable if and only if f is saturated.*

Proof (Sketch). ($\Leftarrow$) To securely realize a saturated f, we use its canonical protocol. The simulator for an adversary $\mathcal{A}$ is a rewinding simulator, which does the following: Without loss of generality, assume $\mathcal{A}$ corrupts Alice. First fix a random tape ω for $\mathcal{A}$, then for every Bob-strategy B, run the canonical protocol against $\mathcal{A}$ (using random tape ω), effecting the strategy B. The choices of $\mathcal{A}$ at each step uniquely determine an Alice-strategy. By the saturation of f, $\mathcal{A}$ is equivalent to some A_x strategy, and we use x as the adversary's input to f. After receiving the output $f(x, y)$, we simulate the unique canonical protocol transcript consistent with $f(x, y)$.

($\Rightarrow$) We shall use the following lemma (proven, in the full version [19], by induction on the number of steps in the protocol):

Lemma 3. *If π is a 2-party protocol whereby both parties agree on the output, then for any way of coloring the possible outputs red and blue, π has one of the 4 properties:*

1. *A malicious Alice can force a red output and can force a blue output.*
2. *A malicious Bob can force a red output and can force a blue output.*
3. *Both a malicious Bob and malicious Alice are able to force a red output.*
4. *Both a malicious Bob and malicious Alice are able to force a blue output.*

By Lemma 1, we have that f is passively realizable and thus decomposable. We can show that if f is not uniquely decomposable, then there is a way to color its outputs red and blue so that each of the 4 conditions of Lemma 3 for any protocol evaluating f is a violation of security. Thus f must be uniquely decomposable.

Now, since f is uniquely decomposable, Theorem 1 implies that f is standalone-realizable if and only if its canonical protocol is standalone secure. Suppose that f is not saturated. Then there is a (without loss of generality) Alice-strategy A that corresponds to no A_x strategy. The strategy A can be trivially effected by a malicious Alice in the canonical protocol. If Bob's input is chosen at random by the environment, then the same outcome can not be achieved by an ideal-world adversary (who must send an input x to f). Thus the canonical protocol is standalone insecure; a contradiction. Therefore f must be saturated. □

6   Characterizing Concurrent Self-composition

Backes et al. [1] showed that even a perfectly secure protocol with rewinding simulator cannot in general be transformed into even a protocol secure under concurrent self-composition. Recall that in concurrent self-composition, the environment initiates several protocol instances, but does not interact with the adversary during their execution. The adversary also corrupts the same party in all protocol instances. We are able to greatly strengthen their negative result to show that concurrent attacks are the rule rather than the exception:

Theorem 6. *If f is standalone-realizable but not UC-realizable, then f has no protocol secure against concurrent self-composition.*

Proof. A function f is UC-realizable if and only if it is decomposable with decomposition depth 1 [5]. Thus an f as in the theorem statement must be uniquely decomposable (by Theorem 5), with decomposition depth at least 2. We show a concurrent attack against two instances of the *canonical protocol* for f, which by Theorem 1 will imply that f is not realizable under concurrent self-composition.

By symmetry, suppose Alice moves first in the canonical protocol, and let x, x' be two inputs which induce different messages in the first round of the protocol. Let y, y' be two inputs which induce different messages in the second round of the protocol when Alice has input x (thus $f(x, y) \neq f(x, y')$). We consider an environment which runs two instances of f, supplies inputs for Alice for the instances, and outputs 1 if Alice reports particular expected outputs for the two instances. The environment chooses randomly from one of the following three cases:

1. Supply inputs x and x'. Expect outputs $f(x, y')$ and $f(x', y)$.
2. Supply inputs x' and x, Expect outputs $f(x', y)$ and $f(x, y')$.
3. Supply inputs x and x. Expect outputs $f(x, y)$ and $f(x, y)$.

A malicious Bob can cause the environment to output 1 with probability 1 in the real world. He waits to receive Alice's first message in *both* protocol instances to determine whether she has x or x' in each instance. Then he can continue the protocols with inputs y or y', as appropriate.

In the ideal world, Bob must send an input to one of the instances of f before the other (i.e., before learning anything about how Alice's inputs have been chosen); suppose he first sends $\hat{y}$ to the first instance of f. If $f(x, \hat{y}) \neq f(x, y)$, then with probability $1/3$, he induces the wrong output in case (3). But if $f(x, \hat{y}) = f(x, y) \neq f(x, y')$, then with probability $1/3$, he induces the wrong output in case (1). Similarly, if he first sends an input to the second instance of f, he violates either case (2) or (3) with probability $1/3$.   □

Put differently, UC-realizability is equivalent to concurrent-self-composition-realizability for SSFE. This is in fact a special case of a theorem by Lindell [18], although our proof is much more direct and requires only 2 instances of the protocol/functionality, as in [1].

7   Finer Complexity Separations

Finally, we develop a new technique for deriving separations among SSFE functions with respect to the $\sqsubseteq$ relation, and apply the technique to concrete functions to inform the landscape of cryptographic complexity.

Theorem 7. *Let $\mathcal{F}$ be any (not necessarily SSFE) UC functionality with a passive-secure, m-round protocol. Let f be an SSFE function with unique decomposition of depth $n > m + 1$.[6] Then there is no UC-secure protocol for f in the $\mathcal{F}$-hybrid world; i.e., $f \not\sqsubseteq \mathcal{F}$.*

Proof (Sketch). We use a modified version of Theorem 1. Suppose for contradiction that π is a protocol for f in the $\mathcal{F}$-hybrid world. By Lemma 1, π is also passive-secure in the same setting. Define $\hat{\pi}$ to be the result of replacing each call to $\mathcal{F}$ in π with the m-round passive-secure protocol for $\mathcal{F}$. $\hat{\pi}$ is a passive-secure protocol for f in the plain setting. Say that an adversary *behaves consistently* for a span of time if it runs the protocol honestly with an input that is fixed throughout the span.

We mark every $\hat{\pi}$ transcript prefix that corresponds to a step in the $\mathcal{F}$ subprotocol, except for the first step of that subprotocol. Intuitively, if a $\hat{\pi}$-adversary behaves consistently during every span of marked transcripts, then that adversary can be mapped to a π adversary that achieves the same effect by interacting with $\mathcal{F}$ appropriately during these points.

---

[6] We remark that this condition can be tightened to $n > m$ via a more delicate analysis.

Since f is uniquely decomposable, we describe a feasible adversary $\mathcal{A}$ attacking the canonical protocol for f, and apply Theorem 1 to obtain an equivalent adversary $\mathcal{S}$ attacking $\widehat{\pi}$. Theorem 1 always constructs an adversary $\mathcal{S}$ that behaves consistently except for a small number of times where it might swap its input. We will construct $\mathcal{A}$ so that $\mathcal{S}$ only needs to swap its input at most once. If we can ensure that $\mathcal{S}$ will always be able to swap its input at an unmarked point in $\widehat{\pi}$, then $\mathcal{S}$ behaves consistently during the marked spans, so we can obtain an adversary successfully attacking π in the $\mathcal{F}$-hybrid world, a contradiction.

Suppose by symmetry that Alice goes first in the canonical protocol for f. Let x_0, y_0 be inputs that cause the canonical protocol to take a maximum number of rounds, and let y_1 be such that the (unique) transcripts for x_0, y_0 and x_0, y_1 agree until Bob's last message; thus $f(x_0, y_0) \neq f(x_0, y_1)$. Let x_1 be an input for which Alice's first message is different than for x_0.

We consider an environment which chooses a random $b \leftarrow \{0, 1\}$ and supplies x_b as Alice's input. It expects the adversary to detect and correctly report its choice of b. If $b = 0$, then the environment chooses a random $c \leftarrow \{0, 1\}$, and gives c to the adversary. The environment expects the adversary to induce Alice to report output $f(x_b, y_c)$. The environment outputs 1 if the adversary succeeds at both phases (guessing b and inducing $f(x_b, y_c)$ when $b = 0$). These conditions are similar to those of a "split adversary" considered by [5]. In the ideal world, an adversary must choose an input for f with no knowledge of Alice's input, and it is easy to see that the adversary fails with probability at least $1/4$. On the other hand, a trivial adversary $\mathcal{A}$ attacking the canonical protocol for f can always succeed.

Applying Theorem 1 to $\mathcal{A}$ will result in a $\mathcal{S}$ that considers n levels of frontiers in $\widehat{\pi}$ – one for each step in the canonical protocol. $\mathcal{S}$ only needs to swap its input at most once (possibly from y_0 to y_1). By the choice of x_0 and x_1, $\mathcal{S}$ can make its decision to swap after visiting the first frontier. Let k be the last round in which Bob moves, then $k \in \{n-1, n\}$. By the choice of y_0 and y_1, $\mathcal{S}$ can safely swap anywhere except after visiting the $(k-1)th$ frontier. It suffices for the transcript to encounter an unmarked step in $\widehat{\pi}$ in this range. This is true with overwhelming probability, since there is an unmarked step in every $m \leq k - 1$ consecutive steps in the protocol, and the frontiers are encountered in strict order with overwhelming probability, $\qquad\qquad\qquad\qquad\qquad\square$

Let $g_n : \{0, 2, \ldots, 2n\} \times \{1, 3, \ldots, 2n + 1\} \to \{0, 1, \ldots, 2n\}$ be defined as $g_n(x, y) = \min\{x, y\}$. It can be seen by inspection that g_n has a unique decomposition of depth $2n$. The corresponding canonical protocol is the one in which Alice first announces whether $x = 0$, then (if necessary) Bob announces whether $y = 1$, and so on — the "Dutch flower auction" protocols from [1].

Corollary 8. *The functions $g_1, g_2, \ldots$ form a strict, infinite hierarchy of increasing $\sqsubseteq$-complexity. That is, $g_i \sqsubseteq g_j$ if and only if $i \leq j$.*

Proof. By Theorem 7, $g_i \not\sqsubseteq g_j$ when $i > j$. It suffices to show that $g_n \sqsubseteq g_{n+1}$, since the $\sqsubseteq$ relation is transitive. It is straight-forward to see that the following is a UC-secure protocol for g_n in the g_{n+1}-hybrid world: both parties to send their g_n-inputs directly to g_{n+1}, and abort if the response is out of bounds for the output of g_n (i.e., $2n$ or $2n + 1$). The simulator for this protocol crucially uses the fact that the adversary can receive the output and then abort.   □

Corollary 9. *There is no function f which is complete (with respect to the $\sqsubseteq$ relation) for the class of passively realizable SSFE functions, or for the class standalone-realizable SSFE functions.*

Proof. This follows from Theorem 7 and by observing that there exist functions of arbitrarily high decomposition depth in both classes in question.   □

Corollary 10. *There exist SSFE functions f, g whose complexities are incomparable; that is, $f \not\sqsubseteq g$ and $g \not\sqsubseteq f$.*

Proof. If f is passively realizable but not standalone realizable, and g is standalone realizable, then $f \not\sqsubseteq g$, since the class of standalone security is closed under composition (at least when restricted to SSFE functions with abort, where the only kind of composition possible is sequential composition). On the other hand, if g has a unique decomposition depth at least 2 larger than the decomposition depth of f, then $g \not\sqsubseteq f$, by Theorem 7.   □

One example of such a pair of functions is $f = \textsc{xor}$ and $g = g_2$ from Corollary 8. In fact, using a more careful analysis, one can choose $g = g_1$ (see [19]). Thus $\textsc{xor}$ is incomparable to the entire $\{g_i\}$ hierarchy.

8   Conclusion and Open Problems

We have gained significant insight into the terrain of SSFE functions. However, there are regions of complexity of SSFE functions that we do not fully understand. In particular, we have not studied the class of incomplete functions which are not passive realizable. Nor have we attempted fully characterizing which functions are reducible to which ones. Going beyond SSFE functions, it remains open to explore similar questions for multi-party functionalities, for reactive functionalities, and (in the case of passive-security) for randomized functionalities. In the computationally bounded setting, the "zero-one conjecture" from [20] — that all functionalities are either realizable or complete — remains unresolved. It is also an intriguing problem to consider cryptographic complexity of multi-party functionalities vis a vis the "complexity" of cryptographic primitives (like one-way functions) that are required to realize them (in different hybrid worlds). In short, our understanding of cryptographic complexity of multi-party functionalities is still quite limited. There are exciting questions that probably call for a fresh set of tools and approaches.

References

1. Backes, M., Müller-Quade, J., Unruh, D.: On the necessity of rewinding in secure multiparty computation. In: Vadhan, S.P. (ed.) TCC 2007. LNCS, vol. 4392, pp. 157–173. Springer, Heidelberg (2007)

2. Beaver, D.: Perfect privacy for two-party protocols. In: Feigenbaum, J., Merritt, M. (eds.) Proceedings of DIMACS Workshop on Distributed Computing and Cryptography, vol. 2, pp. 65–77. American Mathematical Society (1989)
3. Beimel, A., Malkin, T., Micali, S.: The all-or-nothing nature of two-party secure computation. In: Wiener, M. (ed.) CRYPTO 1999. LNCS, vol. 1666, pp. 80–97. Springer, Heidelberg (1999)
4. Canetti, R.: Universally composable security: A new paradigm for cryptographic protocols. Electronic Colloquium on Computational Complexity (ECCC) TR01-016 (2001); Extended abstract in FOCS 2001
5. Canetti, R., Kushilevitz, E., Lindell, Y.: On the limitations of universally composable two-party computation without set-up assumptions. In: Biham, E. (ed.) EUROCRYPT 2003. LNCS, vol. 2656. Springer, Heidelberg (2003)
6. Chor, B., Kushilevitz, E.: A zero-one law for boolean privacy (extended abstract). In: STOC, pp. 62–72. ACM, New York (1989)
7. Goldreich, O., Micali, S., Wigderson, A.: Proofs that yield nothing but their validity or all languages in NP have zero-knowledge proof systems. J. ACM 38(3), 691–729 (1991); Preliminary version in FOCS 1986
8. Impagliazzo, R., Rudich, S.: Limits on the provable consequences of one-way permutations. In: STOC, pp. 44–61. ACM Press, New York (1989)
9. Ishai, Y., Prabhakaran, M., Sahai, A.: Founding cryptography on oblivious transfer - efficiently. In: Wagner, D. (ed.) CRYPTO 2008. LNCS, vol. 5157, pp. 572–591. Springer, Heidelberg (2008)
10. Kilian, J.: Founding cryptography on oblivious transfer. In: STOC, pp. 20–31. ACM, New York (1988)
11. Kilian, J.: Uses of Randomness in Algorithms and Protocols. PhD thesis, Department of Electrical Engineering and Computer Science, Massachusetts Institute of Technology (1989)
12. Kilian, J.: A general completeness theorem for two-party games. In: STOC, pp. 553–560. ACM, New York (1991)
13. Kilian, J.: More general completeness theorems for secure two-party computation. In: Proc. 32nd STOC, pp. 316–324. ACM, New York (2000)
14. Kilian, J., Kushilevitz, E., Micali, S., Ostrovsky, R.: Reducibility and completeness in private computations. SIAM J. Comput. 29(4), 1189–1208 (2000)
15. Kraschewski, D., Müller-Quade, J.: Completeness theorems with constructive proofs for symmetric, asymmetric and general 2-party-functions (unpublished manuscript, 2008)
16. Künzler, R., Müller-Quade, J., Raub, D.: Secure computability of functions in the it setting with dishonest majority and applications to long-term security (in these proceedings)
17. Kushilevitz, E.: Privacy and communication complexity. In: FOCS, pp. 416–421. IEEE, Los Alamitos (1989)
18. Lindell, Y.: Lower bounds for concurrent self composition. In: Naor, M. (ed.) TCC 2004. LNCS, vol. 2951, pp. 203–222. Springer, Heidelberg (2004)
19. Maji, H., Prabhakaran, M., Rosulek, M.: Complexity of multiparty computation problems: The case of 2-party symmetric secure function evaluation. Cryptology ePrint Archive, Report 2008/454 (2008), http://eprint.iacr.org/
20. Prabhakaran, M., Rosulek, M.: Cryptographic complexity of multi-party computation problems: Classifications and separations. In: Wagner, D. (ed.) CRYPTO 2008. LNCS, vol. 5157, pp. 262–279. Springer, Heidelberg (2008)
21. Yao, A.C.: Protocols for secure computation. In: Proc. 23rd FOCS, pp. 160–164. IEEE, Los Alamitos (1982)

Realistic Failures in Secure Multi-party Computation*

Vassilis Zikas, Sarah Hauser, and Ueli Maurer

Department of Computer Science, ETH Zurich, 8092 Zurich, Switzerland
{vzikas,maurer}@inf.ethz.ch,
shauser@student.ethz.ch

Abstract. In secure multi-party computation, the different ways in which the adversary can control the corrupted players are described by different corruption types. The three most common corruption types are active corruption (the adversary has full control over the corrupted player), passive corruption (the adversary sees what the corrupted player sees) and fail-corruption (the adversary can force the corrupted player to crash *irrevocably*). Because fail-corruption is inadequate for modeling recoverable failures, the so-called omission corruption was proposed and studied mainly in the context of Byzantine Agreement (BA). It allows the adversary to selectively block messages sent from and to the corrupted player, but without actually seeing the message.

In this paper we propose a modular study of omission failures in MPC, by introducing the notions of *send-omission* (the adversary can selectively block outgoing messages) and *receive-omission* (the adversary can selectively block incoming messages) corruption. We provide security definitions for protocols tolerating a threshold adversary who can actively, receive-omission, and send-omission corrupt up to t_a, t_ρ, and t_σ players, respectively. We show that the condition $3t_a + t_\rho + t_\sigma < n$ is necessary and sufficient for perfectly secure MPC tolerating such an adversary. Along the way we provide perfectly secure protocols for BA under the same bound. As an implication of our results, we show that an adversary who actively corrupts up to t_a players and omission corrupts (according to the already existing notion) up to t_ω players can be tolerated for perfectly secure MPC if $3t_a + 2t_\omega < n$. This significantly improves a result by Koo in TCC 2006.

1 Introduction

In secure multi-party computation (MPC) n players $p_1, \ldots, p_n$ wish to securely compute a function of their inputs. The computation should be secure, in the sense that the output is correct and the privacy of the players' inputs is not violated. The security should be guaranteed even when some of the players misbehave. The misbehavior of players is modeled by assuming a central adversary who corrupts players. The most typical corruption types are active corruption (the adversary has full control over the corrupted player), passive corruption (the adversary sees whatever the player sees), and fail-corruption (the adversary can make the player crash *irrevocably*).

---

* This research was partially supported by the Swiss National Science Foundation (SNF), project no. 200020-113700/1. The full version of this paper is available at http://www.crypto.ethz.ch/pubs/ZiHaMa09

O. Reingold (Ed.): TCC 2009, LNCS 5444, pp. 274–293, 2009.

The study of MPC was initiated by Yao [Yao82]. The first general solutions were given by Goldreich, Micali, and Wigderson [GMW87]; these protocols are secure under some intractability assumptions. Later solutions [BGW88, CCD88, RB89, Bea91b] provide information-theoretic security.

One of the most studied sub-problems of secure multi-party computation is Byzantine Agreement (BA). BA comes in two flavors, namely *consensus* and *broadcast*. Informally, consensus guarantees that n players, each holding an input, can agree on a common output without destroying pre-agreement. On the other hand, broadcast allows a dedicated player to consistently send his input to every player. BA serves as an important tool for the design of multi-party protocols.

Failures in MPC. For motivating the different corruption-types one typically thinks of MPC as each player running his protocol on his (local) computer, where the computers can communicate over some network (e.g., the Internet). Passive and active corruption correspond, for example, to (the adversary) planting a spyware or a virus, respectively, to the player's computer. Fail-corruption, however, can be criticized as being not so realistic due to the requirement that the crash is irrevocable. Indeed, in real-world scenarios computer-crashes are not irrevocable and are usually fixed soon after they are discovered, e.g., by replacing the computer.

Corruption types modeling more realistic failures than irrevocable computer-crashes have been studied in the literature. An example is the so-called *omission corruption* which allows the adversary to selectively block messages sent or received by the corrupted player, but without seeing the actual message. Omission corruption models failures that are apparent in many real-world applications, e.g., a computer which might lose messages while being restarted due to a hang of the operating system. It also models failures or temporary unavailability of the communication network, e.g., a router's buffer overflow, or instability of the links due to a thunderstorm. Partial asynchrony of the network, i.e., the adversary causing unexpected delays on messages sent from and to certain players, can also be modeled.

Omission corruption has been primarily studied in the context of fault-tolerant consensus [Had85, PT86, Ray02, PR03] and, recently, also in MPC [Koo06].

Summary of known results. In the seminal papers solving the general MPC problem, the adversary is specified by a single corruption type (active or passive) and a threshold t on the tolerated number of corrupted players. Goldreich, Micali, and Wigderson [GMW87] proved that, based on cryptographic intractability assumptions, general secure MPC is possible if and only if $t < n/2$ players are actively corrupted, or, alternatively, if and only if $t < n$ players are passively corrupted. In the information-theoretic model, Ben-Or, Goldwasser, and Wigderson [BGW88] and independently Chaum, Crépeau, and Damgård [CCD88] proved that unconditional security is possible if and only if $t < n/3$ for active corruption, and for passive corruption if and only if $t < n/2$. These results were unified and extended by fail-corruption in [FHM98] by proving that perfectly secure MPC is achievable if and only if $3t_a + 2t_p + t_f < n$, where t_a, t_p, and t_f denote the upper bounds on the number of actively, passively and fail corrupted players, respectively.

A similar development as in MPC can be observed in the area of Byzantine agreement protocols [LSP82, DS82, LF82, MP91, GP92, FM98].

The first to consider omission corruption were Perry and Tueg [PT86]. They considered a threshold adversary who can omission corrupt up to t players and showed that BA tolerating this adversary is possible if and only if $t < n$. However their consistency-guarantee is limited to the outputs of uncorrupted players, i.e., omission corrupted players are allowed to output arbitrary values. Raynal and Parvedy [Ray02, PR03] proved that if we require omission corrupted players to output either the correct value (i.e., consistent with the output of uncorrupted players) or no value, then consensus is possible if and only if $2t < n$.

In the context of general MPC, omission corruption was first studied, in combination with active corruption, by Koo [Koo06]. He considered a threshold adversary who can actively corrupt up to t_a players and, simultaneously, omission corrupt up to t_ω players,[1] and proved that the conditions $3t_a + 2t_\omega < n$ and $3t_a + 4t_\omega < n$ are sufficient for perfectly secure consensus and general MPC, respectively. However, as we show in Section 9, the condition $3t_a + 4t_\omega < n$ is far from optimal.

Our Contributions. We propose a modular study of realistic failures in multi-party computation, by introducing the notions of *send-omission* and *receive-omission* corruption. As the names suggest, send-omission (resp. receive-omission) corruption allows the adversary to selectively block only outgoing (resp. only incoming) messages of the corrupted player, but without seeing the messages (this is consistent with the existent omission-corruption literature). Note that a player who is omission corrupted according to the definitions of [PT86, Ray02, PR03, Koo06] can be thought of as a player who is both send- and receive-omission corrupted at the same time; for clarity we refer to this type of corruption as *full-omission* corruption.

We provide security definitions for the model where the adversary can actively, send-omission, and receive-omission corrupt players, simultaneously. We show that in this model, an adversary who can actively, receive-omission, and send-omission corrupt up to t_a, t_ρ, and t_σ players, respectively, can be tolerated for perfectly secure MPC if and only if $3t_a + t_\rho + t_\sigma < n$. Along the way, we also construct BA primitives for the same bound. Our bound implies that the condition $3t_a + 2t_\omega < n$ is sufficient for perfectly secure MPC.

The novelty of our approach is that, unlike past results on fault-tolerant MPC, we primarily deal with the omissions on the network-level instead of internally in the protocol. In particular, using the paradigm of layered communication (e.g., the OSI-model), first we engineer the actual network to build a new network-layer with better security guarantees, and then we design protocols in which the players communicate over this higher network-layer. This approach leads to simpler and more intuitive protocols. For the construction of our main protocol we also use ideas from the *player-elimination* technique [HMP00].

Outline of this paper. In Section 2 we define the model and introduce some notation. In Section 3 we discuss the security definitions and prove an impossibility result. In Sections 4 and 5 we show how to get an authenticated network with strong security guarantees and then build BA protocols over it. In Section 6 we provide tools that will

---

[1] In [Koo06], omission corrupted players are called *constrained* and actively corrupted are called *corrupted*.

be used as building blocks for the construction of the SFE and MPC protocols;[2] these protocols are described in Sections 7 and 8, respectively. In Section 9 we look at the case of full-omission corruption.

2   The Model

We consider the standard secure-channels model introduced in [BGW88, CCD88]: The players in $\mathcal{P} = \{p_1, \ldots, p_n\}$ are connected by a complete network of bilateral secure channels. The communication is synchronous, i.e., all players have synchronized clocks and there is a known upper bound on the delay of the network. The computation is described as an arithmetic circuit over some finite field $\mathbb{F}$, consisting of addition (or linear) and multiplication gates.

We look at the case of *perfect* security, i.e., information-theoretic without error probability. A protocol is defined to be secure if it realizes a trusted functionality (computing the function f), where the term "realize" is defined via the simulation paradigm [Can00, MR91, Bea91a, DM00, PW01] which, in a nutshell, guarantees that whatever the adversary can achieve in the real world where the protocol is executed, he could also achieve in the ideal setting with the trusted functionality.[3] This security notion implies in particular that the adversary cannot obtain any information about the players' inputs beyond what is implied by the outputs (privacy), and that he cannot influence the outputs other than by choosing the inputs of the corrupted players (correctness).

We consider a rushing[4] threshold adversary who can actively, receive-omission, and send-omission corrupt up to t_a, t_ρ, and t_σ players, respectively. The adversary chooses the players to corrupt non-adaptively, i.e., before the beginning of the protocol.[5]

To simplify the description we adopt the following convention: whenever a player does not receive a message (when expecting one), or receives a message outside of the expected range, then the special symbol $\perp \notin \mathbb{F}$ is taken for this message.

Every $p_i \in \mathcal{P}$ can be in one of the following two internal states: *alive* or *zombie*. At the beginning of the computation every player is alive, which means that he correctly executes all the protocol instructions (unless he is actively corrupted). If p_i realizes that he is receive-omission corrupted, e.g., by receiving fewer messages than what he should in some round, then p_i sets his internal state to zombie (we say that p_i becomes a zombie). Once the state is set to zombie it never switches back. A zombie behaves in the

---

[2] SFE stands for Secure Function Evaluation, i.e., multi-party computation of *non-reactive* functionalities.

[3] While our protocols can be proved secure in any of these simulation-based frameworks, with perfect indistinguishability of the real and the ideal world, we will not give full-fledged simulation-based security proofs in this paper; this is consistent with the previous literature on secure SFE and MPC.

[4] A *rushing* adversary is an adversary who, in each round of the protocol, first sees all the messages sent to actively corrupted players in this round and then decides how the corrupted players should behave in this round.

[5] In contrast, an *adaptive* adversary can corrupt more and more players during the protocol execution, subject only to the constraint that the number of corrupted players of each type is upper-bounded by the corresponding threshold. We do not consider the adaptive setting in this paper, but our results could be generalized to it.

protocols as a player who has crashed, i.e., sends and receives no messages and has no outputs. However, there are two conceptual differences between zombies and crashed players: (1) Being a zombie is a self-imposed state and corresponds to a correct behavior, i.e., players become zombies when the protocol (and not the adversary) instructs them to; (2) zombie-players are "aware of the actual time", as they have clocks which are synchronized with the clocks of the alive players; this will be useful in the context of reactive computation (Section 8) where time plays an important role.

The sets $\mathcal{A}$, $\mathcal{S}$, $\mathcal{R}$, $\mathcal{SR}$, and $\mathcal{H}$. To simplify the description we denote the sets of actively, send-omission only, receive-omission only, and full-omission[6] (but not actively) corrupted players by $\mathcal{A}$, $\mathcal{S}$, $\mathcal{R}$, and $\mathcal{SR}$, respectively, and the set of uncorrupted players by $\mathcal{H}$ ($\mathcal{H}$ stands for "honest"). Note that these sets are a partition of the player set $\mathcal{P}$, they are not known to the players and appear only in the security analysis.

3   Security Definition

Intuitively, the security definition for our model should not allow the adversary to do more with send- and receive-omission corrupted players than to decide which of them give input to and receive output from the computation, respectively. The strongest security one can hope for is to require that the adversary's decision is taken independently of the inputs of non actively corrupted players and before seeing the outputs of actively corrupted players. More precisely one would be interested in securely realizing the functionality STRONG SFE (see below).[7]

STRONG SFE - IDEAL MODEL. *Each $p_i \in \mathcal{P}$ has input x_i. The function to be computed is $f(\cdot)$. The adversary decides which of the send-omission (resp. receive-omission) corrupted players give input to (resp. receive output from) the trusted party* before *seeing the outputs of actively corrupted players.*

1. Every $p_i \in \mathcal{H} \cup \mathcal{R}$ sends his input to the trusted party (TP). Actively corrupted players might send TP arbitrary inputs as instructed by the adversary. For each $p_i \in \mathcal{SR} \cup \mathcal{S}$ the adversary decides (without seeing p_i's input) whether p_i sends TP his input or a default value from $\mathbb{F}$ (e.g., 0). TP denotes the received values by $x'_1, \ldots, x'_n$.
2. TP computes $f(x'_1, \ldots, x'_n) = (y_1, \ldots, y_n)$ (if f is randomized then TP internally generates the necessary random coins). TP asks the adversary which of the players $p_i \in \mathcal{R} \cup \mathcal{SR}$ should receive their output y_i (without revealing any information about y_i).
3. For each $p_i \in \mathcal{H} \cup \mathcal{S} \cup \mathcal{A}$, TP sends y_i to p_i. For each $p_i \in \mathcal{R} \cup \mathcal{SR}$, TP sends y_i to p_i if the adversary allowed that p_i receives output in the previous step, otherwise TP sends nothing to p_i.

---

[6] Recall that a full-omission corrupted player is one who is both send- and receive-omission corrupted at the same time.

[7] We assume that the reader is familiar with the ideal-world/real-world paradigm for defining security of multi-party protocols [Bea91a, MR91, Can00, DM00, BPW03].

We say that a protocol Π *strongly* (t_a, t_ρ, t_σ)-securely evaluates the function f if it securely realizes the functionality STRONG SFE in the presence of an adversary who can actively, receive-omission, and send-omission corrupt up to t_a, t_ρ, and t_σ players, respectively.

Unfortunately, as stated in the following lemma, when the adversary is rushing then for any non-trivial choice for t_a and t_ρ there exist functions which cannot be perfectly strongly (t_a, t_ρ, t_σ)-securely evaluated. In fact our impossibility result is inherent in any setting where we have a threshold adversary with active (or even just passive) and receive-omission corruption, simultaneously. In particular it also applies to the (non-adaptive) case of active and full-omission corruption [Koo06].[8] The idea is the following: the adversary might, with non-zero probability, corrupt the player p_i who is the first (or among the first) to get the output, e.g., by randomly choosing whom to corrupt. In this case, as she is rushing, she can decide, depending on the output, whether the receive-omission corrupted players get full information on the output or not. However, the simulator has to take this decision without seeing the outputs of corrupted players, and hence he is not able to perfectly simulate this behavior. Due to space restrictions the proof of the lemma is deleted from this extended abstract.

Lemma 1. *If $t_a > 0$ and $t_\rho > 0$ and the adversary is rushing, then there exist functions which cannot be perfectly strongly $(t_a, t_\rho, \cdot)$-securely evaluated. The statement holds even when we have passive instead of active corruption.*

We relax the definition of the functionality to allow the adversary to decide which receive-omission corrupted players receive output, even after having seen the outputs of actively corrupted players (and possibly depending on those outputs). Our relaxation is minimal as Lemma 1 suggests. We call the resulting functionality SFE (see below).

SFE – IDEAL MODEL. *Each $p_i \in \mathcal{P}$ has input x_i. The function to be computed is $f(\cdot)$. The adversary decides which of the receive-omission corrupted players receive output from the trusted party* after *receiving the outputs of actively corrupted players.*

1. Every $p_i \in \mathcal{H} \cup \mathcal{R}$ sends his input to the trusted party (TP). Actively corrupted players might send TP arbitrary inputs as instructed by the adversary. For each $p_i \in \mathcal{SR} \cup \mathcal{S}$ the adversary decides (without seeing p_i's input) whether p_i sends TP his input or a default value from $\mathbb{F}$ (e.g., 0). TP denotes the received values by $x'_1, \ldots, x'_n$.
2. TP computes $f(x'_1, \ldots, x'_n) = (y_1, \ldots, y_n)$ (if f is randomized then TP internally generates the necessary random coins). For each $p_i \in \mathcal{H} \cup \mathcal{S} \cup \mathcal{A}$, TP sends y_i to p_i.
3. For $p_i \in \mathcal{R} \cup \mathcal{SR}$, TP asks the adversary if p_i should receive his output y_i (without revealing any information about y_i), if the answer is yes then TP sends y_i to p_i, otherwise it sends nothing to p_i.

---

[8] In [Koo06] the assumed adversary is also rushing and the (non-adaptive) ideal-world functionality requires the adversary to decide which omission corrupted players receive output before seeing the outputs of actively corrupted players.

Definition 1. *We say that a protocol* Π (t_a, t_ρ, t_σ)-securely *evaluates the function* f *if* Π *securely realizes the functionality* SFE *in the presence of an adversary who can actively, receive-omission, and send-omission corrupt up to* t_a, t_ρ, *and* t_σ *players, respectively.*

4   Engineering the Network – Authenticated Channels

A source of difficulties in designing protocols tolerating both active cheaters and omissions is that a player p_j who receives $\perp$ when expecting a message from a player p_i cannot decide whether p_i is send-omission or actively corrupted, or himself (i.e., p_j) is receive-omission corrupted. In [Koo06] the following straight-forward approach was taken in order to overcome this difficulty in the context of p_i sharing a secret: Every player complains when he received no share from the dealer p_i. If more players complain than the number of potentially corrupted players, p_i is disqualified. Otherwise, the players who did not complain pairwise check the consistency of their shares (as in [BGW88, FHM98]), where inconsistencies are publicly reported and resolved by the dealer. This approach, however, leads to thresholds on the number of actively and (full) omission corrupted players which are far from optimal, as discussed in the introduction.

Our approach is different. We deal with this difficulty outside the protocol, on the network level. In particular, using the paradigm of layered communication (e.g., the OSI-model), first we engineer the actual network to get a new network-layer with stronger guarantees, and then we invoke the actual protocol over this layer.

The protocol which is used to build the new network-layer is called FixReceive. It works on the channels of the actual network (the lowest layer), i.e., the secure channels with omissions, and builds on top of them a network of *authenticated* channels (the higher layer), where for any receive-omission corrupted p_i the adversary has to choose *either* to allow p_i to receive all messages that are sent to him *or* to let p_i know that he is receive-omission corrupted. More precisely, FixReceive guarantees that when some p_i sends a message x to a receive-omission corrupted p_j then either p_j receives it, as if he were uncorrupted, or p_j finds out that he is receive-omission corrupted (and becomes a zombie). If p_j becomes a zombie in FixReceive then he notifies every $p_k \in \mathcal{P}$ about this by sending a bilateral message; this information will be used by the players in future invocations of FixReceive. The protocol FixReceive is described in the following. For the proof of the lemma we refer to the full version of this paper.

Protocol FixReceive $(\mathcal{P}, t_a, t_\rho, t_\sigma, p_i, p_j, x)$
 1. p_i sends his input x to every $p_k \in \mathcal{P}$.
 2. Each $p_k \in \mathcal{P}$ forwards x to p_j (if p_k received no value, he sends a special symbol "n/v" to p_j); p_j denotes the received value as x_k (if p_k has become a zombie in the past then p_j sets $x_k =$ "n/v").
 3. If $|\{p_k : x_k \in \mathbb{F} \cup \{\text{"n/v"}\}\}| < n - t_a - t_\sigma$ then p_j becomes zombie (and notifies all players). Otherwise, if there exists $x' \notin \{\perp, \text{"n/v"}\}$ such that $|\{p_k : x_k = x'\}| > t_a$ then p_j outputs x', otherwise p_j outputs $\perp$.

Lemma 2. *If* $3t_a + t_\rho + t_\sigma < |\mathcal{P}|$, *protocol* FixReceive *has the following properties. If* p_j *is alive at the end of the protocol then he outputs a value* x', *where* $x' \in \{x, \perp\}$ *unless* $p_i \in \mathcal{A}$, *and* $x' = x$ *when* $p_i \in \mathcal{H} \cup \mathcal{R}$. *Moreover,* p_j *might become a zombie only when* $p_j \in \mathcal{R} \cup \mathcal{SR}$ *and when he becomes a zombie every player notices.*

5   Byzantine Agreement

In this section we build primitives solving the Byzantine Agreement (BA) problem, which we will later use as tools for constructing the main SFE protocol. BA comes in two flavors, namely *consensus* and *broadcast*. Informally, consensus guarantees that n players, each holding an input, can decide on a common output y, where $y = x$ if all non-actively corrupted players had (the same) input x. On the other hand, broadcast allows a dedicated player p_s holding input x_s, to consistently send x_s to every player.

In our BA protocols, the players communicate over the strengthened authenticated network which is constructed using FixReceive. More precisely, whenever $p_i \in \mathcal{P}$ is instructed to bilaterally send a message to $p_j \in \mathcal{P}$, the protocol FixReceive is invoked. Because alive players might become zombies only within FixReceive, all the designed protocols have the following property: *Only receive-omission corrupted players might become zombies.* The proofs of the lemmas can be found in the full version of the paper.

5.1   Consensus

For constructing a consensus protocol, we use the standard approach [BGP89, FM00]: We construct weaker consensus primitives, and then compose them in a clever way to construct the desired consensus primitive. We construct three such weaker primitives called *Weak Consensus, Graded Consensus*, and *King Consensus*.

Weak Consensus. Informally, weak consensus guarantees that there are no inconsistencies among the outputs of the non-actively corrupted players, but some of them (even alive) might have no output (we say that they output $\perp$). However, we get the guarantee that if the players *pre-agreed* on some value x, i.e., all non-actively corrupted players had input (the same) x, then we get *post-agreement* on x, i.e., all non-actively corrupted players output x.[9] In the following we describe protocol WeakConsensus which achieves weak consensus in our model. The input of each $p_i \in \mathcal{P}$ is denoted as x_i

---

Protocol WeakConsensus $(\mathcal{P}, t_a, t_\rho, t_\sigma, \vec{x} = (x_1, \ldots, x_n))$

1. Each $p_i \in \mathcal{P}$ sends x_i to every $p_j \in \mathcal{P}$, by invoking FixReceive; p_j denotes the received value by $x_j^{(i)}$.
2. Each $p_j \in \mathcal{P}$ sets
$$y_j := \begin{cases} x & , \text{if } (|\{p_i : x_j^{(i)} = x\}| \geq n - t_a - t_\sigma - t_\rho) \ \wedge \\ & \qquad (|\{p_i : x_j^{(i)} \notin \{x, \perp\}\}| \leq t_a) \\ \perp & , \text{otherwise} \end{cases}$$

Lemma 3. *If* $3t_a + t_\rho + t_\sigma < |\mathcal{P}|$, *the protocol* WeakConsensus *has the following properties. Weak Consistency: Every (alive)* $p_j \in \mathcal{P} \setminus \mathcal{A}$ *outputs* $y_j \in \{x', \perp\}$ *for*

---

[9] Recall that the zombies send no values in any protocol and receive no output.

some $x' \in \mathbb{F}$. Correctness: If every $p_i \in \mathcal{P} \setminus \mathcal{A}$ who is alive at the beginning of WeakConsensus *has input $x_i = x$, then $x' = x$.*

Graded Consensus. In Graded Consensus each $p_i \in \mathcal{P}$ outputs a pair (y_i, g_i), where y_i is p_i's actual output-value and $g_i \in \{0, 1\}$ is a bit, called p_i's *grade*. The grade g_i has the meaning of the confidence level of p_i on the fact that all non-actively corrupted players also output y_i. In particular, if $g_i = 1$ for some non-actively corrupted p_i then $y_j = y_i$ for every (alive) non-actively corrupted $p_j \in \mathcal{P}$. Moreover, when the non-actively corrupted players pre-agreed on a value x, then they all output x with grade 1.

In the following we describe the protocol GradedConsensus. The idea is to have the players first invoke the protocol WeakConsensus and then exchange their outputs of WeakConsensus to decide on the actual output and the corresponding grade.

Protocol GradedConsensus $(\mathcal{P}, t_a, t_\rho, t_\sigma, \vec{x} = (x_1, \ldots, x_n))$

1. Invoke WeakConsensus $(\mathcal{P}, t_a, t_\rho, t_\sigma, \vec{x})$; p_i denotes his output by x'_i.
2. Each $p_i \in \mathcal{P}$ sends x'_i to every $p_j \in \mathcal{P}$ by invocation of FixReceive; p_j denotes the received value by $x_j^{(i)}$.
3. Each $p_j \in \mathcal{P}$ sets $y_j := \begin{cases} x & \text{, if there exists } x \in \mathbb{F} \text{ s.t. } |\{p_i : x_j^{(i)} = x\}| > t_a \\ 0 & \text{, otherwise} \end{cases}$

and sets $g_j := \begin{cases} 1 & \text{, if } (|\{p_i : x_j^{(i)} \in \{y_j, \bot\}\}| \geq n - t_a) \; \wedge \\ & \quad (|\{p_i : x_j^{(i)} = y_j\}| \geq n - t_a - t_\rho - t_\sigma) \\ 0 & \text{, otherwise} \end{cases}$

Lemma 4. *If $3t_a + t_\rho + t_\sigma < |\mathcal{P}|$, protocol* GradedConsensus *has the following properties. Graded Consistency: If some $p_j \in \mathcal{P} \setminus \mathcal{A}$ outputs $(y_j, g_j) = (y, 1)$ for some $y \in \mathbb{F}$, then every (alive) $p_k \in \mathcal{P} \setminus \mathcal{A}$ outputs $(y_k, g_k) = (y, g_k)$, where $g_k \in \{0, 1\}$. Graded Correctness: If every $p_i \in \mathcal{P} \setminus \mathcal{A}$ who is alive at the beginning of* GradedConsensus *has input $x_i = x$, then every (alive) $p_j \in \mathcal{P} \setminus \mathcal{A}$ outputs $(y_j, g_j) = (x, 1)$.*

King Consensus. In King Consensus there is a distinguished player $p_k \in \mathcal{P}$, called the *king*. King Consensus guarantees that if the king is uncorrupted, then all non-actively corrupted players output the same value. Additionally, independent of the king's corruption, if the non-actively corrupted players pre-agreed on a value x, then they all output x. The protocol KingConsensus is described in the following.

Protocol KingConsensus $(\mathcal{P}, t_a, t_\rho, t_\sigma, \vec{x} = (x_1, \ldots, x_n), p_k)$

1. Invoke GradedConsensus$(\mathcal{P}, t_a, t_\rho, t_\sigma, \vec{x})$; p_i denotes his output by (x'_i, g_i).
2. The king p_k sends x'_k to every $p_j \in \mathcal{P}$ by invocation of FixReceive.
3. Each $p_j \in \mathcal{P}$ sets

$y_j := \begin{cases} x'_j & \text{, if } (g_j = 1) \text{ or } (p_k \text{ sent } x'_k = \bot) \\ x'_k & \text{, otherwise} \end{cases}$

Lemma 5. *If $3t_a + t_\rho + t_\sigma < |\mathcal{P}|$, the protocol* KingConsensus *has the following properties. King Consistency: If the king p_k is uncorrupted, then every $p_j \in \mathcal{P} \setminus \mathcal{A}$ outputs $y_j = y$. Correctness: If every $p_i \in \mathcal{P} \setminus \mathcal{A}$ who is alive at the beginning of* KingConsensus *has input $x_i = x$ then they all output $y = x$.*

Consensus. Building a consensus protocol from king consensus is straight-forward: Invoke KingConsensus with $t_a + t_\rho + t_\sigma + 1$ different players as king, where the input of the i-th iteration is the output of the $(i-1)$-th iteration. As there are at most $t_a + t_\rho + t_\sigma$ corrupted players, at least one of the kings will be uncorrupted, hence consistency on the output value will be achieved in the corresponding iteration; the correctness of KingConsensus guarantees that this value will not be changed in any future iteration. Note that when we have pre-agreement on some value then consistency on this value is achieved from the first iteration independent of the king.

Lemma 6. *If* $3t_a + t_\rho + t_\sigma < |\mathcal{P}|$*, the protocol* Consensus *has the following properties. Consistency: All (alive)* $p_i \in \mathcal{P} \setminus \mathcal{A}$ *output (the same)* $y \in \mathbb{F}$*. Correctness: If every* $p_i \in \mathcal{P} \setminus \mathcal{A}$ *who is alive at the beginning of* Consensus *has input* $x_i = x$ *then* $y = x$*.*

5.2 Broadcast

The standard approach for achieving broadcast when consensus is given, is to have the sender p_s send his input to every player, and then run consensus on the received values. Unfortunately, this generic approach does not work in our setting, as it provides no guarantees when a send-omission corrupted p_s fails to send his input to some uncorrupted players.

To guarantee that a non actively corrupted p_s never broadcasts a wrong value we extend the above generic protocol by adding the following steps: p_s sends a confirmation bit to every player, i.e., a bit b where $b = 1$ if p_s agrees with the output of the consensus and $b = 0$ otherwise; subsequently, the players invoke consensus on the received bits to establish a consistent view on the confirmation-bit and they accept the output of the generic broadcast protocol only if this bit equals 1, otherwise they output $\bot$. This results in the protocol Broadcast (see below).

Protocol Broadcast $(\mathcal{P}, t_a, t_\rho, t_\sigma, p_s, x_s)$
1. p_s sends x to every $p_j \in \mathcal{P}$ (by FixReceive), who denotes the received value by x_j ($x_j = 0$ if p_j received $\bot$).
2. The players invoke Consensus on the received values. Let y_j denote p_j's output.
3. Each p_j sends y_j to p_s (by FixReceive).
4. p_s sends a confirmation bit b to every $p_i \in \mathcal{P}$ (by FixReceive), where $b = 1$ if p_s received $y_j = x$ from more that t_a players in the previous step and $b = 0$ otherwise; p_i denotes the received bit by b_i ($b_i = 0$ if p_i received $\bot$).
5. Invoke Consensus $(\mathcal{P}, t_a, t_\sigma, t_\rho, (b_1, \dots, b_n))$. For each $p_i \in \mathcal{P}$, if p_i's output in Consensus is 1 then p_i outputs y_i, otherwise he outputs $\bot$.

Lemma 7. *If* $3t_a + t_\rho + t_\sigma < |\mathcal{P}|$*, protocol* Broadcast *has the following properties. Consistency: All (alive)* $p_j \in \mathcal{P} \setminus \mathcal{A}$ *output the (same) value* $y_j = y$*. Correctness:* $y \in \{x, \bot\}$ *when* $p_s \in \mathcal{P} \setminus \mathcal{A}$*, where* $y = x$ *when* $p_s \in \mathcal{H} \cup \mathcal{R}$ *and he is alive at the end of the protocol, and* $y = \bot$ *when* p_s *has been a zombie from the beginning of the protocol.*

6   Tools

In this section we describe sub-protocols that will be used as building-blocks in the construction of the main SFE and MPC protocols. Some of the sub-protocols are non-robust, and might abort with a non-empty set $B \subseteq \mathcal{P}$. When they abort, then all (alive) players in $\mathcal{P}$ notice it and they also learn the set B. As in the case of BA, some alive players might become zombies during the invocation of the sub-protocols, but only when they are receive-omission corrupted.

6.1   Secret Sharing

A secret sharing scheme allows a player, called the *dealer*, to distribute his input among the players in some player set $\mathcal{P}$, so that only qualified sets of players can reconstruct it. As usual in the threshold adversary literature, we use Shamir-sharings for sharing values: With each $p_i \in \mathcal{P}$ a unique publicly known $\alpha_i \in \mathbb{F}$ is associated. A secret s is *t-shared* among the players in $\mathcal{P}$ when there exists a degree-t polynomial $q(\cdot)$ with $q(0) = s$, and every non actively corrupted $p_i \in \mathcal{P}$ holds $s_i \in \{q(\alpha_i), \bot\}$, where $s_i = q(\alpha_i)$ unless p_i is receive-omission corrupted. The value s_i is p_i's *share* of s. We refer to the vector of shares, denoted by $\langle s \rangle = (s_1, \ldots, s_n)$, as a *t-sharing* of s.

We say that $\langle s \rangle$ is a *t-consistent* sharing of s among the players in $\mathcal{P}$ if there exists a degree-t polynomial $q(\cdot)$ such that each non actively corrupted $p_i \in \mathcal{P}$ holds share $s_i \in \{q(\alpha_i), \bot\}$. We say that $\langle s \rangle$ is a *t-valid* sharing of s among the players in $\mathcal{P}$, if $\langle s \rangle$ is t-consistent and for some degree-t polynomial $q(\cdot)$ with $q(0) = s$, each uncorrupted $p_i \in \mathcal{P}$ holds share $s_i = q(\alpha_i)$.

Protocol Share allows a dealer p to t-share his input among the players in any set $\mathcal{P}$. Essentially it is a passive Shamir-sharing protocol: p picks a degree-t uniformly random polynomial $q(\cdot)$ and sends $q(\alpha_i)$ to p_i. The following lemma states the achieved security.

Lemma 8. *Protocol* Share$(\mathcal{P}, t, p, s)$ *has the following properties. Correctness: When $p \in \mathcal{P} \setminus \mathcal{A}$ then* Share *outputs a t-consistent sharing $\langle s \rangle$ of s among the players in $\mathcal{P}$, where $\langle s \rangle$ is even t-valid unless $p \in \mathcal{A} \cup \mathcal{S} \cup \mathcal{SR}$ or unless p is a zombie. Privacy: The players in any set $\mathcal{P}' \subseteq \mathcal{P}$ with $|\mathcal{P}'| \leq t$ get no (joint) information on s.*

In the following we describe the protocols PublicReconstruct and Reconstruct used to reconstruct a shared value publicly and towards some output player p, respectively. The protocols take as input a sharing of a value among the players in some $\mathcal{P}'$ ($\mathcal{P}'$ might be different than $\mathcal{P}$). In protocol Reconstruct (resp. PublicReconstruct) every $p_i \in \mathcal{P}'$ sends his share to p (resp. broadcasts his share to $\mathcal{P}$) and then p (resp. every $p_j \in \mathcal{P}$) reconstructs the shared value using standard error correction. Due to their similarity we only describe protocol Reconstruct and state the security of both protocols in a joint lemma.

Protocol Reconstruct $(\mathcal{P}', t, t', p, \langle s \rangle)$

1. Each $p_i \in \mathcal{P}'$ sends his share s_i to p.
2. p finds, using standard polynomial interpolation techniques, a degree t polynomial $f(\cdot)$ with the property that more than $t + t'$ of the received shares lie on $f(\cdot)$ and outputs $s' = f(0)$. If no such polynomial exists then p_j outputs $\bot$.

Lemma 9. *Assume that there exists t_c such that there are at most t_c corrupted players in $\mathcal{P}'$, of whom at most t' are actively corrupted and the condition $t + t' + t_c < |\mathcal{P}'|$ holds. Then the protocol* Reconstruct *(resp.* PublicReconstruct$)^{10}$ *reconstructs a value s' towards player p (resp. towards every $p_j \in \mathcal{P}$), where $s' \in \{s, \bot\}$ if $\langle s \rangle$ is a t-consistent sharing of s among the players in $\mathcal{P}'$, and $s' = s$ if $\langle s \rangle$ is t-valid.*

6.2 Engineering the Network - Secure Channels

The trick of engineering the network allowed us to reduce the effect of receive-omission corruption. However, because the channels which we achieve provide no privacy guarantees, we cannot use the resulting network directly to build a perfectly secure SFE protocol. In the following, we show how to engineer the initial network of secure channels to get a new network-layer (also of secure channels) with stronger security guarantees.

The new network layer will allow any $p_j \in \mathcal{P}$ who receives $\bot$ instead of a message x from $p_i \in \mathcal{P}$ to decide whether he (i.e., p_j) is receive-omission corrupted or the sender p_i is corrupted. Additionally, when the reception fails because of p_i, then every (alive) player will recognize that p_i is (actively or send-omission) corrupted. Given Broadcast and a uniformly random key $k_{i,j} \in \mathbb{F}$ known exclusively to p_i and p_j, this can be achieved as follows: For p_i to privately send s to p_j, p_i uses $k_{i,j}$ as a one time pad to perfectly blind s, and broadcasts the blinded value $s + k_{i,j}$. Because only p_i and p_j know $k_{i,j}$, only p_j can unblind the broadcasted message and any other player gets no information about it. As syntactic sugar, we denote this protocol as PrivBroadcast.

In the remaining of this section we concentrate on enabling two players p_i and p_j to establish a secret key $k_{i,j}$ (to use in PrivBroadcast). We design two protocols, called WeakExchangeKey and ExchangeKey, which achieve the following: WeakExchangeKey uses the bilateral secure channels and allows any pair $p_i, p_j \in \mathcal{P}$ to exchange a key as long as *one of them* is at most receive-omission corrupted (i.e., is in $\mathcal{H} \cup \mathcal{R}$) and *the other one* is at most send-omission corrupted (i.e., is in $\mathcal{H} \cup \mathcal{S}$). Protocol ExchangeKey uses protocols WeakExchangeKey and Broadcast and allows p_i and p_j to exchange a key, even when *each of them* is *either* at most receive-omission *or* at most send-omission corrupted. Both protocols work in a publicly detectable way, i.e., all (alive) players notice whether or not the key-exchange worked. In the following we describe the protocols WeakExchangeKey and ExchangeKey in more detail.

Protocol WeakExchangeKey is based on the observation that when p_i is at most send-omission and p_j is at most receive-omission corrupted, then p_j can always securely send messages to p_i through the bilateral secure channel. The protocol works as follows: p_i and p_j choose uniformly random values $k_i \in \mathbb{F}$ and $k_j \in \mathbb{F}$, respectively, and exchange them over their bilateral channel. Subsequently, each of them publicly announces, by Broadcast, whether or not he received a value from the other. If any of them confirms reception of a value then this value is used as the secret key and the protocol succeeds; otherwise the protocol fails. WeakExchangeKey is non-robust and might abort with a set $B \in \{\{p_i\}, \{p_j\}\}$, but only when p_i and/or p_j broadcast $\bot$ (if they both broadcast $\bot$ take the one with the smallest index). The detailed description of WeakExchangeKey and the proof of the following lemma can be found in the full version.

---

^{10} For PublicReconstruct we need to assume a broadcast primitive, which when $3t_a + t_\sigma + t_\rho < |\mathcal{P}|$ we can instantiate by Broadcast.

Lemma 10. *If $3t_a + t_\rho + t_\sigma < |\mathcal{P}|$, protocol WeakExchangeKey has the following properties. Correctness: Either it succeeds in p_i and p_j exchanging a uniformly random key k, or it fails, or it aborts with set $B \in \{\{p_i\}, \{p_j\}\}$. It might abort with B only when $B \subseteq \mathcal{R} \cup \mathcal{S} \cup \mathcal{SR} \cup \mathcal{A}$. When it does not abort then the following hold: (1) Every alive $p_k \in \mathcal{P}$ sees whether the protocol succeded or failed, and (2) it always succeeds when $p_i \in \mathcal{H} \cup \mathcal{R}$ and $p_j \in \mathcal{H} \cup \mathcal{S}$ or vice versa (i.e., when $p_i \in \mathcal{H} \cup \mathcal{S}$ and $p_j \in \mathcal{H} \cup \mathcal{R}$). Privacy: The adversary gets no information on k (unless p_i or p_j is actively corrupted).*

We describe the protocol ExchangeKey (see below) and state its achieved security in a lemma. The protocol is non-robust and might abort with set $B \in \{\{p_i\}, \{p_j\}, \{p_i, p_j\}\}$. However, from the fact that it aborted the players can deduce useful information on the corruption of the players in B.

Protocol ExchangeKey $(\mathcal{P}, t_a, p_i, p_j)$

1. For $\ell \in \{i, j\}$: p_ℓ invokes WeakExchangeKey with every $p_r \in \mathcal{P}$. If WeakExchangeKey aborts with B, then ExchangeKey also aborts with B. Denote by $P_{\text{"ok"}}^\ell \subseteq \mathcal{P}$ the set of players who successfully exchanged keys with p_ℓ, and by $P_{\text{"ok"}} := (P_{\text{"ok"}}^i \cap P_{\text{"ok"}}^j)$. If $|P_{\text{"ok"}}| \leq 2t_a$ then ExchangeKey aborts with $B = \{p_i, p_j\}$.

2. For $\ell \in \{i, j\}$: p_ℓ picks a value $k_\ell \in_R \mathbb{F}$ uniformly at random and a degree t_a random polynomial $f_\ell(\cdot)$ with $f_\ell(0) = k_\ell$. For each $p_r \in P_{\text{"ok"}}$, p_ℓ sends, by invoking PrivBroadcast with the exchanged keys, the share $f_\ell(\alpha_r)$ to p_r, who denotes the received value as $s_r^{(\ell)}$. If p_ℓ broadcast $\perp$ then ExchangeKey aborts with $B = \{p_\ell\}$ (if both p_i and p_j broadcast $\perp$ take the one with the smallest index).

3. The players in $P_{\text{"ok"}}$ compute a sharing of the sum $k_i + k_j$, by each player (locally) adding his shares, and then publicly reconstruct it by PublicReconstruct. If PublicReconstruct outputs $\perp$ then ExchangeKey aborts with $B = \{p_i, p_j\}$.

Lemma 11. *If $3t_a + t_\sigma + t_\rho < |\mathcal{P}|$, the protocol ExchangeKey has the following properties. Correctness: Either p_i and p_j succeed in exchanging a uniformly random key k (and all players notice) or the protocol aborts with a set $B \in \{\{p_i\}, \{p_j\}, \{p_i, p_j\}\}$. It might abort with set B only if one of the following two cases holds: (1) $|B| = 1$ and $B \subseteq \mathcal{R} \cup \mathcal{S} \cup \mathcal{SR} \cup \mathcal{A}$ and (2) $|B| = 2$ and $B \cap (\mathcal{SR} \cup \mathcal{A}) \neq \emptyset$. Privacy: The adversary gets no information on k (unless p_i or p_j is actively corrupted).*

6.3   Protocol SFE$^{(\text{BC})}$

The last tool is a protocol, called SFE$^{(\text{BC})}$, which perfectly securely evaluates any given function f without fairness but with unanimous abort [GL02]. In particular, protocol SFE$^{(\text{BC})}$ either perfectly (t_a, t_ρ, t_σ)-securely evaluates the function f, or it aborts with set $B \in \{\{p_i\}, \{p_j\}, \{p_i, p_j\}\}$ for some $p_i, p_j \in \mathcal{P}$. The adversary might force the protocol to abort even after receiving the outputs of actively corrupted players. However, when it aborts every player learns useful information about the corruption of the players in B.

The idea is the following: Let $\Pi_{\mathcal{P},t}(\cdot)$ denote a protocol which perfectly t-securely evaluates any given function, in the presence of an adversary who can (only) actively corrupt up to t players.[11] Such a protocol is known to exist if $3t < n$ [BGW88]. Also, let C_f denote the arithmetic circuit which computes a given function f. To securely evaluate C_f, protocol $\mathsf{SFE}^{(\mathsf{BC})}$ invokes protocol $\Pi_{\mathcal{P},t}(C_f)$ over the engineered network of secure channels. More precisely, each $p_i \in \mathcal{P}$ executes the instructions of $\Pi_{\mathcal{P},t}(C_f)$ with the following modification: whenever p_i is instructed to bilaterally send a message x to some $p_j \in \mathcal{P}$, protocol $\mathsf{ExchangeKey}(\mathcal{P}, p_j, p_j)$ is invoked to have p_i and p_j exchange a uniformly random key, and then the message x is sent using $\mathsf{PrivBroadcast}$ with the established key; whenever p_i is instructed to broadcast a message, he invokes $\mathsf{Broadcast}$. If some invocation of $\mathsf{ExchangeKey}$ aborts with B or some $p_i \in \mathcal{P}$ broadcasts $\perp$ (in this case we set $B = \{p_i\}$) then $\mathsf{SFE}^{(\mathsf{BC})}$ aborts with B.

In the following lemma we state the security of $\mathsf{SFE}^{(\mathsf{BC})}$. The proof follows directly from the perfect t-security of protocol $\Pi_{\mathcal{P},t}(\cdot)$ and the perfect security of protocols $\mathsf{ExchangeKey}$ and $\mathsf{Broadcast}$. $\mathsf{SFE}^{(\mathsf{BC})}$ is parametrized by a single threshold, namely t, but it assumes as given the primitives $\mathsf{Broadcast}$ and $\mathsf{ExchangeKey}$ as specified in Lemmas 7 and 11, respectively.[12]

Lemma 12. *Given* $\mathsf{Broadcast}$ *and* $\mathsf{ExchangeKey}$, *assuming that the condition* $3t < |\mathcal{P}|$ *holds protocol* $\mathsf{SFE}^{(\mathsf{BC})}(\mathcal{P}, t, C_f)$ *has the following properties. Correctness: Either it perfectly* (t, t_σ, t_ρ)-*securely evaluates the circuit* C_f *among the players in* $\mathcal{P}$ *for any* $t_\sigma, t_\rho < n$, *or it aborts with a set* $B \subseteq \mathcal{P}$. *It might abort with set* B *only when one of the following two cases holds:* (1) $|B| = 1$ *and* $B \subseteq \mathcal{R} \cup \mathcal{S} \cup \mathcal{SR} \cup \mathcal{A}$ *and* (2) $|B| = 2$ *and* $B \cap (\mathcal{SR} \cup \mathcal{A}) \neq \emptyset$. *Privacy: The adversary gets no more information than what he can compute from the specified inputs and outputs of actively corrupted players (i.e., from the inputs and outputs she should get when the protocol does not abort).*

7   SFE

In this section we prove the necessary and sufficient condition for perfectly (t_a, t_ρ, t_σ)-securely evaluating any given function $f(\cdot)$, namely we prove the following theorem:

Theorem 1. *Perfectly* (t_a, t_ρ, t_σ)-*secure SFE is possible if and only if* $3t_a + t_\rho + t_\sigma < n$.

The necessity of the condition follows, with some additional work, from the necessity of the conditions $3t_a < n$ for SFE [BGW88]; we state the necessity in the following lemma which is proved in the full version of this paper.

Lemma 13. *If* $3t_a + t_\rho + t_\sigma \geq n$ *then there are functions which cannot be perfectly* (t_a, t_ρ, t_σ)-*securely evaluated.*

---

[11] Here, t-secure evaluation is according to any of the standard security definition (with fairness and guaranteed output delivery) of protocols tolerating an active-only adversary [MR91, Can00, DM00, BPW03].

[12] In slight abuse of notation here, we write $\mathsf{Broadcast}$ and $\mathsf{ExchangeKey}$ to refer *not* to the protocols but to primitives achieving the security specified in Lemmas 7 and 11 (independent of pre-conditions). To be able to instantiate them with our protocols we will have to guarantee that the pre-conditions of the lemmas are satisfied.

The sufficiency is proved by constructing an SFE protocol for computing any given function f. For simplicity, we assume that f takes one input per player and has one global output. Using standard techniques, we can obtain a protocol for computing functions with multiple inputs and/or multiple or even private outputs.

On a high level, the evaluation of the function f proceeds in three stages: In the first stage, called the *input stage*, every $p_i \in \mathcal{P}$ t_a-shares his input to the players in $\mathcal{P}$. Next, in the *computation stage*, the players use $\mathsf{SFE}^{(\mathsf{BC})}$ to compute a random t_a-sharing of the output of the function f. Finally, in the *output stage*, this sharing is reconstructed towards every player using Reconstruct. In the remaining of this section we describe in detail the three stages, and give a detailed description of protocol SFE.

The input stage. In this stage protocol Share is invoked to have each $p_i \in \mathcal{P}$ t_a-share his input $s^{(i)}$ to the players in $\mathcal{P}$. Denote the resulting sharing by $\langle s^{(i)} \rangle$. The security of Share guarantees that for any non actively corrupted p_i $\langle s^{(i)} \rangle$ is a t_a-consistent sharing of $s^{(i)}$, where $\langle s^{(i)} \rangle$ is even t-valid when $p_i \in \mathcal{H} \cup \mathcal{R}$.

The computation stage. The goal is to securely compute, using $\mathsf{SFE}^{(\mathsf{BC})}$, *a uniformly random t_a-valid sharing* of the output of f on input the values that where shared in the input stage. This stage is non-robust and might abort with a player set $B \subseteq \mathcal{P}$, when $\mathsf{SFE}^{(\mathsf{BC})}$ aborts with B. When it aborts, the players use the information about the set B, which is provided by Lemma 12, to repeat this stage in a smaller setting, i.e., among the players in $\mathcal{P}' := \mathcal{P} \setminus B$. The security of $\mathsf{SFE}^{(\mathsf{BC})}$ guarantees that, even when it aborts, the adversary learns at most the outputs of actively corrupted players, which, as they are shares of a (uniformly random) t_a-sharing, give her no information on the input-sharings. Hence, in the successful iteration of $\mathsf{SFE}^{(\mathsf{BC})}$, both the inputs of actively corrupted players and the decision of which send-omission corrupted players give their inputs are independent of the inputs of non actively corrupted players.

Initially $\mathcal{P}' := \mathcal{P}$ and $t'_a := t_a$. Protocol $\mathsf{SFE}^{(\mathsf{BC})}$ is invoked with player set $\mathcal{P}'$ and threshold t'_a, to compute the circuit $C^{t_a}_{\langle f \rangle}$ which does the following: $C^{t_a}_{\langle f \rangle}$ takes as input from each $p_j \in \mathcal{P}'$ his share of each of the input-sharings $\langle s^{(1)} \rangle, \ldots, \langle s^{(n)} \rangle$. For each such sharing $\langle s^{(i)} \rangle$: $C^{t_a}_{\langle f \rangle}$ attempts, exactly as in protocol Reconstruct, to reconstruct the shared value; if the reconstruction succeeds it sets $\hat{s}_i$ to the reconstructed value, otherwise it sets $\hat{s}_i$ to a default value (e.g., $\hat{s}_i := 0$). Note that for $t = t'_a, t' = t'_a$, and $t_c = t'_a + t_\sigma + t_\rho$ all the sufficient conditions for Reconstruct are satisfied; therefore, $C^{t_a}_{\langle f \rangle}$ correctly reconstructs the input of every $p_i \in \mathcal{H} \cup \mathcal{R}$ (which is t-valid), and for every $p_i \in \mathcal{S} \cup \mathcal{SR}$ it either reconstructs p_i's input or it takes a default value (since the sharing of p_i is a t-consistent sharing of his input). Having computed the values $\hat{s}_1, \ldots, \hat{s}_n$, $C^{t_a}_{\langle f \rangle}$ inputs them to the circuit computing f; denote the output by y. Finally, $C^{t_a}_{\langle f \rangle}$ computes and outputs a uniformly random t_a-valid sharing of y among the players in $\mathcal{P}'$. We point out that the circuit $C^{t_a}_{\langle f \rangle}$ can be efficiently computed from the circuit which computes the function f [IKLP06].

To be able to re-invoke $\mathsf{SFE}^{(\mathsf{BC})}$ in $\mathcal{P}' = \mathcal{P}' \setminus B$ when it aborts with B, we need to guarantee that in the updated $\mathcal{P}'$: (1) the condition $3t'_a < |\mathcal{P}'|$, which is sufficient for $\mathsf{SFE}^{(\mathsf{BC})}$, holds and (2) no inputs of non actively corrupted players are lost. To ensure Property (1), we use the idea of *player elimination* [HMP00]:[13] The security of $\mathsf{SFE}^{(\mathsf{BC})}$

---

[13] To our knowledge, this is the first work which uses the idea of player elimination not for improving efficiency but rather for arguing about feasibility of protocols.

guarantees that when it aborts with set B, then either $|B| = 1$ and $B \subseteq \mathcal{R} \cup \mathcal{S} \cup \mathcal{SR} \cup \mathcal{A}$ or $|B| = 2$ and $B \cap (\mathcal{SR} \cup \mathcal{A}) \neq \emptyset$. Therefore, by eliminating the players in B we might only change the ratio of uncorrupted vs. actively corrupted players in $\mathcal{P}'$ in favor of the uncorrupted players. However, as the set $\mathcal{P}'$ becomes smaller, the players might have to reduce the actual threshold t'_a. To be on the safe side, t'_a is reduced only when at least as many players as there can be send-/receive-omission corrupted have been eliminated. Property (2) is guaranteed because, first, the t_a-consistency and t_a-validity of input sharings cannot be destroyed by deleting players and, second, the newly computed t'_a satisfies, as we show, the sufficient condition for Reconstruct.

The output stage. The players invoke Reconstruct with the (latest) t'_a to reconstruct the sharing created in the successful iteration of $\mathsf{SFE}^{(\mathsf{BC})}$. Because the protocol $\mathsf{SFE}^{(\mathsf{BC})}$ outputs a t_a-valid sharing of the output, and, as we will show, t'_a satisfies the sufficient condition for protocol Reconstruct, the reconstruction is robust. For completeness we describe the protocol SFE (see below) and state the achieved security in the following lemma.

Protocol SFE $(\mathcal{P}, t_a, t_\rho, t_\sigma, f)$

0. Set $\mathcal{P}' := \mathcal{P}$, and $t'_a := t_a$.
1. For each $p_i \in \mathcal{P}$ invoke $\mathsf{Share}(\mathcal{P}, t_a, p_i, x_i)$. Each $p_j \in \mathcal{P}$ denotes the vector of all shares he received by $\vec{x}^{\,(j)}$.
2. The players in $\mathcal{P}'$ invoke $\mathsf{SFE}^{(\mathsf{BC})}(\mathcal{P}', t'_a, C^{t_a}_{\langle f \rangle})$, where each $p_i \in \mathcal{P}'$ has input $\vec{x}^{\,(j)}$.[a] If $\mathsf{SFE}^{(\mathsf{BC})}$ aborts with B, then set $\mathcal{P}' = \mathcal{P}' \setminus B$, set $t'_a := t_a - \max\{0, |\mathcal{P} \setminus \mathcal{P}'| - (t_\sigma + t_\rho)\}$ and repeat this step; otherwise denote by $\langle f \rangle$ the output sharing.
3. For each $p_j \in \mathcal{P}$ invoke $\mathsf{Reconstruct}(\mathcal{P}', t_a, t'_a, p_j, \langle f \rangle)$.

---

[a] The required invocations of Broadcast and ExchangeKey are done in the player set $\mathcal{P}$.

Lemma 14. *Protocol SFE is perfectly (t_a, t_ρ, t_σ)-secure if $3t_a + t_\rho + t_\sigma < |\mathcal{P}|$.*

Proof (sketch). Termination is guaranteed because Step 2 is repeated at most $t_a + t_\sigma + t_\rho$ times (in each repetition at least one corrupted player is removed from $\mathcal{P}'$). Correctness follows from the security of the invoked sub-protocols; however one needs to verify that the corresponding sufficient conditions hold whenever they are invoked. This follows from a player-elimination argument, which, due to space restrictions, is deleted from this proceedings version. Privacy follows also from the security of the invoked subprotocols and from the fact that all the sharings that we do are of degree t_a (except of those done internally in $\mathsf{SFE}^{(\mathsf{BC})}$ whose privacy is guaranteed by the security of $\mathsf{SFE}^{(\mathsf{BC})}$), therefore they leak no information to the adversary about the inputs. □

As already mentioned, when the adversary is rushing there are functions that cannot be strongly (t_a, t_ρ, t_σ)-securely evaluated, except in trivial corruption scenarios (i.e., if $t_a = 0$ or $t_\sigma = 0$). However, when the adversary is non-rushing the above protocol can be used to achieve strong security. Indeed, before the output stage, the adversary gains no useful information. As protocol Reconstruct is single round, if, within the output stage, we run it in parallel for every $p_i \in \mathcal{P}$, then a non-rushing adversary has to choose which receive-omission corrupted players do not get enough messages to reconstruct the output before getting any information about the output. This implies strong security. We point out that the necessity of condition $3t_a + t_\rho + t_\sigma < n$ for SFE is independent of whether or not the adversary is rushing.

Corollary 1. *Assuming that the adversary in non-rushing, perfectly strongly (t_a, t_ρ, t_σ)-secure SFE is possible if and only $3t_a + t_\rho + t_\sigma < n$.*

8   Computing Reactive Circuits (MPC)

In this section we show how to compute reactive functionalities, i.e., functionalities that receive inputs from and give outputs to the players several times during the computation (an output can depend on all previous inputs). An important consideration when computing a reactive functionality, is to make sure that the players can keep a secret joint state.

The circuit to be computed consists of input, output, addition, and multiplication gates.[14] We model the reactiveness of the computation by assigning to each gate a point in time in which the gate should be evaluated. The circuit is evaluated in a gate-by-gate fashion, using protocol SFE, where the evaluation of each gate (except for the output gates) yields a uniformly random t_a-valid sharing of the output of the gate among the players in $\mathcal{P}$. Keeping state is guaranteed by the fact that such a sharing is robustly reconstructible, e.g., by using protocol Reconstruct, given that the condition $3t_a + t_\sigma + t_\rho < n$ holds (Lemma 9). The privacy of the state is guaranteed, as there are at most t_a actively corrupted players.

To evaluate addition and multiplication gates, protocol SFE$^{(\text{BC})}$ is invoked to compute the circuits $C_{\langle Mult \rangle}$ and $C_{\langle Add \rangle}$, respectively, which on input t_a-valid sharings of the inputs x_1 and x_2 of the gate output a uniformly random t_a-valid sharing of the sum $x_1 + x_2$ and of the product $x_1 \cdot x_2$, respectively. For an output gate, protocol Reconstruct is invoked (with $\mathcal{P}' = \mathcal{P}$, and $t = t' = t_a$) to reconstruct the shared output towards the output player.

To evaluate an input gate, protocol SFE is invoked to evaluate the circuit $C_{\langle I \rangle}$ which takes as input the input of the corresponding player (and no value from other players) and computes a uniformly random t_a-valid sharing of it among the players in $\mathcal{P}$. *Exceptionally* in the evaluation of input gates, *even* the zombies are required to take part as if they were alive. This is possible as all players (including zombies) hold synchronized clocks, and are aware of when it is time to evaluate an input gate.[15] Instructing the zombies to "wake up" during the evaluation of input gates ensures that every $p_i \in \mathcal{H} \cup \mathcal{R}$, even if he is a zombie, is able to give input to the computation. When the evaluation of the gate finishes, all zombies "sleep" again, i.e., they stop playing (until the next input gate). The security of the MPC protocol follows from the security of protocols SFE and Reconstruct.

---

[14] This does not exclude probabilistic circuits, as a random gate can be simulated by having each player input a random value and taking the sum of the inputs as the output of the gate.

[15] A zombie might re-become zombie during the evaluation of the input gate, in which case he gives up the evaluation of the gate.

Theorem 2. *Perfectly* (t_a, t_ρ, t_σ)*-secure (reactive) MPC is possible if and only if* $3t_a + t_\sigma + t_\rho < n$.

As in the case of SFE, when the adversary is non-rushing, then by evaluating in parallel each tuple of output gates that are due to be evaluated at the same time, we get a strongly perfectly secure MPC protocol.

Corollary 2. *Assuming that the adversary in non-rushing, perfectly strongly* (t_a, t_ρ, t_σ)*-secure (reactive) MPC is possible if an only if* $3t_a + t_\rho + t_\sigma < n$.

9   (Full) Omission Corruption

Our results can be trivially used to obtain sufficient bounds for MPC and SFE in the presence of an adversary who can full-omission corrupt up to t_ω players and, simultaneously, actively corrupted t_a players (as in [Koo06]). Indeed, by setting $t_\sigma = t_\rho = t_\omega$ in our MPC protocols, we get a protocol which perfectly (t_a, t_ω)-securely realized any function when $3t_a + 2t_\omega < n$. Note that this bound is strictly better than the bound $3t_a + 4t_\omega < n$ which was proved sufficient in [Koo06].

Lemma 15. *Perfectly* (t_a, t_ω)*-secure (even reactive) MPC is possible if* $3t_a + 2t_\omega < n$.

10   Extensions

Our results can be extended to deal with adversaries who can, additionally, passively and fail-corrupt players; denote by t_p and t_f the corresponding thresholds. The proof of the following lemma is omitted, but we give some evidence of its validity: Fail-corruption comes almost "for free" as in our protocol a fail-corrupted players behaves exactly as a receive-omission corrupted player with the only difference that, instead of turning him into a zombie the adversary can make him crash. To incorporate passive corruption we need to do the following modifications: (1) the degree of the shares that are computed in SFE is increased by t_p; (2) for SFE$^{(BC)}$, instead of invoking, over the engineered network, the protocol $\Pi_{\mathcal{P},t}(\cdot)$ [BGW88] which tolerates only actively-corruption, we use a protocol which tolerates both active and passive corruption, simultaneously. Such a protocol is known to exist if $3t_a + 2t_p < n$ [FHM98]. These modifications will guarantee privacy of our computation.

Lemma 16. *Perfectly* $(t_a, t_p, t_f, t_\rho, t_\sigma)$*-secure MPC is possible if and only if* $3t_a + 2t_p + t_\sigma + t_\rho + t_f < n$.

Using techniques from Secure Message Transmission [DDWY93], we can extend our results to allow every (even uncorrupted) $p_i \in \mathcal{P}$ to suffer from some message loss, as long as we have the following guarantee: in every round every $p_i \in \mathcal{H} \cup \mathcal{S}$ might lose at most t_a of the messages sent to him by players $p_j \in \mathcal{H} \cup \mathcal{R}$.

Acknowledgements. We would like to thank Martin Hirt for many useful discussions and comments.

References

[Bea91a]   Beaver, D.: Foundations of secure interactive computing. In: Feigenbaum, J. (ed.) CRYPTO 1991. LNCS, vol. 576, pp. 377–391. Springer, Heidelberg (1992)

[Bea91b]   Beaver, D.: Secure multiparty protocols and zero-knowledge proof systems tolerating a faulty minority. Journal of Cryptology 4(2), 370–381 (1991)

[BGP89]   Berman, P.J., Garray, J., Perry, J.: Towards optimal distributed consensus. In: FOCS 1989, pp. 410–415 (1989)

[BGW88]   Ben-Or, M., Goldwasser, S., Wigderson, A.: Completeness theorems for non-cryptographic fault-tolerant distributed computation. In: STOC 1988, pp. 1–10 (1988)

[BPW03]   Backes, M., Pfitzmann, B., Waidner, M.: A universally composable cryptographic library (2003)

[Can00]   Canetti, R.: Security and composition of multiparty cryptographic protocols. Journal of Cryptology 13(1), 143–202 (2000)

[CCD88]   Chaum, D., Crépeau, C., Damgård, I.: Multiparty unconditionally secure protocols (extended abstract). In: STOC 1988, pp. 11–19 (1988)

[DDWY93]   Dolev, D., Dwork, C., Waarts, O., Yung, M.: Perfectly secure message transmission. Journal of the ACM 40(1), 17–47 (1993)

[DM00]   Dodis, Y., Micali, S.: Parallel reducibility for information-theoretically secure computation. In: Bellare, M. (ed.) CRYPTO 2000. LNCS, vol. 1880, pp. 74–92. Springer, Heidelberg (2000)

[DS82]   Dolev, D., Strong, H.R.: Polynomial algorithms for multiple processor agreement. In: STOC 1982, pp. 401–407 (1982)

[FHM98]   Fitzi, M., Hirt, M., Maurer, U.: Trading correctness for privacy in unconditional multi-party computation. In: Krawczyk, H. (ed.) CRYPTO 1998. LNCS, vol. 1462, pp. 121–136. Springer, Heidelberg (1998)

[FM98]   Fitzi, M., Maurer, U.: Efficient byzantine agreement secure against general adversaries. In: Kutten, S. (ed.) DISC 1998. LNCS, vol. 1499, pp. 134–148. Springer, Heidelberg (1998)

[FM00]   Fitzi, M., Maurer, U.: From partial consistency to global broadcast. In: STOC 2000, pp. 494–503 (2000)

[GL02]   Goldwasser, S., Lindell, Y.: Secure computation without agreement. In: Malkhi, D. (ed.) DISC 2002. LNCS, vol. 2508, pp. 17–32. Springer, Heidelberg (2002)

[GMW87]   Goldreich, O., Micali, S., Wigderson, A.: How to play any mental game — a completeness theorem for protocols with honest majority. In: STOC 1987, pp. 218–229 (1987)

[GP92]   Garay, J.A., Perry, K.J.: A continuum of failure models for distributed computing. In: Segall, A., Zaks, S. (eds.) WDAG 1992. LNCS, vol. 647, pp. 153–165. Springer, Heidelberg (1992)

[Had85]   Hadzilacos, V.: Issues of fault tolerance in concurrent computations (databases, reliability, transactions, agreement protocols, distributed computing). PhD thesis, Cambridge, MA, USA (1985)

[HMP00]   Hirt, M., Maurer, U., Przydatek, B.: Efficient secure multi-party computation. In: Okamoto, T. (ed.) ASIACRYPT 2000. LNCS, vol. 1976, pp. 143–161. Springer, Heidelberg (2000)

[IKLP06]   Ishai, Y., Kushilevitz, E., Lindell, Y., Petrank, E.: On combining privacy with guaranteed output delivery in secure multiparty computation. In: Dwork, C. (ed.) CRYPTO 2006. LNCS, vol. 4117, pp. 483–500. Springer, Heidelberg (2006)

[Koo06]   Koo, C.-Y.: Secure computation with partial message loss. In: Halevi, S., Rabin, T. (eds.) TCC 2006. LNCS, vol. 3876, pp. 502–521. Springer, Heidelberg (2006)

[LF82]   Lamport, L., Fischer, M.J.: Byzantine generals and transaction commit protocols. Technical Report Opus 62, SRI International (Menlo Park CA), TR (1982)

[LSP82]   Lamport, L., Shostak, R., Pease, M.: The byzantine generals problem. ACM Transactions on Programming Languages and Systems 4(3), 382–401 (1982)

[MP91]   Meyer, F.J., Pradhan, D.K.: Consensus with dual failure modes. IEEE Transactions on Parallel and Distributed Systems 2(2), 214–222 (1991)

[MR91]   Micali, S., Rogaway, P.: Secure computation. In: Feigenbaum, J. (ed.) CRYPTO 1991. LNCS, vol. 576, pp. 392–404. Springer, Heidelberg (1992)

[PR03]   Parvedy, P.R., Raynal, M.: Uniform agreement despite process omission failures. In: International Symposium on Parallel and Distributed Processing — IPDPS 2003, p. 212.2 (2003)

[PT86]   Perry, K.J., Toueg, S.: Distributed agreement in the presence of processor and communication faults. IEEE Trans. Softw. Eng. 12(3), 477–482 (1986)

[PW01]   Pfitzmann, B., Waidner, M.: A model for asynchronous reactive systems and its application to secure message transmission. In: IEEE Symposium on Security and Privacy, pp. 184–200 (2001)

[Ray02]   Raynal, M.: Consensus in synchronous systems: A concise guided tour. In: Pacific Rim International Symposium on Dependable Computing — PRDC 2002, p. 221 (2002)

[RB89]   Rabin, T., Ben-Or, M.: Verifiable secret sharing and multiparty protocols with honest majority. In: STOC 1989, pp. 73–85 (1989)

[Yao82]   Yao, A.C.: Protocols for secure computations. In: FOCS 1982, pp. 160–164 (1982)

Secure Arithmetic Computation
with No Honest Majority[*]

Yuval Ishai[1,**], Manoj Prabhakaran[2, ***], and Amit Sahai[3,†]

[1] Technion, Israel and University of California, Los Angeles
yuvali@cs.technion.il
[2] University of Illinois, Urbana-Champaign
mmp@cs.uiuc.edu
[3] University of California, Los Angeles
sahai@cs.ucla.edu

Abstract. We study the complexity of securely evaluating arithmetic circuits over finite rings. This question is motivated by natural secure computation tasks. Focusing mainly on the case of *two-party* protocols with security against *malicious* parties, our main goals are to: (1) only make black-box calls to the ring operations and standard cryptographic primitives, and (2) minimize the number of such black-box calls as well as the communication overhead.

We present several solutions which differ in their efficiency, generality, and underlying intractability assumptions. These include:

- An *unconditionally secure* protocol in the OT-hybrid model which makes a black-box use of an arbitrary ring R, but where the number of ring operations grows linearly with (an upper bound on) $\log |R|$.
- Computationally secure protocols in the OT-hybrid model which make a black-box use of an underlying ring, and in which the number of ring operations does not grow with the ring size. The protocols rely on variants of previous intractability assumptions related to linear codes. In the most efficient instance of these protocols, applied to a suitable class of fields, the (amortized) communication cost is a constant number of field elements per multiplication gate and the computational cost is dominated by $O(\log k)$ field operations per gate, where k is a security parameter. These results extend a previous approach of Naor and Pinkas for secure polynomial evaluation (*SIAM J. Comput.*, 2006).
- A protocol for the rings $\mathbb{Z}_m = \mathbb{Z}/m\mathbb{Z}$ which only makes a black-box use of a homomorphic encryption scheme. When m is prime, the

[*] Extended Abstract. Please see full version at Cryptology ePrint Archive: Report 2008/465.
[**] Supported in part by ISF grant 1310/06, BSF grant 2004361, and NSF grants 0205594, 0430254, 0456717, 0627781, 0716389.
[***] Supported in part by NSF grants CNS 07-16626 and CNS 07-47027.
[†] Research supported in part from NSF grants 0627781, 0716389, 0456717, and 0205594, BSF grant 2004361, a subgrant from SRI as part of the Army Cyber-TA program, an equipment grant from Intel, an Alfred P. Sloan Foundation Fellowship, and an Okawa Foundation Research Grant.

(amortized) number of calls to the encryption scheme for each gate of the circuit is constant.

All of our protocols are in fact *UC-secure* in the OT-hybrid model and can be generalized to *multiparty* computation with an arbitrary number of malicious parties.

1   Introduction

This paper studies the complexity of secure multiparty computation (MPC) tasks which involve *arithmetic* computations. Following the general feasibility results from the 1980s [60,34,4,13], much research in this area shifted to efficiency questions, with a major focus on the efficiency of securely distributing natural computational tasks that arise in the "real world". In many of these cases, some inputs, outputs, or intermediate values in the computation are integers, finite-precision reals, matrices, or elements of a big finite ring, and the computation involves arithmetic operations in this ring. To name just a few examples from the MPC literature, such arithmetic computations are useful in the contexts of distributed generation of cryptographic keys [8,28,56,32,2], privacy-preserving data-mining and statistics [48,11], comparing and matching data [50,31,38], auctions and mechanism design [51,21,59,7], and distributed linear algebra computations [15,52,47,20,49].

This motivates the following question:

What is the complexity of securely evaluating a given arithmetic circuit C over a given finite ring R?

Before surveying the state of the art, some clarifications are in place.

Arithmetic circuits. An arithmetic circuit over a ring is defined similarly to a standard boolean circuit, except that the inputs and outputs are ring elements rather than bits and gates are labeled by the ring operations add, subtract, and multiply. (Here and in the following, by "ring" we will refer to a finite ring by default.) In the current context of distributed computations, the inputs and outputs of the circuit are annotated with the parties to which they belong. Thus, the circuit C together with the ring R naturally define a multi-party arithmetic functionality C^R. Note that arithmetic computations over the integers or finite-precision reals can be embedded into a sufficiently large finite ring or field, provided that there is an a-priori upper bound on the bit-length of the output. See Section 1.2 for further discussion of the usefulness of arithmetic circuits and some extensions of this basic model to which our results apply.

Secure computation model. The main focus of this paper is on secure *two-party* computation or, more generally, MPC with an arbitrary number of malicious parties. (In this setting it is generally impossible to guarantee output delivery or even fairness, and one has to settle for allowing the adversary to abort the protocol after learning the output.) Our protocols are described in the "OT-hybrid model," namely in a model that allows parties to invoke an ideal oblivious transfer (OT) oracle [57,27,33]. This has several advantages in generality and efficiency, see [44] and Section 1.2 for discussion.

Ruling out the obvious. An obvious approach for securely realizing an arithmetic computation C^R is by first designing an equivalent *boolean* circuit C' which computes the same function on a binary representation of the inputs, and then using standard MPC protocols for realizing C'. The main disadvantage of such an approach is that it typically becomes very inefficient when R is large. A clean way for ruling out such an approach, which is of independent theoretical interest, is by restricting protocols to only make a *black-box* access to the ring R. That is, Π securely realizes C if Π^R securely realizes C^R for every finite ring R and *every representation of elements in R*. The black-box access to R enables Π to perform ring operations and sample random ring elements, but the correspondence between ring elements and their identifiers (or even the exact size of the ring) will be unknown to the protocol. This automatically ensures that the overhead of Π (compared to an insecure implementation) does not grow with the computational complexity of ring operations. When considering the special case of fields, we allow by default the protocol Π to access an inversion oracle. Most of our protocols will make black-box access to a ring, although we will also consider some protocols outside of this model based on homomorphic encryption (see below).

1.1   Previous Work

In the setting of MPC *with honest majority*, most protocols from the literature can make a black-box use of an arbitrary *field*. An extension to arbitrary black-box rings was given in [19], building on previous black-box secret sharing techniques of [26,18] and previous MPC techniques of [4,16].

In the case of secure two-party computation and MPC with no honest majority, most protocols from the literature apply to boolean circuits. Below we survey some previous approaches from the literature that apply to secure arithmetic computation with no honest majority.

In the semi-honest model, it is easy to employ any homomorphic encryption scheme with plaintext group $\mathbb{Z}_m$ for performing arithmetic MPC over $\mathbb{Z}_m$. (See, e.g., [1,11].) An alternative approach, which relies on oblivious transfer and uses the standard binary representation of elements in $\mathbb{Z}_m$, was employed in [32]. Both of the above protocols make a black-box use of the underlying cryptographic primitives but do not make a black-box use of the underlying ring. Applying the general compilers of [34,12] to these protocols in order to obtain security in the malicious model would result in inefficient protocols which make a non-black-box use of the underlying cryptographic primitives (let alone the ring).

In the *malicious model*, protocols for secure arithmetic computation based on *threshold* homomorphic encryption were given in [17,24][1] (extending a similar protocol for the semi-honest model from [29]). These protocols provide the

---

[1] While [17,24] refer to the case of robust MPC in the presence of an honest majority, these protocols can be easily modified to apply to the case of MPC with no honest majority.

most practical general solutions for secure arithmetic two-party computation we are aware of, requiring a constant number of modular exponentiations for each arithmetic gate. On the down side, these protocols require a nontrivial setup of keys which is expensive to distribute. Moreover, they rely on special-purpose zero-knowledge proofs and specific number-theoretic assumptions and thus do not make a black-box use of the underlying cryptographic primitives, let alone a black-box use of the ring.

The only previous approach which makes a black-box use of an underlying ring (as well as a black-box use of OT) was suggested by Naor and Pinkas [50] in the context of secure polynomial evaluation. Their protocol can make a black-box use of any *field*, and its security is related to the conjectured intractability of decoding Reed-Solomon codes with a sufficiently high level of random noise. The protocol from [50] can be easily used to obtain general secure protocols for arithmetic circuits in the *semi-honest* model. However, extending it to allow full simulation-based security in the malicious model (while still making only a black-box use of the underlying field) is not straightforward. (Even in the special case of secure polynomial evaluation, an extension to the malicious model suggested in [50] only considers *privacy* rather than full simulation-based security.)

Finally, we note that Yao's garbled circuit technique [60], the main known technique for constant-round secure computation of general functionalities, does not have a known arithmetic analogue. Thus, in all general-purpose protocols for secure arithmetic computation (including the ones presented in this work) the round complexity must grow with the multiplicative depth of C – the maximal number of multiplication gates on a path from an input to an output.

1.2 Our Contribution

We study the complexity of general secure arithmetic computation over finite rings in the presence of an arbitrary number of malicious parties. We are motivated by the following two related goals.

- *Black-box feasibility*: only make a black-box use of an underlying ring R or field F and standard cryptographic primitives;
- *Efficiency*: minimize the number of such black-box calls, as well as the communication overhead.

For simplicity, we do not attempt to optimize the dependence of the complexity on the number of parties, and restrict the following discussion to the two-party case.

We present several solutions which differ in their efficiency, generality, and underlying intractability assumptions. All these constructions use the general framework from [44]: one can obtain 2-party UC-secure protocols in the OT-hybrid model, by combining an "outer MPC protocol" secure against active adversaries in the *honest majority setting*, with an "inner two-party protocol" for simple functionalities that need only be secure against *passive* adversaries. The main technical contribution of this work is in designing inner protocols, that can then be combined with appropriate variants of outer protocols from the literature, to obtain secure protocols with desired properties. Below we describe the main protocols we obtain in this way, along with their efficiency and security features.

An unconditionally secure protocol. We present an *unconditionally secure* protocol in the OT-hybrid model which makes a *black-box* use of an *arbitrary* finite ring R, but where the number of ring operations and the number of ring elements being communicated grow linearly with (an upper bound on) $\log |R|$. (We assume for simplicity that an upper bound on $\log |R|$ is given by the ring oracle, though such an upper bound can always be inferred from the length of the strings representing ring elements.) More concretely, the number of ring operations for each gate of C is $poly(k) \cdot \log |R|$, where k is a statistical security parameter. This gives a two-party analogue for the MPC protocol over black-box rings from [19], which requires an honest majority (but does not require the number of ring operations to grow with $\log |R|$).

Protocols based on noisy linear encodings. Motivated by the goal of reducing the overhead of the previous protocol, we present a general approach for deriving secure arithmetic computation protocols over a ring R from linear codes over R. The (computational) security of the protocols relies on intractability assumptions related to the hardness of decoding in the presence of random noise. These protocols generalize and extend in several ways the previous approach of Naor and Pinkas for secure polynomial evaluation [50]. More concretely, we make three main observations: (1) Using [44], secure evaluation of *degree-1* polynomials in the *semi-honest* model can be used in a black-box way for general secure arithmetic computation in the malicious model; (2) In the case of degree-1 polynomials, the approach of [50] can be generalized to rely on *arbitrary* linear codes for which the relevant intractability assumption holds; (3) When using Reed-Solomon codes as in [50], it is possible to significantly improve the efficiency by batching together many instances of secure polynomial evaluation.

Using this approach, we obtain the following types of protocols in the OT-hybrid model.

– A protocol which makes a black-box use of an arbitrary *field F*, in which the number of field operations (and field elements being communicated) does not grow with the field size. More concretely, the number of field operations for each gate of C is bounded by a fixed polynomial in the security parameter k, independently of $|F|$. The underlying assumption is related to the conjectured intractability of decoding a random linear code over F. Our assumption is implied by the assumption that a noisy codeword in a random linear code over F is pseudorandom.

– A variant of the previous protocol which makes a black-box use of an arbitrary *ring R*, and in particular does not rely on inversion. This variant is based on families of linear codes over rings in which decoding in the presence of erasures can be done efficiently, and for which decoding in the presence of (a suitable distribution of) random noise seems intractable.

– The most efficient protocol we present relies on the intractability of decoding Reed-Solomon codes with a (small) constant rate in the presence of a (large) constant fraction of noise.[2] The amortized communication cost is a constant number of field elements per multiplication gate. (Here and in the following, when we refer to "amortized" complexity we ignore an additive term that may depend on the security parameter and the circuit depth, but not on the circuit size. In most natural instances of large circuits this additive term does not form an efficiency bottleneck.)

A careful implementation yields protocols whose amortized *computational* cost is $O(\log k)$ field operations per gate, where k is a security parameter, assuming that the field size is super-polynomial in k. In contrast, protocols which are based on homomorphic encryption schemes (such as [17] or the ones obtained in this work) apply modular exponentiations, which require $\Omega(k + \log |F|)$ ring multiplications per gate, in a ciphertext ring which is larger than F. This is the case even in the semi-honest model.

Protocols making a black-box use of homomorphic encryption. We also consider protocols for the specific rings $\mathbb{Z}_m = \mathbb{Z}/m\mathbb{Z}$ (thus leaving behind the black-box ring model), but which make black-box use of any homomorphic encryption scheme with plaintext group $\mathbb{Z}_m$. Alternatively, the protocol can make a black-box use of homomorphic encryption schemes in which the plaintext group is determined by the key generation algorithm, such as those of Paillier [53] or Damgård-Jurik [23]. In both variants of the protocol, the (amortized) number of communicated ciphertexts and calls to the encryption scheme for each gate of C is constant, assuming that m is prime. This efficiency feature is comparable to the protocols from [17,24] discussed in Section 1.1 above. Our protocols have the advantages of using a more general primitive and only making a *black-box* use of this primitive (rather than relying on special-purpose zero-knowledge protocols). Furthermore, the additive term which we ignore in the above "amortized" complexity measure seems to be considerably smaller than the cost of distributing the setup of the threshold cryptosystem required by [17].

Both variants of the protocol can be naturally extended to the case of matrix rings $\mathbb{Z}_m^{n \times n}$, increasing the communication complexity by a factor of n^2. (Note that emulating matrix operations via basic arithmetic operations over $\mathbb{Z}_m$ would result in a bigger overhead, corresponding to the complexity of matrix multiplication.) Building on the techniques from [49], this protocol can be used to obtain efficient protocols for secure linear algebra which make a black-box use of homomorphic encryption and achieve simulation-based security against malicious parties (improving over similar protocols with security against *covert* adversaries [3] recently presented in [49]).

All of our protocols are in fact *UC-secure* in the OT-hybrid model and can be generalized to *multiparty* computation with an arbitrary number of malicious

---

[2] The precise intractability assumption we use is similar in flavor to an assumption used in [50] for evaluating polynomials of degree $d \geq 2$. With a suitable choice of parameters, our assumption is implied by a natural pseudorandomness variant of the assumption from [50], discussed in [46]. The assumption does not seem to be affected by the recent progress on list-decoding Reed-Solomon codes and their variants [37,14,6,54].

parties. The security of the protocols also holds against *adaptive* adversaries, assuming that honest parties may erase data. (This is weaker than the standard notion of adaptive security [10] which does not rely on data erasure.) The *round complexity* of all the protocols is a constant multiple of the multiplicative depth of C.

From the OT-hybrid model to the plain model. An advantage of presenting our protocols in the OT-hybrid model is that they can be instantiated in a variety of models and under a variety of assumptions. For instance, using UC-secure OT protocols from [55,25], one can obtain efficient UC-secure instances of our protocols in the CRS model. In the stand-alone model, one can implement these OTs by making a black-box use of homomorphic encryption [42]. Thus, our protocols which make a black-box use of homomorphic encryption do not need to employ an additional OT primitive in the stand-alone model.

We finally note that our protocols require only $O(k)$ OTs with security in the malicious model, independently of the circuit size; the remaining OT invocations can all be implemented in the semi-honest model, which can be done very efficiently using the technique of [41]. Furthermore, all the "cryptographic" work for implementing the OTs can be done off-line, before any inputs are available. We expect that in most natural instances of large-scale secure arithmetic computation, the cost of realizing the OTs will not form an efficiency bottleneck.

Extensions. While we explicitly consider here only stateless arithmetic circuits, this model (as well as our results) can be readily generalized to allow stateful, reactive arithmetic computations whose secret state evolves by interacting with the parties.[3]

As it turns out, reactive arithmetic computations are useful not only for the obvious purpose of implementing stateful functionalities, but also, somewhat surprisingly, for enriching the (non-reactive) arithmetic computation model. They can be used to obtain efficient secure realizations of several "non-arithmetic" manipulations of the state, including decomposing a ring element into its bit-representation, equality testing, inversion, comparison, exponentiation, and others [21,59]. These reductions enhance the power of the basic arithmetic model, and allow protocols to efficiently switch from one representation to another in computations that involve both boolean and arithmetic operations.

2   Preliminaries

Black-box rings and fields. A probabilistic oracle R is said to be a valid implementation of a finite ring R if it behaves as follows: it takes as input one of the commands add, subtract, multiply, sample and two m bit "element identifiers" (or none, in the case of sample), and returns a single m bit string. There is a one-to-one mapping label : $R \hookrightarrow \{0,1\}^m$ such that for all $x, y \in R$

---

[3] An ideal functionality which formally captures such general reactive arithmetic computations was defined in [24] (see also [59, Chapter 4]) and referred to as an *arithmetic black-box* (ABB). All of our protocols for arithmetic circuits can be naturally extended to realize the ABB functionality.

R(op, label(x), label(y)) = label($x *_R y$) where op is one of add, subtract and multiply and $*_R$ is the ring operation $+, -$, or $\cdot$ respectively. When an input is not from the range of label, the oracle outputs $\perp$. (In a typical protocol, if a $\perp$ is ever encountered by an honest player, the protocol aborts.) The output of R(sample) is label(x) where x will be drawn uniformly at random from R. We will be interested in oracles of the kind that implements a *family* of rings, of varying sizes. Such a function should take an additional input id to indicate which ring it is implementing.

Definition 1. *A probabilistic oracle $\mathcal{R}$ is said to be a* concrete ring family *(or simply a* ring family*) if, for all strings* id, *the oracle $\mathcal{R}(\mathrm{id}, \cdot)$ (i.e., with first input being fixed to* id*), is an implementation of some ring. This concrete ring will be denoted by $\mathcal{R}_{\mathrm{id}}$.*

Note that so far we have not placed any computability requirement on the oracle; we only require a concrete mapping from ring elements to binary strings. However, when considering computationally secure protocols we will typically restrict the attention to "efficient" families of rings: we say $\mathcal{R}$ is a *computationally efficient ring family* if it is a ring family that can be implemented by a probabilistic polynomial time algorithm.

There are some special cases that we shall refer to:

1. Suppose that for all id, we have that $\mathcal{R}_{\mathrm{id}}$ is a ring with an identity for multiplication, 1. Then, we call $\mathcal{R}$ a *ring family with inverse* if in addition to the other operations, $\mathcal{R}(\mathrm{id}, \mathrm{one})$ returns $\mathrm{label}_{\mathrm{id}}(1)$ and $\mathcal{R}(\mathrm{id}, \mathrm{invert}, \mathrm{label}_{\mathrm{id}}(x))$ returns $\mathrm{label}_{\mathrm{id}}(x^{-1})$ if x is a unit (i.e., has a unique left- and right-inverse) and $\perp$ otherwise.
2. If $\mathcal{R}$ is a ring family with inverse such that for all id the ring $\mathcal{R}_{\mathrm{id}}$ is a field, then we say that $\mathcal{R}$ is a *field family*.
3. We call a ring family with inverse $\mathcal{R}$ a *pseudo-field family*, if for all id, all but negligible fraction of the elements in the ring $\mathcal{R}_{\mathrm{id}}$ are units.

Some special families of rings we will be interested in, other than finite fields, include rings of the form $\mathbb{Z}_m = \mathbb{Z}/m\mathbb{Z}$ for a composite integer m (namely, the ring of residue classes modulo m), and rings of matrices over a finite field or ring. With an appropriate choice of parameters, both of these families are in fact pseudo-fields. Note that a concrete ring family $\mathcal{R}$ for the rings of the form $\mathbb{Z}_m$ could use the binary representation of m as the input id; further the elements in $\mathbb{Z}_m$ could be represented as $\lceil \log m \rceil$-bit strings in a natural way. Of course, a different concrete ring family for the same ring can use a different representation.

Finally, for notational convenience we assume that the length of all element identifiers in $\mathcal{R}_{\mathrm{id}}$ is exactly $|\mathrm{id}|$. In particular, the ring $\mathcal{R}_{\mathrm{id}}$ has at most $2^{|\mathrm{id}|}$ elements.

Arithmetic circuits. We consider arithmetic circuits with gates labeled by add, subtract, or multiply. (In addition, for fields there is an additional constant gate one.) For a concrete ring family $\mathcal{R}$, we denote by $C^{\mathcal{R}}$ the mapping which takes an id and a vector of input identifiers and outputs the corresponding vector of output identifiers. In the context of multi-party computation, each input or output to such a circuit is annotated to indicate which party (or parties) it "belongs" to. Given such an annotated circuit C and a concrete ring family $\mathcal{R}$, we define the functionality $\mathcal{F}_C^{\mathcal{R}}$ to behave as follows:

- The functionality takes id as a common (public) input, and receives (private) inputs to C from each party. It then evaluates the function $C^{\mathcal{R}}(\mathsf{id}, \mathsf{inputs})$ using access to $\mathcal{R}$, and provides the outputs to the parties.[4]

Protocols securely realizing arithmetic computations. We follow the standard UC-security framework [9]. Informally, a protocol π is said to securely realize a functionality $\mathcal{F}$ if there exists a PPT simulator Sim, such that for all (non-uniform PPT) adversaries Adv, and all (non-uniform PPT) environments Env which interact with a set of parties and an adversary, the following two scenarios are indistinguishable: the REAL interaction where the parties run the protocol π and the adversary is Adv; the IDEAL interaction where the parties communicate directly with the ideal functionality $\mathcal{F}$ and the adversary is $\mathsf{Sim}^{\mathsf{Adv}}$. Indistinguishability can either be statistical (in the case of unconditional security) or computational (in the case of computational security). All parties, the adversary, the simulator, the environment and the functionality get the security parameter k as implicit input. Polynomial time computation, computational or statistical indistinguishability and non-uniformity are defined with respect to this security parameter k. However, since we don't impose an a-priori bound on the size of the inputs received from the environment as a function of k, the running time of honest parties is bounded by a fixed polynomial in the total length of their inputs (rather than a fixed polynomial in k).

We distinguish between *static* corruption and *adaptive* corruption. In the latter case it also makes a difference whether the protocols can erase part of their state (so that a subsequent corruption will not have access to the erased information), or no erasure is allowed. Our final protocols will have security against adaptive corruption *with erasures*.

We shall consider protocols which make oracle access to a ring family $\mathcal{R}$. The standard security definition is adapted to this case by giving all algorithms (including the environment) oracle access to $\mathcal{R}$. For such a protocol we define its *arithmetic computation complexity* as the number of oracle calls to $\mathcal{R}$. Similarly the *arithmetic communication complexity* is defined as the number of ring-element labels in the communication transcript. The arithmetic computation (respectively communication) complexity of our protocols will dominate the other computation steps in the protocol execution (respectively, the number of other bits in the transcript). Thus, the arithmetic complexity gives a good measure of efficiency for our protocols.

---

[4] $\mathcal{F}_C^{\mathcal{R}}$ can take id as input from each party, and ensure that all the parties agree on the same id. Alternately, we can restrict to environments which provide the same common input id to all parties.

Note that while any computational implementation of the ring oracle necessarily requires the complexity to grow with the ring size, it is possible that the arithmetic complexity does not depend on the size of the ring at all.

We now define our main notion of secure arithmetic computation.

Definition 2. *Let C be an arithmetic circuit. A protocol π is said to be a* secure black-box realization *of C-evaluation for a given set of ring families if, for each $\mathcal{R}$ in the set,*

1. *$\pi^{\mathcal{R}}$ securely realizes $\mathcal{F}_C^{\mathcal{R}}$, and*
2. *the arithmetic (communication and computation) complexity of $\pi^{\mathcal{R}}$ is bounded by some fixed polynomial in k and $|\mathsf{id}|$ (independently of $\mathcal{R}$).*

In the case of unconditional security we will quantify over the set of *all* ring families, whereas in the case of computational security we will typically quantify only over computationally efficient rings or fields. In both cases, the efficiency requirement on π rules out the option of using a brute-force approach to emulate the ring oracle by a boolean circuit.

We remark that our constructions will achieve a stronger notion of security, as the simulator used to establish the security in item (1) above will not depend on $\mathcal{R}$. A bit more precisely, the stronger definition is quantified as follows: there exists a simulator such that for all adversaries, ring families, and environments, the ideal process and the real process are indistinguishable. For simplicity however we phrase our definition as above which does allow different simulators for different $\mathcal{R}$.

3   Arithmetic Computation with Passive Corruption

To construct a protocol for general arithmetic circuit evaluation over a black-box ring family $\mathcal{R}$, that is secure against passive (adaptive) corruption it is enough to realize the following functionality $\mathcal{F}_{\mathsf{pdt\text{-}shr}}$ (see [45] for more details). Let $R = \mathcal{R}_{\mathsf{id}}$, where id is an implicit common input.

- A sends $a \in R$ and B sends $b \in R$ to $\mathcal{F}_{\mathsf{pdt\text{-}shr}}$.
- $\mathcal{F}_{\mathsf{pdt\text{-}shr}}$ samples two random elements $z^A, z^B \in R$ such that $z^A + z^B = ab$, and gives z^A to A and z^B to B.

A well-known approach for securely realizing this functionality against passive corruption, using a homomorphic encryption scheme (if available), goes as follows:

- Bob generates a public/secret key-pair encryption scheme, and sends an encryption of b along with the public key.
- Alice picks a random element z^A in the ring. She then computes an encryption of $ab - z^A$ from the encryption b (and the public-key) and sends it to Bob.
- Bob decrypts this ciphertext and accepts it as z^B. The encryption scheme should ensure that even with the secret-key, Bob does not learn anything else about (a, z^A) from the message she receives from Alice.

Indeed when such a homomorphic encryption scheme is available this gives a protocol with security against passive corruption for this task. However, such schemes are known only for select families of rings, and further do not meet the goal of making only black-box access to the ring.

This basic approach can be extended to the black-box ring setting with the help of an OT channel to ensure part of the privacy: Instead of an encryption, Alice sends an encoding of a under an appropriate *erasure correcting code*, but with sufficient noise to hide a from Bob. Her "secret-key" is the information about which co-ordinates are noisy. The code should have homomorphic properties to let Bob create a noisy encoding of $ab + z^B$ from this. To ensure that Alice does not learn anything beyond $ab + z^B$, Bob does not send the resulting noisy codeword to Alice, but lets her use an OT channel to pick up only the non-noisy co-ordinates of the codeword.

This high-level description fits the approach taken by Naor and Pinkas [50] for the special case of Reed-Solomon codes. Our protocols in this section provide more general and more efficient instantiations of this approach, to realize the functionality $\mathcal{F}_{\text{pdt-shr}}$ described above. In Section 3.1 we describe our encoding schemes, and in Section 3.2 we show how these encoding schemes can be used in protocols that realize $\mathcal{F}_{\text{pdt-shr}}$ against passive corruption.

3.1   Noisy Encodings

We describe several noisy encoding schemes based on linear codes. All our encoding schemes are specified using a *code generation algorithm* $\mathcal{G}$, over a ring family $\mathcal{R}$. $\mathcal{G}$ is a randomized algorithm such that $\mathcal{G}^{\mathcal{R}}(\text{id}, k)$ outputs (G, L, H), where G is an $n \times k$ generator matrix of a linear code over $\mathcal{R}_{\text{id}}$ of length $n = n(k)$, L is a subset of $[n]$ of size $\ell(k)$ which specifies the set of coordinates which are *not* replaced by noise, and H is another matrix which is used to facilitate efficient decoding. Here k is the security parameter as well as the code dimension, and $n(k)$ (code length) and $\ell(k)$ (number of coordinates *without* noise) are parameters of $\mathcal{G}$. In our instantiations n will be a constant multiple of k and in most cases we will have $\ell = k$.

Let $\mathcal{R}$ be a ring family. Given $\mathcal{G}$, a parameter $t(k) \leq k$ (number of ring elements to be encoded, $t = 1$ by default), and $x \in \mathcal{R}_{\text{id}}^t$, we define a distribution $\mathcal{E}_{(\mathcal{G},t)}^{\mathcal{R}}(\text{id}, k, x)$ as that of the public output in the following encoding process:

- *Encoding* $\text{Encode}_{(\mathcal{G},t)}^{\mathcal{R}}(\text{id}, k, x)$:
  - Input: $x = (x_1, \ldots, x_t) \in R^t$, where $R = \mathcal{R}_{\text{id}}$ and $t = t(k)$.
  - Let $(G, L, H) \leftarrow \mathcal{G}^{\mathcal{R}}(\text{id}, k)$
  - Pick a random vector $u \in R^k$ conditioned on $u_i = x_i$ for $i = 1, \ldots, t$ (i.e., u is x padded with $k - t$ random elements). Compute $Gu \in R^n$.
  - Let $v = Gu + e$, where $e \leftarrow R^n$ is drawn uniformly random conditioned on $e_i := 0$ for $i \in L$.
  - Let the private output be (G, L, H, v) and the public output be (G, v).

The matrix H is not used in the encoding above, but will be useful towards efficient decoding. In our main instantiations H can be readily derived from G

and L. We include H explicitly in the outcome of $\mathcal{G}$, because in some cases it is possible to obtain efficiency gains if (G, H, L) are sampled together.

Below we describe four instantiations of the above encoding scheme. The respective code generation algorithms are denoted by $\mathcal{G}_{\mathsf{Stat}}$, $\mathcal{G}_{\mathsf{Ring}}$, $\mathcal{G}_{\mathsf{Rand}}$, and $\mathcal{G}_{\mathsf{RS}}$. The first three use $t = 1$, i.e., a single ring element is encoded in a noisy codeword, and the last one allows $t(k)$ to be constant fraction of k, say $k/2$. The first three schemes allow homomorphic operations of multiplication and addition of the encoded element with an unencoded element. The last one allows co-ordinate wise multiplication and addition of t-long vectors. The last two require the ring family to be a field family.

The first encoding scheme has a statistical hiding property, whereas the others depend on computational assumptions for their hiding property. The assumption, in these three cases, is as follows:

Assumption 1 (Generic version, for a given $\mathcal{G}$, $\mathcal{R}$ and $t(k)$). *For all sequences $\{(\mathsf{id}_k, x_k, y_k)\}_k$ such that $x_k, y_k \in \mathcal{R}_{\mathsf{id}_k}^{t(k)}$, the ensembles $\{\mathcal{E}_{(\mathcal{G},t)}^{\mathcal{R}}(\mathsf{id}_k, k, x_k)\}_k$ and $\{\mathcal{E}_{(\mathcal{G},t)}^{\mathcal{R}}(\mathsf{id}_k, k, y_k)\}_k$ are computationally indistinguishable (by any $\mathrm{poly}(k)$-size nonuniform distinguisher).*

Statistically hiding encoding. Our statistically hiding encoding mixes an additive secret sharing of x with an equal number of uniformly random ring elements. Following is a more precise description of the encoding algorithm $\mathcal{G}_{\mathsf{Stat}}$ which fits into the above general framework.

- Let $R = \mathcal{R}_{\mathsf{id}}$. Let $n = 2m$ where $m = \log_2 |R| + k$.
- Let A_0 be the $m \times m$ matrix with 1 along the main diagonal and -1 along the rest of the first row.[5] Let G be the fixed $2m \times m$ matrix $G_0 = \begin{bmatrix} A_0 \\ A_0 \end{bmatrix}$.
- Define L as follows. Let $L = \{a_1, \ldots, a_m\}$ where $a_i = i$ or $m + i$ uniformly at random. (That is a_i indices the i-th row in one of the two copies of A_0.)
- Note that $G|_L = A_0$. H has 1 along the main diagonal and the first row, so that $HG|_L = I$.

The encoding of x is the vector $v = G_0 u + e$, where u is a random vector with $u_1 = x$ and e is a random noise vector with $e_i = 0$ for $i \in L$; v is then simply a random vector conditioned on $\sum_{i \in L} v_i = x$. This simple encoding has the useful property that it statistically hides x when the decoding information L is not provided. In the full version [45], we prove this fact using the Leftover Hash Lemma [39] (similarly to previous uses of this lemma in [40,43]).

Lemma 1. *For any $\mathcal{R}$, id, and $x \in \mathcal{R}_{\mathsf{id}}$, the statistical distance between the distribution of $\mathcal{E}_{(\mathcal{G}_{\mathsf{Stat}},1)}^{\mathcal{R}}(\mathsf{id}, k, x)$ and (G_0, v), where v is drawn uniformly from $\mathcal{R}_{\mathsf{id}}^{2m}$, is $2^{-\Omega(k)}$.*

We note that in light of efficient algorithms for low-density instances of subset sum, one cannot hope to obtain significant efficiency improvements by choosing a smaller value of m and settling for computational security.

---

[5] Here it is not necessary to assume that the ring has a multiplicative identity. In computing the matrix product, $1.a$ and $-1.a$ stand for a and $-a$.

Ring code based instantiation. Our next encoding scheme also uses $t = 1$, and works with any arbitrary ring family. It differs from the previous encoding scheme by not requiring n to depend on $|R|$; instead we fix $n = 2k$. The code generation algorithm, denoted by $\mathcal{G}_{\mathsf{Ring}}$, is very similar to $\mathcal{G}_{\mathsf{Stat}}$, except that $G = \begin{bmatrix} A \\ B \end{bmatrix}$, where A and B are two random $k \times k$ upper triangular matrices with 1 along the main diagonal. L is the same as before (using k instead of m). Note that $G|_L$ is an upper triangular matrix with 1 in the main diagonal. It is easy to compute an upper triangular matrix H (also with 1 in the main diagonal) using only the ring operations on elements in $G|_L$ such that $HG|_L = I$.

The hiding property is no longer statistical, but is a consequence of Assumption 1, instantiated with $\mathcal{G}_{\mathsf{Ring}}$ and $t = 1$. $\mathcal{G}_{\mathsf{Ring}}$ could be modified to use more than two matrices A and B, to make the resulting assumption weaker, at the expense of increasing n. In the full version we give an alternative to $\mathcal{G}_{\mathsf{Ring}}$ which relies on a random walk in the special linear group.

Random code based instantiation. Our next instantiation of the generic encoding, again with $t = 1$, uses a code generation algorithm $\mathcal{G}_{\mathsf{Rand}}$ based on a random linear code. It restricts the ring family to be a field family (or a pseudo-field family) $\mathcal{F}$. But this instantiation of Assumption 1 is a more standard assumption, which can be reduced to the hardness of *decoding* a random linear code when the field is small. $\mathcal{G}_{\mathsf{Rand}}^{\mathcal{F}}$ works as follows:

- Let $F = \mathcal{F}_{\mathsf{id}}$ and $n = 2k$.
- Pick a random $n \times k$ matrix $G \leftarrow F^{n \times k}$.
- Pick a random subset $L \subseteq [n]$, $|L| = k$, such that the $k \times k$ submatrix $G|_L$ is non-singular, where $G|_L$ consists of those rows in G whose indices are in L.
- Let $H = G|_L^{-1}$.

These three encoding schemes encode a single element in the ring (or field) R. They allow the following homomorphic operation: given an encoding of $x \in R$, namely $v = Gu + e$ where $u_1 = x$, for any $a, z \in R$, an encoding of $ax + z$ (with the same non-noisy co-ordinates) can be computed as $av + Gw$, where w is a random vector with $w_1 = z$. Further they all have the hiding property that after such a homomorphic operation, the non-noisy co-ordinates of the resulting encoding reveals nothing beyond the value $ax + z$. Finally, a noisy encoding $v \in R^n$ of $x \in R$ can be decoded by taking the first coordinate of $Hv|_L$.

Reed-Solomon code based instantiation. In our final instantiation of the generic encoding, we will let $t(k)$ be a constant fraction of k, say $t = k/2$. This variant of the construction exploits a stronger homomorphic property of Reed-Solomon codes, which was previously exploited in [30]. The code generation algorithm $\mathcal{G}_{\mathsf{RS}}$ uses $n = ck$, for a sufficiently large constant[6] $c > 4$. For a field $F = \mathcal{F}_{\mathsf{id}}$ the $n \times k$ matrix G is a linear transformation that extrapolates a degree $k - 1$ polynomial, given by its value at k randomly chosen points ζ_i in the field

---

[6] We require $c > 4$ so that Assumption 2(c) will not be broken by known list-decoding algorithms for Reed-Solomon codes [37]. Letting $c = 8$ may be a safe choice, with larger values of c being more conservative.

F, to n other randomly chosen evaluation points ϑ_i. (All ζ_i and ϑ_i are distinct,[7] and can be thought of as specifying G.) The non-noisy coordinates $L \subseteq [n]$ are chosen at random, where $|L| = 2k - 1$ to allow reconstructing polynomials of degree $2(k - 1)$.

This encoding allows the following homomorphic operation: given an encoding of $x \in F^t$, namely $v = Gu + e$ where $u_i = x_i$ for $i = 1, \ldots, t$, for any $a, z \in F^t$, an encoding of $ax + z$ (where ax denotes coordinate-wise multiplication) can be computed as $pv + w$, where $p, w \in F^n$ are the values of random polynomials of degree k and $2k$ respectively at the n evaluation points ϑ_i, which evaluate to a and z respectively in the first t of the k points ζ_i. Note that the resulting vector encodes (with noise) a degree $2k$ polynomial.

Instantiations of Assumption 1. Each of the above instantiations of the encoding leads to a corresponding instantiation of Assumption 1. For the sake of clarity we collect these assumptions below.

Assumption 2 *(a)* [**For** $\mathcal{G}_{\mathsf{Rand}}$**, with** $t(k) = 1$]. *For every computationally efficient field family $\mathcal{F}$ and sequence $\{(\mathsf{id}_k, x_k, y_k)\}_k$ such that $x_k, y_k \in \mathcal{F}_{\mathsf{id}_k}$, the ensembles $\{\mathcal{E}^{\mathcal{F}}_{(\mathcal{G}_{\mathsf{Rand}}, 1)}(\mathsf{id}_k, k, x_k)\}_k$ and $\{\mathcal{E}^{\mathcal{F}}_{(\mathcal{G}_{\mathsf{Rand}}, 1)}(\mathsf{id}_k, k, y_k)\}_k$ are computationally indistinguishable.*

(b) [**For** $\mathcal{G}_{\mathsf{Ring}}$**, with** $t(k) = 1$]. *For every computationally efficient ring family $\mathcal{R}$ and sequence $\{(\mathsf{id}_k, x_k, y_k)\}_k$ such that $x_k, y_k \in \mathcal{R}_{\mathsf{id}_k}$, the ensembles $\{\mathcal{E}^{\mathcal{F}}_{(\mathcal{G}_{\mathsf{Ring}}, 1)}(\mathsf{id}_k, k, x_k)\}_k$ and $\{\mathcal{E}^{\mathcal{F}}_{(\mathcal{G}_{\mathsf{Ring}}, 1)}(\mathsf{id}_k, k, y_k)\}_k$ are computationally indistinguishable.*

(c) [**For** $\mathcal{G}_{\mathsf{RS}}$**, with** $t(k) = k/2$].[8] *Let $t(k) = k/2$. For every computationally efficient field family $\mathcal{F}$ and sequence $\{(\mathsf{id}_k, x_k, y_k)\}_k$ such that $x_k, y_k \in \mathcal{F}^{t(k)}_{\mathsf{id}_k}$, the ensembles $\{\mathcal{E}^{\mathcal{F}}_{(\mathcal{G}_{\mathsf{RS}}, t(k))}(\mathsf{id}_k, k, x_k)\}_k$ and $\{\mathcal{E}^{\mathcal{F}}_{(\mathcal{G}_{\mathsf{RS}}, t(k))}(\mathsf{id}_k, k, y_k)\}_k$ are computationally indistinguishable.*

3.2 Product-Sharing Secure against Passive Corruption

Below we list the protocols for securely realizing $\mathcal{F}_{\mathsf{pdt\text{-}shr}}$ that we obtain from the noisy encodings above. They have increasing efficiency, but use stronger assumptions. These protocols use only black-box access to the ring (or field), and are in the OT-hybrid model. The protocols follow the pattern described at the beginning of Section 3. But this achieves security only against *static* corruption. In the full version [45] we show how to transform them into protocols that are secure against adaptive passive corruption, with erasures. In brief, in the new protocol, first the original protocol is run on random inputs, and then its working memory is deleted, and finally the real inputs and the outcome of the original protocol are used to complete the protocol.

---

[7] This requires to ensure that $|F| > n + k$. If id does not satisfy this requirement the algorithm uses a sufficiently large extension field of F.

[8] We can make the assumption weaker by choosing smaller values of t, or larger values of n in $\mathcal{G}_{\mathsf{RS}}$.

The different protocols are as follows:

- Protocol ρ^{OT} (with statistical security): this protocol uses the statistically hiding encoding scheme based on $\mathcal{G}_{Stat}$, and achieves statistical security.
- Protocol σ^{OT}: this protocol uses the computationally hiding encoding scheme based on $\mathcal{G}_{Rand}$ (for fields) or $\mathcal{G}_{Ring}$. Security follows from Assumption 2(a) or (b), respectively.
- Protocol τ^{OT} (using packed encoding): this protocol uses the noisy encoding scheme with the code generation algorithm $\mathcal{G}_{RS}$. It realizes multiple $(t = k/2)$ parallel sessions of $\mathcal{F}_{pdt-shr}$. Security follows from Assumption 2(c).

4   Arithmetic Computation with Active Corruption

In [44] it is shown how to obtain a UC-secure protocol in the OT-hybrid model for any two-party functionality $\mathcal{F}$ against active corruption by making a *black-box* use of the following two ingredients:

1. An "outer protocol" for $\mathcal{F}$ which employs k auxiliary parties (servers); this protocol should be UC-secure against active corruption provided that only some constant fraction of the servers can be (adaptively) corrupted.

2. An "inner protocol" for a reactive two-party functionality corresponding to each server in the outer protocol. In contrast to the outer protocol, this protocol only needs to be secure against *passive* (adaptive) corruption. The inner protocol is allowed to be in the OT-hybrid model and to have memory erasures.

Below we summarize the results we obtain by combining appropriate choices for the outer protocol with the inner protocols from Section 3. All these results can be readily extended to the multi-party setting as well, where the complexity grows polynomially with the number of parties; see [45] for details. All of the protocols provide adaptive security with erasures.

Combining the protocol from [19] (which makes a black-box use of an arbitrary ring) as the outer protocol with ρ^{OT} as the inner protocol, we get:

Theorem 1 (Unconditionally Secure Protocol). *For any arithmetic circuit C, there exists a protocol Π in the OT-hybrid model that is a secure black-box realization of C-evaluation for the set of all ring families. The security holds unconditionally against computationally unbounded adversaries and environments.*

The *arithmetic* communication complexity of the protocol ρ^{OT}, and hence that of the above protocol, grows linearly with (a bound on) $|\log \mathcal{R}_{id}|$. To obtain a computationally secure protocol whose arithmetic communication complexity is independent of the ring, we can replace ρ^{OT} by σ^{OT} (with $\mathcal{G}_{Rand}$ as the code generation scheme) in the previous construction:

Theorem 2. *Suppose that Assumption 2(a) holds. Then, for every arithmetic circuit C, there exists a protocol Π in the OT-hybrid model that is a secure black-box realization of C-evaluation for the set of all computationally efficient field families $\mathcal{F}$. Further, the arithmetic complexity of Π is $\text{poly}(k) \cdot |C|$, independent of $\mathcal{F}$ or id.*

Using $\mathcal{G}_{\text{Ring}}$ instead of $\mathcal{G}_{\text{Rand}}$, this result extends to all computationally efficient ring families:

Theorem 3. *Suppose that Assumption 2(b) holds. Then, for every arithmetic circuit C, there exists a protocol Π in the OT-hybrid model that is a secure black-box realization of C-evaluation for the set of all computationally efficient ring families $\mathcal{R}$. Further, the arithmetic complexity of Π is $\text{poly}(k) \cdot |C|$, independent of $\mathcal{R}$ or id.*

Finally, to obtain our most efficient protocol we use τ^{OT} (with $n = O(k)$ and $t = \Omega(k)$) as the inner protocol. The outer protocol is a variant of the protocol from [22] in which the computational complexity is optimized using an idea from [36] (see [45] for a description). To get the computational complexity specified below, the size of the field should be super-polynomial in the security parameter. (The communication complexity does not depend on this assumption.)

Theorem 4. *Suppose that Assumption 2(c) holds. Then, for every arithmetic circuit C, there exists a protocol Π in the OT-hybrid model with the following properties. The protocol Π is a secure black-box realization of C-evaluation for the set of all computationally efficient field families $\mathcal{F}$, with respect to all computationally bounded environments for which $|\mathcal{F}_{\text{id}}|$ is super-polynomial in k. The arithmetic communication complexity of Π is $O(|C| + k \cdot \text{depth}(C))$, where $\text{depth}(C)$ denotes the depth of C, and its arithmetic computation complexity is $O(\log^2 k) \cdot (|C| + k \cdot \text{depth}(C))$. Its round complexity is $O(\text{depth}(C))$.*

By using a suitable choice of fields and evaluation points for the Reed-Solomon encoding, and under a corresponding specialization of Assumption 2(c), the computational overhead of the above protocol can be reduced from $O(\log^2 k)$ to $O(\log k)$. (In this variant we do not attempt to make a black-box use of the underlying field and rely on the standard representation of field elements.)

4.1 Protocols from Homomorphic Encryption

So far we considered protocols using only black-box access to a ring. If we further assume a black-box access[9] to a homomorphic encryption scheme over the ring, there are simple protocols for $\mathcal{F}_{\text{pdt-shr}}$ secure against passive adversaries. These can then be used instead of the protocols from Section 3.2 in the above constructions.

We are interested in homomorphic encryptions over rings that support addition of two encrypted elements, and multiplication of an encrypted element by an unencrypted element. There are two kinds of such schemes. The more versatile kind — which we shall call a *controlled-ring* scheme — allows one to specify id during the key-generation phase, and then allows operations on elements in $\mathcal{R}_{\text{id}}$, where $\mathcal{R}$ is the ring family associated with the scheme. Candidates for such schemes are the classic Goldwasser-Micali encryption scheme [35] (for which the ring family consists of the single ring $\mathbb{Z}_2$) and Benaloh's scheme [5] (for which

---

[9] When saying that a construction makes a black-box use of a homomorphic encryption primitive, we refer to the notion of a fully black-box reduction as defined in [58].

the ring family consists of rings $\mathbb{Z}_p$ where p is a polynomially bounded prime number). Any such $\mathcal{R}$-homomorphic encryption scheme can be used as a black-box to obtain a homomorphic encryption scheme for the ring family of square matrices over $\mathcal{R}$ (where the matrix size n is specified by id).

Theorem 5. *For every arithmetic circuit C, there exists a protocol Π in the OT-hybrid model, such that for every ring family $\mathcal{R}$, the protocol $\Pi^{\mathcal{R}}$ securely realizes $\mathcal{F}_C^{\mathcal{R}}$ by making a* black-box *use of any controlled-ring homomorphic encryption for $\mathcal{R}$. The number of invocations of the encryption scheme is* $\mathrm{poly}(k) \cdot |C|$, *independent of $\mathcal{R}$ or* id.

Note that the protocol in the above theorem, when instantiated with the ring of $n \times n$ matrices over $\mathbb{Z}_p$, has communication complexity $poly(k) \cdot |C| \cdot n^2$. Combined with [49], this yields constant-round protocols for secure linear algebra which make a black-box use of homomorphic encryption and whose communication complexity is nearly linear in the input size.

For the case of fields, we obtain the following more efficient version of the result by using the same outer protocol as used in Theorem 4:

Theorem 6. *For every arithmetic circuit C, there exists a protocol Π in the OT-hybrid model, such that for every field family $\mathcal{F}$, the protocol $\Pi^{\mathcal{F}}$ securely realizes $\mathcal{F}_C^{\mathcal{F}}$ by making a* black-box *use of any controlled-ring homomorphic encryption for $\mathcal{F}$. The security holds against adaptive corruption with erasures. Further, Π makes $O(|C| + k \cdot \mathsf{depth}(C))$ invocations of the encryption scheme, and the communication complexity is dominated by sending $O(|C| + k \cdot \mathsf{depth}(C))$ ciphertexts.*

Homomorphic encryption schemes like the Paillier cryptosystem [53] are homomorphic with respect to the ring $\mathbb{Z}_N$, where N is a randomly chosen product of two large primes chosen at the time of key generation; N cannot be specified ahead of time. We call such a scheme an "uncontrolled ring" homomorphic encryption scheme. Using standard techniques computation over $\mathbb{Z}_M$ for an *a priori* fixed modulus M can be securely reduced to computation over $\mathbb{Z}_N$ where N is a sufficiently large, dynamically chosen modulus (see [45] for more details). We obtain the following results:

Theorem 7. *Let $\mathcal{R}$ be the ring family where $\mathcal{R}_{\mathsf{id}}$ is the standard representation of the ring $\mathbb{Z}_{\mathsf{id}}$. For every arithmetic circuit C there exists a black-box construction of a protocol Π in the OT-hybrid model from any uncontrolled-ring homomorphic encryption for $\mathcal{R}$, such that Π is a secure realization of C-evaluation for $\mathcal{R}$. The number of invocations of the encryption scheme is $\mathrm{poly}(k) \cdot |C|$, independent of* id, *and the communication complexity is dominated by $\mathrm{poly}(k) \cdot |C|$ ciphertexts. During the protocol, the ring size parameter fed to the encryption scheme by honest parties is limited to $k' = O(k + |\mathsf{id}|)$.*

If, further, the ring over which C should be computed is restricted to be a field, there exists a protocol as above which makes $O(|C| + k \cdot \mathsf{depth}(C))$ invocations of the encryption scheme, and where the communication complexity is dominated by sending $O(|C| + k \cdot \mathsf{depth}(C))$ ciphertexts.

The second part of the above theorem also applies to the case of arithmetic computation over pseudo-fields. Furthermore, it can be generalized to the ring of $n \times n$ matrices, which when used with constructions of uncontrolled-ring $\mathbb{Z}_N$-homomorphic encryption schemes from the literature [53,23] would yield arithmetic protocols for matrices over large rings whose complexity grows quadratically with n.

We finally note that in the *stand-alone* model, the OT oracle in the above protocols can be realized by making a black-box use of the homomorphic encryption primitive without affecting the asymptotic number of calls to the primitive. This relies on the black-box construction from [42] and the fact that only $O(k)$ OTs need to be secure against active corruption. Thus, the above theorems hold also in the plain, stand-alone model (as opposed to the OT-hybrid UC-model).

Acknowledgments. We thank Jens Groth, Venkatesan Guruswami, Farzad Parvaresh, Oded Regev, and Ronny Roth for helpful discussions.

References

1. Abadi, M., Feigenbaum, J.: Secure circuit evaluation. J. Cryptology 2(1), 1–12 (1990)
2. Algesheimer, J., Camenisch, J., Shoup, V.: Efficient computation modulo a shared secret with application to the generation of shared safe-prime products. In: Yung, M. (ed.) CRYPTO 2002. LNCS, vol. 2442, pp. 417–432. Springer, Heidelberg (2002)
3. Aumann, Y., Lindell, Y.: Security against covert adversaries: Efficient protocols for realistic adversaries. In: Vadhan, S.P. (ed.) TCC 2007. LNCS, vol. 4392, pp. 137–156. Springer, Heidelberg (2007)
4. Ben-Or, M., Goldwasser, S., Wigderson, A.: Completeness theorems for non-cryptographic fault-tolerant distributed computation. In: STOC 1988, pp. 1–10 (1988)
5. Benaloh, J.: Verifiable Secret-Ballot Elections. PhD thesis, Department of Computer Science, Yale University (1987)
6. Bleichenbacher, D., Kiayias, A., Yung, M.: Decoding interleaved reed-solomon codes over noisy channels. Theor. Comput. Sci. 379(3), 348–360 (2007)
7. Bogetoft, P., Christensen, D.L., Damgard, I., Geisler, M., Jakobsen, T., Krøigaard, M., Nielsen, J.D., Nielsen, J.B., Nielsen, K., Pagter, J., Schwartzbach, M., Toft, T.: Multiparty computation goes live. Cryptology ePrint Archive, Report 2008/068
8. Boneh, D., Franklin, M.K.: Efficient generation of shared RSA keys. J. ACM 48(4), 702–722 (2001); Earlier version in Crypto 1997
9. Canetti, R.: Universally composable security: A new paradigm for cryptographic protocols. Cryptology ePrint Archive, Report 2000/067 (2005)
10. Canetti, R., Feige, U., Goldreich, O., Naor, M.: Adaptively secure multi-party computation. In: STOC 1996, pp. 639–648 (1996)
11. Canetti, R., Ishai, Y., Kumar, R., Reiter, M.K., Rubinfeld, R., Wright, R.N.: Selective private function evaluation with applications to private statistics. In: PODC 2001, pp. 293–304 (2001)
12. Canetti, R., Lindell, Y., Ostrovsky, R., Sahai, A.: Universally composable two-party computation. In: STOC 2002, pp. 494–503 (2002)

13. Chaum, D., Crépeau, C., Damgård, I.: Multiparty unconditionally secure protocols. In: STOC 1988, pp. 11–19 (1988)
14. Coppersmith, D., Sudan, M.: Reconstructing curves in three (and higher) dimensional space from noisy data. In: STOC 2003, pp. 136–142 (2003)
15. Cramer, R., Damgård, I.: Secure distributed linear algebra in a constant number of rounds. In: Kilian, J. (ed.) CRYPTO 2001. LNCS, vol. 2139, pp. 119–136. Springer, Heidelberg (2001)
16. Cramer, R., Damgård, I.B., Maurer, U.M.: General secure multi-party computation from any linear secret-sharing scheme. In: Preneel, B. (ed.) EUROCRYPT 2000. LNCS, vol. 1807, pp. 316–334. Springer, Heidelberg (2000)
17. Cramer, R., Damgård, I.B., Nielsen, J.B.: Multiparty computation from threshold homomorphic encryption. In: Pfitzmann, B. (ed.) EUROCRYPT 2001. LNCS, vol. 2045, pp. 280–299. Springer, Heidelberg (2001)
18. Cramer, R., Fehr, S.: Optimal black-box secret sharing over arbitrary abelian groups. In: Yung, M. (ed.) CRYPTO 2002. LNCS, vol. 2442, pp. 272–287. Springer, Heidelberg (2002)
19. Cramer, R., Fehr, S., Ishai, Y., Kushilevitz, E.: Efficient multi-party computation over rings. In: Biham, E. (ed.) EUROCRYPT 2003. LNCS, vol. 2656, pp. 596–613. Springer, Heidelberg (2003)
20. Cramer, R., Kiltz, E., Padró, C.: A note on secure computation of the Moore-Penrose pseudoinverse and its application to secure linear algebra. In: Menezes, A. (ed.) CRYPTO 2007. LNCS, vol. 4622, pp. 613–630. Springer, Heidelberg (2007)
21. Damgård, I., Fitzi, M., Kiltz, E., Nielsen, J.B., Toft, T.: Unconditionally secure constant-rounds multi-party computation for equality, comparison, bits and exponentiation. In: Halevi, S., Rabin, T. (eds.) TCC 2006. LNCS, vol. 3876, pp. 285–304. Springer, Heidelberg (2006)
22. Damgård, I., Ishai, Y.: Scalable secure multiparty computation. In: Dwork, C. (ed.) CRYPTO 2006. LNCS, vol. 4117, pp. 501–520. Springer, Heidelberg (2006)
23. Damgård, I., Jurik, M.: A generalisation, a simplification and some applications of paillier's probabilistic public-key system. In: Preneel, B. (ed.) CT-RSA 2002. LNCS, vol. 2271, pp. 79–95. Springer, Heidelberg (2002)
24. Damgård, I., Nielsen, J.B.: Universally composable efficient multiparty computation from threshold homomorphic encryption. In: Boneh, D. (ed.) CRYPTO 2003. LNCS, vol. 2729, pp. 247–264. Springer, Heidelberg (2003)
25. Damgård, I., Nielsen, J.B., Orlandi, C.: Essentially optimal universally composable oblivious transfer. In: ICISC 2008 (2008)
26. Desmedt, Y., Frankel, Y.: Threshold cryptosystems. In: Brassard, G. (ed.) CRYPTO 1989. LNCS, vol. 435, pp. 307–315. Springer, Heidelberg (1990)
27. Even, S., Goldreich, O., Lempel, A.: A randomized protocol for signing contracts. Commun. ACM 28(6), 637–647 (1985)
28. Frankel, Y., MacKenzie, P.D., Yung, M.: Robust efficient distributed rsa-key generation. In: STOC 1998, pp. 663–672 (1998)
29. Franklin, M.K., Haber, S.: Joint encryption and message-efficient secure computation. J. Cryptology 9(4), 217–232 (1996)
30. Franklin, M.K., Yung, M.: Communication complexity of secure computation (extended abstract). In: STOC 1992, pp. 699–710 (1992)
31. Freedman, M.J., Nissim, K., Pinkas, B.: Efficient private matching and set intersection. In: Cachin, C., Camenisch, J.L. (eds.) EUROCRYPT 2004. LNCS, vol. 3027, pp. 1–19. Springer, Heidelberg (2004)
32. Gilboa, N.: Two party RSA key generation. In: Wiener, M. (ed.) CRYPTO 1999. LNCS, vol. 1666, pp. 116–129. Springer, Heidelberg (1999)

33. Goldreich, O.: Foundations of Cryptography: Basic Applications. Cambridge University Press, Cambridge (2004)
34. Goldreich, O., Micali, S., Wigderson, A.: How to play ANY mental game. In: STOC 1987, pp. 218–229 (1987); See [ch. 7] for more details.
35. Goldwasser, S., Micali, S.: Probabilistic encryption. J. Comput. Syst. Sci. 28(2), 270–299 (1984); Preliminary version in STOC 1982
36. Groth, J.: Linear algebra with sub-linear zero-knowledge arguments (manuscript, 2008)
37. Guruswami, V., Sudan, M.: Improved decoding of reed-solomon and algebraic-geometry codes. IEEE Trans. Inf. Theory 45(6), 1757–1767 (1999)
38. Hazay, C., Lindell, Y.: Efficient protocols for set intersection and pattern matching with security against malicious and covert adversaries. In: Canetti, R. (ed.) TCC 2008. LNCS, vol. 4948, pp. 155–175. Springer, Heidelberg (2008)
39. Impagliazzo, R., Levin, L.A., Luby, M.: Pseudo-random generation from one-way functions (extended abstract). In: STOC 1989, pp. 12–24 (1989)
40. Impagliazzo, R., Naor, M.: Efficient cryptographic schemes provably as secure as subset sum. J. Cryptology 9(4), 199–216 (1996)
41. Ishai, Y., Kilian, J., Nissim, K., Petrank, E.: Extending oblivious transfers efficiently. In: Boneh, D. (ed.) CRYPTO 2003. LNCS, vol. 2729, pp. 145–161. Springer, Heidelberg (2003)
42. Ishai, Y., Kushilevitz, E., Lindell, Y., Petrank, E.: Black-box constructions for secure computation. In: STOC 2006, pp. 99–108 (2006)
43. Ishai, Y., Kushilevitz, E., Ostrovsky, R., Sahai, A.: Cryptography from anonymity. In: FOCS 2006, pp. 239–248 (2006)
44. Ishai, Y., Prabhakaran, M., Sahai, A.: Founding cryptography on oblivious transfer - efficiently. In: Wagner, D. (ed.) CRYPTO 2008. LNCS, vol. 5157, pp. 572–591. Springer, Heidelberg (2008)
45. Ishai, Y., Prabhakaran, M., Sahai, A.: Secure arithmetic computation with no honest majority. Cryptology ePrint Archive, Report 2008/465 (2008)
46. Kiayias, A., Yung, M.: Cryptographic hardness based on the decoding of reed-solomon codes. IEEE Transactions on Information Theory 54(6), 2752–2769 (2008)
47. Kiltz, E., Mohassel, P., Weinreb, E., Franklin, M.K.: Secure linear algebra using linearly recurrent sequences. In: Vadhan, S.P. (ed.) TCC 2007. LNCS, vol. 4392, pp. 291–310. Springer, Heidelberg (2007)
48. Lindell, Y., Pinkas, B.: Privacy preserving data mining. J. Cryptology 15(3), 177–206 (2002); Earlier version in Crypto 2000
49. Mohassel, P., Weinreb, E.: Efficient secure linear algebra in the presence of covert or computationally unbounded adversaries. In: Wagner, D. (ed.) CRYPTO 2008. LNCS, vol. 5157, pp. 481–496. Springer, Heidelberg (2008)
50. Naor, M., Pinkas, B.: Oblivious polynomial evaluation. SIAM J. Comput. 35(5), 1254–1281 (2006); Earlier version in STOC 1999
51. Naor, M., Pinkas, B., Sumner, R.: Privacy preserving auctions and mechanism design. In: ACM Conference on Electronic Commerce 1999, pp. 129–139 (1999)
52. Nissim, K., Weinreb, E.: Communication efficient secure linear algebra. In: Halevi, S., Rabin, T. (eds.) TCC 2006. LNCS, vol. 3876, pp. 522–541. Springer, Heidelberg (2006)
53. Paillier, P.: Public-key cryptosystems based on composite degree residuosity classes. In: Stern, J. (ed.) EUROCRYPT 1999. LNCS, vol. 1592, pp. 223–238. Springer, Heidelberg (1999)
54. Parvaresh, F., Vardy, A.: Correcting errors beyond the guruswami-sudan radius in polynomial time. In: FOCS 2005, pp. 285–294 (2005)

55. Peikert, C., Vaikuntanathan, V., Waters, B.: A framework for efficient and composable oblivious transfer. In: Wagner, D. (ed.) CRYPTO 2008. LNCS, vol. 5157, pp. 554–571. Springer, Heidelberg (2008)
56. Poupard, G., Stern, J.: Generation of shared RSA keys by two parties. In: Ohta, K., Pei, D. (eds.) ASIACRYPT 1998. LNCS, vol. 1514, pp. 11–24. Springer, Heidelberg (1998)
57. Rabin, M.: How to exchange secrets by oblivious transfer. Technical Report TR-81, Harvard Aiken Computation Laboratory (1981)
58. Reingold, O., Trevisan, L., Vadhan, S.P.: Notions of reducibility between cryptographic primitives. In: Naor, M. (ed.) TCC 2004. LNCS, vol. 2951, pp. 1–20. Springer, Heidelberg (2004)
59. Toft, T.: Primitives and Applications for Multi-party Computation. PhD thesis, Department of Computer Science, Aarhus University (2007)
60. Yao, A.C.: How to generate and exchange secrets. In: FOCS 1996, pp. 162–167 (1996)

Universally Composable Multiparty Computation with Partially Isolated Parties

Ivan Damgård[1], Jesper Buus Nielsen[1], and Daniel Wichs[2]

[1] University of Aarhus, Denmark
[2] New York University, USA

Abstract. It is well known that universally composable multiparty computation cannot, in general, be achieved in the standard model without setup assumptions when the adversary can corrupt an arbitrary number of players. One way to get around this problem is by having a *trusted third party* generate some global setup such as a *common reference string (CRS)* or a *public key infrastructure (PKI)*. The recent work of Katz shows that we may instead rely on physical assumptions, and in particular *tamper-proof hardware tokens*. In this paper, we consider a similar but *strictly weaker* physical assumption. We assume that a player (Alice) can *partially isolate* another player (Bob) for a brief portion of the computation and prevent Bob from communicating more than some limited number of bits with the environment. For example, isolation might be achieved by asking Bob to put his functionality on a tamper-proof hardware token and assuming that Alice can prevent this token from communicating to the outside world. Alternatively, Alice may interact with Bob directly but in a special office which she administers and where there are no high-bandwidth communication channels to the outside world. We show that, under *standard* cryptographic assumptions, such physical setup can be used to UC-realize any two party and multiparty computation in the presence of an active and *adaptive* adversary corrupting any number of players. We also consider an alternative scenario, in which there are some trusted third parties but no single such party is trusted by all of the players. This compromise allows us to significantly limit the use of the physical set-up and hence might be preferred in practice.

Keywords: universally composable security, multiparty computation, public-key infrastructure.

1 Introduction

Traditionally, the security of cryptographic protocols was considered in the stand-alone setting where a single run of the protocol executes in isolation. In the real world, when many copies of a single protocol and related protocols may be executing concurrently, security in the stand-alone setting becomes insufficient.

The universal composability (UC) framework was introduced by Canetti in [Can01] to fix this problem and allow us to prove the security of protocols in the real-world setting without resorting to intractably complicated proofs. The initial work of Canetti gave hope that UC security is achievable by showing that any multiparty computation (MPC) can be realized in the UC framework, assuming a strict majority of the players

O. Reingold (Ed.): TCC 2009, LNCS 5444, pp. 315–331, 2009.

are honest. Unfortunately, this work was followed by results showing that many natural functionalities cannot be UC realized without an honest majority, including essentially all non-trivial two party computations such as commitments and zero knowledge proofs [CKL03].

To get around these negative results, one can require the existence of additional setup infrastructure available to the parties. For example, such setup can consist of a common reference string (CRS) which is honestly sampled from some pre-defined distribution and given to all the players [CLOS02] or a public key infrastructure (PKI) where a trusted certificate authority (CA) verifies that each player knows the secret key corresponding to his registered public key [BCNP04]. Both of the above setup assumptions require a trusted party to initialize the infrastructure and the protocols become completely insecure if this party is corrupted.

In this paper we rely on a physical assumption instead of a trusted third party. Namely, we assume that a player (Alice) can ensure that another player (Bob) is *partially isolated* for a short portion of the computation. During this time, Bob can only exchange a limited number of bits with the environment but Alice's communication is unrestricted. We show that, under *standard cryptographic assumptions*, the above physical setup allows us to UC realize any two-party and multiparty computation in the presence of an *active and adaptive* adversary corrupting any number of parties. We do *not* assume erasures.

1.1   Related Work

The idea of relying on physical assumptions to achieve universal composability was first proposed by Katz in [K07]. In particular, the work of Katz assumes the existence of *tamper-proof hardware tokens*. A player, Bob, puts some arbitrary functionality inside such a token and sends it to another player, Alice. Alice can then only interact with the token through the intended interface. In addition, it is assumed that Alice can *isolate* the token during this interaction, ensuring that it has no way of communicating with the outside world. In general, there seem to be two ways to take advantage of the fact that Bob's functionality is placed on tamper-proof hardware (rather than having Bob run it remotely):

(1) The tamper-proof hardware token is isolated and cannot communicate with the environment.
(2) The tamper-proof hardware token is a new and *separate* entity from Bob. Bob never sees the content of Alice's interaction with the token.

In [K07], Katz shows how to use tamper-proof tokens to UC realize any multiparty computation in the presence of an active but *static* adversary, under the Diffie Hellman (DH) assumption. We note that this solution only makes use of advantage (1), though this distinction is not explicit and the formal model allows for both advantages.

The work of Chandran et al. [CGS08] extends the result of Katz by considering an adversary who might not necessarily *know* the code of the token he creates. In addition, the adversary may perform *reset attacks* on received tokens, effectively getting the power to rewind tokens at will. The work of Moran and Segev [MS08], on the other hand, presents a protocol for two asymmetrically powerful parties: a powerful Goliath and a limited David. Only Goliath has the ability to create tamper-proof hardware tokens. Moreover, Goliath is not assumed to be computationally bounded (but David is). Both of these works crucially rely on advantage (2) above.

The work of Damgård et al. [DNW08] introduces a new and slightly different physical assumption – namely, that parties can be *partially* isolated so that their communication with the environment is limited. This setting was studied only with regard to *zero knowledge proofs of knowledge* (ZK PoK). Damgård et al. present a witness indistinguishable (WI) PoK protocol for the case where only the prover is partially isolated (while the verifier's communication is unrestricted) and a ZK PoK protocol for the case where both parties are partially isolated.

1.2 Our Contribution

In this paper we consider the partial isolation physical assumption for multiparty computation in general. First, we notice that there is a relationship between the partial isolation model of [DNW08] and the tamper-proof hardware model of [K07]. Namely a party can be (fully) isolated by placing its functionality on a tamper-proof hardware token. However, isolation can also be implemented in many other ways. For example, we may imagine a setting where Bob simply brings his laptop into an office administered by Alice who ensures that there is no wireless or wired internet access available to Bob. Bob then connects his laptop to Alice's machine and they run an interactive protocol between them. During this time, Alice can communicate with the environment as much as she wants, but Bob cannot. In the above example, we see a crucial difference between the isolated parties model and the tamper-proof hardware model: we cannot (in general) assume that some isolated entity is *separate* from Bob – it might be Bob himself who is isolated! Of course, since Bob sees Alice's interaction with himself while he is isolated we do not get advantage (2). Therefore, even the full isolation model is strictly weaker than the tamper proof hardware model and the protocols of [CGS08, MS08] cannot be used in the isolation setting.

Moreover, as a further weakening of our physical assumptions, we only assume that parties can be *partially* isolated from the environment. Specifically, as in [DNW08], we assume the existence of some threshold ℓ, such that Alice can prevent Bob from exchanging more than ℓ bits with the environment. In practice, this might significantly easier to achieve than full isolation. In our example, where Bob meets Alice in her office, it might be significantly easier for Alice to only block *high-bandwidth* communication channels to the outside world than to block *all* such channels. Even if the parties do choose to use tamper-proof hardware tokens, it might not be trivial to fully isolate a token as is required in Katz's model. Partial isolation might be much simpler to achieve. For example, if the tamper-proof token is a smart-card that is too small to have its own power supply, Alice can then observe (and limit) the card's power consumption to limit communication. A study by [BA03] shows that one bit of wireless communication by a smart-card has the same power consumption as 1000 32-bit elementary operations and hence this could be a practical solution in limiting the amount of communication that is possible.

The work of [DNW08] defines partial isolation for the case of zero knowledge proofs of knowledge directly. In this work, we define partial isolation in general as an ideal functionality (similarly to Katz's functionality for tamper-proof hardware tokens). We then construct a protocol for arbitrary two-party and multiparty computation using this functionality. In our protocol, the use of physical assumptions is limited to a short setup phase during which parties register keys with one another while the registrant is partially isolated. In practice, the use of physical setup might be expensive and difficult for individuals. We also propose a hybrid model in which there are some trusted Certificate Authorities (CAs) but no such authority needs to be trusted by all the players. Each player either trusts an external CA (and many players may trust the same one) or can act as his own CA and trust nobody else. This model might be natural in many scenarios where large organizations (countries, companies...) do not trust each other but individuals trust the organization they belong to.

1.3 Overview of Construction

Our basic approach is to set up a public key infrastructure (PKI) between the parties so that each player must know the secret key corresponding to his registered public key. The result of [BCNP04] shows that such a PKI, when created by a trusted third party, can be used to UC-realize the ideal commitment functionality, which in turn allow one to UC-realize arbitrary multiparty computation.

Consider the following naive approach of setting up such a PKI. Each player chooses a public key and registers it with every other player. When a player, Bob, wants to register his key with another player, Alice, he simply sends her his public key and runs a zero knowledge (ZK) proof of knowledge (PoK) to convince her that he knows the corresponding secret key.

Unfortunately, using a standard ZK PoK (secure in the stand-alone setting) does not give us security in the UC-framework. However, if Alice can ensure that Bob is isolated for the duration of the proof, a standard PoK protocol does guarantee that Bob knows his secret key. In fact, the result of [DNW08] shows that for any threshold ℓ there is an ℓ-Isolated Proof of Knowledge (ℓ-IPoK) protocol, which ensures that the prover knows a witness *even* if he can exchange up to ℓ bits of information with the environment during the proof. By using an ℓ-IPoK protocol, Alice will be assured that Bob knows his secret key even if she can only partially isolate Bob and keep him from communicating more than ℓ bits.

We *cannot*, however, guarantee that Alice (who is not isolated during the proof) does not learn anything from such a proof. As is shown in [DNW08], no witness hiding protocol can be zero knowledge simulatable with respect to a verifier that communicates arbitrarily with the environment. Instead, we will only rely on the witness indistinguishability (WI) property of an ℓ-IPoK protocol. This means that our PKI is not perfect and verifying parties might get some limited information about the registered private keys. Nevertheless, we show that an imperfect PKI of this type can be used to implement the ideal commitment functionality. To do so, we modify the commitment scheme of [BCNP04] (which is based on the prior scheme of [CLOS02]) so that it remains secure even if the adversary sees a witness indistinguishable proof of knowledge of the commitment private key. As is shown in [CLOS02], the commitment functionality allows us to implement all other two-party and multiparty computation.

2 The Formal Model of Our Setting

2.1 The $\mathcal{F}_{\text{isolate}}$ Ideal Functionality

We model partial isolation using an ideal functionality $\mathcal{F}_{\text{isolate}}$ described in Fig. 1. It describes a situation where P is partially isolated from the environment during an interaction with P'. This is similar to the ideal functionality $\mathcal{F}_{wrap}$ defined in [K07] to model tamper-proof hardware, but there are several important differences.

The $\mathcal{F}_{\text{isolate}}$ ideal functionality is parametrized by an isolation parameter ℓ, a security parameter κ and a polynomial p.

Isolation of P: Wait until receiving messages $(\texttt{isolate}, sid, P, P')$ from P' and $(\texttt{isolate}, sid, P, P', M)$ from P. If there is already a stored tuple of the form $(P, P', \cdot, \cdot, \cdot, \cdot)$ then ignore the command. Otherwise:
  1. Parse the string M as the description of an ITM with four communication tapes; two tapes ("in" and "out") for regular protocol communication with P' and two tapes for secret communication with P. Let the value $\texttt{state}$ encode the initial state of M (including the value of a work tape and an initialized random tape). Define new values $\texttt{inCom} := 0$, $\texttt{outCom} := 0$ and store the tuple $(P, P', M, \texttt{state}, \texttt{inCom}, \texttt{outCom})$.
  2. Send $(\texttt{isolate}, sid, P)$ to P'.
Interaction with P': On input $(\texttt{run}, sid, P, P', \texttt{msg})$ from P', retrieve the tuple $(P, P', M, \texttt{state}, \texttt{inCom}, \texttt{outCom})$. If there is no such tuple then ignore the command.
  1. Place the string $\texttt{msg}$ on the "in" tape designated for P and run M for $p(\kappa)$ steps.
  2. If there is any value $\texttt{msg}'$ on the output tape for P' then send the message $(\texttt{reply}, sid, P, \texttt{msg}')$ to P'.
  3. If there is any value $\texttt{msg}'$ on the output tape for P and $\texttt{outCom} + |\texttt{msg}'| < \ell$ then send the message $(\texttt{secretCom}, sid, P', P, \texttt{msg}')$ to P and update $\texttt{outCom} := \texttt{outCom} + |\texttt{msg}'|$.
  4. Update the value of $\texttt{state}$ in the stored tuple to encode the updated state of M and the values of its tapes.
Communication: On input $(\texttt{secretCom}, sid, P, P', \texttt{msg})$ from P, if there is no tuple of the form $(P, P', M, \texttt{state}, \texttt{inCom}, \texttt{outCom})$ then ignore. Also if the tuple has $\texttt{inCom} + |\texttt{msg}| > \ell$ then ignore the command. Otherwise
  1. Update $\texttt{inCom} := \texttt{inCom} + |\texttt{msg}|$, place $\texttt{msg}$ on the "in" tape for P and run M for $p(\kappa)$ steps.
  2. Proceed with steps 2,3,4 of the above command.
Release of P: On input $(\texttt{release}, sid, P, P')$ from P', retrieve the tuple $(P, P', M, \texttt{state}, \texttt{inCom}, \texttt{outCom})$ and send $(\texttt{release}, sid, P, P', \texttt{state})$ to P.

Fig. 1. The $\mathcal{F}_{\text{isolate}}$ Ideal Functionality

When Alice wants to isolate Bob, both of them call the $\texttt{isolate}$ command and Bob sends a description of his functionality (modeled as an ITM M) and current state to $\mathcal{F}_{\text{isolate}}$. Alice can then interact with Bob's functionality by issuing $\texttt{run}$ commands to

$\mathcal{F}_{\text{isolate}}$ which internally runs Bob's code to produce replies for Alice. At the conclusion of the interaction, Alice sends a `release` command.

The main differences between Katz's $\mathcal{F}_{wrap}$ functionality and our $\mathcal{F}_{\text{isolate}}$ functionality are as follows. Firstly, we want to capture the fact that it might be Bob himself who is isolated and not some separate token. Therefore, we make a restriction on how honest parties can use this functionality in legitimate protocols. We require that, if Bob is honest, he will be inactive between the time that he issues the `isolate` command and the time that the `release` command is issued by Alice. In addition, when the `release` command is issued, $\mathcal{F}_{\text{isolate}}$ sends Bob the current updated state of his functionality M, which might contain information about the interaction that took place with Alice. Secondly, we want to capture the fact that our isolation is only partial and that there might be some limited secret communication between a partially isolated party and the environment. We parameterize $\mathcal{F}_{\text{isolate}}$ with a communication threshold ℓ. Bob's functionality M can send up to ℓ bits of communication to its creator (and hence the environment) and can receive up to ℓ bits of communication from its creator using `secretCom` commands. We require that only corrupted parties takes advantage of this secret communication — i.e., it describes an allowed flaw of the isolation rather than a useful feature.

2.2 PKI and Certificate Authorities

We use the ideal functionality $\mathcal{F}_{\text{isolate}}$ to setup a public key infrastructure. In the general multiparty computation setting, there are many parties which will register keys and try to implement ideal functionalities among them. We denote these parties by $P_1, \ldots, P_n$. In addition we have parties $CA_1, \ldots, CA_m$ acting as certificate authorities. We allow the case where a player P_i acts as his own certificate authority ($P_i = CA_k$). In general, however, we only require that each party P_i trusts some certificate authority CA_k and many parties may trust the same certificate authority. Any player P_j who wishes to interact with P_i needs to register a key with an authority CA_k trusted by P_i.

We model the certificate authorities as additional players in the game. In the ideal world, the certificate authorities have no inputs and receive no outputs. We define a *certificate authority trust structure* as the mapping of players to the certificate authority they trust, and we assume that each player trusts at least one CA. The group of players who trust a single CA is called the certificate authority's *trust group*. To model the notion of trust, we assume that when an adversary actively corrupts a certificate authority he also actively corrupts all of the players in the authority's trust group. The adversary may actively corrupt an arbitrary number of real players P_i and an arbitrary number of certificate authorities subject to the above restriction. We call any such adversarial corruption strategy a *legal corruption strategy*. An adversary can also passively corrupt any CAs at will. For simplicity, we will just require that an honest CA makes all of its interactions public so such corruptions are unnecessary.

2.3 Statement of Result

We are now ready to state the main theorem of our paper.

Theorem 1. *Assume the existence of one-way permutations and dense public key, IND-CPA secure encryption schemes with pseudorandom ciphertexts. Then any polynomial* *time ideal functionality can be UC realized in the $\mathcal{F}_{\text{isolate}}$-hybrid model under any certificate authority trust structure. We assume that an active and adaptive adversary can corrupt any number of players and certificate authorities using a legal corruption strategy. We do not assume erasures.*

We can instantiate the theorem with the trust structure in which each player acts as his own certificate authority and trusts nobody else. This shows that, as a special case of Theorem 1, any polynomial time ideal functionality can be UC realized in the $\mathcal{F}_{\text{isolate}}$-hybrid model without any additional certificate authority parties and with an adaptive and active adversary corrupting any number of players. The proof of the above theorem spans the remainder of the paper. As in [K07], we will only show how to UC-realize the ideal functionality for multiple commitments and the rest follows from the work of [CLOS02].

3 Proofs of Knowledge and Isolated Proofs of Knowledge

Our construction relies heavily on proofs of knowledge (PoK). Here we review some terminology and results. Recall that an *NP relation* $\mathcal{R}$ is a set of pairs (x, w) where $(x, w) \in^? \mathcal{R}$ can be checked in poly-time in the length of x. For such a relation we define the *witnesses for an instance* $W_{\mathcal{R}}(x) = \{w | (x, w) \in \mathcal{R}\}$ and the *language* $L(\mathcal{R}) = \{x | W_R(x) \neq \emptyset\}$. Given an NP relation $\mathcal{R}$, a PoK is an interactive protocol between two parties called a *Prover P* and a *Verifier V*. The protocol is specified by the PPT ITMs (P, V) where P is given an input $(x, w) \in \mathcal{R}$ and V is given the instance x. The parties run the protocol and, at the end, the verifier outputs a *judgment J* = accept or J = reject. We require *completeness*: when P and V are honest then V outputs the judgment J = accept with all but negligible probability.

In our setting, the prover may communicate with an external adversarial environment during the proof, but this communication is limited to some pre-defined bound of ℓ bits. The verifier, on the other hand, has unbounded communication with the environment. This setting is considered in [DNW08], which defines the notion of an *ℓ-Isolated Proof of Knowledge* (ℓ-IPoK) protocol. Such a protocol ensures that a successful prover knows a witness, even in the above environment.

Formally, knowledge soundness of an ℓ-IPoK protocol is defined by requiring that for any adversarial prover given by a PPT ITM P^*, there exists a *strict* PPT extractor $\mathcal{X}$ which wins the knowledge soundness extraction game outlined in Fig. 2 with all but negligible probability. This should hold for any environment given by a PPT ITM $\mathcal{E}$.

It was shown in [DNW08] that there exists an Isolated Proof of Knowledge compiler (called an IPoK) which, for any NP relation $\mathcal{R}$ and any ℓ polynomial in the security parameter, produces a protocol that is an ℓ-IPoK for R. In addition, the protocol is *witness indistinguishable (WI)*. This means that for any malicious verifier V^*, and any two pairs $(x, w_1) \in \mathcal{R}$, $(x, w_2) \in \mathcal{R}$ the verifier cannot distinguish between a prover that uses the witness w_1 and a prover that uses the witness w_2, even when given w_1 and w_2. Formally, letting $\text{EXEC}(P(x, w), V(x))$ denote the transcript of the execution between P and V where P uses the witness w for the instance x, we require that for any PPT cheating verifier V^*

$$(\text{EXEC}(P(x, w_1), V^*(x)), w_1, w_2) \approx (\text{EXEC}(P(x, w_2), V^*(x)), w_1, w_2)$$

Setup: First the environment $\mathcal{E}$ is run to produce x which it sends to P^* and V. At this stage P^* and $\mathcal{E}$ can communicate arbitrarily.

Execution: Then for $r = 1, \ldots, \rho$ the verifier V is activated to produce a message $v^{(r)}$ that is input to P^* which is activated to produce a message $p^{(r)}$ that is input to V. In addition, P^* can at any point send a message y to $\mathcal{E}$ and receive a response z from $\mathcal{E}$. However, the total number of bits sent and the total number of bits received during the execution stage are both bounded by ℓ. At the conclusion of the ρ rounds, the verifier V produces a judgment $J \in \{\texttt{accept}, \texttt{reject}\}$.

Extraction: If $J = \texttt{reject}$ then the extractor $\mathcal{X}$ wins the extraction game. Otherwise, we construct the view σ to be the description of P^*, its initial random tape, the messages $v^{(r)}, p^{(r)}$ exchanged between P^* and V, and the transcript of the communication between P^* and $\mathcal{E}$. We let $w = \mathcal{X}(\kappa, \sigma)$. If $w \in W_R(x)$, then $\mathcal{X}$ wins the game; otherwise it looses.

Fig. 2. Knowledge soundness extraction game

This notion is significantly weaker than zero knowledge (ZK) but [DNW08] shows that one cannot achieve ZK without isolating the verifier as well and hence we will have to rely on witness indistinguishability only.

4   Construction

We use the results of [CLOS02] which show that one can UC-realize arbitrary MPC given the ideal functionality for multiple commitments $\mathcal{F}_{\mathrm{MCOM}}$ which we review in Fig. 3.

Commit Phase: On input $(\texttt{commit}, sid, ssid, P_j, m)$ from P_i, if there is already a stored tuple of the form $(sid, ssid, P_i, P_j, \cdot)$ then ignore the command. Otherwise store the tuple $(sid, ssid, P_i, P_j, m)$ and send a receipt $(\texttt{receipt}, sid, ssid, P_i)$ to P_j.

Reveal Phase: On input $(\texttt{reveal}, sid, ssid, P_j)$ from P_i, if a tuple $(sid, ssidP_i, P_j, m)$ is stored then send a message $(\texttt{reveal}, sid, ssid, P_i, m)$ to P_j. Otherwise, ignore the command.

Fig. 3. The $\mathcal{F}_{\mathrm{MCOM}}$ Ideal Functionality

There are several challenges in UC realizing the $\mathcal{F}_{\mathrm{MCOM}}$ functionality. Obviously, we need a commitment scheme which is hiding and binding. In addition, the simulator needs to be able to generate commitments for honest parties before knowing the message being committed to and later be able to decommit to any specified message. A scheme with this property is called *equivocal*. For adaptive security, the simulator needs to be able to simulate the corruption of an honest party and thus reveal all of the randomness used to generate such simulated commitments as though they were generated honestly. We call this *strong equivocality*. The simulator also needs to extract the message contained in any valid commitment even if it was adversarially generated. This is called *extractability*.

Luckily, the result of [CLOS02] contains just such a scheme. It relies on two public keys, an extraction key pk_X and an equivocation key pk_E, that are generated randomly and placed in a CRS. The corresponding secret keys, which are known by the simulator but not the players in the real world, give it the power of strong equivocality and extractability. It was already noticed in [BCNP04] that the players can choose these keys themselves. A sender uses his extraction key and the receiver's equivocation key to generate commitment.

We use the basic idea of [BCNP04] but modify it to fit our setting. Firstly, if the honest sender knows his own extraction secret key (and cannot erase it) then the adversary learns this key when the sender is corrupted. This allows the adversary to distinguish if previous commitments sent by the sender were generated honestly (as is done in the real world) or if they were equivocated (as is done by the simulator in the ideal world). To get around this issue, we have the sender and receiver do a coin-flip to generate the extraction public key so that neither party knows the corresponding secret key. To simulate the coin-flip, it is enough to have a strongly equivocal commitment scheme (i.e., no extraction is needed) and so players only register their equivocation public keys. The second problem arises from the fact that the sender's commitments can only be extracted (and in general are only binding) when the sender has no information about the receiver's equivocation key. However, in our setting the adversary gets to run as a verifier in a WI ℓ-IPoK of the equivocation secret key, which might potentially leak useful information. We show how to modify the original scheme so that it remains extractable even with respect to an adversary that sees such proofs.

We begin by formalizing an abstraction which captures the properties achieved by the scheme of [CLOS02]. Then we show how to turn any scheme which has those properties into one that is secure even if the adversary has access to a prover running a WI PoK protocol and using the equivocation secret key as a witness.

4.1 The Commitment Scheme

A *Two-Key Extractable and Strongly Equivocal Commitment Scheme* has two key generation algorithms $(pk_E, sk_E) \leftarrow \mathsf{gen}_E(1^k)$ and $(pk_X, sk_X) \leftarrow \mathsf{gen}_X(1^\kappa)$ for the equivocation and extraction keys respectively. The commitment algorithm takes as input the two public keys and a message m. It produces a commitment $C = \mathsf{commit}^{pk_X}_{pk_E}(m; r)$ using the randomness r. To decommit, the sender simply sends (m, r) and the receiver verifies $C =^? \mathsf{commit}^{pk_X}_{pk_E}(m; r)$.[1] In addition, we need the ability to easily recognize well-formed public key/secret key pairs. For that, we assume that there is an NP relation $\mathcal{R}_E$ which defines well formed equivocation key pairs (pk_E, sk_E), and a relation $\mathcal{R}_X$ that defines well formed extraction key pairs (pk_X, sk_X). We assume that every key pair generated by gen_E (resp. gen_X) is contained in $\mathcal{R}_E$ (resp. $\mathcal{R}_X$) but allow the set of well-formed key pairs to contain other elements.

---

[1] Because we consider adaptive security where the environment can always corrupt the sender to learn all the randomness r used to commit, there is no reason to consider commitment schemes where the decommitment does not consist of sending all this randomness: If the simulator can produce it to simulate a corruption of the sender, it can also produce it to simulate a decommitment.

Extractability. We define an *extraction game* between a challenger and an adversary as follows:

1. The challenger generates random $(pk_E, sk_E) \leftarrow \mathsf{gen}_E(1^\kappa)$ and the adversary is given pk_E.
2. The adversary chooses a pair $(pk_X, sk_X) \in \mathcal{R}_X$, a commitment C and a pair (m', r') and sends these to the challenger.
3. Let $m = \mathsf{extract}_{pk_E}^{(pk_X, sk_X)}(C)$. If $m' \neq m$ and $C = \mathsf{commit}_{pk_E}^{pk_X}(m'; r')$ then the adversary wins the extraction game.

We say that a commitment scheme is extractable if there is a PPT algorithm **extract** such that, for any PPT adversary $\mathcal{A}$, the success probability of $\mathcal{A}$ winning the extraction game is negligible in κ.

Binding. We define a *binding game* between a challenger and an adversary:

1. The challenger generates a random $(pk_E, sk_E) \leftarrow \mathsf{gen}_E(1^\kappa)$ and the adversary is given pk_E.
2. The adversary generates some public key $pk_X \in \{0,1\}^t$. In addition, the adversary specifies a commitment C and two pairs (m, r), (m', r) and sends these to the challenger.
3. The adversary wins the binding game if $m \neq m'$, $C = \mathsf{commit}_{pk_E}^{pk_X}(m; r)$ and $C = \mathsf{commit}_{pk_E}^{pk_X}(m'; r')$.

We say that a commitment scheme is binding if, for any PPT adversary $\mathcal{A}$, the success probability of $\mathcal{A}$ winning the binding game is negligible in κ.

Strong Equivocality/Hiding. We define equivocality by insisting that there is no adversary that can distinguish between the *commitment game* and the *equivocation game* defined below:

The *commitment game* between a challenger and adversary proceeds as follows:

1. The challenger generates a random $(pk_X, sk_X) \leftarrow \mathsf{gen}_X(1^\kappa)$ and gives pk_X to the adversary.
2. The adversary specifies $(pk_E, w_E) \in \mathcal{R}_E$, and a message m.
3. The challenger computes $C = \mathsf{commit}_{pk_E}^{pk_X}(m; r)$ where r is chosen randomly and gives (C, r) to the adversary.

The *equivocation game* between a challenger and adversary as follows:

1. The challenger generates a random $(pk_X, sk_X) \leftarrow \mathsf{gen}_X(1^\kappa)$ and gives pk_X to the adversary.
2. The adversary specifies $(pk_E, w_E) \in \mathcal{R}_E$, and a message m.
3. The challenger computes $(C, \mathsf{aux}) = \mathsf{ecommit}_{pk_E, w_E}^{pk_X}()$, $r \leftarrow \mathsf{equivocate}_{pk_E, w_E}^{pk_X}(C, \mathsf{aux}, m)$ and gives (C, r) to the adversary.

We say that a commitment scheme is *strongly equivocal* if there exists PPT algorithm **ecommit** and PPT algorithm **equivocate** such that no PPT adversary can distinguish between the commitment game and the equivocation game with more than negligible probability.

Fig. 4. Security of a Two Key Extractable and Equivocal Commitment Scheme

Lastly, we require that the extraction keys are dense. More precisely, for $(pk_X, sk_X) \leftarrow \mathsf{gen}_X(1^\kappa)$, the element pk_X is statistically close to a uniformly random element from some $\mathcal{G} = \{0,1\}^t$. We use $\oplus$ to denote bit-wise xor of elements from $\mathcal{G}$. The security properties of the scheme are outlined in Fig. 4. The commitment scheme of [CLOS02] meets our definition.

The observation that the scheme meets the given security requirements was essentially already made in [BCNP04]. For completeness, we include a short description of the scheme in Appendix A.

4.2 Security after WI Proofs

In the security definitions for extractability and binding of the commitment scheme, it is crucial that the adversary has no information about the equivocation secret key sk_E. However, in our protocols the adversary will get to see a witness indistinguishable (WI) proof of knowledge of such a secret key. For this reason, we augment the security definitions for extractability and binding to give the adversary unlimited protocol access to a prover running a WI proof of knowledge for the relation $\mathcal{R}_E$ using the public key pk_E as the instance and the secret key sk_E as a witness. We show how to turn any two-key extractable and strongly equivocal commitment scheme into a scheme that has *security after WI proofs* - i.e. is secure in the above setting.

Assume we have a two-key extractable and strongly equivocal commitment scheme defined by $(\mathsf{gen}_E, \mathsf{gen}_X, \mathsf{commit}, \mathsf{extract}, \mathsf{ecommit}, \mathsf{equivocate})$ and the equivocation key relation $\mathcal{R}_E$. We define a new commitment scheme $(\mathsf{gen}_E', \mathsf{gen}_X', \mathsf{commit}', \mathsf{extract}', \mathsf{ecommit}', \mathsf{equivocate}')$ with equivocation relation $\mathcal{R}_E'$ as follows:

Let gen_E' generate two equivocation keys $(pk_E^{(0)}, sk_E^{(0)}) \leftarrow \mathsf{gen}_E(1^\kappa)$, $(pk_E^{(1)}, sk_E^{(1)}) \leftarrow \mathsf{gen}_E(1^\kappa)$ and let $pk_E' = (pk_E^{(0)}, pk_E^{(1)})$, $sk_E' = sk_E^{(0)}$. We define the relation

$$\mathcal{R}_E' := \left\{ \left(pk_E^{(0)}, pk_E^{(1)}, w_E\right) \ \middle| \ \left(pk_E^{(0)}, w_E\right) \in \mathcal{R}_E \text{ or } \left(pk_E^{(1)}, w_E\right) \in \mathcal{R}_E \right\}.$$

It is clear that this is an NP relation and that $(pk_E', sk_E') = ((pk_E^{(0)}, pk_E^{(1)}), sk_E^{(0)}) \in \mathcal{R}_E'$. We let gen_X' be the same as gen_X so the extraction keys are generated as in the original scheme.

Now assume that the message space is some $\{0,1\}^s$. We use $\oplus$ to denote bitwise xor in $\{0,1\}^s$. To commit to m the sender chooses a uniformly random $m^{(1)}$ and computes $m^{(0)} = m \oplus m^{(1)}$. The sender then computes

$$C^{(0)} = \mathsf{commit}_{pk_E^{(0)}}^{pk_X}(m^{(0)}; r^{(0)}), \quad C^{(1)} = \mathsf{commit}_{pk_E^{(1)}}^{pk_X}(m^{(1)}; r^{(1)}) \qquad (1)$$

and sends the commitment $C = (C^{(0)}, C^{(1)})$.

To open the commitment, the sender sends $(m, r) = (m, (m^{(1)}, r^{(0)}, r^{(1)}))$. The receiver checks that $C^{(0)}, C^{(1)}$ were correctly computed using equation (1).

Equivocality/Hiding. We use a series-of-games argument to show that the above scheme is strongly equivocal. Let us define Game 1 to be the commitment game used in the definition of strong equivocality in Fig. 4.

In Step 2 of the game, the adversary specifies $(pk_E', w_E) = ((pk_E^{(0)}, pk_E^{(1)}), w_E) \in \mathcal{R}_E'$ such that $(pk_E^{(b)}, w_E) \in \mathcal{R}_E$ for $b = 0$ or $b = 1$. In addition, since the relation $\mathcal{R}_E$ is in NP, it is easy to check which is the case (if both, we let $b = 0$). Let $\bar{b} = 1 - b$. We define Game 2 which proceeds as Game 1 except that the challenger computes the commitment by choosing $m^{(\bar{b})}$ randomly and setting $m^{(b)} = m - m^{(\bar{b})}$. Games 1 and 2 have identical distributions and so are indistinguishable.

We define Game 3 which proceeds as Game 2, but the challenger computes $C^{(b)}$ and $r^{(b)}$ using $(C^{(b)}, \text{aux}) = \text{ecommit}^{pk_X}_{pk_E^{(b)}, w_E}()$ and $r^{(b)} = \text{equivocate}^{pk_X}_{pk_E^{(b)}, w_E}(C^{(b)}, m^{(b)}, \text{aux})$. The strong equivocality of the original scheme ensures that Game 2 and 3 are indistinguishable via a simple reduction.

In Game 3, the commitments $C^{(0)}, C^{(1)}$ are computed independently of the message m and hence Game 3 implicitly defines the functions $\text{ecommit}'$ and $\text{equivocate}'$. Since Games 1 and 3 are indistinguishable the equivocality/hiding property holds for the new scheme.

Extractability and Binding after WI Proofs. We show that the extractability property for the new scheme holds even when the adversary has unlimited protocol access to a prover P running a witness indistinguishable proof for the relation $\mathcal{R}_E$. The argument that binding holds as well proceeds in almost exactly the same way and hence we skip it.

Let us assume that there is an adversary $\mathcal{A}'$ which wins the extraction game for the above scheme with non-negligible probability. This time, the adversary is also given protocol access to a prover P running a WI proof for the relation $\mathcal{R}'_E$ using the instance $pk'_E = \left(pk_E^{(0)}, pk_E^{(1)}\right)$ and the witness $sk'_E = sk_E^{(0)}$. We construct an adversary $\mathcal{A}$ which wins the extraction game for the original scheme.

The adversary $\mathcal{A}$ gets a challenge pk_E generated randomly by its challenger. It will pick a bit b at random and choose $(pk_E^{(b)}, sk_E^{(b)}) \leftarrow \text{gen}_E(1^\kappa)$ and set $pk_E^{(1-b)} = pk_E$. Then it sends $pk'_E = (pk_E^{(0)}, pk_E^{(1)})$ as a challenge to $\mathcal{A}'$ and gets back $(pk_X, sk_X) \in \mathcal{R}_X$. Then $\mathcal{A}$ outputs (pk_X, sk_X) to its challenger. (Recall that our construction did not change $\mathcal{R}_X$.) In addition, it will act as a prover for the instance $(pk_E^{(0)}, pk_E^{(1)})$ using the witness $sk_E^{(b)}$. This is different from the original game where the witness $sk_E^{(0)}$ is always used. However, since the proof is WI, the success probability of $\mathcal{A}'$ can be affected at most negligibly.

Next $\mathcal{A}'$ generates some commitment $C = (C^{(0)}, C^{(1)})$ and some decommitment $(m', r') = (m', (m'^{(1)}, r^{(0)}, r^{(1)}))$. Define $m'^{(0)} = m' - m'^{(1)}$. The adversary $\mathcal{A}$ sends $(m'^{(1-b)}, r^{(1-b)})$ to its challenger.

Let
$$m^{(0)} = \text{extract}^{pk_X, sk_X}_{pk_E^{(0)}}(C^{(0)}), \quad m^{(1)} = \text{extract}^{pk_X, sk_X}_{pk_E^{(1)}}(C^{(1)})$$

and $m = m^{(0)} \oplus m^{(1)}$. Then, if $\mathcal{A}'$ wins the extraction game, $m \neq m'$ and so $m^{\hat{b}} \neq m'^{\hat{b}}$ for some $\hat{b} \in \{0, 1\}$. Since b was only used in choosing which witness to use in the WI proof, with probability negligibly close to $1/2$ we have $\hat{b} = 1 - b$. If this is the case, then $\mathcal{A}$ wins the original extraction game. Hence the success probability of $\mathcal{A}$ is negligibly close to half the success probability of $\mathcal{A}'$ which is non-negligible.

4.3 The Protocol

Let $(\text{gen}_S, \text{gen}_R, \text{commit}, \text{extract}, \text{ecommit}, \text{equivocate})$ be a two-key extractable and strongly equivocal commitment scheme secure after WI proofs. Assume the scheme has an equivocation key relation $\mathcal{R}_E$ and that random extraction public keys are statistically close to uniformly random elements from some $\mathcal{G} = \{0,1\}^t$. We use such a scheme to UC realize the $\mathcal{F}_{\text{MCOM}}$ functionality in the $\mathcal{F}_{\text{isolate}}$-hybrid model with isolation parameter ℓ. We label the players involved $P_1, \ldots, P_n$. We also have some certificate authorities $CA_1, \ldots, CA_m$ and some certificate authority trust structure. We specify the protocol in Fig. 5.

In the ideal world, the additional certificate authorities are not involved in the protocol at all. They get no inputs from the environment and receive no outputs. However, in the real world, parties cannot use the commitment functionality without registering their keys first. We model this discrepancy by adding a dummy registration phase to the ideal world functionality. When $\mathcal{F}_{\text{MCOM}}$ gets the input $(\texttt{register}, sid, P_i, CA_k)$ from P_i then, for every P_j in the trust group of CA_k, it sends $(\texttt{certify}, sid, CA_k, P_i)$ to P_j and $(\texttt{registered}, sid, CA_k, P_j)$ to P_i. The adversary decides when these messages are delivered.

The ideal functionality $\mathcal{F}_{\text{MCOM}}$ ignores all request from P_i to commit to P_j until P_i receives the messages $(\texttt{certify}, sid, CA_d, P_j)$ and $(\texttt{registered}, sid, CA_k, P_j)$ for some CA_k, CA_d. This corresponds to the real world where a sender cannot initiate the commitment protocol until he registers a key with some CA_k trusted by the receiver and the receiver registers a key with some CA_d trusted by the sender.

4.4   Outline of Proof of Theorem 1

We now proceed to go over the intuition for how the simulation is performed and why it is indistinguishable. A more complete description and proof appears in the full version of this paper.

We show that for any certificate authority trust structure, any environment $\mathcal{E}$, and any real-world adversary $\mathcal{A}$ attacking the above protocol using a valid corruption strategy, there exists an ideal-world simulator $\mathcal{S}$ such that $\mathcal{E}$ cannot distinguish between interacting with $\mathcal{A}$ in the real-world versus interacting with $\mathcal{S}$ and dummy parties using the ideal functionality $\mathcal{F}_{\text{MCOM}}$. The simulator internally runs a copy of the protocol. The simulator also internally runs a copy of $\mathcal{A}$ and lets $\mathcal{A}$ attack the internal copy of the protocol. It passes messages from $\mathcal{E}$ to its internal copy of $\mathcal{A}$ and outputs from $\mathcal{A}$ to $\mathcal{E}$.

The simulator runs all key registrations honestly, by following the code of **Key Registration** above. In particular, for an honest party P_i, the simulator will pick a key pair $(pk_{(E,i)}, sk_{(E,i)})$ honestly and remember the secret key. For a corrupt P_i, which successfully registers a public keys $pk_{(E,i)}$ with an honest CA_k, the simulator will see the PPT ITM M given by the adversary to the $\mathcal{F}_{\text{isolate}}$ functionality. In addition, M is able to run an ℓ-IPoK for the relation $\mathcal{R}_E$ and the instance $pk_{(E,i)}$ and, by the specification of $\mathcal{F}_{\text{isolate}}$, M communicated at most ℓ bits with its environment during this proof. By the definition of an ℓ-IPoK, this allows $\mathcal{S}$ to extract a witness $w_{(E,i)}$ from M, such that $(pk_{(E,i)}, w_{(E,i)}) \in \mathcal{R}_E$. For public keys $pk_{(E,i)}$ registered by a corrupted P_i at a corrupted CA_k no witness can be computed.

The coin-flipping protocols are simulated in two different ways depending on whether P_j is honest or not. If P_j is honest and accepts the coin-flipping, then it received some message $(\texttt{registered}, sid, CA_k, P_j, P_i, pk_{(E,i)})$ from an authority CA_k trusted by P_j. In addition CA_k was honest since P_j is in the trust group of CA_k

Key Registration: The first step in the protocol is for each party to register a key with every other party. Each party P_i that wants to talk to another party P_j registers a public key with a certificate authority CA_k that is trusted by P_j. Formally, this step happens when P_i gets an input $(\texttt{register}, sid, P_i, CA_k)$. The registration is done as follows:

1. Party P_i chooses an equivocation public/secret key pair $(pk_{(E,i)}, sk_{(E,i)}) \leftarrow \texttt{gen}_E(1^k)$. In addition, P_i generates the PPT ITM M implementing the prover functionality of a WI ℓ-IPoK protocol for the relation $\mathcal{R}_E$ using the instance $pk_{(E,i)}$ and the witness $sk_{(E,i)}$. The random tape of M is initialized with fresh random coins (enough to run one proof). The machine M is set to run a single proof and, at its conclusion, goes into an inactive state in which it produces no further output.
2. The player P_i sends $(\texttt{isolate}, sid, P_i, CA_k, M)$ to the ideal functionality $\mathcal{F}_{\text{isolate}}$ and a key registration request $(\texttt{register}, sid, P_i, CA_k, pk_{(E,i)})$ to CA_k.
3. The authority CA_k, upon receiving $(\texttt{register}, sid, P_i, CA_k, pk_{(E,i)})$ from P_i sends $(\texttt{isolate}, sid, P_i, CA_k)$ to $\mathcal{F}_{\text{isolate}}$. It then runs as a verifier in the ℓ-IPoK protocol by sending challenge messages through the interface provided by $\mathcal{F}_{\text{isolate}}$. At the conclusion of this protocol, CA_k sends $(\texttt{release}, sid, P_i, CA_k)$ to $\mathcal{F}_{\text{isolate}}$.
4. If the conversation is accepting, CA_k sends the message $(\texttt{certify}, sid, CA_k, P_j, P_i, pk_{(E,i)})$ to every player P_j in its trust group.
5. When P_j receives $(\texttt{certify}, sid, CA_k, P_j, P_i, pk_{(E,i)})$ from CA_k it sends $(\texttt{registered}, sid, P_j, P_i, CA_k)$ to P_i.
6. The party P_i ignores all commands instructing it to commit to P_j until it receives a message $(\texttt{registered}, sid, P_j, P_i, \cdot, pk_{(E,i)})$ from P_j and a message $(\texttt{certify}, sid, CA_d, P_i, P_j, pk_{(E,j)})$ from some trusted authority CA_d. Until then, it also ignores all coin-flip requests or commit messages from P_j.

Commitment Setup: The first time that P_i wants to send a commitment to P_j they run a coin-flipping protocol to decide on the extraction key $pk_{(X,i,j)}$. This protocol proceeds as follows:

1. P_i sends a "coin flip request" to P_j.
2. P_j picks a random $g_1 \leftarrow \mathcal{G}$ and a random extraction key $pk_X \leftarrow \mathcal{G}$. It sends pk_X and $C = \texttt{commit}_{pk_{(E,i)}}^{pk_X}(g_1; r)$ to P_i.
3. P_i sends a random $g_2 \leftarrow \mathcal{G}$ to P_j.
4. P_j sends the opening (g_1, r) to P_i and P_i verifies that C was generated correctly as a commitment to g_1. Both parties compute $pk_{(X,i,j)} = g_1 \oplus g_2$.

Commit: Whenever P_i gets input $(\texttt{commit}, sid, ssid, P_j, m)$, it retrieves the key $pk_{(E,j)}$. Then it computes $C = \texttt{commit}_{pk_{(E,j)}}^{pk_{(X,i,j)}}(m; r)$ and sends $(\texttt{commit}, sid, ssid, P_j, C)$ to P_j, which outputs $(\texttt{receipt}, sid, ssid, P_i)$.

Open: If P_i later gets input $(\texttt{reveal}, sid, ssid, P_j)$, then it sends $(\texttt{commit}, sid, ssid, P_j, (m, r))$ to P_j. If $C = \texttt{commit}_{pk_{(E,j)}}^{pk_{(X,i,j)}}(m; r)$, then P_j outputs $(\texttt{reveal}, sid, ssid, P_i, m)$.

Fig. 5. The Commitment Protocol

and P_j is honest. Therefore $\mathcal{S}$ knows $w_{(E,i)}$ such that $(pk_{(E,i)}, w_{(E,i)}) \in \mathcal{R}_E$. The simulator uses $w_{(E,i)}$ to equivocate the commitment sent by P_j. In particular, the simulator uses $\texttt{ecommit}$ in step 2 of the coin-flip protocol. When it receives g_2 from P_i in step 3, it then samples a random key-pair $(pk_{(X,i,j)}, sk_{(X,i,j)})$, lets $g_1 = pk_{(X,i,j)} \oplus g_2$. It then uses the $\texttt{equivocate}$ algorithm in step 4 to open the commitment C to g_2.

This results in a key $pk_{(X,i,j)} = g_1 \oplus g_2$ for which $\mathcal{S}$ knows $sk_{(X,i,j)}$ such that $(pk_{(X,i,j)}, sk_{(X,i,j)}) \in \mathcal{R}_X$. If P_j is corrupted for the coin-flip, then $\mathcal{S}$ simulates P_i by sending a random g_2 as in the protocol. Note that when P_i is honest, then $pk_{(E,i)}$ was picked at random by the simulator and the adversary did not see $sk_{(E,i)}$, except that it saw a WI proof for $sk_{(E,i)}$. Therefore the commitment C sent in step 2 of the coin-flip is computationally binding for P_j, and $pk_{(X,i,j)}$ will have been produced by a Blum coin-flip using a computationally binding commitment scheme. Intuitively it follows that, even if P_j is corrupted, the public key $pk_{(X,i,j)}$ is a random key. Formally, we rely on the "coin tossing lemma" from [CDPW] to argue that strong equivocality/hiding still hold even when the extraction public key $pk_{(X,i,j)}$ is generated using a Blum coin flip protocol as above rather than randomly as in the definition of the commitment and equivocation games in Fig. 4.

The commitments are simulated in two different ways, depending on whether P_j is honest or not. If P_j is honest, then P_j was also honest when $pk_{(X,i,j)}$ was generated. Therefore $\mathcal{S}$ knows a secret key $sk_{(X,i,j)}$ for the key $pk_{(X,i,j)}$ used by P_i. The simulator uses $sk_{(X,i,j)}$ to extract a message m from all commitments C sent by P_i to P_j. By the **Extractability** (Fig. 4) the probability that P_i later opens C to $C = \mathsf{commit}_{pk_{(E,j)}}^{pk_{(X,i,j)}}(m'; r')$ with $m' \neq m$ is negligible. If P_j is corrupted, then $\mathcal{S}$ simulates P_i without knowing m. As above, since P_i is honest and uses the key $pk_{(E,j)}$ to commit to P_j, the simulator knows a witness $w_{(E,j)}$ for the instance $pk_{(E,j)}$ in $\mathcal{R}_E$. The simulator uses $w_{(E,j)}$ to equivocate the commitment. In particular, it computes a commitment without knowing m using $\mathsf{ecommit}$. Later to simulate the opening of the commitment or the corruption of P_i, $\mathcal{S}$ receives m and uses the $\mathsf{equivocate}$ command to compute r which serves as both, an opening of m and an explanation of the randomness used to generate C. By **Strong Equivocality/Hiding** (Fig. 4), it follows that computing r using $\mathsf{equivocate}$ is indistinguishable from the way it is done in the protocol.

References

[BA03]   Barr, K., Asanovic, K.: Energy aware lossless data compression. In: The International Conference on Mobile Systems - MobiSys, San Francisco, CA, USA, pp. 231–244. ACM Press, New York (2003)

[BCNP04]   Barak, B., Canetti, R., Nielsen, J.B., Pass, R.: Universally composable protocols with relaxed set-up assumptions. In: 45th Annual Symposium on Foundations of Computer Science, Rome, Italy, pp. 186–195. IEEE, Los Alamitos (2004)

[BGGL01]   Barak, B., Goldreich, O., Goldwasser, S., Lindell, Y.: Resettably-Sound Zero-Knowledge and its Applications. In: 42nd Annual Symposium on Foundations of Computer Science, Las Vegas, Nevada, pp. 14–17. IEEE, Los Alamitos (2001)

[Can]   Canetti, R.: Universally composable security: A new paradigm for cryptographic protocols. Cryptology ePrint Archive 2000/067

[Can01]   Canetti, R.: Universally composable security: A new paradigm for cryptographic protocols. In: 42nd Annual Symposium on Foundations of Computer Science, Las Vegas, Nevada, pp. 136–145. IEEE, Los Alamitos (2001); Full version in [Can]

[CDPW]   Canetti, R., Dodis, Y., Pass, R., Walfish, S.: Universally Composable Security with Global Setup. Cryptology ePrint Archive 2006/042; Published Version in [CDPW07]

[CDPW07] Canetti, R., Dodis, Y., Pass, R., Walfish, S.: Universally Composable Security with Global Setup. In: Vadhan, S.P. (ed.) TCC 2007. LNCS, vol. 4392, pp. 61–85. Springer, Heidelberg (2007)

[CKL03]   Canetti, R., Kushilevitz, E., Lindell, Y.: On the limitations of universally composable two-party computation without set-up assumptions. In: Biham, E. (ed.) EURO-CRYPT 2003. LNCS, vol. 2656, pp. 68–86. Springer, Heidelberg (2003)

[CLOS02] Canetti, R., Lindell, Y., Ostrovsky, R., Sahai, A.: Universally composable two-party and multi-party secure computation. In: Proceedings of the Thirty-Fourth Annual ACM Symposium on the Theory of Computing, Montreal, Quebec, Canada, pp. 494–503 (2002)

[CGS08]   Chandran, N., Goyal, V., Sahai, A.: New Constructions for UC Secure Computation using Tamper-proof Hardware. In: Smart, N.P. (ed.) EUROCRYPT 2008. LNCS, vol. 4965, pp. 545–562. Springer, Heidelberg (2008)

[DNW08]   Damgård, I., Nielsen, J.B., Wich, D.: Isolated Proofs of Knowledge and Isolated Zero Knowledge. In: Smart, N.P. (ed.) EUROCRYPT 2008. LNCS, vol. 4965, pp. 509–526. Springer, Heidelberg (2008); Full version in Cryptology ePrint Archive 2007/331

[K07]   Katz, J.: Universally composable multi-party computation using tamper-proof hardware. In: Naor, M. (ed.) EUROCRYPT 2007. LNCS, vol. 4515, pp. 115–128. Springer, Heidelberg (2007)

[MS08]   Moran, T., Segev, G.: David and Goliath Commitments: UC Computation for Asymmetric Parties Using Tamper-Proof Hardware. In: Smart, N.P. (ed.) EUROCRYPT 2008. LNCS, vol. 4965, pp. 527–544. Springer, Heidelberg (2008)

A   A Two Key Extractable and Strongly Equivocal Commitment Scheme

In this section we briefly describe the construction of a two-key extractable and strongly equivocal commitment scheme defined in [CLOS02]. Most of the observations here were already made in [BCNP04]. For our purposes we only need to make a slight modification and use a dense public key encryption scheme.

We start with a strongly equivocal, perfectly hiding commitment scheme which is not extractable. For example, we can use the Pedersen commitment scheme which is an efficient scheme based on the DL assumption. Alternatively we can use the Feige-Shamir commitment scheme which is based on the existence of one-way permutations (OWP) alone. This is the approach taken by [CLOS02] where it is shown that a small modification to the Feige-Shamir scheme makes it *strongly* equivocal as well. In the Feige-Shamir scheme, the secret key sk_E is a random string w and the public key pk_E is $f(w)$ where f is some one-way-function. We can define the relation $\mathcal{R}$ as the set of elements $(f(w), w)$ for some one way function f. For *any* such pair, the equivocated commitments and honestly produced commitments have equivalent distributions. The message space of the Feige-Shamir scheme is only 1-bit. The scheme has the property that knowledge of w allows one to create equivocated commitments and openings which are indistinguishable from real commitments and openings even if the adversary knows w as well. However, for an adversary that only sees $f(w)$, the scheme is binding.

To get extractability, we take a strongly equivocal, perfectly hiding commitment scheme and restrict the message space to only 1-bit. Then we use a dense public key

CPA secure encryption scheme ($\mathsf{gen}, \mathsf{Enc}, \mathsf{Dec}$) where the ciphertexts are pseudorandom elements in some easily sampleable range $\mathcal{C}$ and where each ciphertext has only one valid decryption for any public key. To commit to a bit b, the sender computes $C_{com} = \mathsf{commit}_{pk_E}(b; r_{com})$, $C_b = \mathsf{Enc}_{pk_X}(r_{com}; r_{enc})$ and $C_{1-b} \leftarrow \mathcal{C}$ and send the commitment $C = (C_{com}, C_0, C_1)$.

To equivocate using the secret key sk_E we simply compute $(C_{com}, \mathsf{aux}) \leftarrow \mathsf{ecommit}_{pk_E, sk_E}()$, $r_{com}^{(0)} \leftarrow \mathsf{equivocate}_{pk_E, sk_E}(C_{com}, \mathsf{aux}, 0)$, $r_{com}^{(1)} \leftarrow \mathsf{equivocate}_{pk_E, sk_E}(C_{com}, \mathsf{aux}, 1)$ and $C_0 = \mathsf{Enc}(r_{com}^{(0)}; r_{enc}^{(0)})$, $C_1 = \mathsf{Enc}(r_{com}^{(1)}; r_{enc}^{(1)})$. To equivocate to a bit b send $(b, r_{com}^{(b)}, r_{enc}^{(b)})$. It is easy to see that equivocality is preserved because of CPA-security of the encryption scheme and the pseudorandomness of the ciphertexts.

The extractability property holds because the values C_0, C_1 define the encrypted messages $r_{com}^{(0)}, r_{com}^{(1)}$ which can be decrypted using the encryption secret key. If both equations $C_{com} = \mathsf{commit}_{pk_E}(0; r_{com}^{(0)})$ and $C_{com} = \mathsf{commit}_{pk_E}(1; r_{com}^{(1)})$ hold, then the adversary breaks the computational binding property of the original equivocal commitment scheme. If only one such equation holds, say for the bit b, then b is the extracted message and the committer cannot produce a decommitment for $1 - b$. This is true even if the adversary knows the decryption key sk_X. Similarly, binding holds because of the computational binding property of the original strongly equivocal commitment scheme.

Oblivious Transfer from Weak Noisy Channels

Jürg Wullschleger

University of Bristol, UK
j.wullschleger@bristol.ac.uk

Abstract. Various results show that oblivious transfer can be implemented using the assumption of *noisy channels*. Unfortunately, this assumption is not as weak as one might think, because in a cryptographic setting, these noisy channels must satisfy very strong security requirements.

Unfair noisy channels, introduced by Damgård, Kilian and Salvail [Eurocrypt '99], reduce these limitations: They give the adversary an unfair advantage over the honest player, and therefore weaken the security requirements on the noisy channel. However, this model still has many shortcomings: For example, the adversary's advantage is only allowed to have a very special form, and no error is allowed in the implementation.

In this paper we generalize the idea of unfair noisy channels. We introduce two new models of cryptographic noisy channels that we call the *weak erasure channel* and the *weak binary symmetric channel*, and show how they can be used to implement oblivious transfer. Our models are more general and use much weaker assumptions than unfair noisy channels, which makes implementation a more realistic prospect. For example, these are the first models that allow the parameters to come from experimental evidence.

1 Introduction

Secure two-party computation, introduced in [23], allows two mutually distrustful players to calculate a function in a secure way. This means that both players get the correct output, but nothing more than that. Even though secure two-party computation is generally impossible without any further assumption, it has been shown in [11,14] that if a very simple primitive called *oblivious transfer* is available, then any two-party computation can be implemented in an unconditionally secure way.

Oblivious transfer was first defined in [21], however without realizing its connection to cryptography. In the cryptographic context, the two variants of oblivious transfer were defined in [19] and [9], which were shown to be equally powerful in [3]. Throughout this work, we will only consider *chosen one-out-of-two oblivious transfer*, or OT for short. Here, a sender can send two message bits x_0 and x_1, and a receiver can choose which of the two messages he wants to receive by sending a choice bit c. He receives x_c, but does not get to know the other message bit x_{1-c}, and the sender does not get to know the choice bit c. There exist various implementations of OT that are secure against computationally bounded

O. Reingold (Ed.): TCC 2009, LNCS 5444, pp. 332–349, 2009.

adversaries, under various hardness assumptions. Against adversaries with unbounded computational power, OT can only be implemented if the players have
access to an additional (weaker) functionality.

1.1 OT from (Unfair) Noisy Channels

In [5], it has been shown that OT can be implemented from various weaker
forms of OT, as well as *noisy channels*. Therefore, noise is not always a bad
thing; in a cryptographic context it can become a valuable resource. These protocols have later been improved and generalized in [4], [6] and [16]. The basic
idea of all these protocols is very similar: First, they construct some kind of
erasure channel. Then, this erasure channel is used many times to implement
OT. The correctness and the security is guaranteed using error correcting codes
and privacy amplification.

These noisy channels seem to be quite weak primitives and easily implementable, but they have some rather strong requirements: The statistics of the
channel must be *exactly* the same in every instance, and known to both players.
And, apart from the output of the channel, a dishonest player must not get *any*
additional output.

In [8], weaker forms of noisy channels called *unfair noisy channels* were introduced. Unfair noisy channels are binary symmetric noisy channels that let the
dishonest player change the error-rate in the channel by a certain amount. For
example, this makes the protocol secure against an adversary that might use
better transmitters or detectors in order to break the protocol. In this model,
OT must be implemented in a different way, using the following two steps. First,
from only a few instances of the channel, a weak form of OT (called WOT) is
constructed. In the second step, the security is amplified, i.e., many of these
WOTs are used to get one secure instance of OT. The resulting protocol is only
secure in the *semi-honest model*, i.e., under the assumption that the dishonest
player follows the protocol. To make the protocol secure in the *malicious model*,
where the dishonest player may deviate in an arbitrary way from the protocol, a
third step is needed, which uses *bit commitments* and *zero-knowledge proofs* to
force the dishonest player to follow the protocol.

The results from [8] were later improved in [7], and OT amplification was
improved in [22].

1.2 Limitations of Unfair Noisy Channels

Even though unfair noisy channels are much weaker than (fair) noisy channels,
they still have some very strong assumptions, which makes them hard to implement. Let us look at the following example:

A (fair) binary symmetric noisy channel with error ε lets a sender input a
bit $x \in \{0,1\}$. The channel then outputs a value $Y \in \{0,1\}$ to the receiver,
where $\Pr[Y \neq x] = \varepsilon$. Let us assume that neither the sender nor the receiver can
influence ε, but that the dishonest receiver gets an additional value $E \in \{0,1\}$,
where $\Pr[E = 1] = \mu$, and $E = 1 \Rightarrow Y = x$. Therefore, with some probability μ,

the dishonest receiver gets to know that the value Y he received is in fact equal to x. If μ is small, then this channel is very close to a fair binary symmetric noisy channel. However, even then it cannot be modeled by an unfair noisy channel, because there, the receiver can only change the error probability of the channel in a certain range, but he can never be sure that his received bit is the bit sent by the sender. Therefore, unfair noisy channels forbid the adversary to have this kind of advantage.

Now, let us assume that we are given an implementation a noisy channel, and that the statistics of the channel show that the channel behaves like a fair noisy channel. The accuracy of these statistics are only *polynomial*. For example, the channel might as well be the channel from the example above, where μ is only polynomially small. Therefore, we cannot conclude that the channel is really a implementation of a fair noisy channel, and neither can the channel be modeled by an unfair noisy channel. To be able to implement OT in this situation, we need to have a model that allows the implementation to behave *arbitrarily* with some probability.

1.3 Contribution

The goal of this work is to present new, more realistic models for noisy channels (called *weak noisy channels*) and to show that oblivious transfer can be implemented in an unconditionally secure way, assuming that such weak noisy channels exist. Opposed to the unfair noisy channel which is defined as an ideal functionality, our definitions are merely a list of conditions that a implementation of a weak noisy channels should satisfy. For a given implementation, one only needs to check these (quite simple) conditions, and does not need to show a cryptographically secure reduction of an ideal functionality to the implementation. This makes our model easier to apply.

We will introduce the following three models of weak noisy channels:

- *Weak erasure channels in the semi-honest model* [1] *(PassiveWEC)*. These are weak variants of erasure channels[2] (channels that transmit a bit with some probability).
- *Weak binary symmetric channels in the semi-honest model (PassiveWBSC)*. As the unfair noisy channel, these are weak variants of binary symmetric channels.
- *Weak erasure channels in the malicious model (ActiveWEC)*.

We defined these channels such that they fitted well into the protocol proposed in [8], which is the only protocol we know of to implement OT from weak noisy channels.

To show the flexibility and generality of our models, we show that it is very easy to implement a PassiveWEC from *Gaussian channels*, and that the (passive) unfair noisy channel can be seen as an instance of a PassiveWBSC.

---

[1] See Section 2.2 for explanation of the semi-honest and the malicious model.

[2] Note that the original definition of OT by Rabin [19] is in fact also an erasure channel, so a WEC is also a weak form of Rabin OT.

In Sections 3 and 4, we show that PassiveWEC implies WOT, and PassiveWBSC implies PassiveWEC in the semi-honest model. Then, in Section 5, we show that ActiveWEC implies both bit commitment and a committed version of PassiveWEC in the malicious model. This implies that in a certain range of parameters, each of the three weak noisy channels allows for any secure two-party computation to be achieved. For each of the weak noisy channels, we also present a *simulation* of the channel using nothing else than noiseless communication, and in the case of ActiveWEC, shared randomness. Since it is impossible to implement bit commitment or oblivious transfer from noiseless communication and shared randomness, it is also impossible to implement them using the simulated weak noisy channels.

Full proofs are provided in the full version of this work.

2 Preliminaries

We start with some basic definitions and lemmas that we will need later.

We will use the following convention: Lower case letters will denote fixed values and upper case letters will denote random variables. Calligraphic letters will denote sets and domains of random variables. For a random variable X over $\mathcal{X}$, we denote its distribution by $P_X : \mathcal{X} \to [0,1]$ with $\sum_{x \in \mathcal{X}} P_X(x) = 1$. For a given distribution $P_{XY} : \mathcal{X} \times \mathcal{Y} \to [0,1]$, we write for the marginal distribution $P_X(x) := \sum_{y \in \mathcal{Y}} P_{XY}(x,y)$ and, if $P_Y(y) \neq 0$, $P_{X|Y}(x \mid y) := P_{XY}(x,y)/P_Y(y)$ for the conditional distribution. Let $h(x) := -x \log x - (1-x) \log(1-x)$ be the binary entropy function.

2.1 Statistical Distance and Maximal Bit-Prediction Advantage

The *statistical distance* of two distributions P_X and P_Y over the same domain $\mathcal{U}$ is defined as

$$\delta(P_X, P_Y) := \frac{1}{2} \sum_{u \in \mathcal{U}} |P_X(u) - P_Y(u)| .$$

For a distribution P_{XY} over $\{0,1\} \times \mathcal{Y}$, the *maximal bit-prediction advantage* of X from Y for a function f is defined as

$$\mathrm{PredAdv}(X \mid Y) := 2 \cdot \max_f \Pr[f(Y) = X] - 1 .$$

Lemmas 1, 2 and 3 give some intuition about these measures: The random variable B (or C) indicates that an error occurred: If $B = 0$, everything is fine. But if $B = 1$, the adversary may have complete knowledge. (See also [12] and [22].)

Lemma 1. *Let P_{BX} and P_{CY} be distributions over $\{0,1\} \times \mathcal{U}$ such that $\Pr[B = 1] = \Pr[C = 1] = \varepsilon$. Then $\delta(P_X, P_Y) \leq \varepsilon + (1 - \varepsilon) \cdot \delta(P_{X|B=0}, P_{Y|C=0})$.*

Lemma 2. *Let P_{XY} be a distribution over $\{0,1\} \times \mathcal{Y}$. There exists a conditional distribution $P_{B|XY}$ over $\{0,1\} \times \{0,1\} \times \mathcal{Y}$ such that $\Pr[B = 1] \leq \mathrm{PredAdv}(X \mid Y)$ and such that for all functions $f : \mathcal{Y} \to \{0,1\}$, $\Pr[f(Y) = X \mid B = 0] = 1/2$.*

Lemma 3. *Let P_{XY} be a distribution over $\{0,1\} \times \mathcal{Y}$ with $\delta(P_{Y|X=0}, P_{Y|X=1}) \leq \varepsilon$. There exists a random variable B over $\{0,1\}$ such that $\Pr[B = 1 \mid X = 0] = \Pr[B = 1 \mid X = 1] = \varepsilon$, and $P_{Y|X=0,B=0} = P_{Y|X=1,B=0}$.*

Lemma 4. *Let P_{XY} be a distribution over $\{0,1\} \times \mathcal{Y}$ with $\delta(P_{Y|X=0}, P_{Y|X=1}) \leq b$ and $\mathrm{PredAdv}(X) \leq a$. Then $\mathrm{PredAdv}(X \mid Y) \leq 1 - (1-a)(1-b)$.*

We say that X is ε-close to uniform with respect to Y, if $\delta(P_{XY}, P_U P_Y) \leq \varepsilon$, where P_U is the uniform distribution.

2.2   Adversaries

We distinguish between two different models, the *semi-honest model* and the *malicious model*. In the *semi-honest model*, the adversary is *passive*, which means that he follows the protocol, but may try to get additional knowledge from the messages received. In the malicious model, the adversary is *active*, which means that he may change his behavior in an arbitrary way.

2.3   Randomized Functionalities

All our channels are randomized, because we think the security conditions tend to be more intuitive this way. But randomized channels are usually also easier to implement (See for example Protocol ActiveToPassiveWEC in Section 5.3). The results for randomized channels immediately imply similar conditions for non-randomized channel, as they can be converted into a randomized channel simply by requiring the players to choose their inputs at random. Our definitions are weak enough that this works even in the malicious case.

2.4   Oblivious Transfer Amplification

Our work is based on *oblivious transfer amplification* from [8,22], which gives a way to implement oblivious transfer (OT) from weak oblivious transfer (WOT). We will take the definition of WOT from the full version of [22], however we use the weaker requirement of $\mathrm{PredAdv}(C \mid U) \leq p$ instead of $\mathrm{PredAdv}(C \mid U, E) \leq p$. As explained there, the reduction of OT to WOT still works for this weaker definition, as long as the error correction is always done from the sender to the receiver, which is normally the case.

Definition 1 (WOT, semi-honest model). *A weak (randomized) oblivious transfer, denoted by (p, q, ε)-WOT, is a primitive between a sender and a receiver, that outputs (X_0, X_1) to the honest sender and (C, Y) to the honest receiver. Let U be the additional auxiliary output[3] to a dishonest sender and let V be the auxiliary output to a dishonest receiver. Let $E := X_C \oplus Y$. The following conditions must be satisfied:*

---

[3] Or the *view* of the adversary, i.e., everything he knows at the end of the protocol.

- *Correctness:* $\Pr[E = 1] \leq \varepsilon$.
- *Receiver Security:* $\mathrm{PredAdv}(C \mid U) \leq p$.
- *Sender Security:* $\mathrm{PredAdv}(X_{1-C} \mid V, E) \leq q$.

Theorem 1 ([22]). *Let p, q and ε be constants such at least one of the following conditions holds:*

$$p+q+2\varepsilon \leq 0.24 \,, \quad 22q+44\varepsilon < 1-p \,, \quad 22p+44\varepsilon < 1-q \,, \quad 49p+49q < (1-2\varepsilon)^2 \,,$$

$$q = 0 \wedge p < (1 - 2\varepsilon)^2 \,, \qquad p = 0 \wedge q < (1 - 2\varepsilon)^2 \,, \qquad \varepsilon = 0 \wedge p+q < 1 \,.$$

Then there exists a protocol that efficiently implements OT from (p, q, ε)-WOT secure in the semi-honest model.

We will only use the first four bounds, because we assume that $p, q, \varepsilon > 0$.

2.5 Bit Commitment

To achieve oblivious transfer in the malicious model, we will need *bit commitments*. A bit commitment scheme is a pair of protocols, a **Commit** protocol and a **Open** protocol, executed between a committer and a receiver. The players first execute the Commit protocol, where the committer has an input b. Then, they may also execute the Open protocol. After the Open protocol, the receiver either accepts or rejects. If he accepts, he gets a value b'. The protocols are ε-secure, if they satisfy the following properties:

- *Correctness*: If both players follow the protocols, then the receiver rejects with a probability smaller than ε, and if he accepts, he outputs $b' = b$ with probability at least $1 - \varepsilon$.
- *Binding*: If the receiver is honest, then for any malicious sender, with probability $1 - \varepsilon$, there exists at most one value after the commit protocol that the receiver will accept with a probability bigger than ε in the open phase.
- *Hiding*: If the committer is honest, then no malicious receiver gets to know b with a probability bigger than ε.[4]

3 Weak Erasure Channel in the Semi-honest Model

In this section, we present a reduction of WOT to *weak erasure channels (WEC)* in the semi-honest model. A weak erasure channel lets a honest sender send a bit, which is then received by the honest receiver with a certain probability, and gets lost otherwise. Dishonest players are allowed to receive some additional information, so a dishonest receiver may get to know some information about the input even in the case where the channel lost the bit, and a dishonest sender may get information about whether the bit has been lost or not.

---

[4] This means that if $b \in \{0, 1\}$ and V is the receiver's view, then we require that
$\delta(P_{V|B=0}, P_{V|B=0}) \leq \varepsilon$.

Definition 2 (WEC, semi-honest model). $(d_0, d_1, p, q, \varepsilon)$-*PassiveWEC is a primitive where the honest sender has output* $X \in \{0, 1\}$ *and the honest receiver has output* $Y \in \{0, 1, \Delta\}$. *Furthermore, the dishonest sender may receive an additional value* U, *and the dishonest receiver may receive an additional value* V. *These values must satisfy the following conditions:*

- Correctness: $\Pr[Y = \Delta] \in [d_0, d_1]$, $\Pr[Y \neq X \mid Y \neq \Delta] \leq \varepsilon$.
- Receiver Security: $\delta(P_{XU|Y \neq \Delta}, P_{XU|Y=\Delta}) \leq p$.
- Sender Security: $\text{PredAdv}(X \mid V, Y = \Delta) \leq q$.

The parameters can be interpreted as follows: d_0, d_1 and ε are parameters of the honest players. The probability that the output of the channel is Δ is in the interval $[d_0, d_1]$. (Defining this as an interval gives some freedom to the implementation, which may be important, as parameters often cannot be known precisely.) ε is the probability that the output of the honest receiver is wrong, if the output is not Δ. According to Lemma 2, q is the probability that the dishonest receiver gets to know the input of the channel, given that the output of the channel is Δ, and according to Lemma 3, p is the probability that a dishonest sender gets to know whether $Y = \Delta$ or $Y \neq \Delta$.

3.1   Simulation of PassiveWEC

We start by showing for which values a PassiveWEC can be simulated by only using noiseless communication. Since OT cannot be implemented from noiseless communication, such PassiveWEC therefore cannot be used to implement OT. Note that in any simulation that only uses noiseless communication, we always have $d_0 = d_1$, as both players know all the probabilities. In the following simulation, we require that $\varepsilon \in [0, \frac{1}{2}]$, $d, g \in [0, 1]$, and $g \geq (1 - 2\varepsilon)(1 - d)$.

Protocol SimWEC(d, ε, g)

1. The sender chooses x uniformly at random and sends the receiver $m := x$ with probability g, and $m := \Delta$ otherwise. The sender outputs x.
2. If the receiver gets $m \in \{0, 1\}$, he outputs $y := m$ with probability $\frac{(1-2\varepsilon)(1-d)}{g}$, and $y := \Delta$ otherwise.
3. If the receiver gets $m = \Delta$, he outputs y chosen at random with probability $\frac{2\varepsilon(1-d)}{1-g}$, and $y := \Delta$ otherwise.

Theorem 2. *For any* d, ε, p *and* q, *where* $p + q + 2\varepsilon \geq 1$, $(d, d, p, q, \varepsilon)$-*PassiveWEC is simulatable in the semi-honest model.*

3.2   WOT from PassiveWEC

Protocol PassiveWECtoWOT

1. The sender and the receiver execute PassiveWEC twice. The sender receives (x_0, x_1), the receiver (y_0, y_1).

2. If there exists a c, such that $y_c \neq \Delta$ and $y_{1-c} = \Delta$, then the receiver sets $y := y_c$, outputs (c, y), tells the sender to terminate the protocol and terminates.

3. If the sender receives the message to terminate the protocol, he outputs (x_0, x_1) and terminates. Otherwise, they restart the protocol.

Theorem 3. *Protocol* PassiveWECtoWOT *securely implements a*

$$\left(1 - \frac{2d_0(1 - d_1)}{d_1(1 - d_0) + d_0(1 - d_1)}(1 - p)^2, q, \varepsilon\right)\text{-WOT}$$

secure against passive adversaries out of $(d_0, d_1, p, q, \varepsilon)$-PassiveWEC. *The expected number of instances used is at most* $1/\min(2d_0(1 - d_0), 2d_1(1 - d_1))$.

Theorem 3 is not difficult to show using Lemma 3 and Lemma 4. Corollary 1 follows now from Theorem 1, Theorem 3 and

$$1 - 2d_0(1 - d_1)w(1 - p)^2 \leq 2p + (d_1 - d_0)w .$$

Corollary 1. *Let* $d_0 \leq d_1$, p, q *and* ε *be constants, and let* $w = 1/(d_1(1 - d_0) + d_0(1 - d_1))$. *If at least one of the conditions*

$$2p + q + (d_1 - d_0)w + 2\varepsilon \leq 0.24 , \quad 11q + 22\varepsilon < d_0(1 - d_1)w(1 - p)^2 ,$$

$$44p + 22(d_1 - d_0)w + 44\varepsilon < 1 - q , \quad 98p + 49q + 49(d_1 - d_0)w < (1 - 2\varepsilon)^2$$

holds, then there exists a protocol that uses $(d_0, d_1, p, q, \varepsilon)$-PassiveWEC *and efficiently implements OT secure in the semi-honest model.*

3.3   An Example: The Gaussian Channel

The *Gaussian channel* is often used in information theory as a model of a noisy channel, because it models real physical channels quite well. It has been shown that a perfect and fair Gaussian channel implies bit commitment, see [17,18]. A Gaussian channel is a channel where the sender has input $x_g \in \mathbb{R}$ and the receiver has output $Y_g = x_g + E_g$, where $E_g \sim \mathcal{N}(0, 1)$, i.e., the channel has an additive error that is normal distributed.

We can easily implement a PassiveWEC from this channel in the following way: Let $a, b \in \mathbb{R}^+$. The sender chooses $x \in \{0, 1\}$ uniformly at random, sends $x_g := (2x-1)a$ and outputs x. The receiver gets y_g, and outputs $y = \Delta$ if $|y_g| \leq b$, $y = 1$ if $y_g > b$ and $y = 0$ otherwise. With an arbitrary small error, we can make the Gaussian channel discrete. In the limit, we get a $(d, d, p, q, \varepsilon)$-PassiveWEC, where $d = \Phi(b - a) - \Phi(-a - b)$, $\varepsilon = \frac{\Phi(-a-b)}{1-d}$, $p = 0$ and $q = 2\frac{\Phi(b-a)-\Phi(-a)}{d} - 1$. Choosing for example $a = 1$ and $b = 2.5$, we get $d \approx 0.93296$, $\varepsilon \leq 0.0035$, and $q \leq 0.6604$. Since $44 \cdot \varepsilon < 1 - q$, it follows from Corollary 1 that oblivious transfer can be implemented. Together with the bit commitment protocols from [17,18], this implies (using a protocol similar to ActiveToPassiveWEC) that OT can be implemented from (perfect and fair) Gaussian channels in the malicious model.

To the best of our knowledge, this has not been known before, as previous results in [6,16] rely on the fact that the channel is discrete and cannot be applied to the Gaussian channel. Note that in contrast to the reductions from [17,18], our reduction even works for Gaussian channels that are neither perfect nor fair.

4   Weak Binary Symmetric Channel in the Semi-honest Model

Weak Binary Symmetric Channel is a weak form of a binary symmetric channel. The channel transmits the input bit of the sender to the receiver, but flips the bit with some probability. Again, the definition is randomized.

Definition 3 (WBSC, semi-honest model). $(\varepsilon, \varepsilon_0, \varepsilon_1, p, q)$-*PassiveWBSC is defined as follows: The honest sender has output $X \in \{0,1\}$ and the honest receiver has output $Y \in \{0,1\}$. Furthermore, the dishonest sender may receive an additional value $U \in \mathcal{U}$, and the dishonest receiver may receive an additional value $V \in \mathcal{V}$. These values must satisfy the following conditions:*

- Correctness: $\Pr[X = 0] \in [\frac{1-\varepsilon}{2}, \frac{1+\varepsilon}{2}]$, and for $x \in \{0,1\}$, $\Pr[Y \neq x] \in [\varepsilon_0, \varepsilon_1]$.
- Receiver Security: $\delta(P_{UX|Y=X}, P_{UX|Y\neq X}) \leq p$.
- Sender Security: *For all $y \in \{0,1\}$: $\delta(P_{V|X=0,Y=y}, P_{V|X=1,Y=y}) \leq q$.*

The parameters can be interpreted as follows: ε is the bias of X, and ε_0 and ε_1 define the error interval of the honest players. From Lemma 3 it follows that p is the probability that the sender, and q is the probability that the receiver gets to know whether $X = Y$ or not. Note that in order to make our reduction work, the sender security has a slightly different form than the receiver security. If $\varepsilon = 0$ and $\varepsilon_0 = \varepsilon_1$, the sender security implies $\delta(P_{VY|Y=X}, P_{VY|Y\neq X}) \leq q$. So in this case, the sender security is strictly stronger than the receiver security.

4.1   Simulation of PassiveWBSC

The following simulation is basically the same as in [8] for the unfair noisy channel. Let $\varepsilon_A, \varepsilon_B \in [0, \frac{1}{2}]$.

Protocol SimWBSC($\varepsilon_A, \varepsilon_B$)

1. The players toss a uniform coin $M \in \{0,1\}$.
2. The sender calculates $X := 1 - M$ with probability ε_A and $X := M$ otherwise, and outputs X.
3. The receiver calculates $Y := 1 - M$ with probability ε_B and $Y := M$ otherwise, and outputs Y.

Theorem 4. *Let $\varepsilon := \varepsilon_A(1 - \varepsilon_B) + \varepsilon_B(1 - \varepsilon_A)$, $p := \frac{(1-\varepsilon_A)(1-\varepsilon_B)}{1-\varepsilon} - \frac{\varepsilon_A(1-\varepsilon_B)}{\varepsilon}$, and $q := \frac{(1-\varepsilon_A)(1-\varepsilon_B)}{1-\varepsilon} - \frac{(1-\varepsilon_A)\varepsilon_B}{\varepsilon}$. The Protocol SimWBSC($\varepsilon_A, \varepsilon_B$) securely implements a $(0, \varepsilon, \varepsilon, p, q)$-PassiveWBSC in the semi-honest model.*

Theorem 4 implies that $(0, \varepsilon, \varepsilon, p, q)$-PassiveWBSC is simulatable if $p + q > 1$.

4.2 PassiveWEC from PassiveWBSC

We will now give a reduction of PassiveWEC to PassiveWBSC. The protocol itself has already been used in [5] and [4]. The intuition behind the following protocol is simple: The sender sends a bit twice over a binary noisy channel. If the receiver gets twice the same message, he knows (with a small error) what the sender has sent and outputs that. If he receives two different messages, he does not know the input and outputs Δ. Note that since two channels are randomized, the sender cannot choose his input, and therefore has to additionally send $x_0 \oplus x_1$.

Protocol PassiveWBSCtoWEC

1. The players execute PassiveWBSC twice. The sender gets (x_0, x_1), the receiver (y_0, y_1).
2. The sender sends $k := x_0 \oplus x_1$ to the receiver and outputs $x := x_0$.
3. If $y_0 \oplus y_1 = k$, the receiver outputs $y := y_0$. Otherwise, he outputs $y := \Delta$.

Theorem 5. *Let*

$$d_0 := \min(2\varepsilon_0(1 - \varepsilon_0), 2\varepsilon_1(1 - \varepsilon_1)),$$
$$d_1 := \max(2\varepsilon_0(1 - \varepsilon_0), 2\varepsilon_1(1 - \varepsilon_1), \varepsilon_0(1 - \varepsilon_1) + \varepsilon_1(1 - \varepsilon_0)).$$
$$\varepsilon' := \frac{\varepsilon_1 - \varepsilon_0}{\varepsilon_1 + \varepsilon_0 - 2\varepsilon_0\varepsilon_1} - \frac{2\varepsilon}{1 + \varepsilon^2}.$$

Protocol PassiveWBSCtoWEC securely implements a

$$\left(d_0, d_1, 1 - (1 - p)^2, 1 - (1 - \varepsilon')(1 - q)^2, \frac{\varepsilon_1^2}{\varepsilon_1^2 + (1 - \varepsilon_1)^2}\right)\text{-PassiveWEC}$$

in the semi-honest model out of two instances of $(\varepsilon, \varepsilon_0, \varepsilon_1, p, q)$-PassiveWBSC.

Proof (Sketch). It is easy to verify that

$$\Pr[Y \neq X \mid Y \neq \Delta] \leq \frac{\varepsilon_1^2}{\varepsilon_1^2 + (1 - \varepsilon_1)^2}$$

and $\Pr[Y = \Delta] \in [d_0, d_1]$, and the security against a dishonest sender can be shown using Lemma 3 and Lemma 1.

Let V_0 and V_1 be the additional information a dishonest receiver gets in the two executions of the PassiveWBSC. We have $V := (K, V_0, V_1, Y_0, Y_1)$. Using Lemma 3 and Lemma 1 it can be shown that

$$\delta(P_{V_0 V_1 | X=0, K=k, Y_0=y_0, Y_1=y_1}, P_{V_0 V_1 | X=1, K=k, Y_0=y_0, Y_1=y_1}) \leq 1 - (1 - q)^2.$$

We can bound

$$\Pr[X = x \mid Y_0 = y_0, Y_1 = y_1, K = k, Y = \Delta]$$
$$\leq \frac{(1 + \varepsilon)\varepsilon_1 \cdot (1 + \varepsilon)(1 - \varepsilon_0)}{(1 + \varepsilon)\varepsilon_1 \cdot (1 + \varepsilon)(1 - \varepsilon_0) + (1 - \varepsilon)\varepsilon_0 \cdot (1 - \varepsilon)(1 - \varepsilon_1)},$$

from which follows

$$\mathrm{PredAdv}(X \mid Y_0 = y_0, Y_1 = y_1, K = k, Y = \Delta) \leq \frac{\varepsilon_1 - \varepsilon_0}{\varepsilon_1 + \varepsilon_0 - 2\varepsilon_0\varepsilon_1} + \frac{2\varepsilon}{1 + \varepsilon^2} \cdot$$

The statement now follows from Lemma 4. □

4.3   An Example: The Unfair Noisy Channel

The *passive unfair noisy channel* (γ, δ)-PassiveUNC from [8,7] is a special case of a PassiveWBSC, namely a $(0, \delta, \delta, p, p)$-PassiveWBSC, where

$$p := \frac{(1 - \delta)\delta - (1 - \gamma)\gamma}{(1 - 2\gamma)\delta(1 - \delta)} \cdot$$

Note, however, that the bounds that we get using our results are not as good as the bounds from [8,7].

5   WEC in the Malicious Model

The assumption that the adversary is semi-honest and therefore follows the protocol is quite strong and often too strong. As shown in [10], there exist compilers that can convert protocols which are only secure in the semi-honest model into protocols that are also secure in the malicious model. The basic idea is that at the beginning, the players are committed to all the secret data they have, and after every computation step they do, they commit to the newly computed values and show with a zero-knowledge proof that the new committed value contains indeed the correct value, according to the protocol. To implement this in our setting, we need two things: A bit commitment protocol, and a protocol that implements a committed version of the passive weak noisy channel. Hence, for any weak noisy channel in the active model, we need to show that it implies bit commitment and a committed version of either PassiveWEC or PassiveWBSC for parameters that allow us to achieve OT in the semi-honest model. (See also [7] for a more detailed discussion.)

Defining a weak noisy channel in the malicious model turns out to be much more tricky than in the semi-honest model. It is possible to define them in the same way as in the semi-honest model, however we think that this would not give a very realistic model. For example, the dishonest player probably may choose an attack where he does not get the output of the honest player. Therefore, we think that it is preferable to state the security conditions such that the malicious player does not need to get the value of the honest player. In the following we will do this for the WEC. For the WBSC, we were not able to come up with a simple definition.

Definition 4 (WEC, malicious model). $(d_0, d_1, p, g, \varepsilon)$-*ActiveWEC is a primitive with the following properties.*

- *Correctness: If both players are honest, then the sender has output $X \in \{0, 1\}$ and the receiver has output $Y \in \{0, 1, \Delta\}$, where $\Pr[Y = \Delta] \in [d_0, d_1]$ and $\Pr[Y \neq X \mid Y \neq \Delta] \leq \varepsilon$.*

- Receiver Security: *If the receiver is honest, then for all dishonest sender with auxiliary input z and output U, the receiver has output $Y \in \{0, 1, \Delta\}$ where* $\Pr[Y = \Delta] \in [d_0, d_1]$ *and* $\delta(P_{U|Z=z,Y\neq\Delta}, P_{U|Z=z,Y=\Delta}) \leq p$.
- Sender Security: *If the sender is honest, then for all dishonest receiver with auxiliary input z and output V, the sender has output $X \in \{0,1\}$ and* $\mathrm{PredAdv}(X \mid V, Z = z) \leq g$.

Note that the parameter g is different from the parameter q in the semi-honest case, because we do not condition on the event $Y = \Delta$. The honest receiver can guess X using $f(Y) := Y$ if $Y \neq \Delta$, and either 0 or 1 if $Y = \Delta$. We get $\mathrm{PredAdv}(X \mid Y) \geq (1 - 2\varepsilon)(1 - d_1)$. Therefore, an ActiveWEC can only be implemented if $g \geq (1 - 2\varepsilon)(1 - d_1)$.

5.1 Simulation

Using the same simulation as for the semi-honest case, we get

Theorem 6. *For any d, ε, p and g, where*

$$dp + g + 2\varepsilon \geq 1 \qquad \wedge \qquad g \geq (1 - 2\varepsilon)(1 - d) ,$$

$(d, d, p, g, \varepsilon)$-*ActiveWEC is simulatable in the malicious model, given that the players have access to a source of trusted shared randomness.*

5.2 Bit Commitment

Our commitment protocol takes parameters n, c, m, ℓ and κ, where n is the number of instances used, c the error-tolerance of the protocol, ℓ the number of bits committed to, and κ the error. Let $c := n^{-1/3}$, and

$$\kappa := \exp(-2(1 - d_1 - c)nc^2) .$$

Let a be the maximum value that satisfies

$$(1 - d) \cdot a - \sqrt{\frac{a}{2} \cdot \ln \frac{1}{\kappa}} \leq (\varepsilon + c)(1 - d)n$$

for all $d \in [d_0 - c, d_1 + c]$. Let

$$m := (d_1 p + c)n + 2a + 1$$

and let $\mathcal{C} \subset \{0,1\}^n$ be a (n, k, m)-linear code[5], i.e., with 2^k elements and minimal distance m. Let

$$\ell := k - (g + c) \cdot n - 3 \log(1/\kappa)$$

and n be big enough such that $\ell > 0$. Let H be the parity-check matrix of $\mathcal{C}$ and $g : \mathcal{R} \times \{0,1\}^n \rightarrow \{0,1\}^\ell$ be a 2-universal hash function. In the following protocol, the sender is the committer.

Protocol ActiveWECtoBC
  Commit(b).

- The parties execute ActiveWEC n times. The sender gets $x = (x_0, \ldots, x_{n-1})$, and the receiver gets $y = (y_0, \ldots, y_{n-1})$.

---

[5] Since we do not have to decode $\mathcal{C}$, this could be a random linear code.

- The committer chooses $r \in \mathcal{R}$ uniformly at random and sends it to the receiver.
- The committer sends $s := (H(x), b \oplus g(r, x))$ to the receiver.

Open.

- The committer sends (b, x) to the receiver.
- Let n_Δ be the number of y_i equal to Δ. The receiver checks that $n_\Delta/n \in [d_0 - c, d_1 + c]$ and that the number i where $y_i \neq x_i$ and $y_i \neq \Delta$ is smaller than $(n - n_\Delta)(\varepsilon + c)$. He also checks that $s = (H(x), b \oplus g(r, x))$. If this is the case, he accepts, and rejects otherwise.

In the protocol, the committer has to send the receiver the parity-check of a code, because then the committer cannot guess with probability more than ε more than one value x that passes the test of the receiver in the open phase. The committer extracts a string of size ℓ from x, where ℓ is chosen small enough such that the receiver has almost no information about it.

Theorem 7. *Protocol ActiveWECtoBC implements a commitment with an error of 4κ, out of n instances of $(d_0, d_1, p, g, \varepsilon)$-ActiveWEC.*

The correctness of the protocol follows from the Chernoff/Hoeffding bound. It remains to proof that the protocol is also binding and hiding.

Lemma 5. *Protocol ActiveWECtoBC is binding with probability $1 - 4\kappa$.*

Proof. Let $d := n_\Delta/n$. Let B_i be defined as in Lemma 3. If $Y_i \neq \Delta$, let $Y_i' = Y_i$, and let Y_i' be chosen randomly from $\{0, 1\}$ otherwise, such that $\Pr[Y_i' = 1 \mid Y_i = \Delta] = \Pr[Y_i = 1 \mid Y_i \neq \Delta]$. ($Y_i'$ is therefore independent of the event $Y_i = \Delta$.) Let us assume that the sender additionally receives the values B_i and Y_i'.

We divide the n instances into 3 sets. Let S_0 be the set of values where $B_i = 1 \wedge Y_i = \Delta$, S_1 the set of values where $B_i = 1 \wedge Y_i \neq \Delta$, and S_2 the set of values where $B_i = 0$. The sender may choose a subset of S_1 of size a' and a subset of S_2 of size a, where $x_i \neq y_i'$. It follows from the Chernoff/Hoeffding bound that with probability at least κ, the receiver will notice at least $a \cdot (1 - d) - \sqrt{\frac{a}{2} \cdot \ln \frac{1}{\kappa}}$ of these errors in S_2. Therefore, the receiver will only accept with probability at least κ, if

$$a' + a \cdot (1 - d) - \sqrt{\frac{a}{2} \cdot \ln \frac{1}{\kappa}} \leq (\varepsilon + c)(1 - d)n .$$

The sender would only be able to find two values with the same parity-check if

$$(dp + c)n + 2(a' + a) \geq m .$$

The best strategy for the sender is to choose $a' = 0$, and to make a maximal. It follows from the definition of m that the sender cannot find two such values. The statement follows. $\square$

To proof that the protocol is hiding we need some additional lemmas. The *conditional smooth min-entropy of X given Y* [20] is defined as

$$H_{\min}^{\varepsilon}(X \mid Y) := \max_{\Omega:\Pr[\Omega]\geq 1-\varepsilon} \min_{xy}(-\log P_{X\Omega|Y=y}(x)) .$$

Lemma 6 ([2,15]). $H_{\min}^{\varepsilon+\varepsilon'}(X \mid YZ) \geq H_{\min}^{\varepsilon}(XY \mid Z) - \log|\mathcal{Y}| - \log(1/\varepsilon')$.

Lemma 7 (Leftover hash lemma [1,13]). *Let X be a random variable over $\mathcal{X}$ and let $m > 0$. Let $h : \mathcal{R} \times \mathcal{X} \to \{0,1\}^m$ be a 2-universal hash function. If $m \leq H_{\min}^{\varepsilon}(X \mid Y) - 2\log(1/\varepsilon')$, then for R uniform over $\mathcal{R}$, $h(R, X)$ is $(\varepsilon + \varepsilon')$-close to uniform with respect to (R, Y).*

Lemma 8. *Protocol ActiveWECtoBC is hiding with probability $1 - 3\kappa$.*

Proof. The sender holds $X = (X_1, \ldots, X_n)$, and the receiver $V = (V_1, \ldots V_n)$, S and the auxiliary input z. Using Lemma 2, for every pair (X_i, V_i), there exists a random variable B_i, such that $\Pr[B_i = 1] = g$ and X_i is uniform, given $(V_i, B_i = 0, Z = z)$. From the Chernoff/Hoeffding bound follows that with probability $1 - \kappa$, the number of $B_i = 0$ is at least $n(1 - g - c)$ and therefore $H_{\infty}^{\kappa}(X \mid V, Z = z) \geq n(1 - g - c)$. Using Lemma 6, we get $H_{\infty}^{2\kappa}(X \mid V, S, Z = z) \geq n(1 - g - c) - (n - k) - \log(1/\kappa)$. Finally, we can apply Lemma 7, and get that $g(X, R)$ is 3κ-close to uniform, since $\ell \leq H_{\infty}^{2\kappa}(X \mid V, S, Z = z) - 2\log(1/\kappa)$. This implies that the protocol is hiding with probability $1 - 3\kappa$. □

Note that for any $e > 0$, and $k \leq (1 - h(m/n)) n - e$, a random linear (n, k)-code has a minimal distance of at least m with probability at least $1 - 2^{-e}$. If we choose a random linear code and let $n \to \infty$, then $b/n \to \varepsilon$, and hence $m/n \to d_1 p + 2\varepsilon$. From the property of the random linear code, we get $k/n \to 1 - h(d_1 p + 2\varepsilon)$. We need $\ell > 0$, which is equivalent to $g < k/n$. We get the following corollary.

Corollary 2. *For any d_1, d_1, ε, p and q where*

$$d_1 p + 2\varepsilon < \frac{1}{2}, \quad and \quad g + h(d_1 p + 2\varepsilon) < 1 ,$$

$(d_0, d_1, p, g, \varepsilon)$-ActiveWEC implies bit commitment.

Our bound is optimal for $p = 0 \wedge \varepsilon = 0$. Otherwise, it does not reach the simulation bound, since $h(x) > x$ for all $0 < x < \frac{1}{2}$. It would be interesting to know whether this bound can be improved. Note that it is also possible to implement bit commitment in the other direction. We will leave this to the full version of this work.

5.3   Committed PassiveWEC from ActiveWEC

In the following, we present the protocol to implement a committed version of PassiveWEC in the malicious model, using ActiveWEC. It uses a similar idea already used in [7]: The players execute ActiveWEC n times and commit to their output values. Then, they open all except one that is chosen at random, and check if the statistics are fine. If they are, then with high probability, also the statistics of the remaining instance is fine.

The following lemma is essential to the proof, because it can be used to bound the parameter p for any committed value Y produced by the dishonest receiver, if he passes the test by the honest sender. It is easy to verify that the lemma is tight if V is equal to X with probability p and Δ otherwise.

Lemma 9. *Let P_{XV} be a distribution over $\{0,1\} \times \mathcal{V}$. If $\mathrm{PredAdv}(X \mid V) \leq g$, then for any function $Y = f(V) \in \{0, 1, \Delta\}$ where $\Pr[Y = \Delta] \in [d_0, d_1]$ and $\Pr[Y \neq X \mid Y \neq \Delta] \leq \varepsilon$, we have*

$$\delta(P_{V|X=0,Y=\Delta}, P_{V|X=1,Y=\Delta}) \leq \frac{g - (1 - 2\varepsilon)(1 - d_1)}{d_1} .$$

Proof. Let B be the random variable defined by Lemma 2. We have $\Pr[B = 1] = g$ and $P_{V|X=0,B=0} = P_{V|X=1,B=0}$. Given $B = 0$, V does not have any information about X. Hence, for any $Y = f(V)$, we have

$$\Pr[Y \neq X \mid Y \neq \Delta] \geq \frac{1}{2} \cdot \frac{\Pr[Y \neq \Delta \wedge B = 0]}{\Pr[Y \neq \Delta]} .$$

Therefore, it must hold that $2\varepsilon \Pr[Y \neq \Delta] \geq \Pr[Y \neq \Delta \wedge B = 0]$. We get

$$\begin{aligned}
\Pr[B = 1 \mid Y = \Delta] &= \frac{g - \Pr[Y \neq \Delta] + \Pr[Y \neq \Delta \wedge B = 0]}{\Pr[Y = \Delta]} \\
&\leq \frac{g - (1 - 2\varepsilon)(1 - \Pr[Y = \Delta])}{\Pr[Y = \Delta]} \\
&\leq \frac{g - (1 - 2\varepsilon)(1 - d_1)}{d_1} .
\end{aligned}$$

The statement follows now by applying Lemma 1.   □

In addition to ActiveWEC, our protocol needs bit commitments and coin-tosses. Coin-toss can easily be implemented using bit commitments.

Again, c is the error-tolerance, and κ is the error in the protocol. We choose $c := n^{-1/3}$ and $\kappa := \exp(-2(1 - d_1 - c)nc^2)$. Furthermore, let n be big enough such that $c \geq 1/((1 - d_1 - c)n)$.

Protocol ActiveToPassiveWEC

1. The sender and the receiver execute ActiveWEC n times. The sender gets $(x_0, \ldots, x_{n-1})$, and the receiver $(y_0, \ldots, y_{n-1})$.
2. Both players commit to their values.
3. Using coin-toss, they randomly select one instance s of the n instances.
4. They open all commitments, except for instance s. If any of the players does not accept one opening of a commitment, they abort.

5. Let n_Δ be the number of y_i that is equal to Δ. They check if n_Δ is in the interval $[(d_0 - c) \cdot n - 1, (d_1 + c) \cdot n]$, and the number of y_i that is not equal to Δ nor x_i is smaller than $(\varepsilon + c) \cdot (n - n_\Delta)$. If not, they abort.
6. The sender outputs $x := x_s$, the receiver $y := y_s$.

Theorem 8. *Protocol ActiveToPassiveWEC implements a committed version of*

$$\left(d_0 - 2c, d_1 + 2c, p, \frac{g - (1 - 2\varepsilon)(1 - d_1)}{d_1} + \frac{6}{d_1^2}c, \varepsilon + 2c\right)\text{-PassiveWEC}$$

with an error of at most 3κ in the malicious model. It uses coin-toss, bit commitment and n independent instances of $(d_0, d_1, p, g, \varepsilon)$-ActiveWEC.

Theorem 8 can be shown using the Chernoff/Hoeffding bound and Lemma 9. Note that c is only polynomially small and cannot be made negligible. Here we see an advantage of our definition compared to the PassiveUNC in [8,7]: We do not have to introduce the additional error parameter $p(k)$ as it has to be done for the committed version of the PassiveUNC, nor do we have to add an additional amplification step to the reduction to make this additional error negligible. The following corollary follows from Corollary 1 and Theorem 8.

Corollary 3. *Let $d_0 \leq d_1$, p, g and ε be constants, and let $w := 1/(d_1(1 - d_0) + d_0(1 - d_1))$ and $q := \frac{g - (1 - 2\varepsilon)(1 - d_1)}{d_1}$. If at least one of the conditions*

$$p + q + w(d_1 - d_0) + 2\varepsilon < 0.24 , \quad 11q + 22\varepsilon < d_0(1 - d_1)w(1 - p)^2 ,$$

$$44p + 22w(d_1 - d_0) + 44\varepsilon < 1 - q , \quad 98p + 49q + 49w(d_1 - d_0) < (1 - 2\varepsilon)^2$$

holds, then there exists a protocol that uses $(d_0, d_1, p, g, \varepsilon)$-ActiveWEC and bit commitments and efficiently implements OT secure in the malicious model.

To achieve any two party computation from a $(d_0, d_1, p, g, \varepsilon)$-ActiveWEC, the conditions of Corollary 3 and Corollary 2 must be satisfied simultaneously.

6   Conclusions and Open Problems

We gave new, weaker security definitions for the erasure channel and the binary symmetric channel, and showed that they imply oblivious transfer. The advantage of our new definitions is that they allow the use of channels from which the statistics are not known with arbitrary precision, which make it possible to use channels where the parameters come from experimental evidence. Note that together with the computational WOT amplification from [22], our results can also be used in a computational setting.

It seems to be difficult to close the gap between the possibility and the impossibility bounds for OT. But maybe it is possible to get a tight bound for bit commitment. Still missing is a definition of the weak binary symmetric channel in the malicious model. Furthermore, it would be nice to have a bit commitment protocol that works for a weak form of the Gaussian channels.

Acknowledgment

I thank the anonymous referees for many helpful comments. I was supported by the U.K. EPSRC, grant EP/E04297X/1.

References

1. Bennett, C.H., Brassard, G., Robert, J.-M.: Privacy amplification by public discussion. SIAM Journal on Computing 17(2), 210–229 (1988)
2. Cachin, C.: Smooth entropy and rényi entropy. In: Fumy, W. (ed.) EUROCRYPT 1997. LNCS, vol. 1233, pp. 193–208. Springer, Heidelberg (1997)
3. Crépeau, C.: Equivalence between two flavours of oblivious transfers. In: Price, W.L., Chaum, D. (eds.) EUROCRYPT 1987. LNCS, vol. 304, pp. 350–354. Springer, Heidelberg (1988)
4. Crépeau, C.: Efficient cryptographic protocols based on noisy channels. In: Fumy, W. (ed.) EUROCRYPT 1997. LNCS, vol. 1233, pp. 306–317. Springer, Heidelberg (1997)
5. Crépeau, C., Kilian, J.: Achieving oblivious transfer using weakened security assumptions (extended abstract). In: Proceedings of the 29th Annual IEEE Symposium on Foundations of Computer Science (FOCS 1988), pp. 42–52 (1988)
6. Crépeau, C., Morozov, K., Wolf, S.: Efficient unconditional oblivious transfer from almost any noisy channel. In: Blundo, C., Cimato, S. (eds.) SCN 2004. LNCS, vol. 3352, pp. 47–59. Springer, Heidelberg (2005)
7. Damgård, I.B., Fehr, S., Morozov, K., Salvail, L.: Unfair noisy channels and oblivious transfer. In: Naor, M. (ed.) TCC 2004. LNCS, vol. 2951, pp. 355–373. Springer, Heidelberg (2004)
8. Damgård, I., Kilian, J., Salvail, L.: On the (im)possibility of basing oblivious transfer and bit commitment on weakened security assumptions. In: Stern, J. (ed.) EUROCRYPT 1999. LNCS, vol. 1592, pp. 56–73. Springer, Heidelberg (1999)
9. Even, S., Goldreich, O., Lempel, A.: A randomized protocol for signing contracts. Commun. ACM 28(6), 637–647 (1985)
10. Goldreich, O., Micali, S., Wigderson, A.: How to play any mental game. In: Proceedings of the 21st Annual ACM Symposium on Theory of Computing (STOC 1987), pp. 218–229. ACM Press, New York (1987)
11. Goldreich, O., Vainish, R.: How to solve any protocol probleman efficiency improvement. In: Pomerance, C. (ed.) CRYPTO 1987. LNCS, vol. 293, pp. 73–86. Springer, Heidelberg (1988)
12. Holenstein, T.: Strengthening key agreement using hard-core sets. PhD thesis, ETH Zurich, Switzerland, Reprint as vol. 7 of ETH Series in Information Security and Cryptography, Hartung-Gorre Verlag (2006)
13. Impagliazzo, R., Levin, L.A., Luby, M.: Pseudo-random generation from one-way functions. In: Proceedings of the 21st Annual ACM Symposium on Theory of Computing (STOC 1989), pp. 12–24. ACM Press, New York (1989)
14. Kilian, J.: Founding cryptography on oblivious transfer. In: Proceedings of the 20th Annual ACM Symposium on Theory of Computing (STOC 1988), pp. 20–31. ACM Press, New York (1988)
15. Maurer, U., Wolf, S.: Privacy amplification secure against active adversaries. In: Kaliski Jr., B.S. (ed.) CRYPTO 1997. LNCS, vol. 1294, pp. 307–321. Springer, Heidelberg (1997)

16. Nascimento, A., Winter, A.: On the oblivious transfer capacity of noisy correlations. IEEE Trans. on Information Theory 54(6) (2008)
17. Nascimento, A.C.A., Skludarek, S., Barros, J., Imai, H.: The commitment capacity of the gaussian channel is infinite. IEEE Trans. on Information Theory, Special Issue on Information Security (2007)
18. Oggier, F., Morozov, K.: A practical scheme for string commitment based on the gaussian channel. In: Proceedings of 2006 IEEE Information Theory Workshop (ITW 2008) (2008)
19. Rabin, M.O.: How to exchange secrets by oblivious transfer. Technical Report TR-81, Harvard Aiken Computation Laboratory (1981)
20. Renner, R., Wolf, S.: Simple and tight bounds for information reconciliation and privacy amplification. In: Roy, B. (ed.) ASIACRYPT 2005. LNCS, vol. 3788, pp. 199–216. Springer, Heidelberg (2005)
21. Wiesner, S.: Conjugate coding. SIGACT News 15(1), 78–88 (1983)
22. Wullschleger, J.: Oblivious-transfer amplification. In: Naor, M. (ed.) EUROCRYPT 2007. LNCS, vol. 4515, pp. 555–572. Springer, Heidelberg (2007); Full version (PhD Thesis, ETH Zurich), http://arxiv.org/abs/cs.CR/0608076
23. Yao, A.C.: Protocols for secure computations. In: Proceedings of the 23rd Annual IEEE Symposium on Foundations of Computer Science (FOCS 1982), pp. 160–164 (1982)

Composing Quantum Protocols in a Classical Environment

Serge Fehr* and Christian Schaffner**

Centrum Wiskunde & Informatica (CWI)
Amsterdam, The Netherlands
{S.Fehr,C.Schaffner}@cwi.nl

Abstract. We propose a general security definition for cryptographic quantum protocols that implement classical non-reactive two-party tasks. The definition is expressed in terms of simple quantum-information-theoretic conditions which must be satisfied by the protocol to be secure. The conditions are uniquely determined by the ideal functionality $\mathcal{F}$ defining the cryptographic task to be implemented. We then show the following composition result. If quantum protocols $\pi_1, \ldots, \pi_\ell$ securely implement ideal functionalities $\mathcal{F}_1, \ldots, \mathcal{F}_\ell$ according to our security definition, then any purely *classical* two-party protocol, which makes sequential calls to $\mathcal{F}_1, \ldots, \mathcal{F}_\ell$, is equally secure as the protocol obtained by replacing the calls to $\mathcal{F}_1, \ldots, \mathcal{F}_\ell$ with the respective quantum protocols $\pi_1, \ldots, \pi_\ell$. Hence, our approach yields the minimal security requirements which are strong enough for the typical use of quantum protocols as subroutines within larger classical schemes. Finally, we show that recently proposed quantum protocols for secure identification and oblivious transfer in the bounded-quantum-storage model satisfy our security definition, and thus compose in the above sense.

1 Introduction

Background. Finding the right security definition for a cryptographic task is a non-trivial fundamental question in cryptography. From a theoretical point of view, one would like definitions to be as strong as possible in order to obtain strong composability guarantees. However, this often leads to impossibility results or to very complex and inefficient schemes. Therefore, from a practical point of view, one may also consider milder security definitions which allow for efficient schemes, but still offer "good enough" security.

It is fair to say that in computational cryptography, the question of defining security and the trade-offs that come along with these definitions are by now quite well understood. The situation is different in quantum cryptography. For instance, it was realized only recently that the standard security definition of quantum key-agreement does not guarantee the desired kind of security and some

---

* Supported by the Dutch Organization for Scientific Research (NWO).
** Supported by the EU fifth framework project QAP IST 015848 and the Dutch Organization for Scientific Research (NWO).

O. Reingold (Ed.): TCC 2009, LNCS 5444, pp. 350–367, 2009.

work was required to establish the right security definition [13,23,2,22,17]. Security definitions for general quantum protocols have first been proposed in [14] and subsequently been refined for the case of quantum multi-party computation in [26]. In [3,27], strong security definitions for general quantum protocols were proposed by translating Canetti's universal-composability framework and Backes, Pfitzmann and Waidner's reactive-simulatability model, respectively, into the quantum setting. The resulting security definitions are very strong and guarantee full composability. However, they are complex and hard to achieve. Indeed, so far they have been actually used and shown to be achievable only in a couple of isolated cases: quantum key distribution [2] and quantum multi-party computation with dishonest minority [1]. It is still common practice in quantum cryptography that every paper proposes its own security definition of a certain task and proves security with respect to the proposed definition. However, it usually remains unclear whether these definitions are strong enough to guarantee any kind of composability, and thus whether protocols that meet the definition really behave as expected.

Contribution. We propose a general security definition for quantum protocols that implement cryptographic two-party tasks. The definition is in terms of simple quantum-information-theoretic security conditions that must be satisfied for the protocol to be secure. In particular, the definition does not involve additional entities like a "simulator" or an "environment". The security conditions are uniquely determined by the *ideal functionality* that defines the cryptographic task to be realized. Our definition applies to any *non-reactive, classical* ideal functionality $\mathcal{F}$, which obtains classical (in the sense of non-quantum) input from the two parties, processes the provided input according to its specification, and outputs the resulting classical result to the parties. A typical example for such a functionality/task is oblivious transfer (OT). Reactive functionalities, i.e. functionalities that have several phases (like e.g. bit commitment), or functionalities that take quantum input and/or produce quantum output are not the scope of this paper.

We show the following composition result. If quantum protocols $\pi_1, \ldots, \pi_\ell$ securely implement ideal functionalities $\mathcal{F}_1, \ldots, \mathcal{F}_\ell$ according to our security definition, then any purely *classical* two-party protocol, which makes sequential calls to $\mathcal{F}_1, \ldots, \mathcal{F}_\ell$, is equally secure as the protocol obtained by replacing the calls to $\mathcal{F}_1, \ldots, \mathcal{F}_\ell$ with the respective quantum subroutines $\pi_1, \ldots, \pi_\ell$. We stress that our composition theorem, respectively our security definition, only allows for the composition of quantum sub-protocols into a *classical* outer protocol. This is a trade-off which allows for milder security definitions (which in turn allows for simpler and more efficient implementations) but still offers security in realistic situations. Indeed, current technology is far from being able to execute quantum algorithms or protocols which involve complicated quantum operations and/or need to keep a quantum state "alive" for more than a tiny fraction of a second. Thus, the best one can hope for in the near future in terms of practical quantum algorithms is that certain small subroutines, like key-distribution or OT, may be implemented by quantum protocols, while the more complex outer protocol remains classical. From a more theoretical point of view, our general security definition expresses what security properties a quantum protocol must satisfy in order to be able to instantiate a basic cryptographic primitive upon which an information-theoretic cryptographic construction is based. For instance, it expresses the security properties a quantum OT[1] needs to satisfy so that Kilian's classical[2] construction of general secure function evaluation based on OT [15] remains secure when instantiating the OT primitive by a quantum protocol.

Finally, we show that the *ad-hoc* security definitions proposed by Damgård, Fehr, Salvail and Schaffner for their 1-2 OT and secure-identification protocols in the bounded-quantum-storage model [7,9] imply (and are likely to be equivalent) to the corresponding security definitions obtained from our approach.[3] This implies composability in the above sense for these quantum protocols in the bounded-quantum-storage model.

Related work. In the classical setting, Crépeau *et al.* proposed information-theoretic conditions for two-party secure function evaluation [5], though restricted to the *perfect* case, where the protocol is not allowed to make any error. They show equivalence to a simulation-based definition that corresponds to the standard framework of Goldreich [12]. Similar conditions have been subsequently found by Crépeau and Wullschleger for the case of non-perfect classical protocols [6]. Our work can be seen as an extension of [5,6] to the setting where classical subroutines are implemented by quantum protocols.

As pointed out and discussed above, general frameworks for universal composability in the quantum setting have been established in [3,27]. The composability of protocols in the bounded-quantum-storage model has recently been investigated by Wehner and Wullschleger [29]. They propose security definitions that guarantee sequential composability of quantum protocols within *quantum* protocols. This is clearly a stronger composition result than we obtain (though restricted to the bounded-quantum-storage model) but comes at the price of a more demanding security definition. And indeed, whereas we show that the simple definitions used in [8,7] already guarantee composability into classical protocols without any modifications to the original parameters and proofs, [29] need to strengthen the quantum-memory bound (and re-do the security proof) in order to show that the 1-2 OT protocol from [7] meets their strong security definition. As we argued above, this is an overkill in many situations.

---

[1] We are well aware that quantum OT is impossible without any restriction on the adversary, but it becomes possible for instance when restricting the adversary's quantum memory [8,7].

[2] Here, "classical" can be understood as "non-quantum" as well as "being a classic".

[3] Interestingly, this is not true for the definition of Rabin OT given in the first paper in this line of research [8], and indeed in the full version of that paper, it is mentioned that their definition poses some "composability problems" (this problem though has been fixed in the journal version [10]). This supports our claim that failure of satisfying our security definition is strong evidence for a security problem of a quantum protocol (or the definition used).

2   Notation

Quantum States. We assume the reader's familiarity with basic notation and concepts of quantum information processing [21].

Given a bipartite quantum state ρ_{XE}, we say that X is *classical* if ρ_{XE} is of the form $\rho_{XE} = \sum_{x \in \mathcal{X}} P_X(x)|x\rangle\langle x| \otimes \rho_E^x$ for a probability distribution P_X over a finite set $\mathcal{X}$. This can be understood in that the state of the quantum register E depends on the classical random variable X, in the sense that E is in state ρ_E^x exactly if $X = x$. For any event $\mathcal{E}$ defined by $P_{\mathcal{E}|X}(x) = P[\mathcal{E}|X=x]$ for all x, we may then write

$$\rho_{XE|\mathcal{E}} := \sum_x P_{X|\mathcal{E}}(x)|x\rangle\langle x| \otimes \rho_E^x \ . \tag{1}$$

When we omit registers, we mean the partial trace over these register, for instance $\rho_{E|\mathcal{E}} = \mathrm{tr}_X(\rho_{XE|\mathcal{E}}) = \sum_x P_{X|\mathcal{E}}(x)\rho_E^x$, which describes E given that the event $\mathcal{E}$ occurs.

This notation extends naturally to states that depend on several classical random variables X, Y etc., defining the density matrices ρ_{XYE}, $\rho_{XYE|\mathcal{E}}$, $\rho_{YE|X=x}$ etc. We tend to slightly abuse notation and write $\rho_{YE}^x = \rho_{XE|X=x}$ and $\rho_{YE|\mathcal{E}}^x = \rho_{YE|X=x,\mathcal{E}}$, as well as $\rho_E^x = \mathrm{tr}_Y(\rho_{YE}^x)$ and $\rho_{E|\mathcal{E}}^x = \mathrm{tr}_Y(\rho_{YE|\mathcal{E}}^x)$. Given a state ρ_{XE} with classical X, by saying that "there exists a classical random variable Y such that ρ_{XYE} satisfies some condition", we mean that ρ_{XE} can be understood as $\rho_{XE} = \mathrm{tr}_Y(\rho_{XYE})$ for some state ρ_{XYE} with classical X and Y, and that ρ_{XYE} satisfies the required condition.

X is independent of E (in that ρ_E^x does not depend on x) if and only if $\rho_{XE} = \rho_X \otimes \rho_E$, which in particular implies that no information on X can be learned by observing only E. Similarly, X is random and independent of E if and only if $\rho_{XE} = \frac{1}{|\mathcal{X}|}\mathbb{I} \otimes \rho_E$, where $\frac{1}{|\mathcal{X}|}\mathbb{I}$ is the density matrix of the fully mixed state of suitable dimension.

We also need to express that a random variable X is independent of a quantum state E *when given a random variable Y*. This means that when given Y, the state E gives no additional information on X. Yet another way to understand this is that E is obtained from X and Y by solely processing Y. Formally, adopting the notion introduced in [9], this is expressed by requiring that ρ_{XYE} equals $\rho_{X\leftrightarrow Y\leftrightarrow E}$, where the latter is defined as

$$\rho_{X\leftrightarrow Y\leftrightarrow E} := \sum_{x,y} P_{XY}(x,y)|x\rangle\langle x| \otimes |y\rangle\langle y| \otimes \rho_E^y .$$

In other words, $\rho_{XYE} = \rho_{X\leftrightarrow Y\leftrightarrow E}$ precisely if $\rho_E^{x,y} = \rho_E^y$ for all x and y. This notation naturally extends to $\rho_{X\leftrightarrow Y\leftrightarrow E|\mathcal{E}} = \sum_{x,y} P_{XY|\mathcal{E}}(x,y)|x\rangle\langle x| \otimes |y\rangle\langle y| \otimes \rho_{E|\mathcal{E}}^y$.

Full (conditional) independence is often too strong a requirement, and it usually suffices to be "close" to such a situation. Closeness of two states ρ and σ is measured in terms of their trace distance $\delta(\rho,\sigma) = \frac{1}{2}\mathrm{tr}(|\rho - \sigma|)$, where for any operator A, $|A|$ is defined as $|A| := \sqrt{AA^\dagger}$. We write $\rho \approx_\varepsilon \sigma$ to denote that $\delta(\rho,\sigma) \leq \varepsilon$, and we then say that ρ and σ are ε-close. It is known that ε-closeness is preserved under any quantum operation; this in particular implies that if $\rho \approx_\varepsilon \sigma$ then no observer can distinguish ρ from σ with advantage greater than ε [23]. For states ρ_{XE} and $\rho_{X'E'}$ with classical X and X', it is not hard to see that $\delta(\rho_{XE}, \rho_{X'E'}) = \sum_x \delta(P_X(x)\rho_E^x, P_{X'}(x)\rho_{E'}^x)$, and thus $\delta(\rho_{XE}, \rho_{X'E'}) = \sum_x P_X(x)\delta(\rho_E^x, \rho_{E'}^x)$ if $P_X = P_{X'}$. In case of purely classical states ρ_X and $\rho_{X'}$, the trace distance coincides with the statistical distance of the random variables X and X': $\delta(\rho_X, \rho_{X'}) = \frac{1}{2}\sum_x |P_X(x) - P_{X'}(x)|$, and we then write $P_X \approx_\varepsilon P_{X'}$, or $X \approx_\varepsilon X'$, instead of $\rho_X \approx_\varepsilon \rho_{X'}$.

We will make use of the following lemmas whose proofs are given in the full version [11] of this paper.

Lemma 2.1. *1. If $\rho_{XYZE} \approx_\varepsilon \rho_{X\leftrightarrow Y\leftrightarrow ZE}$ then $\rho_{XYZE} \approx_{2\varepsilon} \rho_{X\leftrightarrow YZ\leftrightarrow E}$.*
2. If $\rho_{XZE} \approx_\varepsilon \rho_X \otimes \rho_{ZE}$ then $\rho_{XZE} \approx_{2\varepsilon} \rho_{X\leftrightarrow Z\leftrightarrow E}$.
3. If $\rho_{XZE} \approx_\varepsilon \mathbb{I}/|\mathcal{X}| \otimes \rho_{ZE}$, then $\rho_{XZE} \approx_{4\varepsilon} \rho_{X\leftrightarrow Z\leftrightarrow E}$.

Lemma 2.2. *If $\rho_{XYE} \approx_\varepsilon \rho_{X\leftrightarrow Y\leftrightarrow E}$ then $\rho_{Xf(X,Y)YE} \approx_\varepsilon \rho_{Xf(X,Y)\leftrightarrow Y\leftrightarrow E}$ for any function f.*

Lemma 2.3. *For an event $\mathcal{E}$ which is completely determined by the random variable Y, i.e. for all y, the probability $\Pr[\mathcal{E}|Y = y]$ either vanishes or equals one, we can decompose the density matrix $\rho_{X\leftrightarrow Y\leftrightarrow E}$ into*[4]

$$\rho_{X\leftrightarrow Y\leftrightarrow E} = \Pr[\mathcal{E}] \cdot \rho_{X\leftrightarrow Y\leftrightarrow E|\mathcal{E}} + \Pr[\overline{\mathcal{E}}] \cdot \rho_{X\leftrightarrow Y\leftrightarrow E|\overline{\mathcal{E}}}.$$

3   Protocols and Functionalities

Quantum Protocols. We consider *two-party quantum protocols* $\pi = (\mathsf{A}, \mathsf{B})$, consisting of interactive quantum algorithms A and B. For convenience, we call the two parties who run A and B *Alice* and *Bob*, respectively. There are different approaches to formally define interactive quantum algorithms and thus quantum two-party protocols, in particular when we restrict in- and outputs (of honest participants) to be classical. For instance such a formalization can be done by means of quantum circuits, or by means of a classical Turing machine which outputs unitaries that are applied to a quantum register. For our work, the specific choice of the formalization is immaterial; what is important is that such a two-party quantum protocol, formalized in whatever way, uniquely specifies its input-output behavior. Therefore, in this work, we capture quantum protocols by their input-output behavior, which we formalize by a quantum operation, i.e. a trace-preserving completely-positive map, which maps the common two-partite input state ρ_{UV} to the common two-partite output state ρ_{XY}. We denote this operation by $\rho_{XY} = \pi \rho_{UV}$ or, when we want to emphasize that π is executed by *honest* Alice and Bob, also by $\rho_{XY} = \pi_{\mathsf{A},\mathsf{B}} \rho_{UV}$. If one of the players, say Bob, is *dishonest* and follows a malicious strategy B', then we slightly abuse notation and write $\pi_{\mathsf{A},\mathsf{B}'}$ for the corresponding operator.

---

[4] One is tempted to think that such a decomposition holds for *any* event $\mathcal{E}$; however, this is not true. See Lemma 2.1 of [9] for another special case where the decomposition does hold.

Protocols and Functionalities with Classical In- and Output. In this work, we focus on quantum protocols $\pi = (\mathsf{A}, \mathsf{B})$ with *classical in- and output* for the honest players. This means that we assume the common input state ρ_{UV} to be classical, i.e. of the form $\rho_{UV} = \sum_{u,v} P_{UV}(u,v)|u\rangle\langle u| \otimes |v\rangle\langle v|$ for some probability distribution P_{UV}, and the common output state $\rho_{XY} = \pi_{\mathsf{A},\mathsf{B}}\,\rho_{UV}$ is then guaranteed to be classical as well, i.e., $\rho_{XY} = \sum_{x,y} P_{XY}(x,y)|x\rangle\langle x| \otimes |y\rangle\langle y|$. In this case we may understand U and V as well as X and Y as random variables, and we also write $(X,Y) = \pi(U,V)$. Note that the input-output behavior of the protocol is uniquely determined by the conditional probability distribution $P_{XY|UV}$. If one of the players, say Bob, is dishonest and follows a malicious strategy B', then we may allow his part of the input to be quantum and denote it as V', i.e. $\rho_{UV'} = \sum_u P_U(u)|u\rangle\langle u| \otimes \rho_{V'|U=u}$, and we allow his part Y' of the common output state $\rho_{XY'} = \pi_{\mathsf{A},\mathsf{B}'}\,\rho_{UV'}$ to be quantum, i.e. $\rho_{XY'} = \sum_x P_X(x)|x\rangle\langle x| \otimes \rho_{Y'|X=x}$. We write $\rho_{UV'}$ as $\rho_{U\emptyset} = \rho_U \otimes \rho_\emptyset = \rho_U$ if V' is empty, i.e. if B' has no input at all, and we write it as $\rho_{UZV'}$ if part of his input, Z, is actually classical.

A classical non-reactive two-party *ideal functionality* $\mathcal{F}$ is given by a conditional probability distribution $P_{\mathcal{F}(U,V)|UV}$, inducing a pair of random variables $(X,Y) = \mathcal{F}(U,V)$ for every joint distribution of U and V. We also want to take into account ideal functionalities which allow the dishonest player some additional—though still limited—capabilities (as for instance in Section 6). We do this as follows. We specify $\mathcal{F}$ not only for the "proper" domains $\mathcal{U}$ and $\mathcal{V}$, over which U and V are supposed to be distributed, but we actually specify it for some larger domains $\tilde{\mathcal{U}} \supseteq \mathcal{U}$ and $\tilde{\mathcal{V}} \supseteq \mathcal{V}$. The understanding is that U and V provided by honest players always lie in $\mathcal{U}$ and $\mathcal{V}$, respectively, whereas a dishonest player, say Bob, may select V from $\tilde{\mathcal{V}} \setminus \mathcal{V}$, and this way Bob may cause $\mathcal{F}$, if specified that way, to process its inputs differently and/or to provide a "more informative" output Y to Bob. For simplicity though, we often leave the possibly different domains for honest and dishonest players implicit.

We write $(X,Y) = \mathcal{F}_{\hat{\mathsf{A}},\hat{\mathsf{B}}}(U,V)$ or $\rho_{XY} = \mathcal{F}_{\hat{\mathsf{A}},\hat{\mathsf{B}}}\,\rho_{UV}$ for the execution of the "ideal-life" protocol, where Alice and Bob forward their inputs to $\mathcal{F}$ and output whatever they obtain from $\mathcal{F}$. And we write $\rho_{XY'} = \mathcal{F}_{\hat{\mathsf{A}},\mathsf{B}'}\,\rho_{UV'}$ for the execution of this protocol with a dishonest Bob with strategy B' and quantum input V'. Note that Bob's possibilities are very limited: he can produce some classical input V for $\mathcal{F}$ (distributed over $\tilde{\mathcal{V}}$) from his input quantum state V', and then he can prepare and output a quantum state Y' which might depend on $\mathcal{F}$'s reply Y.

Classical Hybrid Protocols. A two-party *classical hybrid* protocol $\Sigma^{\mathcal{F}_1\cdots\mathcal{F}_\ell} = (\hat{\mathsf{A}}, \hat{\mathsf{B}})$ between Alice and Bob is a protocol which makes a bounded number k of sequential oracle calls to possibly different ideal functionalities $\mathcal{F}_1, \ldots, \mathcal{F}_\ell$. We allow $\hat{\mathsf{A}}$ and $\hat{\mathsf{B}}$ to make several calls to independent copies of the same $\mathcal{F}_i$, but we require from $\Sigma^{\mathcal{F}_1\cdots\mathcal{F}_\ell}$ that for every possible execution, there is always agreement between $\hat{\mathsf{A}}$ and $\hat{\mathsf{B}}$ on when to call which functionality; for instance we may assume that $\hat{\mathsf{A}}$ and $\hat{\mathsf{B}}$ exchange the index i before they call $\mathcal{F}_i$ (and stop if there is disagreement).

Formally, such a classical hybrid protocol is given by a sequence of $k + 1$ quantum protocols formalized by quantum operators with classical in- and output for the honest players, see Figure 1. For an honest player, say Alice, the j-th protocol outputs an index i indicating which functionality is to be called, classical auxiliary (or "state") information information S_j and a classical input U_j for $\mathcal{F}_i$. The $(j+1)$-st protocol expects as input S_j and Alice's classical output X_j from $\mathcal{F}_i$. Furthermore, the first protocol expects Alice's classical input U to the hybrid protocol, and the last produces the classical output X of the hybrid protocol. In case of a dishonest player, say Bob, all in- and outputs may be quantum states V_j' respectively Y_j'. By instantiating the j-th call to a functionality $\mathcal{F}$ (where we from now on omit the index for simpler notation) in the obvious way by the corresponding "ideal-life" proto-

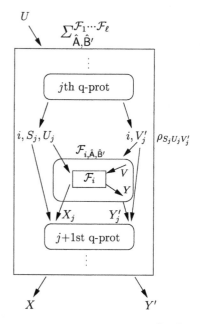

Fig. 1. Hybrid protocol $\Sigma_{\hat{A}, \hat{B}'}^{\mathcal{F}_1 \cdots \mathcal{F}_\ell}$

col $\mathcal{F}_{\hat{A}, \hat{B}}$ (respectively $\mathcal{F}_{\hat{A}', \hat{B}}$ or $\mathcal{F}_{\hat{A}, \hat{B}'}$ in case of a dishonest Alice or Bob), we obtain the instantiated hybrid protocol formally described by quantum operator $\Sigma_{\hat{A}, \hat{B}}^{\mathcal{F}_1 \cdots \mathcal{F}_\ell}$ (respectively $\Sigma_{\hat{A}', \hat{B}}^{\mathcal{F}_1 \cdots \mathcal{F}_\ell}$ or $\Sigma_{\hat{A}, \hat{B}'}^{\mathcal{F}_1 \cdots \mathcal{F}_\ell}$).[5]

For the hybrid protocol to be *classical*, we mean that it has classical in- and output (for the honest players), but also that all communication between Alice and Bob is classical. Since we have not formally modeled the communication within (hybrid) protocols, we need to formalize this property as a property of the quantum operators that describe the hybrid protocol: Consider a dishonest player, say Bob, with no input, and consider the common state $\rho_{S_j U_j V_j'}$ at any point during the execution of the hybrid protocol when a call to functionality $\mathcal{F}_i$ is made. The requirement for the hybrid protocol to be *classical* is now expressed in that there exists a classical Z_j—to be understood as consisting of $\hat{B}'$'s classical communication with $\hat{A}$ and with the $\mathcal{F}_{i'}$'s up to this point—such that given Z_j, Bob's quantum state V_j' is uncorrelated with (i.e. independent of) Alice' classical input and auxiliary information: $\rho_{S_j U_j Z_j V_j'} = \rho_{S_j U_j \leftrightarrow Z_j \leftrightarrow V_j'}$. Furthermore, we require that we may assume Z_j to be part of V_j' in the sense that for any $\hat{B}'$ there exists $\hat{B}''$ such that Z_j is part of V_j'. This definition is motivated by the observation that if Bob can communicate only classically with Alice, then he can correlate his quantum state with information on Alice's side only by means of the classical communication.

---

[5] Note that for simpler notation, we are a bit sloppy and give the same name, like $\hat{A}$ and $\hat{B}'$, to honest Alice's and dishonest Bob's strategy within different (sub)protocols.

We also consider the protocol we obtain by replacing the ideal functionalities by quantum two-party sub-protocols $\pi_1, \ldots, \pi_\ell$ with classical in- and outputs for the honest parties: whenever $\Sigma^{\mathcal{F}_1 \cdots \mathcal{F}_\ell}$ instructs $\hat{\mathsf{A}}$ and $\hat{\mathsf{B}}$ to execute $\mathcal{F}_{i\hat{\mathsf{A}},\hat{\mathsf{B}}}$, they instead execute $\pi_i = (\mathsf{A}_i, \mathsf{B}_i)$ and take the resulting outputs. We write $\Sigma^{\pi_1 \cdots \pi_\ell} = (\mathsf{A}, \mathsf{B})$ for the real-life quantum protocol we obtain this way.

4   Security for Two-Party Quantum Protocols

4.1   The Security Definition

Framework. We use the following framework for defining security of a quantum protocol π with classical in- and output. We distinguish three cases and consider the respective output states obtained by executing π in case of honest Alice and honest Bob, in case of honest Alice and dishonest Bob, and in case of dishonest Alice and honest Bob. For each of these cases we require some security conditions on the output state to hold. More precisely, for honest Alice and Bob, we fix an arbitrary joint probability distribution P_{UV} for the inputs U and V, resulting in outputs $(X, Y) = \pi_{\mathsf{A},\mathsf{B}}(U, V)$ with a well defined joint probability distribution P_{UVXY}. For an honest Alice and a dishonest Bob, we fix an arbitrary distribution P_U for Alice's input and an arbitrary strategy B' *with no input* for Bob, and we consider the resulting joint output state

$$\rho_{UXY'} = \left(\mathrm{id}_U \otimes \pi_{\mathsf{A},\mathsf{B}'}\right)\rho_{UU\emptyset} = \sum_u P_U(u)|u\rangle\langle u| \otimes \pi_{\mathsf{A},\mathsf{B}'}(|u\rangle\langle u|\otimes\rho_\emptyset)$$

augmented with Alice's input U, where U and X are classical and Y' is in general quantum. And, correspondingly, for a dishonest Alice and an honest Bob, we fix an arbitrary distribution P_V for Bob's input and an arbitrary strategy A' *with no input* for Alice, and we consider the resulting joint output state $\rho_{VX'Y} = \left(\mathrm{id}_V \otimes \pi_{\mathsf{A}',\mathsf{B}}\right)\rho_{V\emptyset V}$ augmented with Bob's input V. Then, security is defined by specific information-theoretic conditions on P_{UVXY}, $\rho_{UXY'}$ and $\rho_{VX'Y}$, where the conditions depend on the functionality $\mathcal{F}$ which π is implementing. Definition 4.1 below for a general functionality $\mathcal{F}$, as well as the definitions studied later for specific functionalities (Definitions 6.1), are to be understood in this framework. In particular, the augmented common output states are to be understood as defined above.

We stress once more that the framework assumes that dishonest players have no input at all. This might appear too weak at first glance; one would expect a dishonest player, say Bob, to at least get the input V of the honest Bob. The justification for giving dishonest players no input is that on the one hand, we will show that this "minimalistic approach" is good enough for the level of security we are aiming for (see Theorem 5.1), and on the other hand, our goal is to keep the security definitions as simple as possible.

Restricting the Adversary. Since essentially no interesting two-party task can be implemented securely by a quantum protocol against unbounded quantum attacks [20,19,18,16], one typically has to put some restriction upon the dishonest player's capabilities, like to limit his quantum-storage capabilities [8,7,9,28] or the size of coherent measurements he can do [24]. Throughout, we let $\mathfrak{A}$ and $\mathfrak{B}$ be subfamilies of all possible strategies A' and B' of a dishonest Alice and a dishonest Bob, respectively. In order to circumvent pathological counter examples, we need to assume the following two natural consistency conditions on $\mathfrak{A}$, and correspondingly on $\mathfrak{B}$. If a dishonest strategy A' $\in \mathfrak{A}$ expects as input some state $\rho_{ZU'}$ with classical Z, then for any z and for any $\rho_{U'|Z=z}$, the strategy $A'_{z,\rho_{U'|Z=z}}$, which has z hard-wired and prepares the state $\rho_{U'|Z=z}$ as an initial step but otherwise runs like A', is in $\mathfrak{A}$ as well. And, if A' $\in \mathfrak{A}$ is a dishonest strategy for a protocol Σ^π which makes a call to a sub-protocol π, then the corresponding "sub-strategy" of A', which is active during the execution of π, is in $\mathfrak{A}$ as well. It is for instance clear that bounding the quantum memory leads to a family of strategies that satisfies these conditions.

Defining Security. Following the framework described above, we propose the following security definition for two-party quantum protocols with classical in- and output. The proposed definition implies strong simulation-based security when using quantum protocols as sub-protocols in classical outer protocols (Theorem 5.1), yet it is expressed in a way that is as simple and as weak as seemingly possible, making it as easy as possible to design and prove quantum cryptographic schemes secure according to the definition.

Definition 4.1. *A two-party quantum protocol π ε-securely implements an ideal classical functionality $\mathcal{F}$ against $\mathfrak{A}$ and $\mathfrak{B}$ if the following holds:*

Correctness: *For any joint distribution of the input U and V, the resulting common output $(X,Y) = \pi(U,V)$ satisfies $(U,V,X,Y) \approx_\varepsilon (U,V,\mathcal{F}(U,V))$.*

Security for Alice: *For any B' $\in \mathfrak{B}$ (with no input), and for any distribution of U, the resulting common output state $\rho_{UXY'}$ (augmented with U) is such that there exist[6] classical random variables V,Y such that $P_{UV} \approx_\varepsilon P_U \cdot P_V$, $(U,V,X,Y) \approx_\varepsilon (U,V,\mathcal{F}(U,V))$, and $\rho_{UXVYY'} \approx_\varepsilon \rho_{UX \leftrightarrow VY \leftrightarrow Y'}$.*

Security for Bob: *For any A' $\in \mathfrak{A}$ (with no input), and for any distribution of V, the resulting common output state $\rho_{VX'Y}$ (augmented with V) is such that there exist classical random variables U,X such that $P_{UV} \approx_\varepsilon P_U \cdot P_V$, $(U,V,X,Y) \approx_\varepsilon (U,V,\mathcal{F}(U,V))$, and $\rho_{VYUXX'} \approx_\varepsilon \rho_{VY \leftrightarrow UX \leftrightarrow X'}$.*

The three conditions for dishonest Bob (and similarly for dishonest Alice) express that, up to a small error, V is independent of U, X and Y are obtained by applying $\mathcal{F}$, and the quantum state Y' is obtained by locally processing V and Y.

We would like to point out that Definition 4.1 requires existence of the dishonest party's input, and as such prohibits the dishonest party to execute π in superposition with several inputs and to obtain a superposition of the corresponding outputs. Indeed, it is interesting to note that from a superposition of outputs,

---

[6] As defined in Section 2.

the dishonest party can typically extract "forbidden information" [4,25].This is another way to see that without any restriction on the adversary, non-trivial quantum two-party computation is not possible [18].

4.2   Equivalent Formulations

As already mentioned, Definition 4.1 appears to guarantee security only in a very restricted setting, where the honest player has no information beyond his input, and the dishonest player has no (auxiliary) information at all. Below, we argue that Definition 4.1 actually implies security in a somewhat more general setting, where the dishonest player is allowed as input to have arbitrary classical information Z as well as a quantum state which only depends on Z. For completeness, although this is rather clear, we also argue that not only the honest player's input is protected, but also any classical "side information" S he might additionally have but does not use.

Proposition 4.2. *Let π be a two-party protocol that ε-securely implements $\mathcal{F}$ against $\mathfrak{A}$ and $\mathfrak{B}$. Let $\mathsf{B}' \in \mathfrak{B}$ be a dishonest Bob who takes as input a classical Z and a quantum state V' and outputs (the same) Z and a quantum state Y'. Then, for any $\rho_{SUZV'}$ with $\rho_{SUZV'} = \rho_{SU \leftrightarrow Z \leftrightarrow V'}$, the resulting overall output state (augmented with S and U) $\rho_{SUXZY'} = (\mathrm{id}_{SU} \otimes \pi_{\mathsf{A},\mathsf{B}'})\rho_{SUUZV'}$ is such that there exist classical random variables V, Y such that $P_{SUZV} \approx_\varepsilon P_{SU \leftrightarrow Z \leftrightarrow V}$, $(S, U, V, X, Y, Z) \approx_\varepsilon (S, U, V, \mathcal{F}(U, V), Z)$ and $\rho_{SUXVYZY'} = \rho_{SUX \leftrightarrow VYZ \leftrightarrow Y'}$. The corresponding holds for a dishonest Alice.*

The proof of Proposition 4.2, as well as the proof of Proposition 4.3 below, can be found in the full version [11].

Note the restriction on the adversary's quantum input V', namely that it is only allowed to depend on the honest player's input U (and side information S) "through" Z. It is this limitation which prohibits quantum protocols satisfying Definition 4.1 to securely compose into outer quantum protocols but requires the outer protocol to be classical. Indeed, within a quantum protocol that uses quantum communication, a dishonest player may be able to correlate his quantum state with classical information on the honest player's side; however, within a classical protocol, he can only do so through the classical communication so that his state is still independent when given the classical communication.

The following proposition shows equivalence to a simulation-based definition; this will be a handy formulation in order to prove the composition theorem.

Proposition 4.3. *Let π be a two-party protocol that ε-securely implements $\mathcal{F}$ against $\mathfrak{A}$ and $\mathfrak{B}$. Let $\mathsf{B}' \in \mathfrak{B}$ be a dishonest Bob who takes as input a classical Z and a quantum state V', engages into π with honest Alice and outputs Z and a quantum state Y'. Then, for any $\rho_{SUZV'}$ with $\rho_{SUZV'} = \rho_{SU \leftrightarrow Z \leftrightarrow V'}$ there exists $\hat{\mathsf{B}}'$ such that*

$$(\mathrm{id}_S \otimes \pi_{\mathsf{A},\mathsf{B}'})\rho_{SUZV'} \approx_{3\varepsilon} (\mathrm{id}_S \otimes \mathcal{F}_{\hat{\mathsf{A}},\hat{\mathsf{B}}'})\rho_{SUZV'}.$$

The corresponding holds for a dishonest Alice.

Recall that $\mathcal{F}_{\hat{A},\hat{B}'}$ is the execution of the "ideal-life" protocol, where honest $\hat{A}$ relays in- and outputs, and the only thing dishonest $\hat{B}'$ can do is modify the input and the output. Note that we do not guarantee that $\hat{B}'$ is in $\mathfrak{B}$; we will comment on this after Theorem 5.1.

5   Composability

We show the following composition result. If quantum protocols $\pi_1, \ldots, \pi_\ell$ securely implement ideal functionalities $\mathcal{F}_1, \ldots, \mathcal{F}_\ell$ according to Definition 4.1, then any two-party *classical* hybrid protocol $\Sigma^{\mathcal{F}_1, \ldots, \mathcal{F}_\ell}$ which makes sequential calls to $\mathcal{F}_1, \ldots, \mathcal{F}_\ell$ is essentially equally secure as the protocol obtained by replacing the calls to $\mathcal{F}_1, \ldots, \mathcal{F}_\ell$ by the respective quantum subroutines $\pi_1, \ldots, \pi_\ell$.

We stress that the $\mathcal{F}_i$'s are *classical* functionalities, i.e., even a dishonest player $\hat{A}'$ or $\hat{B}'$ can only input a classical value to $\mathcal{F}_i$, and for instance cannot execute $\mathcal{F}_i$ with several inputs in superposition. This makes our composition result stronger, because we give the adversary less power in the "ideal" (actually hybrid) world.

Theorem 5.1 (Composition Theorem). *Let* $\Sigma^{\mathcal{F}_1 \cdots \mathcal{F}_\ell} = (\hat{A}, \hat{B})$ *be a classical two-party hybrid protocol which makes at most k oracle calls to the functionalities, and for every $i \in \{1, \ldots, \ell\}$, let protocol π_i be an ε-secure implementation of $\mathcal{F}_i$ against $\mathfrak{A}$ and $\mathfrak{B}$. Then, the following holds.*

Correctness: *For every (distribution of) U and V*

$$\delta\left(\Sigma^{\pi_1 \cdots \pi_\ell}_{A,B} \rho_{UV}, \Sigma^{\mathcal{F}_1 \cdots \mathcal{F}_\ell}_{\hat{A},\hat{B}} \rho_{UV} \right) \leq k\varepsilon.$$

Security for Alice: *For every $\mathsf{B}' \in \mathfrak{B}$ there exists $\hat{B}'$ such that for every U*

$$\delta\left(\Sigma^{\pi_1 \cdots \pi_\ell}_{A,B'} \rho_{U\emptyset}, \Sigma^{\mathcal{F}_1 \cdots \mathcal{F}_\ell}_{\hat{A},\hat{B}'} \rho_{U\emptyset} \right) \leq 3k\varepsilon.$$

Security for Bob: *For every $\mathsf{A}' \in \mathfrak{A}$ there exists $\hat{A}'$ such that for every V*

$$\delta\left(\Sigma^{\pi_1 \cdots \pi_\ell}_{A',B} \rho_{\emptyset V}, \Sigma^{\mathcal{F}_1 \cdots \mathcal{F}_\ell}_{\hat{A}',\hat{B}} \rho_{\emptyset V} \right) \leq 3k\varepsilon.$$

Before going into the proof, we would like to point out the following observations. First of all, note that in contrast to typical composition theorems, which per-se guarantee security when replacing *one* functionality by a sub-protocol and where in case of several functionalities security then follows by induction, Theorem 5.1 is stated in such a way that it directly guarantees security when replacing all functionalities by sub-protocols. The reason for this is that the assumption that the outer protocol is classical is not satisfied anymore once the first functionality is replaced by a quantum sub-protocol, and thus the inductive reasoning does not work directly. We stress that our composition theorem nevertheless allows for several levels of compositions (see Corollary 5.2 and the preceding discussion).

Also, note that in Theorem 5.1 we assume the dishonest party to have no input. As in Section 4.2, this can be relaxed to a dishonest party, say Bob, that has an auxiliary input, consisting of a classical part Z and a quantum part V', as long as the quantum part V' depends on Alice' input U only through Z: $\rho_{UZV} = \rho_{U \leftrightarrow Z \leftrightarrow V}$; i.e., dishonest Bob has only classical side-information on Alice' input. This restriction is motivated by our model which captures a classical world except for specific designated quantum sub-protocols, and as such provides dishonest Bob a priori only with classical side-information.

Furthermore, note that we do not guarantee that the hybrid adversary $\hat{B}'$ is in $\mathfrak{B}$ (and similarly for $\hat{A}'$). For instance the specific $\hat{B}'$ we construct in the proof is more involved with respect to classical resources (memory and computation), but less involved with respect to quantum resources: essentially it follows B', except that it remembers all classical communication and except that the actions during the sub-protocols are replaced by sampling a value from some distribution and preparing a quantum state (of a size that also B' has to handle); the descriptions of the distribution and the state have to be computed by $\hat{B}'$ from the stored classical communication. By this, natural restrictions on B' concerning its *quantum capabilities* propagate to $\hat{B}'$. For instance if B' has a quantum memory of bounded size, so has $\hat{B}'$. Furthermore, in many cases the classical hybrid protocol is actually *unconditionally* secure against classical dishonest players and as such in particular secure against unbounded quantum dishonest players (because every dishonest quantum strategy can be simulated by an unbounded classical adversary), so no restriction on $\hat{B}'$ is needed.

Finally, note that we do not specify what it means for the hybrid protocol to be secure; Theorem 5.1 guarantees that *whatever* the hybrid protocol achieves, essentially the same is achieved by the real-life protocol with the oracle calls replaced by protocols. But of course in particular, if the hybrid protocol *is* secure in the sense of Definition 4.1, then so is the real-life protocol, and as such it could itself be used as a quantum sub-protocol in yet another classical outer protocol.

Corollary 5.2. *If $\Sigma^{\mathcal{F}_1 \cdots \mathcal{F}_\ell}$ is a δ-secure implementation of $\mathcal{G}$ against $\mathfrak{A}$ and $\mathfrak{B}$, and if π_i is an ε-secure implementation of $\mathcal{F}_i$ against $\mathfrak{A}$ and $\mathfrak{B}$ for every $i \in \{1, \ldots, \ell\}$, then $\Sigma^{\pi_1 \cdots \pi_\ell}$ is a $(\delta + 3k\varepsilon)$-secure implementation of $\mathcal{G}$.*

Proof (of Theorem 5.1). Correctness is obvious. We show security for Alice; security for Bob can be shown accordingly. Consider a dishonest B'. First we argue that for every distribution for Alice's input U, there exists a $\hat{B}'$ as claimed (which though may depend on P_U). Then, in the end, we show how to make $\hat{B}'$ independent of P_U.

Let A's input U be arbitrarily distributed. We prove the claim by induction on k. The claim holds trivially for protocols that make zero oracle calls. Consider now a protocol $\Sigma^{\mathcal{F}_1 \cdots \mathcal{F}_\ell}$ with at most $k > 0$ oracle calls. For simplicity, we assume that the number of oracle calls equals k, otherwise we instruct the players to makes some "dummy calls". Let $\rho_{S_k U_k V_k'}$ be the common state right before the k-th and thus last call to one of the sub-protocols $\pi_1, \ldots, \pi_\ell$ in the execution of the real protocol $\Sigma^{\pi_1, \ldots, \pi_\ell}$. To simplify notation in the rest of the proof, we omit the index k and write $\rho_{\bar{S}\bar{U}\bar{V}'}$ instead; see Figure 2. We know from the induction hypothesis for $k - 1$ that there exists $\hat{B}'$ such that $\rho_{\bar{S}\bar{U}\bar{V}'} \approx_{3(k-1)\varepsilon} \sigma_{\bar{S}\bar{U}\bar{V}'}$ where

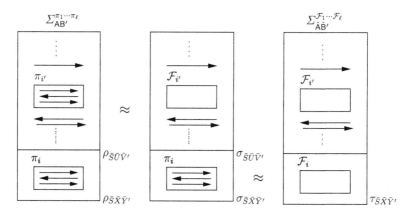

Fig. 2. Steps of the Composability Proof

$\sigma_{\bar{S}\bar{U}\bar{V}'}$ is the common state right before the k-th call to a functionality in the execution of the hybrid protocol $\Sigma_{\hat{A},\hat{B}'}^{\mathcal{F}_1\cdots\mathcal{F}_\ell}\rho_{U\emptyset}$. As described in Section 3, $\bar{S}, \bar{U}$ and $\bar{V}'$ are to be understood as follows. $\bar{S}$ denotes A's (respectively Â's) classical auxiliary information to be "remembered" during the call to the functionality. $\bar{U}$ denotes A's (respectively Â's) input to the sub-protocol (respectively functionality) that is to be called next, and $\bar{V}'$ denotes the dishonest player's current quantum state. For simplicity, we assume that the index i, which determines the sub-protocol π_i (functionality $\mathcal{F}_i$) to be called next, is *fixed* and we just write π and $\mathcal{F}$ for π_i and $\mathcal{F}_i$, respectively. If this is not the case, we consider $\rho_{\bar{S}\bar{U}\bar{V}'|\bar{I}=i}$ and $\sigma_{\bar{S}\bar{U}\bar{V}'|\bar{I}=i}$ instead, and reason as below for any i, where $\bar{I}$ denotes the index of the sub-protocol (functionality) to be called. Note that conditioning on $\bar{I}=i$ means that we allow $\hat{B}'$ to depend on i, but this is legitimate since $\bar{I}$ is known to the dishonest party.

Consider now the evolution of the state $\sigma_{\bar{S}\bar{U}\bar{V}'}$ when executing $\mathcal{F}_{\hat{A},\hat{B}'}$ (as prescribed by the hybrid protocol) with a strategy for $\hat{B}'$ yet to be determined and when executing $\pi_{A,B'}$ instead. Let $\sigma_{\bar{S}\bar{X}\bar{Y}'}$ and $\tau_{\bar{S}\bar{X}\bar{Y}'}$ denote the corresponding states after the execution of respectively $\pi_{A,B'}$ and $\mathcal{F}_{\hat{A},\hat{B}'}$, see Figure 2. We show that $\sigma_{\bar{S}\bar{X}\bar{Y}'}$ and $\tau_{\bar{S}\bar{X}\bar{Y}'}$ are 3ε-close; this then proves the result by the fact that evolution does not increase the trace distance and by the triangle inequality:

$$\rho_{\bar{S}\bar{X}\bar{Y}'} = (\mathrm{id}_{\bar{S}} \otimes \pi_{A,B'})\rho_{\bar{S}\bar{U}\bar{V}'} \approx_{3(k-1)\varepsilon} (\mathrm{id}_{\bar{S}} \otimes \pi_{A,B'})\sigma_{\bar{S}\bar{U}\bar{V}'} = \sigma_{\bar{S}\bar{X}\bar{Y}'}$$
$$\approx_{3\varepsilon} \tau_{\bar{S}\bar{X}\bar{Y}'} = (\mathrm{id}_{\bar{S}} \otimes \mathcal{F}_{\hat{A},\hat{B}'})\sigma_{\bar{S}\bar{U}\bar{V}'}.$$

Let $\sigma_{\bar{S}\bar{U}\bar{Z}\bar{V}'}$, $\sigma_{\bar{S}\bar{X}\bar{Z}\bar{Y}'}$ and $\tau_{\bar{S}\bar{X}\bar{Z}\bar{Y}'}$ be the extensions of the respective states $\sigma_{\bar{S}\bar{U}\bar{V}'}$, $\sigma_{\bar{S}\bar{X}\bar{Y}'}$ and $\tau_{\bar{S}\bar{X}\bar{Y}'}$ when we also consider $\bar{Z}$ (which collects the classical communication dictated by $\Sigma^{\mathcal{F}_1\cdots\mathcal{F}_\ell}$ as well as $\hat{B}'$'s classical inputs to and outputs from the previous oracle calls), which is guaranteed to exist by our formalization of a *classical* hybrid protocol, so that $\bar{Z}$ is without loss of generality contained in $\bar{V}'$ and $\sigma_{\bar{S}\bar{U}\bar{Z}\bar{V}'} = \sigma_{\bar{S}\bar{U}\leftrightarrow\bar{Z}\leftrightarrow\bar{V}'}$. It thus follows from Proposition 4.3 that $\sigma_{\bar{S}\bar{X}\bar{Z}\bar{Y}'}$ and $\tau_{\bar{S}\bar{X}\bar{Z}\bar{Y}'}$ are 3ε-close for a proper strategy of $\hat{B}'$. Note that the strategy of $\hat{B}'$ may depend on the state $\sigma_{\bar{S}\bar{U}\bar{Z}\bar{V}''}$, but since P_U as well as $\hat{A}$'s behavior are fixed, $\sigma_{\bar{S}\bar{U}\bar{Z}\bar{V}'}$ is also fixed.

It remains to argue that we can make $\hat{B}'$ independent of P_U. We use an elegant argument due to Crépeau and Wullschleger [6]. We know that for any P_U there exists a $\hat{B}'$ (though depending on P_U) as required. For any value u that U may take on, let then

$$\varepsilon_u = \delta\left(\Sigma_{\mathsf{A},\mathsf{B}'}^{\pi_1\cdots\pi_\ell}\rho_{U\emptyset|U=u}, \Sigma_{\hat{\mathsf{A}},\hat{\mathsf{B}'}}^{\mathcal{F}_1\cdots\mathcal{F}_\ell}\rho_{U\emptyset|U=u}\right).$$

Then, $\sum_u P_U(u)\varepsilon_u = 3k\varepsilon$. The ε_u's depend on P_U, and thus we also write $\varepsilon_u(P_U)$. Consider now the function F which maps an arbitrary distribution P_U for U to a new distribution defined as $F(P_U)(u) := \frac{1+\varepsilon_u(P_U)}{1+3k\varepsilon}P_U(u)$. Function F is continuous and maps a non-empty, compact, convex set onto itself. Thus, by Brouwer's Fixed Point Theorem, it must have a fixed point: a distribution P_U with $F(P_U) = P_U$, and thus $\varepsilon_u(P_U) = 3k\varepsilon$ for any u. It follows that $\hat{B}'$ which works for that particular distribution P_U in fact works for any specific value for U and so for any distribution of U.   □

6   Example: Secure Identification

We show that the information-theoretic security definition proposed by Damgård *et al.* for their secure-identification quantum protocol in the bounded-quantum-storage model [9] implies security in our sense for a proper functionality $\mathcal{F}_{ID}$; this guarantees composability as in Theorem 5.1 for their protocol. In the full version [11] of this paper, we also show the corresponding for the 1-2 OT scheme [7] and for other variants of OT.

A secure identification scheme allows a user Alice to identify herself to server Bob by securely checking whether the supplied password agrees with the one stored by Bob. Specifically, on respective input strings $W_A, W_B \in \mathcal{W}$ provided by Alice and Bob, the functionality outputs the bit $Y = (W_A \stackrel{?}{=} W_B)$ to Bob. A dishonest server B' should learn essentially no information on W_A beyond that he can come up with a guess W' for W_A and learns whether $W' = W_A$ or not, and similarly a dishonest user A' succeeds in convincing Bob essentially only if she guesses W_B correctly. If her guess is incorrect then the only thing she might learn is that her guess is incorrect. The corresponding ideal functionality is depicted in Figure 3. Note that if dishonest A' provides the "correct" input $W_A = W_B$, then $\mathcal{F}_{ID}$ allows A' to learn this while she may still enforce Bob to reject (by setting the "override bit" D to 0). In [11] we study a slightly stronger variant, which does not allow this somewhat unfair option for A'.[7]

We recall the security definition from [9] for a secure identification scheme. The definition is in the framework described in Section 4.1; thus, it considers a single execution of the protocol with an arbitrary distribution for the honest

---

[7] The reason we study here the weaker version is that this corresponds to the security guaranteed by the definition proposed in [9], as we show.

Functionality $\mathcal{F}_{ID}$: Upon receiving strings W_A and W_B from user Alice and from server Bob, $\mathcal{F}_{ID}$ outputs the bit $W_A \stackrel{?}{=} W_B$ to Bob.

If Alice is dishonest, she may input an additional "override bit" D. Then, $\mathcal{F}_{ID}$ outputs the bit $W_A \stackrel{?}{=} W_B$ to Alice and the bit $(W_A \stackrel{?}{=} W_B) \wedge D$ to Bob.

Fig. 3. The Ideal Password-Based Identification Functionality

players inputs and with no input for dishonest players, and security is defined by information-theoretic conditions on the resulting output states. For consistency with the above notation (and the notation used in [9]), Alice and Bob's inputs are denoted by W_A and W_B, respectively, rather than U and V. Furthermore, note that honest Alice's output X is empty: $X = \emptyset$.

Definition 6.1 (Secure Identification). *A password-based quantum identification scheme is ε-secure (against $\mathfrak{A}$ and $\mathfrak{B}$) if the following properties hold.*

Correctness: *For honest user Alice and honest server Bob, and for any joint input distribution $P_{W_A W_B}$, Bob learns whether their input is equal, except with probability ε.*

Security for Alice: *For any dishonest server $\mathsf{B}' \in \mathfrak{B}$, and for any distribution of W_A, the resulting common output state $\rho_{W_A Y'}$ (augmented with W_A) is such that there exists a classical W' that is independent of W_A and such that*

$$\rho_{W_A W' Y' | W_A \neq W'} \approx_\varepsilon \rho_{W_A \leftrightarrow W' \leftrightarrow Y' | W_A \neq W'} \, ,$$

Security for Bob: *For any dishonest user $\mathsf{A}' \in \mathfrak{A}$, and for any distribution of W_B, the resulting common output state $\rho_{W_B Y X'}$ (augmented with W_B) is such that there exists a classical W' independent of W_B, such that if $W_B \neq W'$ then $Y = 1$ with probability at most ε, and*

$$\rho_{W_B W' X' | W' \neq W_B} \approx_\varepsilon \rho_{W_B \leftrightarrow W' \leftrightarrow X' | W' \neq W_B} \, .$$

Proposition 6.2. *A quantum protocol satisfying Definition 6.1 3ε-securely implements the functionality $\mathcal{F}_{ID}$ from Figure 3 according to Definition 4.1.*

Proof. Correctness follows immediately.

Security for Alice: Consider W' which is guaranteed to exist by Definition 6.1. Let us define $V = W'$ and let Y be the bit $W_A \stackrel{?}{=} W'$. By the requirement of Definition 6.1, W' is independent of Alice's input W_A. Furthermore, we have

$$\bigl(W_A, W', \emptyset, Y\bigr) = \bigl(W_A, W', \mathcal{F}_{ID}(W_A, W')\bigr)$$

by the definition of $\mathcal{F}_{ID}$. Finally, we note that Y completely determines the event $\mathcal{E} := \{W_A \neq W'\}$ and therefore, we conclude using Lemma 2.3 that

$\rho_{W_A \emptyset W'YY'}$

$= \Pr[W_A \neq W'] \cdot \rho_{W_A \emptyset W'YY'|W_A \neq W'} + \Pr[W_A = W'] \cdot \rho_{W_A \emptyset W'YY'|W_A = W'}$

$= \Pr[W_A \neq W'] \cdot \rho_{W_A \emptyset W'YY'|W_A \neq W'} + \Pr[W_A = W'] \cdot \rho_{W_A \leftrightarrow W'Y \leftrightarrow Y'|W_A = W'}$

$\approx_\varepsilon \Pr[W_A \neq W'] \cdot \rho_{W_A \leftrightarrow W'Y \leftrightarrow Y'|W_A \neq W'} + \Pr[W_A = W'] \cdot \rho_{W_A \leftrightarrow W'Y \leftrightarrow Y'|W_A = W'}$

$= \rho_{W_A \leftrightarrow W'Y \leftrightarrow Y'} .$

Security for Bob: Consider the random variable W' which is guaranteed to exist by Definition 6.1. Let us define U and X as follows. We let $U = (W', D)$ where we define $D = Y$ if $W_B = W'$, and else we choose D "freshly" to be 0 with probability $\Pr[Y = 0|W_B = W']$ and to be 1 otherwise. Furthermore, we let $X = (W' \overset{?}{=} W_B)$. Recall that by the requirement of Definition 6.1, W' is independent of Bob's input W_B. Furthermore by construction, $D = 0$ with probability $\Pr[Y = 0|W_B = W']$, independent of the value of W_B (and independent of whether $W_B = W'$ or not). Thus, U is perfectly independent of W_B.

Since by Definition 6.1 the probability for Bob to decide that the inputs are equal, $Y = 1$, does not exceed ε if $W_B \neq W'$, we have that

$P_{UW_BXY} = \Pr[W_B = W'] \cdot P_{UW_BXY|W_B = W'} + \Pr[W_B \neq W'] \cdot P_{UW_BXY|W_B \neq W'}$

$= \Pr[W_B = W'] \cdot P_{UW_B \mathcal{F}_{ID}(U,W_B)|W_B = W'} + \Pr[W_B \neq W'] \cdot P_{UW_BXY|W_B \neq W'}$

$\approx_\varepsilon \Pr[W_B = W'] \cdot P_{UW_B \mathcal{F}_{ID}(U,W_B)|W_B = W'} + \Pr[W_B \neq W'] \cdot P_{UW_B \mathcal{F}_{ID}(U,W_B)|W_B \neq W'}$

$= P_{UW_B \mathcal{F}_{ID}(U,W_B)} .$

Finally, we have

$$\rho_{W_B YUXX'} = \Pr[W_B \neq W'] \cdot \rho_{W_B YW'DXX'|W_B \neq W'} + \Pr[W_B = W'] \cdot \rho_{W_B YW'DXX'|W_B = W'} .$$

In the case $W_B = W'$, we have by construction that $D = Y$ and therefore, we obtain that $\rho_{W_B YW'DXX'|W_B = W'} = \rho_{W_B Y \leftrightarrow W'D \leftrightarrow XX'|W_B = W'}$. If $W_B \neq W'$, it follows from Definition 6.1 and the fact that D is sampled independently that $\rho_{W_B W'DX'|W' \neq W_B} \approx_\varepsilon \rho_{W_B \leftrightarrow W'D \leftrightarrow X'|W' \neq W_B}$. Furthermore, the bit X is fixed to 0 in case $W_B \neq W'$ and we only make an error of at most ε assuming that Bob's output Y is always 0 and therefore,

$$\rho_{W_B YW'DXX'|W_B \neq W'} \approx_\varepsilon \rho_{W_B (Y=0)W'D(X=0)X'|W_B \neq W'}$$
$$\approx_\varepsilon \rho_{W_B (Y=0) \leftrightarrow W'D(X=0) \leftrightarrow X'|W_B \neq W'} \approx_\varepsilon \rho_{W_B Y \leftrightarrow W'DX \leftrightarrow X'|W_B \neq W'}$$

Putting things together, we obtain

$$\rho_{W_B YUXX'} \approx_{3\varepsilon} \Pr[W_B \neq W'] \cdot \rho_{W_B Y \leftrightarrow W'DX \leftrightarrow X'|W_B \neq W'} + \Pr[W_B = W'] \cdot \rho_{W_B Y \leftrightarrow W'D \leftrightarrow XX'|W_B = W'}$$
$$= \rho_{W_B Y \leftrightarrow (W'D)X \leftrightarrow X'} ,$$

where we used Lemma 2.1 and 2.3 in the last step.   □

7   Conclusion

We proposed a general security definition for quantum protocols in terms of simple quantum-information-theoretic conditions and showed that quantum protocols fulfilling the definition do their job as expected when used as subroutines in a larger classical protocol. The restriction to classical "outer" protocols fits our currently limited ability for executing quantum protocols, but can also be appreciated in that our security conditions pose *minimal* requirements for a quantum protocol to be useful beyond running it in isolation.

Acknowledgements

We would like to thank Jürg Wullschleger for sharing a draft of [6] and pointing out how to avoid the dependency of the dishonest player in the ideal model from the honest player's input distribution.

References

1. Ben-Or, M., Crépeau, C., Gottesman, D., Hassidim, A., Smith, A.: Secure multiparty quantum computation with (only) a strict honest majority. In: 46th Annual IEEE Symposium on Foundations of Computer Science (FOCS), pp. 249–260 (2005)
2. Ben-Or, M., Horodecki, M., Leung, D.W., Mayers, D., Oppenheim, J.: The universal composable security of quantum key distribution. In: Kilian, J. (ed.) TCC 2005. LNCS, vol. 3378, pp. 386–406. Springer, Heidelberg (2005)
3. Ben-Or, M., Mayers, D.: General security definition and composability for quantum and classical protocols (September 2004),
   http://arxive.org/abs/quant-ph/0409062
4. Colbeck, R.: The impossibility of secure two-party classical computation (August 2007), http://arxiv.org/abs/0708.2843
5. Crépeau, C., Savvides, G., Schaffner, C., Wullschleger, J.: Information-theoretic conditions for two-party secure function evaluation. In: Vaudenay, S. (ed.) EUROCRYPT 2006. LNCS, vol. 4004, pp. 538–554. Springer, Heidelberg (2006)
6. Crépeau, C., Wullschleger, J.: Statistical security conditions for two-party secure function evaluation. In: Safavi-Naini, R. (ed.) ICITS 2008. LNCS, vol. 5155, pp. 86–99. Springer, Heidelberg (2008)
7. Damgård, I.B., Fehr, S., Renner, R., Salvail, L., Schaffner, C.: A tight high-order entropic quantum uncertainty relation with applications. In: Menezes, A. (ed.) CRYPTO 2007. LNCS, vol. 4622, pp. 360–378. Springer, Heidelberg (2007)
8. Damgård, I.B., Fehr, S., Salvail, L., Schaffner, C.: Cryptography in the bounded quantum-storage model. In: 46th Annual IEEE Symposium on Foundations of Computer Science (FOCS), pp. 449–458 (2005),
   http://arxiv.org/abs/quant-ph/0508222v2
9. Damgård, I.B., Fehr, S., Salvail, L., Schaffner, C.: Secure identification and QKD in the bounded-quantum-storage model. In: Menezes, A. (ed.) CRYPTO 2007. LNCS, vol. 4622, pp. 342–359. Springer, Heidelberg (2007)
10. Damgård, I.B., Fehr, S., Salvail, L., Schaffner, C.: Cryptography in the bounded-quantum-storage model. SIAM Journal on Computing 37(6), 1865–1890 (2008)

11. Fehr, S., Schaffner, C.: Composing quantum protocols in a classical environment (2008), http://arxiv.org/abs/0804.1059
12. Goldreich, O.: Foundations of Cryptography: Basic Applications, vol. II. Cambridge University Press, Cambridge (2004)
13. Gottesman, D., Lo, H.-K.: Proof of security of quantum key distribution with two-way classical communications. IEEE Transactions on Information Theory 49(2), 457–475 (2003), http://arxiv.org/abs/quant-ph/0105121
14. J.: v. d. Graaf. Towards a formal definition of security for quantum protocols. PhD thesis, Université de Montréal (1997)
15. Kilian, J.: Founding cryptography on oblivious transfer. In: 20th Annual ACM Symposium on Theory of Computing (STOC), pp. 20–31 (1988)
16. Kitaev, A.: Quantum coin-flipping. In: QIP 2003 (2003); A review of this technique can be found, http://lightlike.com/~carlosm/publ
17. Koenig, R., Renner, R., Bariska, A., Maurer, U.: Small accessible quantum information does not imply security. Physical Review Letters 98(140502) (April 2007)
18. Lo, H.-K.: Insecurity of quantum secure computations. Physical Review A 56(2), 1154–1162 (1997)
19. Lo, H.-K., Chau, H.F.: Is quantum bit commitment really possible? Physical Review Letters 78(17), 3410–3413 (1997)
20. Mayers, D.: Unconditionally secure quantum bit commitment is impossible. Physical Review Letters 78(17), 3414–3417 (1997)
21. Nielsen, M.A., Chuang, I.L.: Quantum Computation and Quantum Information. Cambridge University Press, Cambridge (2000)
22. Renner, R.: Security of Quantum Key Distribution. PhD thesis, ETH Zürich (Switzerland) (September 2005), http://arxiv.org/abs/quant-ph/0512258
23. Renner, R., König, R.: Universally composable privacy amplification against quantum adversaries. In: Kilian, J. (ed.) TCC 2005. LNCS, vol. 3378, pp. 407–425. Springer, Heidelberg (2005)
24. Salvail, L.: Quantum bit commitment from a physical assumption. In: Krawczyk, H. (ed.) CRYPTO 1998. LNCS, vol. 1462, pp. 338–353. Springer, Heidelberg (1998)
25. Salvail, L., Sotáková, M., Schaffner, C.: On the power of two-party quantum cryptography (submitted, 2008)
26. Smith, A.: Multi-party quantum computation. Master's thesis, MIT (2001)
27. Unruh, D.: Simulatable security for quantum protocols (2004), http://arxiv.org/abs/quant-ph/0409125
28. Wehner, S., Schaffner, C., Terhal, B.M.: Cryptography from noisy storage. Physical Review Letters 100(22), 220502 (2008)
29. Wehner, S., Wullschleger, J.: Composable security in the bounded-quantum-storage model. In: Aceto, L., Damgård, I., Goldberg, L.A., Halldórsson, M.M., Ingólfsdóttir, A., Walukiewicz, I. (eds.) ICALP 2008, Part II. LNCS, vol. 5126, pp. 604–615. Springer, Heidelberg (2008)

LEGO for Two-Party Secure Computation

Jesper Buus Nielsen and Claudio Orlandi

BRICS, Department of Computer Science, Aarhus University
{jbn,orlandi}@cs.au.dk

Abstract. This paper continues the recent line of work of making Yao's garbled circuit approach to two-party computation secure against an active adversary. We propose a new cut-and-choose based approach called LEGO (Large Efficient Garbled-circuit Optimization): It is specifically aimed at large circuits. Asymptotically it obtains a factor $\log |\mathcal{C}|$ improvement in computation and communication over previous cut-and-choose based solutions, where $|\mathcal{C}|$ is the size of the circuit being computed. The protocol is universally composable (UC) in the OT-hybrid model against a static, active adversary.

1 Introduction

In secure two-party computation we have two parties, Alice and Bob, that want to compute a function $f(\cdot, \cdot)$ of their inputs a, b, and learn the result $y = f(a, b)$, without any party learning any other information.

Yao [Yao82, Yao86] was the first to present a solution to this problem. His protocol, presented and proved in [LP04], is only secure against a passive adversary. We give a novel approach to making Yao's idea secure against active adversaries.

In Yao's protocol Alice constructs a garbled circuit and sends it to Bob: a malicious Alice can send Bob a circuit that does not compute the agreed function, as a consequence the computation loses both privacy and correctness. In our protocol instead Alice and Bob both participate in the circuit construction. The main idea of our protocol is to have Alice prepare and send a bunch of garbled NAND gates (together with some other components) to Bob. Bob selects a random subset of the gates and Alice provides Bob with the keys to test them. If they all work correctly he assumes that at most a small fraction of the remaining gates are malfunctioning. Bob shuffles the remaining gates to put the faulty gates in random positions and connects them into a circuit that computes the desired function even in the presence of a few random faults — the scrambled NAND gates are designed such that Bob can, with a limited help from Alice, connect the gates as he likes. Then the circuit is evaluated by Bob as in Yao's protocol: Bob gets his keys running oblivious transfers (OT) with Alice and he evaluates the circuit.

Related Work: In the last years many solutions have been proposed to achieve two-party computation secure against malicious adversaries.

O. Reingold (Ed.): TCC 2009, LNCS 5444, pp. 368–386, 2009.

In [LP07, LPS08], Alice sends s copies of the Yao's garbled circuit to be computed. Bob checks half of them and computes on the remaining circuits. A similar approach was suggested in [MF06]. Due to the circuit replication, they need to introduce some machinery in order to force the parties to provide the same inputs to every circuit, resulting in an overhead of s^2 commitments per input wire for a total of $O(s\kappa|\mathcal{C}| + s^2\kappa|\mathcal{I}|)$, where $|\mathcal{C}|$ is the size of the circuit, $|\mathcal{I}|$ is the size of the input and κ is the length of a hash value. To optimize the cut-and-choose construction, Woodruff [Woo07] proposed a way of proving input consistency using expander graphs: using this construction it is possible to get rid of the dependency on the input size and therefore achieving complexity of $O(s\kappa|\mathcal{C}|)$. More concretely, the protocol in [LP07, LPS08] requires s copies of a circuit of size $|\mathcal{C}| + |D|$, where D is an input decoder, added to the circuit to deal with so-called selective failures. They propose a basic version of D with size $O(s|\mathcal{I}|)$ and a more advanced with size $O(s + |\mathcal{I}|)$. However, because of the $s^2|\mathcal{I}|$ commitments, their optimized encoding gives them just a benefit in the number of OT required. With our construction, we can fully exploit their encoding. In fact we need just to replicate $s/\log(|\mathcal{C}|)$ times a circuit of size $O(|\mathcal{C}| + s + |\mathcal{I}|) = O(|\mathcal{C}|)$, which gives our protocol a complexity of $O((s/\log(|\mathcal{C}|))\kappa|\mathcal{C}|)$, i.e., our replication factor is reduced by the logarithm of the circuit size. The improvement in replication factor from s to $s/\log(|\mathcal{C}|)$ comes from doing cut-and-choose on individual gates instead of doing it on entire circuits.

Another approach to making Yao's idea actively secure is to use generic zero-knowledge proofs to force good behavior [GMW86]. In theory this can be done with just a constant overhead in communication [NN01].

Other related works include: Considering UC security, in [JS07] a solution for two party computation on committed inputs is presented. This construction uses public-key encryption together with efficient zero-knowledge proofs, in order to prove that the circuit was built correctly. Their asymptotic complexity is $O(\kappa'|\mathcal{C}|)$, where κ' is the length of factorization-based public-key cryptosystems. In our protocol the parameter κ can be chosen to be much smaller i.e., the required size for hashing and elliptic curves cryptography.

In [IPS08] a protocol for UC secure two-party computation in the OT-hybrid model is presented with communication complexity $O(|\mathcal{C}|) + \text{poly}(s, d, \log |\mathcal{C}|)$, where s is the security parameter and d is the depth of $\mathcal{C}$. The hidden constant factor and the term $\text{poly}(s, d, \log |\mathcal{C}|)$, however, makes the comparison between the protocols unclear. Moreover, their protocol has non-constant round complexity as opposed to ours. Even the situation for very small or alternatively very large circuits is not clear a priori without trying to optimize and implement both approaches.

Our Contribution: Our scheme is based on three assumptions. We need a UC secure OT scheme. In addition we assume a finite group and an element g of prime order p such that the discrete logarithm problem is hard in $\langle g \rangle$, for instance the group of points over an elliptic curve. Finally we need a function $H : \mathbb{Z}_p \rightarrow \mathbb{Z}_p$ which is collision resistant and which is correlation resistant according to Def. 1.

Definition 1. *Given $F : \mathbb{Z}_p \to \mathbb{Z}_p$, let $\mathcal{O}^F$ be the following oracle: It samples a uniformly random $\Delta \in \mathbb{Z}_p$ and stores Δ. Whenever it is queried on $c \in \{0, 1, 2\}$, it samples a uniformly random $K_0 \in \mathbb{Z}_p$, lets $K_1 = K_0 + \Delta \bmod p$, $K_2 = K_0 + 2\Delta \bmod p$ and returns K_c and $F(K_d)$ for $d \in \{0, 1, 2\} \setminus \{c\}$. We call H correlation resistant if no poly-time adversary can distinguish $\mathcal{O}^H$ from $\mathcal{O}^R$, where R is a uniformly random function from $\mathbb{Z}_p$ to $\mathbb{Z}_p$.*

It is clear that a random function is correlation resistant and it seems reasonable to assume that practical hash functions satisfy this property. The notion of correlation resistance is closely related to the notion of correlation robustness in [IKNP03].

Theorem 1. *If H is collision resistant and correlation resistant according to Def. 1 with output length at most κ, the DL problem is hard in $\langle g \rangle$, elements of $\langle g \rangle$ can be represented with κ bits and 2^{-s} is negligible, then our protocol securely evaluates any function $y = f(a, b)$ computed by a poly-sized Boolean circuit $\mathcal{C}$. The protocol is UC secure against a static, active adversary in the OT-hybrid model. The round complexity is constant, the communication complexity is $O(\kappa s |\mathcal{C}| / \log |\mathcal{C}|)$. Regarding computational complexity, $O(s|\mathcal{C}| / \log |\mathcal{C}|)$ exponentiations in $\langle g \rangle$ are performed and the number of OT calls is $O(|\mathcal{I}| + s)$.*

The version of our protocol presented in this short version of the paper is less efficient than need be, to allow clearer presentation and analysis. In the full version [NO08] some (constant factor) efficiency improvements are presented.

2   Ideal Functionalities

The ideal functionality we implement is described in Fig. 1 (see Fig. 2 for notation). It is "insecure" in the sense that it allows Alice to guess Bob's input bits. This can be solved in a black-box manner by replacing $\mathcal{C}$ with a circuit computing a function of an encoded version of Bob's input. The randomized

---

The ideal functionality $\mathcal{F}_{\text{SCE}}$ is parametrized by a circuit $\mathcal{C}$ and runs as follows:

Inputs: Alice inputs $\{x_w\}_{w \in \mathcal{I}_A}$ to specify an input bit x_w for each of her wires w, and Bob inputs $\{x_w\}_{w \in \mathcal{I}_B}$.

Abort: If a party is corrupted and inputs abort!, then $\mathcal{F}_{\text{SCE}}$ outputs abort! to the other party and terminates.

Evaluation: If no party inputs abort!, then $\mathcal{F}_{\text{SCE}}$ computes $\{y_w\}_{w \in \mathcal{O}} = \mathcal{C}(\{x_w\}_{w \in \mathcal{I}})$ and outputs $\{y_w\}_{w \in \mathcal{O}}$ to Alice.

Guess: If Alice is corrupted, she can after Bob inputs $\{x_w\}_{w \in \mathcal{I}_B}$ give an input $W \subseteq \mathcal{I}_B$ and $\{\beta_w\}_{w \in W}$. If $\beta_w = x_w$ for all $w \in W$, then $\mathcal{F}_{\text{SCE}}$ outputs correct! to Alice and continues as above. Otherwise, it outputs You were busted! to Alice, outputs Alice cheats! to Bob and then terminates.

Fig. 1. Our ideal functionality for secure circuit evaluation

encoding is such that any s bits of a codeword are uniformly random and independent. The security follows from the fact that Alice cannot guess s or more bits with probability more than 2^{-s}, and if she guesses less bits she will learn no information. One method for this is given in [LP07]. The extra number of gates used is $O(|\mathcal{I}_B| + s)$, where $|\mathcal{I}_B|$ is the length of Bob's input and s is a security parameter.

We implement $\mathcal{F}_{\text{SCE}}$ in the OT-hybrid model with an ideal functionality $\mathcal{F}_{\text{OT}}$. We also assume an ideal functionality $\mathcal{F}_{\text{ZK}}$ for zero-knowledge proof of knowledge. This can be implemented by s calls to the OT functionality: The prover offers a reply to challenge $e = 0$ and $e = 1$ and the verifier chooses and verifies one of the replies. The simulator reads the verifier's challenge and sets the corresponding message to be a simulated reply. The extractor can get both replies and compute the witness if both are correct. Repeating s times gives a soundness error of 2^{-s}. We use in total just 2 calls to the proof of knowledge functionality, giving an overhead of $2s$ calls to $\mathcal{F}_{\text{OT}}$.

3   LEGO Circuits

We start by describing our variation of Yao circuits. It is designed to allow Alice to generate garbled gates independently and later let Bob connect the gates in any order. In the description we use the notation in Fig. 2.

We work with garbled circuits where each wire can carry a value $c \in \{0, 1, 2\}$. Input wires to the circuit and output values from the circuit only carry values $c \in \{0, 1\}$ — the value $c = 2$ is only carried by certain internal wires. For a wire with name w we use $\mathcal{V}(w) \in \{0, 1, 2\}$ to denote the *value* carried by the wire. The values on input wires are specified by Alice and Bob.

LEGO circuits consist of wires and so-called bricks. Wires are essentially just names $w \in \{0, 1\}^*$ to which we will associate certain values, in particular a zero-key K_0 and a commitment $[K_0]$. We write $\mathcal{Z}(w) = K_0$ to denote the zero-key associated to wire w and we write $\mathcal{COM}(w) = [K_0]$ to denote the associated commitment to K_0. Alice knows K_0 and $[K_0]$ while Bob knows only $[K_0]$ from the beginning. During the evaluation of the garbled circuit Bob will learn a key $K_c \in \{K_0, K_1, K_2\}$ for each wire w — here $K_c = K_0 + c\Delta \bmod p$. We think of this as the wire w carrying the value $c \in \{0, 1, 2\}$. Bob does not know the value of K_0 and therefore he does not know the value of c.

Bricks are special garbled circuits allowing Bob to compute on keys K_c. As an example we sketch the not-two brick. It has one input wire w_I and one output wire w_O — these are just unique names from $\{0, 1\}^*$. Let $I_0 = \mathcal{Z}(w_I)$ and $O_0 = \mathcal{Z}(w_O)$ be the associated zero-keys. If Bob knows $I_x \in \{I_0, I_1, I_2\}$ then the not-two brick will allow Bob to compute $O_z \in \{O_0, O_1\}$ with $z = \text{nt}(x)$. The not-two brick does not leak either x or z to Bob. We also have an addition brick with two input wires (with zero-keys L_0 and R_0) and one output wire (with zero-key S_0): from $L_x \in \{L_0, L_1\}$ and $R_y \in \{R_0, R_1\}$ it allows Bob to compute $S_{x+y} \in \{S_0, S_1, S_2\}$. Finally we have a key-filter brick with one associated zero-key K_0. Given a possibly large set of keys $\mathcal{K} \subset \mathbb{Z}_p$ it allows to compute $\mathcal{K} \cap \{K_0, K_1\}$. It

- s is the security parameter.
- $\mathcal{C}$ is a Boolean circuit specifying the function to be computed. It consists of NAND gates only.
- $|\mathcal{C}|$ is the number of NAND gates in $\mathcal{C}$.
- $\mathcal{I}, \mathcal{O}$ are the names of input wires respectively output wires in $\mathcal{C}$. $\mathcal{I}_A \subset \mathcal{I}$ are Alice's input wires and $\mathcal{I}_B = \mathcal{I} \setminus \mathcal{I}_A$ are Bob's input wires.
- $\{y_w\}_{w \in \mathcal{O}} = \mathcal{C}\{x_w\}_{w \in \mathcal{I}}$ says that if input wires $w \in \mathcal{I}$ are assigned the bits x_w and $\mathcal{C}$ is evaluated on them, then the output wires $w \in \mathcal{O}$ will hold the bits y_w.
- p is a large prime.
- $[x]$ is a Pedersen commitment to $x \in \mathbb{Z}_p$. It is computed as $[x] = g^x h^r$ for uniformly random $r \in \mathbb{Z}_p$. Here g is a group generator of order p, and $h \in_R \langle g \rangle$ is chosen by Bob. The prime p and the group is picked such that no poly-time algorithm can solve the DL problem in $\langle g \rangle$ with probability better than 2^{-s}.
- If $[x] = g^x h^r$ and $[y] = g^y h^s$ then $[x] \boxplus [y] = [x][y] = g^{x+y} h^{r+s}$ is a commitment to $x + y \bmod p$, and $[x] \boxminus [y] = [x][y]^{-1} = g^{x-y} h^{r-s}$ is a commitment to $x - y \bmod p$.
- $\Delta \in \mathbb{Z}_p$ is the *global difference*, chosen uniformly at random by Alice and unknown by Bob; $[\Delta]$ is a uniformly random commitment to it. Our use of Δ is inspired by [KS08].
- $K_0 \in \mathbb{Z}_p$ denotes a so-called *zero-key*. After Δ is fixed it defines the *one-key* $K_1 = K_0 + \Delta \bmod p$ and the *two-key* $K_2 = K_0 + 2\Delta \bmod p$. The key K_c will be Bob's representation of the "plaintext" $c \in \{0, 1, 2\}$.
- $[K_0]$ is a commitment to a zero-key, always produced by Alice. From $[K_0]$ and $[\Delta]$ Bob can compute commitments $[K_1] = [K_0] \boxplus [\Delta]$ and $[K_2] = [K_1] \boxplus [\Delta]$.
- $\Sigma \in \mathbb{Z}_p$ denotes a so-called *shifting value*: It is a difference $\Sigma = K_0' - K_0 \bmod p$ between two zero-keys K_0 and K_0'. Note that $K_c + \Sigma = K_c'$ for $c = 0, 1, 2$.
- $H : \mathbb{Z}_p \rightarrow \mathbb{Z}_p$ is a hash function which is collision resistant and correlation resistant according to Def. 1. It is picked such that no poly-time adversary can distinguish with probability better than 2^{-s} in Def. 1.
- For $K, K' \in \mathbb{Z}_p$ we let $E_K(K') = H(K) + K' \bmod p$ and think of it as a deterministic encryption of K' using K. We let $D_K(C) = C - H(K) \bmod p = K'$.
- We define nt : $\{0, 1, 2\} \rightarrow \{0, 1\}$ by nt(0) = nt(1) = 1 and nt(2) = 0.
- We define $\bar{\wedge} : \{0, 1\} \times \{0, 1\} \rightarrow \{0, 1\}$ by $a \bar{\wedge} b = 0$ iff $a = 1$ and $b = 1$. Note that if $a, b \in \{0, 1\}$, then $\text{nt}(a + b) = a \bar{\wedge} b$.
- $\mathcal{K} : \{0, 1\}^* \rightarrow 2^{\mathbb{Z}_p}$ maps wire names $w \in \{0, 1\}^*$ to subsets of keys.

Fig. 2. Notation, explained further in main text

is among other things used to ensure that Alice sends valid keys for her input wires.

In the evaluation of the circuit Alice will send $K_a \in \{K_0, K_1\}$ to represent her input $a \in \{0, 1\}$ for each of her input wires w with zero-key K_0. Bob uses a key-filter brick to ensure that $K_a \in \{K_0, K_1\}$. For each of Bob's input wires Alice offers K_0 and K_1 in an OT and Bob inputs b to get the key K_b representing input $b \in \{0, 1\}$. Then the circuit $\mathcal{C}$ is evaluated by Bob on these keys. Each NAND gate in $\mathcal{C}$ is implemented by an addition brick followed by a not-two brick. When Bob knows $K_c = K_0 + c\Delta \bmod p$ for an output wire he sends K_c to Alice who computes $c \in \{0, 1\}$ using K_0 and Δ.

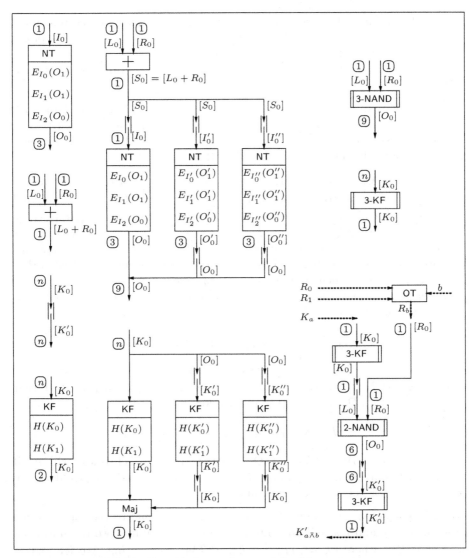

Fig. 3. Graphical notation, explained in main text

The above did not use the associated commitments $[K_0]$. These are used to allow Bob to connect the bricks as he desires, even though they were generated independently. A wire w with zero-key K_0 is connected to a wire w' with zero-key K_0' by Alice opening $[K_0'] \boxminus [K_0]$ to Bob to let him learn $\Sigma = K_0' - K_0 \bmod p$. Given K_c for w, Bob can then compute $K_c' = K_c + \Sigma \bmod p$ for w'. The commitments are use to prevent Alice from sending a wrong value of Σ.

Alice can, however, cheat when she generates the bricks. To deal with this she produces more bricks than needed and a random subset of them is selected by Bob for testing. In the test Bob selects a random input for the brick and Alice sends the appropriate keys by opening the associated commitments. If the selected bricks pass the test Bob will be ensured that most of the remaining bricks will also run correctly. To deal with a small number of incorrect bricks Bob will replicate each of the bricks in the circuit design.

We now give the details of how bricks are generated, evaluated, connected and replicated.

Potential Keys: Above we said that Bob learns $K_c \in \{K_0, K_1, K_2\}$ for each wire. In fact, at some points Bob might hold more than one potential key for a wire. To ease notation we introduce a function $\mathcal{K} : \{0,1\}^* \to 2^{\mathbb{Z}_p}$ mapping a wire name w to the set of potential keys held by Bob.

NT Bricks: An *NT brick* (Fig. 3 top, left) with input wire w_I and output wire w_O contains two commitments $[I_0]$ and $[O_0]$, where $\mathcal{COM}(w_I) \stackrel{\text{def}}{=} [I_0]$ and $\mathcal{COM}(w_O) \stackrel{\text{def}}{=} [O_0]$. Besides it contains three encryptions $C_0 = E_{I_0}(O_1), C_1 = E_{I_1}(O_1), C_2 = E_{I_2}(O_0)$. The zero-keys $I_0, O_0 \in \mathbb{Z}_p$ are chosen uniformly at random by Alice, and Alice computes and sends $(w_I, w_O, [I_0], \{C_0, C_1, C_2\}, [O_0])$ to Bob. The three ciphertexts are sent in a randomly permuted order, so that Bob does not know which ciphertext encrypts which key.

Given potential keys $\mathcal{K}(w_I)$ for the input wire w_I, Bob computes potential keys for the output wire as follows:

$$\mathcal{K}(w_O) \stackrel{\text{def}}{=} \bigcup_{K \in \mathcal{K}(w_I)} \{D_K(C_0), D_K(C_1), D_K(C_2)\} .$$

If $K = I_c$ for $I_c \in \{I_0, I_1, I_2\}$, then it follows from $C_c \in \{C_0, C_1, C_2\}$ that $O_{\text{nt}(c)} \in \{D_K(C_0), D_K(C_1), D_K(C_2)\}$. Therefore

$$I_c \in \mathcal{K}(w_I) \quad \Rightarrow \quad O_{\text{nt}(c)} \in \mathcal{K}(w_O) .$$

Note that $|\mathcal{K}(w_O)| \leq 3|\mathcal{K}(w_I)|$. Our particular use of NT bricks will at all times ensure that $|\mathcal{K}(w_I)| = 1$, and thus $|\mathcal{K}(w_O)| \leq 3$. This is depicted by the numbers in circles next to the wires in Fig. 3.

Addition Bricks: An *addition brick* (Fig. 3 left, second to the top) has two input wires w_L and w_R and one output wire w_S. It contains two commitments $[L_0]$ and $[R_0]$, where $\mathcal{COM}(w_L) \stackrel{\text{def}}{=} [L_0]$, $\mathcal{COM}(w_R) \stackrel{\text{def}}{=} [R_0]$. We let $[S_0] = [L_0] \boxplus [R_0]$ and let $\mathcal{COM}(w_S) \stackrel{\text{def}}{=} [S_0]$. I.e., the zero-key S_0 for the output wire is $S_0 = L_0 + R_0 \bmod p$. Alice picks $L_0, R_0 \in \mathbb{Z}_p$ uniformly at random and sends $(w_L, w_R, w_S, [L_0], [R_0])$ to Bob.

Given potential keys $\mathcal{K}(w_L)$ and $\mathcal{K}(w_R)$ for the input wires, Bob computes potential keys for the output wire as follows:

$$\mathcal{K}(w_S) \stackrel{\text{def}}{=} \mathcal{K}(w_L) + \mathcal{K}(w_R) = \{L + R \bmod p | L \in \mathcal{K}(w_L) \wedge R \in \mathcal{K}(w_R)\} .$$

Note that $L_x + R_y \equiv_p (L_0 + x\Delta) + (R_0 + y\Delta) \equiv_p (L_0 + R_0) + (x + y)\Delta \equiv_p S_{x+y}$. Therefore, if $x, y \in \{0, 1\}$, then

$$L_x \in \mathcal{K}(w_L) \wedge R_y \in \mathcal{K}(w_R) \quad \Rightarrow \quad S_{x+y} \in \mathcal{K}(w_S) .$$

Our particular use of addition bricks will at all times ensure that $|\mathcal{K}(w_L)| = 1$ and $|\mathcal{K}(w_R)| = 1$, and thus $|\mathcal{K}(w_S)| = 1$. This is depicted by the numbers in circles next to the wires in Fig. 3.

Shifts: The next component is the *shift* (Fig. 3 left, second to the bottom). Assume that Bob just evaluated a brick, like an NT brick, to get potential keys $\mathcal{K}(w)$ for its output wire w, and let $[K_0] = \mathcal{COM}(w)$ be the commitment associated to w. Bob needs to use $\mathcal{K}(w)$ as input for the next brick in the LEGO circuit. The next brick was, however, generated independently of the brick producing the potential keys $\mathcal{K}(w)$. This means that it has an input wire $w' \neq w$ and another associated commitment $[K'_0] = \mathcal{COM}(w')$. This is handled as follows: After Bob announces the position of the bricks in the LEGO circuit, Alice will open the commitment $[K'_0] \boxminus [K_0]$ to let Bob learn the *shifting value* $\Sigma = K'_0 - K_0 \bmod p$ for each pair of output wire w and input wire w', where w is to feed w'. We denote such a connection by $w \Longrightarrow w'$. Given Σ, Bob can compute

$$\mathcal{K}(w') \stackrel{\text{def}}{=} \mathcal{K}(w) + \Sigma = \{K + \Sigma \bmod p | K \in \mathcal{K}(w)\} .$$

Note that $K_c + \Sigma \equiv_p (K_0 + c\Delta) + (K'_0 - K_0) \equiv_p K'_0 + c\Delta \equiv_p K'_c$. Therefore

$$K_c \in \mathcal{K}(w) \quad \Rightarrow \quad K'_c \in \mathcal{K}(w') ,$$

unless Alice opens $[K'_0] \boxminus [K_0]$ to $\Sigma \neq K'_0 - K_0 \bmod p$, which would involve breaking the computational binding of the commitment scheme. We make this precise in the formal analysis. Clearly $|\mathcal{K}(w')| = |\mathcal{K}(w)|$.

KF Bricks: Already with the above components one can evaluate a NAND circuit $\mathcal{C}$, using one addition brick and one NT brick to securely evaluate each NAND gate in $\mathcal{C}$. The maximal size of potential-key sets, however, grows by a factor 3 after each NT gate and squares after each addition gate. We deal with this using the key-filter brick.

A *KF brick* (Fig. 3 left, bottom) has one input wire w_I and one output wire w_O. It consists of one commitment $[K_0]$ and $\mathcal{COM}(w_I) \stackrel{\text{def}}{=} [K_0]$ and $\mathcal{COM}(w_O) \stackrel{\text{def}}{=} [K_0]$. It also contains two hash values $T_0 = H(K_0)$ and $T_1 = H(K_1)$. Alice chooses K_0 uniformly at random and sends $([K_0], \{T_0, T_1\})$ to Bob, with T_0 and T_1 in a uniformly random order.

When Bob has potential-key set $\mathcal{K}(w_I)$ for the input wire he computes potential keys for w_O as follows

$$\mathcal{K}(w_O) \stackrel{\text{def}}{=} \{K \in \mathcal{K}(w_I) | H(K) \in \{T_0, T_1\}\} .$$

It clearly holds for $b \in \{0, 1\}$ that $K_b \in \mathcal{K}(w_I) \Rightarrow K_b \in \mathcal{K}(w_O)$. It is also clear that when Alice knows K_0 and K_1 such that $T_0 = H(K_0)$ and $T_1 = H(K_1)$, then unless the collision resistance of the hash function is broken, $\mathcal{K}(w_O) \subseteq \{K_0, K_1\}$. Therefore, except with negligible probability,

$$\mathcal{K}(w_O) = \mathcal{K}(w_I) \cap \{K_0, K_1\} .$$

NAND Composites: Take an addition brick $Add = (w_L, w_R, w_S, [L_0], [R_0])$ and an NT brick $NT = (w_I, w_O, [I_0], \{C_0, C_1, C_2\}, [O_0])$. We call $(Add, w_S \Longrightarrow w_I, NT)$ a NAND brick. I.e., Bob is given $\Sigma = I_0 - S_0 \bmod p$ so that he can shift the potential output keys from the Add brick to the NT brick. If for $x, y \in \{0,1\}$ we have $L_x \in \mathcal{K}(w_L)$ and $R_y \in \mathcal{K}(w_R)$, then $S_{x+y} \in \mathcal{K}(w_S)$ and therefore $O_{\mathrm{nt}(x+y)} \in \mathcal{K}(w_O)$. I.e.,

$$L_x \in \mathcal{K}(w_L) \wedge R_y \in \mathcal{K}(w_R) \quad \Rightarrow \quad O_{x \bar{\wedge} y} \in \mathcal{K}(w_O) \ .$$

The above describes our components and how they work when they are correct. We now proceed to describe how we deal with faulty components using replication. We start with the NT brick.

Consider the NT brick $NT = (w_I, w_O, [I_0], \{C_0, C_1, C_2\}, [O_0])$ connected to an addition brick above. If the encryptions of NT are not correct, it might not output the correct O_z. To deal with this, we use a fresh NT brick $NT' = (w_I', w_O', [I_0'], \{C_0', C_1', C_2'\}, [O_0'])$ and we add a new fresh wire name ν' to the circuit, with $\mathcal{COM}(\nu') = [O_0]$ and $\mathcal{Z}(\nu') = O_0$. We then connect the output wire of Add to the input wire of NT' and the output wire w_O' of NT' to ν' by adding $(w_S \Longrightarrow w_I', NT', w_O' \Longrightarrow \nu')$ to the circuit design.

If for $x, y \in \{0,1\}$ we have $L_x \in \mathcal{K}(w_L)$ and $R_y \in \mathcal{K}(w_R)$, then $S_{x+y} \in \mathcal{K}(w_S)$ and therefore (by $w_S \Longrightarrow w_I'$) $I_{x+y}' \in \mathcal{K}(w_I')$. So, if NT' is correct, then $O_{\mathrm{nt}(x+y)}' \in \mathcal{K}(w_O')$ and thus (by $w_O' \Longrightarrow \nu'$) $O_{\mathrm{nt}(x+y)} \in \mathcal{K}(\nu')$. We already argued that if NT is correct, then $O_{\mathrm{nt}(x+y)} \in \mathcal{K}(w_O)$. This implies that if NT is correct *or* NT' is correct, then $O_{\mathrm{nt}(x+y)} \in \mathcal{K}(w_O) \cup \mathcal{K}(\nu')$. We can therefore add a fresh wire u to the circuit and let Bob compute $\mathcal{K}(u) = \mathcal{K}(w_O) \cup \mathcal{K}(\nu')$. We picture this as the wires w_O and ν' being joined in Fig. 3.

When using a total of ℓ NT bricks we call the structure an ℓ-*NAND composite* or ℓ-*NANDC*, as depicted in the top, center of Fig. 3. When we use a NANDC in a larger construction we depict it as the right, top graphic. It is clear that if at least one of the ℓ NT bricks is correct, then for $x, y \in \{0,1\}$ we have

$$L_x \in \mathcal{K}(w_L) \wedge R_y \in \mathcal{K}(w_R) \quad \Rightarrow \quad O_{x \bar{\wedge} y} \in \mathcal{K}(u) \ .$$

In our use of ℓ-NANDC's we always ensure that $|\mathcal{K}(w_L)| = |\mathcal{K}(w_R)| = 1$, which clearly implies that $|\mathcal{K}(u)| \leq 3\ell$.

KF Composites: Also KF bricks can be incorrect, e.g. by T_0 and T_1 being random values. If both are incorrect it is not a real problem as Bob will end up with $\mathcal{K}(w_O) = \emptyset$ in all cases. Alice might, however, create a brick where $T_0 = H(K_0)$ and T_1 is random. In that case Bob will get $\mathcal{K}(w_O) = \{K_0\}$ if $K_0 \in \mathcal{K}(w_I)$ and $\mathcal{K}(w_O) = \emptyset$ if $K_1 \in \mathcal{K}(w_I)$. In the first case he completes the protocol, but in the second case he has to terminate. Alice detects the termination by not receiving her keys, which allows her to learn the bit on the (possibly) internal wire w_O. To avoid this *leakage via conditional failure* we replicate key filters.

The simple observation is that if we run several KF bricks on the same potential keys, then a correct key will be contained in the output of all correct filters.

If we use $2\ell + 1$ filters under the assumption that that at most ℓ are faulty, this allows us to pick the correct keys by majority voting.

To ease the presentation, we describe just the case $\ell = 1$. Consider three KF bricks $KF = (w_I, w_O, [K_0], \{T_0, T_1\})$, $KF' = (w_I', w_O', [K_0'], \{T_0', T_1'\})$, $KF' = (w_I'', w_O'', [K_0''], \{T_0'', T_1''\})$, add fresh wires ν', ν'', v with $\mathcal{COM}(\nu') \stackrel{\text{def}}{=} [K_0]$, $\mathcal{COM}(\nu'') \stackrel{\text{def}}{=} [K_0]$ and $\mathcal{COM}(v) \stackrel{\text{def}}{=} [K_0]$, and add the shifts $w_I \implies w_I'$, $w_I \implies w_I''$, $w_O' \implies \nu'$, $w_O' \implies \nu''$. Let $\mathcal{K}(v)$ consist of the K which are found in at least two of the sets $\mathcal{K}(w_O), \mathcal{K}(\nu'), \mathcal{K}(\nu'')$. We write

$$\mathcal{K}(v) \stackrel{\text{def}}{=} \text{Maj}(\mathcal{K}(w_O), \mathcal{K}(\nu'), \mathcal{K}(\nu'')) .$$

If $K_c \in \mathcal{K}(w_I)$ and KF (resp KF', KF'') is correct, then $K_c \in \mathcal{K}(w_O)$ (resp. $\mathcal{K}(\nu'), \mathcal{K}(\nu'')$). So, if a majority of the KF bricks are correct, then $K_c \in \mathcal{K}(w_I) \implies K_c \in \mathcal{K}(v)$. Furthermore, since a correct KF brick only output keys in $\{K_0, K_1\}$, it follows that $\mathcal{K}(v) \subseteq \{K_0, K_1\}$. So, except with negligible probability

$$\mathcal{K}(v) = \mathcal{K}(w_I) \cap \{K_0, K_1\} .$$

We depict the KFC in Fig. 3 (center, bottom). When using a KFC in a larger construction we depict it as the right, second from the top graphic.

Overall Architecture: To evaluate a NAND circuit $\mathcal{C}$ with replication parameter ℓ we place one $(\ell + 1)$-NANDC on each NAND gate G in $\mathcal{C}$ and attach a $(2\ell + 1)$-KFC to the output wire of the NANDC — we connect them using the appropriate shift $u \implies w_I$. We consider the two input wires w_L and w_R of the NANDC as the input wires of G and we consider the output wire v of the KFC as the output wire of G. We then use shifts to connect output wires of gates to input wires of other gates according to the architecture of $\mathcal{C}$.

In the evaluation Bob will learn input keys K_{x_w} for his input wires via an OT of K_0 and K_1. By letting Alice send the opening of $[K_{x_w}]$ in the OT, Bob can verify that K_{x_w} is correct. For Alice's input wires, she will simply send K_{x_w} to give input x_w on wire w. To prevent her from sending an incorrect key, $K' \notin \{K_0, K_1\}$ we add an extra $(2\ell + 1)$-KFC and send K' through this filter before feeding it unto w. This ensures that $K' \in \{K_0, K_1\}$, if accepted.

In Fig. 3 (bottom, right) we depict this construction for $\ell = 1$ and $\mathcal{C}$ consisting of one NAND gate where Alice provides the left input and Bob provides the right input.

It is easy to see that if for each $(\ell + 1)$-NANDC at most ℓ NT bricks are incorrect and for each $(2\ell + 1)$-KFC at most ℓ KF bricks are incorrect, then the LEGO circuit will compute the correct result, in the following sense: If $\mathcal{Z}(w) + x_w \Delta \mod p \in \mathcal{K}(w)$ for all input wires $w \in \mathcal{I}$ and $x_w \in \{0, 1\}$, then $\mathcal{Z}(w) + y_w \Delta \mod p \in \mathcal{K}(w)$ for all output wires $w \in \mathcal{O}$ and $\{y_w\}_{w \in \mathcal{O}} = \mathcal{C}(\{x_w\}_{w \in \mathcal{I}})$.

If in addition $\mathcal{K}(w)$ is a singleton set for all $w \in \mathcal{I}$, then $\mathcal{K}(w)$ will be a singleton set for all $w \in \mathcal{O}$, except with negligible probability. As detailed in the analysis, this follows from the use of KFC's and the fact that Bob could not compute both keys K_0 and K_1 for a wire even if he tried.

4   The Protocol

The overall protocol is given in Fig. 4. It depends on the NAND circuit $\mathcal{C}$ and a replication factor $\ell \in \mathbb{N}$.

Bricks Production: Details of how individual bricks are produced and which values are sent to Bob were given in Section 3. Alice will produce $P_{NT} = (1 + \tau_{NT})(\ell+1)|\mathcal{C}|$ NT bricks and Bob will select a uniformly random subset of size $\tau_{NT}(\ell+1)|\mathcal{C}|$ for testing. This leaves $(\ell+1)|\mathcal{C}|$ NT bricks, which is sufficient to construct a $(\ell+1)$-NANDC for each gate in $\mathcal{C}$. She generates $P_{KF} = (1 + \tau_{KF})(2\ell+1)(|\mathcal{C}|+|\mathcal{I}_A|)$ KF bricks and Bob tests a random subset of size $\tau_{KF}(2\ell+1)(|\mathcal{C}|+|\mathcal{I}_A|)$, leaving enough to construct a $(2\ell+1)$-KFC for each gate and each of Alice's input wires.

After sending the components, and *before* the tests, Alice proves to Bob that she knows the openings of all commitments in the bricks, as follows: Let $[K_0^{(1)}], \ldots, [K_0^{(n)}]$ be the set of all commitments contained in the bricks. Bob picks a uniformly random challenge $e \in_R \mathbb{Z}_p^n$ and sends e to Alice. Bob computes $[\sum_{i=1}^{n} e_i K_0^{(i)}] = \boxplus_{i=1}^{n}[K_0^{(i)}]^{e_i}$ and Alice computes the opening of $[\sum_{i=1}^{n} e_i K_0^{(i)}]$. Then Alice uses $\mathcal{F}_{ZK}$ to prove that she knows this opening. Bob terminates if the proof fails.

Addition bricks are by definition correct, as Bob lets $[S_0] = [L_0] \boxplus [R_0]$. For NT bricks and KF bricks Alice can cheat by not providing correct encryptions respectively hash values, which is the reason for the following tests.

NT Bricks: A test of $NT = (w_I, w_O, [I_0], [O_0], \{C_0, C_1, C_2\})$ proceeds as follows:

1. Alice sends a value telling Bob the correct order C_0, C_1, C_2 of $\{C_0, C_1, C_2\}$.
2. Bob computes $[O_1] = [O_0] \boxplus [\Delta]$, $[I_1] = [I_0] \boxplus [\Delta]$, $[I_2] = [I_1] \boxplus [\Delta]$ and sends challenge $e \in_R \{0, 1, 2\}$ to Alice.
3. Alice opens $[I_e]$ and $[O_{nt(e)}]$ to let Bob learn I_e and $O_{nt(e)}$.
4. Bob accepts iff the openings are correct and $E_{I_e}(O_{nt(e)}) = C_e$.

It is clear that if Alice can answer all three challenges correctly and cannot break the computational binding of the commitment scheme, then NT is a correct NT brick, i.e., it would produce the correct output key in $\{O_0, O_1\}$ on all input keys in $\{I_0, I_1, I_2\}$, where I_0 is the key in $[I_0]$ and and O_0 is the key in $[O_0]$. For now, we informally define *the key in* $[I_0], [O_0]$ as the value Alice can open $[I_0], [O_0]$ to. The formal analysis is, of course, more crisp.

KF Bricks: A test of $KF = (w_I, w_O, [K_0], \{T_0, T_1\})$ proceeds as follows:

1. Alice sends a bit telling Bob the correct order T_0, T_1 of $\{T_0, T_1\}$.
2. Bob computes $[K_1] = [K_0] \boxplus [\Delta]$ and sends a one bit challenge $e \in_R \{0, 1\}$ to Alice.
3. Alice opens $[K_e]$ to let Bob learn K_e.
4. Bob accepts iff the opening is correct and $T_e = H(K_e)$.

It is clear that if Alice can answer both challenges, then she can open $[K_0]$ to K_0 such that $H(K_0) = T_0$ and $H(K_0 + \Delta) = T_1$, or she can break the computational binding of the commitment scheme. Therefore KF is correct except with negligible probability. I.e., it would work correctly on both K_0 and K_1, where K_0 is the key in $[K_0]$.

5   Analysis

In the full version of this paper [NO08] we prove Theorem 1. Here we sketch the proof, but assuming that H is a random oracle. By a hybrid argument the random oracle can be replaced by a collision resistant function which is correlation resistant according to Def. 1.

5.1   Corrupted Bob

We first consider the case where Alice is honest and Bob is corrupted. We model H as a non-programmable and non-extractable random oracle.

The UC simulator $\mathcal{S}$ runs Alice's part of the protocol towards Bob completely honestly, except that it uses $x_w = 0$ for each of her input wires. If Bob sends keys in **Result announcement** which makes Alice reject, then $\mathcal{S}$ inputs abort! to $\mathcal{F}_{\text{SCE}}$ on behalf of Bob. Otherwise, $\mathcal{S}$ for each of Bob's wires takes $x_w \in \{0,1\}$ to be the input of Bob to the OT associated to wire $w \in \mathcal{I}_B$ and inputs these x_w to $\mathcal{F}_{\text{SCE}}$ on behalf of Bob. That completes the description of $\mathcal{S}$.

It remains to argue that the views of the environment in the simulation and in the protocol are indistinguishable. When H is random they are in fact statistically close. There are the following differences between the simulation and the protocol:

1. In the simulation $x_w = 0$ for each of Alice's wires. In the protocol the x_w's have the values specified by the environment.
2. In the protocol Alice's output to the environment is computed from the values sent by Bob. In the simulation it is the value output by $\mathcal{F}_{\text{SCE}}$.

To handle the first difference we argue that Bob's view is statistically independent of Alice's input. To handle the second difference we argue that if Alice does not abort she outputs the same y_w values in the two settings, and we argue that she aborts with the same probability in the two settings.

It is straight-forward to verify that when H is a uniformly random function, then the LEGO circuit leaks no information on Alice's input to Bob unless he queries H on two different points $P, P' \in \{K_0, K_0 + \Delta, K_0 + 2\Delta\}$ for some zero-key K_0 of the bricks sent by Alice, and that until he makes such queries, Δ is uniformly random in the view of Bob. It follows from an easy application of the birthday bound that when Δ is uniformly random and independent of the view of Bob, the probability that he makes such queries is less than 2^{-s}, because of our choice of p.

Setup: We assume Alice and Bob agree on a generator g of order p. Bob picks uniformly random $h \in_R \langle g \rangle$ and sends it to Alice. This defines the commitment scheme $[K] = g^K h^r$. Alice picks a global difference $\Delta \in_R \mathbb{Z}_p$ and sends a commitment $[\Delta]$ to Bob and uses $\mathcal{F}_{\text{ZK}}$ to prove knowledge of the opening of $[\Delta]$.

Bricks production: Alice produces P_{NT} uniformly random and independent NT bricks, P_{KF} uniformly random and independent KF bricks and $P_{Add} = |\mathcal{C}|$ uniformly random addition bricks, and sends all these bricks to Bob. All bricks are produced using Δ.

Test: Bob selects some NT bricks and KF bricks for testing. For each such brick he picks a random input and Alice provides Bob with the keys needed to run on those inputs. Bob terminates if any test fails.

Bricks shuffling: For each NAND gate G in $\mathcal{C}$ Bob picks: a random unused addition brick, $\ell + 1$ random unused NT bricks and $2\ell + 1$ random unused KF bricks and assigns them to G. For each of Alice's input wires he picks $2\ell + 1$ random unused KF bricks and assigns them to w. He announce the assignment to Alice.

Bricks connection: The positions of the bricks chosen by Bob define a number of brick output wires w feeding unto brick input wires w'. For each such connection Alice and Bob add $w \Longrightarrow w'$ to the circuit by Alice opening $\mathcal{COM}(w') \boxminus \mathcal{COM}(w)$ to Bob. Bob terminates if any opening is incorrect.

Bob's input: For each input wire $w \in \mathcal{I}_B$ (with $\mathcal{Z}(w) = K_0$ and $\mathcal{COM}(w) = [K_0]$) Bob has an input $x_w \in \{0,1\}$. Alice and Bob run an OT where Alice inputs messages $m_0^{(w)} = (K_0, r_0)$ and $m_1^{(w)} = (K_1, r_1)$ and Bob inputs the selection bit x_w — here (K_0, r_0) is the opening of $[K_0]$ and (K_1, r_1) is the opening of $[K_1] = [K_0] \boxplus [\Delta]$. Bob terminates if $m_{x_w}^{(w)}$ is not an opening of $[K_{x_w}]$. Otherwise he lets $\mathcal{K}(w) \stackrel{\text{def}}{=} \{K_{x_w}\}$ for $w \in \mathcal{I}_B$.

Alice's input: For each input wire $w \in \mathcal{I}_A$ (with $\mathcal{Z}(w) = K_0$ and $\mathcal{COM}(w) = [K_0]$) Alice has an input $x_w \in \{0,1\}$. She sends K_{x_w} to Bob and Bob lets $\mathcal{K}(w) = \{K_{x_w}\}$. Bob evaluates the KFC for w on $\mathcal{K}(w)$ to get $\mathcal{K}(w')$. Since $|\mathcal{K}(w)| = 1$ and $\mathcal{K}(w') = \mathcal{K}(w) \cap \{K_0, K_1\}$, it follows that $|\mathcal{K}(w')| \leq 1$ and $\mathcal{K}(w') \subset \{K_0, K_1\}$. If $|\mathcal{K}(w')| = 1$ for all of Alice's wires w, Bob adopts the sets $\mathcal{K}(w')$ for $w \in \mathcal{I}_A$, otherwise he terminates.

Garbled evaluation: If Bob did not yet terminate, he holds a singleton set $\mathcal{K}(w) \subset \{K_0, K_1\}$ for all input wires $w \in \mathcal{I}$, where $K_0 = \mathcal{Z}(w)$. Now Bob computes singleton sets $\mathcal{K}(w) \subset \{K_0, K_1\}$ for all output wires $w \in \mathcal{O}$, where $K_0 = \mathcal{Z}(w)$. Details of how this is done were given in Section 3.

Result announcement: For each output wire w, Bob sends $(w, \mathcal{K}(w))$ to Alice. Alice terminates if any set $\mathcal{K}(w)$ is not a singleton. Otherwise she picks $K \in \mathcal{K}(w)$ and tries to write K as $K = \mathcal{Z}(w) + y_w \Delta \mod p$ for $y_w \in \{0,1\}$. If this fails, she terminates the protocol; Otherwise she adopts y_w as the output bit for wire w.

Fig. 4. The LEGO protocol

Now note that from the keys $K_{x_w}, w \in \mathcal{I}_A$ sent by Alice and the keys $K_{x_w}, w \in \mathcal{I}_B$ Bob chooses in the OT's, he can compute K_{y_w} for each $w \in \mathcal{O}$, where $\{y_w\}_{w \in \mathcal{O}} = \mathcal{C}(\{x_w\}_{w \in \mathcal{I}})$. So, Bob *knows* the correct output keys K_{y_w}. If Bob sends K_{y_w} for all $w \in \mathcal{O}$, then Alice accepts and outputs the correct values y_w.

If Bob sends $K' \neq K_{y_w}$ for some wire, then Alice rejects unless $K' = K_{1-y_w}$. But then, if $K' = K_{1-y_w}$, Bob could query H on two different points $K_{y_w}, K' \in \{K_0, K_0+\Delta, K_0+2\Delta\}$, and we already argued that this happens with probability less than 2^{-s}. So, the probability that Alice aborts is statistically close to the probability that Bob sends some $K' \neq K_{y_w}$.

5.2 Corrupted Alice

We then consider the case where Bob is honest and Alice is corrupted.

The simulator: The simulator runs Bob honestly in **Setup, Bricks production, Test, Bricks shuffling, Bricks connection, Bob's input** and **Alice's input**, except that:

1. It inspects Alice's input to $\mathcal{F}_{\text{ZK}}$ in **Setup**. If the input to $\mathcal{F}_{\text{ZK}}$ is not an opening of $[\Delta]$, it inputs abort! to $\mathcal{F}_{\text{SCE}}$ on behalf of Alice. Otherwise it records Δ.
2. For each of Bob's input wires $w \in \mathcal{I}_B$ it records both of Alice's inputs $m_0^{(w)}$ and $m_1^{(w)}$ to $\mathcal{F}_{\text{OT}}$ in **Bob's input**. For $c = 0, 1$, if $m_c^{(w)}$ is a valid opening of the $[K_c]$ defined in **Bob's input**, it defines $V_c^{(w)} \overset{\text{def}}{=} K_c$. If $m_c^{(w)}$ is not an opening of $[K_c]$, it defines $V_c^{(w)} \overset{\text{def}}{=} \perp$.

Plain abort: If Alice (formally the adversary) makes Bob abort in any of the steps run so far, e.g. by sending a key for one of Alice's input wires which is not accepted by the corresponding KFC, then $\mathcal{S}$ inputs abort! to $\mathcal{F}_{\text{SCE}}$ to make it output abort! to the environment, exactly as Bob would have done in the protocol. The only step in which $\mathcal{S}$ cannot compute whether Bob would have aborted is in **Bob's input**, as $\mathcal{S}$ does not know the inputs $\{x_w\}_{w \in \mathcal{I}_w}$ of Bob — these were input to $\mathcal{F}_{\text{SCE}}$ by the environment. This is handled as follows.

Handling conditional failures: If $V_0^{(w)} = V_1^{(w)} = \perp$ for some $w \in \mathcal{I}_B$, then $\mathcal{S}$ inputs abort! to $\mathcal{F}_{\text{SCE}}$ on behalf of Alice. Note that Bob would always abort in the protocol in this case, as he would always receive an incorrect opening for the wire w. If not already terminated, then $\mathcal{S}$ computes $W = \{w \in \mathcal{I}_B | V_0^{(w)} = \perp \vee V_1^{(w)} = \perp\}$ and for $w \in W$ sets $\beta_w \in \{0, 1\}$ to be the value where $V_{\beta_w}^{(w)} \neq \perp$. It then inputs $(W, \{\beta_w\}_{w \in W})$ to $\mathcal{F}_{\text{SCE}}$ on behalf of Alice in **Guess**. Note that $\mathcal{F}_{\text{SCE}}$ aborts iff $\beta_w \neq x_w$ for some $w \in W$. By construction of the protocol and definition of the $V_c^{(w)}$, Bob would have aborted in the protocol iff $V_{x_w}^{(w)} \neq K_{x_w}$, if he was running with the inputs $\{x_w\}_{w \in \mathcal{I}_B}$ input to $\mathcal{F}_{\text{SCE}}$ by the environment. Therefore $\mathcal{S}$ makes $\mathcal{F}_{\text{SCE}}$ abort iff Bob would have aborted in the protocol. So, until now the simulation is perfect.

Extracting inputs: If $\mathcal{S}$ did not yet make $\mathcal{F}_{\text{SCE}}$ abort, it must now give an input to $\mathcal{F}_{\text{SCE}}$ on behalf of Alice. Note that if $\mathcal{S}$ did not yet make $\mathcal{F}_{\text{SCE}}$ abort, then it knows a key $K^{(w)}$ for each $w \in \mathcal{I}_A$, namely the keys obtained by running the KFC's on the keys sent by Alice in **Alice's input**. The simulator uses Δ to compute $K_{-1}^{(w)} = K_{x_w}^{(w)} - \Delta$ and $K_{+1}^{(w)} = K_{x_w}^{(w)} + \Delta$.[1] Then it runs the KFC for input wire w on $\mathcal{K} = \{K_{-1}^{(w)}, K^{(w)}, K_{+1}^{(w)}\}$. If the output from the KFC does not consist of two keys K and K' where $K - K' \in \{\Delta, -\Delta\}$ or $K^{(w)} \notin \{K, K'\}$, then $\mathcal{S}$ makes $\mathcal{F}_{\mathrm{SCE}}$ abort. Otherwise it names and orders the two keys as $K_0^{(w)}, K_1^{(w)}$ such that $K_1^{(w)} = K_0^{(w)} + \Delta$ and computes $x_w \in \{0, 1\}$ such that $K^{(w)} = K_{x_w}^{(w)}$. It inputs $\{x_w\}_{w \in \mathcal{I}_A}$ to $\mathcal{F}_{\mathrm{SCE}}$ and gets back $\{y_w\}_{w \in \mathcal{O}} = \mathcal{C}(\{x_w\}_{w \in \mathcal{I}_A} \cup \{x_w\}_{w \in \mathcal{I}_B})$.

Evaluation: By now $\mathcal{S}$ has a key $K^{(w)}$ for each $w \in \mathcal{I}_A$. It lets $\mathcal{K}(w) = \{K^{(w)}\}$. For each $w \in \mathcal{I}_B$ it lets $\mathcal{K}(w) = \{V_0^{(w)}\}$ if $V_0^{(w)} \neq \bot$ and $\mathcal{K}(w) = \{V_1^{(w)}\}$ if $V_0^{(w)} = \bot$. Note that at this point $V_1^{(w)} \neq \bot$ when $V_0^{(w)} = \bot$. It then evaluates the LEGO circuit on these $\mathcal{K}(w)$. If the evaluation fails, then it makes $\mathcal{F}_{\mathrm{SCE}}$ abort. Otherwise it computed $\mathcal{K}(w) = \{K^{(w)}\}$ for $w \in \mathcal{O}$. It computes $\mathcal{K} = \{K_{-1}^{(w)}, K^{(w)}, K_{+1}^{(w)}\}$ as above and runs the KFC from the gate computing $K^{(w)}$ on $\mathcal{K}$. If the output does not consist of two keys which can be named and ordered such that $K_1^{(w)} = K_0^{(w)} + \Delta$ it makes $\mathcal{F}_{\mathrm{SCE}}$ abort. Otherwise, it sends $\{K_{y_w}^{(w)}\}_{w \in \mathcal{O}}$ to Alice, where the y_w are the values received from $\mathcal{F}_{\mathrm{SCE}}$.

That completes the description of the simulator.

Analysis: We argued that the simulation is perfect up until the evaluation. What remains is to argue that the simulator aborts in the evaluation with the same probability as Bob would in the protocol and that when it does not abort, then the keys $\{K_{y_w}^{(w)}\}_{w \in \mathcal{O}}$ sent to Alice have the same distribution in the protocol and the simulation. This boils down to arguing that the LEGO circuit is correct if Bob does not abort before the evaluation phase.

Defining good and bad: We start the analysis by dividing bricks into *good* and *bad* and use this to define good and bad composites.

Consider the entire state of an execution after Alice sent all bricks to Bob and before Bob sends the challenge $e \in \mathbb{Z}_p^n$. Define q to be the probability that Bob accepts Alice's proof for $[K_e] \stackrel{\mathrm{def}}{=} [\sum_{i=1}^n e_i K_0^{(i)}]$, when run from this fixed state (formally we fix the random tape of the environment and the adversary). If $q \leq 2/p$ we define all bricks to be *bad*. Otherwise we extract the openings of all commitments and use these to define the goodness of the bricks, as described now.

For a fixed state of Alice she will for a fixed $e \in E \stackrel{\mathrm{def}}{=} \mathbb{Z}_p^n$ input a correct opening of $[K_e]$ to $\mathcal{F}_{\mathrm{ZK}}$ with probability 0 or 1. Let E_g denote the $e \in E$ where the probability is 1. Then $q = |E_g|/|E|$, and we have a poly-time oracle for computing $K_e \stackrel{\mathrm{def}}{=} \sum_{i=1}^n e_i K_0^{(i)}$ for $e \in E_g$: Given any $e \in E$ we can run the execution from the fixed state until Alice inputs to $\mathcal{F}_{\mathrm{ZK}}$. If $e \in E_g$, this gives us K_e. Let n denote the number of commitments $[K_i]$. It is clear that if we get K_{e^i} for n linear independent values $e^i \in \mathbb{Z}_p^n$, then we can efficiently solve for openings of all $[K_i]$. We get the e^i by querying the oracle on several uniformly

---

[1] Both additions are modulo p. Below we will continue not explicitly mentioning the mod p when computing on elements from $\mathbb{Z}_p$.

random $e \in E$. At any point, let $\{e^i\}$ denote the e^i on which we got an opening of $[K_{e^i}]$. We continue querying as long as $\text{span}\{e^i\} \neq \mathbb{Z}_p^n$. If we query on a fresh uniformly random $e \in E$, then the probability that $e \in \text{span}\{e^i\}$ is at most $1/p$. The probability that $e \in E_g$ is q. Therefore the probability that $e \in E_g \backslash \text{span}\{e^i\}$ is at least $q - 1/p \geq q/2$. It follows that the expected number of queries to get $\text{span}\{e^i\} = \mathbb{Z}_p^n$ is at most $n(q/2)^{-1} = 2nq^{-1}$. The expected running time is therefore $\text{poly}(s)q^{-1}$. We only need to run the above extraction when Bob accepts the proof in the simulation, which he does with probability q. Therefore the expected running time of running the simulation once and extracting if Bob accepts is $q \, \text{poly}(s)q^{-1} = \text{poly}(\kappa)$.

After having extracted openings of all commitments we let $\mathcal{Z}(w) = K_0$ for all wires, where $[K_0] = \mathcal{COM}(w)$ and K_0 is the key extracted from $[K_0]$.

We then define a KF brick $(w_I, w_O, [K_0], \{T_0, T_1\})$ to be *good* if there is an ordering T_0, T_1 of $\{T_0, T_1\}$ such that $T_0 = H(K_0)$ and $T_1 = H(K_0 + \Delta)$. We define an NT brick $(w_L, w_R, w_O, [L_0], [R_0], [O_0], \{C_0, C_1, C_2\})$ to be *good* if there is an ordering C_0, C_1, C_2 of $\{C_0, C_1, C_2\}$ such that $C_0 = H(I_0) + O_0 + \Delta$, $C_1 = H(I_0 + \Delta) + O_0 + \Delta$ and $C_2 = H(I_0 + 2\Delta) + O_0$. All other KF and NT bricks are defined to be *bad*.

We already now note that Alice cannot later open shifting values incorrectly: Assume that Alice opens $[K_0'] \boxminus [K_0]$ to $\Sigma \neq K_0' - K_0$ as part of implementing $w \implies w'$ (here $[K_0] = \mathcal{COM}(w), K_0 = \mathcal{Z}(w), [K_0'] = \mathcal{COM}(w'), K_0' = \mathcal{Z}(w')$). We can then use the opening of $[K_0']$ to K_0', the opening of $[K_0]$ to K_0 and the opening of $[K_0'] \boxminus [K_0]$ to Σ to compute an opening of e.g. $[K_0]$ to two different values. Since all three openings originate from Alice[2] and were computed in expected poly-time, we broke the computational binding of the commitment scheme. We also note the easy fact that if a KF brick is defined to be bad, then there exists at least one challenge $e \in \{0, 1\}$ on which Alice cannot make Bob accept a test, and if an NT brick is defined to be bad, then there exists at least one challenge $e \in \{0, 1, 2\}$ on which Alice cannot make Bob accept a test.[3]

We call a composite *bad* if it consists of more than ℓ bad composites. Otherwise it is *good*. An NANDC consists of $\ell + 1$ bricks and a KFC consists of $2\ell + 1$ bricks. As a consequence they will work correctly if they are good. It then follows from Section 3 that a LEGO circuit consisting of only good composites will compute correct $\mathcal{K}(w)$ for all $w \in \mathcal{O}$.

We first analyze the probability that any NANDC is bad. Let $P = (1 + \tau)(\ell + 1)|\mathcal{C}|$ be the number of NT bricks produced by Alice, let B be the number of NT bricks being defined as bad, let $T = \tau(\ell + 1)|\mathcal{C}|$ be the number of NT bricks being tested, and let $U = (\ell + 1)|\mathcal{C}|$ be the number of NT bricks used in the LEGO circuit.

Split each good NT brick into three green balls and split each bad NT brick into two green balls and a red ball. A ball represents one of the values a specific brick can be tested on. Each bad brick has at least one value which will catch

---

[2] The two first ones from extraction and the last because she sent it.

[3] In both case, unless she breaks the computational binding of the commitment scheme.

Alice if Bob tests on that value — this is the red ball. There are $3P$ balls and B are red. We analyze the game where Bob chooses T balls at random and accepts if they are all green. This upper bounds his probability of accepting in the protocol, except for a negligible amount coming from Alice's possible breaking of the commitment scheme. The probability that Bob accepts in the game is upper bounded by $\left(\frac{3P-B}{3P}\right)^T$. The probability that a given NANDC is bad is upper bounded by $\left(\frac{B}{U}\right)^{\ell+1}$, so by a union bound, the probability that Bob accepts and there is at least one bad NANDC is upper bounded by

$$\left(\frac{3P-B}{3P}\right)^T |\mathcal{C}| \left(\frac{B}{U}\right)^{\ell+1} = |\mathcal{C}| \left(\left(\frac{3P-B}{3P}\right)^{\tau|\mathcal{C}|} \frac{B}{U}\right)^{\ell+1}. \tag{1}$$

This is maximal in B when $(3P-B)^{\tau|\mathcal{C}|} B$ is maximal in B, which it is when $B = 3P(1+\tau|\mathcal{C}|)^{-1} = 3(1+\tau)(\ell+1)|\mathcal{C}|(1+\tau|\mathcal{C}|)^{-1} \approx 3\frac{1+\tau}{\tau}(\ell+1) = 3(1+\tau^{-1})(\ell+1)$. We use $B = 3(1+\tau^{-1})(\ell+1)$. Then $B/U = 3(1+\tau^{-1})|\mathcal{C}|^{-1}$, and $(3P-B)(3P)^{-1} = 1 - B(3P)^{-1} = 1 - (1+\tau^{-1})(1+\tau)|\mathcal{C}|^{-1} = 1 - \tau^{-1}|\mathcal{C}|^{-1}$. So, $\left(\frac{3P-B}{3P}\right)^{\tau|\mathcal{C}|} = (1-\tau^{-1}|\mathcal{C}|^{-1})^{\tau|\mathcal{C}|} \approx e^{-1}$. Plugging this into (1), we get an error probability of $|\mathcal{C}|^{-\ell}\left(\frac{3(\tau+1)}{e\tau}\right)^{\ell+1}$. Isolating for a probability of 2^{-s} that there is a bad NT composite we get

$$\ell \approx \frac{s + 0.1423 + \log\frac{\tau+1}{\tau}}{\log|\mathcal{C}| - 0.1423 - \log\frac{\tau+1}{\tau}} = O(s/\log|\mathcal{C}|).$$

We round up this value to the nearest integer to get the replication factor for the protocol. A more careful analysis without the approximations shows that this actually gives a probability of $O(2^{-s})$ that Bob accepts and yet there is a NANDC being defined as bad.

Correctness: In the full version [NO08] we give a similar analysis for KFCs and recommend optimal values of τ for a wide range of circuit sizes and desired security levels. Here it suffices to note that for $\ell = O(s/\log|\mathcal{C}|)$ also the probability that Bob accepts and there is a bad KFC is $O(2^{-s})$. Therefore, if Bob accepts (in the protocol or simulation) then all composites are good, except with negligible probability. As described in Section 3, this implies that Bob will be able to evaluate the LEGO circuit in both settings, or at least abort with negligible probability. Furthermore, in the protocol the keys sent to Alice are $\{K_{y_w}^{(w)}\}_{w\in\mathcal{O}}$, where $K_0^{(w)}$ is the key she can open $\mathcal{COM}(w)$ to (by the extraction argument) and $\{y_w\}_{w\in\mathcal{O}} = \mathcal{C}(\{x_w\}_{w\in\mathcal{I}_A} \cup \{x_w\}_{w\in\mathcal{I}_B})$ for Bob's inputs $\{x_w\}_{w\in\mathcal{I}_B}$ and inputs $\{x_w\}_{w\in\mathcal{I}_A}$ defined by the keys $\{K_{x_w}^{(w)}\}_{w\in\mathcal{I}_A}$ sent by Alice and the keys $K_0^{(w)}$ she can open the commitment of the input KFCs to. By construction of the simulator the keys sent to Alice are $\{K_{y_w}^{(w)}\}_{w\in\mathcal{O}}$, where $K_0^{(w)}$ is the key she can open $\mathcal{COM}(w)$ to and $\{y_w\}_{w\in\mathcal{O}} = \mathcal{C}(\{x_w\}_{w\in\mathcal{I}_A} \cup \{x_w\}_{w\in\mathcal{I}_B})$ for Bob's inputs $\{x_w\}_{w\in\mathcal{I}_B}$ (held by the $\mathcal{F}_{\text{SCE}}$) and the inputs $\{x_w\}_{w\in\mathcal{I}_A}$ extracted from the keys sent by Alice. So, it suffices to argue that the extraction produces the correct values of x_w for $w \in \mathcal{I}_A$: Since Bob did not abort, the KFC for w produced a unique output. Since the KFC is correct, the output is of the form $K^{(w)} = K^{(0)} + c\Delta$, defined relative to the $K^{(0)}$ Alice can open $\mathcal{COM}(w)$ to and $c \in \{0, 1\}$. Therefore $\{K_0^{(w)}, K_1^{(w)}\} \subset \{K_{-1}^{(w)}, K^{(w)}, K_{+1}^{(w)}\}$. Therefore the KFC outputs $\mathcal{K} \cap \{K_0^{(w)}, K_1^{(w)}\} = \{K_0^{(w)}, K_1^{(w)}\}$. By construction this leads to $\mathcal{S}$ computing the $x_w \in \{0, 1\}$ for which $K^{(w)} = K_0^{(w)} + x_w \Delta$.

Acknowledgments

We thank the TCC reviewers for their suggestions on improving the presentation and in particular Yuval Ishai who provided valuable feedback during our writing of the conference version.

References

[GMW86] Goldreich, O., Micali, S., Wigderson, A.: Proofs that yield nothing but their validity and a methodology of cryptographic protocol design (extended abstract). In: FOCS (1986)

[IKNP03] Ishai, Y., Kilian, J., Nissim, K., Petrank, E.: Extending oblivious transfers efficiently. In: Boneh, D. (ed.) CRYPTO 2003. LNCS, vol. 2729, pp. 145–161. Springer, Heidelberg (2003)

[IPS08] Ishai, Y., Prabhakaran, M., Sahai, A.: Founding cryptography on oblivious transfer – efficiently. In: Wagner, D. (ed.) CRYPTO 2008. LNCS, vol. 5157, pp. 572–591. Springer, Heidelberg (2008)

[JS07] Jarecki, S., Shmatikov, V.: Efficient two-party secure computation on committed inputs. In: Naor, M. (ed.) EUROCRYPT 2007. LNCS, vol. 4515, pp. 97–114. Springer, Heidelberg (2007)

[KS08] Kolesnikov, V., Schneider, T.: Improved garbled circuit: Free XOR gates and applications. In: Aceto, L., Damgård, I., Goldberg, L.A., Halldórsson, M.M., Ingólfsdóttir, A., Walukiewicz, I. (eds.) ICALP 2008, Part II. LNCS, vol. 5126, pp. 486–498. Springer, Heidelberg (2008)

[LP04] Lindell, Y., Pinkas, B.: A proof of Yao's protocol for secure two-party computation. Electronic Colloquium on Computational Complexity (2004)

[LP07] Lindell, Y., Pinkas, B.: An efficient protocol for secure two-party computation in the presence of malicious adversaries. In: Naor, M. (ed.) EUROCRYPT 2007. LNCS, vol. 4515, pp. 52–78. Springer, Heidelberg (2007)

[LPS08] Lindell, Y., Pinkas, B., Smart, N.P.: Implementing two-party computation efficiently with security against malicious adversaries. In: Ostrovsky, R., De Prisco, R., Visconti, I. (eds.) SCN 2008. LNCS, vol. 5229, pp. 2–20. Springer, Heidelberg (2008)

[MF06] Mohassel, P., Franklin, M.K.: Efficiency tradeoffs for malicious two-party computation. In: Yung, M., Dodis, Y., Kiayias, A., Malkin, T.G. (eds.) PKC 2006. LNCS, vol. 3958, pp. 458–473. Springer, Heidelberg (2006)

[NN01] Naor, M., Nissim, K.: Communication preserving protocols for secure function evaluation. In: STOC (2001)

[NO08] Nielsen, J.B., Orlandi, C.: Lego for two party secure computation. Cryptology ePrint Archive, Report 2008/427 (2008), http://eprint.iacr.org/

[Woo07] Woodruff, D.P.: Revisiting the efficiency of malicious two-party computation. In: Naor, M. (ed.) EUROCRYPT 2007. LNCS, vol. 4515, pp. 79–96. Springer, Heidelberg (2007)

[Yao82] Yao, A.C.: Protocols for secure computations (extended abstract). In: FOCS (1982)

[Yao86] Yao, A.C.: How to generate and exchange secrets (extended abstract). In: FOCS (1986)

Simple, Black-Box Constructions of Adaptively Secure Protocols*

Seung Geol Choi[1], Dana Dachman-Soled[1], Tal Malkin[1], and Hoeteck Wee[2,**]

[1] Columbia University
{sgchoi,dglasner,tal}@cs.columbia.edu
[2] Queens College, CUNY
hoeteck@cs.qc.cuny.edu

Abstract. We present a compiler for transforming an oblivious transfer (OT) protocol secure against an adaptive semi-honest adversary into one that is secure against an adaptive malicious adversary. Our compiler achieves security in the universal composability framework, assuming access to an ideal commitment functionality, and improves over previous work achieving the same security guarantee in two ways: it uses black-box access to the underlying protocol and achieves a constant multiplicative overhead in the round complexity. As a corollary, we obtain the first constructions of adaptively secure protocols in the stand-alone model using black-box access to a low-level primitive.

1 Introduction

Secure multi-party computation (MPC) allows several mutually distrustful parties to perform a joint computation without compromising, to the greatest extent possible, the privacy of their inputs or the correctness of the outputs. An important criterion in evaluating the security guarantee is *how many* parties an adversary is allowed to corrupt and *when* the adversary determines which parties to corrupt. In this work, we focus on MPC protocols secure against an adversary that corrupts an arbitrary number of parties, and in addition, *adaptively* determines who and when to corrupt during the course of the computation. Even though the latter is a very natural and realistic assumption about the adversary, most of the MPC literature only addresses security against a static adversary, namely one that chooses (and fixes) which parties to corrupt before the protocol starts executing.

In the absence of an honest majority, secure MPC protocols can only be realized under computational assumptions. From both a theoretical and practical stand-point, it is desirable for these protocols to be based on general hardness assumptions, and in addition, to require only black-box access to the primitive guaranteed by the assumption (that is, the protocol refers only to the input/output behavior of the primitive). Indeed, the first MPC protocols achieving security without an honest majority [GMW87] require non-black-box access to the underlying cryptographic primitives: the first step in the construction is to obtain protocols that are secure against semi-honest adversaries, and the second handles malicious behavior by having the parties prove in zero knowledge

---

* Supported in part by NSF Grants CNS-0716245, CCF-0347839, and SBE-0245014.
** Part of this work was done while a post-doc at Columbia University.

O. Reingold (Ed.): TCC 2009, LNCS 5444, pp. 387–402, 2009.

that they are adhering to the protocol constructions. It is the second step that requires the code of the underlying primitive with the use of general NP reductions to prove statements in zero knowledge. This aversely affects both computational complexity and communication complexity of the resulting protocol as well as the complexity of implementing the protocol.

In a recent work, Ishai et al. [IKLP06] exhibited MPC protocols that are secure against a static adversary corrupting any number of parties and that rely only on black-box access to a low-level primitive, such as (enhanced) trapdoor permutations and homomorphic encryption schemes. This, along with the follow-up work of Haitner [H08], resolves the theoretical question of the minimal assumptions under which we may obtain black-box constructions of secure MPC protocols against a static adversary. The main technical contribution in both works is to construct a secure protocol for a specific two-party functionality, that of oblivious transfer (OT). The general result then follows from a classic result of Kilian's [K88] showing that any multi-party functionality can be securely computed using black-box access to a secure OT protocol. However, none of these works addresses security against an adaptive adversary, which begs the following question:

> Is it possible to construct MPC protocols secure against a malicious, adaptive adversary that may corrupt any number of parties, given only black-box access to a low-level primitive?

Towards resolving this question, Ishai, Prabhakaran and Sahai [IPS08] established an analogue of Kilian's result for an adaptive adversary. While there has been fairly extensive work on secure OT protocols against a static malicious adversary (e.g. [NP01, K05, PVW08]), very few - namely [B98, CLOS02, KO04] - provide security against an adaptive adversary; moreover, all of those which do follow [GMW87] paradigm and exploit non-black-box access to the underlying primitive.

1.1 Our Results

Our main technical contribution is the construction of efficient OT protocols that achieve security against an adaptive adversary, while relying only upon black-box access to some low-level primitive. Specifically, we provide a compiler that transforms an OT protocol secure against a semi-honest, adaptive adversary into one that is secure against a malicious, adaptive adversary, given only black-box access to the underlying OT protocol and an "ideal" commitment scheme. In addition, we achieve security in the universal composability (UC) model, where a protocol may be executed concurrently with an unknown number of other protocols [C01]. This is a notable improvement over afore-mentioned works of Ishai et al. [IKLP06, H08] which provide a compiler for semi-honest OT to malicious OT, but only for static adversaries in the stand-alone model.[1]

Theorem 1 (informal). *There exists a black-box construction of a protocol that UC realizes OT against a malicious, adaptive adversary in the $\mathcal{F}_{COM}$-hybrid model, starting*

---

[1] We note that our construction does not improve on the computational complexity of the previous compiler, as measured by the number of invocations of the underlying semi-honest OT protocol. However, we believe our construction may be combined with the OT extension protocol in [IPS08, Section 5.3] to achieve better efficiency.

from any protocol that UC realizes OT against a semi-honest, adaptive adversary.[2] *Moreover, the construction achieves a constant multiplicative blow-up in the number of rounds.*

Our construction also improves upon the earlier work of Canetti et. al [CLOS02] achieving the same guarantee; their construction is non-black-box and incurs a blow-up in round complexity proportional to the depth of the circuit computing the semi-honest OT protocol. Combined with the 2-round semi-honest OT protocol in [CLOS02, CDMW08], we obtain the first constant-round protocol for OT in the $\mathcal{F}_{COM}$-hybrid model secure against a malicious, adaptive adversary.[3] Moreover, the protocol uses black-box access to a low-level primitive, that of trapdoor simulatable cryptosystems[4], which may in turn be based on the RSA, DDH, worst-case lattice assumptions or hardness of factoring.

The key conceptual insight underlying the construction is to view the [IKLP06, H08] compiler as an instantiation of the [GMW87] paradigm in the $\mathcal{F}_{COM}$-hybrid model, except enforcing consistency via cut-and-choose techniques instead of using zero-knowledge proofs. This perspective leads naturally to a simpler, more modular, and more direct analysis of the previous compiler for static adversaries. In addition, we immediately obtain a string OT protocol, which is important for obtaining round-efficient MPC protocols [LP07, IPS08]. Showing that the modified compiler achieves UC security against an adaptive adversary requires new insight in constructing a simulator and in the analysis. We defer a more detailed discussion of the construction to Section 2, and instead focus here on the applications to secure MPC derived by combining our OT protocol with various MPC protocols in the $\mathcal{F}_{OT}$-hybrid model in [IPS08].

MPC in the $\mathcal{F}_{COM}$-Hybrid Model. Combining our OT protocol with [IPS08, Theorem 2], we obtain UC-secure MPC protocols in the $\mathcal{F}_{COM}$-hybrid model against a malicious, adaptive adversary, which improves upon [CLOS02] in that we only require black-box access to the underlying primitive:

Theorem 2 (informal). *There exists a protocol in the $\mathcal{F}_{COM}$-hybrid model that uses black-box access to a (trapdoor) simulatable cryptosystem and UC realizes any well-formed multi-party functionality against a malicious, adaptive adversary that may corrupt any number of parties.*

The round complexity of the protocol is proportional to the depth of the circuit computing the functionality. By combining our OT protocol with [IPS08, Theorem 3],

---

[2] In both the semi-honest and the malicious OT protocols, we allow the adaptive adversary to corrupt both the sender and the receiver.

[3] In an independent work [GWZ08], Garay, Wichs and Zhou also constructed a constant-round protocol for OT in the common reference string model, secure against a malicious, adaptive adversary. Their underlying assumptions are comparatively more restrictive.

[4] Trapdoor simulatable cryptosystems are introduced in [CDMW08], as a relaxation of simulatable cryptosystems [DN00]. These are semantically secure encryption schemes with special algorithms for "obliviously" sampling public keys and ciphertexts without learning the respective secret keys and plaintexts. In addition, both of these oblivious sampling algorithms are efficiently invertible given the corresponding secret key.

we obtain a constant-round MPC protocol in the $\mathcal{F}_{\text{COM}}$ with the same guarantees, except that the adversary is limited to corrupting up to $m-1$ parties for a m-party functionality. The advantage of constructing UC-secure MPC protocols in the $\mathcal{F}_{\text{COM}}$-hybrid model is that they may be combined with many of the existing UC feasibility results under various set-up or modeling assumptions e.g. [CLOS02, BCNP04, CDPW07, K07], almost all of which start by showing how to UC realize $\mathcal{F}_{\text{COM}}$ in some new security model[5]. Moreover, if the protocol realizing $\mathcal{F}_{\text{COM}}$ uses black-box access to a low-level primitive, so will the combined protocol.

MPC in the Stand-Alone Model. Next, we present our results for the stand-alone model with adaptive post-execution corruptions [C00], which is a weaker notion of security than UC security with adaptive corruptions (and in particular, our protocols in the $\mathcal{F}_{\text{COM}}$-hybrid model achieve this notion of security). Here, there is a constant-round two-party protocol that uses black-box access to a one-way function and securely realizes $\mathcal{F}_{\text{COM}}$ in the plain model without any set-up assumptions [PW09]. This immediately yields the following corollaries (via the composition theorem in [C00]):

Corollary 1 (informal). *There exists a constant-round string OT protocol that uses black-box access to a (trapdoor) simulatable cryptosystem and is secure in the stand-alone model against a malicious, adaptive adversary.*

Corollary 2 (informal). *There exists a protocol that uses black-box access to a (trapdoor) simulatable cryptosystem and securely computes any well-formed multi-party functionality in the stand-alone model against a malicious, adaptive adversary that may corrupt any number of parties.*

Both of these results improve on the work of Beaver's [B98] which achieve similar security guarantees but with non-black-box access to the underlying primitive.

Corollary 3 (informal). *For any constant $m \geq 2$, there exists a constant-round protocol that uses black-box access to a (trapdoor) simulatable cryptosystem and securely computes any well-formed m-party functionality in the stand-alone model against a malicious, adaptive adversary that may corrupt up to $m-1$ parties.*

This extends a result of Katz and Ostrovsky [KO04] which presents a 4-round protocol achieving the same security guarantee for two parties but relying on non-black-box access to the underlying primitive.

2   Construction

High-Level Description. We provide an overview of the [IKLP06, H08] compiler. Our presentation is slightly different from, and simpler than, that in the original works, and is closer in spirit to the [GMW87] compiler. Our presentation is easier to adapt to the UC setting and the adaptive setting (and also OT with strings instead of bits) since we do not need to rely on the intermediate notion and construction of a defensible

---

[5] This is because it is impossible to UC realize any non-trivial 2-party functionality in the plain model (even against static adversaries) [CKL06, C01].

OT protocol.[6] We focus on the main transformation Comp (shown in Fig 3), which "boosts" the security guarantee of an OT protocol Π from security against semi-honest receivers to security against malicious receivers while preserving the security guarantee for corrupted senders.

Phase I: Random tape generation. The sender and the receiver engage in a coin-tossing (in the well) protocol to determine a collection of $2n$ random strings for the receiver.

Phase II: Basic execution. The sender and the receiver engage in $2n$ parallel executions of Π with random inputs: the sender will choose its own inputs randomly and independently for each of the $2n$ executions, whereas the inputs and randomness for the receiver are determined by the preceding coin-tossing protocol (one random string for each execution of Π).

Phase III: Cut-and-choose. The sender and the receiver engage in a coin-tossing protocol to pick a random subset Q of n executions, and the receiver proves that it acted accordingly to Π for the n executions in Q by revealing the inputs and randomness used for those executions. The sender verifies that the inputs and randomness are indeed consistent with both the n executions of Π and the coin-tossing protocol, and if so, we are guaranteed that the receiver must have behaved honestly in at least one of the n executions of Π not in Q (except with negligible probability).

Phase IV: Combiner. We may then apply a combiner that (essentially) yields a single secure OT protocol, starting a collection of n OT protocols all of which guarantee security against a malicious sender, but only one of which guarantee security against a malicious receiver.

To obtain a full-fledged string-OT protocol secure against both a malicious sender and a malicious receiver starting from a semi-honest bit-OT protocol, we proceed as in [IKLP06], with the addition of Step 3 to directly obtain a string-OT protocol and with references to semi-honest instead of defensible adversaries:

1. Use Comp to obtain a bit-OT protocol secure against a semi-honest sender and a malicious receiver.
2. Use OT reversal [WW06] (shown in Fig 4) to obtain a bit-OT protocol secure against a malicious sender and a semi-honest receiver.
3. Repeat in parallel to obtain a string-OT protocol secure against a malicious sender and a semi-honest receiver.
4. Use Comp again to obtain a string-OT protocol secure against malicious sender and receiver.

In this work, we will view the above construction in the $\mathcal{F}_{\text{COM}}$-hybrid model, where the $\mathcal{F}_{\text{COM}}$ functionality is used to implement the coin-tossing protocol in Phases I

---

[6] Specifically, the previous compiler proceeds in two phases. The first [H08] transforms any semi-honest OT protocol into defensible OT protocols. A defensible OT protocol provides an intermediate level of security interpolating semi-honest and malicious OT. The second [IKLP06] transforms any defensible OT protocol into a malicious one.

Functionality $\mathcal{F}_{\text{COM}}$

1. Upon receiving input (commit, sid, P_j, x) from P_i where $x \in \{0,1\}^m$, internally record the tuple (P_i, P_j, x) and send the message (sid, P_i, P_j) to the adversary; When receiving (ok) from the adversary, output (receipt, sid, P_i) to P_j. Ignore all subsequent (commit, ...) inputs.

2. Upon receiving a value (reveal, sid) from P_i, where a tuple (P_i, P_j, x) is recorded, send (x) to the adversary; When receiving (ok) from the adversary, output (reveal, sid, x) to P_j.

Fig. 1. String Commitment Functionality

Functionality $\mathcal{F}_{\text{OT}}$

1. Upon receiving input (sender, sid, s_0, s_1) from **S** where $s_0, s_1 \in \{0,1\}^\ell$, record the pair (sid, s_0, s_1).

2. Upon receiving input (receiver, sid, r) from **R** where $r \in \{0,1\}$, send (sid, s_r) to **R** and (sid) to the adversary, and halt. If no (sender, $sid, \ldots$) message was previously sent, send nothing to **R**.

Fig. 2. Oblivious Transfer Functionality

and III [B81, CR03]. To instantiate the protocol in the plain stand-alone model, we will need to replace $\mathcal{F}_{\text{COM}}$ with an extractable trapdoor commitment scheme. This is different from the original [IKLP06, H08] compiler, where a standard commitment scheme is used in the Phase I commitments. We will use the same construction for an adaptive adversary except a minor simplification to Comp: the sender picks the challenge Q in Phase III (even when it may be malicious) We note here that we will exploit the extractability of the Phase I commitments in a crucial way when handling an adaptive adversary.

Improved Analysis for Static Adversaries. We sketch an improved analysis of Comp for static adversaries in the stand-alone $\mathcal{F}_{\text{COM}}$-hybrid model (i.e. showing that if Π is secure against a semi-honest receiver, then Comp(Π) is secure against a malicious receiver). Our simulator for a malicious receiver $\mathbf{R}^*$ is similar to that in [IKLP06], except we extract $\mathbf{R}^*$'s input to $\mathcal{F}_{\text{OT}}$ (in the ideal model) from Phase I instead of Phase III. This way, we achieve straight-line and strict polynomial time simulation. In the reduction[7], we will need to use repeated sampling to estimate for each $r = 1, 2, \ldots, n$, a quantity related to the probability that an honest sender interacting with a malicious receiver $\mathbf{R}^*$ in Comp(Π) does not abort at the end of Phase III, and amongst the remaining executions not in Q, exactly r are consistent with the random tape determined in Phase I. This is reminiscent of the analysis in [H08, Lemma 3] but much simpler. Putting everything together, we obtain the following result[8]:

Proposition 1. *There exists a black-box construction of a string-OT protocol secure against a static, malicious adversary in the stand-alone $\mathcal{F}_{\text{COM}}$-hybrid model, starting*

---

[7] This step is not necessary if we use a non-uniform reduction.

[8] We believe our analysis also extends to static adversaries in the UC model and we plan to look into that in the full version of this paper.

INITIALIZATION.

    Sender input: (sender, sid, s_0, s_1) where $s_0, s_1 \in \{0,1\}^\ell$.

    Receiver input: (receiver, sid, r) where $r \in \{0,1\}$.

PHASE I: RANDOM TAPE GENERATION.

1. **R** chooses $2n$ random strings $(r_1^{\mathrm{R}}, \tau_1^{\mathrm{R}}), \ldots, (r_{2n}^{\mathrm{R}}, \tau_{2n}^{\mathrm{R}})$ and sends (commit, sid_i, r_i^{R}, τ_i^{R}) to $\mathcal{F}_{\mathrm{COM}}$, for $i = 1, 2, \ldots, 2n$.
2. Upon receiving (receipt, sid_1), $\ldots$, (receipt, sid_{2n}) from $\mathcal{F}_{\mathrm{COM}}$, **S** sends $2n$ random strings $(r_1^{\mathrm{S}}, \tau_1^{\mathrm{S}}), \ldots, (r_{2n}^{\mathrm{S}}, \tau_{2n}^{\mathrm{S}})$.
3. **R** sets $r_i = r_i^{\mathrm{R}} \oplus r_i^{\mathrm{S}}$ and $\tau_i = \tau_i^{\mathrm{R}} \oplus \tau_i^{\mathrm{S}}$, for $i = 1, 2, \ldots, 2n$.

PHASE II: BASIC EXECUTION.

1. **S** chooses $2n$ pairs of random inputs $(s_1^0, s_1^1), \ldots, (s_{2n}^0, s_{2n}^1)$.
2. **S** and **R** engages in $2n$ parallel executions of the protocol Π. In the ith execution, **S** inputs (s_i^0, s_i^1) and **R** inputs r_i with randomness τ_i and obtains output $s_i^{r_i}$.

PHASE III: CUT-AND-CHOOSE.

1. **S** chooses a random $q = (q_1, \ldots, q_n) \in \{0,1\}^n$. The string q is used to define a set of indices $Q \subset \{1, 2, \ldots, 2n\}$ of size n in the following way: $Q = \{2i - q_i\}_{i=1}^n$.
2. For every $i \in Q$, **R** sends (reveal, sid_i) to $\mathcal{F}_{\mathrm{COM}}$ and upon receiving (reveal, sid_i, r_i^{R}, τ_i^{R}) from $\mathcal{F}_{\mathrm{COM}}$, **S** computes (r_i, τ_i).
3. **S** checks that for all $i \in Q$, (r_i, τ_i) is consistent with **R**'s messages in the i'th execution of Π. If not, **S** aborts and halts.

PHASE IV: COMBINER.

1. For every $j \notin Q$, **R** computes $\alpha_j = r \oplus r_j$ and sends $\{\alpha_j\}_{j \notin Q}$ to **S**.
2. **S** computes $\sigma_0 = s_0 \oplus (\bigoplus_{j \notin Q} s_j^{\alpha_j})$ and $\sigma_1 = s_1 \oplus (\bigoplus_{j \notin Q} s_j^{1-\alpha_j})$ and sends (σ_0, σ_1).
3. **R** computes and outputs $s_r = \sigma_r \oplus (\bigoplus_{j \notin Q} s_j^{r_j})$.

Fig. 3. THE COMPILER Comp(Π)

from any bit-OT protocol secure against a static, semi-honest adversary. Moreover, the construction achieves a constant multiplicative blow-up in the number of rounds, and has a strict polynomial-time and straight-line simulator.

Our result and analysis even for static adversaries offers several improvements over that in [IKLP06, H08]:

- The simulator in [IKLP06] uses rewinding and runs in expected polynomial time, even in the $\mathcal{F}_{\mathrm{COM}}$-hybrid model.
- Our result immediately yields string-OT protocols and in a constant number of rounds.
- We eliminate many of the tedious steps in the analysis in both [IKLP06] and [H08], most notably verifying that the OT reversal protocol in [WW06] works for defensible adversaries [IKLP06, Claim 5.2]. The overall analysis is simpler, more modular, and more intuitive.

As shown in [PW09], there exists a constant-round protocol that securely realizes $\mathcal{F}_{\mathrm{COM}}$ against a static adversary in the stand-alone model and that uses black-box access to a one-way function. Combined with Proposition 1, we obtain a constant-round string-OT protocol secure against a static, malicious adversary, relying only on black-box access to a constant-round bit-OT protocol secure against a static, semi-honest adversary.

Achieving Security against Adaptive Adversaries. In order to cope with adaptive adversaries in $\mathsf{Comp}(\Pi)$, we will further modify our simulator for static adversaries. The main difference lies in how we simulate the sender messages in Phase II of $\mathsf{Comp}(\Pi)$. For static adversaries, we may simply follow the honest sender strategy for $\mathsf{Comp}(\Pi)$ (i.e., run the honest sender in Π on random inputs for all $2n$ parallel executions of Π). This simulation strategy fails against an adaptive adversary because we will then be unable to present random tapes that are consistent with different sender's inputs and the protocol transcript if the sender is corrupted at the end of the protocol. Instead, we will simulate honest sender messages in Phase II against a malicious receiver as follows:

1. For each i, extract the receiver's input and randomness for the i'th execution of Π from the commitment in Phase I.
2. Upon receiving a message from the receiver in the i'th execution of Π, check if all of the receiver's messages so far are consistent with its input and randomness. If so, generate the sender's response by using the simulator for Π. Otherwise, corrupt the sender in the i'th execution of Π to obtain its input and random tape and complete the simulation of the sender's messages using the honest sender strategy.

Organization. We present our analysis of Comp and OT reversal for adaptive adversaries in the UC model in Sections 3 and 4 respectively. We defer the proof of Proposition 1 for static adversaries to the full version of the paper. Henceforth, we will always refer to adaptive adversaries.

3   Achieving Security against a Malicious Receiver

In this section, we show that Comp boosts the security guarantee from security against semi-honest receivers to security against malicious receivers.

Proposition 2. *Suppose Π is a protocol that UC realizes $\mathcal{F}_{\mathrm{OT}}$ against a semi-honest, adaptive adversary, and let $\mathsf{Comp}(\Pi)$ be the protocol obtained by applying the compiler in Fig 3 to Π. Then, $\mathsf{Comp}(\Pi)$ UC realizes $\mathcal{F}_{\mathrm{OT}}$ against an adaptive adversary with a semi-honest sender and a malicious receiver. Moreover, if Π is in addition secure against a malicious, adaptive sender, then $\mathsf{Comp}(\Pi)$ UC realizes $\mathcal{F}_{\mathrm{OT}}$ against an adaptive adversary with malicious sender and receiver.*

A Hybrid Execution. To facilitate the analysis, we introduce an intermediate setting (inspired by [IKOS07]) in which the protocol $\mathsf{Comp}(\Pi)$ is executed, where there is again a sender $\mathbf{S}$ and a receiver $\mathbf{R}$ and in addition $2n$ pairs of "virtual" parties $(\mathbf{S}_1, \mathbf{R}_1), \ldots, (\mathbf{S}_{2n}, \mathbf{R}_{2n})$. The i'th execution of Π in $\mathsf{Comp}(\Pi)$ will be executed by $\mathbf{S}_i$ and $\mathbf{R}_i$ with inputs from $\mathbf{S}$ and $\mathbf{R}$ respectively. We will require that $\mathbf{R}_1, \ldots, \mathbf{R}_{2n}$ are always semi-honest; i.e. they use a truly random tape for Π. Moreover, the environment is not aware of the "virtual parties".

PHASE I/II: BASIC EXECUTION.[9] **S** chooses $2n$ pairs of random inputs $(s_1^0, s_1^1), \ldots,$
(s_{2n}^0, s_{2n}^1) and **R** chooses $2n$ random inputs $r_1, \ldots, r_{2n}$. For each $i = 1, \ldots, 2n$, **S**
activates $\mathbf{S}_i$ with (sender, sid_i, s_i^0, s_i^1) and **R** activates $\mathbf{R}_i$ with (receiver, sid_i, r_i).
In HYBRID$_{\Pi, \mathcal{A}, \mathcal{Z}}$, the parties $\mathbf{S}_i$ and $\mathbf{R}_i$ execute Π in parallel. In HYBRID$_{\mathcal{F}_{\text{OT}}, \mathcal{A}, \mathcal{Z}}$,
the parties $\mathbf{S}_i$ and $\mathbf{R}_i$ call the ideal functionality $\mathcal{F}_{\text{OT}}$.

PHASE III: CUT-AND-CHOOSE. **S** chooses a random $q \in \{0, 1\}^n$ which identifies
$Q \subset \{1, 2, \ldots, 2n\}$ as in Comp(Π) and sends q to **R**. **S** checks that for all $i \in Q$,
$\mathbf{S}_i$ is not corrupted. Otherwise, abort.

PHASE IV: COMBINER. Proceed as in Phase IV of Comp(Π).

We say that an adversary $\mathcal{A}$ in the hybrid execution is well-formed if it satisfies the
following properties:

- When $\mathcal{A}$ corrupts **S**, it also corrupts each of $\mathbf{S}_1, \ldots, \mathbf{S}_{2n}$. Moreover, if **S** is semi-honest, then $\mathbf{S}_1, \ldots, \mathbf{S}_{2n}$ are semi-honest.
- When $\mathcal{A}$ corrupts **R**, it also corrupts each of $\mathbf{R}_1, \ldots, \mathbf{R}_{2n}$. Moreover, $\mathbf{R}_1, \ldots, \mathbf{R}_{2n}$ are always semi-honest, even if **R** is malicious.
- If **R** is corrupted, then $\mathcal{A}$ may corrupt any of $\mathbf{S}_1, \ldots, \mathbf{S}_{2n}$ with semi-honest behavior, without corrupting **S**.
- Upon receiving the set Q in Phase III from **S**, $\mathcal{A}$ may corrupt all of $\mathbf{R}_j, j \in Q$ with semi-honest behavior, even if neither **R** nor **S** is corrupted. However, if **R** is not corrupted, then $\mathbf{R}_j, j \notin Q$ are also not corrupted.

We will also stipulate that the communication channels between **S** and each of
$\mathbf{S}_1, \ldots, \mathbf{S}_{2n}$ are private and authenticated. The same holds for the communication
channels between **R** and each of $\mathbf{R}_1, \ldots, \mathbf{R}_{2n}$. In addition, **S** learns whether each of
$\mathbf{S}_1, \ldots, \mathbf{S}_{2n}$ is corrupted.

Lemma 1. *For every adversary $\mathcal{A}$ that interacts with* Comp(Π) *in the* $\mathcal{F}_{\text{COM}}$*-hybrid
model, there exists a well-formed adversary $\mathcal{A}'$ that interacts in the hybrid execution
running Π, such that for every environment $\mathcal{Z}$,*

$$\text{EXEC}_{\text{Comp}(\Pi), \mathcal{A}, \mathcal{Z}}^{\mathcal{F}_{\text{COM}}} \equiv \text{HYBRID}_{\Pi, \mathcal{A}', \mathcal{Z}}$$

In the first step, we show how to enforce semi-honest behavior of $\mathbf{R}_1, \ldots, \mathbf{R}_{2n}$
in HYBRID$_{\Pi, \mathcal{A}', \mathcal{Z}}$. The high-level strategy is as follows: if a corrupted receiver in
Comp(Π) deviates from semi-honest behavior in the i'th execution of Π in Phase III,
we corrupt $\mathbf{S}_i$ in HYBRID$_{\Pi, \mathcal{A}', \mathcal{Z}}$ to obtain its input and randomness, and continue the
simulation by running the honest sender strategy.

Proof. As usual, $\mathcal{A}'$ works by invoking a copy of $\mathcal{A}$ and running a simulated interaction
of $\mathcal{A}$ with $\mathcal{Z}$ and the parties **S** and **R**. We will refer to the communication of $\mathcal{A}'$ with
$\mathcal{Z}$ and Comp(Π) as external communication, and that with the simulated $\mathcal{A}$ as internal
communication. More precisely, $\mathcal{A}'$ works as follows:

---

[9] The choice of notation is so that Phase III always corresponds to cut-and-choose and Phase IV corresponds to combiner in both Comp(Π) and in the hybrid executions.

Simulating the communication with $\mathcal{Z}$: Every input value that $\mathcal{A}'$ receives from $\mathcal{Z}$ externally is written into the adversary $\mathcal{A}$'s input tape (as if coming from $\mathcal{A}$'s environment). Every output value written by $\mathcal{A}$ on its output tape is copied to $\mathcal{A}'$'s own output tape (to be read by the external $\mathcal{Z}$).

Simulating the case when neither party is corrupted:

PHASE I. $\mathcal{A}'$ internally passes $\mathcal{A}$ the message (receipt, sid_1), (receipt, sid_2), ..., (receipt, sid_{2n}) as if sent from $\mathcal{F}_{\text{COM}}$ to $\mathbf{S}$. Then, $\mathcal{A}'$ chooses $2n$ random strings $(r_1^{\mathrm{S}}, \tau_1^{\mathrm{S}}), \ldots, (r_{2n}^{\mathrm{S}}, \tau_{2n}^{\mathrm{S}})$, and simulates $\mathbf{S}$ sending $\mathbf{R}$ those $2n$ strings.

PHASE II. For each round in the protocol Π, if it is the receiver's turn, then for each $i = 1, \ldots, 2n$, $\mathcal{A}'$ obtains β_i from $\mathbf{R}_i$ for the corresponding round. Next, $\mathcal{A}'$ internally passes $\mathcal{A}$ the message $(\beta_1, \ldots, \beta_{2n})$, as if sent from $\mathbf{R}$ to $\mathbf{S}$. The sender's turn is handled analogously.

PHASE III. When $\mathbf{S}$ externally sends q which determines Q, then for each $i \in Q$: corrupt $\mathbf{R}_i$ to obtain (r_i, τ_i) and compute $r_i^{\mathrm{R}} = r_i \oplus r_i^{\mathrm{S}}$ and $\tau_i^{\mathrm{R}} = \tau_i \oplus \tau_i^{\mathrm{S}}$. Send (reveal, $sid_i, r_i^{\mathrm{R}}, \tau_i^{\mathrm{R}}$) to $\mathcal{A}$ as if coming from $\mathcal{F}_{\text{COM}}$.

PHASE IV. Just forward all the messages between $\mathbf{S}$ and $\mathbf{R}$.

Simulating the case when only the sender is corrupted: This is essentially the same as when neither party is corrupted, except the values $(r_1^{\mathrm{S}}, \tau_1^{\mathrm{S}}), \ldots, (r_{2n}^{\mathrm{S}}, \tau_{2n}^{\mathrm{S}})$ in Phase I and the value q in Phase III are chosen by $\mathcal{A}$.

Simulating the case when only the receiver is corrupted:

PHASE I. $\mathcal{A}'$ externally corrupts $(\mathbf{R}_1, \ldots, \mathbf{R}_{2n})$ to obtain $(\tau_1, \ldots, \tau_{2n})$ and picks $2n$ random values $r_1, \ldots, r_{2n}$. Next, $\mathcal{A}'$ obtains from $\mathcal{A}$ the messages (commit, $sid_i, r_i^{\mathrm{R}}, \tau_i^{\mathrm{R}}$) as sent by $\mathbf{R}$ to $\mathcal{F}_{\text{COM}}$. Then, $\mathcal{A}'$ sets $r_i^{\mathrm{S}} = r_i \oplus r_i^{\mathrm{R}}$ and $\tau_i^{\mathrm{S}} = \tau_i \oplus \tau_i^{\mathrm{R}}$ for $i = 1, 2, \ldots, 2n$ and internally passes $(r_1^{\mathrm{S}}, \tau_1^{\mathrm{S}}), \ldots, (r_{2n}^{\mathrm{S}}, \tau_{2n}^{\mathrm{S}})$ to $\mathcal{A}$ as if sent by $\mathbf{S}$ to $\mathbf{R}$.

PHASE II. We need to simulate the external messages sent by $\mathbf{S}$ in $\text{Comp}(\Pi)$ (with the "help" of $\mathbf{S}_1, \ldots, \mathbf{S}_{2n}$). If $\mathbf{R}$ behaves consistently in the ith execution of Π, we will just obtain the corresponding message from $\mathbf{S}_i$; otherwise, we will corrupt $\mathbf{S}_i$ so that we may compute those messages.

First, we handle receiver messages in $\text{Comp}(\Pi)$. Whenever $\mathcal{A}$ sends a message $(\beta_1, \ldots, \beta_{2n})$ from $\mathbf{R}$ where β_i is the message in the ith parallel execution of Π, do the following for each $i = 1, \ldots, 2n$:

- If $\mathbf{R}_i$ has not aborted and β_i is consistent with (r_i, τ_i), deliver the corresponding message from $\mathbf{R}_i$ to $\mathbf{S}_i$.
- If $\mathbf{R}_i$ has not aborted and β_i is not consistent with (r_i, τ_i), $\mathcal{A}'$ tells $\mathbf{R}_i$ to abort. In addition, $\mathcal{A}'$ corrupts $\mathbf{S}_i$ to obtain its input (s_i^0, s_i^1) and its randomness.
- If $\mathbf{R}_i$ has aborted, then record β_i and do nothing.

Next, we handle sender messages in $\text{Comp}(\Pi)$. Whenever $\mathcal{A}$ expects a message $(\gamma_1, \ldots, \gamma_{2n})$ from $\mathbf{S}$, where γ_i is the message in the ith parallel execution of Π, do the following for each $i = 1, \ldots, 2n$:

- If $\mathbf{S}_i$ is corrupted, then $\mathcal{A}'$ computes γ_i according to $\mathbf{S}_i$'s input and randomness and the previous messages from $\mathbf{R}_i$.
- If $\mathbf{S}_i$ is not corrupted, then set γ_i to be the corresponding message sent from $\mathbf{S}_i$ to $\mathbf{R}_i$.

$\mathcal{A}'$ then sends $(\gamma_1, \ldots, \gamma_{2n})$ to $\mathcal{A}$ as if sent by $\mathbf{S}$ to $\mathbf{R}$.

PHASE III. Deliver q sent externally by $\mathbf{S}$ to $\mathbf{R}$. Check that for all $i \in Q$, $\mathbf{S}_i$ is not corrupted. Otherwise, abort.

PHASE IV. Just forward all the messages between $\mathbf{S}$ and $\mathbf{R}$.

Dealing with corruption of parties: When the simulated $\mathcal{A}$ internally corrupts $\mathbf{R}$, $\mathcal{A}'$ externally corrupts $\mathbf{R}$ and thus $\mathbf{R}_1, \ldots, \mathbf{R}_{2n}$, and learns the values $r_1, \ldots, r_{2n}$ and $\tau_1, \ldots, \tau_{2n}$ (in addition to the input r). $\mathcal{A}'$ then sets $r_i^{\mathrm{R}} = r_i \oplus r_i^{\mathrm{S}}$ and $\tau_i^{\mathrm{R}} = \tau_i \oplus \tau_i^{\mathrm{S}}$ for $i = 1, 2, \ldots, 2n$ and internally passes $(r_1^{\mathrm{R}}, \tau_1^{\mathrm{R}}), \ldots, (r_{2n}^{\mathrm{R}}, \tau_{2n}^{\mathrm{R}})$ to $\mathcal{A}$ as the randomness for $\mathbf{R}$ in $\mathrm{Comp}(\Pi)$. Similarly, when the simulated $\mathcal{A}$ internally corrupts $\mathbf{S}$, $\mathcal{A}'$ externally corrupts $\mathbf{S}$ and thus $\mathbf{S}_1, \ldots, \mathbf{S}_{2n}$ and learns the values $(s_1^0, s_1^1), \ldots, (s_{2n}^0, s_{2n}^1)$ along with the randomness used by $\mathbf{S}_1, \ldots, \mathbf{S}_{2n}$ in the $2n$ executions of Π. $\mathcal{A}'$ then internally passes all of these values to $\mathcal{A}$ as the randomness for $\mathbf{S}$ in $\mathrm{Comp}(\Pi)$. In addition, for all $i \in Q$, $\mathcal{A}'$ passes the value $(r_i^{\mathrm{R}}, \tau_i^{\mathrm{R}})$ to $\mathcal{A}$ as the value sent from $\mathcal{F}_{\mathrm{COM}}$ to $\mathbf{S}$ in Phase III.

It is straight-forward to verify that in Phase III, checking $\mathbf{S}_i$ is not corrupted in $\mathrm{HYBRID}_{\Pi, \mathcal{A}', \mathcal{Z}}$ is identical to $\mathbf{R}$ behaving consistently in the ith execution of Π in $\mathrm{Comp}(\Pi)$. Thus, the abort condition at the end of Phase III are identical. We may therefore conclude that the ensembles EXEC and HYBRID are identical. $\square$

Lemma 2. *For every well-formed adversary $\mathcal{A}'$ that interacts in the hybrid execution running Π, there exists a well-formed adversary $\mathcal{A}''$ that interacts in the hybrid execution running $\mathcal{F}_{\mathrm{OT}}$, such that for every environment $\mathcal{Z}$,*

$$\mathrm{HYBRID}_{\Pi, \mathcal{A}', \mathcal{Z}} \overset{c}{\approx} \mathrm{HYBRID}_{\mathcal{F}_{\mathrm{OT}}, \mathcal{A}'', \mathcal{Z}}$$

Proof (sketch). The idea is that we may interpret $\mathrm{HYBRID}_{\Pi, \mathcal{A}', \mathcal{Z}}$ as an execution involving $4n+2$ parties $\mathbf{S}, \mathbf{R}, \mathbf{S}_1, \ldots, \mathbf{S}_{2n}, \mathbf{R}_1, \ldots, \mathbf{R}_{2n}$ jointly running some protocol that uses Π as a sub-routine, and $\mathrm{HYBRID}_{\mathcal{F}_{\mathrm{OT}}, \mathcal{A}'', \mathcal{Z}}$ as an execution involving the same $4n+2$ parties running the same protocol except with an ideal $\mathcal{F}_{\mathrm{OT}}$ functionality instead of Π. The claim then follows from the UC composition [C01]. $\square$

Lemma 3. *For every well-formed adversary $\mathcal{A}''$ that interacts in the hybrid execution running $\mathcal{F}_{\mathrm{OT}}$, there exists an ideal-process adversary $\mathcal{S}$, such that for every environment $\mathcal{Z}$,*

$$\mathrm{HYBRID}_{\mathcal{F}_{\mathrm{OT}}, \mathcal{A}'', \mathcal{Z}} \overset{s}{\approx} \mathrm{IDEAL}_{\mathcal{F}_{\mathrm{OT}}, \mathcal{S}, \mathcal{Z}}$$

Proof. Again, we first specify $\mathcal{S}$ depending on the corruption pattern:

Simulating the communication with $\mathcal{Z}$: Every input value that $\mathcal{S}$ receives from $\mathcal{Z}$ externally is written into the adversary $\mathcal{A}''$'s input tape (as if coming from $\mathcal{A}''$'s environment). Every output value written by $\mathcal{A}''$ on its output tape is copied to $\mathcal{S}$'s own output tape (to be read by the external $\mathcal{Z}$).

Simulating the case when neither party is corrupted:

PHASE I/II. Send $(sid_1, \ldots, sid_{2n})$ internally to $\mathcal{A}''$ as if sent from $\mathcal{F}_{\mathrm{OT}}$.

PHASE III. Send a random $q \in \{0,1\}^n$ as if sent from $\mathbf{S}$ to $\mathbf{R}$. For each $i \in Q$, when $\mathcal{A}''$ corrupts $\mathbf{R}_i$, pick a random $r_i \in \{0,1\}$ and a random $s_i^{r_i} \in \{0,1\}^\ell$.

PHASE IV. Send random $\{\alpha_j\}_{j \notin Q}$ as if sent from $\mathbf{R}$ and random (σ_0, σ_1) as if sent from $\mathbf{S}$.

Simulating the case when only the sender is corrupted:

PHASE I/II. When $\mathcal{A}''$ sends (sender, sid_i, s_i^0, s_i^1) to $\mathcal{F}_{\mathrm{OT}}$ as $\mathbf{S}_i$, then $\mathcal{S}$ records (s_i^0, s_i^1). Then, send $(sid_1, \ldots, sid_{2n})$ internally to $\mathcal{A}''$ as if sent from $\mathcal{F}_{\mathrm{OT}}$.

PHASE III. Proceed as in the case neither party is corrupted, except q is chosen by $\mathcal{A}''$.

PHASE IV. Send random $\{\alpha_j\}_{j \notin Q}$ to $\mathcal{A}''$ as if sent from $\mathbf{R}$. When $\mathcal{A}''$ sends (σ_0, σ_1) as $\mathbf{S}$, compute $s_0 = \sigma_0 \oplus (\bigoplus_{j \notin Q} s_j^{\alpha_j})$ and $s_1 = \sigma_1 \oplus (\bigoplus_{j \notin Q} s_j^{1-\alpha_j})$. Next, send (sender, s_0, s_1) to $\mathcal{F}_{\mathrm{OT}}$ as if sent from $\mathbf{S}$.

Simulating the case when only the receiver is corrupted:

PHASE I/II. $\mathcal{S}$ picks $2n$ pairs of random inputs $(s_1^0, s_1^1), \ldots, (s_{2n}^0, s_{2n}^1)$. If $\mathcal{A}''$ sends (receiver, sid_i, r_i) to $\mathcal{F}_{\mathrm{OT}}$ as $\mathbf{R}_i$, record r_i and pass $(sid_i, s_i^{r_i})$ to $\mathcal{A}''$ as if sent by $\mathcal{F}_{\mathrm{OT}}$ to $\mathbf{R}_i$. If $\mathcal{A}''$ corrupts $\mathbf{S}_i$, then $\mathcal{S}$ presents (s_i^0, s_i^1) as $\mathbf{S}_i$'s input to $\mathcal{A}''$.

PHASE III. Pick a random $q \in \{0,1\}^n$ and send q to $\mathcal{A}''$ as if coming from $\mathbf{S}$. Compute $Q \subset \{1, 2, \ldots, 2n\}$ as in $\mathrm{Comp}(\Pi)$. Check that for all $i \in Q$, $\mathbf{S}_i$ is not corrupted. Otherwise, $\mathcal{S}$ simulates an abort from $\mathbf{S}$.

PHASE IV. Compute $j^* \notin Q$ where $\mathbf{S}_{j^*}$ is not corrupted; output failure if such a j^* does not exist. When $\mathcal{A}''$ sends $\{\alpha_j\}_{j \notin Q}$ as $\mathbf{R}$, compute $r = \alpha_{j^*} \oplus r_{j^*}$ and send (receiver, sid, r) to $\mathcal{F}_{\mathrm{OT}}$. Upon receiving (sid, s_r) from $\mathcal{F}_{\mathrm{OT}}$, compute (σ_0, σ_1) so that σ_r is consistent with s_r as follows:
  – If $r = 0$, then $\sigma_0 = s_0 \oplus (\bigoplus_{j \notin Q} s_j^{\alpha_j})$ and σ_1 is a random string in $\{0,1\}^\ell$.
  – If $r = 1$, then σ_0 is a random string in $\{0,1\}^\ell$ and $\sigma_1 = s_1 \oplus (\bigoplus_{j \notin Q} s_j^{1-\alpha_j})$.
$\mathcal{S}$ then sends (σ_0, σ_1) to $\mathcal{A}''$ as if sent by $\mathbf{S}$ to $\mathbf{R}$.

Dealing with corruptions: Corruptions of $\mathbf{R}_1, \ldots, \mathbf{R}_{2n}, \mathbf{S}_1, \ldots, \mathbf{S}_{2n}$ may be handled as above. For corruptions of $\mathbf{R}$ and $\mathbf{S}$, we will consider two cases depending on the corruption schedule. In the first case, at least one of the parties is corrupted before the message (σ_0, σ_1) is sent.
  – Once $\mathbf{S}$ is corrupted, $\mathcal{S}$ learns the actual input (s_0, s_1). If $\mathbf{S}$ is corrupted before the messages (σ_0, σ_1) are computed, then $\mathcal{S}$ may simply present $(s_1^0, s_1^1), \ldots, (s_{2n}^0, s_{2n}^1)$ (as chosen in Phase I) as the randomness of $\mathbf{S}$. Otherwise, $\mathcal{S}$ modifies $s_{j^*}^{1-r_{j^*}}$ (if necessary) so that both relations $\sigma_0 = s_0 \oplus (\bigoplus_{j \notin Q} s_j^{\alpha_j})$ and $\sigma_1 = s_1 \oplus (\bigoplus_{j \notin Q} s_j^{1-\alpha_j})$ are satisfied.

- Once $\mathbf{R}$ is corrupted, S learns the actual input r. If $\mathbf{R}$ is corrupted before the messages $\{\alpha_j\}_{j \notin Q}$ are computed, then S may simply present $(r_1, \ldots, r_{2n})$ (as chosen in Phase I) as the randomness of $\mathbf{R}$. Otherwise, S modifies $\{r_j\}_{j \notin Q}$ so that $r_j = r \oplus \alpha_j$. In addition, S presents $s_i^{r_i}$ as the output of $\mathbf{R}_i, i = 1, 2, \ldots, 2n$.

In the other case, neither party is corrupted when the message (σ_0, σ_1) is sent.

- Once $\mathbf{S}$ is corrupted, we will modify both s_{j*}^0 and s_{j*}^1 so that (σ_0, σ_1) is consistent with (s_0, s_1).
- Once $\mathbf{R}$ is corrupted, we will first modify $\{r_j\}_{j \notin Q}$ as in the previous case and then modify $s_{j*}^{r_{j*}}$ so that σ_r is consistent with s_r.

We claim that if S does not output failure, then the ensembles $\text{HYBRID}_{\mathcal{F}_{\text{OT}}, \mathcal{A}'', \mathcal{Z}}$ and $\text{IDEAL}_{\mathcal{F}_{\text{OT}}, S, \mathcal{Z}}$ are identical. This is clear up to the end of Phase III. For Phase IV, observe that if $\mathbf{S}$ and $\mathbf{S}_{j*}$ are not corrupted, then from the view of $\mathcal{A}''$ and $\mathcal{Z}$ in $\text{HYBRID}_{\mathcal{F}_{\text{OT}}, \mathcal{A}'', \mathcal{Z}}$, the string $s_{j*}^{1-r_{j*}}$ is truly random. As such, σ_{1-r} is also truly random. Similarly, if $\mathbf{R}$ is not corrupted, then from the view of $\mathcal{A}''$ and $\mathcal{Z}$, the n values $\{r_j\}_{j \notin Q}$ are truly random and thus $\{\alpha_j\}_{j \notin Q}$ are also truly random. Furthermore, if neither $\mathbf{S}$ nor $\mathbf{R}$ is corrupted just before the message (σ_0, σ_1) is sent, then from the view of $\mathcal{A}''$ and $\mathcal{Z}$, both s_{j*}^0 and s_{j*}^1 are truly random, and thus both σ_0 and σ_1 are truly random.

It remains to show that S outputs failure with negligible probability. Observe that S only outputs failure if at the start of Phase IV, all of the following conditions hold:

- Neither party has aborted. In addition, the sender is honest at the start of Phase IV, so the challenge q is chosen at random.
- Amongst the n pairs of parties $(\mathbf{S}_1, \mathbf{S}_2), \ldots, (\mathbf{S}_{2n-1}, \mathbf{S}_{2n})$, exactly one party in each pair is corrupted. Otherwise, if there is a pair where both parties are corrupted, then $\mathbf{S}$ will abort at the end of Phase III; and if there is a pair where neither party is corrupted, then there is an uncorrupted $\mathbf{S}_{j*}$.
- The set Q corresponding to the challenge q is exactly the set of n uncorrupted parties (one in each pair).

Clearly, the last condition only holds with probability 2^{-n} over a random choice of q. □

4 Malicious Sender and Semi-honest Receiver

In this section, we reverse the OT protocol from the previous section to obtain one that is secure for a malicious sender and a semi-honest receiver. The construction (shown in Fig 4) is the same as that in [WW06], the novelty lies in the analysis which establishes security against an adaptive adversary. We note that the analysis though tedious, is fairly straight-forward.

Proposition 3. *For every adaptive adversary $\mathcal{A}$ that interacts with the protocol ψ in the $\mathcal{F}_{\text{OT}}$-hybrid model, there exists an adaptive adversary S that interacts with $\mathcal{F}_{\text{OT}}$, such that for every environment $\mathcal{Z}$,*

$$\text{EXEC}_{\psi, \mathcal{A}, \mathcal{Z}}^{\mathcal{F}_{\text{OT}}} \equiv \text{IDEAL}_{\mathcal{F}_{\text{OT}}, S, \mathcal{Z}}.$$

Moreover, the corruption pattern in S is the reverse of that in $\mathcal{A}$.

INITIALIZATION.

    Sender input: (sender, sid, s_0, s_1) where s_0, $s_1 \in \{0, 1\}$.
    Receiver input: (receiver, sid, r) where $r \in \{0, 1\}$.

PHASE I: CALL $\mathcal{F}_{\mathrm{OT}}$.

    1. **R** chooses a bit $\rho \in \{0, 1\}$ and sends (sender, sid, ρ, $\rho \oplus r$) to $\mathcal{F}_{\mathrm{OT}}$.
    2. **S** sends (receiver, sid, $s_0 \oplus s_1$) to $\mathcal{F}_{\mathrm{OT}}$.

PHASE II: REVERSE.

    1. Upon receiving (sid, a) from $\mathcal{F}_{\mathrm{OT}}$, **S** computes $\alpha = s_0 \oplus a$ and sends α to **R**.
    2. Upon receiving α, **R** computes and outputs $\rho \oplus \alpha$.

Fig. 4. The OT-Reversal Protocol ψ

Proof (sketch). As usual, S works by invoking a copy of $\mathcal{A}$ and running a simulated interaction of $\mathcal{A}$ with $\mathcal{Z}$ and the parties **S** and **R** in the $\mathcal{F}_{\mathrm{OT}}$-hybrid model. We will refer to the communication of S with $\mathcal{Z}$ and ψ as external communication, and that with the simulated $\mathcal{A}$ as internal communication. In addition, we will refer to the $\mathcal{F}_{\mathrm{OT}}$ functionality in the real execution as the internal $\mathcal{F}_{\mathrm{OT}}$, and that in the ideal execution as the external $\mathcal{F}_{\mathrm{OT}}$. S works as follows:

Simulating the communication with $\mathcal{Z}$: Every input value that S externally receives from $\mathcal{Z}$ is written into the adversary $\mathcal{A}$'s input tape (as if coming from $\mathcal{A}$'s environment). Every output value written by $\mathcal{A}$ on its output tape is copied to S's own output tape (to be read by the external $\mathcal{Z}$).

Simulating the case when neither party is corrupted: S internally passes (sid) to $\mathcal{A}$ as if coming from the internal $\mathcal{F}_{\mathrm{OT}}$. When S receives (sid) from the external $\mathcal{F}_{\mathrm{OT}}$, S chooses $\alpha \in \{0, 1\}$ at random and sends it to $\mathcal{A}$ as if coming from **S**.

Simulating the case when only the sender is corrupted: When $\mathcal{A}$ sends (receiver, sid, d) to the internal $\mathcal{F}_{\mathrm{OT}}$ as **S**, S chooses $a \in \{0, 1\}$ at random and sends (sid, a) to $\mathcal{A}$ the output from the internal $\mathcal{F}_{\mathrm{OT}}$. When $\mathcal{A}$ sends α as **S**, S sends (sender, sid, $a \oplus \alpha$, $a \oplus \alpha \oplus d$) to the external $\mathcal{F}_{\mathrm{OT}}$.

Simulating the case when only the receiver is corrupted: When $\mathcal{A}$ internally sends (sender, sid, a_0, a_1) to $\mathcal{F}_{\mathrm{OT}}$ as **R**, S sets $\rho = a_0$, $r = a_0 \oplus a_1$ and externally sends (receiver, sid, r) to $\mathcal{F}_{\mathrm{OT}}$. Upon receiving (sid, s_r) externally from $\mathcal{F}_{\mathrm{OT}}$, S internally sends $\alpha = s_r \oplus \rho$ to $\mathcal{A}$ as if coming from **S**.

Dealing with corruptions: When **R** is corrupted, S needs to present $\mathcal{A}$ with a consistent random tape comprising of a single bit ρ. When **S** is corrupted, S needs to present $\mathcal{A}$ with the output bit a which **S** receives from the internal $\mathcal{F}_{\mathrm{OT}}$. We consider four cases depending on the corruption schedule:

    – Case 1: **R** is corrupted before it sends its input to the internal $\mathcal{F}_{\mathrm{OT}}$. In this case, S proceeds as in the case when only the receiver is corrupted to compute ρ and r. If and when **S** is corrupted, S computes $a = \rho \oplus rd$ where d is **S**'s input to the internal $\mathcal{F}_{\mathrm{OT}}$ (set to $s_0 \oplus s_1$ if **S** is honest when it submits its input to the internal $\mathcal{F}_{\mathrm{OT}}$).

– Case 2: Neither party is corrupted when α is sent. In this case, $\mathcal{S}$ picks a random $\alpha \in \{0, 1\}$. Then, when $\mathbf{R}$ is corrupted, $\mathcal{S}$ learns both its input r and its output s_r, and computes $\rho = \alpha \oplus s_r$. When $\mathbf{S}$ is corrupted, $\mathcal{S}$ learns its input s_0, s_1 and computes $a = \alpha \oplus s_0$.

If neither Case 1 nor Case 2 holds, then the adversary $\mathcal{A}$ corrupts either $\mathbf{R}$ or $\mathbf{S}$ (or both) and learns at least one of ρ and a before seeing α.

– Case 3: $\mathcal{A}$ learns a first. This means $\mathcal{A}$ corrupts $\mathbf{S}$ first and corrupts $\mathbf{R}$ (if at all) after $\mathbf{S}$ receives a from the internal $\mathcal{F}_{\mathrm{OT}}$. Then, $\mathcal{S}$ proceeds as in the case where only the sender is corrupted and picks a random $a \in \{0, 1\}$. When $\mathbf{R}$ is corrupted, $\mathcal{S}$ learns r and computes $\rho = a \oplus rd$ (where d is again $\mathbf{S}$'s input to the internal $\mathcal{F}_{\mathrm{OT}}$).
– Case 4: $\mathcal{A}$ learns ρ first. This means either $\mathcal{A}$ corrupts $\mathbf{R}$ first, or $\mathcal{A}$ corrupts $\mathbf{R}$ before $\mathbf{S}$ receives a from the internal $\mathcal{F}_{\mathrm{OT}}$.[10] In this case, $\mathcal{S}$ picks $\rho \in \{0, 1\}$ at random when $\mathbf{R}$ is corrupted, and subsequently (if and when $\mathcal{A}$ corrupts $\mathbf{S}$) computes $a = \rho \oplus rd$.

Finally, we need to check that $\mathrm{EXEC}^{\mathcal{F}_{\mathrm{OT}}}_{\psi, \mathcal{A}, \mathcal{Z}} \equiv \mathrm{IDEAL}_{\mathcal{F}_{\mathrm{OT}}, \mathcal{S}, \mathcal{Z}}$, which is similar to that in [ww06] which addresses static corruptions.   □

Acknowledgments. We thank Ran Canetti, Yehuda Lindell and Manoj Prabhakaran for helpful discussions.

References

[B81]   Blum, M.: Coin flipping by telephone. In: CRYPTO (1981)
[B98]   Beaver, D.: Adaptively secure oblivious transfer. In: Ohta, K., Pei, D. (eds.) ASIACRYPT 1998. LNCS, vol. 1514, pp. 300–314. Springer, Heidelberg (1998)
[BCNP04]   Barak, B., Canetti, R., Nielsen, J.B., Pass, R.: Universally composable protocols with relaxed set-up assumptions. In: FOCS (2004)
[C00]   Canetti, R.: Security and composition of multiparty cryptographic protocols. J. Cryptology 13(1), 143–202 (2000)
[C01]   Canetti, R.: Universally composable security: A new paradigm for cryptographic protocols. In: FOCS (2001)
[CDMW08]   Choi, S.G., Dachman-Soled, D., Malkin, T., Wee, H.: Non-committing encryption and adaptively secure protocols from weaker assumptions (manuscript, 2008)
[CDPW07]   Canetti, R., Dodis, Y., Pass, R., Walfish, S.: Universally composable security with global setup. In: Vadhan, S.P. (ed.) TCC 2007. LNCS, vol. 4392, pp. 61–85. Springer, Heidelberg (2007)
[CKL06]   Canetti, R., Kushilevitz, E., Lindell, Y.: On the limitations of universally composable two-party computation without set-up assumptions. J. Cryptology 19(2), 135–167 (2006)
[CLOS02]   Canetti, R., Lindell, Y., Ostrovsky, R., Sahai, A.: Universally composable two-party and multi-party secure computation. In: STOC (2002)

---

[10] In particular, it could be that $\mathcal{A}$ corrupts $\mathbf{S}$ at the start of the protocol (learning nothing at this point), and then corrupts $\mathbf{R}$ immediately after it sends its input to the internal $\mathcal{F}_{\mathrm{OT}}$.

[CR03]   Canetti, R., Rabin, T.: Universal composition with joint state. In: Boneh, D. (ed.)
           CRYPTO 2003. LNCS, vol. 2729, pp. 265–281. Springer, Heidelberg (2003)
[DN00]   Damgård, I.B., Nielsen, J.B.: Improved non-committing encryption schemes based
           on a general complexity assumption. In: Bellare, M. (ed.) CRYPTO 2000. LNCS,
           vol. 1880, p. 432. Springer, Heidelberg (2000)
[GMW87]   Goldreich, O., Micali, S., Wigderson, A.: How to play any mental game or a
           completeness theorem for protocols with honest majority. In: STOC (1987)
[GWZ08]   Garay, J.A., Wichs, D., Zhou, H.-S.: Somewhat non-committing encryption and
           efficient adaptively secure oblivious transfer. Cryptology ePrint 2008/534 (2008)
[H08]   Haitner, I.: Semi-honest to malicious oblivious transfer—the black-box way. In:
           Canetti, R. (ed.) TCC 2008. LNCS, vol. 4948, pp. 412–426. Springer, Heidelberg
           (2008)
[IKLP06]   Ishai, Y., Kushilevitz, E., Lindell, Y., Petrank, E.: Black-box constructions for
           secure computation. In: STOC (2006)
[IKOS07]   Ishai, Y., Kushilevitz, E., Ostrovsky, R., Sahai, A.: Zero-knowledge from secure
           multiparty computation. In: STOC (2007)
[IPS08]   Ishai, Y., Prabhakaran, M., Sahai, A.: Founding cryptography on oblivious transfer
           – efficiently. In: Wagner, D. (ed.) CRYPTO 2008. LNCS, vol. 5157, pp. 572–591.
           Springer, Heidelberg (2008)
[K88]   Kilian, J.: Founding cryptography on oblivious transfer. In: STOC (1988)
[K05]   Kalai, Y.T.: Smooth projective hashing and two-message oblivious transfer. In:
           Cramer, R. (ed.) EUROCRYPT 2005. LNCS, vol. 3494, pp. 78–95. Springer,
           Heidelberg (2005)
[K07]   Katz, J.: Universally composable multi-party computation using tamper-proof
           hardware. In: Naor, M. (ed.) EUROCRYPT 2007. LNCS, vol. 4515, pp. 115–128.
           Springer, Heidelberg (2007)
[KO04]   Katz, J., Ostrovsky, R.: Round-optimal secure two-party computation. In: Franklin,
           M. (ed.) CRYPTO 2004. LNCS, vol. 3152, pp. 335–354. Springer, Heidelberg
           (2004)
[LP07]   Lindell, Y., Pinkas, B.: An efficient protocol for secure two-party computation
           in the presence of malicious adversaries. In: Naor, M. (ed.) EUROCRYPT 2007.
           LNCS, vol. 4515, pp. 52–78. Springer, Heidelberg (2007)
[NP01]   Naor, M., Pinkas, B.: Efficient oblivious transfer protocols. In: SODA (2001)
[PVW08]   Peikert, C., Vaikuntanathan, V., Waters, B.: A framework for efficient and
           composable oblivious transfer. In: Wagner, D. (ed.) CRYPTO 2008. LNCS,
           vol. 5157, pp. 554–571. Springer, Heidelberg (2008)
[PW09]   Pass, R., Wee, H.: Black-box constructions of two-party protocols from one-
           way functions. In: Reingold, O. (ed.) TCC 2009. LNCS, vol. 5444, pp. 403–418.
           Springer, Heidelberg (2009)
[WW06]   Wolf, S., Wullschleger, J.: Oblivious transfer is symmetric. In: Vaudenay, S. (ed.)
           EUROCRYPT 2006. LNCS, vol. 4004, pp. 222–232. Springer, Heidelberg (2006)

Black-Box Constructions of Two-Party Protocols from One-Way Functions

Rafael Pass[1,*] and Hoeteck Wee[2,**]

[1] Cornell University
rafael@cs.cornell.edu
[2] Queens College, CUNY
hoeteck@cs.qc.cuny.edu

Abstract. We exhibit constructions of the following two-party cryptographic protocols given only black-box access to a one-way function:

- constant-round zero-knowledge arguments (of knowledge) for any language in NP;
- constant-round trapdoor commitment schemes;
- constant-round parallel coin-tossing.

Previous constructions either require stronger computational assumptions (e.g. collision-resistant hash functions), non-black-box access to a one-way function, or a super-constant number of rounds. As an immediate corollary, we obtain a constant-round black-box construction of secure two-party computation protocols starting from only semi-honest oblivious transfer. In addition, by combining our techniques with recent constructions of concurrent zero-knowledge and non-malleable primitives, we obtain black-box constructions of concurrent zero-knowledge arguments for NP and non-malleable commitments starting from only one-way functions.

Keywords: black-box constructions, zero-knowledge arguments, trapdoor commitments, parallel coin-tossing, secure two-party computation, non-malleable commitments.

1 Introduction

Much of the modern work in foundations of cryptography rests on general cryptographic assumptions like the existence of one-way functions and trapdoor permutations. General assumptions provide an abstraction of the functionalities and hardness we exploit in specific assumptions such as hardness of factoring and discrete log without referring to any specific underlying algebraic structure. The expressive nature of general assumptions means that we could then derive constructions based on a large number of concrete assumptions of our choice, even ones that may not have been considered at the time of designing the protocols. Constructions based on general assumptions may use the primitive guaranteed by the assumption in one of two ways:

---

* Supported in part by NSF CAREER Award CCF-0746990, AFOSR Award FA9550-08-1-0197, BSF Grant 2006317 and I3P grant 2006CS-001-0000001-02.
** Most of this work was done while a post-doc at Columbia University, supported in part by NSF Grants CNS-0716245 and SBE-0245014.

O. Reingold (Ed.): TCC 2009, LNCS 5444, pp. 403–418, 2009.

Black-box usage: A construction is black-box if it refers only to the input/output behavior of the underlying primitive; we would typically also require that in the proof of security, we can use an adversary breaking the security of the construction as an oracle to break the underlying primitive (c.f. [39,27]).

Non-black-box usage: A construction is non-black box if it uses the code computing the functionality of the primitive.

Motivated by the fact that the vast majority of constructions in cryptography indeed are black-box, a rich and fruitful body of work initiated in [27] seeks to understand the power and limitations of black-box constructions in cryptography, resulting in a fairly complete picture of the relations amongst most cryptographic primitives with respect to black-box constructions. We stress that the general question of whether we can securely realize tasks via black-box access to a general primitive is not merely of theoretical interest. A practical reason is related to efficiency, as non-black box constructions tend to be less efficient due to the use of general NP reductions to order to prove statements in zero knowledge; this impacts both computational complexity as well as communication complexity. As such, non-black box constructions traditionally only serve as "feasibility" results; moreover, the constructions underlying such feasibility results often do not translate readily into "practical" black-box constructions without a recourse to the use of either specific assumptions or additional general assumptions.

Fortunately, a recent line of work has narrowed - and in several cases even closed - the gap between black-box and non-black-box constructions for many cryptographic tasks. A notable example is the work of Ishai et al. [28,24], which building on the early work of Kilian's [30], provides a black-box construction of secure multi-party protocols that can tolerate any number of static malicious adversaries, assuming only the existence a semi-honest oblivious transfer protocol (which can in turn be based on homomorphic encryption schemes or enhanced trapdoor permutations); the corresponding non-black-box feasibility result was known since the 1980s [21]. Several other works address improvements in efficiency for two-party protocols and multi-party protocols with an honest majority e.g. [12,34], as well as public-key encryption schemes secure against chosen-ciphertext attacks and variants thereof [9,37]. In fact, the recent success we have had with black-box constructions of secure protocols seems to hint that there is perhaps no inherent gap between non-black-box "feasibility" results and black-box "practical" constructions for natural cryptographic tasks: that is, any feasibility result may also be realized by a practical black-box construction under the *same* assumptions. If so, this would be a stark contrast to black-box versus non-black-box use of the adversary's code in the proof of security in simulation-based notions of security, for which a gap has been established [20,1].

Upon closer examination, one notices that while the afore-mentioned black-box constructions of secure protocols do improve on the efficiency of previous non-black-box constructions as measured in terms of computational and communication complexity, most (except for [12]) do not match the round complexity of existing non-black-box constructions. Indeed, there are several fundamental constant-round two-party cryptographic tasks, notably zero-knowledge arguments for NP for which we do know how to realize via non-black-box usage of a one-way function [16], but existing

cryptographic task	black-box OWF	non-black-box OWF	black-box SHC
zero-knowledge argument	$\omega(1)$ [22]	$O(1)$ [16]	$O(1)$ [19]
trapdoor commitments (m bits)	$\tilde{O}(m)$ [13]	$O(1)$ [16]	$O(1)$ [13]
coin-tossing (m coins)	$\tilde{O}(m)$ [3]	$O(1)$ [33]	$O(1)$ [33]

Fig. 1. Round complexity of existing constructions of several cryptographic tasks from one-way functions (OWF) and constant-round statistically hiding commitments (SHC). Here $\tilde{O}(\cdot)$ hides savings of multiplicative factors that are logarithmic in the security parameter. The trapdoor commitment with $\tilde{O}(m)$ rounds from black-box OWF is obtained by combining the [13] protocol with Blum's coin-tossing protocol [3].

black-box constructions either require a super-constant number of rounds or stronger assumptions [21,19]. This raises the following intriguing question:

Is there an inherent trade-off between round complexity and either efficiency or computational assumptions in realizing these two-party cryptographic tasks?

Put differently, if we require a constant-round zero-knowledge argument system, must we necessarily turn to a non-black-box construction (thereby incurring a loss in efficiency) or use collision-resistant hash functions (a stronger assumption)? Interestingly, the Feige-Shamir zero-knowledge arguments [16] constitute one of the earliest examples of non-black-box constructions; the same work also presents a non-black-box construction of constant-round trapdoor commitments from one-way functions, for which there is again a gap with respect to existing black-box constructions. Other related tasks with a similar gap include parallel coin-tossing from one-way functions, and secure two-party computation from semi-honest oblivious transfer. In each of these cases, constant-round non-black-box constructions are known [33,43,21], whereas existing black-box constructions either *additionally* assuming collision-resistant hash functions or constant-round statistically hiding commitments [19,13,33,34] (which we know cannot be realized via black-box access to a one-way function [25,42][1]) or by considering protocols with super-constant number of rounds. We summarize these prior results in Figure 1.

Our Results. In this work, we answer the afore-mentioned question negatively: we present black-box constructions of constant-round zero-knowledge arguments for NP and several other two-party functionalities under the minimal assumption of one-way functions:

Theorem 1 (informal). *There exist black-box constructions of constant-round zero-knowledge arguments (of knowledge) for* NP, *constant-round trapdoor commitments and constant-round parallel coin-tossing, starting from any one-way function.*

---

[1] This is true even if we allow black-box access to semi-honest oblivious transfer, by observing that the impossibility result in [25] extends to enhanced trapdoor permutations with which we could realize constant-round semi-honest oblivious transfer [15].

We stress that reducing the computational assumptions for these cryptographic protocols from collision-resistant hash functions to one-way functions is important also in practice; recent attacks on the popular MD4, MD5 and SHA1 hash functions demonstrate that achieving collision-resistance in the heuristic sense is much harder than achieving one-way'ness.

The above constructions may be modified to achieve security against adaptive corruptions in the stand-alone model (c.f. [5]) while maintaining constant round complexity. This improves on the early work of Beaver [2], who provided constructions assuming hardness of factoring. The idea is to have the receiver in the commitment scheme from [13] (which we observe to be adaptively secure) commit to its challenge using our trapdoor commitment scheme.

Secure Two-Party Computation. A series of recent work [28,34,24,10,29] (building on [30]) provided a black-box construction of secure two-party protocols starting from semi-honest oblivious transfer. The resulting protocol has constant round complexity assuming a constant-round parallel coin-tossing protocol. The following result then follows as an immediate corollary of our coin-tossing protocol:

Theorem 2 (informal). *There exists a black-box construction of constant-round secure two-party computation protocol with respect to static malicious adversaries, starting from any constant-round oblivious transfer protocol secure against static semi-honest adversaries.*

This result also extends to any constant number of parties, while preserving constant round complexity. We point out that in concurrent work, Choi et al. [10] established an analogous statement for adaptive corruptions, using as a building block our trapdoor commitment schemes tolerating adaptive corruptions.

Additional Constructions. Combining our techniques with previous work, we also obtain black-box constructions of concurrent zero-knowledge arguments and non-malleable commitments from one-way functions:

Theorem 3 (informal). *There exist black-box constructions of the following cryptographic protocols starting from any one-way function:*

- *concurrent zero-knowledge arguments for* NP *with* $c \log n$ *rounds for any super-constant function* $c(\cdot)$;
- *non-malleable commitments with* $O(\log n)$ *rounds, and concurrent non-malleable commitments with* $O(n)$ *rounds.*

The concurrent zero-knowledge argument system follows readily from modifying the challenge-response preamble in our stand-alone zero-knowledge argument system in the manner of [38,36]. The non-malleable commitment scheme requires substantially more work, combining ideas from our stand-alone zero-knowledge argument system, an encoding scheme from [9] along the messaging scheduling and analysis from [14,32].

2   Overview of Our Constructions

We begin with our overview of our constant-round zero-knowledge arguments and trap-door commitment schemes, which are obtained by applying a compiler to challenge-response protocols with a certain structure.

Challenge-Response Protocol. Consider a 3-round challenge-response protocol, say between a prover and a verifier with possibly a common input, with the following structure: In the first round, the prover commits to values $v_1, \ldots, v_k$ (bit by bit, in parallel). The verifier responds with a random challenge $e \in \{0,1\}^k$, and the prover responds by opening to some subset of bits in each value $v_1, \ldots, v_k$. Then verifier then decides whether to accept or reject.

SPECIAL SOUNDNESS: For every message in the first round, there exists at most one "easy challenge" $\tilde{e}$ that allows the prover to cheat. For a language, cheating means convincing the verifier to accept a NO instance; for a commitment scheme, cheating means generating an accepting commit phase transcript that can be opened to two different values. Moreover, we require that the "easy challenge" is efficiently recoverable in the following sense: there is an efficient procedure that given the values $v_1, \ldots, v_k$ (along with the common input), outputs a string $\tilde{e}$ such that if an easy challenge exists, it must equal $\tilde{e}$.

LOOK-AHEAD SIMULATION: Roughly speaking, this condition says simulation is easy if we can look ahead and obtain the verifier's challenge e. For a language, this condition stipulates that the protocol is special honest-verifier zero-knowledge [11]: we require that the simulator on input any fixed verifier's challenge e generates an "honest-looking" transcript. Here, honest-looking means computationally indistinguishable from an honest prover-verifier interaction wherein the verifier always sends e. For a commitment scheme, this condition stipulates that there exists a simulator that on input any fixed verifier's challenge e generates an "honest-looking" transcript of the commit phase that can be later opened to any value v. Here, honest-looking means computationally indistinguishable from an honest commitment and opening to the value v wherein the verifier always sends e in the commit phase.

The Compiler. We have the verifier commits to its challenge e in advance before running the challenge-response protocol. Indeed, this approach was adopted in [19,17] for zero knowledge, and in [13] for trapdoor commitments. The difficulty is that we do not know how to guarantee soundness as there could be a malleability attack (specifically, we do not know how to rule out the possibility that after seeing the verifier's commitment to e, the cheating prover could send some carefully crafted commitments that can be open to a valid accepting response once the verifier opens the commitment to e). This problem can be circumvented in one of three ways:

– Have the verifier commit using a *perfectly hiding* commitment scheme and the prover use a *statistically binding* commitment scheme [19,13].

– Have the verifier commit using a *trapdoor* commitment scheme and the prover use a *statistically binding* commitment scheme (implicit in [33,38,41,23][2]).
– Have both the prover and verifier commit using a *computationally hiding* commitment scheme, but have the prover prove that it "knows" the values underlying its commitments (e.g., by using a zero-knowledge proof of knowledge) before the verifier opens the commitment to its challenge [17, Sec 4.9.2.2].

We adopt the third approach in this paper. Specifically, we use an extractable commitment scheme, which is informally a commitment scheme with a proof of knowledge property. Such a commitment scheme can be constructed via black-box access to any commitment scheme using cut-and-choose techniques [38,14]. Note that the first approach cannot work in our setting because there is no black-box construction of constant-round perfectly hiding commitment schemes from one-way functions [25], whereas the second requires a functionality that we are trying to construct.

Towards Trapdoor Commitments and Parallel Coin-Tossing. For zero-knowledge arguments, Blum's challenge-response protocol for the NP-complete problem Graph Hamiltonicity [4] suffices. On the other hand, for trapdoor commitments, we need to design a new challenge-response protocol because we do not know how to efficiently recover the easy challenge in the [13] protocol. Next, we show how to derive an extractable trapdoor commitment scheme starting from any trapdoor commitment scheme (such as ours), and from there, we obtain a constant-round parallel coin-tossing protocol from the works of [3,8].

3   Preliminaries

We will use 1^k to denote the security parameter. We refer the reader to [17] for definitions of various cryptographic notions, such as zero knowledge.

Commitment Schemes. Recall that a commitment scheme Com is a 2-party protocol between a sender $\mathcal{C}$ and a receiver $\mathcal{R}$. In this paper, we always refer to computationally hiding commitment schemes. The binding property however, may be either statistical or computational. A commitment scheme has a commit phase and an open phase; we only consider commitment schemes where the open phase consists of a single message from the sender to the receiver. We know that there is a black-box construction of a 2-round statistically binding commitment scheme from any one-way function [35,26].

Trapdoor Commitment Schemes. Let $(\mathcal{C}, \mathcal{R})$ be a (computationally hiding) commitment scheme. We say that $(\mathcal{C}, \mathcal{R})$ is a trapdoor commitment scheme if there exists an expected polynomial-time probabilistic oracle machine $\mathcal{S} = (\mathcal{S}_1, \mathcal{S}_2)$ such that for any PPT $\mathcal{R}^*$ and all $v \in \{0, 1\}^n$, the output (τ, w) of the following experiments are computationally indistinguishable:

---

[2] In these works (specifically, the protocols based on one-way functions), the verifier commits to its challenge e, and to reveal the challenge, it sends the string e, along with a zero-knowledge proof that the value in the commitment is e; the verifier is effectively using a trapdoor commitment.

- an honest sender $\mathcal{C}$ interacts with $\mathcal{R}^*$ to commit to v, and then opens the commitment: τ is the view of $\mathcal{R}^*$ in the commit phase, and w is the message $\mathcal{C}$ sends in the open phase.
- the simulator $\mathcal{S}$ generates a simulated view τ for the commit phase, and then opens the commitment to v in the open phase: formally, $\mathcal{S}_1^{\mathcal{R}^*}(1^n, 1^k) \rightarrow (\tau, \text{STATE})$, $\mathcal{S}_2(\text{STATE}, v) \rightarrow w$.

Extractable Commitment Schemes. Let $(\mathcal{C}, \mathcal{R})$ be a statistically binding commitment scheme. We say that $(\mathcal{C}, \mathcal{R})$ is an extractable commitment scheme if there exists an expected polynomial-time probabilistic oracle machine (the extractor) E that given oracle access to any PPT cheating sender $\mathcal{C}^*$ outputs a pair (τ, σ^*) such that:

- (simulation) τ is identically distributed to the view of $\mathcal{C}^*$ at the end of interacting with an honest receiver $\mathcal{R}$ in commit phase.
- (extraction) the probability that τ is accepting and $\sigma^* = \perp$ is negligible.
- (binding) if $\sigma^* \neq \perp$, then it is statistically impossible to open τ to any value other than σ^*.

We will also consider extractable commitment schemes that are computationally binding; the definition is as above, except if $\sigma^* \neq \perp$, we only require that it is computationally infeasible to open τ to any value other than σ^*.

4   Extractable Commitment Schemes

The Basic Construction. The following protocol used in the works of [14,38,40] (also [30]) yields an extractable commitment scheme, starting from any commitment scheme Com:

PROTOCOL ExtCom.
- Common input: security parameter 1^k.
- Sender's input: a string $\sigma \in \{0,1\}^m$.

COMMIT PHASE.
- The sender commits (using Com) to k pairs of strings $(v_1^0, v_1^1), \ldots, (v_k^0, v_k^1)$ where $(v_i^0, v_i^1) = (\eta_i, \sigma \oplus \eta_i)$ and $\eta_1, \ldots, \eta_k$ are random strings in $\{0,1\}^m$.
- Upon receiving a challenge $e = (e_1, \ldots, e_k)$ from the receiver, the sender opens the commitments to $v_1^{e_1}, \ldots, v_k^{e_k}$.
- The receiver checks that the openings are valid.

OPEN PHASE.
- The sender sends σ and opens the commitments to all k pairs of strings.
- The receiver checks that all the openings are valid, and also that $\sigma = v_1^0 \oplus v_1^1 = \cdots = v_k^0 \oplus v_k^1$.

We sketch the proof (implicit in [14,38,40]) that ExtCom is an extractable commitment scheme.

Computationally hiding. The proof proceeds by a hybrid argument. Fix a cheating receiver, σ, σ' and suppose we want to show that $\text{ExtCom}(\sigma)$ and $\text{ExtCom}(\sigma')$ are computationally indistinguishable. In the i'th hybrid distribution, the first i pairs of strings are random shares of σ and the last $k - i$ pairs of strings are random shares of σ'. Suppose we have a distinguisher for the i'th and $i + 1$'th hybrids. If the distribution of the bit e_i is noticeably biased, then we can break the hiding property of the underlying commitment Com right away. Otherwise, we can guess e_i with probability roughly $1/2$ and obtain a distinguisher for $\mathsf{Com}(\sigma \oplus \eta_i)$ and $\mathsf{Com}(\sigma' \oplus \eta_i)$.

Extractable. We start with the easier case where Com is statistically binding, upon which ExtCom is also statistically binding. Fix a cheating sender C^*. We construct the extractor E as follows:

1. First, simulate an execution of C^* by internally emulating an honest receiver $\mathcal{R}$ to obtain a transcript τ of the commit phase. If τ is rejecting, then output $(\tau, \perp)$ and halt.
2. If τ is accepting with some challenge e, then keep rewinding C^* with random challenges until we receive another accepting response from C^* with some challenge e'. If $e = e'$, then output $(\tau, \perp)$ and halt. Otherwise, extract a value σ^* from the C^*'s responses to distinct challenges e, e' (by combining the appropriate shares), and output (τ, σ^*).

Now, suppose the probability over e that we obtain an accepting transcript τ is p. Then, the expected number of queries E makes to C^* is $(1 - p) + p \cdot \frac{1}{p} \leq 2$. Also, the failure probability, i.e., the probability that τ is accepting and $e = e'$ is at most $p \cdot \frac{2^{-k}}{p} = 2^{-k}$.

We can still use the same extractor E in the case where Com is computationally binding. Now, if there is a cheating sender that can open the commitment in τ to a different value from σ^*, then we can combine this with the opening to σ^* obtained by E to derive an efficient adversary that breaks the binding property of Com.

The Parallel Variant. For our compiler, we will actually need an extractable commitment scheme to a string σ for which we can open any subset of the bits in σ without compromising the security (i.e. hiding) of the remaining bits. We may obtain such a scheme PExtCom by running ExtCom to commit to each bit of σ in parallel. That PExtCom is hiding follows from the more general fact that the hiding property of commitment schemes is preserved under parallel composition. To show that PExtCom is extractable, we may use the same extractor E as before, except for a modification in step 2. Note that the receiver's challenge in PExtCom is a k-tuple of m-bit strings, which again we denote by $e \in (\{0, 1\}^m)^k$. Once we obtain responses to two challenges e, e' in Step 2, we proceed as follows: if e' agrees with e in any of the k components, we output $(\tau, \perp)$ and halt. Otherwise, we will be able to extract each of the m bits in the m parallel executions of ExtCom. As before, the expected number of queries E makes to C^* is at most 2. The failure probability in this case is now at most $m \cdot 2^{-k}$.

5   Zero-Knowledge Arguments for NP

Look-Ahead Zero-Knowledge Proof System. We use as our look-ahead zero-knowledge proof system the parallel repetition variant of Blum's Hamiltonicity protocol [4], which we already know to be special honest-verifier zero-knowledge.

HAMILTONICITY PROTOCOL Π_{HAM}.
- Common input: a graph G on n vertices.
- Prover's input, a cycle h in G
1. The prover picks random permutations π_i over $[n]$ and commits to $v_i = (\pi_i, A_i)$, where A_i denotes the adjacency matrix of the graph $\pi_i(G)$.
2. Upon receiving the verifier's challenge $e = (e_1, \ldots, e_k)$, the prover responds as follows for each $i = 1, \ldots, k$: if $e_i = 0$, it opens the commitment to (π_i, A_i); if $e_i = 1$, it opens the commitment to entries in A_i corresponding to the edges of the cycle $\pi_i(h)$.
3. The verifier checks that the openings are valid and in addition, that $A_i = \pi_i(G_i)$ if $e_i = 0$ and that the open entries correspond to edges of a Hamiltonian cycle if $e_i = 1$.

We just need to verify that the easy challenge is efficiently recoverable:

Special soundness. Given a non-Hamiltonian graph G and the values $v_i = (\pi_i, A_i)$, we can compute $\tilde{e} = (\tilde{e}_1, \ldots, \tilde{e}_k)$ as follows: $\tilde{e}_i$ equals 0 if $\pi_i(G) = A_i$ and 1 otherwise. It is easy to see that if an easy challenge (that allows the prover to cheat) exists, then it must equal $\tilde{e}$.[3]

The Zero-Knowledge Argument System. The zero-knowledge protocol is as follows:

1. The verifier picks a random $e \in \{0, 1\}^k$ and commits to e using Com, a statistically-binding commitment scheme.
2. The prover commits to $v_1, \ldots, v_k$ as in Π_{HAM} using PExtCom.
3. The verifier opens the commitment to e.
4. The prover aborts if the opening to e is not valid. Otherwise, it responds to the challenge e according to Π_{HAM}.
5. The verifier runs the final verification step as in Π_{HAM}.

The Analysis. Completeness is straight-forward.

Computational Soundness. Suppose there exists a cheating prover P^* (WLOG deterministic) that convinces the verifier to accept a non-Hamiltonian graph G with probability $\epsilon = 1/\text{poly}(k)$. Intuitively this means that P^* on input Com(e) predicts e with probability roughly $\epsilon \gg 2^{-k}$, which must contradict the hiding property of Com. More formally, fix the graph G, and we know that with probability $\epsilon/2$ over e, P^* succeeds with probability $\epsilon/2$. Let Γ denote the set of such challenges e, so $|\Gamma| \geq \frac{\epsilon}{2} \cdot 2^k$, and consider the procedure A that on input a commitment Com(e):

1. sends Com(e) to P^*;
2. uses the extractor for PExtCom with P^* as the cheating sender to obtain commitments to $v_1, \ldots, v_k$ along with the values $v_1, \ldots, v_k$.
3. computes a candidate easy challenge $\tilde{e}$ from $v_1, \ldots, v_k$ and outputs $\tilde{e}$.

It is easy to see that for all $e \in \Gamma$, $\Pr[A(\text{Com}(e)) \to e] \geq \frac{\epsilon}{2} - \text{neg}(k) \geq \frac{\epsilon}{4}$. By using a non-uniform reduction, we may WLOG assume that $0^k \in \Gamma$. Now, the sets of strings Γ'

---

[3] Note that determining whether an easy challenge exists is NP-hard, since we must determine whether A_i contains a cycle.

in the output of $A(\mathsf{Com}(0^k))$ that occurs with probability at least $\frac{\epsilon}{8}$ is at most $\frac{8}{\epsilon}$. Since $|\Gamma'| < |\Gamma|$, there must exist a string, say 1^k, that lies in Γ but outside Γ'. Now,

$$1^k \notin \Gamma' \;\Rightarrow\; \Pr[A(\mathsf{Com}(0^k)) \to 1^k] \le \tfrac{\epsilon}{8}$$
$$1^k \in \Gamma \;\Rightarrow\; \Pr[A(\mathsf{Com}(1^k)) \to 1^k] \ge \tfrac{\epsilon}{4}$$

This yields a distinguisher for $\mathsf{Com}(0^k)$ and $\mathsf{Com}(1^k)$, which contradicts the hiding property of Com.

Zero-Knowledge. The zero-knowledge simulator is virtually identically to that in the Goldreich-Kahan protocol [19]. Roughly speaking, upon receiving the verifier's commitment to e, the prover sends the cheating verifier V^* dummy commitments. If the verifier aborts, we are basically done. Otherwise, we learn the challenge e and then we could use the honest-verifier zero-knowledge simulator to complete the simulation. As in [19], we will need to estimate the probability that V^* aborts on dummy commitments.

Argument of Knowledge. We may obtain a zero-knowledge argument of knowledge for NP by instantiating the Feige-Shamir protocol [16] with the trapdoor commitment scheme, which we present in the next section.

6   Trapdoor Commitments

We construct a "look-ahead trapdoor commitment". This is a statistically binding commitment scheme wherein the commit phase comprises a 3-round challenge-response protocol. In addition, the scheme will be "look-ahead trapdoor" in the following sense: if we fix the receiver's challenge in the challenge-response phase, then we may generate a simulated transcript for the commit phase which we may later open to both a 0 and a 1. Moreover, the transcript together with either bit b is computationally indistinguishable from a legitimate commitment to b followed by an opening to b. We note similar constructions appear in [31,30]. In addition, we stress that we cannot use the challenge-response protocol in [13] because we do not know how to efficient compute the easy challenge in that protocol.[4]

Look-Ahead Trapdoor Bit Commitment. To commit to a bit σ. Again, we fix some statistically binding commitment scheme Com.

COMMIT PHASE.
  – Each v_i is a 2×2 0,1-matrix given by

$$\begin{pmatrix} v_i^{00} & v_i^{01} \\ v_i^{10} & v_i^{11} \end{pmatrix} = \begin{pmatrix} \eta_i & \sigma \oplus \eta_i \\ \eta_i & \sigma \oplus \eta_i \end{pmatrix}$$

---

[4] Roughly speaking, easy challenges in [13] are the first-round messages in Naor's commitment scheme [35] that allow a computationally unbounded sender to cheat, i.e. strings of the form $G(a) \oplus G(b) \in \{0,1\}^k$ ranging over all $a, b \in \{0,1\}^k$, and where $G : \{0,1\}^{k/3} \to \{0,1\}^k$ is a pseudorandom generator.

where η_i is a random bit. The sender commits to $v_1, \ldots, v_k$ using Com (bit by bit in parallel). In addition, the sender prepares $(a_1^0, a_1^1), \ldots, (a_k^0, a_k^1)$ where a_i^β is the opening of the commitment to $v_i^{0\beta}, v_i^{1\beta}$ (i.e., either the left or right column of v_i).

- Upon receiving a challenge $e = (e_1, \ldots, e_k)$ from the receiver, the sender responds with $a_1^{e_1}, \ldots, a_k^{e_k}$.
- The receiver checks that the openings are valid and that $v_i^{0e_i} = v_i^{1e_i}$ for $i = 1, 2, \ldots, k$ (i.e., every column that is open contains two equal bits).

OPEN PHASE.
- The sender sends σ. In addition, it chooses a random $\gamma \in \{0, 1\}$, sends γ, opens the commitments to $v_i^{\gamma 0}, v_i^{\gamma 1}$ for $i = 1, 2, \ldots, k$ (i.e., either the top rows or the bottom rows of all the matrices).
- The receiver checks that all the openings are valid, and also that $\sigma = v_1^{\gamma 0} \oplus v_1^{\gamma 1} = \cdots = v_k^{\gamma 0} \oplus v_k^{\gamma 1}$.

Analysis. It is straight-forward to show that the commitment scheme is computationally hiding.

Special soundness. Suppose we have a cheating sender that generates a transcript for the commit phase that can be successfully open to both a 0 and a 1. It must be the case that every matrix v_i contains at least one column with two unequal bits; call that column $\tilde{e}_i$. Then, the cheating sender will get caught in the commit phase unless $e = \tilde{e} = (\tilde{e}_1, \ldots, \tilde{e}_k)$. Moreover, given $v_1, \ldots, v_k$ it is easy to compute $\tilde{e}$.

Look-ahead trapdoor. We construct a simulator as follows:
- Given the challenge e, pick a random $\beta \in \{0, 1\}$, and prepare the matrices v_i as follows:

$$\begin{pmatrix} v_i^{00} & v_i^{01} \\ v_i^{10} & v_i^{11} \end{pmatrix} = \begin{pmatrix} \eta_i & \beta \oplus \eta_i \\ \bar{\eta}_i & \bar{\beta} \oplus \eta_i \end{pmatrix} \quad \text{if } e_i = 0; \text{ and}$$

$$\begin{pmatrix} v_i^{00} & v_i^{01} \\ v_i^{10} & v_i^{11} \end{pmatrix} = \begin{pmatrix} \beta \oplus \eta_i & \eta_i \\ \bar{\beta} \oplus \eta_i & \bar{\eta}_i \end{pmatrix} \quad \text{otherwise;}$$

where η_i is a random bit. When the receiver sends e, open the commitments to $v_i^{0e_i}$ and $v_i^{1e_i}$ like the honest sender.
- To open to σ, send $\gamma = \beta \oplus \sigma$, and open the commitments to $v_i^{\gamma 0}, v_i^{\gamma 1}$ for $i = 1, 2, \ldots, k$.

The Trapdoor Bit Commitment Scheme. The construction and the analysis is completely analogous to the zero-knowledge protocol. The verifier begins by committing to a random challenge $e \in \{0, 1\}^k$ using a statistically-binding commitment Com, and then we proceed according to the look-ahead scheme except the prover commits using ExtCom. Completeness is again straight-forward. Establishing computational binding is analogous to establishing computational soundness for the zero-knowledge protocol; we transform any cheating sender a distinguisher for Com by arguing that it must on input $\mathsf{Com}(e)$ predict e with noticeable probability. Trapdoor simulation is again based on the Goldreich-Kahan simulation strategy [19].

Extension to Multiple Bits. We claim that by running the trapdoor bit commitment scheme in parallel, we obtain a trapdoor commitment scheme for multiple bits, with the additional property that we can open the commitment to any subset of the bits without compromising the security of the remaining bits. We know that parallel repetition preserves the hiding and binding properties of commitment schemes. To see that the parallel version is still trapdoor, observe that we may still use the Goldreich-Kahan simulation strategy and that the look-ahead simulation property is preserved under parallel repetition.

7   Parallel Coin-Tossing

We present a constant-round parallel coin-tossing protocol in this section. Using the composition theorem in [6] and the results of [3,8], it is sufficient to implement the ideal string commitment functionality $\mathcal{F}_{\mathsf{Com}}$ (shown in Fig 2) with stand-alone security a la [21,5,18] in constant rounds. Moreover, by the results of [7], it suffices to construct a constant-round extractable trapdoor commitment scheme.

Functionality $\mathcal{F}_{\mathsf{Com}}$

1. Upon receiving input $(\mathtt{Commit}, sid, P_j, x)$ from P_i where $x \in \{0,1\}^m$, internally record the tuple (P_i, P_j, x) and send the message (sid, P_i, P_j) to the adversary; When receiving $(\mathtt{ok})$ from the adversary, output $(\mathtt{Receipt}, sid, P_i)$ to P_j. Ignore all subsequent $(\mathtt{Commit}, ...)$ inputs.
2. Upon receiving a value $(\mathtt{Open}, sid)$ from P_i, where a tuple (P_i, P_j, x) is recorded, send (x) to the adversary; When receiving $(\mathtt{ok})$ from the adversary, output $(\mathtt{Open}, sid, x)$ to P_j.

Fig. 2. Ideal String Commitment Functionality

Extractable Trapdoor Commitment Scheme. We provide a general construction of an extractable trapdoor commitment scheme ExtTDCom starting from any trapdoor commitment scheme TDCom: simply instantiate the protocol ExtCom with the trapdoor commitment scheme TDCom. Specifically, the sender in ExtTDCom on input a string $\sigma \in \{0,1\}^m$, commits to k pairs of strings $(v_1^0, v_1^1), \ldots, (v_k^0, v_k^1)$ (with $v_1^0 \oplus v_1^1 = \cdots = v_k^0 \oplus v_k^1 = \sigma$) using TDCom, by treating the k pairs of strings as a single string of length $2km$. The trapdoor property is straight-forward: if we could equivocate on the commitment to the string $(v_1^0, v_1^1), \ldots, (v_k^0, v_k^1)$, then we could easily equivocate on the commitment to σ. The extractable property is already established in Section 4.

The Coin-Tossing Protocol. For self-containment, we present the coin-tossing protocol based directly on ExtTDCom.

1. Party 1 chooses a random $s_1 \in \{0,1\}^m$ and commits to s_1 using ExtTDCom. Party 2 aborts with output $\bot$ if the commitment protocol fails.
2. Party 2 chooses $s_2 \in \{0,1\}^m$ and sends s_2 to Party 1.

3. If Party 1 receives an invalid message from Party 2, then Party 1 aborts. Otherwise, Party 1 opens the commitment to s_1. Party 2 aborts with output $\perp$ if the opening is invalid.
4. Output: both parties output $s_1 \oplus s_2$.

The high level proof strategy is as follows.

- If Party 1 is corrupted, we will use the extractor for ExtTDCom to extract s_1 and then set $s_2 = s_1 \oplus s$ (where s is the string chosen by the trusted party).
- If Party 2 is corrupted, we will use the trapdoor commitment property so that upon receiving s_2 from Party 2, the simulator can open the commitment to $s_1 = s \oplus s_2$.

8 Non-malleable Commitments

We begin by describing a commitment scheme satisfying some strong notions of extractability and hiding, based on an encoding scheme from [9].

An Intermediate Construction. To commit to a string v with parameter 1^ℓ (ℓ is the length of the identities):

COMMIT PHASE.
1. The receiver commits to a random subset $S \subset [10k]$ of size k using Com.
2. The sender picks random $\alpha_1, \ldots, \alpha_k \in \mathrm{GF}(2^n)$ and set $s_j = p(j), j \in [10k]$ where $p(x) = v + \alpha_1 x + \ldots + \alpha_k x^k$. (Note that $(s_1, \ldots, s_{10k})$ encodes v under the Reed-Solomon code.) The sender then commits to $(s_1, \ldots, s_{10k})$ a total of 2ℓ times using PExtCom sequentially.
3. The receiver opens the commitment to S.
4. The sender opens the 2ℓ commitments to the value s_j for all $j \in S$.
5. The receiver checks that for each $j \in S$, the 2ℓ commitments to s_j open to the same value.

OPEN PHASE.
1. The sender sends v and opens the commitments to $(s_1, \ldots, s_{10k})$ in the first execution of PExtCom.
2. The receiver computes the codeword $w = (w_1, \ldots, w_{10k})$ that agrees with $(s_1, \ldots, s_{10k})$ in at least $9k$ positions, and checks that $(s_1, \ldots, s_{10k})$ is a codeword corresponding to v and that for all $j \in S$, $s_j = w_j$.

We sketch the properties satisfied by this commitment scheme, and defer the analysis to the full version of this paper.

Extractability. There exists expected polynomial-time probabilistic oracle machines $E_1, E_2, \ldots, E_{2\ell}$ such that for all $i = 1, 2, \ldots, 2\ell$, the machine E_i given oracle access to any PPT cheating sender C^* outputs a pair (τ, σ^*) such that
- (simulation) τ is identically distributed to the view of C^* at the end of interacting with an honest receiver $\mathcal{R}$ in commit phase.
- (strong extraction) the pair (τ, σ^*) is computationally indistinguishable from the view of C^* at the end of interacting with an honest receiver $\mathcal{R}$ in commit phase, together with the committed value implicitly specified by the view.

We will also require that the machine E_i extracts from the i'th execution of PExtCom, for $i = 1, \ldots, 2\ell$.

Hiding. We require that the commitment scheme is (computationally) hiding even against a PPT cheating receiver $\mathcal{R}^*$ that may request for an arbitrary number of additional commitments to $(s_1, \ldots, s_{10k})$ using PExtCom, along with the openings to s_j for each $j \in S$ in these additional commitments.

We stress that the notion of extractability above is stronger than that in Section 3. In particular, it guarantees that if there is no valid opening for the commit phase transcript τ, then the extractor must output $\sigma^* = \bot$.

Achieving Non-malleability. To obtain a non-malleable commitment scheme from the previous construction, we just need to schedule the messages in the 2ℓ copies of PExtCom according to the message scheduling in [14]. It follows from the analysis in [32] that the resulting $O(n)$-round commitment scheme is one-many non-malleable. By further applying the results in [14,32], we obtain a $O(\log n)$-round non-malleable commitment and a $O(n)$-round concurrent non-malleable commitment.

References

1. Barak, B.: How to go beyond the black-box simulation barrier. In: FOCS, pp. 106–115 (2001)
2. Beaver, D.: Adaptive zero knowledge and computational equivocation. In: STOC, pp. 629–638 (1996)
3. Blum, M.: Coin flipping by telephone. In: CRYPTO, pp.11–15 (1981)
4. Blum, M.: How to prove a theorem so no one else can claim it. In: Proc. ICM (1986)
5. Canetti, R.: Security and composition of multiparty cryptographic protocols. J. Cryptology 13(1), 143–202 (2000)
6. Canetti, R.: Universally composable security: A new paradigm for cryptographic protocols. In: FOCS, pp. 136–145 (2001)
7. Canetti, R., Fischlin, M.: Universally composable commitments. In: Kilian, J. (ed.) CRYPTO 2001. LNCS, vol. 2139, pp. 19–40. Springer, Heidelberg (2001)
8. Canetti, R., Rabin, T.: Universal composition with joint state. In: Boneh, D. (ed.) CRYPTO 2003. LNCS, vol. 2729, pp. 265–281. Springer, Heidelberg (2003)
9. Choi, S.G., Dachman-Soled, D., Malkin, T.G., Wee, H.M.: Black-box construction of a non-malleable encryption scheme from any semantically secure one. In: Canetti, R. (ed.) TCC 2008. LNCS, vol. 4948, pp. 427–444. Springer, Heidelberg (2008)
10. Choi, S.G., Dachman-Soled, D., Malkin, T., Wee, H.: Simple, black-box constructions of adaptively secure protocols. In: TCC (to appear, 2009)
11. Cramer, R., Damgård, I.B., Schoenmakers, B.: Proof of partial knowledge and simplified design of witness hiding protocols. In: Desmedt, Y.G. (ed.) CRYPTO 1994. LNCS, vol. 839, pp. 174–187. Springer, Heidelberg (1994)
12. Damgård, I.B., Ishai, Y.: Constant-round multiparty computation using a black-box pseudorandom generator. In: Shoup, V. (ed.) CRYPTO 2005. LNCS, vol. 3621, pp. 378–394. Springer, Heidelberg (2005)
13. Di Crescenzo, G., Ostrovsky, R.: On concurrent zero-knowledge with pre-processing (Extended abstract). In: Wiener, M. (ed.) CRYPTO 1999. LNCS, vol. 1666, pp. 485–502. Springer, Heidelberg (1999)

14. Dolev, D., Dwork, C., Naor, M.: Nonmalleable cryptography. SIAM J. Comput. 30(2), 391–437 (2000)
15. Even, S., Goldreich, O., and Lempel, A. A randomized protocol for signing contracts. In: CRYPTO 1982, pp. 205–210 (1982)
16. Feige, U., Shamir, A.: Zero knowledge proofs of knowledge in two rounds. In: Brassard, G. (ed.) CRYPTO 1989. LNCS, vol. 435, pp. 526–544. Springer, Heidelberg (1990)
17. Goldreich, O.: Foundations of Cryptography: Basic Tools. Cambridge University Press, Cambridge (2001)
18. Goldreich, O.: Foundations of Cryptography: Basic Applications, vol. II. Cambridge University Press, Cambridge (2004)
19. Goldreich, O., Kahan, A.: How to construct constant-round zero-knowledge proof systems for NP. J. Cryptology 9(3), 167–190 (1996)
20. Goldreich, O., Krawczyk, H.: On the composition of zero-knowledge proof systems. SIAM J. Comput. 25(1), 169–192 (1996)
21. Goldreich, O., Micali, S., Wigderson, A.: How to play any mental game or a completeness theorem for protocols with honest majority. In: STOC, pp. 218–229 (1987)
22. Goldreich, O., Micali, S., Wigderson, A.: Proofs that yield nothing but their validity for all languages in NP have zero-knowledge proof systems. J. ACM 38(3), 691–729 (1991); Prelim. version in FOCS 1986
23. Goyal, V., Moriarty, R., Ostrovsky, R., Sahai, A.: Concurrent statistical zero-knowledge arguments for NP from one way functions. In: Kurosawa, K. (ed.) ASIACRYPT 2007. LNCS, vol. 4833, pp. 444–459. Springer, Heidelberg (2007)
24. Haitner, I.: Semi-honest to malicious oblivious transfer—the black-box way. In: Canetti, R. (ed.) TCC 2008. LNCS, vol. 4948, pp. 412–426. Springer, Heidelberg (2008)
25. Haitner, I., Hoch, J., Reingold, O., Segev, G.: Finding collisions in interactive protocols – a tight lower bound on the round complexity of statistically-hiding commitments. In: FOCS, pp. 669–679 (2007)
26. Håstad, J., Impagliazzo, R., Levin, L.A., Luby, M.: A pseudorandom generator from any one-way function. SIAM J. Comput. 28(4), 1364–1396 (1999)
27. Impagliazzo, R., Rudich, S.: Limits on the provable consequences of one-way permutations. In: STOC, pp. 44–61 (1989)
28. Ishai, Y., Kushilevitz, E., Lindell, Y., Petrank, E.: Black-box constructions for secure computation. In: STOC, pp. 99–108 (2006)
29. Ishai, Y., Prabhakaran, M., Sahai, A.: Founding cryptography on oblivious transfer – efficiently. In: Wagner, D. (ed.) CRYPTO 2008. LNCS, vol. 5157, pp. 572–591. Springer, Heidelberg (2008)
30. Kilian, J.: Founding cryptography on oblivious transfer. In: STOC, pp. 20–31 (1988)
31. Kilian, J.: On the complexity of bounded-interaction and noninteractive zero-knowledge proofs. In: FOCS, pp. 466–477 (1994)
32. Lin, H., Pass, R., Venkitasubramaniam, M.: Concurrent non-malleable commitments from any one-way function. In: Canetti, R. (ed.) TCC 2008. LNCS, vol. 4948, pp. 571–588. Springer, Heidelberg (2008)
33. Lindell, Y.: Parallel coin-tossing and constant-round secure two-party computation. J. Cryptology 16(3), 143–184 (2003)
34. Lindell, Y., Pinkas, B.: An efficient protocol for secure two-party computation in the presence of malicious adversaries. In: Naor, M. (ed.) EUROCRYPT 2007. LNCS, vol. 4515, pp. 52–78. Springer, Heidelberg (2007)
35. Naor, M.: Bit commitment using pseudorandomness. J. Cryptology 4(2), 151–158 (1991)
36. Pass, R., Tseng, W., Venkitasubramaniam, M.: Unconditional characterizations of concurrent zero knowledge (manuscript, 2008)

37. Peikert, C., Waters, B.: Lossy trapdoor functions and their applications. In: STOC (to appear, 2008); Cryptology ePrint Archive, Report 2007/279
38. Prabhakaran, M., Rosen, A., Sahai, A.: Concurrent zero knowledge with logarithmic round-complexity. In: FOCS, pp. 366–375 (2002)
39. Reingold, O., Trevisan, L., Vadhan, S.P.: Notions of reducibility between cryptographic primitives. In: Naor, M. (ed.) TCC 2004. LNCS, vol. 2951, pp. 1–20. Springer, Heidelberg (2004)
40. Rosen, A.: A note on constant-round zero-knowledge proofs for NP. In: Naor, M. (ed.) TCC 2004. LNCS, vol. 2951, pp. 191–202. Springer, Heidelberg (2004)
41. Rosen, A.: The Round-Complexity of Black-Box Concurrent Zero-Knowledge. Ph.D., Weizmann Institute of Science (May 2004)
42. Simon, D.R.: Findings collisions on a one-way street: Can secure hash functions be based on general assumptions? In: Nyberg, K. (ed.) EUROCRYPT 1998. LNCS, vol. 1403, pp. 334–345. Springer, Heidelberg (1998)
43. Yao, A.C.-C.: How to generate and exchange secrets (extended abstract). In: FOCS, pp. 162–167 (1986)

Chosen-Ciphertext Security
via Correlated Products

Alon Rosen[1,*] and Gil Segev[2]

[1] Efi Arazi School of Computer Science,
Herzliya Interdisciplinary Center (IDC), Herzliya 46150, Israel
`alon.rosen@idc.ac.il`
[2] Department of Computer Science and Applied Mathematics,
Weizmann Institute of Science, Rehovot 76100, Israel
`gil.segev@weizmann.ac.il`

Abstract. We initiate the study of one-wayness under *correlated products*. We are interested in identifying necessary and sufficient conditions for a function f and a distribution on inputs $(x_1, \ldots, x_k)$, so that the function $(f(x_1), \ldots, f(x_k))$ is one-way. The main motivation of this study is the construction of public-key encryption schemes that are secure against chosen-ciphertext attacks (CCA). We show that any collection of injective trapdoor functions that is secure under a very natural correlated product can be used to construct a CCA-secure encryption scheme. The construction is simple, black-box, and admits a direct proof of security.

We provide evidence that security under correlated products is achievable by demonstrating that lossy trapdoor functions (Peikert and Waters, STOC '08) yield injective trapdoor functions that are secure under the above mentioned correlated product. Although we currently base security under correlated products on existing constructions of lossy trapdoor functions, we argue that the former notion is potentially weaker as a general assumption. Specifically, there is no fully-black-box construction of lossy trapdoor functions from trapdoor functions that are secure under correlated products.

1   Introduction

The construction of secure public-key encryption schemes lies at the heart of cryptography. Following the seminal work of Goldwasser and Micali [20], increasingly strong security definitions have been formulated. The strongest notion to date is that of semantic security against a chosen-ciphertext attack (CCA) [27,32], which protects against an adversary that is given access to decryptions of ciphertexts of her choice.

Constructions of CCA-secure public-key encryption schemes have followed several structural approaches. These approaches, however, either result in rather complicated schemes, or rely only on specific number-theoretic assumptions. Our

---

* Research supported in part by BSF grant 2006317.

O. Reingold (Ed.): TCC 2009, LNCS 5444, pp. 419–436, 2009.

goal in this paper is to construct a simple CCA-secure public-key encryption scheme based on general computational assumptions.

The first approach for constructing a CCA-secure encryption scheme was put forward by Naor and Yung [27], and relies on any semantically secure public-key encryption scheme and non-interactive zero-knowledge (NIZK) proof system for $\mathcal{NP}$. Their approach was later extended by Dolev, Dwork and Naor [11] for a more general notion of chosen-ciphertext attack, and subsequently simplified by Sahai [35] and by Lindell [26]. Schemes resulting from this approach, however, are somewhat complicated and impractical due to the use of generic NIZK proofs.

An additional approach was introduced by Cramer and Shoup [10], and is based on "smooth hash proof systems", which were shown to exist based on several number-theoretic assumptions. Elkind and Sahai [12] observed that both the above approaches can be viewed as special cases of a single paradigm in which ciphertexts include "proofs of well-formedness". Even though in some cases this paradigm leads to elegant and efficient constructions [9], the complexity of the underlying notions makes the general framework somewhat cumbersome.

Recently, Peikert and Waters [31] introduced the intriguing notion of lossy trapdoor functions, and demonstrated that such functions can be used to construct a CCA-secure public-key encryption scheme in a black-box manner. Their construction can be viewed as an efficient and elegant realization of the "proofs of well-formedness" paradigm. Lossy trapdoor functions seem to be a very powerful primitive. In particular, they were shown to also imply oblivious transfer protocols and collision-resistant hash functions[1]. It is thus conceivable that CCA-secure encryption can be realized based on weaker primitives.

A different approach was suggested by Canetti, Halevi and Katz [6] (followed by [3,4,5]) who constructed a CCA-secure public-key encryption scheme based on any identity-based encryption (IBE) scheme. Their construction is elegant, black-box, and essentially preserves the efficiency of the underlying IBE scheme. However, IBE is a rather strong cryptographic primitive, which is currently realized only based on a small number of specific number-theoretic assumptions.

1.1   Our Contributions

Motivated by the task of constructing a simple CCA-secure public-key encryption scheme, we initiate the study of one-wayness under *correlated products*. The main question in this context is to identify necessary and sufficient conditions for a collection of functions $\mathcal{F}$ and a distribution on inputs $(x_1, \ldots, x_k)$ so that the function $(f_1(x_1), \ldots, f_k(x_k))$ is one-way, where $f_1, \ldots, f_k$ are independently chosen from $\mathcal{F}$. Our results are as follows:

1. We show that any collection of injective trapdoor functions that is secure under a very natural correlated product can be used to construct a CCA-secure public-key encryption scheme. The construction is simple, black-box,

---

[1] We note that the constructions of CCA-secure encryption and collision-resistant hash functions presented in [31] require lossy trapdoor functions that are "sufficiently lossy" (i.e., they rely on lossy trapdoor functions with sufficiently good parameters).

and admits a direct proof of security. Arguably, both the underlying assumption and the proof of security are simple enough to be taught in an undergraduate course in cryptography.

2. We demonstrate that any collection of lossy trapdoor functions (with appropriately chosen parameters) yields a collection of injective trapdoor functions that is secure under the correlated product that is required by our encryption scheme. In turn, existing constructions of lossy trapdoor functions [1,31,34] imply that our encryption scheme can be based on the hardness of the decisional Diffie-Hellman problem, and of Paillier's decisional composite residuosity problem.

3. We argue that security under correlated products is potentially weaker than lossy trapdoor functions as a general computational assumption. Specifically, we prove that there is no fully-black-box construction of lossy trapdoor functions from trapdoor functions (and even from enhanced trapdoor permutations) that are secure under correlated products.

Following our work Peikert [30] and Goldwasser and Vaikuntanathan [21] recently showed that security under correlated products is achievable also under the worst-case hardness of lattice problems (although these assumptions are currently not known to imply lossy trapdoor functions with the appropriately chosen parameters that are required for our transformation). Their constructions result in new CCA-secure public-key encryption schemes that are based on lattices, and this demonstrates that the correlated products approach for chosen-ciphertext security is fruitful, and that security under correlated products is achievable under a variety of number-theoretic assumptions.

In the remainder of this section we provide a high-level overview of our contributions, and then turn to describe the related work.

1.2   Security under Correlated Products

It is well known that for every collection of one-way functions $\mathcal{F} = \{f_s\}_{s \in S}$ and polynomially-bounded $k \in \mathbb{N}$, the collection $\mathcal{F}_k = \{f_{s_1,\dots,s_k}\}_{(s_1,\dots,s_k) \in S^k}$, whose members are defined as

$$f_{s_1,\dots,s_k}(x_1,\dots,x_k) = (f_{s_1}(x_1),\dots,f_{s_k}(x_k))$$

is also one-way. Moreover, such a direct product amplifies the one-wayness of $\mathcal{F}$ [19,37], and this holds even when considering a single function (i.e., when $s_1 = \cdots = s_k$).

In general, however, the one-wayness of $\mathcal{F}_k$ is guaranteed only when the inputs are independently chosen, and when the inputs are correlated no such guarantee can exist. A well-known example for insecurity under correlated products is Håstad's attack [2,23] on plain-broadcast RSA: there is an efficient algorithm that is given as input $x^3 \bmod N_1$, $x^3 \bmod N_2$, and $x^3 \bmod N_3$, and outputs x. More generally, it is rather easy to show that if collections of one-way functions exist, then there exists a collection of one-way functions $\mathcal{F} = \{f_s\}_{s \in S}$ such that $f_{s_1,s_2}(x,x) = (f_{s_1}(x), f_{s_2}(x))$ is not one-way. However, this does not rule out the possibility of constructing a collection of one-way functions whose product remains one-way even when the inputs are correlated.

Informally, given a collection $\mathcal{F}$ of functions and a distribution $\mathcal{C}_k$ of inputs $(x_1, \ldots, x_k)$, we say that $\mathcal{F}$ is *secure under a $\mathcal{C}_k$-correlated product* if $\mathcal{F}_k$ is one-way when the inputs $(x_1, \ldots, x_k)$ are distributed according to $\mathcal{C}_k$ (a formal definition is provided in Section 2). The main goal in this setting is to characterize the class of collections $\mathcal{F}$ and distributions $\mathcal{C}_k$ that satisfy this notion.

We motivate the study of security under correlated products by relating it to the study of chosen-ciphertext security. Specifically, we show that any collection of injective trapdoor functions that is secure under a very natural correlated product can be used to construct a CCA-secure public-key encryption scheme. The simplest form of distribution $\mathcal{C}_k$ on inputs that is sufficient for our construction is the *uniform k-repetition distribution* that outputs k copies of a uniformly chosen input x. We note that although this seems to be a strong requirement, we demonstrate that it can be based on various number-theoretic assumptions.

More generally, our construction can rely on any distribution $\mathcal{C}_k$ with the property that any $(x_1, \ldots, x_k)$ in the support of $\mathcal{C}_k$ can be reconstructed given any $t = (1 - \epsilon)k$ entries from $(x_1, \ldots, x_k)$, for some constant $0 < \epsilon < 1$. For example, $\mathcal{C}_k$ may be a distribution that evaluates a random polynomial of degree at most $t-1$ on a set of k points (in this case the x_i's are t-wise independent, but other choices which do not guarantee such a strong property are also possible).

1.3   Chosen-Ciphertext Security via Correlated Products

Consider the following, very simple, public-key encryption scheme. The public-key consists of an injective trapdoor function f, and the secret-key consists of its trapdoor td. Given a message $m \in \{0, 1\}$, the encryption algorithm chooses a random input x and outputs the ciphertext $(f(x), m \oplus h(x))$, where h is a hard-core predicate of f. The decryption algorithm uses the trapdoor to retrieve x and then extracts m. In what follows we frame our approach as a generalization of this fundamental scheme.

The above scheme is easily proven secure against a chosen-plaintext attack. Any adversary $\mathcal{A}$ that distinguishes between an encryption of 0 and an encryption of 1 can be used to construct an adversary $\mathcal{A}'$ that distinguishes between $h(x)$ and a randomly chosen bit with exactly the same probability. Specifically, $\mathcal{A}'$ receives a function f, a value $y = f(x)$, and a bit w (which is either $h(x)$ or a uniformly chosen bit), and emulates $\mathcal{A}$ with f as the public-key and $(y, m \oplus w)$ as the challenge ciphertext for a random message m. This scheme, however, fails to be proven secure against a chosen-ciphertext attack (even when considering only CCA1 security). There is a conflict between the fact that $\mathcal{A}'$ is required to answer decryption queries, and the fact that $\mathcal{A}'$ does not have the trapdoor for inverting f.

The following simplified variant of our scheme is designed to resolve this conflict. The public-key consists of k pairs of functions $(f_1^0, f_1^1), \ldots, (f_k^0, f_k^1)$, where each function is sampled independently from a collection $\mathcal{F}$ of injective trapdoor functions[2]. The secret-key consists of the trapdoors $(td_1^0, td_1^1), \ldots,$ (td_k^0, td_k^1), where each td_i^b is the trapdoor of the function f_i^b. Given a message $m \in \{0, 1\}$, the encryption algorithm chooses a random $v = v_1 \cdots v_k \in \{0, 1\}^k$, a random input x, and outputs the ciphertext

$$E_{PK}(m; v, x) = (v, f_1^{v_1}(x), \ldots, f_k^{v_k}(x), m \oplus h(x)) ,$$

where h is a hard-core predicate of $\mathcal{F}_k$ with respect to the uniform k-repetition distribution. The decryption algorithm acts as follows: given a ciphertext of the form $(v, y_1, \ldots, y_k, z)$ it inverts $y_1, \ldots, y_k$ to obtain $x_1, \ldots, x_k$, and if $x_1 = \cdots = x_k$ then it outputs $h(x_1) \oplus z$ (otherwise it outputs $\perp$).

In order to prove the CCA1 security of this scheme, we show that any adversary $\mathcal{A}$ that breaks the CCA1 security of the scheme can be used to construct an adversary $\mathcal{A}'$ that distinguishes between $h(x)$ and a randomly chosen bit with exactly the same probability. The adversary $\mathcal{A}'$ receives as input k functions $f_1, \ldots, f_k \in \mathcal{F}$, k values $y_1 = f_1(x), \ldots, y_k = f_k(x)$, and a bit w (which is either $h(x)$ or a uniformly chosen bit). $\mathcal{A}'$ simulates the CCA1 interaction to $\mathcal{A}$ by choosing a random value $v^* = v_1^* \cdots v_k^* \in \{0, 1\}^k$, and for each pair (f_i^0, f_i^1) it sets $f_i^{vk_i^*} = f_i$ and samples $f_i^{1-vk_i^*}$ together with its trapdoor from $\mathcal{F}$. Note that now $\mathcal{A}'$ is able to answer decryption queries as long as none of them contain the value v^*, and in this case we claim that essentially no information on v^* is revealed. The challenge ciphertext is then computed as $(v^*, y_1, \ldots, y_k, m \oplus w)$ for a random message m. If $\mathcal{A}$ guesses the bit m correctly then $\mathcal{A}'$ outputs that $w = h(x)$, and otherwise $\mathcal{A}'$ outputs that w is a random bit.

Our scheme can be viewed as a realization of the Naor-Yung paradigm [27] in which a message is encrypted using several independently chosen keys, and ciphertexts include "proofs of well-formedness". In our scheme, however, the decryption algorithm can verify "well-formedness" of ciphertexts without any additional "proof": given any one of the trapdoors it is possible to verify that the remaining values are consistent with the same input x.

Our scheme is inspired also by the one based on lossy trapdoor functions [31], and specifically, by the generic construction of *all-but-one* lossy trapdoor functions from lossy trapdoor functions. However, the proof security of our construction is simpler than that of [31] due to the additional hybrids resulting from using both lossy trapdoor functions and all-but-one trapdoor functions. In addition, our construction only relies on *computational* hardness, whereas the construction of [31] relies on the *statistical* properties of lossy trapdoor functions.

Finally, we note that our proof of security is rather similar to that of the IBE-based schemes [4,5,6]. The value v^* can be viewed as the challenge identity, for which $\mathcal{A}'$ does not have the secret key, and is therefore not able to decrypt ciphertexts for this identity. For any other identity $v \neq v^*$, $\mathcal{A}'$ has sufficient information to decrypt ciphertexts.

---

[2] For CCA1 security any $k = \omega(\log n)$ is sufficient, where n is the security parameter. For our more generalized construction that guarantees CCA2 security, any $k = n^\epsilon$ for some constant $0 < \epsilon < 1$ is sufficient.

In some sense, our approach enjoys "the best of both worlds" in that both the underlying assumption and the proof of security are simpler than those of previous approaches.

1.4   A Black-Box Separation

Although we currently base security under correlated products on lossy trapdoor functions, we argue that security under correlated products is potentially weaker than lossy trapdoor functions as a general computational assumption. Specifically, we prove that there is no fully-black-box construction of lossy trapdoor functions from trapdoor functions that are secure under correlated products. We present an oracle relative to which there exists a collection of injective trapdoor functions (and even of enhanced trapdoor permutations) that is secure under a correlated product with respect to the above mentioned uniform k-repetition distribution, but there is no collection of lossy trapdoor functions. The oracle is essentially the collision-finding oracle due to Simon [36], and the proof follows the approach of Haitner et al. [22] while overcoming several technical difficulties.

Informally, consider a circuit A which is given as input $(f_1(x), \ldots, f_k(x))$, and whose goal is to retrieve x. The circuit A is provided access to an oracle Sam that receives as input a circuit C and outputs random w and w' such that $C(w) = C(w')$. As in the approach of Haitner et al. the idea underlying the proof is to distinguish between two cases: one in which A obtains information on x via one of its Sam-queries, and the other in which none of A's Sam-queries provides information on x. The proof consists of two modular parts dealing with these two cases separately. In first part we generalize an argument of Haitner et al. (who in turn generalized the reconstruction lemma of Gennaro and Trevisan [14]) to deal with the product of several functions. We show that the probability that $\mathcal{A}$ retrieves x in the first case is exponentially small. In the second part we show that the second case can essentially be reduced to the first case. This part of the proof is simpler than the corresponding argument of Haitner et al. that considers a more interactive setting.

1.5   Related Work

Much research has been devoted for the construction of CCA-secure public-key encryption schemes. A significant part of this research was already mentioned in the previous sections, and here we mainly focus on results regarding the possibility and limitations of basing such schemes on general assumptions.

Pass, shelat and Vaikuntanathan [28] constructed a public-key encryption scheme that is non-malleable against a chosen-plaintext attack from any semantically secure one (building on the scheme of Dolev, Dwork and Naor [11]). Their technique was later shown by Cramer et al. [8] to also imply non-malleability against a weak notion of chosen-ciphertext attack, in which the number of decryption queries is bounded. These approaches, however, are rather impractical due to the use of generic (designated verifier) NIZK proofs. Very recently, Choi et al. [7] showed that the latter notions of security can in fact be elegantly realized in a black-box manner based on the same assumptions. The reader is referred to [11,29] for classifications of the different notions of security.

Impagliazzo and Rudich [24] introduced a paradigm for proving impossibility results for cryptographic constructions. They showed that there are no black-box constructions of key-agreement protocols from one-way permutations, and substantial additional work in this line followed (see, for example [13,15,17,25,36] and many more). The reader is referred to [33] for a comprehensive discussion and taxonomy of black-box constructions. In the context of public-key encryption schemes, most relevant to our result is the work of Gertner, Malkin and Myers [16], who addressed the question of whether or not semantically secure public-key encryption schemes imply the existence of CCA-secure schemes. They showed that there are no black-box constructions in which the decryption algorithm of the proposed CCA-secure scheme does not query the encryption algorithm of the semantically secure one.

1.6   Paper Organization

The remainder of the paper is organized as follows. In Section 2 we provide a formal treatment of security under correlated products, which is shown to be satisfied by lossy trapdoor functions. In Section 3 we describe a simplified version of our encryption scheme which already illustrates the main ideas underlying our approach. In Section 4 we prove that there is no fully-black-box construction of lossy trapdoor functions from trapdoor functions secure under correlated products. Due to space limitation we refer the reader to the full version for a more generalized version of the encryption scheme, and for a complete proof of the black-box separation.

2   Security under Correlated Products

In this section we formally define the notion of security under correlated products, and demonstrate that the notion is satisfied by any collection of lossy trapdoor functions (with appropriately chosen parameters) for a very natural and useful correlation. We then discuss the exact parameters that are required for our encryption scheme, and the number-theoretic assumptions that are currently known to guarantee such parameters.

A collection of functions is represented as a pair of algorithms $\mathcal{F} = (G, F)$, where G is a generation algorithm used for sampling a description of a function, and F is an evaluation algorithm used for evaluating a function on a given input. The following definition formalizes the notion of a k-wise product which introduces a collection $\mathcal{F}_k$ consisting of all k-tuples of functions from $\mathcal{F}$.

Definition 2.1 (k-wise product). *Let $\mathcal{F} = (G, F)$ be a collection of efficiently computable functions. For any integer k, we define the k-wise product $\mathcal{F}_k = (G_k, F_k)$ as follows:*

- *The generation algorithm G_k on input 1^n invokes $G(1^n)$ for k times independently and outputs $(s_1, \ldots, s_k)$. That is, a function is sampled from $\mathcal{F}_k$ by independently sampling k functions from $\mathcal{F}$.*

- *The evaluation algorithm F_k on input $(s_1, \ldots, s_k, x_1, \ldots, x_k)$ invokes F to evaluate each function s_i on x_i. That is,*

$$F_k(s_1, \ldots, s_k, x_1, \ldots, x_k) = (F(s_1, x_1), \ldots, F(s_k, x_k)) .$$

The notion of a one-way function asks for a function that is efficiently computable but is hard to invert given the image of a uniformly chosen input. More generally, one can naturally extend this notion to consider one-wayness under any specified input distribution, not necessarily the uniform distribution. That is, informally, we say that a function is one-way with respect to an input distribution $\mathcal{I}$ if it is efficiently computable but hard to invert given the image of a random input sampled according to $\mathcal{I}$.

In the context of k-wise products, a standard argument shows that for any collection $\mathcal{F}$ which is one-way with respect to some input distribution $\mathcal{I}$, the k-wise product $\mathcal{F}_k$ is one-way with respect to the input distribution which samples k independent inputs from $\mathcal{I}$. The following definition formalizes the notion of security under correlated products, where the inputs for $\mathcal{F}_k$ may be correlated.

Definition 2.2 (Security under correlated products). *Let $\mathcal{F} = (G, F)$ be a collection of efficiently computable functions, and let $\mathcal{C}_k$ be a distribution where $\mathcal{C}_k(1^n)$ is distributed over $\{0,1\}^{k \cdot n}$ for some integer $k = k(n)$. We say that $\mathcal{F}$ is secure under a $\mathcal{C}_k$-correlated product if $\mathcal{F}_k$ is one-way with respect to the input distribution $\mathcal{C}_k$.*

Correlated products security based on lossy trapdoor functions. We conclude this section by demonstrating that, for an appropriate choice of parameters, any collection of lossy trapdoor functions yields a collection of injective trapdoor functions that is secure under a $\mathcal{C}_k$-correlated product. The input distribution under consideration, $\mathcal{C}_k$, samples a uniformly random input x and outputs k copies of x. We refer to this distribution as the *uniform k-repetition distribution*, and this distribution is the one required for the simplified variant of our encryption scheme, presented in Section 3.

Specifically, given a collection of lossy trapdoor functions $\mathcal{F} = (G, F, F^{-1})$ we define a collection $\mathcal{F}_{\text{inj}}$ of injective trapdoor functions by restricting $\mathcal{F}$ to its injective functions. That is, $\mathcal{F}_{\text{inj}} = (G_{\text{inj}}, F, F^{-1})$ where $G_{\text{inj}}(1^n) = G(1^n, \text{injective})$. We prove the following theorem:

Theorem 2.1. *Let $\mathcal{F} = (G, F, F^{-1})$ be a collection of (n, ℓ)-lossy trapdoor functions. Then, for any integer $k < \frac{n - \omega(\log n)}{n - \ell}$, for any probabilistic polynomial-time algorithm $\mathcal{A}$ and polynomial $p(\cdot)$, it holds that*

$$\Pr\left[\mathcal{A}(1^n, s_1, \ldots, s_k, F(s_1, x), \ldots, F(s_k, x)) = x\right] < \frac{1}{p(n)} ,$$

for all sufficiently large n, where the probability is taken over the choices of $s_1 \leftarrow G_{\text{inj}}(1^n), \ldots, s_k \leftarrow G_{\text{inj}}(1^n)$, $x \leftarrow \{0,1\}^n$, and over the internal coin tosses of $\mathcal{A}$.

Proof. Peikert and Waters [31] proved that any collection of $(n, \omega(\log n))$-lossy trapdoor functions is in particular a collection of one-way functions. Thus, it is sufficient to prove that $\mathcal{F}_k$ is a collection of $(n, \omega(\log n))$-lossy trapdoor functions. For any k functions $s_1, \ldots, s_k$ sampled according to $G_{\mathsf{inj}}(1^n)$, the function $F_k(s_1, \ldots, s_k, x_1, \ldots, x_k) = (F(s_1, x_1), \ldots, F(s_k, x_k))$ is clearly injective. For any k functions $s_1, \ldots, s_k$ sampled according to $G_{\mathsf{lossy}}(1^n)$, the function $F_k(s_1, \ldots, s_k, x_1, \ldots, x_k) = (F(s_1, x_1), \ldots, F(s_k, x_k))$ obtains at most $2^{k(n-\ell)}$ values, which is upper bounded by $2^{n-\omega(\log n)}$ for any $k < \frac{n-\omega(\log n)}{n-\ell}$. Finally, note that a standard hybrid argument shows that the distribution obtained by independently sampling k functions according to $G_{\mathsf{inj}}(1^n)$ is computationally indistinguishable from the distribution obtained by independently sampling k functions according to $G_{\mathsf{lossy}}(1^n)$. Thus, $\mathcal{F}_k$ is a collection of $(n, \omega(\log n))$-lossy trapdoor functions. □

The required parameters for our scheme. The assumption underlying our encryption scheme asks for $k(n) = \omega(\log n)$ for CCA1 security, and for $k(n) = n^\epsilon$ (for some constant $0 < \epsilon < 1$) for CCA2 security. In turn, existing constructions of lossy trapdoor functions guaranteing these parameters [1,31,34] imply that our encryption scheme can be realized under the hardness of the decisional Diffie-Hellman problem, and of Paillier's decisional composite residuosity problem. We note that the lattice-based construction of Peikert and Waters [31] guarantees only a constant $k(n)$ that is not sufficient for our encryption scheme. However, Peikert [30] and Goldwasser and Vaikuntanathan [21] recently showed that security under correlated products (with sufficiently large $k(n)$) is nevertheless achievable under the worst-case hardness of lattice problems, although these are currently known to imply lossy trapdoor functions with only a relatively small amount of loss.

3   A Simplified Construction

In this section we describe a simplified version of our construction which already illustrates the main ideas underlying our approach. The encryption scheme presented in the current section is a simplification in the sense that it relies on a seemingly stronger computational assumption than the more generalized construction which is presented in the full version. In addition, we first present the scheme as encrypting only one bit messages, and then demonstrate that it naturally extends to multi-bit messages. In what follows we state the computational assumption, describe the encryption scheme, prove its security, and describe the extension to multi-bit messages.

The underlying computational assumption. The computational assumption underlying the simplified scheme is that there exists a collection $\mathcal{F}$ of injective trapdoor functions and an integer function $k = k(n)$ such that $\mathcal{F}$ is secure under a $\mathcal{C}_k$-correlated product, where $\mathcal{C}_k$ is the uniform k-repetition distribution (i.e., outputs k copies of a uniformly distributed input x). Specifically, our scheme uses a hard-core predicate $h : \{0,1\}^* \to \{0,1\}$ for $\mathcal{F}_k$ with respect to $\mathcal{C}_k$. That is, the underlying computational assumption is that for any probabilistic polynomial-time predictor $\mathcal{P}$ it holds that

$$\left| \Pr\left[\mathcal{P}(1^n, s_1, \ldots, s_k, F(s_1, x), \ldots, F(s_k, x)) = h(s_1, \ldots, s_k, x)\right] - \frac{1}{2} \right|$$

is negligible in n, where the probability is taken over the choices of $s_1 \leftarrow G(1^n), \ldots, s_k \leftarrow G(1^n)$, $x \leftarrow \{0,1\}^n$, and over the internal coin tosses of $\mathcal{P}$.

The integer function $k(n)$ should correspond to the bit-length of verification keys of some one-time strongly-unforgeable signature scheme $(\mathsf{KG}_{\mathsf{sig}}, \mathsf{Sign}, \mathsf{Ver})$. By applying a universal one-way hash function to the verification keys (as in [11]) it suffices that the above assumption holds for $k(n) = n^\epsilon$ for a constant $0 < \epsilon < 1$. For simplicity, however, when describing our scheme we do not apply a universal one-way hash function to the verification keys. We also note that for an even more simplified version which is only CCA1-secure (the one described in Section 1.3), any $k(n) = \omega(\log n)$ suffices.

The construction. The following describes our simplified encryption scheme given by the triplet (KG, E, D).

- **Key generation:** On input 1^n the key generation algorithm invokes $G(1^n)$ for $2k$ times independently to obtain $2k$ descriptions of functions denoted $(s_1^0, s_1^1), \ldots, (s_k^0, s_k^1)$ with trapdoors $(td_1^0, td_1^1), \ldots, (td_k^0, td_k^1)$. The public-key and secret-key are defined as follows:

$$PK = \left((s_1^0, s_1^1), \ldots, (s_k^0, s_k^1)\right)$$
$$SK = \left((td_1^0, td_1^1), \ldots, (td_k^0, td_k^1)\right) .$$

- **Encryption:** On input a message $m \in \{0,1\}$ and a public key PK, the algorithm samples $(vk, sk) \leftarrow \mathsf{KG}_{\mathsf{sig}}(1^n)$ where $vk = vk_1 \circ \cdots \circ vk_k \in \{0,1\}^k$, chooses a uniformly distributed $x \in \{0,1\}^n$, and outputs the ciphertext

$$(vk, y_1, \ldots, y_k, c_1, c_2) ,$$

where

$$y_i = F\left(s_i^{vk_i}, x\right) \ \forall i \in [k]$$
$$c_1 = m \oplus h\left(s_1^{vk_1}, \ldots, s_k^{vk_k}, x\right)$$
$$c_2 = \mathsf{Sign}(sk, (y_1, \ldots, y_k, c_1)) .$$

- **Decryption:** On input a ciphertext $(vk, y_1, \ldots, y_k, c_1, c_2)$ and a secret-key SK, the algorithm acts as follows. If $\mathsf{Ver}(vk, (y_1, \ldots, y_k, c_1), c_2) = 0$, it outputs $\perp$. Otherwise, for every $i \in [k]$ it computes $x_i = F^{-1}\left(td_i^{vk_i}, y_i\right)$. If $x_1 = \cdots = x_k$ then it outputs $c_1 \oplus h\left(s_1^{vk_1}, \ldots, s_k^{vk_k}, x_1\right)$, and otherwise it outputs $\perp$.

The following theorem establishes the security of the scheme.

Theorem 3.1. *Assuming that $\mathcal{F}$ is secure under a $\mathcal{C}_k$-correlated product, where $\mathcal{C}_k$ is the uniform k-repetition distribution, and that $(\mathsf{KG}_{\mathsf{sig}}, \mathsf{Sign}, \mathsf{Ver})$ is one-time strongly unforgeable, the encryption scheme (KG, E, D) is CCA2-secure.*

Proof. Let $\mathcal{A}$ be a probabilistic polynomial-time CCA2-adversary. We denote by Forge the event in which for one of $\mathcal{A}$'s decryption queries $(vk, y_1, \ldots, y_k, c_1, c_2)$ during the CCA2 interaction it holds that $vk = vk^*$ (where vk^* is given in the secret key) and $\mathsf{Ver}(vk, (y_1, \ldots, y_k, c_1), c_2) = 1$. We first argue that the event Forge has a negligible probability due to the security of the one-time signature scheme. Then, assuming that the event Forge does not occur, we construct a probabilistic polynomial-time algorithm $\mathcal{P}$ that predicts the hard-core predicate h while preserving the advantage of $\mathcal{A}$.

More formally, we denote by Success the event in which $\mathcal{A}$ successfully guesses the bit b used for encrypting the challenge ciphertext. Then, the advantage of $\mathcal{A}$ in the CCA2 interaction is bounded as follows:

$$\left| \Pr\left[\mathsf{Success}\right] - \frac{1}{2} \right| = \left| \Pr\left[\mathsf{Success} \wedge \mathsf{Forge}\right] + \Pr\left[\mathsf{Success} \wedge \overline{\mathsf{Forge}}\right] - \frac{1}{2} \right|$$

$$\leq \Pr\left[\mathsf{Forge}\right] + \left| \Pr\left[\mathsf{Success} \wedge \overline{\mathsf{Forge}}\right] - \frac{1}{2} \right| \, .$$

The theorem follows from the following two claims:

Claim 3.2. $\Pr\left[\mathsf{Forge}\right]$ *is negligible.*

Proof. We show that any probabilistic polynomial-time adversary $\mathcal{A}$ for which $\Pr\left[\mathsf{Forge}\right]$ is non-negligible, can be used to construct a probabilistic polynomial-time adversary $\mathcal{A}'$ that breaks the security of the one-time signature with the same probability. The adversary $\mathcal{A}'$ is given a verification key vk^* sampled using $\mathsf{KG}_{\mathsf{sig}}(1^n)$ and simulates the CCA2 interaction to $\mathcal{A}$ as follows. $\mathcal{A}'$ begins by invoking the key generation algorithm on input 1^n and using vk^* for forming the public and secret keys. In the decryption phases, whenever $\mathcal{A}$ submits a decryption query $(vk, y_1, \ldots, y_k, c_1, c_2)$, $\mathcal{A}'$ acts as follows. If $vk = vk^*$ and $\mathsf{Ver}(vk, (y_1, \ldots, y_k, c_1), c_2) = 1$, then $\mathcal{A}'$ outputs $((y_1, \ldots, y_k, c_1), c_2)$ as the forgery and halts. Otherwise, $\mathcal{A}'$ invokes the decryption procedure. In the challenge phase, upon receiving two message m_0 and m_1, $\mathcal{A}'$ chooses $b \in \{0, 1\}$ and $x \in \{0, 1\}^n$ uniformly at random, and computes

$$y_i = F\left(s_i^{vk_i^*}, x\right) \ \forall i \in [k]$$

$$c_1 = m_b \oplus h\left(s_1^{vk_1^*}, \ldots, s_k^{vk_k^*}, x\right) \, .$$

Then, it obtains a signature c_2 on $(y_1, \ldots, y_k, c_1)$ with respect to vk^* (recall that $\mathcal{A}'$ is allowed to ask for a signature on one message). Finally, it sends $(vk^*, y_1, \ldots, y_k, c_1, c_2)$ to $\mathcal{A}$. We note that during the second decryption phase, if $\mathcal{A}$ submits the challenge ciphertext as a decryption query, then $\mathcal{A}'$ responds with $\bot$.

Note that prior to the first decryption query in which Forge occurs (assuming that Forge indeed occurs), the simulation of the CCA2 interaction is perfect. Therefore, the probability that $\mathcal{A}'$ breaks the security of the one-time signature scheme is exactly $\Pr[\text{forge}]$. The security of the signature scheme implies that this probability is negligible.   □

Claim 3.3. $\left|\Pr\left[\text{Success} \wedge \overline{\text{Forge}}\right] - \frac{1}{2}\right|$ *is negligible.*

Proof. Given any efficient adversary $\mathcal{A}$ for which $\left|\Pr\left[\text{Success} \wedge \overline{\text{Forge}}\right] - \frac{1}{2}\right|$ is non-negligible, we construct a predictor $\mathcal{P}$ that breaks the security of the hard-core predicate h. That is,

$$\left|\Pr\left[\mathcal{P}(1^n, s_1, \ldots, s_k, F(s_1, x), \ldots, F(s_k, x)) = h(s_1, \ldots, s_k, x)\right] - \frac{1}{2}\right|$$

is non-negligible, where $s_1 \leftarrow G(1^n), \ldots, s_k \leftarrow G(1^n)$ independently, and the probability is taken over the uniform choice of $x \in \{0,1\}^n$, and over the internal coin tosses of both G and $\mathcal{P}$.

For simplicity, we first construct an efficient distinguisher $\mathcal{A}'$ which receives input of the form $(1^n, s_1, \ldots, s_k, F(s_1, x), \ldots, F(s_k, x))$ and a bit $w \in \{0,1\}$ which is either $h(s_1, \ldots, s_k, x)$ or a uniformly random bit, and is able to distinguish between the two cases with non-negligible probability. The distinguisher $\mathcal{A}'$ acts by simulating the CCA2 interaction to $\mathcal{A}$. More specifically, on input $(1^n, s_1, \ldots, s_k, y_1, \ldots, y_k)$ and a bit w, the distinguisher $\mathcal{A}'$ first creates a pair (PK, SK) as follows. It samples $(vk^*, sk^*) \leftarrow \text{KG}_{\text{sig}}(1^n)$, where $vk^* = vk_1^* \circ \cdots \circ vk_k^* \in \{0,1\}^k$, and for every $i \in [k]$ sets $s_i^{vk_i^*} = s_i$ and samples $\left(s_i^{1-vk_i^*}, td_i^{1-vk_i^*}\right) \leftarrow G(1^n)$. Then, $\mathcal{A}'$ outputs the public-key

$$PK = \left(\left(s_1^0, s_1^1\right), \ldots, \left(s_k^0, s_k^1\right)\right) \ .$$

Whenever $\mathcal{A}$ submits a decryption query of the form $(vk, y_1, \ldots, y_k, c_1, c_2)$, $\mathcal{A}'$ acts as follows. If $vk = vk^*$ or $\text{Ver}(vk, (y_1, \ldots, y_k, c_1), c_2) = 0$, it outputs $\bot$ and halts. Otherwise, it picks some $i \in [k]$ for which $vk_i \neq vk_i^*$ and computes $x = F^{-1}\left(td_i^{vk_i}, y_i\right)$. If for every $j \in [k]$ it holds that $y_j = F\left(s_j^{vk_j}, x\right)$, it outputs $c_1 \oplus h\left(s_1^{vk_1}, \ldots, s_k^{vk_k}, x\right)$, and otherwise it outputs $\bot$.

In the challenge phase, given two messages m_0 and m_1, $\mathcal{A}'$ chooses a random bit $b \in \{0,1\}$ and replies with the challenge ciphertext

$$c = (vk^*, y_1, \ldots, y_k, c_1, c_2) \ ,$$

where $c_1 = m_b \oplus w$, and $c_2 = \text{Sign}(sk^*, (y_1, \ldots, y_k, c_1))$. We note that during the second decryption phase, if $\mathcal{A}$ submits the challenge ciphertext as a decryption query, then $\mathcal{A}'$ responds with $\bot$. At the end of this interaction $\mathcal{A}$ outputs a bit b'. If $b' = b$ then $\mathcal{A}'$ outputs 1, and otherwise $\mathcal{A}'$ outputs 0.

In order to compute the advantage of $\mathcal{A}'$ we observe the following:

1. If w is a uniformly random bit, then the challenge ciphertext in the simulated interaction is independent of b. Therefore, the probability that $\mathcal{A}'$ outputs 1 in this case is exactly $1/2$.
2. If $w = h(s_1, \ldots, s_k, x)$, then as long as the event Forge does not occur, the simulated interaction is identical to the CCA2 interaction (a formal argument follows). Therefore, the probability that $\mathcal{A}'$ outputs 1 in this case is exactly $\Pr\left[\text{Success} \wedge \overline{\text{Forge}}\right]$.

   Note that the only difference between the CCA2 interaction and the simulated interaction is the distribution of the challenge ciphertext: In the CCA2 interaction the value vk in the challenge ciphertext is a randomly chosen verification key, and in the simulated interaction the value vk is chosen ahead of time by $\mathcal{A}$. In what follows we claim that as long as the event Forge does not occur, the distribution of vk in the challenge ciphertext is identical in the two cases.

   Formally, denote by $vk_1, \ldots, vk_q$ the random variables corresponding to the value of vk in $\mathcal{A}$'s decryption queries (without loss of generality we assume that $\mathcal{A}$ always submits q queries, and that the signature verification never fails on these queries). In the CCA2 interaction, as long as the event Forge does not occur, it holds that the verification key used for the challenge ciphertext is a random verification key with the only restriction that it is different than $vk_1, \ldots, vk_q$. In the simulated interaction, given that $vk^* \notin \{vk_1, \ldots, vk_q\}$, we claim that from $\mathcal{A}$'s point of view, the value vk^* is also a random verification key which is different than $vk_1, \ldots, vk_q$. That is, each $vk^* \notin \{vk_1, \ldots vk_q\}$ produces exactly the same transcript. Indeed, first note that the public key is independent of vk^*. Now consider a decryption query $(vk, y_1, \ldots, y_k, c_1, c_2)$ for some $vk \in \{vk_1, \ldots, vk_q\}$. For any $vk^* \neq vk$, if $y_1, \ldots, y_k$ have the same preimage x, then the decryption algorithm will always output $c_1 \oplus h\left(s_1^{vk_1}, \ldots, s_k^{vk_k}, x\right)$. In addition, for any $vk^* \neq vk$, if $y_1, \ldots, y_k$ do not have the same preimage, then the decryption algorithm will always output $\bot$.

The above observations imply that

$$\left|\Pr\left[\mathcal{A}' \text{ outputs } 1 \mid w = h(s_1, \ldots, s_k, x)\right] - \Pr\left[\mathcal{A}' \text{ outputs } 1 \mid w \text{ is random}\right]\right|$$
$$= \left|\Pr\left[\text{Success} \wedge \overline{\text{Forge}}\right] - \frac{1}{2}\right| .$$

A standard argument (see, for example, [18, Chapter 3.4]) can be applied to efficiently transform $\mathcal{A}'$ into a predictor $\mathcal{P}$ that predicts $h(s_1, \ldots, s_k, x)$ with the same probability.   □

Encrypting any polynomial number of bits. For simplicity we presented the encryption scheme above for one-bit plaintexts. We now demonstrate that our approach extends to plaintexts of any polynomial length while relying on the same computational assumption[3].

Recall that the underlying computational assumption is the existence of a collection $\mathcal{F}$ of injective trapdoor functions such that $\mathcal{F}_k$ is one-way under the uniform k-repetition distribution (i.e., $x_1 = \cdots = x_k$ where x_1 is chosen uniformly at random). Specifically, the scheme uses a hard-core predicate $h : \{0,1\}^* \rightarrow \{0,1\}$ for $\mathcal{F}_k$ to mask the plaintext bit. This assumption clearly implies that for any polynomial $T = T(n)$ there exists a collection $\mathcal{F}'$ of injective trapdoor functions such that $\mathcal{F}'$ is one-way under the uniform k-repetition distribution, and has a hard-core *function* $h' : \{0,1\}^* \rightarrow \{0,1\}^T$ that can be used in our scheme to mask T-bit plaintexts. Specifically, the collection $\mathcal{F}'$ is defined as follows: for every function $f : \{0,1\}^n \rightarrow \{0,1\}^m$ in $\mathcal{F}$ define a function $f' : \{0,1\}^{Tn} \rightarrow \{0,1\}^{Tm}$ by $f'(x_1, \ldots x_T) = (f(x_1), \ldots, f(x_T))$. The security proof of the T-bit encryption scheme is essentially identical to the proof of Theorem 3.1 by showing that any successful CCA-adversary can be used to either break the one-time signature scheme or to break the pseudorandomness of h'.

4   A Black-Box Separation

In this section we show that there is no fully-black-box construction of lossy trapdoor functions (with even a single bit of lossiness) from injective trapdoor functions that are secure under correlated products. We show that this holds for the seemingly strongest form of correlated product, where independently chosen functions are evaluated on the same input (i.e., we consider the uniform k-repetition distribution).

Our proof consists of constructing an oracle $\mathcal{O}$ relative to which there exists a collection of injective trapdoor functions that are permutations secure under a correlated product[4], but there are no collections of lossy trapdoor functions. In what follows, we describe the oracle $\mathcal{O}$, and show that it breaks the security of any collection of lossy trapdoor functions.

The oracle. The oracle $\mathcal{O}$ is of the form $(\tau, \mathsf{Sam}^\tau)$, where τ is a collection of trapdoor permutations, and Sam^τ is an oracle that samples random collision. Specifically, Sam receives as input a description of a circuit C (which may contain τ-gates), chooses a random input w, and then samples a uniformly distributed $w' \in C^{-1}(C(w))$.

We now explain how exactly Sam samples w and w'. We provide Sam with a collection of permutations $\mathcal{F}$, where for every possible circuit C the collection $\mathcal{F}$ contains two permutations f_C^1 and f_C^2 over the domain of C. Given a circuit

---

[3] It is well-known that for semantic security under a chosen-plaintext attack it is straightforward to construct a multi-bit encryption scheme from any one-bit encryption scheme by independently encrypting the individual bits of the plaintext. For semantic security under a chosen-ciphertext attack, however, this approach fails in general.

[4] These functions are in fact enhanced trapdoor permutations, but we note that this is not essential for our result.

On input a circuit $C : \{0,1\}^m \rightarrow \{0,1\}^{\ell(m)}$, **the oracle Sam**$^{\tau, \mathcal{F}}$ **acts as follows:**

1. Compute $w = f_C^1(0^m)$.
2. Compute $w' = f_C^2(t)$ for the lexicographically smallest $t \in \{0,1\}^m$ such that $C(f_C^2(t)) = C(w)$.
3. Output (w, w')

Fig. 1. The oracle Sam

$C : \{0,1\}^m \rightarrow \{0,1\}^{\ell(m)}$, for some m and $\ell(m)$, the oracle Sam uses f_C^1 to compute $w = f_C^1(0^m)$. Then, it computes $w' = f_C^2(t)$ for the lexicographically smallest $t \in \{0,1\}^m$ such that $C(f_C^2(t)) = C(w)$. Note that whenever the permutations f_C^1 and f_C^2 are chosen uniformly at random, and independently of all other permutations in $\mathcal{F}$, then w is uniformly distributed over $\{0,1\}^m$, and w' is uniformly distributed over $C^{-1}(C(w))$. In the remainder of the proof, whenever we consider the probability of an event over the choice of the collection $\mathcal{F}$, we mean that for each circuit C, two permutations f_C^1 and f_C^2 are chosen uniformly at random and independently of all other permutations. A complete and formal description of the oracle is provided in Figure 1.

Distinguishing between injective functions and lossy functions. The oracle Sam can be easily used to distinguish between the injective mode and the lossy mode of any collection of $(n, 1)$-lossy functions. Consider the following distinguisher A: given a circuit C (which may contain τ-gates[5]), which is a description of either an injective function or a lossy function (with image size at most 2^{n-1}), A queries Sam with C. If Sam returns (w, w') such that $w = w'$, then A outputs 1, and otherwise A outputs 0. Clearly, if C corresponds to an injective function, then always $w = w'$ and A outputs 1. In addition, if C corresponds to a lossy function, then with probability at least $1/4$ it holds that $w \neq w'$, where the probability is taken over the randomness of Sam (i.e., over the collection $\mathcal{F}$).

Outline of the proof. For simplicity we first consider only two permutations. Then, we extend our argument to more than two permutations, and to trapdoor permutations. Our goal is to upper bound the success probability of circuits having oracle access to Sam in the task of inverting $(\pi_1(x), \pi_2(x))$ for random permutations $\pi_1, \pi_2 \in \Pi_n$ and a random $x \in \{0,1\}^n$ (where Π_n is the set of all permutations over $\{0,1\}^n$). We prove the following theorem:

Theorem 4.1. *For any circuit A of size at most $2^{n/40}$ and for all sufficiently large n, it holds that*

$$\Pr_{\substack{\pi_1, \pi_2, \mathcal{F} \\ x \leftarrow \{0,1\}^n}} \left[A^{\pi_1, \pi_2, \mathsf{Sam}^{\pi_1, \pi_2, \mathcal{F}}}(\pi_1(x), \pi_2(x)) = x \right] \leq \frac{1}{2^{n/40}} .$$

---

[5] We allow the circuits given as input to Sam to contain τ-gates, but we do not allow them to contain Sam-gates. This suffices, however, for ruling out fully-black-box constructions.

Consider a circuit A which is given as input $(\pi_1(x), \pi_2(x))$, and whose goal is to retrieve x. The idea underlying the proof is to distinguish between two cases: one in which A obtains information on x via one of its Sam-queries, and the other in which none of A's Sam-queries provides information on x. More specifically, we define:

Definition 4.1. *A* Sam-*query C produces a x-hit if* Sam *outputs (w, w') such that some π_1-gate or π_2-gate in the computations of $C(w)$ or $C(w')$ has input x.*

Given π_1, π_2, $\mathcal{F}$, a circuit A, and a pair $(\pi_1(x), \pi_2(x))$, we denote by SamHIT_x the event in which one of the Sam-queries made by A produces a x-hit. From this point on, the proof proceeds in two modular parts. In the first part of the proof, we consider the case that the event SamHIT_x does not occur, and generalize an argument of Haitner et al. [22] (who in turn generalized the reconstruction lemma of Gennaro and Trevisan [14]). We show that if a circuit A manages to invert $(\pi_1(x), \pi_2(x))$ for many x's, then π_1 and π_2 have a short representation given A. This enables us to prove the following lemma:

Lemma 4.1. *For any circuit A of size at most $2^{n/7}$ and for all sufficiently large n, it holds that*

$$\Pr_{\substack{\pi_1, \pi_2, \mathcal{F} \\ x \leftarrow \{0,1\}^n}} \left[A^{\pi_1, \pi_2, \mathsf{Sam}^{\pi_1, \pi_2, \mathcal{F}}}(\pi_1(x), \pi_2(x)) = x \ \wedge \ \overline{\mathsf{SamHIT}_x} \right] \leq 2^{-n/8} \ .$$

In the second part of the proof, we show that the case where the event SamHIT_x does occur can be reduced to the case where the event SamHIT_x does not occur. Given a circuit A that tries to invert $(\pi_1(x), \pi_2(x))$, we construct a circuit M that succeeds almost as well as A, without M's Sam-queries producing any x-hits. This proof is a simpler case of a similar argument due to Haitner et al. [22]. The following theorem is proved:

Lemma 4.2. *For any circuit A of size $s(n)$, if*

$$\Pr_{\substack{\pi_1, \pi_2, \mathcal{F} \\ x \leftarrow \{0,1\}^n}} \left[A^{\pi_1, \pi_2, \mathsf{Sam}^{\pi_1, \pi_2, \mathcal{F}}}((\pi_1(x), \pi_2(x))) = x \right] \geq \frac{1}{s(n)}$$

for infinitely many values of n, then there exists a circuit M of size $O(s(n))$ such that

$$\Pr_{\substack{\pi_1, \pi_2, \mathcal{F} \\ x \leftarrow \{0,1\}^n}} \left[M^{\pi_1, \pi_2, \mathsf{Sam}^{\pi_1, \pi_2, \mathcal{F}}}((\pi_1(x), \pi_2(x))) = x \ \wedge \ \overline{\mathsf{SamHIT}_x} \right] \geq \frac{1}{s(n)^5}$$

for infinitely many values of n.

Due to space limitations the remainder of the proof is provided in the full version.

Acknowledgments

We thank Oded Goldreich, Moni Naor, Chris Peikert, and Omer Reingold for useful discussions. In particular, we thank Oded for suggesting the relaxation that led to the more generalized scheme.

References

1. Boldyreva, A., Fehr, S., O'Neill, A.: On notions of security for deterministic encryption, and efficient constructions without random oracles. In: Wagner, D. (ed.) CRYPTO 2008. LNCS, vol. 5157, pp. 335–359. Springer, Heidelberg (2008)
2. Boneh, D.: Twenty years of attacks on the RSA cryptosystem. Notices of the American Mathematical Society 46(2), 203–213 (1999)
3. Boneh, D., Canetti, R., Halevi, S., Katz, J.: Chosen-ciphertext security from identity-based encryption. SIAM J. Comput. 36(5), 1301–1328 (2007)
4. Boneh, D., Katz, J.: Improved efficiency for CCA-secure cryptosystems built using identity-based encryption. In: Menezes, A. (ed.) CT-RSA 2005. LNCS, vol. 3376, pp. 87–103. Springer, Heidelberg (2005)
5. Boyen, X., Mei, Q., Waters, B.: Direct chosen ciphertext security from identity-based techniques. In: 12th ACM CCS, pp. 320–329 (2005)
6. Canetti, R., Halevi, S., Katz, J.: Chosen-ciphertext security from identity-based encryption. In: Cachin, C., Camenisch, J.L. (eds.) EUROCRYPT 2004. LNCS, vol. 3027, pp. 207–222. Springer, Heidelberg (2004)
7. Choi, S.G., Dachman-Soled, D., Malkin, T.G., Wee, H.M.: Black-box construction of a non-malleable encryption scheme from any semantically secure one. In: Canetti, R. (ed.) TCC 2008. LNCS, vol. 4948, pp. 427–444. Springer, Heidelberg (2008)
8. Cramer, R., Hanaoka, G., Hofheinz, D., Imai, H., Kiltz, E., Pass, R., Shelat, A., Vaikuntanathan, V.: Bounded CCA2-secure encryption. In: Kurosawa, K. (ed.) ASIACRYPT 2007. LNCS, vol. 4833, pp. 502–518. Springer, Heidelberg (2007)
9. Cramer, R., Shoup, V.: A practical public key cryptosystem provably secure against adaptive chosen ciphertext attack. In: Krawczyk, H. (ed.) CRYPTO 1998. LNCS, vol. 1462, pp. 13–25. Springer, Heidelberg (1998)
10. Cramer, R., Shoup, V.: Design and analysis of practical public-key encryption schemes secure against adaptive chosen-ciphertext attack. SIAM J. Comput. 33(1), 167–226 (2003)
11. Dolev, D., Dwork, C., Naor, M.: Non-malleable cryptography. SIAM J. Comput. 30(2), 391–437 (2000)
12. Elkind, E., Sahai, A.: A unified methodology for constructing public-key encryption schemes secure against adaptive chosen-ciphertext attack. Cryptology ePrint Archive, Report 2002/042 (2002)
13. Gennaro, R., Gertner, Y., Katz, J., Trevisan, L.: Bounds on the efficiency of generic cryptographic constructions. SIAM J. Comput. 35(1), 217–246 (2005)
14. Gennaro, R., Trevisan, L.: Lower bounds on the efficiency of generic cryptographic constructions. In: 41st FOCS, pp. 305–313 (2000)
15. Gertner, Y., Kannan, S., Malkin, T., Reingold, O., Viswanathan, M.: The relationship between public key encryption and oblivious transfer. In: 41st FOCS, pp. 325–335 (2000)
16. Gertner, Y., Malkin, T.G., Myers, S.: Towards a separation of semantic and CCA security for public key encryption. In: Vadhan, S.P. (ed.) TCC 2007. LNCS, vol. 4392, pp. 434–455. Springer, Heidelberg (2007)
17. Gertner, Y., Malkin, T., Reingold, O.: On the impossibility of basing trapdoor functions on trapdoor predicates. In: 42nd FOCS, pp. 126–135 (2001)
18. Goldreich, O.: Foundations of Cryptography: Basic Tools, vol. 1. Cambridge University Press, Cambridge (2001)
19. Goldreich, O., Impagliazzo, R., Levin, L.A., Venkatesan, R., Zuckerman, D.: Security preserving amplification of hardness. In: 31st FOCS, pp. 318–326 (1990)

20. Goldwasser, S., Micali, S.: Probabilistic encryption. J. Comput. Syst. Sci. 28(2), 270–299 (1984)

21. Goldwasser, S., Vaikuntanathan, V.: New constructions of correlation-secure trapdoor functions and CCA-secure encryption schemes (manuscript, 2008)

22. Haitner, I., Hoch, J.J., Reingold, O., Segev, G.: Finding collisions in interactive protocols – A tight lower bound on the round complexity of statistically-hiding commitments. In: 48th FOCS, pp. 669–679 (2007)

23. Håstad, J.: Solving simultaneous modular equations of low degree. SIAM J. Comput. 17(2), 336–341 (1988)

24. Impagliazzo, R., Rudich, S.: Limits on the provable consequences of one-way permutations. In: 21st STOC, pp. 44–61 (1989)

25. Kim, J.H., Simon, D.R., Tetali, P.: Limits on the efficiency of one-way permutation-based hash functions. In: 40th FOCS, pp. 535–542 (1999)

26. Lindell, Y.: A simpler construction of CCA2-secure public-key encryption under general assumptions. J. Cryptology 19(3), 359–377 (2006)

27. Naor, M., Yung, M.: Public-key cryptosystems provably secure against chosen ciphertext attacks. In: 22nd STOC, pp. 427–437 (1990)

28. Pass, R., Shelat, A., Vaikuntanathan, V.: Construction of a non-malleable encryption scheme from any semantically secure one. In: Dwork, C. (ed.) CRYPTO 2006. LNCS, vol. 4117, pp. 271–289. Springer, Heidelberg (2006)

29. Pass, R., Shelat, A., Vaikuntanathan, V.: Relations among notions of non-malleability for encryption. In: Kurosawa, K. (ed.) ASIACRYPT 2007. LNCS, vol. 4833, pp. 519–535. Springer, Heidelberg (2007)

30. Peikert, C.: Public-key cryptosystems from the worst-case shortest vector problem. Cryptology ePrint Archive, Report 2008/481 (2008)

31. Peikert, C., Waters, B.: Lossy trapdoor functions and their applications. In: 40th STOC, pp. 187–196 (2008)

32. Rackoff, C., Simon, D.R.: Non-interactive zero-knowledge proof of knowledge and chosen ciphertext attack. In: Feigenbaum, J. (ed.) CRYPTO 1991. LNCS, vol. 576, pp. 433–444. Springer, Heidelberg (1992)

33. Reingold, O., Trevisan, L., Vadhan, S.P.: Notions of reducibility between cryptographic primitives. In: Naor, M. (ed.) TCC 2004. LNCS, vol. 2951, pp. 1–20. Springer, Heidelberg (2004)

34. Rosen, A., Segev, G.: Efficient lossy trapdoor functions based on the composite residuosity assumption. Cryptology ePrint Archive, Report 2008/134 (2008)

35. Sahai, A.: Non-malleable non-interactive zero knowledge and adaptive chosen-ciphertext security. In: 40th FOCS, pp. 543–553 (1999)

36. Simon, D.R.: Finding collisions on a one-way street: Can secure hash functions be based on general assumptions? In: Nyberg, K. (ed.) EUROCRYPT 1998. LNCS, vol. 1403, pp. 334–345. Springer, Heidelberg (1998)

37. Yao, A.C.: Theory and applications of trapdoor functions. In: 23rd FOCS, pp. 80–91 (1982)

Hierarchical Identity Based Encryption with Polynomially Many Levels

Craig Gentry[1] and Shai Halevi[2,*]

[1] Stanford & IBM
[2] IBM

Abstract. We present the first hierarchical identity based encryption (HIBE) system that has full security for more than a constant number of levels. In all prior HIBE systems in the literature, the security reductions suffered from exponential degradation in the depth of the hierarchy, so these systems were only proven fully secure for identity hierarchies of constant depth. (For deep hierarchies, previous work could only prove the weaker notion of selective-ID security.) In contrast, we offer a tight proof of security, regardless of the number of levels; hence our system is secure for polynomially many levels.

Our result can very roughly be viewed as an application of Boyen's framework for constructing HIBE systems from exponent-inversion IBE systems to a (dramatically souped-up) version of Gentry's IBE system, which has a tight reduction. In more detail, we first describe a generic transformation from "identity based broadcast encryption with key randomization" (KR-IBBE) to a HIBE, and then construct KR-IBBE by modifying a recent construction of IBBE of Gentry and Waters, which is itself an extension of Gentry's IBE system. Our hardness assumption is similar to that underlying Gentry's IBE system.

1 Introduction

Identity-Based Encryption (IBE) is a public-key encryption scheme where one's public key can be freely set to any value (such as one's identity): An authority that holds a master secret key can take any arbitrary identifier and extract a secret key corresponding to this identifier. Anyone can then encrypt messages using the identifier as a public encryption key, and only the holder of the corresponding secret key can decrypt these messages. This concept was introduced by Shamir [19], a partial solution was proposed by Maurer and Yacobi [18], and

---

* Research was sponsored by US Army Research laboratory and the UK Ministry of Defense and was accomplished under Agreement Number W911NF-06-3-0001. The views and conclusions contained in this document are those of the authors and should not be interpreted as representing the official policies, either expressed or implied, of the US Army Research Laboratory, the U.S. Government, the UK Ministry of Defense, or the UK Government. The US and UK Governments are authorized to reproduce and distribute reprints for Government purposes notwithstanding any copyright notation hereon.

O. Reingold (Ed.): TCC 2009, LNCS 5444, pp. 437–456, 2009.

the first fully functional IBE systems were described by Boneh and Franklin [5] and Cocks [11].

IBE systems can greatly simplify the public-key infrastructure for encryption solutions, but they are still not as general as one would like. Many organizations have an hierarchical structure, perhaps with one central authority, several sub-authorities and sub-sub-authorities and many individual users, each belonging to a small part of the organization tree. We would like to have a solution where each authority can delegate keys to its sub-authorities, who in turn can keep delegating keys further down the hierarchy to the users. The depth of the hierarchy can range from two or three in small organizations, up to ten or more in large ones. An IBE system that allows delegation as above is called Hierarchical Identity-Based Encryption (HIBE). In HIBE, messages are encrypted for *identity-vectors*, representing nodes in the identity hierarchy. This concept was introduced by Horwitz and Lynn [17], who also described a partial solution to it, and the first fully functional HIBE system was described by Gentry and Silverberg [15].

The security model for IBE and HIBE systems postulates an attacker that can adaptively make "key-reveal" queries, thereby revealing the decryption keys of identities of its choice. The required security property asserts that such an attacker still cannot break the encryption at any identity other than those for which it issued key-reveal queries. (Or in the case of HIBE, other than those for which it issued key-reveal queries or their descendants.)

For the first IBE and HIBE systems, the only known proofs of security are carried out in the random-oracle model. Canetti et al. [9] introduced a weaker notion of security called *selective-ID*, where the attacker must choose the identity to attack before the system parameters are chosen (but can still make adaptive key-reveal queries afterward). They proved that a variant of the Gentry-Silverberg system is secure in this model even without random oracles. Boneh and Boyen described a more efficient selective-ID HIBE [1], and later described a fully secure IBE system without a random oracle [2]. Waters [22] described what is currently the most practical adaptively-secure HIBE system without random oracles.

All currently known fully-secure HIBE systems, however, suffer from loose security reductions (whether they use random oracles or not). Specifically, they lose a multiplicative factor of $\Omega(q/\ell)^\ell$ in the success probability, where q is the number of key-reveal queries and ℓ is the depth of the identity hierarchy. This means that asymptotically these reductions can only be used for hierarchies of constant depth. When considering concrete parameters, these reductions are only meaningful for hierarchies of depth two or three.

Gentry [13] proposed the first adaptively-secure IBE system without random oracles that has a tight reduction to its underlying hard problem. Recently Gentry and Waters extended Gentry's IBE to construct an adaptively-secure identity based broadcast encryption (IBBE) system without random oracles [16], whose security is tightly based on a related hard problem. Our HIBE system builds on the Gentry-Waters system.

Boyen [8] proposed a framework for constructing HIBE systems from exponent-inversion IBE systems. Specifically, Boyen described some properties of pairing-based IBE systems (called parallel IBE and linear IBE), and proved that an IBE system with these properties can be transformed to HIBE with comparable security. Boyen noted that Gentry's IBE does not quite fit within this template, and left it as an open problem to construct a HIBE system from Gentry's IBE system. Our system, which solves this problem, does not quite fit within Boyen's framework, yet our approach owes much to Boyen's idea.

We construct the first fully-secure HIBE with a tight proof of security. Namely, ours is the first HIBE system that can be proven fully secure for more than a small constant number of levels. This solves an open problem posed in [15,1,2,22,3,13,8]. Similarly to the systems of Gentry [13] and Gentry-Waters [16], we exhibit a tight reduction, albeit to a problem whose instances are of size linear in $q + \ell$.

1.1 Loose and Tight Reductions

On a high level, the reason that most IBE systems have loose reductions is that those reductions involve the following trade-off: For each identity ID, either the simulator knows a decryption key for ID, or it doesn't. If it knows a key for ID then it does not learn anything new if the adversary chooses ID as the target identity to attack, since it could have used the decryption key to learn the same information. And if the simulator does not know a decryption key for ID then it must abort if the adversary makes a key-reveal query for this identity.

The crucial difference in the security proof of Gentry's IBE [13] is that there are many different decryption keys for each identity, and the simulator knows a *small subset* of these keys. Thus, the simulator can answer every key-reveal query without aborting, but still learn something when the adversary choses that identity for the challenge ciphertext. In this sense, Gentry's IBE system follows the universal hash proof paradigm of Cramer and Shoup [12]: Given a well-formed ciphertext, all the decryption keys recover the same message, but they recover different messages when the ciphertext is mal-formed (in a certain sense). The adversary is assumed to have a non-negligible advantage when the challenge ciphertext is well-formed, but has essentially no advantage (statistically) when it is mal-formed; the adversary's different behavior in these cases allows the simulator to solve the underlying decision problem. Gentry's reduction uses an underlying hard problem that has a large problem instance (size $\theta(q)$), to ensure that the adversary cannot use its q key-reveal queries to determine what keys the simulator possesses for the target identity. In this work we extend Gentry's IBE system and proof to the case of a HIBE.

1.2 Constructing HIBE, Step 1: From IBBE to HIBE

In our quest to construct HIBE, we use as an intermediate step a specific type of identity-based broadcast encryption (IBBE). An IBBE system can be seen as somewhere in between regular IBE and HIBE: It allows a sender to encrypt a message to a set identities, and each member of this set can use its own key to decrypt the message. This is somewhat similar to HIBE, in that encryption is targeted at a group of identities (similarly to the identity vector in HIBE).[1] However, IBBE is simpler than HIBE since decryption keys correspond only to single identities (see Section 2.2).

As a first step in constructing HIBE systems, we provide a generic transformation from IBBE to HIBE. This transformation, however, requires an "augmented IBBE system" that also has decryption keys corresponding to sets of identities (for decrypting ciphertexts that were encrypted for these sets). Specifically, we require a *key-randomizable identity based broadcast encryption* (KR-IBBE), where it is possible to generate a *uniformly random* decryption key K_S for a set of identities S from any decryption key $K_{S'}$ for $S' \subset S$ (see Section 2.3). KR-IBBE is rather close to HIBE, but a major difference is that security is defined with respect to an adversary that can only ask for decryption keys corresponding to single identities, not for sets of identities. Hence it is still simpler to design KR-IBBE and use our transformation than to design a HIBE "from scratch."

1.3   Constructing HIBE, Step 2: Constructing KR-IBBE

Even with the simplification of KR-IBBE, our construction and its proof are still rather complex. Part of the reason for the complexity of our system and proof stems from the inherent tension between the key-randomization requirement and the Cramer-Shoup proof paradigm: On one hand, key-randomization implies in particular that one can generate a *random* decryption key for an identity set S from any fixed valid encryption key for the same set. On the other hand, the Cramer-Shoup paradigm require that the simulator be able to generate only a small subset of the decryption keys for the target identity set.

Our proof resolves this tension by going through an intermediate step in which we replace the full-randomization requirement with "pseudo-randomization": Namely, from each fixed valid encryption key we can only generate a small subset of the decryption keys, but this small subset still looks random. In our case, the difference between "fully-random" and "pseudo-random" keys is that "fully random" keys are taken from some linear space and "pseudo-random" keys are taken from a proper subspace of this linear space. These being linear spaces of group elements, they are indistinguishable under the Decision Linear Assumption [4].

We prove the security of the "pseudo-random" system using techniques and hard problems analogous to those used by Gentry and Waters in [16], but we we need to make rather substantial modifications to the system given in [16]. Most notably, the randomization requirement seems to imply that we cannot have scalars in the decryption key, so we must convert everything into vectors of group elements.

---

[1] We use IBBE as a tool for constructing HIBE, so we consider a variant where the intended recipients must be enumerated explicitly by the encryption procedure. Note that it is more common for IBBE to have the "revoked" recipients enumerated on encryption. Arguably, our variant should have been called *multicast* encryption.

2   HIBE and IBBE: Definitions

For simplicity, we define our encryption systems as key encapsulation mechanisms (KEM). The standard transformation from KEM to encryption is ignored here.

2.1   Hierarchical Identity-Based Encryption

A HIBE system consists of the following five procedures:

Setup(λ, ℓ) Takes as input a security parameter λ and the hierarchy depth ℓ. It outputs a public key PK and a master secret key SK. The public key implies also a key space $\mathcal{K}(PK)$ and an identity space $\mathcal{ID}(PK)$, and hierarchical identities are (ordered) tuples in $\mathcal{ID}(PK)^{\leq \ell}$.

KeyGen(PK, SK, ID) Takes as input the public key PK and master secret key SK, and an identity vector $\mathsf{ID} = [\mathsf{ID}_1, \ldots, \mathsf{ID}_t] \in \mathcal{ID}(PK)^{\leq \ell}$. It outputs a decryption key K_{ID} for ID.

KeyDerive$(PK, \mathsf{ID}, K_{\mathsf{ID}}, \mathsf{ID}')$ Takes as input the public key PK, the identity vector ID and corresponding decryption key K_{ID}, and another vector ID' such that ID is a prefix of ID'. It outputs a decryption key $K_{\mathsf{ID}'}$ for ID'.

KEM(PK, ID) Takes as input the public key PK and identity vector ID. It outputs a pair (K, C), where K is the KEM key (from the key space $\mathcal{K}(PK)$) and C is the ciphertext.

Decrypt$(PK, C, \mathsf{ID}, K_{\mathsf{ID}})$ On input the public key PK, ciphertext C, identity vector ID and corresponding decryption key K_{ID}. It outputs the corresponding KEM key K (or an error message $\perp$).

We require the usual "completeness", namely that decryption with the correct decryption key always recovers the correct KEM key. In particular, setting $(PK, SK) \leftarrow$ Setup(λ, ℓ) and fixing any chain of identity vectors $\mathsf{ID}_1, \mathsf{ID}_2, \ldots, \mathsf{ID}_t$ with each ID_i a prefix of ID_{i+1}, if we set $K_{\mathsf{ID}_1} \leftarrow$ KeyGen(PK, SK, ID_1) and then $K_{\mathsf{ID}_i} \leftarrow$ KeyDerive$(PK, \mathsf{ID}_{i-1}, K_{\mathsf{ID}_{i-1}}, \mathsf{ID}_i)$ for $i = 2, \ldots, t$ and $(K, C) \leftarrow$ KEM(PK, ID_t), then we have Decrypt$(PK, C, \mathsf{ID}_t, K_{\mathsf{ID}_t}) = K$ (with probability one).

Security.[2] Chosen-plaintext security for a HIBE system $\mathcal{E}$ against an adversary A is defined by the following game between A and a "challenger" (both given the parameters λ, ℓ as input):

Setup: The challenger runs $(PK, SK) \leftarrow \mathcal{E}.$Setup$(\lambda, \ell)$ and gives PK to A.

Key-Reveal: The adversary A makes adaptive key-reveal queries to the challenger, each consisting of an identity vector $\mathsf{ID} = [\mathsf{ID}_1, \ldots, \mathsf{ID}_t] \in \mathcal{ID}(PK)^{\leq \ell}$. If the adversary already made the challenge query and ID is a prefix of the target identity ID^* then the challenger ignores this query, and otherwise it returns to the adversary the decryption key $K_{\mathsf{ID}} \leftarrow \mathcal{E}.$KeyGen$(PK, SK, \mathsf{ID})$.

---

[2] Our security definition below ignores the delegation issue that was noted by Shi and Waters [20], see discussions later in this section.

Challenge: The adversary queries the challenger with the target identity vector $\mathsf{ID}^* = [\mathsf{ID}_1^*, \ldots, \mathsf{ID}_t^*] \in \mathcal{ID}(PK)^{\leq \ell}$. If the adversary already made a challenge query before, or if it made a key-reveal query for any prefix of the target identity ID^* then the challenger ignores this query. Otherwise the challenger sets $(K_1, C) \leftarrow \mathcal{E}.\mathsf{KEM}(PK, \mathsf{ID}^*)$, chooses another random key $K_0 \in_R \mathcal{K}(PK)$ and a "challenge bit" $\sigma \in_R \{0, 1\}$, and returns (K_σ, C) to the adversary.

The adversary can make many Key-Reveal queries and one Challenge query, in whatever order. Then it halts, outputting a guess σ' for the challenge bit σ. The HIBE advantage of A is

$$\mathsf{AdvHIBE}_A^{\mathcal{E}}(\lambda, \ell) \;=\; \Pr[A \Rightarrow 1 | \sigma = 1] \;-\; \Pr[A \Rightarrow 1 | \sigma = 0]$$

Definition 1 (CPA-secure HIBE). *The system $\mathcal{E}$ is CPA-secure if for any efficient adversary A and any $\ell = poly(\lambda)$ it holds that $\mathsf{AdvHIBE}_A^{\mathcal{E}}(\lambda, \ell(\lambda))$ is negligible in λ.*

CCA-security is defined similarly, where the adversary can also make decryption queries (except for decrypting the target ciphertext by the target identity vector).

Key delegation. Shi and Waters observed recently [20] that definitions such as the one above are incomplete model of the real world. In the definition above the adversary only sees decryption keys that were generated by KeyGen, whereas compromised nodes in the real world have keys that were generated by KeyDerive. This could be significant, since different delegation paths could result in different distributions of secret keys. Shi wand Waters presented a more elaborate definition in which the adversary is allowed to specify a delegation path and obtain a key that was generated using this delegation path.

   For our construction, the key-randomization property ensures that the distribution of keys is nearly identical, whether they are generated by KeyGen or by KeyDerive. Hence, we only prove security with respect to this simplified definition.

2.2   Identity-Based Broadcast Encryption

An IBBE system consists of the procedures (Setup, KeyGen, KEM, Decrypt). Setup, KeyGen, and KEM are similar to HIBE, except that KeyGen can only be used for single identities (not identity vectors), and KEM gets a set of identities S instead of an ordered vector. Decrypt is defined as follows:

Decrypt$(PK, C, S, \mathsf{ID}, K_{\mathsf{ID}})$ On input the public key PK, ciphertext C, identity set $S = \{\mathsf{ID}_1, \ldots, \mathsf{ID}_t\}$ (with $t \leq \ell$) and the decryption key K_{ID} for some $\mathsf{ID} \in S$. It outputs the corresponding KEM key K (or an error message $\perp$).

The security definition for IBBE is similar to the one for HIBE, the difference being that the adversary can only make key-reveal queries for single identities rather than identity-vectors. See the long version [14] for the formal definitions.

2.3 Key-Randomizable IBBE

To construct HIBE systems, we will use "augmented IBBE systems" that also have decryption keys corresponding to sets of identities: A decryption key corresponding to an identity-set S makes it possible to decrypt ciphertexts that were created with respect to this set. A *Key-Randomizable Identity-Based Broadcast Encryption* system (KR-IBBE) is an IBBE system with extended key generation KeyGen*, extended decryption Decrypt*, and key-derivation KeyDerive, as follows:

KeyGen$^*(PK, SK, S)$ Takes as input the public key PK, master secret key SK, and an identity set $S = \{ID_1, \ldots, ID_t\} \in \mathcal{ID}(PK)^{\leq \ell}$, and outputs a decryption key K_S for S. We require that KeyGen$^*(PK, SK, S)$ degenerates to the original KeyGen when S is a singleton set $S = \{ID\}$.

KeyDerive(PK, S, K_S, S') Takes as input the public key PK, an identity set S and corresponding decryption key K_S, and a superset $S' \supseteq S$, and outputs a decryption key $K_{S'}$ for S'.[3]

Decrypt$^*(PK, C, S, K_S)$ Takes as input the public key PK, an identity set S, ciphertext C that was generated with respect to S, and the decryption key K_S for S. It outputs the KEM key K (or an error message $\perp$).

We stress that *we make no security requirements* regarding these additional procedures: the CPA-security game is still defined with respect to the original four procedures Setup, KeyGen, KEM, Decrypt. However, we do make some functionality requirements, specifically the standard "completeness" requirement on Decrypt* and a distribution requirement on KeyDerive.

The "completeness" requirement says that for any $(PK, SK) \leftarrow$ Setup(λ, ℓ) and any set of identities S, if we set $K_S \leftarrow$ KeyGen$^*(PK, SK, S)$ and $(K, C) \leftarrow$ KEM(PK, S), then we get Decrypt$^*(PK, C, S, K_S) = K$ (with probability one).

The distribution requirement says for any $(PK, SK) \leftarrow$ Setup(λ, ℓ), any two sets of identities $S \subseteq S'$, and any decryption key $K_S \leftarrow$ KeyGen$^*(PK, SK, S)$, the output distributions of KeyGen$^*(PK, SK, S')$ and KeyDerive(PK, S, K_S, S') are almost identical. (That is, their statistical distance is negligible in λ.)

Remark. Due to the distribution requirement above, our transformation from key-randomizable IBBE to HIBE in Section 3 results in a system where the decryption keys generated by KeyDerive have the same distribution as the ones generated by KeyGen. As we pointed out before, this property allows us to ignore the delegation issue of Shi and Waters [20].

3   From Key-Randomizable IBBE to HIBE

The transformation from key-randomizable IBBE to HIBE is quite straightforward: we use collision-resistant hashing to map identity-vectors to identity-sets, and then just use each of the procedures Setup, KeyGen*, KeyDerive, KEM,

---

[3] Note that in this setting of broadcast encryption, keys corresponding to smaller sets are "more powerful" than ones corresponding to larger sets: one can derive a key for the superset S' from any key for a subset S, but not the other way around.

Decrypt* as-is. The only non-trivial aspect of this transformation is the security reduction, since the HIBE adversary can make key-reveal queries on identity-vectors whereas the IBBE adversary can only ask for keys of "top level" single identities. We handle this difference by having the reduction algorithm generate decryption keys differently than is done in the system, which is where we need the distribution requirement of key randomization.

3.1   The Transformation

Let $\mathcal{E} = (\mathsf{Setup}, \mathsf{KeyGen}^*, \mathsf{KeyDerive}, \mathsf{KEM}, \mathsf{Decrypt}^*)$ be a key-randomizable IBBE system, and we assume that we have a "matching" collision resistant hash function H that can hash identity-vectors into the identity space of $\mathcal{E}$.[4] We use H to hash identity vectors in the HIBE system into identity sets for $\mathcal{E}$ by setting:

$$H(\mathsf{ID}_1, \ldots, \mathsf{ID}_i) \stackrel{\text{def}}{=} \{H(\mathsf{ID}_1),\ H(\mathsf{ID}_1, \mathsf{ID}_2),\ \ldots,\ H(\mathsf{ID}_1, \mathsf{ID}_2, \ldots, \mathsf{ID}_i)\}$$

Note that short of finding collisions in H, we can only get $H(\mathsf{ID}_1, \mathsf{ID}_2, \ldots, \mathsf{ID}_i) \in H(\mathbf{ID'})$ if $(\mathsf{ID}_1, \mathsf{ID}_2, \ldots, \mathsf{ID}_i)$ is a prefix of $\mathbf{ID'}$. Then we construct a HIBE system as follows:[5]

$\mathsf{HIBE.Setup}(\lambda, \ell)$: Set $(SK_0, PK_0) \leftarrow \mathcal{E}.\mathsf{Setup}(\lambda, \ell)$. Output SK and PK, which are the same as SK_0 and PK_0, except that each includes a description of the hash function H as above.

$\mathsf{HIBE.KeyGen}(PK, SK, \mathbf{ID})$: Set $S \leftarrow H(\mathbf{ID})$ (as above), $K_S \leftarrow \mathcal{E}.\mathsf{KeyGen}^*(PK_0, SK_0, S)$ and output $K_{\mathbf{ID}} = K_S$.

$\mathsf{HIBE.KeyDerive}(PK, \mathbf{ID}, K_{\mathbf{ID}}, \mathbf{ID'})$: Set $S \leftarrow H(\mathbf{ID})$ and $S' \leftarrow H(\mathbf{ID'})$, and note that $S \subseteq S'$ since $\mathbf{ID}$ is a prefix of $\mathbf{ID'}$. Also let $K_S = K_{\mathbf{ID}}$, compute $K_{S'} \leftarrow \mathcal{E}.\mathsf{KeyDerive}(PK_0, S, K_S, S')$ and output $K_{\mathbf{ID'}} = K_{S'}$.

$\mathsf{HIBE.KEM}(PK, S)$: Set $S \leftarrow H(\mathbf{ID})$, compute $(K, C) \leftarrow \mathcal{E}.\mathsf{KEM}(PK_0, S)$ and output (K, C).

$\mathsf{HIBE.Decrypt}(PK, C, \mathbf{ID}, K_{\mathbf{ID}})$: Set $S \leftarrow H(\mathbf{ID})$ and $K_S = K_{\mathbf{ID}}$, and return $\mathcal{E}.\mathsf{Decrypt}^*(PK_0, C, S, K_S)$.

Theorem 1. *Suppose that there exists a HIBE adversary $\mathcal{A}$ that breaks CPA security (resp. CCA security) of the HIBE construction with advantage ϵ. Then, there exists an IBBE adversary $\mathcal{B}$ and a collision finder $\mathcal{B'}$, both running in about the same time as $\mathcal{A}$, such that $\mathcal{B'}$ finds a hash function collision with some probability ϵ' and $\mathcal{B}$ breaks the CPA security (resp. CCA security) of the underlying KR-IBBE system $\mathcal{E}$ with advantage $\epsilon - \epsilon'$.*

---

[4] The identity space in our IBBE system from Section 5 is $\mathbb{Z}_q$ for a large q, so "matching" a hash function is easy.

[5] Note that this transformation is completely black box; in particular, it does not depend on whether or not the IBBE system uses a bilinear map.

The proof is in the long version [14]. The only non-trivial aspect of the proof is that to get a key for the set $S \leftarrow \boldsymbol{H}(\mathsf{ID}_1, \ldots, \mathsf{ID}_t)$, the simulator makes a query for the singleton key of the identity $\mathsf{ID}'_t = H(\mathsf{ID}_1, \ldots, \mathsf{ID}_t) \in S$, and then uses key-derivation to get the key for S.

4   Notations and Preliminaries

We now introduce notations and hardness assumption that are used to establish our key-randomizable IBBE in Section 5. We denote the set of integers from m to n (inclusive) by $[m, n]$. We denote polynomials by uppercase letters in San-serif font, for example P, Q, T, etc. We use the following simple fact about polynomials:

Lemma 1. *For any polynomial* $\mathsf{P}(x)$ *and any scalar* a, $\mathsf{P}(x) - \mathsf{P}(a)$ *is divisible by* $x - a$. *In other words,* $\frac{\mathsf{P}(x) - \mathsf{P}(a)}{x - a}$ *is a polynomial (without denominator) of degree* $\deg(\mathsf{P}) - 1$.

4.1   Bilinear Maps and Our Additive Notations

Our system and its security proof make heavy use of linear algebra. We therefore use additive notations for all the groups that are involved in the system. Specifically, we use $\mathbb{Z}_q$ — the field of integers modulo a prime q — as our base scalar field, and we have two order-q groups that we call the *source group* $\mathbb{G}$ and *target group* $\mathbb{G}_T$, both of which can be viewed as vector spaces over $\mathbb{Z}_q$.

Throughout the writeup we denote elements of the source group with a hat over lowercase letters (e.g., $\hat{a}, \hat{b}$, etc.) and elements of the target group with a tilde ($\tilde{a}, \tilde{b}$, etc.). Scalars will be denoted with no decorations (e.g., a, b, and sometimes also τ, ρ, etc.)

We will make use of an efficiently computable bilinear map from the source group to the target group $\mathsf{e} : \mathbb{G} \times \mathbb{G} \to \mathbb{G}_T$,[6] such that for any two source-group elements $\hat{a}, \hat{b} \in \mathbb{G}$ and any two scalars $u, v \in \mathbb{Z}_q$ it holds that

$$\mathsf{e}(u \cdot \hat{a}, v \cdot \hat{b}) = uv \cdot \mathsf{e}(\hat{a}, \hat{b})$$

The neutral elements in the groups $\mathbb{G}, \mathbb{G}_T$ are denoted by $\hat{0}, \tilde{0}$, respectively. We also denote by $\hat{1}$ some fixed generator in $\mathbb{G}$, which we consider to be part of the description of $\mathbb{G}$. We require that the mapping e is non-trivial, which means that $\mathsf{e}(\hat{1}, \hat{1})$ is a generator in $\mathbb{G}_T$, and we denote this generator by $\tilde{1} = \mathsf{e}(\hat{1}, \hat{1})$.

More generally, for a scalar $a \in \mathbb{Z}_q$, we denote the source-group element $a \cdot \hat{1}$ by $\hat{a}$, and the target-group element $a \cdot \tilde{1} = \mathsf{e}(\hat{a}, \hat{1})$ by $\tilde{a}$. Conversely, for an element $\hat{a} \in \mathbb{G}$, its discrete-logarithm based $\hat{1}$ is denoted $a \in \mathbb{Z}_q$. (Readers who are used to multiplicative notations may find it easier to think of $\hat{a}, \tilde{a}$ as denoting

---

[6] Our system can just as well use a-symmetric bilinear maps where you have two different source groups, $\mathsf{e} : \mathbb{G}_1 \times \mathbb{G}_2 \to \mathbb{G}_T$. We chose to describe it for the symmetric case $\mathbb{G}_1 = \mathbb{G}_2$ in order to avoid introducing even more notations.

"a in the exponent" in the appropriate groups.) Note also that in these notations, the discrete-logarithm of $\hat{a}$ with respect to $\hat{b}$ is just their "ratio" $\hat{a}/\hat{b}$, which is a scalar.

With these notations, we usually omit the map e altogether, and simply denote it as a "product" of two source-group elements:

$$\hat{a} \cdot \hat{b} \overset{\text{def}}{=} \mathsf{e}(\hat{a}, \hat{b}) = \tilde{ab} \in \mathbb{G}_T$$

Note that the bi-linearity of e looks in these notations just like the natural commutative property of products $u\hat{a} \cdot v\hat{b} = uv \cdot \tilde{ab}$.

Below we slightly abuse notations to denote "powers of group elements": If $\hat{a}$ is a group element with discrete-logarithm a, then we denote $\hat{a}^i \overset{\text{def}}{=} a^i \cdot \hat{1}$ and we call $\hat{a}^i$ the i'th power of $\hat{a}$. [7]

Vectors and Matrices. We extend our notations to vectors and matrices: A vector of scalars is denoted with no decoration $\boldsymbol{a} = [a_1, a_2, \ldots, a_n]$, a vector of source-group elements denoted with a hat, $\hat{\boldsymbol{a}} = [\hat{a}_1, \hat{a}_2, \ldots, \hat{a}_n]$, and a vector of target-group elements denoted with a tilde $\tilde{\boldsymbol{a}} = [\tilde{a}_1, \tilde{a}_2, \ldots, \tilde{a}_n]$. All these vectors are considered row vectors.

Matrices are denoted by uppercase letters, e.g., A for a matrix of scalars, $\hat{A}$ for a matrix of source-group elements, and $\tilde{A}$ for a matrix of target-group elements. We denote the i'th row of A by A_i, the sub-matrix consisting of rows i, j, k by $A_{i,j,k}$, and the sub-matrix consisting of rows i through j is denoted $A_{i..j}$. As usual, the transposed matrix of A is denoted A^t.

We denote by $\mathsf{span}(\boldsymbol{x}, \boldsymbol{y}, \boldsymbol{z})$ the linear space that is spanned by the vectors $\boldsymbol{x}, \boldsymbol{y}, \boldsymbol{z}$, and also use the same notation to denote the uniform distribution over this space. For example, we use $\hat{\boldsymbol{u}} \leftarrow \hat{\boldsymbol{w}} + \mathsf{span}(\hat{A}_{1,2,4})$ as a shorthand for the process of choosing three random scalars $a, b, c \in_R \mathbb{Z}_p$ and setting $\hat{\boldsymbol{u}} \leftarrow \hat{\boldsymbol{w}} + a\hat{A}_1 + b\hat{A}_2 + c\hat{A}_4$.

Inner and Outer-Products. For vectors $\boldsymbol{a}, \boldsymbol{b}$, we denote their inner product by $\langle \boldsymbol{a}, \boldsymbol{b} \rangle \overset{\text{def}}{=} \sum_i a_i b_i$. We use the same inner-product notations also for vectors of source-group elements, namely:

$$\left\langle \boldsymbol{a}, \hat{\boldsymbol{b}} \right\rangle = \left\langle \hat{\boldsymbol{b}}, \boldsymbol{a} \right\rangle \overset{\text{def}}{=} \sum_i a_i \hat{b}_i = \langle \boldsymbol{a}, \boldsymbol{b} \rangle \cdot \hat{1}, \quad \text{and} \quad \left\langle \hat{\boldsymbol{a}}, \hat{\boldsymbol{b}} \right\rangle \overset{\text{def}}{=} \sum_i \mathsf{e}(\hat{a}_i, \hat{b}_i) = \langle \boldsymbol{a}, \boldsymbol{b} \rangle \cdot \tilde{1}$$

It is easy to check that all the commutative, associative, and distributive properties of inner products hold for both scalars and group elements.

Similar notations apply to matrix multiplication, for either scalar matrices or group-element matrices. For example, if A is an $\ell \times m$ scalar matrix and $\hat{B}$ is an $m \times n$ matrix of source-group elements, then $A\hat{B} \in \mathbb{G}[\ell \times n]$ is a matrix of source-group elements whose i, j element is the inner product of the i'th row

---

[7] This abuse of notation may take some getting used to: notice that the a's themselves should be thought of as being "in the exponent." In multiplicative notation, this power of $\hat{a}$ would be denoted as something like $g^{(a^i)}$.

of A by the j'th column of $\hat{B}$. We also use $\boldsymbol{a} \times \boldsymbol{b}$ to denote the outer product of two vectors. Namely, the outer product of the m-vector $\boldsymbol{a}$ by the n-vector $\boldsymbol{b}$ is the $m \times n$ matrix obtained as the matrix product of the $m \times 1$ matrix $\boldsymbol{a}^t$ by the $1 \times n$ matrix $\boldsymbol{b}$. The same notation applies to vectors of group elements.

Linear Algebra. All the standard concepts from linear algebra behave just the same with either scalars or group elements. For example, if $\hat{A} \in \mathbb{G}[n \times n]$ is a square matrix of source-group elements and A is the matrix of the discrete logarithm of all the elements in $\hat{A}$ (with respect to the fixed generator $\hat{1}$), then the inverse of $\hat{A}$ is $\hat{A}^{-1} = A^{-1} \cdot \hat{1}$. (Equivalently, the inverse of $\hat{A}$ is the unique matrix $\hat{B}$ such that $\hat{A} \cdot \hat{B} = \hat{I}$.) Similarly, the rank of a scalar matrix is defined as usual, and the rank of a matrix of group elements is defined as the rank of their discrete-logarithm matrix.

4.2   The BDHE-Set Assumption

The BDHE-Set assumption (used also in [16]) is a parameterized generalization of the t-BDHI problem from [1].[8] Recall that a t-BDHI adversary is given $t+1$ powers of a random source-group element, $\hat{1}, \hat{a}, \hat{a}^2, \ldots \hat{a}^t$, and it needs to distinguish the target-group element $\tilde{a}^{-1}$ from random.

An instance of the BDHE-Set assumption is parameterized by a set of integers $\mathcal{S} \subset \mathbb{Z}$ and another "target integer" m. The BDHE-Set adversary is given some powers of a random source-group element, $\{\hat{a}^i : i \in \mathcal{S}\}$, and it (roughly) needs to distinguish the target-group element $\tilde{a}^m$ from random. Denoting $\mathcal{S} +_q \mathcal{S} \overset{\text{def}}{=} \{i + j \bmod \lambda(q) : i, j \in \mathcal{S}\}$, where $\lambda(q)$ is the order of elements modulo q, it is easy to see that if $\mathbb{G}$ is an order-q bilinear-map group and $m \in \mathcal{S} +_q \mathcal{S}$ then the problem is easy: Just choose some $i, j \in \mathcal{S}$ such that $i + j = m \bmod \lambda(q)$ and compute the bilinear map

$$\mathsf{e}(\hat{a}^i, \hat{a}^j) = \hat{a}^i \cdot \hat{a}^j = \tilde{a}^{i+j} = \tilde{a}^m$$

However, when $m \notin \mathcal{S} +_q \mathcal{S}$ then there does not seem to be an easy way of distinguishing $\tilde{a}^m$ from random given the source-group elements $\{\hat{a}^i : i \in \mathcal{S}\}$. The formal BDHE-Set assumption below is somewhat stronger, however, giving the adversary not the target group element $\tilde{a}^m$ itself, but rather two random source group elements whose product is $\tilde{a}^m$. Even so, this may be a reasonable assumption to make.

Definition 2 (Decision BDHE-Set). *Fix a prime number q, a set of integers $\mathcal{S}$ and another integer $m \notin \mathcal{S} +_q \mathcal{S}$. Also fix two order-$q$ groups $\mathbb{G}$ and $\mathbb{G}_T$, admitting a non-trivial, efficiently computable bilinear map $\mathsf{e} : \mathbb{G} \times \mathbb{G} \to \mathbb{G}_T$.*

The $(\mathcal{S}, m)$-BDHE-Set problem with respect to $\mathbb{G}$ and $\mathbb{G}_T$ consists of the following experiment: Choose at random a scalar $a \in_R \mathbb{Z}_q^*$ and a bit $\sigma \in_R \{0, 1\}$. If $\sigma = 0$ then choose two random scalars $z_1, z_2 \in_R \mathbb{Z}_q^*$, and if $\sigma = 1$ then choose a random scalar $z_1 \in_R \mathbb{Z}_q^*$ and set $z_2 \leftarrow a/z_1 \bmod q$. The BDHE-Set adversary

---

[8] This assumption is called q-BDHI in [1], but we use the letter q as our group order.

gets as input $\hat{a}^i = a^i \cdot \hat{1}$ *for all* $i \in \mathcal{S}$ *and also* $\hat{z}_1, \hat{z}_2$, *and its goal is to guess the bit* σ. *The advantage of an adversary* A *is defined as*

$$\mathsf{AdvBDHE}_A^{\mathcal{S},m}(\mathbb{G}, \mathbb{G}_T) \overset{\text{def}}{=} \Pr\left[a, z_1 \in_R \mathbb{Z}_q^*, \; z_2 \leftarrow \frac{a}{z_1}, A\left(\{\hat{a}^i : i \in \mathcal{S}\}, \hat{z}_1, \hat{z}_2\right) \Rightarrow 1\right]$$
$$- \Pr\left[a, z_1, z_2 \in_R \mathbb{Z}_q^*, A\left(\{\hat{a}^i : i \in \mathcal{S}\}, \hat{z}_1, \hat{z}_2\right) \Rightarrow 1\right]$$

Informally, the asymptotic Decision BDHE-Set assumption states that for any $m \notin \mathcal{S} + \mathcal{S}$ and a large enough prime q, efficient adversaries (that work in time $\mathrm{poly}(|\mathcal{S}|, \log q)$ only have insignificant advantage in the experiment from above. Making this formal is rather straightforward (though getting the quantification right takes some care).

Jumping ahead, for our system we use the assumption above with the target integer $m = -1$ and the set $\mathcal{S}$ defined as:

$$\begin{aligned} \mathcal{S} \;=\; & [-2h - 2\ell, \; -2h - \ell - 2] \;\cup\; [-h - \ell, \; -\ell - 1] \\ & \cup\; [0, \ell - 1] \;\cup\; [h + \ell, 2h + \ell] \;\cup\; [2h + 2\ell, 3h + 2\ell + 1] \end{aligned} \tag{1}$$

where ℓ is the depth of the identity-hierarchy of the system and $h > \ell$ is some other parameter. (Specifically, if q^* is a bound on the number of queries then $h = q^* + \ell + 2$.) It is easy to check that indeed $m = -1 \notin \mathcal{S} + \mathcal{S}$.

The Linear Assumption. The decision linear assumption, first defined in [4], states (in our additive notations) that given the six source group elements $\hat{a}, \hat{b}, \hat{c}, \hat{d}, \hat{e}, \hat{f}$, it is hard to distinguish the case where these elements are completely random from the case where they are chosen at random subject to the condition $\hat{f}/\hat{c} = \hat{e}/\hat{b} + \hat{d}/\hat{a}$. (I.e., the discrete logarithm of f relative to c is the sum of the discrete logarithm of e relative to b and the discrete logarithm of d relative to a.) Note that this assumption is equivalent to saying that given the matrix of group elements

$$M \;=\; \begin{pmatrix} \hat{a} & \hat{0} & \hat{c} \\ \hat{0} & \hat{b} & \hat{c} \\ \hat{d} & \hat{e} & \hat{f} \end{pmatrix}$$

it is hard to decide if this matrix is invertible or has rank two. In this work we use a slightly weaker variant of this assumption: Specifically, we assume that given a 3×3 matrix of source-group elements, it is hard to distinguish the case where this is a random invertible matrix from the case where it is a random rank-two matrix. (The advantage of an adversary in distinguishing these cases is denoted $\mathsf{AdvLinear}_A(\mathbb{G}, \mathbb{G}_T)$.) This assumption is implied both by the standard linear assumption from [4] and by our BDHE-Set assumption, but we make it a separate assumption just to make the exposition of our security-proof easier.

5   A Key-Randomizable IBBE System

Our system operates in prime-order bilinear-map groups. In the description below we assume that these order-q groups are fixed "once and for all" and everyone knows their description. (An alternative description will include the group-generation as part of the Setup procedure.) We also fix the hierarchy-depth of the system to some integer ℓ.

The identity space of the system is the scalar field $\mathbb{Z}_q$, except that we have ℓ "forbidden identities" within this range: $\ell - 1$ of them are arbitrary (and we set them to be $0, 1, \ldots, \ell - 2$), and the last one is a random scalar a that is chosen during Setup (see below).

Setup: Choose three random scalars $a, b, s \in Z_q$ and a random invertible matrix $A \in G[7 \times 7]$, and set $\hat{B} = (\hat{A}^{-1})^t$. We note that the system only uses the top four rows of $\hat{A}$ and five rows of $\hat{B}$. The seventh dimension is only used in the security proof. Below we denote by $\boldsymbol{a}_i$ the vector $\boldsymbol{a}_i \stackrel{\text{def}}{=} [1\ a\ a^2 \ldots a^i]$.

- The master secret key is $SK = (\hat{B}_{1..6}, s, \boldsymbol{a}_\ell)$.
- The public key consists of three parts, $PK = (PK_1, PK_2, PK_3)$ with PK_1 consisting of a target-group element that is used to compute the KEM key, PK_2 consisting of multiples of the rows of $\hat{A}$ that are used to compute the ciphertext, and PK_3 consisting of multiples of the rows of $\hat{B}$ that are used only for key randomization. Specifically we have $PK_1 = a^{\ell-1}\tilde{s}$ and

$$PK_2 = \left\{ \underbrace{\{a^i\hat{A}_1 : i = 0, \ldots, \ell\}}_{\boldsymbol{a}_\ell \times \hat{A}_1}, s\hat{A}_2, \underbrace{\{a^i\hat{A}_3 : i = 0, \ldots, \ell - 1\}}_{\boldsymbol{a}_{\ell-1} \times \hat{A}_3}, \hat{A}_4 \right\} \quad (2)$$

$$PK_3 = \left\{ bs\hat{B}_1,\ abs\hat{B}_1,\ \hat{B}_5,\ \hat{B}_6,,\ \underbrace{\{a^ib\hat{B}_1 : i = 0, \ldots, \ell\}}_{b(\boldsymbol{a}_\ell \times \hat{B}_1)}, \right.$$

$$\left. \underbrace{\{a^ib\hat{B}_2 : i = 0, \ldots, \ell\}}_{b(\boldsymbol{a}_\ell \times \hat{B}_2)},\ \underbrace{\{a^ib\hat{B}_3 : i = 0, \ldots, \ell + 1\}}_{b(\boldsymbol{a}_{\ell+1} \times \hat{B}_3)} \right\}$$

KeyGen(PK, SK, ID): Choose a key of $3\ell - 3$ seven-dimensional vectors of source-group elements as follows: Pick at random $r_{\mathsf{ID}} \in Z_q$ and set $\hat{K}_{\mathsf{ID}} = (\hat{\boldsymbol{u}}_{\mathsf{ID}}, \hat{V}_{\mathsf{ID}}, \hat{W}_{\mathsf{ID}}, \hat{X}_{\mathsf{ID}}, \hat{\boldsymbol{y}}_{\mathsf{ID}})$, where

$$\hat{\boldsymbol{u}}_{\mathsf{ID}} = \frac{s - r_{\mathsf{ID}}}{a - \mathsf{ID}} \hat{B}_1 \qquad \hat{V}_{\mathsf{ID}} = r_{\mathsf{ID}}(\boldsymbol{a}_{\ell-2} \times \hat{B}_1)\ \left(= \{r_{\mathsf{ID}}a^i\hat{B}_1 : i = 0, \ldots, \ell - 2\}\right)$$

$$\hat{W}_{\mathsf{ID}} = \boldsymbol{a}_{\ell-2} \times \hat{B}_2\ \left(= \{a^i\hat{B}_2 : i = 0, \ldots, \ell - 2\}\right)$$

$$\hat{\boldsymbol{y}}_{\mathsf{ID}} = r_{\mathsf{ID}}a^{\ell-1}\hat{B}_3 + \mathrm{span}(\hat{B}_{5,6})\ \hat{X}_{\mathsf{ID}} = r_{\mathsf{ID}}(\boldsymbol{a}_{\ell-2} \times \hat{B}_3)\ \left(= \{r_{\mathsf{ID}}a^i\hat{B}_3 : i = 0, \ldots, \ell - 2\}\right) \tag{3}$$

Note that the $\hat{W}_{\mathsf{ID}}$ component is the same for all identities (so it really belongs in the public key). It is included in the secret key only for the purpose of the key-randomization procedure below.

KEM(PK, S): If $|S| < \ell$ then add to S the first $\ell - |S|$ of the "forbidden identities" $0, 1, \ldots$. Denote the resulting ℓ identities by $\{\mathsf{ID}_1, \mathsf{ID}_2, \ldots, \mathsf{ID}_\ell\}$.

- Set the monic degree-ℓ polynomial $\mathsf{P}(x) \stackrel{\text{def}}{=} \prod_{i=1}^{\ell}(x - \mathsf{ID}_i)$, let $p_0, \ldots, p_\ell$ be the coefficients of P and denote $\boldsymbol{p} \stackrel{\text{def}}{=} [p_0 \; \cdots \; p_\ell]$ (so $\mathsf{P}(a) = \langle \boldsymbol{p}, \boldsymbol{a}_\ell \rangle$).
- Choose at random $f_0, \ldots, f_{\ell-1} \in Z_q$ and denote $\boldsymbol{f} \stackrel{\text{def}}{=} [f_0 \; f_1 \; \cdots \; f_{\ell-1}]$ and $\mathsf{F}(x) \stackrel{\text{def}}{=} \sum_{i=0}^{\ell-1} f_i x^i$. Make sure that $\mathsf{F}(\mathsf{ID}_i) \neq 0$ for all $i = 1, \ldots, \ell$ (otherwise re-choose F until this condition holds).
- Choose a random scalar $t \in Z_q$.
- Output the ciphertext containing the polynomial F and the vector

$$\hat{c} = t\bigg(\underbrace{\mathsf{P}(a)\hat{A}_1}_{\boldsymbol{p}(\boldsymbol{a}_\ell \times \hat{A}_1)} + s\hat{A}_2 + \underbrace{\mathsf{F}(a)\hat{A}_3}_{\boldsymbol{f}(\boldsymbol{a}_{\ell-1} \times \hat{A}_3)} \bigg) + \mathsf{span}(\hat{A}_4) \qquad (4)$$

The implied KEM key is the target-group element $\tilde{k} = t \cdot PK_1 = a^{\ell-1} t \tilde{s}$.

Remark. Note that the ciphertext include seven source group elements and ℓ scalars (to specify F). The ciphertext size can be reduced in a particular way, so that when encrypting to a set S of size $m < \ell$ we only have m scalars in the ciphertext: Instead of choosing F completely at random, we impose the condition that $\mathsf{F}(ID) = 1$ for each of the "forbidden identities" that were added to S. This way, the encryptor can specify F using only the m scalars $\mathsf{F}(\mathsf{ID}_i)$ for all $\mathsf{ID}_i \in S$. This optimization requires a small change to the proof of security, see remark at the end of Section 6. We also note that we can get a constant-size ciphertext by moving to the random-oracle model: the encryptor just sends some nonce, and F is determined by applying the random oracle to this nonce.

Decrypt$(PK, (\mathsf{F}, \hat{c}), S, \mathsf{ID}, \hat{K}_{\mathsf{ID}})$, where $\mathsf{ID} \in S$. If $|S| < \ell$ then add to S the first $\ell - |S|$ of the "forbidden identities" $0, 1, \ldots$. Denote the resulting ℓ identities by $\{\mathsf{ID}_1, \mathsf{ID}_2, \ldots, \mathsf{ID}_\ell\}$. Parse the key as $\hat{K}_{\mathsf{ID}} = (\hat{u}_{\mathsf{ID}}, \hat{V}_{\mathsf{ID}}, \hat{W}_{\mathsf{ID}}, \hat{X}_{\mathsf{ID}}, \hat{y}_{\mathsf{ID}})$, recalculate the monic ℓ-degree polynomial $\mathsf{P}(x) = \prod_{i=1}^{\ell}(x - \mathsf{ID}_i)$, and do the following:

- Set $\mathsf{Q}_{\mathsf{ID}}(x) \stackrel{\text{def}}{=} \dfrac{\mathsf{P}(x)}{x - \mathsf{ID}}$ and $\mathsf{Q}'_{\mathsf{ID}}(x) = \mathsf{Q}_{\mathsf{ID}}(x) - a^{\ell-1}$. (That is, Q' is the polynomial Q without the top coefficient of $1 \cdot x^{\ell-1}$.) Denote the coefficient vector of $\mathsf{Q}'_{\mathsf{ID}}$ by $\boldsymbol{q}'_{\mathsf{ID}} = [q_0 \; q_1 \; \cdots \; q_{\ell-2}]$.
- Set $\mathsf{G}_{\mathsf{ID}}(x) \stackrel{\text{def}}{=} \dfrac{\mathsf{F}(x) - \mathsf{F}(\mathsf{ID})}{x - \mathsf{ID}}$ and denote the coefficient vector of G_{ID} by $\boldsymbol{g}_{\mathsf{ID}} = [g_0 \; g_1 \; \cdots \; g_{\ell-2}]$.
- Set

$$\hat{d}_{\mathsf{ID}} = \hat{u}_{\mathsf{ID}} - \boldsymbol{q}'_{\mathsf{ID}} \cdot \hat{W}_{\mathsf{ID}} - \frac{\boldsymbol{g}_{\mathsf{ID}} \cdot \hat{V}_{\mathsf{ID}} - \boldsymbol{q}'_{\mathsf{ID}} \cdot \hat{X}_{\mathsf{ID}} - \hat{y}_{\mathsf{ID}}}{\mathsf{F}(\mathsf{ID})} \qquad (5)$$

Finally, recover the KEM key as $\tilde{k} = \left\langle \hat{c}, \hat{d}_{\mathsf{ID}} \right\rangle$.

5.1   Correctness

To argue correctness, we can rewrite

$$
\hat{d}_{\mathsf{ID}} = \overbrace{\hat{u}_{\mathsf{ID}}}^{\frac{s-r_{\mathsf{ID}}}{a-\mathsf{ID}}\hat{B}_1} - q'_{\mathsf{ID}} \cdot \overbrace{\hat{W}_{\mathsf{ID}}}^{a_{\ell-2}\times\hat{B}_2} \frac{g_{\mathsf{ID}} \cdot \overbrace{\hat{V}_{\mathsf{ID}}}^{r_{\mathsf{ID}}\,a_{\ell-2}\times\hat{B}_1} - q'_{\mathsf{ID}} \cdot \overbrace{\hat{X}_{\mathsf{ID}}}^{r_{\mathsf{ID}}\,a_{\ell-2}\times\hat{B}_3} - \overbrace{\hat{y}_{\mathsf{ID}}}^{r_{\mathsf{ID}}\,a^{\ell-1}\hat{B}_3+\mathsf{span}(\hat{B}_{5,6})}}{\mathsf{F(ID)}}
$$

$$
= \frac{s-r_{\mathsf{ID}}}{a-\mathsf{ID}}\hat{B}_1 - \langle q'_{\mathsf{ID}}, a_{\ell-2}\rangle\,\hat{B}_2
$$

$$
-\frac{r_{\mathsf{ID}}}{\mathsf{F(ID)}}\Big(\langle g_{\mathsf{ID}}, a_{\ell-2}\rangle\,\hat{B}_1 - \big(\langle q'_{\mathsf{ID}}, a_{\ell-2}\rangle + a^{\ell-1}\big)\,\hat{B}_3 - \mathsf{span}(\hat{B}_{5,6})\Big)
$$

$$
=\Big(\frac{s-r_{\mathsf{ID}}}{a-\mathsf{ID}} - \frac{r_{\mathsf{ID}}\,\mathsf{G}_{\mathsf{ID}}(a)}{\mathsf{F(ID)}}\Big)\,\hat{B}_1 - (\mathsf{Q}_{\mathsf{ID}}(a)-a^{\ell-1})\hat{B}_2 + \frac{r_{\mathsf{ID}}}{\mathsf{F(ID)}}\Big(\mathsf{Q}_{\mathsf{ID}}(a)\hat{B}_3 + \mathsf{span}(\hat{B}_{5,6})\Big)
$$

Further developing the coefficient of $\hat{B}_1$ (using $\mathsf{G}_{\mathsf{ID}}(a)(a-\mathsf{ID}) = \mathsf{F}(a) - \mathsf{F(ID)}$), we get

$$
\Big(\frac{s-r_{\mathsf{ID}}}{a-\mathsf{ID}} - \frac{r_{\mathsf{ID}}\,\mathsf{G}_{\mathsf{ID}}(a)}{\mathsf{F(ID)}}\Big) = \frac{\mathsf{F(ID)}(s-r_{\mathsf{ID}}) - r_{\mathsf{ID}}\,\mathsf{G}_{\mathsf{ID}}(a)(a-\mathsf{ID})}{\mathsf{F(ID)}(a-\mathsf{ID})} = \frac{s\cdot\mathsf{F(ID)} - r_{\mathsf{ID}}\cdot\mathsf{F}(a)}{\mathsf{F(ID)}(a-\mathsf{ID})}
$$

Examining the inner-product of $\hat{c}$ with $\hat{d}_{\mathsf{ID}}$, we use the fact that $\langle\hat{A}_i,\hat{B}_j\rangle$ is either 0 (when $i \neq j$) or $\tilde{1}$ (when $i = j$). Hence the span's of $\hat{A}_4$ and of $\hat{B}_{5,6}$ drop out completely, and we are left with the product of the matching coefficients only:

$$
\langle\hat{c},\hat{d}_{\mathsf{ID}}\rangle = \Big(\underbrace{t\mathsf{P}(a)\frac{s\cdot\mathsf{F(ID)} - r_{\mathsf{ID}}\cdot\mathsf{F}(a)}{\mathsf{F(ID)}(a-\mathsf{ID})}}_{\text{coefficients of }\hat{A}_1,\hat{B}_1} \underbrace{- ts(\mathsf{Q}_{\mathsf{ID}}(a)-a^{\ell-1})}_{\text{coefficients of }\hat{A}_2,\hat{B}_2} + \underbrace{t\mathsf{F}(a)\frac{r_{\mathsf{ID}}}{\mathsf{F(ID)}}\mathsf{Q}_{\mathsf{ID}}(a)}_{\text{coefficients of }\hat{A}_3,\hat{B}_3} \Big) \cdot \tilde{1}
$$

The first term in the parenthesis can be simplified using $\mathsf{Q}_{\mathsf{ID}}(a) = \mathsf{P}(a)/(a-\mathsf{ID})$, so we get

$$
\langle\hat{c},\hat{d}_{\mathsf{ID}}\rangle = t\Big(\mathsf{Q}_{\mathsf{ID}}(a)\frac{s\cdot\mathsf{F(ID)} - r_{\mathsf{ID}}\cdot\mathsf{F}(a)}{\mathsf{F(ID)}} - s(\mathsf{Q}_{\mathsf{ID}}(a)-a^{\ell-1}) + \mathsf{F}(a)\frac{r_{\mathsf{ID}}}{\mathsf{F(ID)}}\mathsf{Q}_{\mathsf{ID}}(a)\Big)\cdot\tilde{1}
$$

$$
= t\Big(\mathsf{Q}_{\mathsf{ID}}(a)s - \frac{r_{\mathsf{ID}}\,\mathsf{Q}_{\mathsf{ID}}(a)\mathsf{F}(a)}{\mathsf{F(ID)}} - \mathsf{Q}_{\mathsf{ID}}(a)s + a^{\ell-1}s + \frac{r_{\mathsf{ID}}\,\mathsf{Q}_{\mathsf{ID}}(a)\mathsf{F}(a)}{\mathsf{F(ID)}}\Big)\cdot\tilde{1}
$$

$$
= t\cdot a^{\ell-1}s\cdot\tilde{1} \;=\; \tilde{k} \qquad\qquad \square
$$

5.2   Key Randomization

Our key-randomization follows Boyen's idea from [8], where the key for identity-set $S = \{\mathsf{ID}_1,\ldots,\mathsf{ID}_m\}$ consists of m "shifted versions" of the keys, $r'_{\mathsf{ID}_1}\hat{K}_{\mathsf{ID}_1}$, $\ldots, r'_{\mathsf{ID}_n}\hat{K}_{\mathsf{ID}_m}$, such that $\sum_i r'_{\mathsf{ID}_i} = 1 \pmod q$. Namely, the augmented procedure

KeyGen$^*(PK, SK, S)$ uses the same KeyGen procedure from above m times to get $\hat{K}_{\mathsf{ID}_i} \leftarrow$ KeyGen(PK, SK, ID_i). Then for $i = 1 \ldots m$ it chooses $r'_{\mathsf{ID}_i} \in Z_q$ at random subject to the constraint $\sum_i r'_{\mathsf{ID}_i} = 1 \pmod{q}$, and outputs the secret key

$$\hat{K}_S = [r'_{\mathsf{ID}_1} \hat{K}_{\mathsf{ID}_1}, \ldots, r'_{\mathsf{ID}_m} \hat{K}_{\mathsf{ID}_m}]$$

where $r'_{\mathsf{ID}_i} \hat{K}_{\mathsf{ID}_i}$ means multiplying all the elements in $\hat{K}_{\mathsf{ID}_i}$ by the scalar r'_{ID_i}. Below we call $\hat{K}_{\mathsf{ID}_i}$ the *singleton key* corresponding to ID_i, and $r'_{\mathsf{ID}_i} \hat{K}_{\mathsf{ID}_i}$ is the *shifted singleton key* for ID_i. Note that for the special case $m = 1$, we have $r'_{\mathsf{ID}} = 1$, so KeyGen* degenerates to the original KeyGen.

Extended Decryption. The extended decryption procedure Decrypt* is given a ciphertext $(F, \hat{c})$ together with a set of identities $S = \{\mathsf{ID}_1, \ldots, \mathsf{ID}_m\}$ ($m \leq \ell$) and a matching decryption key $\hat{K}_S$. It parses the decryption key as $\hat{K}_S = [\hat{K}'_{\mathsf{ID}_1}, \ldots, \hat{K}'_{\mathsf{ID}_m}]$ where the $\hat{K}'_{\mathsf{ID}_i}$'s are shifted singleton keys. Namely we have $\hat{K}'_{\mathsf{ID}_i} = r'_{\mathsf{ID}_i} \hat{K}_{\mathsf{ID}_i}$ where the $\hat{K}_{\mathsf{ID}_i}$'s are singleton keys and $\sum_i r'_{\mathsf{ID}_i} = 1 \pmod{q}$. Then we use each shifted singleton key to produce $\hat{d}'_{\mathsf{ID}_i}$ just as in Eq. (5), sets $\hat{d}_S = \sum_i \hat{d}'_{\mathsf{ID}_i}$, and recover $\tilde{k} = \left\langle \hat{c}, \hat{d}_S \right\rangle$.

Correctness holds since the decryption process in linear: Denote by $\hat{d}_{\mathsf{ID}_i}$ the vector that would have been obtained from the singleton key $\hat{K}_{\mathsf{ID}_i}$ using Eq. (5). Then on one hand decryption is linear so we have $\hat{d}'_{\mathsf{ID}_i} = r'_{\mathsf{ID}_i} \hat{d}'_{\mathsf{ID}_i}$. On the other hand by correctness of the basic decryption procedure we know that $\left\langle \hat{c}, \hat{d}_{\mathsf{ID}_i} \right\rangle = \tilde{k}$. We therefore get

$$\left\langle \hat{c}, \hat{d}_S \right\rangle = \sum_i \left\langle \hat{c}, \hat{d}'_{\mathsf{ID}_i} \right\rangle = \sum_i \left\langle \hat{c}, r'_{\mathsf{ID}_i} \hat{d}_{\mathsf{ID}_i} \right\rangle = \sum_i r'_{\mathsf{ID}_i} \left\langle \hat{c}, \hat{d}_{\mathsf{ID}_i} \right\rangle = \sum_i r'_{\mathsf{ID}_i} \tilde{k} = \tilde{k}$$

Key Derivation. Key-derivation uses Boyen's idea of reciprocal keys [8]. Namely, given the public key and any two identities ID_1 and ID_2, anyone can compute a pair of shifted singleton keys $\delta \hat{K}_{\mathsf{ID}_1}$ and $\delta \hat{K}_{\mathsf{ID}_2}$ for the same (unknown) scalar factor δ. The procedure for generating these reciprocal keys (which is used as a subroutine for key derivation) is as follows:

ReciprocalKeys$(PK, \mathsf{ID}_1, \mathsf{ID}_2)$: Recall that the public key PK depends on the unknown scalars a, b, s (among other things).

- Choose at random $z \in Z_q$. The shifted singleton keys $\delta \hat{K}_{\mathsf{ID}_i}$ will have $\delta = bz(a - \mathsf{ID}_1)(a - \mathsf{ID}_2)$.
- Choose at random $r_1, r_2 \in Z_q$ (which will play the role of r_{ID_1} and r_{ID_2} in the reciprocal keys).

– Compute $\delta \hat{K}_{\mathsf{ID}_1}$ as

$$\delta \cdot \frac{s-r_1}{a-\mathsf{ID}_1} \hat{B}_1 = (abs - bs\mathsf{ID}_2 - abr_1 + br_1\mathsf{ID}_2)z\hat{B}_1$$
$$\delta \cdot r_1(\boldsymbol{a}_{\ell-2} \times \hat{B}_1) = \{br_1z(a^{i+2} - a^{i+1}(\mathsf{ID}_1 + \mathsf{ID}_2) + a^i\mathsf{ID}_1\mathsf{ID}_2)\hat{B}_1$$
$$: i = 0, \ldots, \ell - 2\}$$
$$\delta \cdot (\boldsymbol{a}_{\ell-2} \times \hat{B}_2) = \{bz(a^{i+2} - a^{i+1}(\mathsf{ID}_1 + \mathsf{ID}_2) + a^i\mathsf{ID}_1\mathsf{ID}_2)\hat{B}_2$$
$$: i = 0, \ldots, \ell - 2\}$$
$$\delta \cdot r_1(\boldsymbol{a}_{\ell-2} \times \hat{B}_3) = \{br_1z(a^{i+2} - a^{i+1}(\mathsf{ID}_1 + \mathsf{ID}_2) + a^i\mathsf{ID}_1\mathsf{ID}_2)\hat{B}_3$$
$$: i = 0, \ldots, \ell - 2\}$$
$$\delta \left(r_1a^{\ell-1}\hat{B}_3 + \mathsf{span}(\hat{B}_{5,6})\right) = br_1z(a^{\ell+1} - a^\ell(\mathsf{ID}_1 + \mathsf{ID}_2) + a^{\ell-1}\mathsf{ID}_1\mathsf{ID}_2)\hat{B}_3$$
$$+ \mathsf{span}(\hat{B}_{5,6})$$

and similarly for $\delta\hat{K}_{\mathsf{ID}_2}$ (using r_2 instead of r_1 and swapping the roles of $\mathsf{ID}_1, \mathsf{ID}_2$). Notice that the terms $a^ib\hat{B}_j$ for $i \in [0,\ell], j = 1,2,3$, as well as $bs\hat{B}_1$, $abs\hat{B}_1$, $a^{\ell+1}b\hat{B}_3$, and $\hat{B}_{5,6}$, are all part of the PK_3 component of the public key.

From the description above it is clear that when $\mathsf{ID}_1, \mathsf{ID}_2 \neq a$, then **ReciprocalKeys** indeed returns the correct distribution, namely two shifted singleton keys $\delta\hat{K}_{\mathsf{ID}_1}$, $\delta\hat{K}_{\mathsf{ID}_2}$ where each $\hat{K}_{\mathsf{ID}}$ is drawn from the same distribution as the singleton keys for ID in **KeyGen** and δ is chosen at random in Z_q (and independently of $\hat{K}_{\mathsf{ID}_1}, \hat{K}_{\mathsf{ID}_2}$).

KeyDerive$(PK, S, \hat{K}_S, S')$ (where $S' = \{\mathsf{ID}_1, \ldots, \mathsf{ID}_m\}$ and $S \subseteq S'$). Assume (w.l.o.g.) that S consists of the first n identities in S', namely $S = \{\mathsf{ID}_1, \ldots, \mathsf{ID}_n\}$ with $n \leq m$. Denote $\hat{K}_S = \{\hat{K}'_{\mathsf{ID}_1}, \ldots, \hat{K}'_{\mathsf{ID}_n}\}$, where $\hat{K}'_{\mathsf{ID}_i}$ is the shifted singleton key for ID_i (consisting of $3\ell - 3$ 7-dimensional vectors of source-group elements).

For $i = 1, \ldots, m$, run the ReciprocalKeys procedure from above with identities ID_i and ID_{i+1} (indexing mod m) to get two shifted singleton keys for these ID's, which we denote by $\hat{L}_{\mathsf{ID}_i}, \hat{M}_{\mathsf{ID}_{i+1}}$, respectively. Namely, set

$$(\hat{L}_{\mathsf{ID}_i}, \hat{M}_{\mathsf{ID}_{i+1}}) \leftarrow \mathsf{ReciprocalKeys}(PK, \mathsf{ID}_i, \mathsf{ID}_{i+1})$$

Then for $i \in [1, n]$ set $\hat{K}^*_{\mathsf{ID}_i} = \hat{K}'_{\mathsf{ID}_i} + \hat{L}_{\mathsf{ID}_i} - \hat{M}_{\mathsf{ID}_i}$, and for $i \in [n+1, m]$ set $\hat{K}^*_{\mathsf{ID}_i} = \hat{L}_{\mathsf{ID}_i} - \hat{M}_{\mathsf{ID}_i}$ (where addition and subtraction is element-wise). The new key is $\hat{K}_S = [\hat{K}^*_{\mathsf{ID}_1}, \ldots, \hat{K}^*_{\mathsf{ID}_m}]$. In Lemma 2 below we show that this KeyDerive procedure induces almost the same distribution as KeyGen over the decryption key $\hat{K}_{S'}$.

Lemma 2. *For every $S \subseteq S'$ (with $|S'| = m$) and every secret key $\hat{K}_S$ corresponding to S, the procedure* KeyDerivation$(PK, S, \hat{K}_S, S')$ *draws from a distribution at most $O(m/q)$ away from that of* KeyGen(PK, SK, S').

Proof. Observe that every 7-vector in a singleton key $\hat{K}_{\mathsf{ID}}$ corresponding to identity ID (as computed by KeyGen) is of the form

$$(\mathsf{expr}(a, s, \mathsf{ID}) + r_{\mathsf{ID}}\mathsf{expr}'(a, s, \mathsf{ID})) \cdot \hat{B}_k$$

where r_{ID} is the scalar that was chosen for this singleton key, $\hat{B}_k$ is one specific row of the matrix $\hat{B}$, and $\text{expr}(a, s, \text{ID})$, $\text{expr}'(a, s, \text{ID})$ are two fixed scalar-valued expressions that depend only on the scalars a, s from the master secret key and on the identity ID. (Note that either $\text{expr}(a, s, \text{ID})$ or $\text{expr}'(a, s, \text{ID})$ can be zero, but not both.)

Considering the same vector in all the shifted singleton keys in $\hat{K}_S$, we have a collection of n vectors, $\hat{x}_1, \ldots, \hat{x}_n$, where $\hat{x}_i = r'_{\text{ID}_i}(\text{expr}(a, s, \text{ID}_i) + r_{\text{ID}_i}\text{expr}'(a, s, \text{ID}_i)) \cdot \hat{B}_k$, and the scalars r'_{ID_i} satisfy $\sum_i r'_{\text{ID}_i} = 1$. For notational convenience, for $i \in [n+1, m]$ we denote $r_{\text{ID}_i} = r'_{\text{ID}_i} = 0$ and $\hat{x}_i = \hat{0}$ (so we still have $\hat{x}_i$'s of the right format with $\sum_i r'_{\text{ID}_i} = 1$, even when we consider all m elements). Similarly considering the same vector in all the shifted singleton keys that are generated by ReciprocalKeys, we have vectors $\hat{y}_1 \ldots \hat{y}_m$ (from the $\hat{L}_{\text{ID}_i}$'s) and $\hat{z}_1 \ldots \hat{z}_m$ (from the $\hat{M}_{\text{ID}_i}$'s) of the form

$$\hat{y}_i = \delta_i(\text{expr}(a, s, \text{ID}_i) + \rho_i \text{ expr}'(a, s, \text{ID}_i)) \cdot \hat{B}_k$$
$$\text{and } \hat{z}_i = \delta_{i-1}(\text{expr}(a, s, \text{ID}_i) + \tau_i \text{ expr}'(a, s, \text{ID}_i)) \cdot \hat{B}_k$$

where all the scalars δ_i, ρ_i, τ_i, $i = 1 \ldots m$, are chosen at random in Z_q (and indexing is mod m, so $\delta_0 = \delta_m$). Hence the corresponding element in the shifted singleton key $\hat{K}^*_{\text{ID}_i}$ is

$$\hat{x}_i + \hat{y}_i - \hat{z}_i = \left((r'_{\text{ID}_i} + \delta_i - \delta_{i-1})\text{expr}(a, s, \text{ID}_i) + (r'_{\text{ID}_i}r_{\text{ID}_i} + \delta_i\rho_i - \delta_{i-1}\tau_i)\text{expr}'(a, s, \text{ID}_i)\right) \hat{B}_k$$

Assuming that $r'_{\text{ID}_i} + \delta_i - \delta_{i-1} \neq 0$, we can denote

$$r^{**}_{\text{ID}_i} \stackrel{\text{def}}{=} r'_{\text{ID}_i} + \delta_i - \delta_{i-1} \quad \text{and} \quad r^*_{\text{ID}_i} \stackrel{\text{def}}{=} \frac{r'_{\text{ID}_i}r_{\text{ID}_i} + \delta_i\rho_i - \delta_{i-1}\tau_i}{r'_{\text{ID}_i} + \delta_i - \delta_{i-1}}$$

and then we have $\hat{x}_i + \hat{y}_i - \hat{z}_i = r^{**}_{\text{ID}_i}(\text{expr}(a, s, \text{ID}_i) + r^*_{\text{ID}_i}\text{expr}'(a, s, \text{ID}_i))\hat{B}_k$, which is of the right form, and indeed the scalars $r^{**}_{\text{ID}_i}$ satisfy

$$\sum_{i=1}^m r^{**}_{\text{ID}_i} = \sum_{i=1}^m r'_{\text{ID}_i} + \sum_{i=1}^m \delta_i - \sum_{i=1}^m \delta_{i-1} = \sum_{i=1}^m r'_{\text{ID}_i} = 1$$

Since the δ_i's are random and independent then the $r^{**}_{\text{ID}_i}$'s are also random and independent subject to the constraint that their sum is one. Finally, assuming that none of the $r^{**}_{\text{ID}_i}$'s is zero and also none of the δ_i's are zero (which happens with probability at least $1 - O(m/q)$) then all the $r^*_{\text{ID}_i}$'s are random and independent (since the τ_i's and ρ_i's are). $\qquad\square$

6 Security of Our System

Theorem 2. *The IBBE system from Section 5 is secure under the BDHE-Set assumption and the decision Linear assumption. Specifically, for an ℓ-depth* *hierarchy, groups* $\mathbb{G}, \mathbb{G}_T$ *of order* q, *and an adversary that makes upto* q^* *key-extraction queries, we have* $\mathsf{AdvIBBE}^{\mathcal{E}}(\log q, \ell) \leq \mathsf{AdvBDHE}(\mathbb{G}, \mathbb{G}_T) + \mathsf{AdvLinear}$ $(\mathbb{G}, \mathbb{G}_T)$, *where the BDHE-Set instances are of size* $O(\ell + q^*)$.

The proof is found in the long version [14]. On a very high level, the proof consists of four games: Game 0 is the actual interaction of the adversary with our system, in Game 1 we use decryption rather than encryption to compute the KEM key corresponding to the challenge ciphertext (which has no effect on the outcome), in Game 2 we add a component of $\hat{B}_7$ to the secret keys (which is indistinguishable by the Linear assumption), and in Game 3 we add a component of $\hat{A}_7$ to the challenge ciphertext vector (thus making the KEM key statistically independent of the adversary's view).

The main reduction then proves indistinguishability of Game 3 from Game 2 based on the BDHE-Set assumption. That reduction follows the hash-proof approach: The simulator generates the challenge ciphertext so that this is either a valid ciphertext or an invalid one, depending on whether the input of the simulator is a YES instance or a NO instance of the decision BDHE-Set problem. In our case, a valid ciphertext is spanned by the rows $\hat{A}_{1,2,3,4}$, and an invalid ciphertext also has a component of $\hat{A}_7$. The secret keys have a random $\hat{B}_7$ component in them, so an invalid ciphertext is decrypted to a random KEM key (while a valid ciphertext are always decrypted to the "right KEM key").

In the reduction itself, the simulator gets as input source-group elements $\hat{a^i} = a^i \cdot \hat{1}$ for all $i \in \mathcal{S}$ and two additional source-group elements $\hat{z}_1, \hat{z}_2$, and uses these elements to answer all the queries of the adversary: Very roughly, it chooses a random polynomial $\mathsf{H}(x)$ of high-enough degree over Z_q, sets $s = \mathsf{H}(a)$ for the master secret key, $r_{\mathsf{ID}} = H(\mathsf{ID})$ in all the key-reveal queries, and $\mathsf{F} = \mathsf{H} \bmod \mathsf{P}$ for the challenge ciphertext. To compute the appropriate terms, the simulator uses the powers $\hat{a^i}$ from its input. The main challenge is to make the set $\mathcal{S}$ "large enough" so the simulator can produce the entire view of the adversary from the elements $\hat{a^i}$ that it knows, while at the same time ensuring that $\mathcal{S}$ is "small enough" so that the target integer m is not in $\mathcal{S} + \mathcal{S}$ (since otherwise the problem becomes easy).

References

1. Boneh, D., Boyen, X.: Efficient Selective-ID Secure Identity Based Encryption Without Random Oracles. In: Cachin, C., Camenisch, J.L. (eds.) EUROCRYPT 2004. LNCS, vol. 3027, pp. 223–238. Springer, Heidelberg (2004)
2. Boneh, D., Boyen, X.: Secure Identity Based Encryption Without Random Oracles. In: Franklin, M. (ed.) CRYPTO 2004. LNCS, vol. 3152, pp. 443–459. Springer, Heidelberg (2004)
3. Boneh, D., Boyen, X., Goh, E.-J.: Hierarchical Identity Based Encryption with Constant Size Ciphertexts. In: Cramer, R. (ed.) EUROCRYPT 2005. LNCS, vol. 3494, pp. 440–456. Springer, Heidelberg (2005)
4. Boneh, D., Boyen, X., Shacham, H.: Short group signatures. In: Franklin, M. (ed.) CRYPTO 2004. LNCS, vol. 3152, pp. 41–55. Springer, Heidelberg (2004)

5. Boneh, D., Franklin, M.: Identity Based Encryption from the Weil Pairing. In: Kilian, J. (ed.) CRYPTO 2001. LNCS, vol. 2139, pp. 213–229. Springer, Heidelberg (2001)
6. Boneh, D., Gentry, C., Hamburg, M.: Space Efficient Identity Based Encryption without Pairings. In: Proceedings of FOCS 2007, pp. 647–657. IEEE, Los Alamitos (2007)
7. Boneh, D., Gentry, C., Waters, B.: Collusion Resistant Broadcast Encryption with Short Ciphertexts and Private Keys. In: Shoup, V. (ed.) CRYPTO 2005. LNCS, vol. 3621, pp. 258–275. Springer, Heidelberg (2005)
8. Boyen, X.: General Ad Hoc Encryption from Exponent Inversion IBE. In: Naor, M. (ed.) EUROCRYPT 2007. LNCS, vol. 4515, pp. 394–411. Springer, Heidelberg (2007)
9. Canetti, R., Halevi, S., Katz, J.: A Forward-Secure Public-Key Encryption Scheme. In: Biham, E. (ed.) EUROCRYPT 2003. LNCS, vol. 2656, pp. 255–271. Springer, Heidelberg (2003)
10. Canetti, R., Halevi, S., Katz, J.: Chosen-Ciphertext Security from Identity-Based Encryption. In: Cachin, C., Camenisch, J.L. (eds.) EUROCRYPT 2004. LNCS, vol. 3027, pp. 207–222. Springer, Heidelberg (2004)
11. Cocks, C.: An Identity Based Encryption Scheme Based on Quadratic Residues. In: IMA Int. Conf. 2001 (2001)
12. Cramer, R., Shoup, V.: Universal Hash Proofs and a Paradigm for Adaptive Chosen Ciphertext Secure Public-Key Encryption. In: Knudsen, L.R. (ed.) EUROCRYPT 2002. LNCS, vol. 2332, pp. 45–64. Springer, Heidelberg (2002)
13. Gentry, C.: Practical Identity-Based Encryption without Random Oracles. In: Vaudenay, S. (ed.) EUROCRYPT 2006. LNCS, vol. 4004, pp. 445–464. Springer, Heidelberg (2006)
14. Gentry, C., Halevi, S.: Hierarchical Identity Based Encryption with Polynomially Many Levels, http://eprint.iacr.org/2008/383
15. Gentry, C., Silverberg, A.: Hierarchical ID-Based Cryptography. In: Zheng, Y. (ed.) ASIACRYPT 2002. LNCS, vol. 2501, pp. 548–566. Springer, Heidelberg (2002)
16. Gentry, C., Waters, B.: Adaptive Security in Broadcast Encryption Systems (manuscript, 2008), http://eprint.iacr.org/2008/268
17. Horwitz, J., Lynn, B.: Toward Hierarchical Identity-Based Encryption. In: Knudsen, L.R. (ed.) EUROCRYPT 2002. LNCS, vol. 2332, pp. 466–481. Springer, Heidelberg (2002)
18. Maurer, U.M., Yacobi, Y.: Non-interative Public-Key Cryptography. In: Davies, D.W. (ed.) EUROCRYPT 1991. LNCS, vol. 547, pp. 498–507. Springer, Heidelberg (1991)
19. Shamir, A.: Identity-Based Cryptosystems and Signature Schemes. In: Blakely, G.R., Chaum, D. (eds.) CRYPTO 1984. LNCS, vol. 196, pp. 47–53. Springer, Heidelberg (1985)
20. Shi, E., Waters, B.: Delegating Capabilities in Predicate Encryption Systems. In: Aceto, L., Damgård, I., Goldberg, L.A., Halldórsson, M.M., Ingólfsdóttir, A., Walukiewicz, I. (eds.) ICALP 2008, Part II. LNCS, vol. 5126, pp. 560–578. Springer, Heidelberg (2008)
21. Weisstein, E.W.: Sylvester Matrix. From MathWorld, a Wolfram Web Resource, http://mathworld.wolfram.com/SylvesterMatrix.html
22. Waters, B.: Efficient Identity Based Encryption without Random Oracles. In: Cramer, R. (ed.) EUROCRYPT 2005. LNCS, vol. 3494, pp. 114–127. Springer, Heidelberg (2005)

Predicate Privacy in Encryption Systems

Emily Shen[1], Elaine Shi[2], and Brent Waters[3],[*]

[1] MIT
eshen@csail.mit.edu
[2] CMU/PARC
eshi@parc.com
[3] UT Austin
bwaters@cs.utexas.edu

Abstract. *Predicate encryption* is a new encryption paradigm which gives a master secret key owner fine-grained control over access to encrypted data. The master secret key owner can generate secret key tokens corresponding to predicates. An encryption of data x can be evaluated using a secret token corresponding to a predicate f; the user learns whether the data satisfies the predicate, i.e., whether $f(x) = 1$.

Prior work on public-key predicate encryption has focused on the notion of data or plaintext privacy, the property that ciphertexts reveal no information about the encrypted data to an attacker other than what is inherently revealed by the tokens the attacker possesses. In this paper, we consider a new notion called *predicate privacy*, the property that tokens reveal no information about the encoded query predicate. Predicate privacy is inherently impossible to achieve in the public-key setting and has therefore received little attention in prior work. In this work, we consider predicate encryption in the symmetric-key setting and present a symmetric-key predicate encryption scheme which supports inner product queries. We prove that our scheme achieves both plaintext privacy and predicate privacy.

1 Introduction

In traditional public-key encryption, a user encrypts a message under a public key, and only the owner of the corresponding secret key can decrypt the ciphertext. In some applications, however, the user may wish to have more fine-grained control over what is revealed about the encrypted data. For example, in a medical context an administrative assistant might only be able to learn whether an encrypted record was generated at a certain clinic. *Predicate encryption* is a new encryption paradigm which allows for such fine-grained control over access to encrypted data. In a predicate encryption scheme, the owner of a master secret key can create and issue secret key *tokens* to other users. Tokens are associated with

[*] Supported by NSF CNS-0524252, CNS-0716199; the U.S. Department of Homeland Security under Grant Award Number 2006-CS-001-000001.

predicates which can be evaluated over encrypted data. Specifically, an encryption of a data x can be evaluated using a token TK_f associated with a predicate f to learn whether $f(x) = 1$.

Prior work on public-key predicate encryption [7,12,1,9,19,11,28,27] has focused on the security property that ciphertexts reveal no information about the underlying plaintext or data other than what is implied by the tokens in one's possession. More specifically, an adversary in possession of tokens $TK_{f_1}, \ldots, TK_{f_\ell}$ for predicates $f_1, \ldots, f_\ell$ learns no information about the underlying plaintext x other than the values of $f_1(x), \ldots, f_\ell(x)$[1]. We refer to the above property as plaintext or data privacy.

In this work, we consider a different dimension of predicate encryption – *predicate privacy*. In addition to protecting the privacy of plaintexts, we would like to protect the description of the predicates encoded by tokens. Informally, predicate privacy says that a token hides all information about the encoded predicate other than what is implied by the ciphertexts in one's possession. Unfortunately, predicate privacy is inherently impossible to achieve in the public-key setting. Since encryption does not require a secret key, an adversary can encrypt any plaintext of his choice and evaluate a token on the resulting ciphertext to learn whether the plaintext satisfies the predicate associated with the token. In this way, an adversary can gain information about the predicate encoded in a token. Therefore, it does not make sense to consider the notion of predicate privacy for predicate encryption in the public-key setting.

However, it is interesting to consider predicate privacy in the symmetric-key setting, in applications where we want to hide information about the predicate being tested from the party evaluating a token. For example, suppose a user Alice uses a remote storage service to back up her files. Alice wishes to protect the privacy of her files by encrypting them using her secret key before sending them to the server. (Only Alice possesses her secret key.) Later on, Alice may wish to retrieve all files satisfying a certain predicate. To do this, Alice can create a token (using her secret key) for this predicate and issue the token to the server. The server can then evaluate the predicate on the encrypted files and return those files which satisfy the predicate. We want to guarantee that the server learns nothing about the predicate it evaluates on Alice's behalf.

1.1   Our Results

In this paper, we present formal definitions of security for predicate encryption in the symmetric-key setting, for general classes of predicates. We present a symmetric-key predicate encryption scheme that achieves both plaintext privacy and predicate privacy. Our construction supports the class of predicates corresponding to the evaluation of inner products. We take the set of plaintexts to be $\Sigma = \mathbb{Z}_N^n$ and the class of predicates to be $\mathcal{F} = \{f_{\boldsymbol{v}} | \boldsymbol{v} \in \mathbb{Z}_N^n\}$ where $f_{\boldsymbol{v}}(\boldsymbol{x}) = 1$

---

[1] In some works the authors also distinguish an extra "payload message" M such that in the case that one of $f_1(x), \ldots, f_\ell(x)$ evaluates to 1, the adversary learns the payload message M. In our work we solely consider the predicate encryption system property where the evaluation reveals $f(x)$.

iff $\langle v, x \rangle = 0$, where $\langle v, x \rangle$ denotes the inner product $\sum_{i=1}^{n} v_i \cdot x_i \mod N$ of vectors v and x. Our construction is based on the KSW construction [21], which uses bilinear groups whose order is the product of three primes. Our construction uses groups whose order is the product of four primes. Our complexity assumptions have all been introduced in prior work but for the case of groups whose order is the product of fewer than four primes.

Why Inner Product Queries? An important goal in predicate encryption is to support complex, expressive queries. Prior work has focused on achieving more expressive schemes, with the most expressive scheme to date being that of Katz, Sahai and Waters [21]. The KSW scheme supports inner product queries, which are strictly more expressive than conjunctive queries and, as shown in [21], imply conjunctions, disjunctions, CNF/DNF formulas, polynomial evaluation, and exact thresholds. Therefore, our goal in this work is to construct a symmetric-key predicate encryption scheme that supports inner product queries.

1.2   Related Work

Public-Key Predicate Encryption. The earliest examples of public-key predicate encryption are *anonymous identity-based encryption* (A-IBE) schemes with *keyword search* (which corresponds to an equality predicate) [7,12,1,9]. Since then, more expressive schemes such as those supporting conjunctive queries [19,11,28] and multi-dimensional range queries [27] have been proposed. The most expressive scheme known to date is due to Katz, Sahai and Waters [21] and supports inner-product queries. As explained above, the KSW scheme is strictly more expressive than previously proposed predicate encryption schemes.

Searchable Encrypted Databases. A related line of research is secure searching on outsourced encrypted databases. The problem was considered by Goldreich and Ostrovsky [22,18] when cast as a problem of oblivious RAM, and they provided general solutions. Song, Wagner, and Perrig [29] later gave more efficient solutions for equality searches, but made a tradeoff of letting a storage server learn the access pattern of a user. Curtmola et al. [17] considered stronger security definitions in a similar setting. While we do not directly address searchable encrypted databases in this work, our predicate encryption solution allows for more complex queries to be made in this particular application.

Identity-Based Encryption and Attribute-Based Encryption. Identity-based encryption (IBE) [26,8,16] can be viewed as a special, more limited, case of predicate encryption for the class of equality tests. In attribute-based encryption (ABE) [25,20,3,24,15,23], a user can receive a capability representing an access control policy over the attributes of an encrypted record.

In both IBE and ABE schemes, the identity or attributes are not hidden in the ciphertext. In fact, access to the encrypted data itself is inherently "all-or-nothing." The important distinction between these systems and the ones we consider is that they only hide a "payload message" M. In particular, the ciphertext is associated with a payload message M and some extra structure x

(e.g., the "identity" or set of attributes associated with the ciphertext). The security guarantee of these systems is that M remains hidden as along as the attacker does not have a secret key associated with a predicate function f such that $f(x) = 1$; however, there is no guarantee about hiding the structure of x, which in general might be leaked to the attacker. One advantage, however, is that this relaxation might allow for more expressive access predicates.

2   Definitions

In this section, we formally define symmetric-key predicate encryption and its security. For simplicity, we consider the *predicate-only* variant, in which evaluating a token on a ciphertext outputs a bit indicating whether the encrypted plaintext satisfies the predicate corresponding to the token. We note that a predicate-only scheme can easily be extended to obtain a full-fledged predicate encryption scheme, in which evaluating a token on a ciphertext outputs the encrypted plaintext if the plaintext satisfies the predicate corresponding to the token, using techniques from prior work such as [11,27,21].

We give definitions for the general case of an arbitrary set of plaintexts Σ and an arbitrary set of predicates $\mathcal{F}$. Our construction in Section 4 will be for the specific case of $\Sigma = \mathbb{Z}_N^n$ and $\mathcal{F} = \{f_v | v \in \mathbb{Z}_N^n\}$ with $f_x(v) = 1$ iff $\langle x, v \rangle = 0$ mod N, where $\langle x, v \rangle$ denotes the inner product $\sum_{i=1}^n x_i \cdot v_i \mod N$ of vectors x and v. We follow the notation of [21].

2.1   Symmetric-Key Predicate-Only Encryption

Let Σ denote a finite set of plaintexts, and let $\mathcal{F}$ denote a finite set of predicates $f : \Sigma \to \{0, 1\}$. We say that $x \in \Sigma$ satisfies a predicate f if $f(x) = 1$.

Definition 1 (Symmetric-Key Predicate-Only Encryption Scheme). *A symmetric-key predicate-only encryption scheme for the class of predicates $\mathcal{F}$ over the set of attributes Σ consists of the following probabilistic polynomial time (PPT) algorithms.*

Setup(1^λ): *Takes as input a security parameter 1^λ and outputs a secret key SK.*
Encrypt(SK, x): *Takes as input a secret key SK and a plaintext $x \in \Sigma$ and outputs a ciphertext CT.*
GenToken(SK, f): *Takes as input a secret key SK and a description of a predicate $f \in \mathcal{F}$ and outputs a token TK_f.*
Query(TK_f, CT): *Takes as input a token TK_f for a predicate f and a ciphertext CT. It outputs either 0 or 1, indicating the value of the predicate f evaluated on the underlying plaintext.*

Correctness. For correctness, we require the following condition. For all λ, all $x \in \Sigma$, and all $f \in \mathcal{F}$, letting $SK \leftarrow Setup(1^\lambda)$, $TK_f \leftarrow GenToken(SK, f)$, and $CT \leftarrow Encrypt(SK, x)$,

- *If $f(x) = 1$, then $Query(TK_f, CT) = 1$.*
- *If $f(x) = 0$, then $\Pr[Query(TK_f, CT) = 0] > 1 - \epsilon(\lambda)$ where ϵ is a negligible function.*

2.2   Security Definitions

We now give formal definitions of security for a symmetric-key predicate-only encryption scheme. We first define *full security*, which, roughly speaking, says that given a set of tokens for predicates $f_1, \ldots, f_k$ and a set of encryptions of plaintexts $x_1, \ldots, x_\ell$, an adversary $\mathcal{A}$ gains no information about any of the predicates $f_1, \ldots, f_k$ or the plaintexts $x_1, \ldots, x_\ell$ (other than the value of each of the predicates evaluated on each of the plaintexts).

However, the full security notion turns out to be difficult to work with in our proofs of security. Therefore, we introduce a second security notion called *single challenge security*, which resembles the security notions used in previous work such as [11,21]. As we show later, full security implies single challenge security, and, for the specific case of inner product predicates, single challenge security implies full security in the sense that, given a single challenge secure scheme for inner product predicates over $\Sigma = \mathbb{Z}_N^{2n}$, we can construct a fully secure scheme for inner product predicates over $\Sigma = \mathbb{Z}_N^n$. Therefore, for our construction it suffices to consider the single challenge security definition. To prove the security of our construction, we will use the *selective* relaxation of single challenge security. The notion of selective security was first introduced by [13] and has been used widely in the literature [13,14,5,11,12,27].

Full Security. We define full security of a symmetric-key predicate-only encryption scheme using the following game between an adversary $\mathcal{A}$ and a challenger.

Setup: The challenger runs $Setup(1^\lambda)$ and keeps SK to itself. The challenger picks a random bit b.

Queries: $\mathcal{A}$ adaptively issues queries, where each query is of one of two types:
  - Ciphertext query. On the jth ciphertext query, $\mathcal{A}$ outputs a bit $t = 0$ (indicating a ciphertext query) and two plaintexts $x_{j,0}, x_{j,1} \in \Sigma$. The challenger responds with $Encrypt(SK, x_{j,\mathsf{b}})$.
  - Token query. On the ith token query, $\mathcal{A}$ outputs a bit $t = 1$ (indicating a token query) and descriptions of two predicates $f_{i,0}, f_{i,1} \in \mathcal{F}$. The challenger responds with $GenToken(SK, f_{i,\mathsf{b}})$.

$\mathcal{A}$'s queries are subject to the restriction that, for all ciphertext queries $(x_{j,0}, x_{j,1})$ and all predicate queries $(f_{i,0}, f_{i,1})$, $f_{i,0}(x_{j,0}) = f_{i,1}(x_{j,1})$.

Guess: $\mathcal{A}$ outputs a guess b' of b.

The advantage of $\mathcal{A}$ is defined as $\mathrm{Adv}_{\mathcal{A}} = \left| \Pr[\mathsf{b}' = \mathsf{b}] - \frac{1}{2} \right|$.

Definition 2 (Full Security). *A symmetric-key predicate-only encryption scheme is* fully secure *if, for all PPT adversaries $\mathcal{A}$, the advantage of $\mathcal{A}$ in winning the above game is negligible in λ.*

Single Challenge Security. In order to prove the security of our construction, we will need to introduce a second security definition, which we refer to as *single challenge security*. Whereas in the full security game, each of the adversary's queries is considered part of the challenge, in the single challenge security game, the challenge consists of only one pair of plaintexts or predicates. The single challenge security game resembles security games used previously in the IBE and predicate encryption literature. The game proceeds as follows.

Setup: The challenger runs $Setup(1^\lambda)$ and keeps SK to itself.

Query Phase 1: $\mathcal{A}$ adaptively issues queries, where each query is of one of two types:

- Ciphertext query. On the jth ciphertext query, $\mathcal{A}$ outputs a bit $t = 0$ (indicating a ciphertext query) and a plaintext x_j. The challenger responds with $Encrypt(SK, x_j)$.
- Token query. On the jth token query, $\mathcal{A}$ outputs a bit $t = 1$ (indicating a token query) and a description of a predicate f_j. The challenger responds with $GenToken(SK, f_j)$.

Challenge: $\mathcal{A}$ outputs a request for one of the following:

- Ciphertext challenge. $\mathcal{A}$ outputs a bit $t = 0$ (indicating a ciphertext challenge) and two plaintexts x_0^* and x_1^* such that, for all previous token queries f_j, $f_j(x_0^*) = f_j(x_1^*)$. The challenger picks a random bit b and responds with $Encrypt(SK, x_\mathsf{b}^*)$.
- Token challenge. $\mathcal{A}$ outputs a bit $t = 1$ (indicating a token challenge) and descriptions of two predicates f_0^* and f_1^* such that, for all previous ciphertext queries x_j, $f_0^*(x_j) = f_1^*(x_j)$. The challenger picks a random bit b and responds with $GenToken(SK, f_\mathsf{b}^*)$.

Query Phase 2: $\mathcal{A}$ adaptively issues additional queries as in Query Phase 1, subject to the same restriction with respect to the challenge as above.

Guess: $\mathcal{A}$ outputs a guess b' of b.

The advantage of $\mathcal{A}$ is defined as $\mathrm{Adv}_\mathcal{A} = \left| \Pr[\mathsf{b}' = \mathsf{b}] - \frac{1}{2} \right|$.

Definition 3 (Single Challenge Security). *A symmetric-key predicate-only encryption scheme is* single challenge secure *if, for all PPT adversaries $\mathcal{A}$, the advantage of $\mathcal{A}$ in winning the above game is negligible in λ.*

Selective Single Challenge Security. We will need to use the *selective* variant of single challenge security, defined below. The notion of selective security was first introduced by [13] and has been used previously by [13,14,5,11,12,27].

Definition 4 (Selective Single Challenge Security). *In the selective single challenge security game, the adversary $\mathcal{A}$ outputs the challenge strings at the start of the game during an **Init** phase (instead of during a **Challenge** phase). The rest of the game proceeds in the same way as in the single challenge security game. We say that a symmetric-key predicate-only encryption scheme is* selective single challenge secure *if, for all PPT adversaries $\mathcal{A}$, the advantage of $\mathcal{A}$ in winning the selective single challenge game is negligible in λ.*

For our proofs of security, it will be useful to define separate notions of plaintext privacy and predicate privacy, which correspond to a ciphertext challenge and a token challenge, respectively, in the selective single challenge security game.

Definition 5 (Plaintext Privacy). *A symmetric-key predicate-only encryption scheme has* selective single challenge plaintext privacy *(plaintext privacy, for short) if, for all PPT adversaries $\mathcal{A}$, the advantage of $\mathcal{A}$ in winning the selective single challenge game for a ciphertext challenge is negligible in λ.*

Definition 6 (Predicate Privacy). *A symmetric-key predicate-only encryption scheme has* selective single challenge predicate privacy *(predicate privacy, for short) if, for all PPT adversaries $\mathcal{A}$, the advantage of $\mathcal{A}$ in winning the selective single challenge game for a token challenge is negligible in λ.*

We note that plaintext privacy and predicate privacy, together, are equivalent to selective single challenge security.

Relationship Between Single Challenge Security and Full Security. It is useful to consider the relationship between the security definitions introduced above. The full security notion implies single challenge security. For the specific case of inner product query predicates, a single challenge secure scheme for vectors of length $2n$ can be used to construct a fully secure scheme for vectors of length n. Therefore, we consider single challenge security to be a sufficiently strong notion of security for our construction.

These relationships are stated formally in the following theorems.

Theorem 1. *If a symmetric-key predicate-only encryption scheme is fully secure, then it is single challenge secure.*

Proof. Suppose an adversary $\mathcal{A}$ wins the single challenge security game with advantage ϵ. We can define an adversary $\mathcal{B}$ that wins the full security game with advantage ϵ as follows. When $\mathcal{A}$ makes a ciphertext query $\boldsymbol{x}$, $\mathcal{B}$ in turn makes the ciphertext query $(\boldsymbol{x}, \boldsymbol{x})$ to $\mathcal{B}$'s challenger and responds to $\mathcal{A}$ with the ciphertext it receives. Similarly, when $\mathcal{A}$ makes a token query $\boldsymbol{v}$, $\mathcal{B}$ in turn makes the token query $(\boldsymbol{v}, \boldsymbol{v})$ to $\mathcal{B}$'s challenger and responds to $\mathcal{A}$ with the token it receives. When $\mathcal{A}$ issues its challenge request, $\mathcal{B}$ outputs the challenge request as a query to its challenger and responds to $\mathcal{A}$ with the answer it receives. $\mathcal{B}$ outputs the same guess b' as $\mathcal{A}$ does. It is clear that all of $\mathcal{B}$'s responses to $\mathcal{A}$ are properly constructed, and $\mathcal{B}$ wins the full security game with the same advantage ϵ with which $\mathcal{A}$ wins the single challenge security game.

Theorem 2. *Let SCHEME_{2n} denote a single challenge secure symmetric-key predicate-only encryption scheme for inner product queries, where plaintext and predicate vectors have length $2n$. Then SCHEME_{2n} can be used to construct a fully secure symmetric-key predicate-only encryption scheme SCHEME_n for inner product queries, where plaintext and predicate vectors have length n.*

The proof of this theorem is deferred to Appendix A.

3   Background and Complexity Assumptions

Our symmetric-key predicate encryption scheme uses bilinear groups of composite order, first introduced by [10]. While the public-key predicate encryption scheme of [21] uses bilinear groups whose order is the product of three distinct primes, we use bilinear groups whose order is the product of four distinct primes.

We briefly review some facts about bilinear groups and then state the assumptions we use to prove security of our construction.

3.1   Bilinear Groups of Composite Order

Let $\mathcal{G}$ denote a group generator algorithm that takes as input a security parameter 1^λ and outputs a tuple $(p, q, r, s, \mathbb{G}, \mathbb{G}_T, e)$ where p, q, r, s are distinct primes, $\mathbb{G}$ and $\mathbb{G}_T$ are two cyclic groups of order $N = pqrs$, and $e : \mathbb{G} \times \mathbb{G} \to \mathbb{G}_T$ satisfies the following properties:

- (Bilinear) $\forall u, v \in \mathbb{G}, \forall a, b \in \mathbb{Z}, e(u^a, v^b) = e(u, v)^{ab}$.
- (Non-degenerate) $\exists g \in \mathbb{G}$ such that $e(g, g)$ has order N in $\mathbb{G}_T$.

We assume that group operations in $\mathbb{G}$ and $\mathbb{G}_T$ as well as the bilinear map e can be computed in time polynomial in λ.

We use the notation $\mathbb{G}_p, \mathbb{G}_q, \mathbb{G}_r, \mathbb{G}_s$ to denote the subgroups of $\mathbb{G}$ having order p, q, r, s, respectively.

We will use the following facts about bilinear groups of composite order. Although these facts are stated in terms of $\mathbb{G}_p$ and $\mathbb{G}_q$, similar facts hold in general for distinct subgroups of a composite order bilinear group.

- Let $a_p \in \mathbb{G}_p, b_q \in \mathbb{G}_q$ denote two elements from distinct subgroups. Then $e(a_p, b_q) = 1$.
- Let $\mathbb{G}_{pq} = \mathbb{G}_p \times \mathbb{G}_q$, $a, b \in \mathbb{G}_{pq}$. a and b can be rewritten (uniquely) as $a = a_p a_q$, $b = b_p b_q$, where $a_p, b_q \in \mathbb{G}_p$, and $a_q, b_q \in \mathbb{G}_q$. Then $e(a, b) = e(a_p, b_p)e(a_q, b_q)$.

3.2   Our Assumptions

The security of our symmetric-key predicate-only encryption scheme relies on three assumptions. All of these assumptions have been introduced previously but in groups whose order is the product of at most three distinct primes. Specifically, Assumption 1 involves 3 subgroups, C3DH involves 2 subgroups, and DL involves 1 subgroup. We assume that these assumptions hold when the relevant subgroups are contained in a larger group whose order is the product of four distinct primes. Note that the naming of subgroups is not significant in our assumptions; that is, the assumptions are the same if the subgroups are renamed.

Assumption 1. We use Assumption 1 of KSW [21], which was used for bilinear groups whose order is the product of three distinct primes. We restate the assumption in the context of a bilinear group whose order is the product of four distinct primes.

Let $\mathcal{G}$ be a group generator algorithm as above. Run $\mathcal{G}(1^\lambda)$ to obtain $(p, q, r, s, \mathbb{G}, \mathbb{G}_T, e)$. Let $N = pqrs$ and let g_p, g_q, g_r, g_s be random generators of $\mathbb{G}_p, \mathbb{G}_q, \mathbb{G}_r, \mathbb{G}_s$, respectively. Choose random $Q_1, Q_2, Q_3 \in \mathbb{G}_q$, random $R_1, R_2, R_3 \in \mathbb{G}_r$, random $a, b, c \in \mathbb{Z}_p$, and a random bit b. If b = 0, let $T = g_p^{b^2 c} R_3$; if b = 1, let $T = g_p^{b^2 c} Q_3 R_3$. Give the adversary $\mathcal{A}$ the description of the bilinear group, $(N, \mathbb{G}, \mathbb{G}_T, e)$, along with the following values:

$$\left(g_p, \quad g_r, \quad g_s, \quad g_q R_1, \quad g_p^b, \quad g_p^{b^2}, \quad g_p^a g_q, \quad g_p^{ab} Q_1, \quad g_p^c, \quad g_p^{bc} Q_2 R_2, \quad T \right)$$

The adversary $\mathcal{A}$ outputs a guess b' of b. The advantage of $\mathcal{A}$ is defined as $\text{Adv}_{\mathcal{A}} = \left| \Pr[\text{b}' = \text{b}] - \frac{1}{2} \right|$.

Definition 7. *We say that $\mathcal{G}$ satisfies Assumption 1 if, for all PPT algorithms $\mathcal{A}$, the advantage of $\mathcal{A}$ in winning the above game is negligible in the security parameter λ.*

We note that Assumption 1 implies the hardness of finding a non-trivial factor of N.

Generalized 3-Party Diffie-Hellman Assumption (C3DH). We use the composite 3-party Diffie-Hellman assumption first introduced by [11]. We restate the assumption in the context of a bilinear group whose order is the product of four distinct primes.

Let $\mathcal{G}$ be a group generator algorithm as above. Run $\mathcal{G}(1^{\lambda})$ to obtain $(p, q, r, s, \mathbb{G}, \mathbb{G}_T, e)$. Let $N = pqrs$ and let g_p, g_q, g_r, g_s be random generators of $\mathbb{G}_p, \mathbb{G}_q, \mathbb{G}_r, \mathbb{G}_s$, respectively. Choose random $R_1, R_2, R_3 \in \mathbb{G}_r$, random $a, b, c \in \mathbb{Z}_N$, and a random bit b. If b $= 0$, let $T = g_p^c \cdot R_3$; if b $= 1$, let T be a random element in $\mathbb{G}_{pr} = \mathbb{G}_p \times \mathbb{G}_r$. Give the adversary $\mathcal{A}$ the description of the bilinear group, $(N, \mathbb{G}, \mathbb{G}_T, e)$, along with the following values:

$$\left(g_p, \quad g_q, \quad g_r, \quad g_s, \quad g_p^a, \quad g_p^b, \quad g_p^{ab} \cdot R_1, \quad g_p^{abc} \cdot R_2, \quad T \right)$$

The adversary $\mathcal{A}$ outputs a guess b' of b. The advantage of $\mathcal{A}$ is defined as $\text{Adv}_{\mathcal{A}} = \left| \Pr[\text{b}' = \text{b}] - \frac{1}{2} \right|$.

Definition 8. *We say that $\mathcal{G}$ satisfies the C3DH assumption if for all PPT algorithms $\mathcal{A}$, the advantage of $\mathcal{A}$ in winning the above game is negligible in the security parameter λ.*

We note that the C3DH assumption implies the hardness of finding a non-trivial factor N.

Decisional Linear assumption (DLinear). We use the Decisional Linear assumption introduced by [6]. We restate the assumption in the context of a bilinear group whose order is the product of four distinct primes.

Let $\mathcal{G}$ be a group generator algorithm as above. Run $\mathcal{G}(1^{\lambda})$ to obtain $(p, q, r, s, \mathbb{G}, \mathbb{G}_T, e)$. Let $N = pqrs$ and let g_p, g_q, g_r, g_s be random generators of $\mathbb{G}_p, \mathbb{G}_q, \mathbb{G}_r, \mathbb{G}_s$, respectively. Choose random $z_1, z_2, z_3, z_4 \in \mathbb{Z}_p$ and a random bit b. If b $= 0$, let $Z = g_p^{z_3 + z_4}$; if b $= 1$, let Z be a random element in $\mathbb{G}_p$. Give the adversary $\mathcal{A}$ the description of the bilinear group, $(N, \mathbb{G}, \mathbb{G}_T, e)$, along with the following values:

$$\left(g_p, \quad g_q, \quad g_r, \quad g_s, \quad g_p^{z_1}, \quad g_p^{z_2}, \quad g_p^{z_1 z_3}, \quad g_p^{z_2 z_4}, \quad Z \right)$$

The adversary $\mathcal{A}$ outputs a guess b' of b. The advantage of $\mathcal{A}$ is defined as $\text{Adv}_{\mathcal{A}} = \left| \Pr[\text{b}' = \text{b}] - \frac{1}{2} \right|$.

Definition 9. *We say that $\mathcal{G}$ satisfies the DLinear assumption if for all PPT algorithms $\mathcal{A}$, the advantage of $\mathcal{A}$ in winning the above game is negligible in the security parameter n.*

4   Construction

Our goal is to construct a symmetric-key predicate encryption scheme supporting inner product queries that has both plaintext privacy and predicate privacy. The KSW construction [21] is a public-key predicate encryption scheme supporting inner product queries that has plaintext privacy. A natural first attempt might be to convert the KSW scheme into a symmetric-key scheme simply by withholding the public key. Such a scheme would immediately inherit plaintext privacy from the KSW construction. However, it is difficult to prove the predicate privacy of such a scheme. Our primary challenge is to achieve predicate privacy.

To achieve predicate privacy, we use the observation that, for inner product queries, ciphertexts and tokens play symmetric roles in the scheme and the security definitions. In particular, a token and a ciphertext each encode a vector in $\mathbb{Z}_N^n$, and the inner product $\langle x, v \rangle$ is commutative. Furthermore, for inner products, ciphertexts and tokens have symmetric roles in the security definitions. One way to interpret this observation is to view a ciphertext as an encryption of a plaintext vector and a token as an encryption of a predicate vector.

Based on this observation, our general approach is to start from a construction that resembles the KSW construction, so that we can prove plaintext privacy in a relatively straightforward manner. We then show through a series of modifications to our construction that it is indistinguishable from one in which ciphertexts and tokens are formed symmetrically. Using this symmetry, we can leverage the plaintext privacy proven for our main construction to achieve predicate privacy as well. Taken all together, the "native" formation of our system gives us plaintext privacy by a KSW type of approach, and the indistinguishability of our construction from one in which ciphertexts and tokens are symmetrically formed implies that our construction also has predicate privacy.

4.1   A Symmetric-Key Predicate Encryption Scheme

Our main construction is a symmetric-key predicate-only encryption scheme supporting inner product queries. We take the class of plaintexts to be $\Sigma = \mathbb{Z}_N^n$ and the class of predicates to be $\mathcal{F} = \{f_v | v \in \mathbb{Z}_N^n\}$ with $f_x(v) = 1$ iff $\langle x, v \rangle = 0$ mod N.

We now describe our construction in detail.

Setup(1^λ): The setup algorithm runs $\mathcal{G}(1^\lambda)$ to obtain $(p, q, r, s, \mathbb{G}, \mathbb{G}_T, e)$ with $\mathbb{G} = \mathbb{G}_p \times \mathbb{G}_q \times \mathbb{G}_r \times \mathbb{G}_s$. Next it picks generators g_p, g_q, g_r, g_s of $\mathbb{G}_p, \mathbb{G}_q, \mathbb{G}_r, \mathbb{G}_s$, respectively. It chooses $h_{1,i}, h_{2,i}, u_{1,i}, u_{2,i} \in \mathbb{G}_p$ uniformly at random for $i = 1$ to n. The secret key is

$$SK = \left(g_p, \ g_q, \ g_r, \ g_s, \ \{h_{1,i}, h_{2,i}, u_{1,i}, u_{2,i}\}_{i=1}^n \right).$$

Encrypt(SK, x): Let $x = (x_1, \ldots, x_n) \in \mathbb{Z}_N^n$. The encryption algorithm chooses random $y, z, \alpha, \beta \in \mathbb{Z}_N$, random $S, S_0 \in \mathbb{G}_s$, and random $R_{1,i}, R_{2,i} \in \mathbb{G}_r$ for $i = 1$ to n. It outputs the ciphertext

$$CT = \begin{pmatrix} C = S \cdot g_p^y, \quad C_0 = S_0 \cdot g_p^z, \\ \{C_{1,i} = h_{1,i}^y \cdot u_{1,i}^z \cdot g_q^{\alpha x_i} \cdot R_{1,i}, \quad C_{2,i} = h_{2,i}^y \cdot u_{2,i}^z \cdot g_q^{\beta x_i} \cdot R_{2,i}\}_{i=1}^n \end{pmatrix}.$$

GenToken(SK, v): Let $v = (v_1, \ldots, v_n) \in \mathbb{Z}_N^n$. The token generation algorithm chooses random $f_1, f_2 \in \mathbb{Z}_N$, random $r_{1,i}, r_{2,i} \in \mathbb{Z}_N$ for $i = 1$ to n, random $R, R_0 \in \mathbb{G}_r$, and random $S_{1,i}, S_{2,i} \in \mathbb{G}_s$ for $i = 1$ to n. It outputs the token

$$
TK_v = \begin{pmatrix} K = R \cdot \prod_{i=1}^n h_{1,i}^{-r_{1,i}} \cdot h_{2,i}^{-r_{2,i}}, & K_0 = R_0 \cdot \prod_{i=1}^n u_{1,i}^{-r_{1,i}} \cdot u_{2,i}^{-r_{2,i}}, \\ \{K_{1,i} = g_p^{r_{1,i}} \cdot g_q^{f_1 v_i} \cdot S_{1,i}, & K_{2,i} = g_p^{r_{2,i}} \cdot g_q^{f_2 v_i} \cdot S_{2,i}\}_{i=1}^n \end{pmatrix}.
$$

Query(TK_v, CT) : Let $CT = (C, C_0, \{C_{1,i}, C_{2,i}\}_{i=1}^n)$ and $TK_v = (K, K_0, \{K_{1,i}, K_{2,i}\}_{i=1}^n)$ as above. The query algorithm outputs 1 iff

$$
e(C, K) \cdot e(C_0, K_0) \cdot \prod_{i=1}^n e(C_{1,i}, K_{1,i}) \cdot e(C_{2,i}, K_{2,i}) \overset{?}{=} 1.
$$

Correctness. Let CT and TK_v be as above. Then

$$
e(C, K) \cdot e(C_0, K_0) \cdot \prod_{i=1}^n e(C_{1,i}, K_{1,i}) \cdot e(C_{2,i}, K_{2,i})
$$

$$
= e(S \cdot g_p^y, \; R \cdot \prod_{i=1}^n h_{1,i}^{-r_{1,i}} \cdot h_{2,i}^{-r_{2,i}}) \cdot e(S_0 \cdot g_p^z, \; R_0 \cdot \prod_{i=1}^n u_{1,i}^{-r_{1,i}} \cdot u_{2,i}^{-r_{2,i}})
$$

$$
\cdot \prod_{i=1}^n e(h_{1,i}^y \cdot u_{1,i}^z \cdot g_q^{\alpha x_i} \cdot R_{1,i}, \; g_p^{r_{1,i}} \cdot g_q^{f_1 v_i} \cdot S_{1,i})
$$

$$
\cdot e(h_{2,i}^y \cdot u_{2,i}^z \cdot g_q^{\beta x_i} \cdot R_{2,i}, \; g_p^{r_{2,i}} \cdot g_q^{f_2 v_i} \cdot S_{2,i})
$$

$$
= e(g_p^y, \; \prod_{i=1}^n h_{1,i}^{-r_{1,i}} \cdot h_{2,i}^{-r_{2,i}}) \cdot e(g_p^z, \; \prod_{i=1}^n u_{1,i}^{-r_{1,i}} \cdot u_{2,i}^{-r_{2,i}})
$$

$$
\cdot \prod_{i=1}^n e(h_{1,i}^y \cdot u_{1,i}^z \cdot g_q^{\alpha x_i}, \; g_p^{r_{1,i}} \cdot g_q^{f_1 v_i}) \cdot e(h_{2,i}^y \cdot u_{2,i}^z \cdot g_q^{\beta x_i}, \; g_p^{r_{2,i}} \cdot g_q^{f_2 v_i})
$$

$$
= \prod_{i=1}^n e(g_q, g_q)^{(\alpha f_1 + \beta f_2) x_i v_i} = e(g_q, g_q)^{(\alpha f_1 + \beta f_2 \mod q)\langle x, v \rangle}
$$

If $\langle x, v \rangle = 0 \mod N$, then the above expression evaluates to 1. If $\langle x, v \rangle \neq 0 \mod N$, then there are two cases. If $\langle x, v \rangle = 0 \mod q$, then the above expression evaluates to 1; however, this case would reveal a non-trivial factor of N and, therefore, this case occurs with negligible probability. If $\langle x, v \rangle \neq 0 \mod q$, then with all but negligible probability $\alpha f_1 + \beta f_2 \neq 0 \mod q$ and the above expression does not evaluate to 1.

4.2   Discussion

To understand our construction, it is useful to examine the role of each of the subgroups $\mathbb{G}_p, \mathbb{G}_q, \mathbb{G}_r, \mathbb{G}_s$.

The $\mathbb{G}_q$ subgroup is used to encode the plaintext vector x in the $C_{1,i}$ and $C_{2,i}$ terms of the ciphertext and the predicate vector v in the $K_{1,i}$ and $K_{2,i}$ terms of the token. When a token for v is applied to an encryption of x, the computation of the inner product $\langle x, v \rangle$ is evaluated in the exponent of the $\mathbb{G}_q$ subgroup.

The $\mathbb{G}_p$ subgroup is used to prevent an adversary from manipulating components of either a ciphertext or a token and then evaluating a query on the improperly formed inputs. The $\mathbb{G}_p$ subgroup encodes an equation which will evaluate to 0 in the exponent if the inputs to the query algorithm are properly formed.

The $\mathbb{G}_r$ subgroup is used for to hide factors from other subgroups and ensure plaintext privacy. In an analogous manner, the $\mathbb{G}_s$ subgroup is used to ensure predicate privacy. Also, the additional subgroup $\mathbb{G}_s$ allows us to construct our scheme in a slightly different manner from KSW. For example, the $\mathbb{G}_s$ subgroup allows us to eliminate the factor Q from the $\mathbb{G}_q$ subgroup in the K term of the token.

As discussed earlier, in our proofs of security we will need to show that our main construction is computationally indistinguishable from a scheme in which ciphertexts and tokens are formed symmetrically. In the KSW construction, all terms in the ciphertext have the same exponent y in the $\mathbb{G}_p$ subgroup. In our construction, we introduce an additional degree of randomness using the exponent z. Terms in the ciphertext now contain two degrees of randomness, y and z, in the $\mathbb{G}_p$ subgroup. This change is necessary to ensure symmetry of the ciphertext and the token in the $\mathbb{G}_p$ subgroup.

To see why this is the case, recall that Decisional Diffie-Hellman is easy in bilinear groups. That is, for a random vector $g_p^{\alpha_1}, g_p^{\alpha_2}, \ldots, g_p^{\alpha_k}$, it is easy to decide whether the exponents $(\alpha_1, \alpha_2, \ldots, \alpha_k)$ are picked independently at random or picked from a prescribed one-dimensional subspace. On the other hand, an informal interpretation of the Decisional Linear assumption tells us that it is computationally hard to decide whether the exponents $(\alpha_1, \alpha_2, \ldots, \alpha_k)$ are picked independently at random or picked randomly from a prescribed 2-dimensional subspace. The reason for introducing the extra randomness z in the ciphertext is to ensure that the exponents in the $\mathbb{G}_p$ subgroup are picked from a 2-dimensional subspace instead of a 1-dimensional subspace.

Similarly to [11,12,21], our construction consists of two parallel sub-systems. Note that $C_{1,i}$ and $C_{2,i}$ (similarly, $K_{1,i}$ and $K_{2,i}$) play identical roles. Our proof of security will rely on having these two parallel sub-systems.

For comparison, we provide a review of the KSW construction in Appendix B.

4.3 Proof Overview

Our main security statement is the following theorem.

Theorem 3. *Under the generalized Assumption 1 of the KSW construction, the generalized C3DH assumption, and the Decisional Linear assumption, the symmetric-key predicate-only encryption scheme presented in Section 4.1 is selectively single challenge secure.*

Our proof technique consists of two steps. First, we prove that our construction achieves plaintext privacy. Second, we prove that, for our construction, plaintext privacy implies predicate privacy. Taken together, these results imply the security of our scheme.

Our construction defined above, which we call SCHEMEREAL, does not immediately yield a proof of these two properties. In order to argue these properties, we define two different schemes that are computationally indistinguishable from our original construction. That is, no adversary can tell whether tokens and ciphertexts are generated from our actual system or from one of the two defined for the purposes of the proof.

We first define a system that we call SCHEMEQ, which very closely follows the KSW construction. We reduce the plaintext privacy of SCHEMEQ to the plaintext privacy of the KSW construction.

Next we define a system that we call SCHEMESYM, in which ciphertexts and tokens are formed symmetrically. For this system it is straightforward to argue that plaintext privacy implies predicate privacy.

We observe that since our main construction and the two variants defined are all computationally indistinguishable (from an adversary's view), it is actually possible to define any of them as the "real" construction that we will actually use. We chose the variant described above due to (relative) notational simplicity and slight efficiency advantages. Details of our proof and further discussion are given in the full version of our paper.

5 Conclusions

We examined the idea of protecting the privacy of predicates in predicate encryption systems. While this turns out to be an inherently unachievable in a public-key system, we showed that there exist solutions in the symmetric-key setting. We first provided security definitions for predicate encryption schemes in the symmetric-key setting and then presented a construction supporting inner product queries, which are the most expressive queries supported by currently known schemes.

While semantic security of predicates is inherently impossible in the public-key setting, in the future we might wish to consider relaxations of public-key encryption. For example, is it possible to find interesting systems where predicates are drawn from a high entropy distribution, in a fashion similar to recent work on deterministic encryption [4,2]? Another open direction is to consider "partial public-key encryption," in which a public key might allow a user to generate only a subset of valid ciphertexts. (The rest may be generated from a secret key or other public keys kept hidden from an attacker.) Thus, certain predicates might be indistinguishable given the partial public keys published.

Acknowledgments

We thank Philippe Golle for helpful discussions. The second author thanks Adrian Perrig for his support while part of this research was conducted.

References

1. Abdalla, M., Bellare, M., Catalano, D., Kiltz, E., Kohno, T., Lange, T., Malone-Lee, J., Neven, G., Paillier, P., Shi, H.: Searchable encryption revisited: Consistency properties, relation to anonymous IBE, and extensions. In: Shoup, V. (ed.) CRYPTO 2005. LNCS, vol. 3621, pp. 205–222. Springer, Heidelberg (2005)
2. Bellare, M., Fischlin, M., O'Neill, A., Ristenpart, T.: Deterministic encryption: Definitional equivalences and constructions without random oracles. In: Wagner, D. (ed.) CRYPTO 2008. LNCS, vol. 5157, pp. 360–378. Springer, Heidelberg (2008)
3. Bethencourt, J., Sahai, A., Waters, B.: Ciphertext-policy attribute-based encryption. In: SP 2007: Proceedings of the 2007 IEEE Symposium on Security and Privacy, Washington, DC, USA, pp. 321–334. IEEE Computer Society Press, Los Alamitos (2007)
4. Boldyreva, A., Fehr, S., O'Neill, A.: On notions of security for deterministic encryption, and efficient constructions without random oracles. In: Wagner, D. (ed.) CRYPTO 2008. LNCS, vol. 5157, pp. 335–359. Springer, Heidelberg (2008)
5. Boneh, D., Boyen, X.: Efficient selective-ID secure identity based encryption without random oracles. In: Cachin, C., Camenisch, J.L. (eds.) EUROCRYPT 2004. LNCS, vol. 3027, pp. 223–238. Springer, Heidelberg (2004), http://www.cs.stanford.edu/~xb/eurocrypt04b/
6. Boneh, D., Boyen, X., Shacham, H.: Short group signatures. In: Franklin, M. (ed.) CRYPTO 2004. LNCS, vol. 3152, pp. 41–55. Springer, Heidelberg (2004)
7. Boneh, D., Di Crescenzo, G., Ostrovsky, R., Persiano, G.: Public key encryption with keyword search. In: Cachin, C., Camenisch, J.L. (eds.) EUROCRYPT 2004. LNCS, vol. 3027, pp. 506–522. Springer, Heidelberg (2004)
8. Boneh, D., Franklin, M.: Identity-based encryption from the Weil pairing. In: Kilian, J. (ed.) CRYPTO 2001. LNCS, vol. 2139, pp. 213–229. Springer, Heidelberg (2001)
9. Boneh, D., Gentry, C., Hamburg, M.: Space-efficient identity based encryption without pairings. In: FOCS (2007)
10. Boneh, D., Goh, E.-J., Nissim, K.: Evaluating 2-DNF Formulas on Ciphertexts. In: Kilian, J. (ed.) TCC 2005. LNCS, vol. 3378, pp. 325–342. Springer, Heidelberg (2005)
11. Boneh, D., Waters, B.: A fully collusion resistant broadcast trace and revoke system with public traceability. In: ACM Conference on Computer and Communication Security (CCS) (2006)
12. Boyen, X., Waters, B.: Anonymous hierarchical identity-based encryption (Without random oracles). In: Dwork, C. (ed.) CRYPTO 2006. LNCS, vol. 4117, pp. 290–307. Springer, Heidelberg (2006)
13. Canetti, R., Halevi, S., Katz, J.: A forward-secure public-key encryption scheme. In: Biham, E. (ed.) EUROCRYPT 2003. LNCS, vol. 2656, pp. 255–271. Springer, Heidelberg (2003)
14. Canetti, R., Halevi, S., Katz, J.: Chosen-ciphertext security from identity-based encryption. In: Cachin, C., Camenisch, J.L. (eds.) EUROCRYPT 2004. LNCS, vol. 3027, pp. 207–222. Springer, Heidelberg (2004)
15. Chase, M.: Multi-authority attribute based encryption. In: Vadhan, S.P. (ed.) TCC 2007. LNCS, vol. 4392, pp. 515–534. Springer, Heidelberg (2007)
16. Cocks, C.: An identity based encryption scheme based on quadratic residues. In: Honary, B. (ed.) Cryptography and Coding 2001. LNCS, vol. 2260, pp. 360–363. Springer, Heidelberg (2001)

17. Curtmola, R., Garay, J., Kamara, S., Ostrovsky, R.: Searchable symmetric encryption: improved definitions and efficient constructions. In: CCS 2006: Proceedings of the 13th ACM conference on Computer and communications security (2006)
18. Goldreich, O., Ostrovsky, R.: Software protection and simulation by oblivious rams. JACM (1996)
19. Golle, P., Staddon, J., Waters, B.: Secure conjunctive keyword search over encrypted data. In: Proc. of the 2004 Applied Cryptography and Network Security Conference (2004)
20. Goyal, V., Pandey, O., Sahai, A., Waters, B.: Attribute-based encryption for fine-grained access control of encrypted data. In: CCS 2006: Proceedings of the 13th ACM conference on Computer and communications security, pp. 89–98. ACM Press, New York (2006)
21. Katz, J., Sahai, A., Waters, B.: Predicate encryption supporting disjunctions, polynomial equations, and inner products. In: Smart, N.P. (ed.) EUROCRYPT 2008. LNCS, vol. 4965, pp. 146–162. Springer, Heidelberg (2008)
22. Ostrovsky, R.: Software protection and simulation on oblivious RAMs. PhD thesis, M.I.T (1992); Preliminary version in STOC 1990
23. Ostrovsky, R., Sahai, A., Waters, B.: Attribute-based encryption with non-monotonic access structures. In: CCS 2007: Proceedings of the 14th ACM conference on Computer and communications security (2007)
24. Pirretti, M., Traynor, P., McDaniel, P., Waters, B.: Secure attribute-based systems. In: CCS 2006: Proceedings of the 13th ACM conference on Computer and communications security, New York, NY, USA, pp. 99–112 (2006)
25. Sahai, A., Waters, B.: Fuzzy identity-based encryption. In: Cramer, R. (ed.) EUROCRYPT 2005. LNCS, vol. 3494, pp. 457–473. Springer, Heidelberg (2005)
26. Shamir, A.: Identity-based cryptosystems and signature schemes. In: Blakely, G.R., Chaum, D. (eds.) CRYPTO 1984. LNCS, vol. 196, pp. 47–53. Springer, Heidelberg (1985)
27. Shi, E., Bethencourt, J., Chan, T.-H.H., Song, D., Perrig, A.: Multi-dimensional range query over encrypted data. In: IEEE Symposium on Security and Privacy (May 2007)
28. Shi, E., Waters, B.: Delegating capabilities in predicate encryption systems. In: Aceto, L., Damgård, I., Goldberg, L.A., Halldórsson, M.M., Ingólfsdóttir, A., Walukiewicz, I. (eds.) ICALP 2008, Part II. LNCS, vol. 5126, pp. 560–578. Springer, Heidelberg (2008),
http://sparrow.ece.cmu.edu/~elaine/docs/delegation.pdf
29. Song, D., Wagner, D., Perrig, A.: Practical techniques for searches on encrypted data. In: Proceedings of the 2000 IEEE symposium on Security and Privacy, S&P 2000 (2000)

A   Proof of Theorem 2

Here, we prove that a single challenge secure symmetric-key predicate-only encryption scheme supporting inner product queries for vectors of length $2n$ can be used to construct a fully secure symmetric-key predicate-only encryption scheme supporting inner product queries for vectors length n. Our proof is inspired by the hybrid argument used by [21].

Proof. Let SCHEME$_{2n}$ be a single challenge secure symmetric-key predicate-only encryption scheme supporting inner product queries over $\mathbb{Z}_N^{2n}$. We construct a fully secure symmetric-key predicate-only encryption scheme SCHEME_n supporting inner product queries over $\mathbb{Z}_N^n$.

For any two vectors $\boldsymbol{x} = (x_1, \ldots x_n), \boldsymbol{y} = (y_1, \ldots, y_n) \in \mathbb{Z}_N^n$, define $\boldsymbol{x} \| \boldsymbol{y} = (x_1, \ldots, x_n, y_1, \ldots, y_n)$ to be the vector obtained by concatenating $\boldsymbol{x}$ and $\boldsymbol{y}$.

Informally, SCHEME_n works as follows. To encrypt a vector $\boldsymbol{x} \in \mathbb{Z}_N^n$, encrypt the vector $\boldsymbol{x} \| \boldsymbol{x} \in \mathbb{Z}_N^{2n}$ using SCHEME_{2n}. Similarly, to construct a token for the vector $\boldsymbol{v} \in \mathbb{Z}_N^n$, use SCHEME_{2n} to construct a token for the vector $\boldsymbol{v} \| \boldsymbol{v} \in \mathbb{Z}_N^{2n}$. The algorithms of Scheme_n are defined as follows.

$\text{SCHEME}_n.Setup(1^\lambda)$: Run $\text{SCHEME}_{2n}.Setup(1^\lambda)$. The secret key SK is the same as that generated by SCHEME_{2n}.
$\text{SCHEME}_n.Encrypt(SK, \boldsymbol{x})$: Output $\text{SCHEME}_{2n}.Encrypt(SK, \boldsymbol{x} \| \boldsymbol{x})$.
$\text{SCHEME}_n.GenToken(SK, \boldsymbol{v})$: Output $\text{SCHEME}_{2n}.GenToken(SK, \boldsymbol{v} \| \boldsymbol{v})$.
$\text{SCHEME}_n.Query(TK_{\boldsymbol{v}}, CT)$: Output $\text{SCHEME}_{2n}.Query(TK_{\boldsymbol{v}}, CT)$.

The correctness of SCHEME_n results from the fact that for vectors $\boldsymbol{x}, \boldsymbol{v} \in \mathbb{Z}_N^n$,

$$\langle \boldsymbol{x}, \boldsymbol{v} \rangle = 0 \quad \textit{iff} \quad \langle \boldsymbol{x} \| \boldsymbol{x}, \ \boldsymbol{v} \| \boldsymbol{v} \rangle = 0.$$

We now show that SCHEME_n is fully secure. Recall the full security game defined in Section 2.2. First, the challenger picks a random bit b. Next, the adversary $\mathcal{A}$ adaptively issues queries to the challenger. If a query is a ciphertext query $(\boldsymbol{x}_{j,0}, \boldsymbol{x}_{j,1})$, the challenger responds with an encryption of $\boldsymbol{x}_{j,\text{b}}$. If a query is a token query $(\boldsymbol{v}_{i,0}, \boldsymbol{v}_{i,1})$, the challenger responds with a token for $\boldsymbol{v}_{i,\text{b}}$. $\mathcal{A}$'s queries are subject to the restriction that, for all ciphertext queries $(x_{j,0}, x_{j,1})$ and all predicate queries $(f_{i,0}, f_{i,1})$, $f_{i,0}(x_{j,0}) = f_{i,1}(x_{j,1})$. At the end of the game, $\mathcal{A}$ outputs a guess b' of b and wins if b' = b.

Suppose that the adversary $\mathcal{A}$ makes c ciphertext queries, $(\boldsymbol{x}_{1,0}, \boldsymbol{x}_{1,1})$, $\ldots$, $(\boldsymbol{x}_{c,0}, \boldsymbol{x}_{c,1})$, and t token queries, $(\boldsymbol{v}_{1,0}, \boldsymbol{v}_{1,1}), \ldots, (\boldsymbol{v}_{t,0}, \boldsymbol{v}_{t,1})$.

Our task is to show that $\mathcal{A}$ cannot distinguish between two experiments: one where the challenger constructs ciphertexts for $\boldsymbol{x}_{1,0}, \ldots, \boldsymbol{x}_{c,0}$ and tokens for $\boldsymbol{v}_{1,0}, \ldots, \boldsymbol{v}_{t,0}$ (call this Game 0), and one where the challenger constructs ciphertexts for $\boldsymbol{x}_{1,1}, \ldots, \boldsymbol{x}_{c,1}$ and tokens for $\boldsymbol{v}_{1,1}, \ldots, \boldsymbol{v}_{t,1}$ (call this Game 1). To do this, we construct a series of hybrid games as follows.

Game 0: The challenger calls SCHEME_{2n} and computes ciphertexts for $\boldsymbol{x}_{1,0} \| \boldsymbol{x}_{1,0}, \ \boldsymbol{x}_{2,0} \| \boldsymbol{x}_{2,0}, \ \ldots, \ \boldsymbol{x}_{c,0} \| \boldsymbol{x}_{c,0}$ and tokens for $\boldsymbol{v}_{1,0} \| \boldsymbol{v}_{1,0}, \ \boldsymbol{v}_{2,0} \| \boldsymbol{v}_{2,0}, \ \ldots, \ \boldsymbol{v}_{t,0} \| \boldsymbol{v}_{t,0}$.

Game A: The challenger calls SCHEME_{2n} and computes ciphertexts for $\boldsymbol{x}_{1,0} \| \boldsymbol{0}$, $\boldsymbol{x}_{2,0} \| \boldsymbol{0}, \ \ldots, \ \boldsymbol{x}_{c,0} \| \boldsymbol{0}$ and tokens for $\boldsymbol{v}_{1,0} \| \boldsymbol{v}_{1,0}, \ \boldsymbol{v}_{2,0} \| \boldsymbol{v}_{2,0}, \ \ldots, \ \boldsymbol{v}_{t,0} \| \boldsymbol{v}_{t,0}$.

Game B: The challenger calls SCHEME_{2n} and computes ciphertexts for $\boldsymbol{x}_{1,0} \| \boldsymbol{0}$, $\boldsymbol{x}_{2,0} \| \boldsymbol{0}, \ \ldots, \ \boldsymbol{x}_{c,0} \| \boldsymbol{0}$ and tokens for $\boldsymbol{v}_{1,0} \| \boldsymbol{v}_{1,1}, \ \boldsymbol{v}_{2,0} \| \boldsymbol{v}_{2,1}, \ \ldots, \ \boldsymbol{v}_{t,0} \| \boldsymbol{v}_{t,1}$.

Game M: The challenger picks a random $\alpha \overset{\text{R}}{\leftarrow} \mathbb{Z}_N$, calls SCHEME_{2n} and computes ciphertexts for $\boldsymbol{x}_{1,0} \| \alpha \boldsymbol{x}_{1,1}, \ \boldsymbol{x}_{2,0} \| \alpha \boldsymbol{x}_{2,1}, \ \ldots, \ \boldsymbol{x}_{c,0} \| \alpha \boldsymbol{x}_{c,1}$ and tokens for $\boldsymbol{v}_{1,0} \| \boldsymbol{v}_{1,1}, \ \boldsymbol{v}_{2,0} \| \boldsymbol{v}_{2,1}, \ \ldots, \ \boldsymbol{v}_{t,0} \| \boldsymbol{v}_{t,1}$.

Notice that in the above sequence of hybrid games, the outcomes of the predicates corresponding to the generated tokens on the plaintexts in $\mathbb{Z}_N^{2n}$ encrypted by the challenger remain the same between all pairs of adjacent games, except with negligible probability.

Claim. If SCHEME_{2n} is single challenge secure, then no PPT adversary $\mathcal{A}$ has more than negligible advantage in distinguishing between any pair of adjacent games in the above sequence of games.

Proof. By a hybrid argument.

Similarly, we can construct a sequence of hybrid games connecting Game M and Game 1. Using a hybrid argument, we conclude that no PPT adversary has more than negligible advantage in distinguishing between Game 0 and Game 1.

B KSW Predicate Encryption Scheme

To aid in the understanding of our construction and the proof of security, we review the KSW public key predicate-only encryption scheme for inner product queries [21].

Let $\mathcal{G}'$ denote a group generator algorithm for a bilinear group whose order is the product of three distinct primes.

Setup(1^λ): The setup algorithm runs $\mathcal{G}'(1^\lambda)$ to obtain $(p, q, r, \mathbb{G}, \mathbb{G}_T, e)$ with $\mathbb{G} = \mathbb{G}_p \times \mathbb{G}_q \times \mathbb{G}_r$. Next it picks generators g_p, g_q, g_r from subgroups $\mathbb{G}_p, \mathbb{G}_q, \mathbb{G}_r$, respectively. It then chooses, uniformly at random, $h_{1,i}, h_{2,i} \in \mathbb{G}_p, R_{1,i}, R_{2,i} \in \mathbb{G}_r$ for $i = 1$ to n, and $R_0 \in \mathbb{G}_r$.
The public key consists of:

$$PK = (g_p, \ g_r, \ Q = g_q \cdot R_0, \ \{H_{1,i} = h_{1,i} \cdot R_{1,i}, \ H_{2,i} = h_{2,i} \cdot R_{2,i}\}_{i=1}^n)$$

The secret key is set to:

$$SK = \left(p, q, r, g_q, \{h_{1,i}, h_{2,i}\}_{i=1}^n\right).$$

Encrypt($PK, \boldsymbol{x}$): Let $\boldsymbol{x} = (x_1, \ldots, x_n) \in \mathbb{Z}_N^n$. The encryption algorithm first picks random exponents y, α, β from $\mathbb{Z}_N$, and it chooses random $R_{3,i}, R_{4,i} \in \mathbb{G}_r$ for $i = 1$ to n. It outputs the ciphertext

$$CT = \left(C = g_p^y, \ \{C_{1,i} = H_{1,i}^y \cdot Q^{\alpha x_i} \cdot R_{3,i}, \ C_{2,i} = H_{2,i}^y \cdot Q^{\beta x_i} \cdot R_{4,i}\}_{i=1}^n\right).$$

GenToken($SK, \boldsymbol{v}$): Let $\boldsymbol{v} = (v_1, \ldots, v_n) \in \mathbb{Z}_N^n$. The token generation algorithm chooses random $f_1, f_2, \{r_{1,i}, r_{2,i}\}_{i=1}^n$ from $\mathbb{Z}_N$, random $R_5 \in \mathbb{G}_r$, and random $Q_6 \in \mathbb{G}_q$. It outputs the token

$$TK_{\boldsymbol{v}} = \left(\begin{array}{c} K = R_5 \cdot Q_6 \cdot \prod_{i=1}^n h_{1,i}^{-r_{1,i}} \cdot h_{2,i}^{-r_{2,i}}, \\ \{K_{1,i} = g_p^{r_{1,i}} \cdot g_q^{f_1 v_i}, \ K_{2,i} = g_p^{r_{2,i}} \cdot g_q^{f_2 v_i}\}_{i=1}^n \end{array} \right).$$

Query($TK_{\boldsymbol{v}}, CT$): Let $CT = (C, \{C_{1,i}, C_{2,i}\}_{i=1}^n)$ and $TK_{\boldsymbol{v}} = (K, \{K_{1,i}, K_{2,i}\}_{i=1}^n)$ as above. The query algorithm outputs 1 iff

$$e(C, K) \cdot \prod_{i=1}^n e(C_{1,i}, K_{1,i}) \cdot e(C_{2,i}, K_{2,i}) \stackrel{?}{=} 1.$$

Simultaneous Hardcore Bits and Cryptography against Memory Attacks

Adi Akavia[1,*], Shafi Goldwasser[2,**], and Vinod Vaikuntanathan[3,* * *]

[1] IAS and DIMACS
[2] MIT and Weizmann Insitute
[3] MIT and IBM Research

Abstract. This paper considers two questions in cryptography.

Cryptography Secure Against Memory Attacks. A particularly devastating side-channel attack against cryptosystems, termed the "memory attack", was proposed recently. In this attack, a significant fraction of the bits of a secret key of a cryptographic algorithm can be measured by an adversary if the secret key is ever *stored* in a part of memory which can be accessed even after power has been turned off for a short amount of time. Such an attack has been shown to completely compromise the security of various cryptosystems in use, including the RSA cryptosystem and AES.

We show that the public-key encryption scheme of Regev (STOC 2005), and the identity-based encryption scheme of Gentry, Peikert and Vaikuntanathan (STOC 2008) are remarkably robust against memory attacks where the adversary can measure a large fraction of the bits of the secret-key, or more generally, can compute an arbitrary function of the secret-key of bounded output length. This is done without increasing the size of the secret-key, and without introducing any complication of the natural encryption and decryption routines.

Simultaneous Hardcore Bits. We say that a block of bits of x are *simultaneously hard-core* for a one-way function $f(x)$, if given $f(x)$ they cannot be distinguished from a random string of the same length. Although any candidate one-way function can be shown to hide one hardcore bit and even a logarithmic number of simultaneously hardcore bits, there are few examples of one-way or trapdoor functions for which a linear number of the input bits have been proved simultaneously hardcore; the ones that are known relate the simultaneous security to the difficulty of factoring integers.

We show that for a lattice-based (injective) trapdoor function which is a variant of function proposed earlier by Gentry, Peikert and Vaikuntanathan, an $N - o(N)$ number of input bits are simultaneously hardcore, where N is the total length of the input.

These two results rely on similar proof techniques.

_____

* Supported in part by NSF grant CCF-0514167, by NSF grant CCF-0832797, and by Israel Science Foundation 700/08.
** Supported in part by NSF grants CCF-0514167, CCF-0635297, NSF-0729011, the Israel Science Foundation 700/08 and the Chais Family Fellows Program.
* * * Supported in part by NSF grant CCF-0635297 and Israel Science Foundation 700/08.

O. Reingold (Ed.): TCC 2009, LNCS 5444, pp. 474–495, 2009.

1   Introduction

The contribution of this paper is two-fold.

First, we define a new class of strong side-channel attacks that we call "memory attacks", generalizing the "cold-boot attack" recently introduced by Halderman et al. [22]. We show that the public-key encryption scheme proposed by Regev [39], and the identity-based encryption scheme proposed by Gentry, Peikert, and Vaikuntanathan [16] can provably withstand these side channel attacks under essentially the same intractability assumptions as the original systems[1].

Second, we study how many bits are simultaneously hardcore for the candidate trapdoor one-way function proposed by [16]. This function family has been proven one-way under the assumption that the learning with error problem (LWE) for certain parameter settings is intractable, or alternatively the assumption that approximating the length of the shortest vector in an integer lattice to within a polynomial factor is hard for quantum algorithms [39]. We first show that for the set of parameters considered by [16], the function family has $O(\frac{N}{\log N})$ simultaneously hardcore bits (where N is the length of the input to the function). Next, we introduce a new parameter regime for which we prove that the function family is still trapdoor one-way and has upto $N - o(N)$ simultaneously hardcore bits[2], under the assumption that approximating the length of the shortest vector in an integer lattice to within a *quasi-polynomial* factor in the worst-case is hard for quantum algorithms running in quasi-polynomial time.

The techniques used to solve both problems are closely related. We elaborate on the two results below.

1.1   Security against Memory Attacks

The absolute privacy of the secret-keys associated with cryptographic algorithms has been the corner-stone of modern cryptography. Still, in practice, keys do get compromised at times for a variety of reasons.

A particularly disturbing loss of secrecy is as a result of side-channel attacks. These attacks exploit the fact that every cryptographic algorithm is ultimately implemented on a physical device and such implementations typically enable 'observations' which can be made and measured, such as the amount of power consumption or the time taken by a particular implementation of a cryptographic algorithm. These side-channel observations lead to information leakage about secret-keys which can (and have) lead to complete breaks of systems which have been proved mathematically secure, without violating any of the underlying mathematical principles or assumptions (see, for example, [28, 29, 12, 1, 2]). Traditionally, such attacks have been followed by ad-hoc 'fixes' which make particular implementations invulnerable to particular attacks, only to potentially be broken anew by new examples of side-channel attacks.

In their pioneering paper on *physically observable cryptography* [33], Micali and Reyzin set forth the goal of building a general theory of physical security against a

---

[1] Technically, the assumptions are the same except that they are required to hold for problems of a smaller size, or dimension. See Informal Theorems 1 and 2 for the exact statements.

[2] The statement holds for a particular $o(N)$ function. See Informal Theorem 3.

large class of side channel attacks which one may call *computational* side-channel attacks. These include *any* side channel attack in which leakage of information on secrets occurs as a result of performing a *computation* on secrets. Some well-known examples of such attacks include Kocher's timing attacks [28], power analysis attacks [29], and electromagnetic radiation attacks [1] (see [32] for a glossary of examples.) A basic defining feature of a computational side-channel attack, as put forth by [33] is that *computation and only computation leaks information*. Namely, the portions of memory which are not involved in computation do not leak any information.

Recently, several works [33, 26, 37, 20, 15] have proposed cryptographic algorithms provably robust against computational side-channel attacks, by limiting in various ways the portions of the secret key which are involved in each step of the computation [26, 37, 20, 15].

In this paper, we consider an entirely different family of side-channel attacks that are not included in the computational side-channel attack family, as they violate the basic premise (or axiom, as they refer to it) of Micali-Reyzin [33] that *only computation* leaks information. The new class of attacks, which we call "memory attacks", are inspired by (although not restricted to) the "cold-boot attack" introduced recently by Halderman et al. [22]. The Halderman et al. paper shows how to measure a significant fraction of the bits of secret keys if the keys were *ever stored* in a part of memory which could be accessed by an adversary (e.g. DRAM), even after the power of the machine has been turned off. They show that uncovering half of the bits of the secret key that is stored in the natural way completely compromises the security of cryptosystems, such as the RSA and Rabin cryptosystems.[3]

A New Family of Side Channel Attacks. Generalizing from [22], we define the family of memory attacks to leak a bounded number of bits computed as a result of applying *an arbitrary function* whose output length is bounded by $\alpha(N)$ to the content of the secret-key of the cryptographic algorithm (where N is the size of the the secret-key).[4] Naturally, this family of attacks is inherently parameterized and quantitative in nature. If $\alpha(N) = N$, then the attack could uncover the entire secret key at the outset, and there is no hope for any cryptography. However, it seems that in practice, only a fraction of the secret key is recovered [22]. The question that emerges is how large a fraction of the secret-key can leak without compromising the security of the cryptosystems.

For the public-key case (which is the focus of this paper), we differentiate between two flavors of memory attacks.

The first is *non-adaptive α-memory attacks*. Intuitively, in this case, a function h with output-length $\alpha(N)$ (where N is the length of the secret-key in the system) is first chosen by the adversary, and then the adversary is given $(PK, h(SK))$, where (PK, SK) is a random key-pair produced by the key-generation algorithm. Thus, h is chosen independently of the system parameters and in particular, PK. This definition captures the attack specified in [22] where the bits measured were only a function of the hardware or the storage medium used. In principle, in this case, one could design

---

[3] This follows from the work of Rivest and Shamir, and later Coppersmith [40, 13], and has been demonstrated in practice by [22]: their experiments successfuly recovered RSA and AES keys.

[4] The special case considered in [22] corresponds to a function that outputs a subset of its input bits.

the decryption algorithm to protect against the particular h which was fixed a-priori. However, this would require the design of new software (i.e, the decryption algorithm) for every possible piece of hardware (e.g, a smart-card implementing the decryption algorithm) which is highly impractical. Moreover, it seems that such a solution will involve artificially expanding the secret-key, which one may wish to avoid. We avoid the aforementioned disadvantages by showing an encryption scheme that protects against all leakage functions h (with output of length at most $\alpha(N)$).

The second, stronger, attack is the *adaptive α-memory attacks*. In this case, a key-pair (PK, SK) is first chosen by running the key generation algorithm with security parameter n, and then the adversary on input PK chooses functions h_i adaptively (depending on the PK and the outputs of $h_j(SK)$, for $j < i$) and the adversary receives $h_i(SK)$. The total number of bits output by $h_i(SK)$ for all i, is bounded by $\alpha(N)$.

Since we deal with public-key encryption (PKE) and identity-based encryption (IBE) schemes in this paper, we tailor our definitions to the case of encryption. However, we remark that similar definitions can be made for other cryptographic tasks such as digital signatures, identification protocols, commitment schemes etc. We defer these to the full version of the paper.

New Results on PKE Security. There are two natural directions to take in desiging schemes which are secure against memory attacks. The first is to look for redundant representations of secret-keys which will enable battling memory attacks. The works of [26, 25, 10] can be construed in this light. Naturally, this entails expansion of the storage required for secret keys and data. The second approach would be to examine natural and existing cryptosystems, and see how vulnerable they are to memory attacks. We take the second approach here.

Following Regev [39], we define the learning with error problem (LWE) in dimension n, to be the task of learning a vector $\mathbf{s} \in \mathbb{Z}_q^n$ (where q is a prime), given m pairs of the form $(\mathbf{a}_i, \langle \mathbf{a}_i, \mathbf{s} \rangle + x_i \bmod q)$ where $\mathbf{a}_i \in \mathbb{Z}_q^n$ are chosen uniformly and independently and the x_i are chosen from some "error distribution" $\overline{\Psi}_\beta$ (Throughout, we one may think of x_i's as being small in magnitude. See section 2 for precise definition of this error distribution.). We denote the above parameterization by $\mathsf{LWE}_{n,m,q,\beta}$. The hardness of the LWE problem is chiefly parametrized by the dimension n: we say that $\mathsf{LWE}_{n,m,q,\beta}$ is t-hard if no probabilistic algorithm running in time t can solve it.

We prove the following two main theorems.

Informal Theorem 1. *Let the parameters m, q and β be polynomial in the security parameter n. There exist public key encryption schemes with secret-key length $N = n \log q = O(n \log n)$ that are:*

1. *semantically secure against a non-adaptive $(N - k)$-memory attack, assuming the $\mathsf{poly}(n)$-hardness of $\mathsf{LWE}_{O(k/\log n),m,q,\beta}$, for any $k > 0$. The encryption scheme corresponds to a slight variant of the public key encryption scheme of [39].*
2. *semantically secure against an adaptive $O(N/\mathsf{polylog}(N))$-memory attack, assuming the $\mathsf{poly}(n)$-hardness of $\mathsf{LWE}_{k,m,q,\beta}$ for $k = O(n)$. The encryption scheme is the public-key scheme proposed by [39].*

Informal Theorem 2. *Let the parameters m, q and β be polynomial in the security parameter n. The GPV identity-based encryption scheme [16] with secret-key length $N = n \log q = O(n \log n)$ is:*

1. *semantically secure against a* non-adaptive $(N - k)$*-memory attack, assuming the* poly(n)*-hardness of* $\mathsf{LWE}_{O(k/\log n), m, q, \beta}$ *for any $k > 0$.*
2. *semantically secure against an* adaptive $O(N/\text{polylog}(N))$*-memory attack, assuming the* poly(n)*-hardness of* $\mathsf{LWE}_{k, m, q, \beta}$ *for $k = O(n)$.*

The parameter settings for these theorems require some elaboration. First, the theorem for the non-adaptive case is fully parametrized. That is, for any k, we prove security in the presence of leakage of $N - k$ bits of information about the secret-key, under a corresponding hardness assumption. The more the leakage we would like to tolerate, the stronger the hardness assumption. In particular, setting the parameter k to be $O(N)$, we prove security against leakage of a constant fraction of the secret-key bits assuming the hardness of LWE for $O(N/\log n) = O(n)$ dimensions. If we set $k = N^\epsilon$ (for some $\epsilon > 0$) we prove security against a leakage of *all but N^ϵ bits of the secret-key*, assuming the hardness of LWE for a polynomially smaller dimension $O(N^\epsilon/\log n) = O((n \log n)^\epsilon / \log n)$.

For the adaptive case, we prove security against a leakage of $O(N/\text{polylog}(N))$ bits, assuming the hardness of LWE for $O(n)$ dimensions, where n is the security parameter of the encryption scheme.

Due to lack of space, we describe only the public-key encryption result in this paper, and defer the identity-based encryption result to the full version.

Idea of the Proof. The main idea of the proof is *dimension reduction*. To illustrate the idea, let us outline the proof of the non-adaptive case in which this idea is central.

The hardness of the encryption schemes under a non-adaptive memory attack relies on the hardness of computing $\mathbf{s}$ given $m = \text{poly}(n)$ LWE samples $(\mathbf{a}_i, \langle \mathbf{a}_i, \mathbf{s} \rangle + x_i \bmod q)$ and the leakage $h(\mathbf{s})$. Let us represent these m samples compactly as $(\mathbf{A}, \mathbf{As} + \mathbf{x})$, where the $\mathbf{a}_i$ are the rows of the matrix $\mathbf{A}$. This is exactly the LWE problem except that the adversary also gets to see $h(\mathbf{s})$. Consider now the mental experiment where $\mathbf{A} = \mathbf{BC}$, where $\mathbf{C} \in \mathbb{Z}_q^{m \times l}$ for some $l < n$. The key observations are that (a) since $h(\mathbf{s})$ is small, $\mathbf{s}$ still has considerable min-entropy given $h(\mathbf{s})$, and (b) matrix multiplication is a strong randomness extractor. In particular, these two observations together mean that $\mathbf{t} = \mathbf{Cs}$ is (statistically close to) random, even given $h(\mathbf{s})$. The resulting expression now looks like $\mathbf{Bt} + \mathbf{x}$, which is exactly the LWE distribution with secret $\mathbf{t}$ (a vector in $l < n$ dimensions). The proof of the adaptive case uses similar ideas in a more complex way: we refer the reader to Section 3.1 for the proof.

A few remarks are in order.

(Arbitrary) Polynomial number of measurements. We find it extremely interesting to construct encryption schemes secure against *repeated* memory attacks, where the combined number of bits leaked can be larger than the size of the secret-key (although any single measurement leaks only a small number of bits). Of course, if the secret-key is unchanged, this is impossible. It seems that to achieve this goal, some off-line (randomized) refreshing of the secret key must be done periodically. We do not deal with these further issues in this paper.

Leaking the content of the entire secret memory. The secret-memory may include more than the secret-keys. For example, results of intermediate computations produced during the execution of the decryption algorithm may compromise the security of the scheme even more than a carefully stored secret-key. Given this, why not allow the definition of memory attacks to measure the entire content of the secret-memory? We have two answers to this issue. First, in the case of the adaptive definition, when the decryption algorithm is deterministic (as is the case for the scheme in question and all schemes in use today), there is no loss of generality in restricting the adversary to measure the leakage from just the secret-key. This is the case because the decryption algorithm is itself only a function of the secret and public keys as well as the ciphertext that it receives, and this can be captured by a leakage function h that the adversary chooses to apply. In the non-adaptive case, the definition does not necessarily generalize this way; however, the constructions we give are secure under a stronger definition which allows leakage from the entire secret-memory. Roughly, the reason is that the decryption algorithm in question can be implemented using a small amount of extra memory, and thus the intermediate computations are an insignificant fraction of memory at any time.

1.2 Simultaneous Hard-Core Bits

The notion of hard-core bits for one-way functions was introduced very early in the developement of the theory of cryptography [42, 21, 8]. Indeed, the existence of hard-core bits for particular proposals of one-way functions (see, for example [8, 4, 23, 27]) and later for any one-way function [17], has been central to the constructions of secure public-key (and private-key) encryption schemes, and strong pseudo-random bit generators, the cornerstones of modern cryptography.

The main questions which remain open in this area concern the generalized notion of "simultaneous hard-core bit security" loosely defined as follows. Let f be a one-way function and h an easy to compute function. We say that h is a *simultaneously hard-core function* for f if given $f(x)$, $h(x)$ is computationally indistinguishable from random. In particular, we say that a block of bits of x are simultaneously hard-core for $f(x)$ if given $f(x)$, they cannot be distinguished from a random string of the same length (this corresponds to a function h that outputs a subset of its input bits).

The question of how many bits of x can be proved simultaneously hard-core has been studied for general one-way functions as well as for particular candidates in [41, 4, 31, 24, 18, 17], but the results obtained are far from satisfactory. For a general one-way function (modified in a similar manner as in their hard-core result), [17] showed the existence of an h that outputs $O(\log N)$ bits (where we let N denote the length of the input to the one-way function throughout) which is a simultaneous hard-core function for f. For particular candidate one-way functions such as the exponentiation function (modulo a prime p), the RSA function and the Rabin function, [41, 31] have pointed to particular blocks of $O(\log N)$ input bits which are simultaneously hard-core given $f(x)$.

The first example of a one-way function candidate that hides more than $O(\log N)$ simultaneous hardcore bits was shown by Hastad, Schrift and Shamir [24, 18] who proved that the modular exponentiation function $f(x) = g^x \bmod M$ hides *half* the bits of x under the intractability of factoring the modulus M. The first example of a trapdoor function for which many bits were shown simultaneous hardcore was the Pallier function. In particular, Catalano, Gennaro and Howgrave-Graham [11] showed that $N - o(N)$ bits are simulatenously hard-core for the Paillier function, under a stronger assumption than the standard Paillier assumption.

A question raised by [11] was whether it is possible to construct other natural and efficient trapdoor functions with many simultaneous hardcore bits and in particular, functions whose conjectured one-wayness is not based on the difficulty of the factoring problem. In this paper, we present two lattice-based trapdoor functions for which is the case.

First, we consider the following trapdoor function family proposed in [16]. A function $f_{\mathbf{A}}$ in the family is described by a matrix $\mathbf{A} \in \mathbb{Z}_q^{m \times n}$, where $q = \text{poly}(n)$ is prime and $m = \text{poly}(n)$. $f_{\mathbf{A}}$ takes two inputs $\mathbf{s} \in \mathbb{Z}_q^n$ and a sequence of random bits $\mathbf{r}$; it first uses $\mathbf{r}$ to sample a vector $\mathbf{x}$ from (a discretized form of) the Gaussian distribution over $\mathbb{Z}_q^m$. $f_{\mathbf{A}}$ then outputs $\mathbf{As} + \mathbf{x}$. The one-wayness of this function is based on the learning with error (LWE) problem $\text{LWE}_{n,m,q,\beta}$. Alternatively, the one-wayness can also be based on the worst-case quantum hardness of $\text{poly}(n)$-approximate shortest vector problem ($\text{gapSVP}_{\text{poly}(n)}$), by a reduction of Regev [39] from gapSVP to LWE. We prove that $O(N/\log N)$ bits (where N is the total number of input bits) of $f_{\mathbf{A}}$ are simultaneously hardcore.

Second, for a new setting of the parameters in $f_{\mathbf{A}}$, we show that $N - N/\text{polylog}(N)$ bits (out of the N input bits) are simultaneously hardcore. The new parameter setting is a much larger modulus $q = n^{\text{polylog}(n)}$, a much smaller $m = O(n)$ and a Gaussian noise with a much smaller (inverse superpolynomial) standard deviation. At first glance, it is unclear whether for these new parameter setting, the function is still a trapdoor (injective) function. To this end, we show that the function is injective, is sampleable with an appropriate trapdoor (which can be used to invert the function) and that it is one-way. The one-wayness is based on a much stronger (yet plausible) assumption, namely the quantum hardness of gapSVP with approximation factor $n^{\text{polylog}(n)}$ (For details, see Section 4.2).

We stress that our results (as well as the results of [24, 18, 11]) show that particular sets of *input bits* of these functions are simultaneously hardcore (as opposed to arbitrary hardcore functions that output many bits).

Informal Theorem 3

1. *Let m and q be polynomial in n and let $\beta = 4\sqrt{n}/q$. There exists an injective trapdoor function $\mathcal{F}_{n,m,q,\beta}$ with input length N for which a $1/\log N$ fraction of the input bits are simultaneously hardcore, assuming the $\text{poly}(n)$-hardness of $\text{LWE}_{O(n),m,q,\beta}$.*
2. *Let $m = O(n)$, $q = n^{\text{polylog}(n)}$ and $\beta = 4\sqrt{n}/q$. There exists an injective trapdoor function $\mathcal{F}_{n,m,q,\beta}$ with input length N for which a $1-1/\text{polylog}(N)$ fraction of input bits are simultaneously hardcore, assuming the hardness of $\text{LWE}_{n/\text{polylog}(n),m,q,\beta}$.*

Our proof is simple and general: one of the consequences of the proof is that a related one-way function based on the well-studied learning parity with noise problem (LPN) [7] also has $N - o(N)$ simultaneous hardcore bits. We defer the proof of this result to the full version due to lack of space.

Idea of the Proof. In the case of security against non-adaptive memory attacks, the statement we showed (see Section 1.1) is that given $\mathbf{A}$ and $h(\mathbf{s})$, $\mathbf{As} + \mathbf{x}$ looks random. The statement of hardcore bits is that given $\mathbf{A}$ and $\mathbf{As} + \mathbf{x}$, $h(\mathbf{s})$ (where h is the particular function that outputs a subset of bits of $\mathbf{s}$) looks random. Though the statements look different, the main idea in the proof of security against non-adaptive memory attacks, namely dimension reduction, carries over and can be used to prove the simultaneous hardcore bits result also. For details, see Section 4.

1.3   Other Related Work

Brent Waters, in a personal communication, has suggested a possible connection between the recently proposed notion of deterministic encryption [9,6], and simultaneous hardcore bits. In particular, his observation is that deterministic encryption schemes (which are, informally speaking, trapdoor functions that are uninvertible even if the input comes from a min-entropy source) satisfying the definition of [9] imply trapdoor functions with many simultaneous hardcore bits. Together with the construction of deterministic encryption schemes from lossy trapdoor functions [36] (based on DDH and LWE), this gives us trapdoor functions based on DDH and LWE with many simultaneous hardcore bits. However, it seems that using this approach applied to the LWE instantiation, it is possible to get only $o(N)$ hardcore bits (where N is the total number of input bits); roughly speaking, the bottleneck is the "quality" of lossy trapdoor functions based on LWE. In contrast, in this work, we achieve $N - o(N)$ hardcore bits.

Recently, Peikert [34] has shown a classical reduction from a variant of the worst-case shortest vector problem (with appropriate approximation factors) to the average-case LWE problem. This, in turn, means that our results can be based on the *classical* worst-case hardness of this variant shortest-vector problem as well.

A recent observation of [38] surprisingly shows that any public-key encryption scheme is secure against an adaptive $\alpha(N)$-memory attack, *under (sub-)exponential hardness assumptions* on the security of the public-key encryption scheme. Slightly more precisely, the observation is that any semantically secure public-key encryption scheme that cannot be broken in time roughly $2^{\alpha(N)}$ is secure against an adaptive $\alpha(N)$-memory attack. In contrast, the schemes in this paper make only *polynomial hardness assumptions*. (See Section 3.1 for more details).

2   Preliminaries and Definitions

We will let bold capitals such as $\mathbf{A}$ denote matrices, and bold small letters such as $\mathbf{a}$ denote vectors. $\mathbf{x} \cdot \mathbf{y}$ denotes the inner product of $\mathbf{x}$ and $\mathbf{y}$. If $\mathbf{A}$ is an $m \times n$ matrix and $S \subseteq [n]$ represents a subset of the columns of $\mathbf{A}$, we let $\mathbf{A}_S$ denote the restriction of $\mathbf{A}$ to the columns in S, namely the $m \times |S|$ matrix consisting of the columns with indices in S. In this case, we will write $\mathbf{A}$ as $[\mathbf{A}_S, \mathbf{A}_{\overline{S}}]$.

A problem is t-hard if no (probabilistic) algorithm running in time t can solve it. When we say that a problem is hard without further qualification, we mean that it is poly(n)-hard, where n is the security parameter of the system (which is usually explicitly specified).

2.1   Cryptographic Assumptions

The cryptographic assumptions we make are related to the hardness of learning-type problems. In particular, we will consider the hardness of learning with error (LWE); this problem was introduced by Regev [39] where he showed a relation between the hardness of LWE and the *worst-case hardness* of certain problems on lattices (see Proposition 1).

We now define a probability distribution $A_{\mathbf{s},\chi}$ that is later used to specify this problem. For positive integers n and $q \geq 2$, a vector $\mathbf{s} \in \mathbb{Z}_q^n$ and a probability distribution χ on $\mathbb{Z}_q$, define $A_{\mathbf{s},\chi}$ to be the distribution obtained by choosing a vector $\mathbf{a}_i \in \mathbb{Z}_q^n$ uniformly at random, a noise-term $x_i \in \mathbb{Z}_q$ according to χ and outputting $(\mathbf{a}_i, \langle \mathbf{a}_i, \mathbf{s} \rangle + x_i)$, where addition is performed in $\mathbb{Z}_q$.[5]

Learning With Error (LWE). Our notation here follows [39, 35]. The normal (or the Gaussian) distribution with mean 0 and variance σ^2 (or standard deviation σ) is the distribution on $\mathbb{R}$ with density function $\frac{1}{\sigma \cdot \sqrt{2\pi}} \exp(-x^2/2\sigma^2)$.

For $\beta \in \mathbb{R}^+$ we define Ψ_β to be the distribution on $\mathbb{T} = [0, 1)$ of a normal variable with mean 0 and standard deviation $\beta/\sqrt{2\pi}$, reduced modulo 1.[6] For any probability distribution $\phi : \mathbb{T} \to \mathbb{R}^+$ and an integer $q \in \mathbb{Z}^+$ (often implicit) we define its *discretization* $\bar{\phi} : \mathbb{Z}_q \to \mathbb{R}^+$ to be the distribution over $\mathbb{Z}_q$ of the random variable $\lfloor q \cdot X_\phi \rceil \bmod q$, where X_ϕ has distribution ϕ.[7] In our case, the distribution $\overline{\Psi}_\beta$ over $\mathbb{Z}_q$ is defined by choosing a number in $[0, 1)$ from the distribution Ψ_β, multiplying it by q, and rounding the result.

Definition 1. *Let* $\mathbf{s} \in \mathbb{Z}_q^n$ *be uniformly random. Let* $q = q(n)$ *and* $m = m(n)$ *be integers, and let* $\chi(n)$ *be the distribution* $\overline{\Psi}_\beta$ *with parameter* $\beta = \beta(n)$. *The goal of the learning with error problem in* n *dimensions, denoted* $\mathsf{LWE}_{n,m,q,\beta}$, *is to find* $\mathbf{s}$ *(with overwhelming probability) given access to an oracle that outputs* m *samples from the distribution* $A_{\mathbf{s},\chi}$. *The goal of the decision variant* $\mathsf{LWE\text{-}Dist}_{n,m,q,\beta}$ *is to distinguish (with non-negligible probability) between* m *samples from the distribution* $A_{\mathbf{s},\chi}$ *and* m *uniform samples over* $\mathbb{Z}_q^n \times \mathbb{Z}_q$. *We say that* $\mathsf{LWE}_{n,m,q,\beta}$ *(resp.* $\mathsf{LWE\text{-}Dist}_{n,m,q,\beta}$) *is* t-*hard if no (probabilistic) algorithm running in time* t *can solve it.*

The LWE problem was introduced by Regev [39], where he demonstrated a connection between the LWE problem for certain moduli q and error distributions χ, and worst-case lattice problems. In essence, he showed that LWE is as hard as solving several standard *worst-case* lattice problems *using a quantum algorithm*. We state a version of his result here. Informally, $\mathsf{gapSVP}_{c(n)}$ refers to the (worst-case) promise problem of distinguishing between lattices that have a vector of length at most 1 from ones that have no vector shorter than $c(n)$ (by scaling, this is equivalent to distinguishing between lattices with a vector of length at most k from ones with no vector shorter than $k \cdot c(n)$).

Proposition 1 ([39]). *Let* $q = q(n)$ *be a prime and* $\beta = \beta(n) \in [0, 1]$ *be such that* $\beta q > 2\sqrt{n}$. *Assume that we have access to an oracle that solves* $\mathsf{LWE}_{n,m,q,\beta}$. *Then,*

---

[5] Here, we think of n as the security parameter, and $q = q(n)$ and $\chi = \chi(n)$ as functions of n. We will sometimes omit the explicit dependence of q and χ on n.

[6] For $x \in \mathbb{R}$, $x \bmod 1$ is simply the fractional part of x.

[7] For a real x, $\lfloor x \rceil$ is the result of rounding x to the nearest integer.

there is a polynomial (in n and m) time quantum algorithm to solve $\mathsf{gapSVP}_{200n/\beta}$ *for any n-dimensional lattice.*

We will use Proposition 1 as a guideline for which parameters are hard for LWE. In particular, the (reasonable) assumption that $\mathsf{gapSVP}_{n^{\mathrm{polylog}(n)}}$ is hard to solve in quasi-polynomial (quantum) time implies that $\mathsf{LWE}_{n,m,q,\beta}$ (as well as $\mathsf{LWE\text{-}Dist}_{n,m,q,\beta}$) where $q = n^{\mathrm{polylog}(n)}$ and $\beta = 2\sqrt{n}/q$ is hard to solve in polynomial time.

Regev [39] also showed that an algorithm that solves the decision version $\mathsf{LWE\text{-}Dist}$ with m samples implies an algorithm that solves the search version LWE in time $\mathrm{poly}(n, q)$.

Proposition 2. *There is a polynomial (in n and q) time reduction from the search version* $\mathsf{LWE}_{n,m,q,\beta}$ *to the decision version* $\mathsf{LWE\text{-}Dist}_{n,m\cdot\mathrm{poly}(n,q),q,\beta}$, *and vice versa (for some polynomial* poly*).*

Sampling $\overline{\varPsi}_\beta$*.* The following proposition gives a way to sample from the distribution $\overline{\varPsi}_\beta$ using few random bits. This is done by a simple rejection sampling routine (see, for example, [16]).

Proposition 3. *There is a PPT algorithm that outputs a vector* $\mathbf{x}$ *whose distribution is statistically close to* $\overline{\varPsi}_\beta^m$ *(namely, m independent samples from* $\overline{\varPsi}_\beta$*) using $O(m \cdot \log(q\beta) \cdot \log^2 n)$ uniformly random bits.*

2.2 Defining Memory Attacks

In this section, we define the semantic security of public-key encryption schemes against memory attacks. The definitions in this section can be extended to other cryptographic primitives as well; these extensions are deferred to the full version. We proceed to define semantic security against two flavors of memory attacks, (the stronger) adaptive memory attacks and (the weaker) non-adaptive memory attacks.

Semantic Security Against Adaptive Memory Attacks. In an adaptive memory attack against a public-key encryption scheme, the adversary, upon seeing the public-key PK, chooses (efficiently computable) functions h_i adaptively (depending on PK and the outputs of $h_j(SK)$ for $j < i$) and receives $h_i(SK)$. This is called the probing phase. The definition is parametrized by a function $\alpha(\cdot)$, and requires that *the total number of bits output by $h_i(SK)$ for all i is bounded by $\alpha(N)$* (where N is the length of the secret-key).

After the probing phase, the adversary plays the semantic security game, namely he chooses two messages (m_0, m_1) of the same length and gets $\mathrm{ENC}_{PK}(m_b)$ for a random $b \in \{0, 1\}$ and he tries to guess b. We require that the adversary guesses the bit b with probability at most $\frac{1}{2} + \mathrm{negl}(n)$, where n is the security parameter and negl is a negligible function. We stress that the adversary is allowed to get the measurements $h_i(SK)$ only *before* he sees the challenge ciphertext. The formal definition follows.

Definition 2 (Adaptive Memory Attacks). *Let $\alpha : \mathbb{N} \to \mathbb{N}$ be a function, and let N be the size of the secret-key output by* $\mathrm{GEN}(1^n)$*. Let H_{SK} be an oracle that takes as input* *a polynomial-size circuit h and outputs $h(SK)$. A PPT adversary $A = (A_1^{H_{SK}}, A_2)$ is called admissible if the total number of bits that A gets as a result of oracle queries to H_{SK} is at most $\alpha(N)$.*

A public-key encryption scheme PKE $=$ (GEN, ENC, DEC) *is semantically secure against* adaptive $\alpha(N)$-*memory attacks if for any* admissible *PPT adversary* $A = (A_1, A_2)$, *the probability that A wins in the following experiment differs from $\frac{1}{2}$ by a negligible function in n.*

$(\text{PK}, \text{SK}) \leftarrow \text{GEN}(1^n)$
$(m_0, m_1, \text{state}) \leftarrow A_1^{H_{SK}}(\text{PK})$ *s.t.* $|m_0| = |m_1|$
$y \leftarrow \text{ENC}_{PK}(m_b)$ *where $b \in \{0, 1\}$ is a random bit*
$b' \leftarrow A_2(y, \text{state})$

The adversary A wins the experiment if $b' = b$.

The definitions of security for identity-based encryption schemes against memory attacks is similar in spirit, and is deferred to the full version.

Semantic Security Against Non-Adaptive Memory Attacks. Non-adaptive memory attacks capture the scenario in which a polynomial-time computable leakage function h whose output length is bounded by $\alpha(N)$ is fixed in advance (possibly as a function of the encryption scheme, and the underlying hardware). We require that the encryption scheme be semantically secure even if the adversary is given the auxiliary input $h(SK)$. We stress that h is chosen *independently of the public-key PK*. Even though this is much weaker than the adaptive definition, schemes satisfying the non-adaptive definition could be much easier to design and prove (as we will see in Section 3). Moreover, in some practical scenarios, the leakage function is just a characteristic of the hardware and is independent of the parameters of the system, including the public-key. The formal definition follows.

Definition 3 (Non-adaptive Memory Attacks). *Let $\alpha : \mathbb{N} \to \mathbb{N}$ be a function, and let N be the size of the secret-key output by* GEN(1^n). *A public-key encryption scheme* PKE $=$ (GEN, ENC, DEC) *is semantically secure against* non-adaptive $\alpha(N)$-*memory attacks if for any function $h : \{0, 1\}^N \to \{0, 1\}^{\alpha(N)}$, and any PPT adversary $A = (A_1, A_2)$, the probability that A wins in the following experiment differs from $\frac{1}{2}$ by a negligible function in n:*

$(\text{PK}, \text{SK}) \leftarrow \text{GEN}(1^n)$
$(m_0, m_1, \text{state}) \leftarrow A_1(\text{PK}, h(\text{SK}))$ *s.t.* $|m_0| = |m_1|$
$y \leftarrow \text{ENC}_{PK}(m_b)$ *where $b \in \{0, 1\}$ is a random bit*
$b' \leftarrow A_2(y, \text{state})$

The adversary A wins the experiment if $b' = b$.

Remarks about the Definitions

A Simpler Definition that is Equivalent to the adaptive definition. We observe that without loss of generality, we can restrict our attention to an adversary that outputs a single function h (whose output length is bounded by $\alpha(N)$) and gets $(PK, h(PK, SK))$

(where $(PK, SK) \leftarrow \text{GEN}(1^n)$) as a result. Informally, the equivalence holds because the adversary can encode all the functions h_i (that depend on PK as well as $h_j(SK)$ for $j < i$) into a *single polynomial-size circuit* h that takes PK as well as SK as inputs. We will use this formulation of Definition 2 later in the paper.

The Dependence of the Leakage Function on the Challenge Ciphertext. In the adaptive definition, the adversary is not allowed to obtain $h(SK)$ *after* he sees the challenge ciphertext. This restriction is necessary: if we allow the adversary to choose h depending on the challenge ciphertext, he can use this ability to decrypt it (by letting h be the decryption circuit and encoding the ciphertext into h), and thus the definition would be unachievable.

A similar issue arises in the definition of CCA2-security of encryption schemes, where the adversary should be prohibited from querying the decryption oracle on the challenge ciphertext. Unfortunately, whereas the solution to this issue in the CCA2-secure encryption case is straightforward (namely, explicitly disallow querying the decryption oracle on the challenge ciphertext), it seems far less clear in our case.

The Adaptive Definition and Bounded CCA1-security. It is easy to see that a bit-encryption scheme secure against an adaptive $\alpha(N)$-memory attack is also secure against a CCA1 attack where adversary can make at most $\alpha(N)$ decryption queries (also called an $\alpha(N)$-bounded CCA1 attack).

3   Public-Key Encryption Secure against Memory Attacks

In this section, we construct a public-key encryption scheme that is secure against memory attacks. In Section 3.1, we show that the Regev encryption scheme [39] is secure against *adaptive α-memory attacks*, for $\alpha(N) = O(\frac{N}{\log N})$, under the assumption that $\text{LWE}_{O(n),m,q,\beta}$ is poly(n)-hard (where n is the security parameter and $N = 3n \log q$ is the length of the secret-key). The parameters q, m and β are just as in Regev's encryption scheme, described below.

In Section 3.2, we show that a slight variant of Regev's encryption scheme is secure against *non-adaptive $(N - k)$-memory attacks*, assuming the poly(n)-hardness of $\text{LWE}_{O(k/\log n),m,q,\beta}$. On the one hand, this allows the adversary to obtain more information about the secret-key but on the other hand, achieves a much weaker (namely, non-adaptive) definition of security.

The Regev Encryption Scheme. First, we describe the public-key encryption scheme of Regev, namely $\text{RPKE} = (\text{RGEN}, \text{RENC}, \text{RDEC})$ which works as follows. Let n be the security parameter and let $m(n), q(n), \beta(n) \in \mathbb{N}$ be parameters of the system. For concreteness, we will set $q(n)$ be a prime between n^3 and $2n^3$, $m(n) = 3n \log q$ and $\beta(n) = 4\sqrt{n}/q$.

- RGEN(1^n) picks a random matrix $\mathbf{A} \in \mathbb{Z}_q^{m \times n}$, a random vector $\mathbf{s} \in \mathbb{Z}_q^n$ and a vector $\mathbf{x} \leftarrow \overline{\Psi}_\beta^m$ (that is, where each entry x_i is chosen independently from the probability distribution $\overline{\Psi}_\beta$). Output $PK = (\mathbf{A}, \mathbf{As} + \mathbf{x})$ and $SK = \mathbf{s}$.
- RENC(PK, b), where b is a bit, works as follows. First, pick a vector $\mathbf{r}$ at random from $\{0, 1\}^m$. Output $(\mathbf{rA}, \mathbf{r}(\mathbf{As} + \mathbf{x}) + b\lfloor \frac{q}{2} \rfloor)$ as the ciphertext.

- RDEC(SK, c) first parses $c = (\mathbf{c}_0, c_1)$, computes $b' = c_1 - \mathbf{c}_0 \cdot \mathbf{s}$ and outputs 0 if b' is closer to 0 than to $\frac{q}{2}$, and 1 otherwise.

Decryption is correct because the value $b' = \mathbf{r} \cdot \mathbf{x} + b\lfloor q/2 \rfloor$ computed by the decryption algorithm is very close to $b\lfloor q/2 \rfloor$: this is because the absolute value of $\mathbf{r} \cdot \mathbf{x}$ is much smaller than $q/4$. In particular, since $||\mathbf{r}||_2 \leq \sqrt{m}$ and $||\mathbf{x}||_2 \leq mq\beta = 4m\sqrt{n}$ with high probability, $|\mathbf{r} \cdot \mathbf{x}| \leq ||\mathbf{r}||_2||\mathbf{x}||_2 \leq 4m\sqrt{mn} \ll q/4$.

3.1   Security against Adaptive Memory Attacks

Let $N = 3n \log q$ be the length of the secret-key in the Regev encryption scheme. In this section, we show that the scheme is secure against $\alpha(N)$-adaptive memory attacks for any $\alpha(N) = O(\frac{N}{\log N})$, assuming that $\mathsf{LWE}_{O(n),m,q,\beta}$ is poly(n)-hard, where m, q and β are as in encryption scheme described above.

Theorem 1. *Let the parameters m, q and β be as in* RPKE. *Assuming that* $\mathsf{LWE}_{O(n),m,q,\beta}$ *is* poly(n)-*hard, the scheme is semantically secure against adaptive $\alpha(N)$-memory attacks for $\alpha(N) \leq N/10 \log N$.*

Proof. *(Sketch.)* First, we observe that without loss of generality, we can restrict our attention to an adversary that outputs single function h (whose output length is bounded by $\alpha(N)$) and the adversary gets $(PK, h(PK, SK))$ as a result. Informally, the equivalence holds because the adversary can encode all the functions h_i (that depend on PK as well as $h_j(SK)$ for $j < i$) into a *single polynomial (in n) size circuit* h that takes PK as well as SK as inputs.

Thus, it suffices to show that for any polynomial-size circuit h,

$$(PK, \mathrm{ENC}_{PK}(0), h(PK, SK)) \approx_c (PK, \mathrm{ENC}_{PK}(1), h(PK, SK))$$

In our case, it suffices to show the following statement (which states that the encryption of 0 is computationally indistinguishable from uniform)

$$(\mathbf{A}, \mathbf{As} + \mathbf{x}, \mathbf{rA}, \mathbf{r}(\mathbf{As} + \mathbf{x}), h(\mathbf{A}, \mathbf{s}, \mathbf{x})) \approx_c (\mathbf{A}, \mathbf{As} + \mathbf{x}, \mathbf{u}, u', h(\mathbf{A}, \mathbf{s}, \mathbf{x})) \quad (1)$$

where $\mathbf{u} \in \mathbb{Z}_q^n$ and $u' \in \mathbb{Z}_q$ are uniformly random and independent of all other components. That is, the ciphertext is computationally indistinguishable from uniformly random, given the public-key and the leakage $h(PK, SK)$.

We will in fact show a stronger statement, namely that

$$(\mathbf{A}, \mathbf{As} + \mathbf{x}, \mathbf{rA}, \mathbf{rAs}, h(\mathbf{A}, \mathbf{s}, \mathbf{x}), \mathbf{rx}) \approx_c (\mathbf{A}, \mathbf{As} + \mathbf{x}, \mathbf{u}, u', h(\mathbf{A}, \mathbf{s}, \mathbf{x}), \mathbf{rx}) \quad (2)$$

The difference between (1) and (2) is that in the latter, the distributions also contain the additional information $\mathbf{r} \cdot \mathbf{x}$. Clearly, this is stronger than (1). We show (2) in four steps.

Step 1.   We show that $\mathbf{rA}$ can be replaced with a uniformly random vector in $\mathbb{Z}_q^n$ while maintaining statistical indistinguishability, even given $\mathbf{A}, \mathbf{As} + \mathbf{x}$, the leakage $h(\mathbf{A}, \mathbf{s}, \mathbf{x})$ and $\mathbf{r} \cdot \mathbf{x}$. More precisely,

$$(\mathbf{A}, \mathbf{As} + \mathbf{x}, \mathbf{rA}, \mathbf{rAs}, h(\mathbf{A}, \mathbf{s}, \mathbf{x}), \mathbf{r} \cdot \mathbf{x}) \approx_s (\mathbf{A}, \mathbf{As} + \mathbf{x}, \mathbf{u}, \mathbf{u} \cdot \mathbf{s}, h(\mathbf{A}, \mathbf{s}, \mathbf{x}), \mathbf{r} \cdot \mathbf{x}) \quad (3)$$

where $\mathbf{u} \in \mathbb{Z}_q^n$ is uniformly random.

Informally, 3 is true because of the leftover hash lemma. (A variant of) leftover hash lemma states that if (a) $\mathbf{r}$ is chosen from a distribution over $\mathbb{Z}_q^n$ with min-entropy $k \geq 2n \log q + \omega(\log n)$, (b) $\mathbf{A}$ is a uniformly random matrix in $\mathbb{Z}_q^{m \times n}$, and (c) the distributions of $\mathbf{r}$ and $\mathbf{A}$ are statistically independent, then $(\mathbf{A}, \mathbf{rA}) \approx_s (\mathbf{A}, \mathbf{u})$ where $\mathbf{u}$ is a uniformly random vector in $\mathbb{Z}_q^n$. Given $\mathbf{r} \cdot \mathbf{x}$ (which has length $\log q = O(\log n)$), the residual min-entropy of $\mathbf{r}$ is at least $m - \log q \geq 2n \log q + \omega(\log n)$. Moreover, the distribution of $\mathbf{r}$ given $\mathbf{r} \cdot \mathbf{x}$ depends only on $\mathbf{x}$, and is statistically independent of $\mathbf{A}$. Thus, leftover hash lemma applies and $\mathbf{rA}$ can be replaced with a random vector $\mathbf{u}$.

Step 2. This is the crucial step in the proof. Here, we replace the (uniformly random) matrix $\mathbf{A}$ with a matrix $\mathbf{A}'$ drawn from another distribution $\mathcal{D}$. Informally, the (efficiently sampleable) distribution $\mathcal{D}$ satisfies two properties: (1) a random matrix drawn from $\mathcal{D}$ is computationally indistinguishable from a uniformly random matrix, assuming the poly(n)-hardness of $\mathsf{LWE}_{O(n),m,q,\beta}$, and (2) given $\mathbf{A}' \leftarrow \mathcal{D}$ and $\mathbf{y} = \mathbf{A}'\mathbf{s} + \mathbf{x}$, the min-entropy of $\mathbf{s}$ is at least n. The existence of such a distribution follows from Lemma 1 below.

The intuition behind this step is the following: Clearly, $\mathbf{As} + \mathbf{x}$ is computationally indistinguishable from $\mathbf{A}'\mathbf{s} + \mathbf{x}$. Moreover, given $\mathbf{A}'\mathbf{s} + \mathbf{x}$, $\mathbf{s}$ has high (information-theoretic) min-entropy. Thus, in some informal sense, $\mathbf{s}$ has high "computational entropy" given $\mathbf{As} + \mathbf{x}$. This is the intuition for the next step.

Summing up, the claim in this step is that

$$(\mathbf{A}, \mathbf{As} + \mathbf{x}, \mathbf{u}, \mathbf{u} \cdot \mathbf{s}, h(\mathbf{A}, \mathbf{s}, \mathbf{x}), \mathbf{r} \cdot \mathbf{x}) \approx_c (\mathbf{A}', \mathbf{A}'\mathbf{s} + \mathbf{x}, \mathbf{u}, \mathbf{u} \cdot \mathbf{s}, h(\mathbf{A}', \mathbf{s}, \mathbf{x}), \mathbf{r} \cdot \mathbf{x}) \quad (4)$$

where $\mathbf{A}' \leftarrow \mathcal{D}$. This follows directly from Lemma 1 below.

Step 3. By Lemma 1, $\mathbf{s}$ has min-entropy at least $n \geq \frac{N}{9 \log N}$ given $\mathbf{A}'\mathbf{s} + \mathbf{x}$. Since the output length of h is at most $\frac{N}{10 \log N}$ and the length of $\mathbf{r} \cdot \mathbf{x}$ is $\log q = O(\log n)$, $\mathbf{s}$ still has residual min-entropy $\omega(\log n)$ given $\mathbf{A}', \mathbf{A}'\mathbf{s} + \mathbf{x}, h(\mathbf{A}', \mathbf{s}, \mathbf{x})$ and $\mathbf{r} \cdot \mathbf{x}$. Note also that the vector $\mathbf{u}$ on the left-hand side distribution is independent of $(\mathbf{A}, \mathbf{As} + \mathbf{x}, h(\mathbf{A}, \mathbf{s}, \mathbf{x}), \mathbf{r} \cdot \mathbf{x})$. This allows us to apply leftover hash lemma again (with $\mathbf{u}$ as the "seed" and $\mathbf{s}$ as the min-entropy source). Thus,

$$(\mathbf{A}', \mathbf{A}'\mathbf{s} + \mathbf{x}, \mathbf{u}, \mathbf{u} \cdot \mathbf{s}, h(\mathbf{A}', \mathbf{s}, \mathbf{x}), \mathbf{r} \cdot \mathbf{x}) \approx_s (\mathbf{A}', \mathbf{A}'\mathbf{s} + \mathbf{x}, \mathbf{u}, u', h(\mathbf{A}', \mathbf{s}, \mathbf{x}), \mathbf{r} \cdot \mathbf{x}) \quad (5)$$

where $u' \leftarrow \mathbb{Z}_q$ is uniformly random and independent of all the other components in the distribution.

Step 4. In the last step, we switch back to a uniform matrix $\mathbf{A}$. That is,

$$(\mathbf{A}', \mathbf{A}'\mathbf{s} + \mathbf{x}, \mathbf{u}, u', h(\mathbf{A}', \mathbf{s}, \mathbf{x}), \mathbf{r} \cdot \mathbf{x}) \approx_c (\mathbf{A}, \mathbf{As} + \mathbf{x}, \mathbf{u}, u', h(\mathbf{A}, \mathbf{s}, \mathbf{x}), \mathbf{r} \cdot \mathbf{x}) \quad (6)$$

Putting the four steps together proves (2). $\qquad\qquad\square$

Lemma 1. *There is a distribution $\mathcal{D}$ such that*

- *$\mathbf{A} \leftarrow U_{\mathbb{Z}_q^{m \times n}} \approx_c \mathbf{A}' \leftarrow \mathcal{D}$, assuming the poly$(n)$-hardness of $\mathsf{LWE}_{O(n),m,q,\beta}$, where m, q, β are as in Regev's encryption scheme.*
- *The min-entropy of $\mathbf{s}$ given $\mathbf{A}'\mathbf{s} + \mathbf{x}$ is at least n. That is, $\mathsf{H}_\infty(\mathbf{s} \mid \mathbf{A}'\mathbf{s} + \mathbf{x}) \geq n$ [8].*

---

[8] The precise statement uses the notion of average min-entropy due to Dodis, Reyzin and Smith [14].

Remark: The above lemma is a new lemma proved in [19]; it has other consequences such as security under auxiliary input, which is beyond the scope of this paper.

A Different Proof of Adaptive Security under (Sub-)Exponential Assumptions. Interestingly, [38] observed that *any* public-key encryption scheme that is $2^{\alpha(N)}$-hard can be proven to be secure against $\alpha(N)$ adaptive memory attacks. In contrast, our result (Theorem 1) holds under a standard, polynomial (in the security parameter n) hardness assumption (for a reduced dimension, namely $O(n)$). We sketch the idea of the [38] proof here.

The proof follows from the existence of a simulator that breaks the standard semantic security with probability $\frac{1}{2} + \frac{\epsilon}{2^{\alpha(N)}}$ given an adversary that breaks the adaptive $\alpha(N)$-memory security with probability $\frac{1}{2} + \epsilon$. The simulator simply guesses the (at most $\alpha(N)$) bits of the output of h and runs the adversary with the guess; if the guess is correct, the adversary succeeds in guessing the encrypted bit with probability $\frac{1}{2} + \epsilon$. The key observation that makes this idea work is that there is indeed a way for the simulator to "test" if its guess is correct or wrong: simply produce many encryptions of random bits and check if the adversary succeeds on more than $1/2 + \epsilon$ fraction of these encryptions. We remark that this proof idea carries over to the case of symmetric encryption schemes secure against a chosen plaintext attack (that is, CPA-secure) as well.

3.2 Security against Non-adaptive Memory Attacks

In this section, we show that a variant of Regev's encryption scheme is secure against non-adaptive $N - o(N)$ memory attacks (where N is the length of the secret-key), assuming that $\mathsf{LWE}_{o(n),m,q,\beta}$ is $\mathsf{poly}(n)$-hard. The variant encryption scheme differs from Regev's encryption scheme only in the way the public-key is generated.

The key generation algorithm picks the matrix $\mathbf{A}$ as $\mathbf{BC}$ where $\mathbf{B}$ is uniformly random in $\mathbb{Z}_q^{m \times k}$ and $\mathbf{C}$ is uniformly random in $\mathbb{Z}_q^{k \times n}$ (as opposed to uniformly random in $\mathbb{Z}_q^{n \times m}$). We will let $k = n - \frac{\alpha(N)}{3 \log q}$ (note that $k < n$). For this modified key-generation procedure, it is easy to show that the decryption algorithm is still correct. We show:

Theorem 2. *The variant public-key encryption scheme outlined above is secure against a non-adaptive α-memory attack, where $\alpha(N) \leq N - o(N)$ for some $o(N)$ function, assuming that $\mathsf{LWE}_{o(n),m,q,\beta}$ is $\mathsf{poly}(n)$-hard, where the parameters m, q and β are exactly as in Regev's encryption scheme.*

We sketch a proof of this theorem below. The proof of semantic security of Regev's encryption is based on the fact that the public-key $(\mathbf{A}, \mathbf{As} + \mathbf{x})$ is computationally indistinguishable from uniform. In order to show security against non-adaptive memory attacks, it is sufficient to show that this computational indistinguishability holds even given $h(\mathbf{s})$, where h is an arbitrary (polynomial-time computable) function whose output length is at most $\alpha(N)$.

The proof of this essentially follows from the leftover hash lemma. First of all, observe that $\mathbf{s}$ has min-entropy at least $N - \alpha(N)$, given $h(\mathbf{s})$ (this is because the output length of h is at most $\alpha(N)$). Furthermore, the distribution of $\mathbf{s}$ given $h(\mathbf{s})$ is independent of $\mathbf{A}$ (since h depends only on $\mathbf{s}$ and is chosen independent of $\mathbf{A}$). By our choice of parameters, $N - \alpha(N) \geq 3k \log q$. Thus, leftover hash lemma implies that $\mathbf{Cs}$ is a vector $\mathbf{t}$ whose distribution is statistically close to uniform (even given $\mathbf{C}$ and $h(\mathbf{s})$). Thus, $\mathbf{As} + \mathbf{x} = \mathbf{BCs} + \mathbf{x} = \mathbf{Bt} + \mathbf{x}$ is distributed exactly like the output of an LWE distribution with dimension k (since $\mathbf{t} \in \mathbb{Z}_q^k$). This is computationally indistinguishable from random, assuming $\mathsf{LWE}_{k,m,q,\beta} = \mathsf{LWE}_{o(n),m,q,\beta}$ (since $k = o(n)$ by our choice).

4   Simultaneous Hardcore Bits

In this section, we show that variants of the trapdoor one-way function proposed by Gentry et al. [16] (the GPV trapdoor function) has many simultaneous hardcore bits. For the parameters of [16], we show that a $1/\mathsf{polylog}(N)$ fraction of the input bits are simultaneously hardcore, assuming the $\mathsf{poly}(n)$-hardness of $\mathsf{LWE}_{O(n),m,q,\beta}$ (here, m and q are polynomial in n and β is inverse-polynomial in n, the GPV parameter regime).

More significantly, we show a different (and non-standard) choice of parameters for which the function has $N - N/\mathsf{polylog}(N)$ hardcore bits. The choice of parameters is $m = O(n)$, a modulus $q = n^{\mathsf{polylog}(n)}$ and $\beta = 4\sqrt{n}/q$. This result assumes the $\mathsf{poly}(n)$-hardness of $\mathsf{LWE}_{n/\mathsf{polylog}(n),m,q,\beta}$ for these parameters m, q and β. The parameters are non-standard in two respects: first, the modulus is superpolynomial, and the noise rate is very small (i.e, inverse super-polynomial) which makes the hardness assumption stronger. Secondly, the number of samples m is linear in n (as opposed to roughly $n \log n$ in [16]): this affects the trapdoor properties of the function (for more details, see Section 4.2). Also, note that the hardness assumption here refers to a reduced dimension (namely, $n/\mathsf{polylog}(n)$).

We remark that for any sufficiently large $o(N)$ function, we can show that the GPV function is a trapdoor function with $N - o(N)$ hardcore bits for different choices of parameters. We defer the details to the full version.

4.1   Hardcore Bits for the GPV Trapdoor Function

In this section, we show simultaneous hardcore bits for the GPV trapdoor function. First, we show a general result about hardcore bits that applies to a wide class of parameter settings: then, we show how to apply it to get $O(N/\mathsf{polylog}(N))$ hardcore bits for the GPV parameters, and in Section 4.2, $N - N/\mathsf{polylog}(N)$ hardcore bits for our new setting of parameters.

The collection of (injective) trapdoor functions $\mathcal{F}_{n,m,q,\beta}$ is defined as follows. Let $m = m(n)$ be polynomial in n. Each function $f_{\mathbf{A}} : \mathbb{Z}_q^n \times \{0, 1\}^r \to \mathbb{Z}_q^m$ is indexed by a matrix $\mathbf{A} \in \mathbb{Z}_q^{m \times n}$. It takes as input $(\mathbf{s}, \mathbf{r})$ where $\mathbf{s} \in \mathbb{Z}_q^n$ and $\mathbf{r} \in \{0, 1\}^r$, first uses $\mathbf{r}$ to sample a vector $\mathbf{x} \leftarrow \overline{\Psi}_\beta^m$ (that is, a vector each of whose components is independently drawn from the Gaussian error-distribution $\overline{\Psi}_\beta$), and outputs $\mathbf{As} + \mathbf{x}$. Clearly, the one-wayness of this function is equivalent to solving $\mathsf{LWE}_{n,m,q,\beta}$. Gentry et al. [16] show that $\mathcal{F}_{n,m,q,\beta}$ is a *trapdoor* one-way function for the parameters $q = O(n^3)$, $m = 3n \log q$ and $\beta = 4\sqrt{n}/q$ (assuming the hardness of $\mathsf{LWE}_{n,m,q,\beta}$).

Lemma 2. *For any integer $n > 0$, integer $q \geq 2$, an error-distribution $\chi = \overline{\Psi}_\beta$ over $\mathbb{Z}_q$ and any subset $S \subseteq [n]$, the two distributions $(\mathbf{A}, \mathbf{As} + \mathbf{x}, \mathbf{s}|_S)$ and $(\mathbf{A}, \mathbf{As} + \mathbf{x}, U_{\mathbb{Z}_q^{|S|}})$*

are computationally indistinguishable assuming the hardness of the decision version $\mathsf{LWE\text{-}Dist}_{n-|S|,m,q,\beta}$.

Proof. We will show this in two steps.

Step 1. The first and the main step is to show that $(\mathbf{A}, \mathbf{As}+\mathbf{x}, \mathbf{s}|_S) \approx_c (\mathbf{A}, U_{\mathbb{Z}_q^m}, U_{\mathbb{Z}_q^{|S|}})$. The distribution on the right consists of uniformly random and independent elements. This statement is shown by contradiction: Suppose a PPT algorithm D distinguishes between the two distributions. Then, we construct a PPT algorithm E that breaks the decision version $\mathsf{LWE\text{-}Dist}_{n-|S|,m,q,\beta}$. E gets as input $(\mathbf{A}', \mathbf{y}')$ such that $\mathbf{A}' \in \mathbb{Z}_q^{m\times(n-|S|)}$ is uniformly random and $\mathbf{y}'$ is either drawn from the LWE distribution (with dimension $n - |S|$) or is uniformly random. E does the following:

1. Let $\mathbf{A}_{\bar{S}} = \mathbf{A}'$. Choose $\mathbf{A}_S$ uniformly at random from $\mathbb{Z}_q^{m\times|S|}$ and set $\mathbf{A} = [\mathbf{A}_S, \mathbf{A}_{\bar{S}}]$.
2. Choose $\mathbf{s}_S \leftarrow \mathbb{Z}_q^{|S|}$ uniformly at random and compute $\mathbf{y} = \mathbf{y}' + \mathbf{A}_S\mathbf{s}_S$.
3. Run D with input $(\mathbf{A}, \mathbf{y}, \mathbf{s}_S)$, and output whatever D outputs.

First, suppose $(\mathbf{A}', \mathbf{y}')$ is drawn from the LWE distribution $A_{\mathbf{s}',\chi}$ for some $\mathbf{s}'$. Let $\mathbf{s}_{\bar{S}} = \mathbf{s}'$ and let $\mathbf{s} = [\mathbf{s}_S, \mathbf{s}_{\bar{S}}]$. Then, $(\mathbf{A}, \mathbf{y})$ constructed by E is distributed identical to $A_{\mathbf{s},\chi}$. On the other hand, if $(\mathbf{A}', \mathbf{y}')$ is drawn from the uniform distribution, then $(\mathbf{A}, \mathbf{y})$ is uniformly distributed, and independent of $\mathbf{s}|_S$. Thus, if D distinguishes between the two distributions, then E solves $\mathsf{LWE\text{-}Dist}_{n-|S|,m,q,\beta}$.

Step 2. The second step is to show that $(\mathbf{A}, U_{\mathbb{Z}_q^m}, U_{\mathbb{Z}_q^{|S|}}) \approx_c (\mathbf{A}, \mathbf{As} + \mathbf{x}, U_{\mathbb{Z}_q^{|S|}})$. This is equivalent to the hardness of $\mathsf{LWE\text{-}Dist}_{n,m,q,\beta}$. $\qquad\square$

The theorem below shows that for the GPV parameter settings, a $1/\mathsf{polylog}(N)$ fraction of the bits are simultaneously hardcore.

Theorem 3. *Let* $\gamma = m\log(q\beta)\log^2 n/n\log q$. *For any* $k > 0$, *assuming that* $\mathsf{LWE}_{k,m,q,\beta}$ *is* $\mathsf{poly}(n,q)$-*hard, the fraction of simultaneous hardcore bits for the family* $\mathcal{F}_{n,m,q,\beta}$ *is* $\frac{1}{1+\gamma}(1 - \frac{k}{n})$. *In particular, for the GPV parameters as above, the number of hardcore bits is* $O(N/\mathsf{polylog}(N))$.

Proof. We first bound the total input length of a function in $\mathcal{F}_{n,m,q,\beta}$, in terms of n, m, q and β. The number of bits r needed to sample $\mathbf{x}$ from $\overline{\Psi}_\beta^m$ is $mH(\beta) = O(m\log(q\beta)\log^2 n)$, by Proposition 3. Thus, the total input length is $n\log q + r = n\log q + O(m\log(q\beta)\log^2 n) = n\log q(1+\gamma)$.

By Lemma 2, assuming the hardness of the decision problem $\mathsf{LWE\text{-}Dist}_{k,m,q,\beta}$ (or, by Proposition 2, assuming the $\mathsf{poly}(n,q)$-hardness of the search problem $\mathsf{LWE}_{k,m,q,\beta}$), the number of simultaneously hardcore bits is at least $(n - k)\log q$. The fraction of hardcore bits, then, is $\frac{(n-k)\log q}{n\log q(1+\gamma)} = \frac{1}{1+\gamma}(1 - \frac{k}{n})$.

For the GPV parameters $\gamma = \mathsf{polylog}(N)$, and with $k = O(n)$, the number of hardcore bits is $O(N/\mathsf{polylog}(N))$ assuming the hardness of $\mathsf{LWE}_{O(n),m,q,\beta}$. $\qquad\square$

4.2 A New Setting of Parameters for the GPV Function

In this section, we show a choice of the parameters for the GPV function for which the function remains trapdoor one-way and an $1 - o(1)$ fraction of the input bits are simultaneously hardcore. Although the number of hardcore bits remains the same as in the GPV parametrization (as a function of n and q), namely $(n - k) \log q$ bits assuming the hardness of $\mathsf{LWE}_{k,m,q,\beta}$, the length of the input relative to this number will be much smaller. Overall, this means that the *fraction* of input bits that are simultaneously hardcore is larger.

We choose the parameters so that r (the number of random bits needed to sample the error-vector $\mathbf{x}$) is a subconstant fraction of $n \log q$. This could be done in one (or both) of the following ways. (a) Reduce m relative to n: note that m cannot be too small relative to n, otherwise the function ceases to be injective. (b) Reduce the standard deviation β of the Gaussian noise relative to the modulus q: as β/q gets smaller and smaller, it becomes easier to invert the function and consequently, the one-wayness of the function has to be based on progressively stronger assumptions. Indeed, we will employ both these methods (a) and (b) to achieve our goal.

In addition, we have to show that for our choice of parameters, it is possible to sample a random function in $\mathcal{F}_{n,m,q,\beta}$ (that is, the trapdoor sampling property) and that given the trapdoor, it is possible to invert the function (that is, the trapdoor inversion property). See the proof of Theorem 4 below for more details.

Our choice of parameters is $m(n) = 6n$, $q(n) = n^{\log^3 n}$ and $\beta = 4\sqrt{n}/q$.

Theorem 4. *Let $m(n) = 6n$, $q(n) = n^{\log^3 n}$ and $\beta = 4\sqrt{n}/q$. Then, the family of functions $\mathcal{F}_{n,m,q,\beta}$ is a family of trapdoor injective one-way functions with an $1 - 1/\mathsf{polylog}(N)$ fraction of hardcore bits, assuming the $n^{\mathsf{polylog}(n)}$-hardness of the search problem $\mathsf{LWE}_{n/\mathsf{polylog}(n),m,q,\beta}$. Using Regev's worst-case to average-case connection for LWE, the one-wayness of this function family can also be based on the worst-case $n^{\mathsf{polylog}(n)}$-hardness of $\mathsf{gapSVP}_{n^{\mathsf{polylog}(n)}}$.*

Proof. (*Sketch.*) Let us first compute the fraction of hardcore bits. By Theorem 3 applied to our parameters, we get a $1 - \frac{1}{\log n}$ fraction of hardcore bits assuming the hardness of $\mathsf{LWE\text{-}Dist}_{O(n/\log n),m,q,\beta}$. By Propositions 2 and 1, this translates to the assumptions claimed in the theorem.

We now outline the proof that for this choice of parameters, $\mathcal{F}_{n,m,q,\beta}$ is an injective trapdoor one-way function. Injectivity[9] follows from the fact that for all but an exponentially small fraction of $\mathbf{A}$, the minimum distance (in the ℓ_2 norm) of the lattice defined by $\mathbf{A}$ is very large; the proof is by a simple probabilistic argument and is omitted due to lack of space. Inverting the function is identical to solving $\mathsf{LWE}_{n,m,q,\beta}$. By Proposition 1, this implies that inverting the function on the average is as hard as solving $\mathsf{gapSVP}_{n^{\log^3 n}}$ in the worst-case.

---

[9] In fact, what we prove is a slightly weaker statement. More precisely, we show that for all but an exponentially small fraction of $\mathbf{A}$, there are no two pairs $(\mathbf{s}, \mathbf{x})$ and $(\mathbf{s}', \mathbf{x}')$ such that $\mathbf{As} + \mathbf{x} = \mathbf{As}' + \mathbf{x}'$ where $\mathbf{s}, \mathbf{s}' \in \mathbb{Z}_q^m$ and $||\mathbf{x}||_2, ||\mathbf{x}'||_2 \leq \beta\sqrt{mn}$. This does not affect the applications of injective one-way and trapdoor functions such as commitment and encryption schemes.

Trapdoor Sampling. The trapdoor for the function indexed by $\mathbf{A}$ is a short basis for the lattice $\Lambda^{\perp}(\mathbf{A}) = \{\mathbf{y} \in \mathbb{Z}^m : \mathbf{y}\mathbf{A} = 0 \bmod q\}$ defined by $\mathbf{A}$ (in a sense described below). We use here a modification of the procedure due to Ajtai [3] (and its recent improvement due to Alwen and Peikert [5]) which generates a pair $(\mathbf{A}, \mathbf{S})$ such that $\mathbf{A} \in \mathbb{Z}_q^{m \times n}$ is statistically close to uniform and $\mathbf{S} \in \mathbb{Z}^{m \times m}$ is a short basis for $\Lambda^{\perp}(\mathbf{A})$.

We outline the main distinction between [3, 5] and our theorem. Both [3] and [5] aim to construct bases for $\Lambda^{\perp}(\mathbf{A})$ that is as short as possible (namely, where each basis vector has length $\text{poly}(n)$). Their proof works for the GPV parameter choices, that is $q = \text{poly}(n)$ and $m = \Omega(n \log q) = \Omega(n \log n)$, for which they construct a basis $\mathbf{S}$ such that each basis vector has length $O(m^3)$ (this was recently improved to $m^{0.5}$ by [5]). In contrast, we deal with a much smaller m (linear in n) and a much larger q (superpolynomial in n). For this choice of parameters, the shortest vectors in $\Lambda^{\perp}(\mathbf{A})$ are quite long: indeed, they are unlikely to be much shorter than $q^{n/m} = q^{O(1)}$ (this follows by a simpler probabilistic argument). What we do is to construct a basis that is nearly as short; it turns out that this suffices for our purposes. Reworking the result of Ajtai for our parameters, we get the following theorem. The proof is omitted from this extended abstract.

Theorem 5. *Let $m = 6n$ and $q = n^{\log^3 n}$. There is a polynomial (in n) time algorithm that outputs a pair $(\mathbf{A}, \mathbf{S})$ such that (a) The distribution of $\mathbf{A}$ is statistically close to the uniform distribution in $\mathbb{Z}_q^{m \times n}$. (b) $\mathbf{S} \in \mathbb{Z}^{m \times m}$ is a full-rank matrix and is a short basis for $\Lambda^{\perp}(\mathbf{A})$. In particular, $\mathbf{S}\mathbf{A} = 0 \bmod q$. (c) Each entry of $\mathbf{S}$ has absolute value at most $q' = q/m^4$.*

Trapdoor Inversion. As in GPV, we use the procedure of Liu, Lyubashevsky and Micciancio [30] for trapdoor inversion. In particular, we show a procedure that, given the basis $\mathbf{S}$ for the lattice $\Lambda^{\perp}(\mathbf{A})$ from above, outputs $(\mathbf{s}, \mathbf{x})$ given $f_{\mathbf{A}}(\mathbf{s}, \mathbf{r})$ (if such a pair $(\mathbf{s}, \mathbf{x})$ exists, and $\perp$ otherwise). Formally, they show the following:

Lemma 3. *Let n, m, q, β be as above, and let L be the length of the basis $\mathbf{S}$ of $\Lambda^{\perp}(\mathbf{A})$ (namely, the sum of the lengths of all the basis vectors). If $\beta \leq 1/Lm$, then there is an algorithm that, with overwhelming probability over the choice of $(\mathbf{A}, \mathbf{S})$ output by the trapdoor sampling algorithm, efficiently computes $\mathbf{s}$ from $f_{\mathbf{A}}(\mathbf{s}, \mathbf{r})$.*

The length L of the basis output by the trapdoor sampling algorithm is at most $m^2 q' \leq q/m^2$. For our choice of parameters, namely $\beta = 4\sqrt{n}/q$, and $m = 6n$, clearly $\beta \leq 1/Lm$. Thus, the inversion algorithm guaranteed by Lemma 3 succeeds with overwhelming probability over the choice of inputs. Note that once we compute $\mathbf{s}$, we can also compute the unique value of $\mathbf{x}$.   □

5   Open Questions

In this paper, we design public-key and identity-based encryption schemes that are secure against memory attacks. The first question that arises from our work is whether it is possible to (define and) construct other cryptographic primitives such as signature schemes, identification schemes and even protocol tasks that are secure against memory attacks. The second question is whether it is possible to protect against memory attacks that measure an arbitrary polynomial number of bits. Clearly, this requires some form of (randomized) refreshing of the secret-key, and it would be interesting to construct such a mechanism. Finally, it would be interesting to improve the parameters of our construction, as well as the complexity assumptions, and also to design encryption schemes against memory attacks under other cryptographic assumptions.

Acknowledgments. We thank Yael Kalai, Chris Peikert, Omer Reingold, Brent Waters and the TCC program committee for their excellent comments. The third author would like to acknowledge delightful discussions with Rafael Pass about the simultaneous hardcore bits problem in the initial stages of this work.

References

1. Agrawal, D., Archambeault, B., Rao, J.R., Rohatgi, P.: The EM side-channel(s). In: Kaliski Jr., B.S., Koç, Ç.K., Paar, C. (eds.) CHES 2002. LNCS, vol. 2523, pp. 29–45. Springer, Heidelberg (2003)
2. Agrawal, D., Rao, J.R., Rohatgi, P.: Multi-channel attacks. In: Walter, C.D., Koç, Ç.K., Paar, C. (eds.) CHES 2003. LNCS, vol. 2779, pp. 2–16. Springer, Heidelberg (2003)
3. Ajtai, M.: Generating hard instances of the short basis problem. In: Wiedermann, J., Van Emde Boas, P., Nielsen, M. (eds.) ICALP 1999. LNCS, vol. 1644, pp. 1–9. Springer, Heidelberg (1999)
4. Alexi, W., Chor, B., Goldreich, O., Schnorr, C.-P.: Rsa and rabin functions: Certain parts are as hard as the whole. SIAM J. Comput. 17(2), 194–209 (1988)
5. Alwen, J., Peikert, C.: Generating shorter bases for hard random lattices (manuscript, 2008)
6. Bellare, M., Fischlin, M., O'Neill, A., Ristenpart, T.: Deterministic encryption: Definitional equivalences and constructions without random oracles. In: Wagner, D. (ed.) CRYPTO 2008. LNCS, vol. 5157, pp. 360–378. Springer, Heidelberg (2008)
7. Blum, A., Furst, M., Kearns, M., Lipton, R.J.: Cryptographic primitives based on hard learning problems. In: Stinson, D.R. (ed.) CRYPTO 1993. LNCS, vol. 773, pp. 278–291. Springer, Heidelberg (1994)
8. Blum, M., Micali, S.: How to generate cryptographically strong sequences of pseudo-random bits. SIAM J. Comput. 13(4), 850–864 (1984)
9. Boldyreva, A., Fehr, S., O'Neill, A.: On notions of security for deterministic encryption, and efficient constructions without random oracles. In: Wagner, D. (ed.) CRYPTO 2008. LNCS, vol. 5157, pp. 335–359. Springer, Heidelberg (2008)
10. Canetti, R., Eiger, D., Goldwasser, S., Lim, D.-Y.: How to protect yourself without perfect shredding. In: Aceto, L., Damgård, I., Goldberg, L.A., Halldórsson, M.M., Ingólfsdóttir, A., Walukiewicz, I. (eds.) ICALP 2008, Part II. LNCS, vol. 5126, pp. 511–523. Springer, Heidelberg (2008)
11. Catalano, D., Gennaro, R., Howgrave-Graham, N.: Paillier's trapdoor function hides up to $O(n)$ bits. J. Cryptology 15(4), 251–269 (2002)
12. Chari, S., Rao, J.R., Rohatgi, P.: Template attacks. In: Kaliski Jr., B.S., Koç, Ç.K., Paar, C. (eds.) CHES 2002. LNCS, vol. 2523, pp. 13–28. Springer, Heidelberg (2003)
13. Coppersmith, D.: Small solutions to polynomial equations, and low exponent rsa vulnerabilities. J. Cryptology 10(4), 233–260 (1997)
14. Dodis, Y., Reyzin, L., Smith, A.: Fuzzy extractors: How to generate strong keys from biometrics and other noisy data. In: Cachin, C., Camenisch, J.L. (eds.) EUROCRYPT 2004. LNCS, vol. 3027, pp. 523–540. Springer, Heidelberg (2004)

15. Dziembowski, S., Pietrzak, K.: Leakage-resilient stream ciphers. In: IEEE Foundations of Computer Science (to appear, 2008)
16. Gentry, C., Peikert, C., Vaikuntanathan, V.: Trapdoors for hard lattices and new cryptographic constructions. In: STOC, pp. 197–206 (2008)
17. Goldreich, O., Levin, L.A.: A hard-core predicate for all one-way functions. In: STOC, pp. 25–32 (1989)
18. Goldreich, O., Rosen, V.: On the security of modular exponentiation with application to the construction of pseudorandom generators. Journal of Cryptology 16, 2003 (2000)
19. Goldwasser, S., Kalai, Y., Peikert, C., Vaikuntanathan, V (manuscript in preparation, 2008)
20. Goldwasser, S., Kalai, Y.T., Rothblum, G.N.: One-time programs. In: Wagner, D. (ed.) CRYPTO 2008. LNCS, vol. 5157, pp. 39–56. Springer, Heidelberg (2008)
21. Goldwasser, S., Micali, S.: Probabilistic encryption. J. Comput. Syst. Sci. 28(2), 270–299 (1984)
22. Halderman, A., Schoen, S., Heninger, N., Clarkson, W., Paul, W., Calandrino, J., Feldman, A., Appelbaum, J., Felten, E.: Lest we remember: Cold boot attacks on encryption keys. In: Usenix Security Symposium (2008)
23. Håstad, J., Näslund, M.: The security of individual rsa bits. In: FOCS, pp. 510–521 (1998)
24. Håstad, J., Schrift, A.W., Shamir, A.: The discrete logarithm modulo a composite hides o(n) bits. J. Comput. Syst. Sci. 47(3), 376–404 (1993)
25. Ishai, Y., Prabhakaran, M., Sahai, A., Wagner, D.: Private circuits II: Keeping secrets in tamperable circuits. In: Vaudenay, S. (ed.) EUROCRYPT 2006. LNCS, vol. 4004, pp. 308–327. Springer, Heidelberg (2006)
26. Ishai, Y., Sahai, A., Wagner, D.: Private circuits: Securing hardware against probing attacks. In: Boneh, D. (ed.) CRYPTO 2003. LNCS, vol. 2729, pp. 463–481. Springer, Heidelberg (2003)
27. Kaliski Jr., B.S.: A pseudo-random bit generator based on elliptic logarithms. In: Odlyzko, A.M. (ed.) CRYPTO 1986. LNCS, vol. 263, pp. 84–103. Springer, Heidelberg (1987)
28. Kocher, P.C.: Timing attacks on implementations of diffie-hellman, RSA, DSS, and other systems. In: Koblitz, N. (ed.) CRYPTO 1996. LNCS, vol. 1109, pp. 104–113. Springer, Heidelberg (1996)
29. Kocher, P.C., Jaffe, J., Jun, B.: Differential power analysis. In: Wiener, M. (ed.) CRYPTO 1999. LNCS, vol. 1666, pp. 388–397. Springer, Heidelberg (1999)
30. Liu, Y.-K., Lyubashevsky, V., Micciancio, D.: On bounded distance decoding for general lattices. In: APPROX-RANDOM, pp. 450–461 (2006)
31. Long, D.L., Wigderson, A.: The discrete logarithm hides o(log n) bits. SIAM J. Comput. 17(2), 363–372 (1988)
32. Side-Channel Cryptanalysis Lounge (2008),
    http://www.crypto.rub.de/en_sclounge.html.
33. Micali, S., Reyzin, L.: Physically observable cryptography. In: Naor, M. (ed.) TCC 2004. LNCS, vol. 2951, pp. 278–296. Springer, Heidelberg (2004)
34. Peikert, C.: Public-key cryptosystems from the worst-case shortest vector problem. Cryptology ePrint Archive, Report 2008/481 (2008), http://eprint.iacr.org/
35. Peikert, C., Vaikuntanathan, V., Waters, B.: A framework for efficient and composable oblivious transfer. In: Wagner, D. (ed.) CRYPTO 2008. LNCS, vol. 5157, pp. 554–571. Springer, Heidelberg (2008)
36. Peikert, C., Waters, B.: Lossy trapdoor functions and their applications. In: STOC, pp. 187–196 (2008)
37. Petit, C., Standaert, F.-X., Pereira, O., Malkin, T., Yung, M.: A block cipher based pseudo random number generator secure against side-channel key recovery. In: ASIACCS, pp. 56–65 (2008)

38. Pietrzak, K., Vaikuntanathan, V.: Personal Communication (2009)
39. Regev, O.: On lattices, learning with errors, random linear codes, and cryptography. In: STOC, pp. 84–93 (2005)
40. Rosen, A., Segev, G.: Chosen-ciphertext security via correlated products. Cryptology ePrint Archive, Report 2008/116 (2008)
41. Vazirani, U.V., Vazirani, V.V.: Efficient and secure pseudo-random number generation. In: Blakely, G.R., Chaum, D. (eds.) CRYPTO 1984. LNCS, vol. 196, pp. 193–202. Springer, Heidelberg (1985)
42. Yao, A.C.: Theory and application of trapdoor functions. In: Symposium on Foundations of Computer Science, pp. 80–91 (1982)

The Differential Privacy Frontier
(Extended Abstract)

Cynthia Dwork

Microsoft Research

Abstract. We review the definition of *differential privacy* and briefly survey a handful of very recent contributions to the differential privacy frontier.

1 Background

Differential privacy is a strong privacy guarantee for an individual's input to a (randomized) function or sequence of functions, which we call a *privacy mechanism*. Informally, the guarantee says that the behavior of the mechanism is essentially unchanged independent of whether any individual opts into or opts out of the data set. Designed for statistical analysis, for example, of health or census data, the definition protects the privacy of individuals, and small groups of individuals, while permitting very different outcomes in the case of very different data sets.

We begin by recalling some differential privacy basics. While the frontier of a vibrant area is always in flux, we will endeavor to give an impression of the state of the art by surveying a handful of extremely recent advances in the field.

Formally, The degree of privacy offered is described by a parameter, ε.

Definition 1. *A randomized function $\mathcal{K}$ gives ε-differential privacy if for all data sets D and D' of Hamming distance $d(D, D') \leq 1$ and all $S \subseteq Range(\mathcal{K})$,*

$$\Pr[\mathcal{K}(D) \in S] \leq e^{\varepsilon} \times \Pr[\mathcal{K}(D') \in S] \tag{1}$$

The probability is taken is over the coin tosses of $\mathcal{K}$.

The definition represents a paradigm shift: instead of a simulation-style definition, in which we compare what an adversary can learn about an individual with, versus without, access to the outputs of the privacy mechanism, differential privacy focuses on limiting the additional risk – of anything! – incurred by an individual as a consequence of opting into (or opting out of) a data set. This is no accident, as any "with vs. without access" definition is doomed to fail [5,9]. The definition is suited to the real world because it is a property of the mechanism alone, and has no bearing on what the consumer of information produced in a differentially private fashion might or might not know. In consequence the

O. Reingold (Ed.): TCC 2009, LNCS 5444, pp. 496–502, 2009.

outputs of a differentially private mechanism preserve differential privacy independent of the information and computational power available to an adversary, now or in the future.

Two principal techniques for ensuring differential privacy have appeared, one for the case of (vectors of) real-valued outputs and the other for outputs of arbitrary types [7,20]; the former is efficient, the latter may not be [10]. These positive results and a key precursor [11] (which used a cumbersome definition now known to imply a natural, mild relaxation of pure differential privacy and which showed that if the number of queries is sublinear in the size of the data set then privacy can be obtained "for free," *i.e.*, with noise smaller than the sampling error) have been used to obtain highly accurate differentially private solutions to a host of problems in datamining, statistics, and learning (see, *e.g.*, [2,1,19,3]). A central concept is the *sensitivity* of a real-valued function mapping data sets to (vectors of) reals:

Definition 2. *Let $\mathcal{D}$ denote the space of all databases. For $f : \mathcal{D} \rightarrow R^d$, the sensitivity of f is*

$$\Delta f = \max_{D,D'} \|f(D) - f(D)\|_1$$

for all D, D' of distance at most 1.

Roughly speaking, real-valued data analyses that have low sensitivity permit highly accurate differentially private mechanisms [7]. The true answer is computed and Lapalacian (symmetric exponential) noise is added with variance depending on ε and the sensitivity of the query. For analyses whose outcome need not be real (it might be the choice of a color, or a set of locations, or a string), or in cases where the output is real-valued but adding noise makes no sense (the output might be a price when the data set is a collection of bids in an auction), if there is an insensitive function for evaluating the quality of an output (for example, revenue, in the case of an auction), then again high-quality outputs can be obtained in a differentially private fashion [20]. This is done using the *exponential mechanism* which, roughly speaking, weights each possible answer with a density that falls exponentially with its (in)utility, again depending on ε and also the sensitivity, this time, of the utility function[1].

Very recently Ghosh *et al.* considered the question of what it means for a privacy mechanism to be optimal [14]. Intuitively, different users may have different preconceptions before seeing the output of a privacy mechanism, and therefore two users might place different values on the same piece of information. In such a setting what sort of utility function should the mechanism employ? Using a very general notion of utility, and permitting each user to have *her own* utility function, Ghosh *et al.* show that a discretized version of the Laplace distribution used in [7] simultaneously maximizes utility for all users for the case of counting functions ("How many rows in the data set satisfy predicate P?").

---

[1] The addition of Laplacian noise to a real-valued output is a special case of the exponential mechanism: the (in)utility of an output is its L_1 distance from the true answer.

2   Differentially Private Synthetic Data Sets and Coresets

A series of negative results concerning privacy, says, roughly, that there is a class of queries with the property that it is blatantly non-private (allowing almost full reconstruction) if "too many" queries receive "overly accurate" responses [4,8,12]. These results have been viewed as saying that, in contrast to the sublinear queries work discussed above, one cannot privately answer a small polynomial number of queries, say, n^3 or even n^2, with reasonably small noise (here, n is the number of elements in the data set).

The idea of creating a *synthetic* data set whose statistics closely mirror those of the original data set, but which preserves privacy of individuals, was proposed in the statistics community as far back as 1993 [24]. However, the negative results imply that no such data set can safely provide very accurate answers to too many questions, motivating the interactive approach to private data analysis ([11] *et sequelae*). Intuitively, the advantage of the interactive approach is that only the questions actually asked receive responses, while to offer the same utility in the non-interactive approach all, or at least most, questions must receive very accurate responses, leading to blatant non-privacy.

Against this backdrop, Blum, *et al.* revisited the non-interactive case from a learning theory perspective, and challenged the above interpretation about the necessity of limiting the number of queries [3]. Let X be a universe of data items and C be a "concept" class consisting of efficiently computable functions $c : X \rightarrow \{0, 1\}$. Given a sufficiently large database $x \in X^n$, Blum *et al.* inefficiently, but with differential privacy, obtain a *synthetic database* that maintains approximately correct fractional counts for *all* concepts in C. That is, letting y denote the synthetic database produced, with high probability over the choices made by the privacy mechanism, for every concept $c \in C$, the fraction of elements in y that satisfy c is approximately the same as the fraction of elements in x that satisfy c.[2]

This remarkable result has rekindled interest in synthetic databases in particular and non-interactive solutions in general. When can differentially private synthetic databases be constructed efficiently? Very roughly, if either the universe X of data items or the concept class C is of size superpolynomial in a computation parameter κ, then, under standard computational assumptions, there exists a distribution on databases and a concept class C for which there is no efficient (in κ) mechanism for privately generating synthetic databases. In contrast, if both the concept class and the data universe are of size polynomial in κ then not only is there an efficient mechanism, but the size of the input database can be surprisingly small, namely $|C|^{o(1)} \cdot \log |X|$ (or even $O(2^{\sqrt{\log |C|}} \log |X|))$ [10]. Thus C can be very large, as a function of n (while still polynomial in κ).

Interestingly, for the potentially easier problem of privately generating a data structure (as opposed to a synthetic data set) from which it is possible to

---

[2] This does not contradict the negative results because of the size of the error in the case of attacks using a polynomial number of queries, or the size of the input database in the case of attacks using an exponential number of queries.

approximate counts, there is a tight "if and only if" connection between hardness of sanitization and the existence of *traitor tracing* schemes in cryptography [10].

2.1   Coresets

In computational geometry a coreset for a point set P is a small, weighted, point set C that is useful in computing approximate solutions of problems for P. For example, the queries might consist of a set of k points (not necessarily related to P), and the exact answer to the query might be the sum of the distances from each point $p \in P$ to its closest point in the query set Q; this is a k-*median* query. Coresets enjoy an extensive literature; different techniques are used for creating coresets appropriate for different sorts of queries.

Feldman *et al.* define *private coresets*. These are coresets in the traditional sense, but they are generated from P in a differentially private fashion [13]. Thus, the private coreset problem is similar to the problem of private generation of a synthetic data set, where the class of queries to be handled by the coreset plays a role analogous to the fractional concept class counts. Using similar techniques to those in [3], Feldman *et al.* show how any coreset construction can (ineffeciently) be modified to yield differentially private coresets, and using new techniques they obtain an efficient construction of coresets for k-median queries.

3   Connections to Other Fields of Study

As the study of privacy broadens, differential privacy productively blends, Zelig-like, with a surprising variety of concepts[3]. We have already seen this in the connection between traitor-tracing and non-interactive sanitization. Here we offer four additional examples.

Truthful Mechanisms for Strategic Agents. In a truthful mechanism, reporting one's true value is a dominant strategy. Designing mechanisms to be truthful simplifies their analysis, making truthful mechanisms a widely studied solution concept in economics. One way of ensuring truthfulness is to arrange that the price paid by an individual is *independent* of his or her reported value. Analogously, if a price is set by a differentially private mechanism, then the price paid by an individual is "almost" independent of her bid. This intuition has been validated: differential privacy can be used to obtain "approximate truthfulness" [20], yielding the first collusion-resilient mechanism; it can also be used to better approximately solve combinatorial public project problems than can be done with any efficient truthful solution (unless $NP \subseteq BPP$) [15]. In each case an agent can gain only slightly by lying.

Additive Combinatorics and Dense Model Theorems. Reingold *et al.* [23] give (almost) the following definition of density: Consider distributions X and

---

[3] The Internet Movie Database summarizes Woody Allen's *Zelig*: "Fictional documentary about the life of human chameleon Leonard Zelig, a man who becomes a celebrity in the 1920s due to his ability to look and act like whoever is around him."

Y over a set R. X is e^ε-dense in Y if for all $x \in R$, $\Pr[X = x] \leq e^\varepsilon \Pr[Y = x]$. Thus, a randomized mechanism f is ε-differentially private if and only if $f(D)$ is e^ε-dense in $f(D')$ for all D, D' such that $d(D, D') \leq 1$. This connection between differential privacy and (mutually) dense distributions has been exploited in an investigation of *computational* differential privacy, *i.e.*, differential privacy against a computationally bounded adversary. [21], which extends the dense model theorem in [23] to demonstrate equivalence between two definitions (indistinguishability-based and simulatability-based, respectively) of computational differential privacy.

Robust Statistics and the Influence Function. *Robust statistics* is the subfield of statistics that attempts to cope with outliers. In consequence, in a robust analysis the specific data for any one individual should not greatly affect the outcome of the analysis, suggesting a connection to differential privacy. Indeed, independently of our community and unknown to us, as early as 2005 Heitzig [17] proposed adapting, for the sake of privacy, a specific robust technique for reducing bias and estimating variance, known as the Jackknife [22,25].

The Jackknife is related to the the *influence function* IF$(x, T; F)$, which describes how an estimator T applied to samples from distribution F changes if we replace F by a distribution G with an infinitesimal contamination at x: $G = (1 - t)F + t\Delta_x$, for very small t. (See [18,16].) This, in turn, is related to sensitivity "in a statistical setting" (that is, *whp* over samples from the distribution F). Typically, robust estimators are designed to have bounded influence function, implying *vanishing* sensitivity in a statistical setting. Heitzig's intuition, supported by detailed statistical insight but not made rigorous, was that it should be possible to ensure privacy by reporting an interval for the results of an analysis, rather than the exact value, where the size of the interval is determined by his (randomized) Jackknife-like procedure. Independently of Heitzig, but later, Dwork and Lei were also inspired by the implications of vanishing sensitivity offered by bounded influence functions. They adapted several robust algorithms, for varying statistical tasks, to provably (and always) yield differential privacy, with excellent accuracy whenever certain mild statistical assumptions hold [6].

4   Concluding Remarks

We have surveyed at least six very recent contributions on the differential privacy frontier. In several cases the work has forged links with other fields and communities: statistics, cryptography, complexity, geometry, mechanism design, and optimization. The plethora of new techniques, the formulation of new problems, and the fruitful interplay with other fields provides fertile ground for ebullient growth in an intellectually exciting and socially valuable endeavor.

Acknowledgements. I am grateful to all the authors of the new works for sharing their results with me before publication. Thanks also to Guy Rothblum for his helpful comments on an early draft of this extended abstract.

References

1. Barak, B., Chaudhuri, K., Dwork, C., Kale, S., McSherry, F., Talwar, K.: Privacy, accuracy, and consistency too: A holistic solution to contingency table release. In: Proceedings of the 26th Symposium on Principles of Database Systems, pp. 273–282 (2007)
2. Blum, A., Dwork, C., McSherry, F., Nissim, K.: Practical privacy: The SuLQ framework. In: Proceedings of the 24th ACM SIGMOD-SIGACT-SIGART Symposium on Principles of Database Systems (June 2005)
3. Blum, A., Ligett, K., Roth, A.: A learning theory approach to non-interactive database privacy. In: Proceedings of the 40th ACM SIGACT Symposium on Thoery of Computing (2008)
4. Dinur, I., Nissim, K.: Revealing information while preserving privacy. In: Proceedings of the Twenty-Second ACM SIGACT-SIGMOD-SIGART Symposium on Principles of Database Systems, pp. 202–210 (2003)
5. Dwork, C.: Differential privacy. In: Bugliesi, M., Preneel, B., Sassone, V., Wegener, I. (eds.) ICALP 2006. LNCS, vol. 4052, pp. 1–12. Springer, Heidelberg (2006)
6. Dwork, C., Lei, J.: Differential privacy and robust statistics (manuscript) (November 2008)
7. Dwork, C., McSherry, F., Nissim, K., Smith, A.: Calibrating noise to sensitivity in private data analysis. In: Proceedings of the 3rd Theory of Cryptography Conference, pp. 265–284 (2006)
8. Dwork, C., McSherry, F., Talwar, K.: The price of privacy and the limits of lp decoding. In: Proceedings of the 39th ACM Symposium on Theory of Computing, pp. 85–94 (2007)
9. Dwork, C., Naor, M.: On the difficulties of disclosure prevention in statistical databases or the case for differential privacy (manuscript, 2008)
10. Dwork, C., Naor, M., Reingold, O., Rothblum, G., Vadhan, S.: When and how can privacy-preserving data release be done efficiently? (November 2008) (manuscript)
11. Dwork, C., Nissim, K.: Privacy-preserving datamining on vertically partitioned databases. In: Franklin, M. (ed.) CRYPTO 2004. LNCS, vol. 3152, pp. 528–544. Springer, Heidelberg (2004)
12. Dwork, C., Yekhanin, S.: New efficient attacks on statistical disclosure control mechanisms. In: Wagner, D. (ed.) CRYPTO 2008. LNCS, vol. 5157, pp. 468–480. Springer, Heidelberg (2008)
13. Feldman, D., Fiat, A., Kaplan, H., Nissim, K.: Private coresets (November 2008) (manuscript)
14. Ghosh, A., Roughgarden, T., Sundarajan, M.: Universally utility-maximizing privacy mechanisms (November 2008) (manuscript)
15. Gupta, A., Ligett, K., McSherry, F., Roth, A., Talwar, K.: Differentially private approximation algorithms (November 2008) (manuscript)
16. Hampel, F., Ronchetti, E., Rousseeuw, P., Stahel, W.: Robust Statistics: The Approach Based on Influence Functions. John Wiley, New York (1986)
17. Heitzig, J.: The "jackknife" method: Confidentiality protection for complex statistical analyses. In: Proceedings of the Joint UNECE/Eurostat work session on statistical data confidentiality (2005)
18. Huber, P.: Robust statistics. John Wiley & Sons, Chichester (1981)
19. Kasiviswanathan, S., Lee, H., Nissim, K., Raskhodnikova, S., Smith, A.: What can we learn privately? In: Proceedings of FOCS 2008 (2008)

20. McSherry, F., Talwar, K.: Mechanism design via differential privacy. In: Proceedings of the 48th Annual Symposium on Foundations of Computer Science (2007)
21. Mironov, I., Pandey, O., Reingold, O., Vadhan, S.: Computational differential privacy (November 2008) (manuscript)
22. Quenouille, M.: Notes on bias in estimation. Biometrika 43, 353–360 (1956)
23. Reingold, O., Trevisan, L., Tulsiani, M., Vadhan, S.: Dense subsets of pseudorandom sets. In: Proceedings 49th Annual IEEE Symposium on Foundations of Computing (2008)
24. Rubin, D.: Discussion: Statistical disclosure limitation. Journal of Official Statistics 9, 462–468 (1993)
25. Tukey, J.: Bias and confidence in not-quite large samples (abstract). Ann. Math. Statist., 29 (1958)

How Efficient Can Memory Checking Be?

Cynthia Dwork[1], Moni Naor[2,*],
Guy N. Rothblum[3,**], and Vinod Vaikuntanathan[4,***]

[1] Microsoft Research
[2] The Weizmann Institute of Science
[3] MIT
[4] IBM Research

Abstract. We consider the problem of memory checking, where a user wants to maintain a large database on a remote server but has only limited local storage. The user wants to use the small (but trusted and secret) local storage to detect faults in the large (but public and untrusted) remote storage. A memory checker receives from the user store and retrieve operations to the large database. The checker makes its own requests to the (untrusted) remote storage and receives answers to these requests. It then uses these responses, together with its small private and reliable local memory, to ascertain that all requests were answered correctly, or to report faults in the remote storage (the public memory).

A fruitful line of research investigates the complexity of memory checking in terms of the number of queries the checker issues per user request (query complexity) and the size of the reliable local memory (space complexity). Blum et al., who first formalized the question, distinguished between online checkers (that report faults as soon as they occur) and offline checkers (that report faults only at the end of a long sequence of operations). In this work we revisit the question of memory checking, asking *how efficient can memory checking be?*

For online checkers, Blum et al. provided a checker with logarithmic query complexity in n, the database size. Our main result is a lower bound: we show that for checkers that access the remote storage in a deterministic and non-adaptive manner (as do all known memory checkers), their query complexity must be at least $\Omega(\log n/\log\log n)$. To cope with this negative result, we show how to trade off the read and write complexity of online memory checkers: for any desired logarithm base d, we construct an online checker where either reading *or* writing is inexpensive and has query complexity $O(\log_d n)$. The price for this is that the other operation (write or read respectively) has query complexity $O(d \cdot \log_d n)$. Finally, if even this performance is unacceptable, *offline* memory checking may be an inexpensive alternative. We provide a scheme with $O(1)$ amortized query complexity, improving Blum et al.'s construction, which only had such performance for long sequences of at least n operations.

* Incumbent of the Judith Kleeman Professorial Chair; Research supported in part by a grant from the Israel Science Foundation.
** Research supported by NSF Grants CCF-0635297, NSF-0729011, CNS-0430336, Israel Science Foundation Grant 700/08 and by a Symantec Graduate Fellowship.
*** Supported in part by NSF CCF-0635297 and Israel Science Foundation 700/08.

1   Introduction

Consider a user who wants to maintain a large database but has only limited local storage. A natural approach is for the user to store the database on a remote storage server. This solution, however, requires that the user trust the remote storage server to store the information reliably. It is natural to ask whether the user can use his or her small (but trusted and secret) local storage to detect faults in the large (but public and untrusted) remote storage. This is the problem of *memory checking*, as introduced by Blum, Evans, Gemmel, Kannan and Naor [6] in 1991. Since then, this problem has gained even more importance for real-world applications, see for example the more recent works of Clarke *et al.* [8], Ateniese *et al.* [4], Juels and Kaliski [13], Oprea and Reiter [16] and Shacham and Waters [17]. Large databases are increasingly being outsourced to untrusted storage providers, and this is happening even with medical or other databases where reliability is crucial. Another wide-spread and growing phenomenon are services that offer individual users huge and growing remote storage capacities (e.g. webmail providers, social networks, repositories of digital photographs, etc.). In all of these applications it is important to guarantee the integrity of the remotely stored data.

Blum *et al.* formalized the above problem as the problem of memory checking. A memory checker can be thought of as a layer between the user and the remote storage. The checker receives from its user a sequence of "store" and "retrieve" operations to a large unreliable memory. Based on these "store" and "retrieve" requests, it makes its own requests to the (untrusted) remote storage and receives answers to these requests. The checker then uses these responses, together with a small private and reliable "local" memory, to ascertain that all requests were answered correctly, or to report that the remote storage (the *public memory*) was faulty. The checker's assertion should be correct with high probability (a small two-sided error is permitted). Blum *et al.* made the distinction between online and offline memory checking. An *online* checker verifies the correctness of each answer it gives to the user. An *offline* checker gives only the relaxed guarantee that after a (long) sequence of operations a user can verify whether or not there was an error *somewhere* in the sequence of operations. Two important complexity measures of a memory checker are its *space complexity*, the size of the secret reliable "local" memory, and its *query complexity*, the number of queries made to the unreliable memory per user request. One may consider additional complexity measures such as the alphabet size (the size of words in the public memory), and more measures such the checker's and public memory's running times, the amount of public storage, etc. See Section 2 for formal definitions and a fuller discussion.

In this work we revisit the question of designing efficient memory checkers. Our main result is a lower bound on the query complexity of deterministic and non-adaptive online memory checkers. We also present new upper bounds for both online and off-line memory checking.

Online Memory Checkers. The strong verification guarantee given by online memory checkers makes them particularly appealing for a wide variety of applications. Blum *et al.* construct efficient online memory checkers with space complexity that is proportional to the size of a cryptographic key, and *logarithmic* query complexity. Their construction(s) assume that a one-way function exists and that the adversary who controls the public memory is efficient and cannot invert the function. In fact, this assumption was shown to be *essential* by Naor and Rothblum [15], who showed that any online memory checker with a non-trivial query-space tradeoff can only be computationally secure and must be based on the existence of a one-way function. Even in the computational setting, the space complexity of Blum *et al.*'s online memory checkers is intuitively optimal, since if the secret memory is s bits long, an (efficient) adversary can guess it (and fool the memory checker) with probability at least 2^{-s}. What is less clear, however, is whether the logarithmic query complexity is essential (in a computational setting). This is an important question, since while this logarithmic overhead is reasonable, in many applications it remains a significant price to have to pay for data verification.

Where then does this overhead come from? The logarithmic query complexity is needed to avoid *replay attacks*, in which the correct public memory is swapped for some older version of it. In most applications replay attacks are a serious threat, and Blum *et al.* (and all other solutions we know of) use a tree structure to overcome this devastating class of attacks. This tree structure incurs a logarithmic overhead which is basically the depth of the tree. We begin by asking whether it is possible to avoid the logarithmic overhead and construct memory checkers with lower query complexity. We show that the answer is negative (even in the cryptographic setting!) for all known and/or practical methods of designing memory checkers.

A Query Complexity Lower Bound. Consider online memory checkers, where for each store or retrieve request made by the user, the locations that the checker accesses in the public memory are fixed and known. We call such a checker a deterministic and non-adaptive checker. Known checker constructions are all deterministic and non-adaptive, indeed tree authentication structures all have this property. Our main result is a new lower bound, showing that any deterministic non-adaptive memory checker must have query complexity $\Omega(\log n / \log \log n)$. Thus the logarithmic query complexity overhead is (almost) unavoidable for online memory checking. This is stated more fully (but still informally) below, see Section 3 for the full details.

Theorem 1. Let $\mathcal{C}$ be a non-adaptive and deterministic memory checker for an n-index boolean database, with space complexity $s \leq n^{1-\varepsilon}$ for some $\varepsilon > 0$, query complexity q and a polylog-length alphabet (public memory word size). It must be that $q = \Omega(\frac{\log n}{\log \log n})$.

Let us examine the above theorem more closely. Considering only checkers that are deterministic and non-adaptive may seem at first glance to be quite restrictive. We argue, however, that *practical* checkers will likely have to conform to this restriction:

An *adaptive* checker is one that chooses sequentially which locations in the remote storage it reads and writes, and chooses these locations based on the contents of earlier read locations. This means that the checker needs to conduct, for every user request, several rounds of communication with the remote storage (the checker needs to know the contents of a location before deciding which location it accesses next). Since this communication happens over a network, it may very well lead to latency which results in more of an overhead than the logarithmic query complexity of non-adaptive checkers. In addition, in cases where the memory contents are encrypted non-adaptive memory access may be especially desirable, as the set of locations accessed reveals nothing about the (decrypted) contents of the memory.

Another problem with adaptive checkers is that they make *caching* the results much more difficult, since the actual locations needed to be stored in the faster memory change between accesses.

A *non-deterministic* checker may also result in worse performance. Such a checker strategy, with queries that are either significantly randomized or hard to predict (depending on the secret memory), destroys locality in the user's queries and makes it hard to utilize caching mechanisms. In particular, user accesses to neighboring database indices would not necessarily be mapped to checker accesses to neighboring locations in the remote storage, and repeated user accesses to the same database index would not necessarily be mapped to the same locations in the remote storage. For many of the applications of memory checking, this will result in an unacceptable overhead for the remote storage server. We note that Blum *et al.*'s constructions, as well as all of the constructions we present in this work, have the important property that they do preserve (to a large extent) the locality of a user's data accesses.

Finally, we note that the restriction on sub-linear space is essential, as the problem of memory checking makes very little sense with linear secret memory; the checker can simply store the entire database in reliable memory! Finally, it is interesting to ask whether the lower bound can be extended to larger alphabets (we focus on polylog word lengths or quasi-polynomial alphabet size). We do note that the best parameters attained both in our work and in [6] can be attained with words of poly-logarithmic length.

Trading Off Reads and Writes. Is all hope of improving the performance of online memory checkers lost in light of Theorem 1? We argue that this is not the case. While we cannot improve the query complexity of online checkers beyond logarithmic, we observe that in many applications read operations are far more numerous than write, and vice versa. One example for frequent read operation is a database that is read frequently but updated only periodically. An example for frequent write is a repository of observed data (say climate measurements) that is constantly updated but polled much less frequently.

For these settings we show how to trade off the query complexity of read and write operations. For any desired logarithm base d, we show how to build an online checker where the frequent operation (read or write) is inexpensive and has query complexity $O(\log_d n)$, and the infrequent operation (write or read respectively) has query complexity $O(d \cdot \log_d n)$. The space complexity is proportional to a security parameter (it can be poly-logarithmic under an exponential hardness assumption), and the alphabet size is the logarithm of the desired soundness. The construction uses a pseudo-random function (see [12]), and can thus be based on the existence of any one-way function. This means, for example, that if one is willing to have a polynomial (n^ε) write complexity, then we can get a constant $(O(1/\varepsilon))$ read complexity (and vice versa). This may be very useful for a database that is read frequently but only updated infrequently.

To achieve this tradeoff, we provide two constructions: one for efficient write and one for efficient read. Both of these use a tree-based authentication structure, where the tree's depth is $\log_d n$. The efficient-write construction can viewed as a generalization of Blum *et al.*'s tree-based online memory checker. The efficient-read construction is different in the way it stores authentication information. Intriguingly, we do not know how to get a good read-write trade-off based on UOWHFs where the checker's memory only needs to be reliable (and not necessarily private). Blum *et al.* were able to present such a construction (albeit with a nearly exponential-size alphabet) with logarithmic query complexity, but their construction does not easily yield itself to a read-write tradeoff. See Section 4 for the full details.

While we believe that these trade-offs are very useful for many applications, we still cannot beat the lower bound of Theorem 1: the *sum* of read and write complexities is still at least logarithmic in n (not surprisingly, since the above checkers are deterministic and non-adaptive). For many other applications this may still be prohibitively expensive. This leads us then to revisit Blum *et al.*'s notion of *offline memory checking*, where the verification guarantee of the checker is weaker, but it is possible to achieve better performance.

An Off-Line Alternative. Blum *et al.* suggested the notion of an *offline* memory checker. Such a memory checker gives the relaxed guarantee that after a (long) sequence of operations it can be used to check whether there was an error. In other words, whether any value retrieved from public memory was different from the last value stored at that location. The advantage of offline memory checkers is that they allow much better parameters. Specifically, Blum *et al.* gave a construction where for any *long* sequence of user operations (at least n operations) the *amortized* query complexity is $O(1)$ and the space complexity is logarithmic in n and in the soundness parameter. Remarkably, the security of their checker is *information theoretic*, and does not rely on any cryptographic assumptions.

We conclude that for applications in which the offline guarantee suffices, say when the user does not mind that some of the data may be retrieved incorrectly as long as this is eventually detected, the query complexity of both read and write can be reduced to $O(1)$. It is natural to ask what can possibly be improved in the above construction, as the (amortized) query and space complexity seem optimal. One place for improvement is that Blum *et al.*'s construction is highly *invasive*: the checker stores a significant amount of additional information in the public memory on top of the database. Ajtai [2] showed that this invasiveness cannot be avoided (see the full version for an overview of Ajtai's results).

We focus on a different parameter. The above off-line checker only guarantees good amortized performance for *long sequences* of at least n operations. We observe that for shorter operation sequences, the amortized performance will be quite bad, as their checker needs to always scan the *entire public memory* before deciding whether there were any errors. So for a k operation sequence, the amortized query complexity will be $O(n/k)$. In Section 5 we overcome this obstacle. We present a simple and inexpensive offline memory checker where the amortized query complexity for *any sequence of operations* (even a short one) is $O(1)$. Moreover, we show that similar ideas can be used to decrease the invasiveness of the checker, and that the invasiveness (the amount of extra information stored in public memory on top of the database) only needs to be proportional to the number of database locations that the checker actually accesses (instead of always being proportional to the entire database size as in Blum *at al.*). We note that we can overcome Ajtai's invasiveness lower bound in this setting because the proof of that lower bound considers sequences of operations that access every location in the database (again, see the full version for the details).

Organization. We begin in **Section 2** with definitions of memory checkers (we refer the reader to Goldreich [10,11] for standard cryptographic definitions). In **Section 3** we state and prove our lower bound for the query complexity of online memory checkers. Constructions of read-write tradeoffs are presented in **Section 4**. Finally, in **Section 5** we present a new and improved construction of offline checkers.

2   Memory Checkers: Definitions

A memory checker is a probabilistic Turing machine C with five tapes: a read-only input tape for receiving read/write requests from the user U to the RAM or database, a write-only output tape for sending responses back to the user, a read-write work tape (the secret reliable memory), a write-only tape for sending read/write requests to the memory M and a read only input tape for receiving M's responses.

Let n be the size of the database (the RAM) U is interested in using. A checker is presented with "store" (write) and "retrieve" (read) requests made by U to M. After each "retrieve" request C returns an answer or outputs that M's operation is BUGGY. C's operation should be both correct and complete for all polynomial (in n) length request sequences. Formally, we say that a checker has completeness c (2/3 by default) and soundness s (1/3 by default) if:

– Completeness. For any polynomial-length sequence of U-requests, as long as M answers all of C's "retrieve" requests correctly (with the last value that C stored at that location), C also answers all of U's "retrieve" requests correctly with probability at least c.[1]

---

[1] In fact in all our constructions we get *perfect completeness*; the checker answers all requests correctly with probability 1.

– Soundness. For any polynomial-length sequence of $\mathcal{U}$-requests, for *any* (even incorrect or malicious) answers returned by $\mathcal{M}$, the probability that C answers a user request incorrectly is at most s. C may either recover the correct answer independently or answer that $\mathcal{M}$ is "BUGGY", but it may not answer a request incorrectly (beyond probability s).

Note that the completeness and soundness requirements are for *any* request sequence and for *any* behavior of the unreliable memory. Thus we think of $\mathcal{U}$ and $\mathcal{M}$ as being controlled by a malicious adversary. A memory checker is secure in the computational setting if the soundness property holds versus any PPTM adversary. In this setting, if one-way functions exist, then they can be used to construct very good online memory checkers (see [6]).

As previously noted, [6] make the distinction between memory checkers that are online and offline. An *offline* checker is notified before it receives the last "retrieve" request in a sequence of requests. It is only required that if at some point in the sequence a user retrieve request was answered incorrectly, then the checker outputs BUGGY (except with probability s). The task of an *online* checker is more difficult: if $\mathcal{M}$'s response to some request was incorrect, C must immediately detect the error or recover from it (with high probability). C is not allowed (beyond a small probability) to ever return an erroneous answer to $\mathcal{U}$. Note that after the memory checker informs the user that $\mathcal{M}$'s operation was BUGGY, there are no guarantees about the checker's answers to future queries.

Recall that the two important measures of the complexity of a memory checker are the size of its secret memory (*space complexity*) and the number of requests it makes per request made by the user (*query complexity*). The query complexity bounds the number of locations in public memory accessed (read or written) per user request. We would prefer memory checkers to have small space complexity and small query complexity. A memory checker is *polynomial time* if C is a PPTM (in n).

A **deterministic and non-adaptive** memory checker is a checker C where the locations it queries in public memory are set and depend (deterministically) only on the database index being stored or retrieved. We call such a checker non-adaptive because it chooses the entire list of locations to access in public memory without knowing the value of the public (or secret) memory at any location. We note, though, that even a non-adaptive checker can decide which *values* to write into those (non-adaptively chosen) locations in an adaptive manner, based on values it reads and the secret memory. One way to think of a deterministic non-adaptive checker is by associating with each index in the database a static set of locations that the checker accesses when storing or retrieving that index.

Similarly, for a deterministic and non-adaptive checker, each location in the public memory can be associated with the set of database indices that "access" it. We say that a location in public memory is *t-heavy* if there are at least t database indices that access it (for store or retrieve requests).

We say that C is a $(\mathbf{\Sigma}, \mathbf{n}, \mathbf{q}, \mathbf{s})$-**checker** if it can be used to store a (binary) database of n indices with query complexity q and space complexity s, where the secret and public memory are over the alphabet Σ (we allow this alphabet to be non-binary).

3 Lower Bounds

Throughout this section we obtain a lower bound for memory checking by using **restrictions of memory checkers**. When we talk about restricting a memory checker to a subset of database indices, we start with a checker $\mathcal{C}$ say for databases with n indices, and obtain from it a checker $\mathcal{C}'$ for databases with $n' < n$ indices. This is done simply by selecting a subset I of the indices that $\mathcal{C}$ works on ($|I| = n'$) and ignoring all of the rest. Naturally, the completeness and soundness of $\mathcal{C}$ carry over to $\mathcal{C}'$. Intuitively, this may also mean that we can ignore some of the locations in public memory or some of the secret memory, but we make no such assumptions in this work. It may seem that this is a bad bargain: the number of indices is decreased without gaining anything. However, when performing the restrictions below we gain (reduce) something in other complexity measures such as the query complexity. Sometimes this will require making additional changes to the checker, such as moving some locations from public to secret memory.

We will assume without loss of generality that the read and the write operations access the same locations. This involves at most doubling the number of accesses per operation.

We now present our lower bound for non-adaptive and deterministic checkers.

Theorem 1. Let $\mathcal{C}$ be a (Σ, n, q, s) deterministic and non-adaptive online memory checker, with $s \leq n^{1-\varepsilon}$ for some $\varepsilon > 0$ and $|\Sigma| \leq n^{\text{poly} \log n}$. It must be that $q = \Omega(\frac{\log n}{\log \log n})$.

Proof (of Theorem 1). Let $q_0 = q$ be the query complexity of the checker $\mathcal{C}$. The proof proceeds by iteratively restricting the checker, gradually lowering its query complexity until a lower bound can be obtained. This is done by examining the memory checker and determining whether there is a relatively large set of "heavily queried" locations in the public memory. I.e. whether there is a polynomial size set of locations in the public memory, each of which is queried when reading or writing many database indices. Recall that we call such heavily-queried locations in the public memory "heavy locations".[2] If there is such a set of heavy locations, then those public memory locations are moved into the secret memory and the query complexity of the checker is reduced significantly. In this case we advance towards our goal of lower bounding the query complexity. This intuition is formalized by Lemma 1 (the proof appears below):

Lemma 1. Let $\mathcal{C}$ be a (Σ, n, q, s) deterministic and non-adaptive online memory checker. For every threshold $t \in \mathbb{N}$ such that $n > t$ the following holds: If there exists $m \in \mathbb{N}$ such that there are m or more t/m-heavy locations in public

---

[2] I.e. locations accessed by many indices - more formally a location is t-*heavy* if there are t different queries $i \in [n]$ that access it.

memory, then for some $i \in [q]$ *the memory checker* $\mathcal{C}$ *can be restricted to a* $(\Sigma, t/2^{i+2}, q - i, s + m)$-*checker.*

Lemma 1 is used iteratively as long as there are heavy public memory locations, restricting the memory checker to only a (large) subset of its indices while lowering its query complexity (q). This comes at the cost of only a modest drop in the number of indices (n) and a moderate increase in the space complexity (s). We repeat this iteratively, reducing the query complexity until there is no set of "heavy" locations in the public memory. If we can apply the lemma many times, then we get a lower bound on the checker's query complexity: each application of the lemma reduces the query complexity, so if we applied the lemma many times the initial query complexity had to have been high.

The reason that we can apply Lemma 1 many times is that otherwise we are left with a checker on many indices with no set of "heavy" locations. If there is no set of "heavy" public memory locations, then the (possibly reduced) public memory can be partitioned into relatively many parts that are *disjoint* in the sense that each part is queried only by a single index of the database. We can restrict the checker again, but this time to obtain a checker with many indices, relatively small secret memory and query complexity 1. This is formalized in Lemma 2 (proof below):

Lemma 2. *Let* $\mathcal{C}$ *be a* (Σ, n, q, s) *deterministic and non-adaptive online memory checker. Then, for every* $\alpha \in \mathbb{N}$ *such that* $\alpha < n$, *and for every threshold* $t \in \mathbb{N}$ *such that* $n > 4t \cdot q \cdot \log n$, *the following holds:*

If for every integer $m \in \{1, \ldots, \alpha\}$, *there are fewer than* m *locations in public memory that are* t/m-*heavy, then the memory checker* $\mathcal{C}$ *can be restricted to a* $(\Sigma^q, n \cdot \alpha/(2q \cdot t), 1, s/q)$-*checker.*

Finally, we show that such a "disjoint" checker implies a contradiction. In particular, it must have space complexity that is more or less proportional to the number of disjoint parts. Unless the memory checker has already been restricted to very few indices (in which case we have a query complexity lower bound), this results in a contradiction, since the checker's space complexity is bounded (by a small polynomial in n). The intuition that a disjoint checker must have large space complexity is formalized in Lemma 3 (proof below):

Lemma 3. *Let* $\mathcal{C}$ *be a* $(\Sigma, n, q = 1, s)$ *deterministic and non-adaptive online memory checker, i.e. a checker that makes only a single query, where the location that each index queries in public memory is different. Then,* $s \geq \frac{n}{\log |\Sigma|} - 1$.

We postpone proving the lemmas and proceed with a formal analysis. We take $\alpha = n^d$, for a constant $0 < d < 1$ to be specified later. We iteratively examine and restrict the memory checker. Let $\mathcal{C}_i$ be the checker obtained after the i-th iteration ($\mathcal{C}_0 = \mathcal{C}$ is the original checker), let n_i be the number of indices in its database and s_i its space complexity. Taking a threshold $t_i = \frac{n_i}{\log^c n}$, where $c > 1$ is a constant specified below, we check whether or not the "new" checker $\mathcal{C}_i$ has a set of heavy indices in its public memory. We only iterate as long as $n_i > \alpha$. Formally, there are two possible cases:

1. If $\mathcal{C}_i$ has a set of $m \leq \alpha$ public memory locations that are at least t_i/m-heavy, then by Lemma 1:
   For some $j \in \{1, \ldots, q\}$, we can build from C_i a $(\Sigma, t_i/2^{j+2}, q - j, s_i + \alpha)$ deterministic and non-adaptive online memory checker $\mathcal{C}_{i+1}$.
2. If for every integer $m \leq \alpha$ the checker $\mathcal{C}_i$ does not have a set of m public memory locations that are t_i/m-heavy, and choosing c, d such that $n_i > 4t_i \cdot q \cdot \log \alpha$, by Lemma 2:
   We can build from C_i a $(\Sigma^q, n_i \cdot \alpha/(2q \cdot t_i), 1, s_i/q)$ deterministic and non-adaptive online memory checker. If n_i is reasonably large, i.e. has not been reduced by repeated iterations of Case 1, then this will imply a contradiction.

Recall that q_0 denotes the query complexity of the initial checker $\mathcal{C}$, before any application of Lemmas 1 and 2. Assume for a contradiction that $q_0 \leq \log n/(3c \cdot \log \log n)$. Let $j \in [q + 1]$ be the *total* number of queries reduced by the iterative applications of Lemma 1, i.e., the number of queries reduced by the iterations in which Case 1 occurred. Since we assumed $q \leq \log n/(3c \cdot \log \log n)$, we know that $j < \log n/(3c \cdot \log \log n)$. Thus, in the first iteration in which Case 2 applies (say the i-th iteration in total), it must be the case that

$$n_i \geq n/(\log^{c \cdot j} n \cdot 2^{3 \log n/3c \cdot \log \log n}) = n/(\log^{c \cdot j} n \cdot 2^{\log n/c \log \log n}) > n^{1-\varepsilon/2}.$$

Recall that we only iterate so long as $n_i > \alpha$, so we can choose any $\alpha < n^{1-\varepsilon/2}$. The space s_i used by this restricted checker is at most $s + i \cdot \alpha \leq s + \log n \cdot \alpha$. As usual, $t_i = n_i/\log^c n$, and choosing $c > 2$ we get that

$$4t_i \cdot q \cdot \log \alpha \leq n_i/(\log^c n \cdot \log n \cdot d \log n) < n_i$$

Applying Lemma 2, we obtain a $(\Sigma^q, n_i \cdot \alpha/(2q \cdot t_i), 1, s_i/q)$-checker. Now, by Lemma 3, which bounds the space complexity of one-query checkers, we get that it must be the case that:

$$s_i \geq n_i \cdot \alpha/(2q \cdot t_i \cdot \log |\Sigma|) \geq \log^{c-1} n \cdot \alpha/(2 \log |\Sigma|)$$

But on the other hand we know that

$$s_i \leq s + \log n \cdot \alpha.$$

We know $|\Sigma| \leq 2^{\text{poly} \log n}$, and choose c such that $\log^{c-1} n/(2 \log |\Sigma|) > 2 \log n$. We also set $\alpha > 2s = 2n^{1-\varepsilon}$. Recall that we also needed $\alpha < n_i$, but this is fine since $n_i > n^{1-\varepsilon/2}$. In conclusion, we set α by choosing d such that $1 - \varepsilon < d < 1 - \varepsilon/2$, i.e. such that

$$2s = 2n^{1-\varepsilon} < \alpha = n^d < n_i = n^{1-\varepsilon/2}$$

We get that

$$s > \log^{c-1} n \cdot \alpha/(2 \cdot \log |\Sigma|) - \log n \cdot \alpha > \log n \cdot \alpha > 2s$$

This is a contradiction!

Proof (of Lemma 1). If there is a set M of m locations in public memory that are all t/m-heavy (i.e. each accessed by at least t/m indices), then we "restrict" the memory checker to only work for some of the indices that access one or more of the heavy locations. Let $I \subseteq [n]$ be the set of database indices that access at least one of the locations in M (the "heavy" locations).

We claim that for some $i \in \{1, \ldots, q\}$, there are at least $t/2^{i+2}$ indices in I that each access at least i locations in M. To see this, assume for a contradiction that this is not the case. Then the sum of the number of locations in M that are accessed by each database index (and in particular by the indices in I) is less than:

$$\sum_{i=1}^{q} i \cdot t/2^{i+2} = t \cdot \sum_{i=1}^{q} i/2^{i+2} < t$$

On the other hand, since there are m locations in M that are at least t/m-heavy, the sum of locations in M read by database indices must be at least t and we get a contradiction.

We restrict the checker to the indices in I that read at least i locations in M, and move these locations to the secret memory. This increases the space complexity (size of the secret memory) from s to $s + m$. By the above, there are at least $t/2^{i+2}$ such indices. For each of them, we have reduced their query complexity from q to $q - i$. The alphabet size remains unchanged.

Proof (of Lemma 2). If there are only a few relatively heavy locations in the public memory, then we eliminate indices and split the public memory in "disjoint chunks": subsets of the public memory that are disjoint in the sense that no location in any chunk is accessed by two different indices. This is done in a greedy iterative manner. We go over the locations in public memory one by one; for each of them we choose one index (say j) that accesses them and eliminate any other index that accesses a location in public memory also accessed by j. This is repeated iteratively (for the analysis, we think of this as being done from the heavy public memory locations to the lighter ones). After the checker cannot be restricted any more we are left with a checker for which no two indices access the same location in public memory, and we will show that the number of remaining indices is reasonably high.

More concretely, for any value $i \in [1 \ldots \log \alpha]$, we know that there are at most $2^i - 1$ locations that are between $t/2^i$-heavy and $2t/2^i$-heavy. In fact, in the iterative restriction process, when we consider i we have already restricted the memory checker so that no location in the public memory is more than $2t/2^i$-heavy.

We go over these (at most $2^i - 1$) locations one by one, say in lexicographic order. For each of them, we examine one index that accesses that location, say index j. We restrict the checker by eliminating all "intersecting" indices: indices k such that there is a public memory location queried by both j and k. Index j queries at most q locations in the public memory, and these in turn are queried by at most $2t/2^i$ indices each (since we have already restricted the checker so that there is no $2t/2^i$-heavy location in the public memory). Thus, we eliminate at most $2t \cdot q/2^i$ indices per heavy location in the public memory, or at most $2t \cdot q$ indices in all.

Repeating this for $i \leftarrow 1 \ldots \log \alpha$, in the i-th iteration there are at most 2^i locations that are at least $t/2^i$-heavy, and none of these locations can be more than $2t/2^i$-heavy. We go over these locations one by one, and if they have an index accessing them that has not been eliminated yet we restrict the checker as above. This eliminates at most $2t \cdot q/2^i$ indices per heavy public memory location, or $2t \cdot q$ indices in all.

In total, in all of these $\log \alpha$ iterations, with their restrictions, the number of indices eliminated is at most:

$$\sum_{i=1}^{\log \alpha} 2t \cdot q = 2t \cdot q \cdot \log \alpha$$

If $n > 4t \cdot q \cdot \log \alpha$ then we have only eliminated at most $n/2$ indices. Now, after all the restrictions, there are no locations in the public memory that are t/α-heavy. We go over the remaining indices in lexicographic order, and for each of them we restrict the checker by eliminating all other indices that intersect its public memory accesses. Since there are no more t/α-heavy locations in the public memory, each such restriction eliminates at most $q \cdot t/\alpha$ indices.

In the end, we are left with a memory checker on at least $n \cdot \alpha/(2q \cdot t)$ indices, with the property that no two indices access the same location in public memory. We can thus re-order the public memory into "chunks", of q symbols each, such that each chunk is queried only by a single index and each index queries only that chunk. If we enlarge the alphabet to be comprised of these q-symbol chunks, we get a checker with query complexity 1. The "price" is restricting the checker to only $n \cdot \alpha/(2q \cdot t)$ indices and increasing the alphabet size to Σ^q. Since we have increased the alphabet size, we can represent the secret memory as fewer symbols of the new larger alphabet, so the secret memory is of size s/q new alphabet symbols.

Proof (of Lemma 3). The intuition is that the public memory has a single location for storing information about each database index. When reading or writing the value of the database at that index, the only information read from public memory is the information held in that index's location. Further, for two different database indices, their locations in public memory are different. To achieve soundness the checker must (intuitively) store, for every index in the database, separate "authentication information" in the secret memory about the value at that index's location. There are n indices (say holding boolean data base values), and only $s \cdot \log |\Sigma|$ bits of secret memory, and thus s should be at least on the order of $\frac{n}{\log |\Sigma|}$.

To prove this we examine an adversary $\mathcal{A}$, who begins by storing the all 0 database into the memory checker. This yields some public memory p_1. $\mathcal{A}$ then picks a random database $r \in \{0,1\}^n$ and stores it into the checker: for every index in r which has value 1, $\mathcal{A}$ uses the checker to store the value 1 into that index. Say now that at the end of this operation sequence, the public memory is

p_2 and the secret memory is s_2. The important thing to note is that for indices of r whose values are 0, the value of their locations in the public memory has not changed between p_1 and p_2 (since each index has a unique location in public memory that it accesses).

The adversary $\mathcal{A}$ now replaces the public memory p_2 with the "older" information p_1.[3] Now the adversary tries to retrieve some index of the database, say the i-th ($i \in [n]$). The checker runs with secret memory s_2 and public memory p_1 to retrieve the i-th bit of r. Note that if $r[i] = 0$, then the value of the i-th index's location in public memory is *unchanged* between p_1 and p_2. By *completeness*, the checker should w.h.p. output 0 (the correct value of $r[i]$). On the other hand, if $r[i] = 1$, then by its *soundness* guarantee the memory checker should w.h.p. output either 1 or $\perp$ - we take either of these answers as an indication that $r[i] = 1$. We conclude that for each index $i \in [n]$, the checker can be used to retrieve the i-th bit of r w.h.p. The checker achieves this using only the public memory p_1, which is completely independent of r, and the secret memory s_2. Intuitively, s_2 holds nearly all the information about the (randomly chosen) vector r, and thus s_2 cannot be much smaller than r, an n-bit vector.

More formally, suppose that $s < \frac{n}{\log |\Sigma|} - 1$. We can view the above procedure as allowing us to transmit a random n-bit string using only $s \log |\Sigma|$ bits and succeeding with high probability: the sender and the receiver share the initial assignment to the secret memory s_1 and the public memory p_1 resulting from writing the all 0 vector (all this is independent of r). Given the string $r \in \{0,1\}^n$ the sender simulates writing r to the memory as above and the resulting secret memory at the end is s_2. This is the only message it sends to the receiver. The receiver runs the above reconstructing procedure for each $1 \leq i \leq n$, i.e. using secret memory s_2 and public memory p_1 tries to read location i and decides that $r[i] = 0$ iff it gets as an answer a 0 (1 or $\perp$ are interpreted that $r[i] = 1$). Since for each i the procedure the receiver is running is just what the memory checker will run with the above adversary, the probability of error in any of the i's is small. Therefore we get that the receiver reconstructs all of r correctly with high probability. But by simple counting this should happen with probability at most $\frac{2^{s \log |\Sigma|}}{2^n} < 1/2$.

4   Read-Write Tradeoffs for Online Checking

In this section we present two read-write tradeoffs for the query complexity of online memory checking. These can be viewed as counterparts to the lower bound of Theorem 1 (all of the memory checkers in this section are deterministic and non-adaptive). While Theorem 1 states that the *sum* of the query complexities of read and write operations cannot be low, in this section we show that the query complexity of either read or write *can* be made significantly lower, at the cost of increasing the query complexity of the other operation (write or read respectively).

---

[3] Note that this is a "replay attack". As noted above, the Lemma and this section's query complexity lower bounds do not hold for checkers that are not required to work against replay attacks.

We present two trade-offs. The first gives an memory checker with efficient write operations but expensive read operations. The second is a checker with efficient read but expensive write. In particular, in both these tradeoffs, for any well-behaved function $d(n) : \mathbb{N} \to \mathbb{N}$, the "efficient" operation (write or read) has query complexity $O(\log_{d(n)} n)$, and the "inefficient" operation (read or write respectively) has query complexity $O(d(n) \cdot \log_{d(n)} n)$. In both cases the space complexity is polynomial in the security parameter, and the checker uses a pseudo-random function. For desired soundness ε the length of alphabet symbols is $O(\log(1/\varepsilon) + \log n)$.

Overview of the Constructions. We proceed with an overview of the common elements of both constructions, the details are in the full version. Following Blum *et al.* (Section 5.1.2), we construct a tree structure "on top" of the memory. Where they constructed a binary tree, we construct instead a $d(n)$-ary tree. Each internal node has $d(n)$ children, so the depth of the tree is $\log_{d(n)} n$. The n leaves of the tree correspond to the n database indices. We assume for convenience w.l.o.g that n is a power of $d(n)$.

In both constructions we associate a time-stamp with each node in the tree. The time-stamp of a leaf is the number of times that the user wrote to the database index that the leaf represents. The time-stamp of an internal node is the sum of its children's time-stamps, and thus the time-stamp of the root is the total number of times that the user has written to the database. We use t_u to denote the current time-stamp of tree node u. The time-stamps are used to defeat *replay attacks* (where the adversary "replays" an old version of the public memory). If the adversary replays old information, then the replayed time-stamps will have smaller values than they should.

For each tree node u, we store in public memory its value $v_u \in V$ and its time-stamp $t_u \in [T]$. For an internal node u, its value is simply 0, for a leaf ℓ, its value represents the value that the user stored in the database index associated with that leaf. The root's time-stamp is stored in the secret reliable memory, together with the seed of a pseudo-random function (PRF). This simply a generalization of Blum *et al.*'s construction (the tree is $d(n)$-ary and not binary).

Our two construction differ from each other and from [6] in their use of authentication tags to authenticate different nodes' values and time-stamps. In the **first construction** (efficient write), we store for each node u an authentication tag which is the PRF evaluated on (u, t_u, v_u). When writing a new value to a leaf, we verify the tags of all the nodes on the path from the root to that leaf and then update the leaf's value and the time-stamps of all the nodes on the path to the leaf. Thus the write complexity is proportional to the tree depth, or $O(\log_{d(n)} n)$. To read the value from some leaf, we read the values, time-stamps and tags of that leaf, all nodes on the path from the root to the leaf and all their children, a total of $O(d(n) \cdot \log_{d(n)} n)$ public memory locations. We verify the consistency of all the tags, and that the time-stamp of every internal node is the sum of its children's time-stamps. This prevents replay attacks, as the root's time-stamp is in the reliable memory and thus always correct. The **second construction** (efficient read) is different. For each tree *edge* connecting a node u and one of its $d(n)$ children w, we store in public memory a tag which is the PRF evaluated on $(u, t_u, v_u, w, t_w, v_w)$. Now, to read the value from a leaf we read the values and time-stamps of all nodes on the path from the root, and the tags of the edges. For each edge we verify that the tag is consistent. This requires making $O(\log_{d(n)} n)$ queries to public memory. To write a new value to a leaf, read and write the values and time-stamps at the leaf and all nodes on the path from the root to the leaf, as well as all their children and edge tags, a total of $O(d(n) \cdot \log_{d(n)} n)$ queries. Verify that all tags are consistent and that the time-stamp of each internal node is the sum of its children's time-stamps. If all checks pass, update the proper time-stamps and the leaf's value. See the full version for details.

5 Offline Checking of RAMs

In this section we describe how to check "offline" the operation of a RAM, that is a sequence of read and write (or store and retrieve) operations. To check that a RAM operates correctly we must verify that the value we obtain from reading an address in public memory is equal to the last value we wrote to that address. Blum *et al.* [6] showed an (invasive) scheme, where if one scans the whole memory at the end of the sequence of operations, then it is possible to detect (with hight probability) any malfunction. The cost (in query complexity) is $O(1)$ per operation, plus the final scan. Thus, for sequences of n operations or more, the *amortized* query complexity is $O(1)$. As discussed in the introduction, our goal is to improve upon that, by not running a final scan of all the memory. Instead, we scan only the locations that were changed. This implies that at any point, after t operations, we can check that the memory worked appropriately by investing time $O(t)$, so for *any* sequence of operations (not only for long ones) the amortized query complexity is $O(1)$. This result can be viewed as a generalization of those in Amato and Loui [3].

Our ideas follow closely those of Blum *et al.* [6]. First, add to each memory address a slot for the time it was written - a "timestamp". The "time" can be any discrete variable that is incremented whenever a write or read operation is performed. The timestamp of each location is actually updated after either read or write. So one can view each operation as read followed by write. The offline checker needs to verify that the set of (value, address, time) triples which are written equals the set of (value, address, time) triples which are read. More precisely, consider the following two sets:

$$R = \{(v, a, t) | \text{location } a \text{ was read with value } v \text{ and timestamp } t\}$$

$$W = \{(v, a, t) | \text{location } a \text{ was written with value } v \text{ and timestamp } t\}$$

Suppose that at no point in time did a read operation return a timestamp larger than the current time (call this the timestamp condition), a clear malfunction, and suppose that the memory is scanned (i.e. read completely) at the end of the sequence of operations. Then Blum *et al* [6] showed

Claim. $W = R$ iff the memory functioned properly.

In other words, a procedure that checks online for the timestamp condition plus an offline test for $W = R$ results in an offline checker for the RAM. It is useful to note that the proof actually implies that if the timestamp condition was not violated, then actually $W \nsubseteq R$.

We modify slightly the above and note that if we scan *only those locations that were actually modified*, then we can similarly say that $W = R$ iff the memory functioned properly. This is true, since the locations that were not written do not affect W and hence whether we access them or not does not make $R = W$.

Now the question is, how do we scan only the locations that were modified? For this we keep a linked list of all locations that were accessed. When a new location is read it is added to the end of the list. The starting and ending locations of the list are stored in the secure memory. To scan the locations accessed we trace the list, see below on possible implementations of the list, the important thing is that adding a memory location to the list and checking whether a memory location is already in the list can be done with $O(1)$ queries (possibly amortized).

A natural question now is how to authenticate the list to ensure that an adversary did not tamper with it (i.e. *who guards the guard?*). The point here is that *the list itself need not be authenticated.* To address the issue of faults in the linked list, observe that as indicated above to make the checker accept the adversary needs to "cover" W by R. If the adversary tampers with the list, and a wrong set of locations is accessed in the final scan, then it will not cover W. Since we do not authenticate the list, the one remaining concern is that a faulty memory can even lead the scanning process into loops (by putting loops into the list). To prevent this, we use a simple counter that bound the number of locations we traverse in the list.

To check whether W and R are the same or not, we can use the same methods as described in Blum et al. The problem can be thought of as one in *streaming*, since the sets are accessed one value at a time. We assume there is a secret hash function h mapping sets to some range and we can compute on the fly $h(R)$ and $h(W)$ and compare the results. That is, h can be updated incrementally in each operation (read or write).

Specifically, we require that where for every k there exists a family H where: (i) representing a member $h \in H$ of the family takes $O(k + log n)$ bits (ii) the range of $h \in H$ can be represented by $O(k)$ bits (iii) the probability that two different sets hash to the same value is at most $1/2^k$ (the probability is over the choice of the function in the family) and (iv) given $h \in H$ an element x and the value of $H(S)$, the value of $h(S \cup \{x\})$ can be computed in $O(1)$ operations. There are constructions of such functions (see e.g. Naor and Naor [14] and the analysis in [6]). The procedures for reading and writing are now as follows:

Write of value v to address a

- read the value v' and time t' stored in address a.
- verify that t' is less than the current time.
- update the hash $h(R)$ of set R with (v', a, t').

- write the new value v and current time t to address a.
- update the hash $h(W)$ of set W with (v, a, t).
- if location a is not in the linked list add it to the end and update the endpoint in the secure memory.

Read of address a

- read the value v' and time t' from address a.
- verify that t' is less than the current time t.
- update the hash $h(R)$ of set R with (v', a, t').
- write v' and t to address a.
- update the hash $h(W)$ of set W with (v', a, t).
- if location a is not in the linked list add it to the end and update the endpoint in the secure memory.

To check the functioning of the RAM at the end of any sequence of operations, the checker reads all the memory locations in the linked list, starting from the first location in the list, which is stored in the secure memory. As the scan proceeds $h(R)$ is updated accordingly. Assuming initially $W = R = 0$ and the RAM is empty, $h(W)$ should equal $h(R)$ if the memory functioned correctly, and should be different from $h(R)$ with high probability if the memory was faulty. To maintain the list of modified locations, we can use a simple linked list (see below for a more efficient alternative). It is enough to add a pointer to each address in public memory (together with the value and timestamp of that address). The pointer is initially NULL (or 0), and whenever we access a public memory location for the first time we modify the pointer of the current list tail to point to the new list end and update the list end (there is no need to update R and W for list maintenance operations, faults in the list will be detected).

Note that we do not have to assume that the memory is initialized to be all 0 before the beginning of the operations, since it is possible to use the "uninitialized memory trick", where one keeps a list of pointers to the modified locations and all other locations are 0. See [1], exercise 2.12 or [5,9,7].

Since the scheme is invasive (has to change the memory), it makes the most sense when the basic unit we read is relatively large. Suppose the length of a database word is μ, then the additional timestamp takes $\log n$ bits and the pointer to the linked list takes another $\log n$ bits. We summarize the results in the following theorem.

Theorem 2. *For a RAM with n words of size μ there exists an invasive, offline memory checker using n memory locations storing $\mu + 2 \log n$-bit words, which uses $O(\log n + \log 1/\varepsilon)$ private memory. Each read or write operation takes $O(1)$ queries, and a procedure for detecting error can be executed after any sequence of t steps at the cost of $O(m)$ where m is the actual number of locations that were used. An error is detected with probability at least $1 - \varepsilon$.*

Finally, we re-examine the issue of invasiveness. We note that in fact we do not need to store time-stamps and list-pointers for *all* of the database indices, just for those that are accessed. This leads to a method for reducing the invasiveness of the checker (the total number of non-database bits that it stores in public memory). We can maintain the timestamps and the list itself as a separate data structure, whose size is proportional (say linear) to the number of database indices which have been accessed. Any data structure that supports insertion and membership queries in amortized $O(1)$ time work. We note once more that Ajtai [2] proved a lower bound on the invasiveness of offline memory checkers, but his proof uses long sequences of operations that access every database index, and thus it does not apply to our setting of short sequences of operations that access only a few locations in the database.

References

1. Aho, A.V., Hopcroft, J.E., Ullman, J.D.: The Design and Analysis of Computer Algorithms. Addison-Wesley Series in Computer Science and Information Processing. Addison Wesley, Reading (1974)
2. Ajtai, M.: The invasiveness of off-line memory checking. In: STOC, pp. 504–513 (2002)
3. Amato, N.M., Loui, M.C.: Checking linked data structures. In: Proceedings of the 24th Annual International Symposium on Fault-Tolerant Computing (FTCS), pp. 164–173 (1994)
4. Ateniese, G., Burns, R., Curtmola, R., Herring, J., Kissner, L., Peterson, Z., Song, D.: Provable data possession at untrusted stores. Cryptology ePrint Archive, Report 2007/202 (2007)
5. Bentley, J.: Programming Pearls. ACM, New York (1986)
6. Blum, M., Evans, W.S., Gemmell, P., Kannan, S., Naor, M.: Checking the correctness of memories. Algorithmica 12(2/3), 225–244 (1994)
7. Briggs, P., Torczon, L.: An efficient representation for sparse sets. ACM Letters on Programming Languages and Systems 2, 59–69 (1993)
8. Clarke, D.E., Suh, G.E., Gassend, B., Sudan, A., van Dijk, M., Devadas, S.: Towards constant bandwidth overhead integrity checking of untrusted data. In: IEEE Symposium on Security and Privacy, pp. 139–153 (2005)
9. Cox, R.: http://research.swtch.com/2008/03/using-uninitialized-memory-for-fun-and.html
10. Goldreich, O.: The Foundations of Cryptography, vol. 1. Cambridge University Press, Cambridge (2001)
11. Goldreich, O.: The Foundations of Cryptography, vol. 2. Cambridge University Press, Cambridge (2004)
12. Goldreich, O., Goldwasser, S., Micali, S.: How to construct pseudorandom functions. Journal of the ACM 33(2), 792–807 (1986)
13. Juels, A., Kaliski, B.: Pors: proofs of retrievability for large files. In: CCS 2007: Proceedings of the 14th ACM conference on Computer and communications security, pp. 584–597. ACM, New York (2007)
14. Naor, J., Naor, M.: Small-bias probability spaces: Efficient constructions and applications. SIAM J. Comput. 22(4), 838–856 (1993)
15. Naor, M., Rothblum, G.N.: The complexity of online memory checking. In: FOCS, pp. 573–584 (2005)
16. Oprea, A., Reiter, M.K.: Integrity checking in cryptographic file systems with constant trusted storage. In: USENIX Security Symposium (2007)
17. Shacham, H., Waters, B.: Compact proofs of retrievability. In: Pieprzyk, J. (ed.) ASIACRYPT 2008. LNCS, vol. 5350, pp. 90–107. Springer, Heidelberg (2008)

Goldreich's One-Way Function Candidate and Myopic Backtracking Algorithms

James Cook[1,*], Omid Etesami[1,**], Rachel Miller[2,***], and Luca Trevisan[1,†]

[1] Computer Science Division, U.C. Berkeley
{jcook,etesami,luca}@cs.berkeley.edu
[2] University of Virginia
rachel.an.miller@gmail.com

Abstract. Goldreich (ECCC 2000) proposed a candidate one-way function construction which is parameterized by the choice of a small predicate (over $d = O(1)$ variables) and of a bipartite expanding graph of right-degree d. The function is computed by labeling the n vertices on the left with the bits of the input, labeling each of the n vertices on the right with the value of the predicate applied to the neighbors, and outputting the n-bit string of labels of the vertices on the right.

Inverting Goldreich's one-way function is equivalent to finding solutions to a certain constraint satisfaction problem (which easily reduces to SAT) having a "planted solution," and so the use of SAT solvers constitutes a natural class of attacks.

We perform an experimental analysis using MiniSat, which is one of the best publicly available algorithms for SAT. Our experiment shows that the running time required to invert the function grows exponentially with the length of the input, and that such an attack becomes infeasible already with small input length (a few hundred bits).

Motivated by these encouraging experiments, we initiate a rigorous study of the limitations of back-tracking based SAT solvers as attacks against Goldreich's function. Results by Alekhnovich, Hirsch and Itsykson imply that Goldreich's function is secure against "myopic" backtracking algorithms (an interesting subclass) if the 3-ary parity predicate $P(x_1, x_2, x_3) = x_1 \oplus x_2 \oplus x_3$ is used. One must, however, use non-linear predicates in the construction, which otherwise succumbs to a trivial attack via Gaussian elimination.

We generalized the work of Alekhnovich et al. to handle a more general class of predicates, and we present a lower bound for the construction that uses the predicate $P_d(x_1, \ldots, x_d) := x_1 \oplus x_2 \oplus \cdots \oplus x_{d-2} \oplus (x_{d-1} \wedge x_d)$ and a random graph.

* Work supported by the National Science Foundation under grant No. CCF-0729137 and by the National Sciences and Engineering Research Council of Canada under a PGS award.
** Work supported by the National Science Foundation under grant No. CCF-0729137.
*** Work done at U.C. Berkeley, supported by an NSF SUPERB fellowship.
† Work supported by the National Science Foundation under grant No. CCF-0729137 and by the BSF under grant 2002246.

O. Reingold (Ed.): TCC 2009, LNCS 5444, pp. 521–538, 2009.

1   Introduction

Goldreich [11] proposed in 2000 a candidate one-way function construction based on expanding graphs. His construction is parameterized by the choice of a bipartite graph with n vertices per side and right-degree d (where d is either a constant independent of n, or grows very moderately as $O(\log n)$) and of a boolean predicate $P : \{0,1\}^d \to \{0,1\}$. To compute the function, on input $x \in \{0,1\}^n$ we label the vertices on the left by the bits of x, and we label each vertex on the right by the value of P applied to the label of the neighbors. The output of the function is the sequence of n labels of the vertices on the right.

Goldreich's Function and Cryptography in NC_0. A function is computable in NC_0 if every bit of the output depends only on a constant number of bits of the input. One can see any NC_0-computable function as a generalization of Goldreich's function in which the graph is allowed to be arbitrary, subject to having bounded right-degree, and in which different predicates can be used for different bits of the output.

Cryan and Miltersen [7] first raised the question of whether cryptographic primitives (their work focused on pseudorandom generators) can be computed in NC_0. Mossel, Shpilka and Trevisan [13] construct, for arbitrarily large constant c, a function $f : \{0,1\}^n \to \{0,1\}^{cn}$ based on a bipartite graph of right-degree 5 and the fixed predicate $P(x_1, \cdots, x_5) := x_1 \oplus x_2 \oplus x_3 \oplus (x_4 \wedge x_5)$, and show that the function computes a small-bias generator. Such a construction may in fact be a pseudorandom generator, and hence a one-way function.[1]

Applebaum, Ishai and Kushilevtiz [4,5] show that, under standard assumptions, there are one-way functions and pseudorandom generators that can be computed in NC_0; their one-way function is computable with right-degree 3.[2] In their construction, the graph encodes the computation of a log-space machine computing a one-way function that is used as a primitive.

In this paper, we are interested in the security of Goldreich's original proposal, implemented using a random graph and a fixed predicate.

Goldreich's Function and DPLL Algorithms. Inverting Goldreich's one-way function (and, indeed, inverting any one-way function that is computable in NC_0) can be seen as the task of finding a solution to a constraint satisfaction problem with a planted solution. A plausible line of attack against such a construction is thus to employ a general-purpose SAT solver to solve the constraint satisfaction problem. We performed an experimental study using MiniSat, which is one of the best publicly available SAT solvers, and has solved instances with several thousand variables. Using a random graph of right-degree 5, and the

---

[1] The graph used in this construction, however, is not a random graph or a strong expander graph of right-degree 5, so this is not an instantiation of Goldreich's proposal.

[2] This is the best possible, because it is easy to show that no function based on a bipartite graph of right-degree 2 can be one-way, by reducing the problem of finding the inverse to a 2SAT instance.

predicate $(x_1 \oplus x_2 \oplus x_3 \oplus (x_4 \wedge x_5))$, we observed an exponential increase of the running time as a function of the input length, and an attack with MiniSat appears infeasible already for moderate input lengths (a few hundred bits). See Appendix A.

Our goal in this paper is to provide a rigorous justification for these experimental results, and to show that "DPLL-style" algorithms based on backtracking (such as most general SAT solvers) cannot break Goldreich's construction in subexponential time. We restrict ourselves to algorithms that instantiate variables one at a time, in an order chosen adaptively by a "scheduler" procedure, and then recurse on the instance obtained by by fixing the variable to zero and then to the instance obtained by fixing the variable to one, or viceversa (the scheduler decides which assignment to try first). The recursion stops if the current partial assignment contradicts one of the constraints in the instance, or if we find a satisfying assignment.

When such an algorithm runs on an unsatisfiable instance, then a transcript of the algorithm gives a "tree-like resolution proof" of unsatisfiability; a number of techniques are known to prove exponential lower bounds on the size of tree-like resolutions proofs of unsatisfiability, and so such proofs give lower bounds to the running time of any such algorithm, regardless of how the scheduler is designed.

When dealing with *satisfiable* instances, however, one cannot prove lower bounds without putting some restriction on the scheduler. (If unrestricted in complexity, the scheduler could compute a satisfying assignment, and then assign the variables accordingly, making the algorithm converge in a linear number of steps.)

The Lower Bound of Alekhnovich et al. Alekhnovich, Hirsch and Itsykson [3] consider two such restrictions: they consider (i) "myopic" algorithms in which the scheduler chooses which variable to assign based on only a bounded number of variables and clauses of the current formula, and (ii) "drunken" algorithms in which the order of variables is chosen arbitrarily by the scheduler, but the choice of whether to assign first zero or one to the next chosen variable is made randomly with equal probability. The result of the second type is proven for carefully designed instances, and it remains an open question to prove a lower bound for drunken algorithms on a random satisfiable constraint satisfaction problem. Lower bounds of the first type are proven for random instances, and they are proved via a reduction to the problem of certifying unsatisfiability: Alekhnovich et al. show that a myopic algorithm, with high probability, after assigning a certain number of variables will be left with an instance that is unsatisfiable, but for which there is no sub-exponential size tree-like resolution proof of unsatisfiability. Hence the algorithm will take an exponential amount of time before it realizes it has chosen a bad partial assignment.

Our Results. The result of Alekhnovich et al. applies to myopic algorithms for random instances of 3XOR with a planted solution, and provided a lower bound for myopic DPLL inversion algorithms for the instantiation of Goldreich's proposal using the 3XOR predicate.

Unfortunately, the use of 3XOR as a predicate in Goldreich's construction leads to a total break via Gaussian elimination, so our goal is to extend the result of Alekhnovich et al. to a setting in which we have either a random predicate or the predicate $(x_1 \oplus \cdots \oplus x_{d-2} \oplus (x_{d-1} \wedge x_d))$ which is inspired by the work of Mossell et al.

In order to extend the work of Alekhnovich et al. to the setting of Goldreich's one way functions, we need to make the following changes:

- The proof in [3] uses the fact that all constraints have arity 3. It is not difficult to adapt it to handle linear constraints of larger constant arity, by relying on the strong expansion properties which are true of random constraint graphs.
- The proof in [3] uses the linearity of the constraints. We show that it is sufficient for the predicate to be such that it remains nearly balanced even after many variables have been fixed to arbitrary values. For example, a d-ary parity remains perfectly balanced even after $d - 1$ variables are assigned arbitrary variables. The predicate $(x_1 \oplus \cdots \oplus x_{d-2} \oplus (x_{d-1} \wedge x_d))$ remains perfectly balanced even after $d - 3$ variables are assigned arbitrary variables, and a random predicate remains ϵ-close to balanced after any $d - O(\log d/\epsilon)$ variables are fixed to arbitrary values. (Those parameters are sufficient for our proof to go through.)
- The proof in [3] assumes that there is a unique solution, and this is not true in our setting. We show that the proof carries over if one assumes that the total number of pairs x, y such that $f(x) = f(y)$ is at most $2^{(1+\epsilon)n}$ for small ϵ. We are able to show that such a condition is satisfied by the predicate $(x_1 \oplus \cdots \oplus x_{d-2} \oplus (x_{d-1} \wedge x_d))$ and by the choice of a highly-expanding graph, with $\epsilon = 2^{-\Omega(d)}$. We believe that the same result holds with high probability if we choose a random d-ary predicate, but we have not been able to prove it.

With such results, we are able to show an exponential lower bound for myopic algorithms in a construction that uses a random graph and the predicate $(x_1 \oplus \cdots \oplus x_{d-2} \oplus (x_{d-1} \wedge x_d))$. If we consider the construction that uses a random graph and a random predicate, then we have a conditional exponential lower bound under the assumption that the resulting function is nearly injective.

Goldreich's Analysis. Goldreich [11] considered the following algorithm for computing x given $y = f(x)$. The algorithm proceeds in n steps, revealing the output bits one at a time. Let R_i be the set of inputs connected to the first i outputs. Then in the ith step, the algorithm computes the list L_i of all strings in $\{0, 1\}^{R_i}$ which are consistent with the first i bits of y. Goldreich proves that if the graph satisfies an expansion condition, then for a random input x, the expected size of one of the sets L_i is exponentially large.

Since Goldreich's algorithm is forced to consider all consistent assignments to the bits in each set R_i, it takes no less time than a (myopic) backtracking algorithm that chooses the input bits in the same order, and possibly much more time. For this reason, our new lower bounds are more general.

Open Questions. We believe that there is motivation for further experimental and rigorous analysis of Goldreich's construction.

The main limitation of the present work is the somewhat artificial setup of myopic algorithms, which fails to capture certain natural "global" heuristics used in SAT solvers. Since the algorithm is required to work only with partial information on the object given as an input, negative results for myopic algorithms are similar in spirit (but very different technically) to results on "space bounded cryptography." It would be very interesting to have a lower bounds for drunken algorithms, which are restricted in a way that is more computational than information-theoretic. As a first step, it would be interesting to show that drunken algorithms take exponential time to find planted solutions in a random 3XOR instance.

It would also be interesting to show that no "variation of Gaussian elimination" can invert Goldreich's function when non-linear predicates are used. Unfortunately it is not clear how to even formalize such a statement.

2   Preliminaries

2.1   Goldreich's Function

Goldreich [11] constructs a function $f : \{0,1\}^n \to \{0,1\}^n$ parameterized by a d-ary predicate P and a bipartite graph $G = (V, E)$ connecting n *input nodes* u_i on the left to n *output nodes* v_i on the right. The output nodes all have degree d. To compute the function, on input $x \in \{0,1\}^n$, we label the input nodes with the bits of x, and label each output node by the value of P applied to the labels of its neighbors. The output of the function is the sequence of n labels of the output nodes. For example, if the neighborhood of v_i is $\{u_{j_1}, u_{j_2}, \ldots, u_{j_d}\}$, then

$$(f(x))_i = P(x_{j_1}, x_{j_2}, \ldots, x_{j_d}).$$

We denote by A the $n \times n$ matrix adjacency matrix of G, whose columns correspond to input nodes and whose rows correspond to output nodes:

$$A_{ij} = \begin{cases} 1 & (u_j, v_i) \in E \\ 0 & (u_j, v_i) \notin E \end{cases}.$$

Goldreich suggests using a random predicate P, and a graph G with expansion properties.

2.2   Myopic Backtracking Algorithms

We consider the class of algorithms that might invert Goldreich's function by backtracking.

First, we need a notion of a *partial truth assignment*.

Definition 2.1 (partial assignment). Taken from [2]. *A partial assignment is a function* $\rho : [n] \to \{0, 1, *\}$. *Its set of* fixed variables *is* $\text{Vars}(\rho) = \rho^{-1}(\{0, 1\})$.

Its size *is defined to be* $|\rho| = |\mathrm{Vars}(\rho)|$. *Given* $f : \{0,1\}^n \to \{0,1\}^n$, *the* restriction *of* f *by* ρ, *denoted* $f|_\rho$, *is the function obtained by fixing the variables in* $\mathrm{Vars}(\rho)$ *and allowing the rest to vary.*

Definition 2.2. *A* backtracking *algorithm for solving an equation* $f(x) = b$ *for* x *is defined by two procedures* **N** *and* **T**. **N** *takes a partial assignment* ρ *and returns the index of a new variable* $\mathbf{N}(\rho) \in [n]$ *to assign, and* **T** *chooses a truth value* $\mathbf{T}(\rho) \in \{0,1\}$ *for* $x_{\mathbf{N}(\rho)}$. *More precisely, the algorithm has the form:*

- *Initialize* ρ *to the empty truth assignment* $(*,*,\ldots,*)$.
- *While not all variables in* ρ *are fixed,*
    - $j \leftarrow \mathbf{N}(\rho)$.
    - *Update* ρ *by assigning* x_j *the truth value* $\mathbf{T}(\rho)$.
    - *If there is row* i *such that* $f(\rho)_i$ *is determined by* ρ *but* $f(\rho)_i \neq b_i$ *then backtrack.*

We study a special class of backtracking algorithms which we call *myopic* backtracking algorithms, after [1].

Definition 2.3. *A* myopic *backtracking algorithm for* $f(x) = b$ *is a backtracking algorithm where procedures* **N** *and* **T** *are restricted in that they are not allowed to see all the output bits in vector* b. *More precisely, myopic backtracking algorithm of parameter* K *have the following properties:*

- *In the beginning of the algorithm, the algorithm does not have the value of any of* b.
- *At each step of fixing a new variable, the algorithm is allowed to ask the value of* K *output bits corresponding to* K *equations chosen by the algorithm.*
- *When we backtrack from a step we have already taken, we lose the value of the output bits that were revealed to us at that step.*

Thus, in the middle of the algorithm, when the partial assignment is ρ, the algorithm sees the values of $K|\mathrm{Vars}(\rho)|$ output bits, and the outputs of procedures **N** and **T** are allowed to depend only on these $K|\mathrm{Vars}(\rho)|$ output bits. But notice that procedures **N** and **T** can use the structure of the function f; they have restricted access to only b.

Notice that in the above definitions, there is no restriction on the computational complexity of procedures **N** and **T**. Therefore without the *myopic* constraint, there is no way to prevent **T** from guiding the algorithm immediately towards the correct solution.

The work in [3] gives a lower bound for myopic backtracking algorithms for SAT instances. They translate a system of linear equations $Ax = b$ into a CNF formula. Similarly, for inverting Goldreich's function $f(x) = b$ for a fixed $b \in \{0,1\}^n$, we can define a d-CNF formula $\Phi_b(x)$ which is logically equivalent to the statement $f(x) = b$. The i-th bit of b translates to a set of at most 2^d clauses that enforce the constraint $P(x_{S_i}) = b_i$. Then the problem of inverting f can be reduced to finding a solution to the SAT instance Φ_b. Notice that (myopic) backtracking algorithms for solving Φ_b are similar to (myopic) backtracking algorithms for solving $f(x) = b$.

In [3] the authors consider a notion of myopic backtracking algorithms that is slightly more powerful, called myopic DPLL algorithms after [12,8,1], which might get more information about b using two new rules called Unit Clause Propagation and Pure Literal Elimination. It can be seen that when the equations of $f(x) = b$ are linear, these two rules do not give an advantage to the backtracking algorithm. However, the same reduction from DPLL to ordinary backtracking does not apply to the more general case $f(x) = b$ which we consider. Therefore, in this paper we restrict ourselves to backtracking algorithms.

2.3 Random Predicates

We follow Goldreich's suggestion in choosing $P : \{0,1\}^d \to \{0,1\}$ uniformly at random. Here we define two useful properties that most random predicates have.

Definition 2.4 (robust predicate). $P : \{0,1\}^d \to \{0,1\}$ *is* h-robust *iff every restriction* ρ *such that* $f|_\rho$ *is constant satisfies* $d - |\rho| \le h$ *[2, Definition 2.2]. For example, the predicate that sums all its inputs modulo 2 is 0-robust.*

Definition 2.5 (balanced predicate). $P : \{0,1\}^d \to \{0,1\}$ *is* (h, ϵ)-balanced *if, after fixing all variables but* $h + 1$ *of them,*

$$\left| \Pr[P(x) = 0] - \frac{1}{2} \right| \le \epsilon.$$

For example, predicates of the form $P_d(x) = x_1 \oplus \cdots \oplus x_{d-2} \oplus (x_{d-1} \wedge x_d)$ are $(2, 0)$-balanced and $(1, \frac{1}{4})$-balanced. The predicate that sums all its inputs is $(0, 0)$-balanced.

Lemma 2.6. *A random predicate on d variables is* $(\Theta(\log \frac{d}{\epsilon}), \epsilon)$-balanced *with probability* $1 - \exp[-\text{poly}(d/\epsilon)]$.

(We omit the proof to save space.)

Corollary 2.7. *A random predicate on d variables is* $\Theta(\log d)$-robust *with probability* $1 - \exp[-\text{poly}(d)]$.

2.4 Expansion Properties

Let G be a bipartite graph with n nodes on each side and right-degree d. Equivalently, let A be an $n \times n$ matrix with d ones and $n - d$ zeros in each row.

Definition 2.8 (Boundary and Neighborhood). *Taken from* [3, Definition 2.1]. *Let I be a set of output nodes. Its* boundary, *denoted* ∂I, *is the set of all nodes* $j \in U$ *such that there is exactly one edge from j to I. The* neighborhood *of I, $\Gamma(I) \subseteq U$ is the set of all nodes connected to I.*

Definition 2.9 (Expansion). *Taken from* [3, Definition 2.1]. *G (or A) is an* (r, d, c)-boundary expander *if for all $I \subseteq V, (|I| \le r \Rightarrow |\partial I| \ge c|I|)$. G (or A) is an* (r, d, c)-expander *if* $\forall I \subseteq V, (|I| \le r \Rightarrow |\Gamma(I)| \ge c|I|)$.

Lemma 2.10. Analogous to [3, Lemma 2.1]. *Every (r, d, c)-expander is an $(r, d, 2c - d)$-boundary expander.*

Throughout our paper, we will use c to denote neighborhood expansion, and c' to denote boundary expansion, with $c' = 2c - d$.

2.5   Closure Operation

We define the *closure* of a set of input nodes, or columns of A.

Definition 2.11 (closure). Analogous to [3, Definition 3.2]. *For a set of columns $J \subseteq [n]$, define the following relation on $2^{[n]}$:*

$$I \vdash_J I_1 \iff I \cap I_1 = \varnothing \wedge |I_1| \leq \frac{r}{2} \wedge \left| \partial(I_1) \setminus \left[\bigcup_{i \in I} A_i \cup J \right] \right| < c/2 |I_1|.$$

Define the closure *of J, $\mathrm{Cl}(J)$, as follows. Let $G_0 = \varnothing$. Having defined G_k, choose a non-empty I_k such that $G_k \vdash_J I_k$, set $G_{k+1} = G_k \cup I_k$, and remove equations I_k from matrix A. (Fix an ordering on $2^{[n]}$ to ensure a deterministic choice of I_k.) When k is large enough that no non-empty I_k can be found, set $\mathrm{Cl}(J) = G_k$.*

We omit the proofs in this section, since similar facts are proved in Section 3 of [3].

Lemma 2.12. Analogous to [3, Lemma 3.5]. *If $|J| < \frac{cr}{4}$, then $|\mathrm{Cl}(J)| < 2c^{-1}|J|$.*

Lemma 2.13. Analogous to [3, Lemma 3.4]. *Assume that A is an arbitrary matrix and J is a set of its columns. Denote by $\hat{A}$ the matrix that results from A by removing the rows in $\mathrm{Cl}(J)$ and the columns in $\bigcup_{i \in \mathrm{Cl}(J)} A_i$. If $\hat{A}$ is non-empty then it is an $(r/2, d, c/2)$-boundary expander.*

Definition 2.14. From [3, Definition 3.4]. *A substitution ρ is said to be* locally consistent *w.r.t. the function $f(x) = b$ if and only if ρ can be extended to an assignment on X which satisfies the equations corresponding to $\mathrm{Cl}(\mathrm{Vars}(\rho))$:*

$$(f(x))_{\mathrm{Cl}(\mathrm{Vars}(\rho))} = b_{\mathrm{Cl}(\mathrm{Vars}(\rho))}$$

Lemma 2.15. Analogous to [3, Lemma 3.6]. *Assume that f employs a (r, d, c)-boundary expander and a h-robust predicate with $c > 2h$. Let $b \in \{0, 1\}^n$ and ρ be a locally consistent partial assignment. Then for any set $I \subseteq [n]$ with $|I| \leq r/2$, ρ can be extended to an assignment x which satisfies the subsystem $(f(x))_I = b_I$.*

3   Myopic Algorithms Use Exponential Time in the Average Case

Theorem 3.1. *Assume A is an $n \times n$ (r, d, c)-boundary expander with left and right degree d and that P is an (h, ϵ)-balanced predicate. Let f be Goldreich's* *function for A and P, and assume f is M-to-one-on-average, in the sense that the number of pairs (x, y) such that $f(x) = f(y)$ is at most $M2^n$. Let A be any myopic backtracking algorithm. Choose $x \in \{0, 1\}^n$ uniformly at random and let $b = f(x)$. Let $F = \lceil 2c - d - h \rceil - 1$, and $s = F/(F + d(d - 1))$. Then the probability that A solves $f(x) = b$ in time $2^{O(r(c-h))}$ is at most*

$$M2^{-s\lfloor \frac{cr}{4dK} \rfloor} \left(\frac{1 + 2\epsilon}{1 - 2\epsilon} \right)^{r/2}. \tag{1}$$

(We can relax the degree requirement to say that A has right degree d_{right}, and the nodes in every set of $s\lfloor \frac{cr}{4dK} \rfloor$ input nodes have average degree at most d_{left}, where in this case $s = F/(F + d_{\text{left}}(d_{\text{right}} - 1))$.)

Applications of Theorem 3.1

1. Use the predicate $P_d(x) = x_1 \oplus \cdots \oplus x_{d-2} \oplus (x_{d-1} \wedge x_d)$ and a random graph of right-degree d. Then $h = \Theta(1)$, $\epsilon = 0$, $c = d/2 + \Theta(d)$ and $r = \Theta(n/d)$. This gives $F = \Theta(d)$ and $s = \Theta(1/d)$. In Section 4, we show that with high probability $M = 2^{n2^{-\Omega(d)}}$. Furthermore, the average degree of any set of $s\lfloor \frac{cr}{4dK} \rfloor$ input nodes is at most $3d$ with high probability. With these parameters, Theorem 3.1 says that for constant K, the myopic algorithm takes time $2^{\Theta(n)}$ with probability $1 - 2^{-Cn}$, where C depends on d and K.
2. Use a random predicate P and a random graph of right-degree d. Then with high probability, P is (h, ϵ) balanced, with $h = \Theta(\log d)$ and $\epsilon = 1/\text{poly}(d)$. Conditioned on the assumption $M = 2^{nO(d^{-C_0})}$ for $C_0 > 2$, Theorem 3.1 says as before that for constant K, the myopic algorithm takes time $2^{\Theta(n)}$ with probability $1 - 2^{-C_1 n}$, where C_1 depends on d and K.

The rest of this section is devoted to proving Theorem 3.1. First in Section 3.1 we show how it is possible to assume that after a fixed number of steps, the partial truth assignment ρ made by the algorithm will be locally consistent. Then in Section 3.2 we show that the algorithm can only have selected one of many possible locally consistent partial truth assignments – and for any fixed $b \in \{0, 1\}^n$, most of these partial assignments will be wrong. Thus, with high probability, the algorithm will have selected globally inconsistent values that lead to an unsatisfiable formula. We then show in Section 3.3 that any resolution proof showing that this new formula is unsatisfiable has size $2^{\Omega(r(c-h))}$, so the algorithm must take that many steps before correcting its mistake.

3.1   Clever Myopic Algorithms

Without loss of generality, we allow our algorithm to be a "clever" myopic algorithm in the sense that, as defined in [3], it satisfies these two properties.

1. Let J be the set of indices of all variables x_j that appear in equations whose output bit b_i has been revealed. Then the algorithm may also read all clauses in $\text{Cl}(J)$ for free and reveal the corresponding new variables.

2. The algorithm never makes stupid guesses: whenever the equations corresponding to the revealed output bits b_i determine the value of a variable x_j, the algorithm will never make the wrong assignment for x_j.

Property 2 can only reduce the number of backtracking steps taken. Property 1 is justified by the following proposition.

Proposition 3.2. Analogous to [3, Proposition 3.1]. *After the first $\lfloor \frac{cr}{4dK} \rfloor$ steps a clever myopic algorithm reads at most $r/2$ bits of b.*

Proof. At each step, the algorithm makes K clause-queries, asking for dK variable entries. This sums to at most $dK(cr/4dK) = cr/4$ variables, which by Lemma 2.12 will result in at most $r/2$ bits of b.   □

Once we have assumed that our algorithm is clever, the following proposition shows that we can further assume the algorithm only makes locally consistent assignments in its first $\lfloor cr/4dK \rfloor$ steps.

Proposition 3.3. Analogous to [3, Proposition 3.2]. *During the first $\lfloor \frac{cr}{4dK} \rfloor$ steps the current partial assignment made by a clever myopic algorithm is locally consistent, and so the algorithm will not backtrack.*

Proof. This statement follows by repeated application of Lemma 2.15. Note that clever myopic algorithms are required to select a locally consistent choice of variables if one is available. The proof is accomplished through induction. Initially, the partial assignment is empty, and so is locally consistent. For each step t (with $t < \frac{cr}{4dK}$) with a locally consistent partial assignment ρ_t, a clever myopic algorithm will extend this assignment to ρ_{t+1} which is also locally consistent if possible. By Lemma 2.15 it can always do so as long as $|Cl(Vars(\rho_t)) \cup \{x_j\}| \leq r/2$ for the newly chosen x_j.   □

3.2   The Probability of a Correct Guess Is Small

Now choose b randomly from the set of attainable outputs of $f(x)$; more formally, let $x \sim \mathrm{Unif}(\{0, 1\}^n)$ and $b = f(x)$. Initially, the value of b should be hidden from the algorithm. Whenever the algorithm reveals a clause corresponding to the ith row of A, the ith-bit of b should be revealed to the algorithm. We consider the situation after $\lfloor \frac{cr}{4dK} \rfloor$ steps of the algorithm. By Proposition 3.3, the current partial assignment must be locally consistent, and no backtracking will have occurred. Thus, at this point in time we observe the algorithm in the $\lfloor \frac{cr}{4dK} \rfloor$-th vertex v in the leftmost branch of its backtracking tree. By Proposition 3.2, the algorithm has revealed at most $r/2$ bits of b.

Define random variable $I_v \subseteq [n]$ to be the set of output bits revealed after $\lfloor \frac{cr}{4dK} \rfloor$ steps. Similarly define random variable ρ_v to be the the partial truth assignment given by the algorithm at that time. Define $R_v = Vars(\rho_v)$. Hence $|R_v| = \lfloor \frac{cr}{4dK} \rfloor$.

Definition 3.4. *Let $I \subseteq [n]$, $R \subseteq [n]$, $\iota \in \{0,1\}^I$ and $\rho \in \{0,1\}^R$. We say (I, R, ι, ρ) is a* consistent state *if*

$$\Pr[I_v = I \wedge R_v = R \wedge b_{I_v} = \iota \wedge \rho_v = \rho] > 0.$$

Put another way, (I, R, ι, ρ) is a consistent state iff there exists some $x \in \{0,1\}^n$ such that after $\lfloor \frac{cr}{4dK} \rfloor$ steps, I is exactly the set of revealed bits, R is exactly the set of assigned variables, ρ is the values assigned those variables, and $b_I = \iota$.

Our first attempt at a proof is to show that there are many choices for ρ_v that are locally consistent. Intuitively, if the number of those possible choices for ρ is large compared to M, we expect the algorithm to make the wrong choice with high probability. This line of reasoning would need a result of the following form.

Lemma 3.5. *Analogous to* [3, Lemma 3.10]. *Assume that an $n \times n$ matrix A is an (r, d, c)-boundary expander and P is an (h, ϵ)-balanced predicate, and let f be Goldreich's function corresponding to A and P. Let $\hat{X} \subseteq [n]$ be a set of input variables with $|\hat{X}| < r$. Choose x uniformly at random from $\{0,1\}^n$ and let $b = f(x)$. Let $Y \subseteq [n]$ be a set of output variables, $|Y| < r$. For $i \in Y$ let ℓ_i be the constraint $(f(x))_i = b_i$, and let $\mathcal{L} = \{\ell_i : i \in Y\}$. Denote by L the set of partial assignments who assign values to the variables in $\hat{X}$ and can be extended to complete truth assignments that satisfy $\mathcal{L}$.*

Let s be defined as in Theorem 3.1. Then $\log |L| \geq s|\hat{X}|$.

In fact, our proof requires the following stronger lemma instead of Lemma 3.5, which states that conditioned on what the algorithm has seen, the minimum entropy of x_{R_v} is high.

Lemma 3.6. *Let x, $\hat{X}$, $\mathcal{L}$, and s be as in Lemma 3.5. Then for any $\hat{x} \in \{0,1\}^{|\hat{X}|}$,*

$$\Pr[x_{|\hat{X}|} = \hat{x}|\mathcal{L}] \leq 2^{-s|\hat{X}|} \left(\frac{1 + 2\epsilon}{1 - 2\epsilon}\right)^{|\mathcal{L}|}.$$

We postpone the proof of Lemma 3.6 (and do not prove Lemma 3.5, since we do not use it). We are now prepared to complete the proof of the main theorem.

Proof (Theorem 3.1). Our goal is to bound the probability of the following event:

$$E = \{\rho_v \in (f^{-1}(b))_{R_v}\}.$$

We first condition on the state of the algorithm, considering all consistent states in the sense of Definition 3.4:

$$\Pr[E]$$

$$= \sum_{(I,R,\iota,\rho) \text{ consistent}} \Pr[E|I_v = I \wedge R_v = R \wedge b_{I_v} = \iota \wedge \rho_v = \rho] \cdot$$

$$\Pr[I_v = I \wedge R_v = R \wedge b_{I_v} = \iota \wedge \rho_v = \rho]$$

$$= \mathbf{E}[\Pr[E|I_v, R_v, b_{I_v}, \rho_v]].$$

Since the algorithm is deterministic and only observes the bits in b_{I_v}, the event $[I_v = I \wedge R_v = R \wedge \rho_v = \rho]$ is implied by the event $[b_{I_v} = \iota]$ – put another way, if bits of b outside the set of observed bits I_v are changed, the behavior of the algorithm will not be affected, so the values of I_v, R_v and ρ_v will not change. This gives us:

$$\Pr[E]$$
$$= \mathbf{E}[\Pr[E|b_{I_v}]]$$
$$= \mathbf{E}[\Pr[\rho_v \in (f^{-1}(b))_{R_v}|b_{I_v}]]$$
$$\leq \mathbf{E}\left[\max_{\rho_v^* \in \{0,1\}^{R_v}} |f^{-1}(b)| \Pr[\rho_v^* = x_v|b_{I_v}]\right]$$
$$\leq \mathbf{E}[|f^{-1}(b)|] \max_{I_v,b_{I_v},\rho_v^*} \Pr[\rho_v^* = x_v|b_{I_v}]$$
$$\leq M 2^{-s}\left(\frac{1+2\epsilon}{1-2\epsilon}\right)^{r/2}.$$

In the last step, we applied Lemma 3.6, replacing $\hat{X}$ by R_v, $\mathcal{L}$ by I_v and $\hat{x}$ by ρ. Note that $|I_v| < r/2$, so $|\mathcal{L}| < r/2$ in the hypothesis of the lemma.

We have shown that it will be likely that ρ_v, though locally consistent, can not be extended to satisfy b, and an unsatisfiable instance will occur. In Section 3.3, we explore the running time of backtracking algorithms on unsatisfiable cases to show if E does not occur, the algorithm will take time $2^{\Omega(r(c-h))}$.

Proof of Lemma 3.6

Lemma 3.7. *Fix any $g \subseteq \hat{X}$. If each output in $\mathcal{L}$ has at most $F = \lceil 2c-d-h \rceil - 1$ of its d inputs in g, then $\forall I \subseteq \mathcal{L}$, $|\partial I \setminus g| > h|I|$.*

Proof. Consider any subset $I \subseteq \mathcal{L}$. By Lemma 2.10, $|\partial I| \geq (2c-d)|I|$, so $|\partial I \setminus g| \geq (2c-d-F)|I| > h|I|$.

Lemma 3.8. *Fix any $g \subseteq \hat{X}$. If $\forall I \subseteq \mathcal{L}$, $|\partial I \setminus g| > h|I|$, then any partial assignment $\rho : g \to \{0,1\}$ can be extended to a complete assignment that satisfies $\mathcal{L}$.*

Proof. We make our proof by contradiction; assume ρ cannot be extended to satisfy the equations in $\mathcal{L}$. Let k be a minimal set of unsatisfiable equations. We assume our predicate is h-robust. $\forall I \subseteq \mathcal{L}$, $|\partial I \setminus g| > h|I|$ implies that some equation in I must have at least $h+1$ boundary elements outside of g. However, no equation in k should have more than h boundary variables; otherwise, those $h+1$ boundary variables could be set to a value that satisfies that equation, and it should not be in the minimal set k.

Lemma 3.9. *Let s and F be as in Theorem 3.1. We can find $g \subseteq \hat{X}$ with $|g| \geq s|\hat{X}|$, such that no output has more than F inputs in g.*

Proof. Construct g using the following algorithm:

- $g \leftarrow \varnothing$.
- $n_i \leftarrow \begin{cases} F & i \in \hat{X} \\ 0 & i \notin \hat{X} \end{cases}$.
- **while** $\exists i,\ n_i > 0$,
  - *Invariant:* If an output has $F - a$ inputs in g, then for every input i connected to it, $n_i \leq a$.
  - $g \leftarrow g \cup \{i\}$.
  - $n_i \leftarrow n_i - F$.
  - $\forall j$, if $\text{dist}(i,j) = 2$, then $n_j \leftarrow n_j - 1$.

We start with $F|\hat{X}|$ counters, and remove on average $F + d_{\text{left}}(d_{\text{right}} - 1)$ counters at every step. (Recall that output nodes have degree d_{right}, and as long as $|g| \leq s\lfloor \frac{cr}{4dK} \rfloor$, the average degree of g is at most d_{left}.) In the end,

$$|g| \geq \frac{F|\hat{X}|}{F + d_{\text{left}}(d_{\text{right}} - 1)}.$$

$\square$

Proof (Lemma 3.6). Choose $g \subseteq \hat{X}$ with $|g| \geq s|\hat{X}|$ as in Lemma 3.9. By Lemma 3.7, every subset of $\mathcal{L}$ has a row with at least $h + 1$ boundary variables that are not in g. Therefore we can order the rows of $\mathcal{L}$ as $\ell_1, \ldots, \ell_{|\mathcal{L}|}$ such that setting $L_i = \{\ell_1, \ldots, \ell_i\}$, for all i we have $|\Gamma(\ell_i) \setminus (\Gamma(L_{i-1}) \cup g)| \geq h + 1$. Then

$$\frac{\Pr[x_{|g} = g_1 | L_{i+1}]}{\Pr[x_{|g} = g_2 | L_{i+1}]} = \frac{\Pr[L_{i+1} | x_{|g} = g_1] \Pr[x_{|g} = g_2]}{\Pr[L_{i+1} | x_{|g} = g_2] \Pr[x_{|g} = g_1]}$$

$$= \frac{\Pr[L_i | x_{|g} = g_1] \Pr[\ell_{i+1} | L_i, x_{|g} = g_1]}{\Pr[L_i | x_{|g} = g_2] \Pr[\ell_{i+1} | L_i, x_{|g} = g_2]}$$

(Use the fact that the predicate is (h, ϵ)-balanced.)

$$\leq \left(\frac{\frac{1}{2} + \epsilon}{\frac{1}{2} - \epsilon} \right) \frac{\Pr[L_i | x_{|g} = g_1]}{\Pr[L_i | x_{|g} = g_2]}.$$

Observe that

$$\frac{\Pr[x_{|g} = g_1 | L_0]}{\Pr[x_{|g} = g_2 | L_0]} = 1.$$

It follows that

$$\frac{\Pr[x_{|g} = g_1 | \mathcal{L}]}{\Pr[x_{|g} = g_2 | \mathcal{L}]} \leq \left(\frac{1 + 2\epsilon}{1 - 2\epsilon} \right)^{|\mathcal{L}|}.$$

Take $g_1 \in \{0, 1\}^g$ that minimizes $\Pr[x_{|g} = g_1 | \mathcal{L}]$. There are $2^{|g|}$ possible values for g_1, so $\Pr[x_{|g} = g_1 | \mathcal{L}] \leq 2^{-|g|} \leq 2^{-s|\hat{X}|}$. For any $\hat{x} \in \{0, 1\}^{\hat{X}}$,

$$\Pr[x_{|\hat{X}} = \hat{x} | \mathcal{L}] \leq \Pr[x_{|g} = \hat{x}_{|g} | \mathcal{L}]$$

$$= \Pr[x_{|g} = g_1 | \mathcal{L}] \frac{\Pr[x_{|g} = \hat{x}_{|g} | \mathcal{L}]}{\Pr[x_{|g} = g_1 | \mathcal{L}]} \leq 2^{-s|\hat{X}|} \left(\frac{\frac{1}{2} + \epsilon}{\frac{1}{2} - \epsilon} \right)^{|\mathcal{L}|}.$$

$\square$

3.3 Backtracking Algorithms Use Exponential Running Time on Unsatisfiable Formulas

In Section 3.2, we showed that with high probability a myopic backtracking algorithm will choose a partial assignment to x that cannot be extended to satisfy $f(x) = b$. We now prove that once this happens, the algorithm must run for exponential time:

Theorem 3.10. Analogous to [3, Lemma 3.9]. *Let f be Goldreich's function for predicate P and graph G, where G is an $n \times n$ (r, d, c)-boundary expander with right-degree d and P is h-robust with $h < c/2$. Fix $b \in \{0, 1\}^n$. Let ρ be a locally consistent partial assignment such that $|Vars(\rho)| \leq cr/4$. If ρ cannot be extended to any input x satisfying $f(x) = b$, then every backtracking algorithm that makes the partial assignment ρ will run for time $2^{\Omega(r(c-h))}$.*

We will make use of the following lemma from [6, Corollary 3.4]. The *width* of a resolution proof is the greatest width of any clause that occurs in it, and the width of a clause is the number of variables in it.

Lemma 3.11. *The size of any tree-like resolution refutation of a formula Ψ is at least 2^{w-w_Ψ}, where w is the minimal width of a resolution refutation of Ψ, and w_Ψ is the maximal width of a clause in Ψ.*

Our setup and proof strategy are similar to those found in [2, Section 3] and [3]. [2] measures robustness in terms of ℓ, where $\ell = d - h$.

Proof (Theorem 3.10). Let $I = \text{Cl}(\rho)$ and $J = \Gamma(I)$. By Lemma 2.12 $|I| \leq r/2$. By Lemma 2.15, ρ can be extended to another partial assignment ρ' on variables x_J, such that ρ' satisfies every equation in $(f(x))_I = b_I$. The restricted formula $(f(x) = b)|_{\rho'}$ still encodes an unsatisfiable system $f'(x) = b'$. The underlying graph G' of f' is produced from G by removing every output node in I and every input node in J. By Lemma 2.13, G' is an $(r/2, d, c/2)$-boundary expander.

We can express the equation $f'(x) = b'$ using a CNF formula Φ, by representing each equation $(f'(x))_i = b'_i$ by at most 2^d clauses. The computation of a backtracking algorithm as it discovers that $f'(x) = b'$ is unsatisfiable can be translated to a tree-like resolution refutation of the formula Φ, such that the size of the refutation is the working time of the algorithm. Thus it is sufficient to show that every tree-like resolution refutation of Φ is large.

We say a set of equations $(f'(x))_I = b'_I$ *semantically implies* a clause C iff every truth assignment satisfying $(f(x))_I = b_I$ also satisfies C. Following [2, Section 3], we define the *measure* of C to be

$$\mu(C) = \min_{I:(f'(x))_I = b'_I \models C} |I|.$$

We omit the proofs of the following facts; similar facts are proved in [2].

1. For any $D \in \Phi$, $\mu(D) = 1$.
2. $\mu(\varnothing) > r$.

3. μ is subadditive: if C_2 is the resolution of C_0 and C_1, then $\mu(C_2) \leq \mu(C_0) + \mu(C_1)$.

4. For any clause C, if $\frac{r}{2} \leq \mu(C) \leq r$, then C has width at least $\frac{(c/2-h)r}{4}$.

1, 2 and 3 together imply that any resulation proof will result in a clause C whose width is between $\frac{r}{2}$ and r. By 4, C has width at least $\frac{(c/2-h)r}{4}$, so by Lemma 3.11, the resolution proof has size $2^{\Omega(r(c-h))}$. $\qquad\square$

4   The Size of Pre-images of Goldreich's Function

In this section we prove that Goldreich's function has pre-images sufficiently small for Theorem 3.1 to work.

Theorem 4.1. *For every degree d, let $P_d(x_1, \ldots, x_d) = x_1 \oplus \cdots \oplus x_{d-2} \oplus (x_{d-1} \wedge x_d)$. Choose a random graph for Goldreich's function by connecting each output to d inputs chosen uniformly at random (with replacement). Then*

$$\mathbf{E}[\#(x,y) : f(x) = f(y)] = 2^{(1+2^{-\Omega(d)})n},$$

where the expectation is over the choice of graph.

Proof. For $x, y \in \{0,1\}^n$ and $i, j \in \{0,1\}$, let $n_{ij}(x,y)$ be the number of indices k where $x_k = i$ and $y_k = j$. We have $n_{00}(x,y) + n_{01}(x,y) + n_{10}(x,y) + n_{11}(x,y) = n$.

Since the input indices to the predicate are selected uniformly at random, the probability that a single output bit will be equal for inputs x and y is only a function of $\alpha_{ij} \stackrel{\text{def}}{=} n_{ij}(x,y)/n$. We call this function the *probability of equality*, $\mathrm{PE}(\alpha_{00}, \alpha_{01}, \alpha_{10}, \alpha_{11})$.

Then

$$\mathbf{E}[\#(x,y) : f(x) = f(y)]$$

$$= \sum_{x,y \in \{0,1\}^n} \Pr[f(x) = f(y)]$$

$$= \sum_{n_{00}+n_{01}+n_{10}+n_{11}=n} \binom{n}{n_{00}, n_{01}, n_{10}, n_{11}} \Pr[f(x) = f(y) | n_{ij}]$$

$$\leq n^4 \max_{\alpha_{00}+\alpha_{01}+\alpha_{10}+\alpha_{11}=1} \binom{n}{n\alpha_{00}, n\alpha_{01}, n\alpha_{10}, n\alpha_{11}} \mathrm{PE}(\alpha_{00}, \alpha_{01}, \alpha_{10}, \alpha_{11})^n$$

$$= \max_{\alpha_{00}+\alpha_{01}+\alpha_{10}+\alpha_{11}=1} (2^{H(\alpha_{00}, \alpha_{01}, \alpha_{10}, \alpha_{11})} \mathrm{PE}(\alpha_{00}, \alpha_{01}, \alpha_{10}, \alpha_{11}))^{n(1+o(1))}$$

where $H(\alpha_{ij})$ is the base-2 entropy of the distribution defined by α_{ij}. Thus, it suffices to show that there is a constant $\epsilon > 0$ such that for sufficiently large d,

$$\forall \alpha_{ij} \; H(\alpha_{ij}) + \log_2 \mathrm{PE}(\alpha_{ij}) \leq 1 + 2^{-\epsilon d}.$$

It can be shown that for the predicate P_d which we have defined,

$$\mathrm{PE}(\alpha_{ij}) \leq \frac{1 + (\alpha_{00} + \alpha_{11} - \alpha_{01} - \alpha_{10})^{d-2}}{2}.$$

Now, let $p = \alpha_{00} + \alpha_{11}$ and $q = \alpha_{01} + \alpha_{10} = 1 - p$. Forcing $\alpha_{00} = \alpha_{11}$ and $\alpha_{01} = \alpha_{10}$ can only increase $H(\alpha_{ij})$, so without loss of generality, we assume $\alpha_{00} = \alpha_{11} = p/2$ and $\alpha_{01} = \alpha_{10} = q/2$, and prove

$$\forall p \in [0,1]: \ H(p/2, p/2, q/2, q/2) + \log_2 \left(\frac{1 + (p-q)^{d-2}}{2} \right) \leq 1 + 2^{-\epsilon d},$$

or equivalently,

$$\forall q \in [0, \tfrac{1}{2}]: \ H(q) + \log_2(1 + (1 - 2q)^{d-2}) \leq 1 + 2^{-\epsilon d}.$$

We consider four cases for the value of q. (We will choose positive constants $\epsilon, \epsilon_1, \epsilon_2, \epsilon_3$ suitably as we go along.)

Case 1: $q > \epsilon_1$

$$H(q) + \log_2(1 + (1 - 2q)^{d-2}) \leq 1 + (1 - 2\epsilon_1)^{d-2} \log_2 e \leq 1 + 2^{-\epsilon d},$$

for $\epsilon < -\log_2(1 - 2\epsilon_1)$ and sufficiently large d.

For the remaining three cases, q is small. Using the Taylor expansion of $\log_2$ around 2, we get

$$\log_2(1 + (1 - 2q)^{d-2}) \leq 1 + \frac{(1 - 2q)^{d-2} - 1}{2 \ln 2} \leq 1 + \frac{e^{-2qd} - 1}{2 \ln 2}.$$

Case 2: $\epsilon_1 \geq q > \epsilon_2/d$

$$H(q) + \log_2(1 + (1 - 2q)^{d-2}) \leq H(\epsilon_1) + 1 + \frac{e^{-2\epsilon_2} - 1}{2 \ln 2} \leq 1$$

if we choose ϵ_1 small enough that $H(\epsilon_1) < \frac{1 - e^{-2\epsilon_2}}{2 \ln 2}$.

For the remaining two cases we fix ϵ_2, say $\epsilon_2 = \tfrac{1}{2}$. Now, $qd \leq \tfrac{1}{2}$, and we have the approximation

$$H(q) + 1 + \frac{e^{-2qd} - 1}{2 \ln 2} \leq (q \log_2(1/q) + 2q) + 1 - \frac{qd}{2 \ln 2} = q(\log_2(1/q) - \Theta(d)) + 1.$$

Case 3: $\epsilon_2/d \geq q > 2^{-\epsilon_3 d}$

For $\epsilon_3 < \frac{1}{\ln 2}$ and sufficiently large d: $\log_2(1/q) - \Theta(d) < 0$.

Case 4: $2^{-\epsilon_3 d} \geq q$

For $\epsilon < \epsilon_3$ and sufficiently large d: $q \log(1/q) \leq \epsilon_3 d 2^{-\epsilon_3 d} \leq 2^{-\epsilon d}$.
    This completes our proof.   □

References

1. Achlioptas, D., Sorkin, G.B.: Optimal myopic algorithms for random 3-SAT. In: FOCS, pp. 590–600 (2000)
2. Alekhnovich, M., Ben-Sasson, E., Razborov, A.A., Wigderson, A.: Pseudorandom generators in propositional proof complexity. SIAM Journal on Computing 34(1), 67–88 (2004)
3. Alekhnovich, M., Hirsch, E.A., Itsykson, D.: Exponential lower bounds for the running time of DPLL algorithms on satisfiable formulas. J. Autom. Reasoning 35, 51–72 (2005)
4. Applebaum, B., Ishai, Y., Kushilevitz, E.: Cryptography in NC0. SIAM J. on Computing 36(4), 845–888 (2006)
5. Applebaum, B., Ishai, Y., Kushilevitz, E.: On pseudorandom generators with linear stretch in NC^0 . In: Díaz, J., Jansen, K., Rolim, J.D.P., Zwick, U. (eds.) APPROX 2006 and RANDOM 2006. LNCS, vol. 4110, pp. 260–271. Springer, Heidelberg (2006)
6. Ben-Sasson, Wigderson: Short proofs are narrow–resolution made simple. J. ACM: Journal of the ACM 48 (2001)
7. Cryan, M., Miltersen, P.B.: On pseudorandom generators in NC. In: Sgall, J., Pultr, A., Kolman, P. (eds.) MFCS 2001. LNCS, vol. 2136, p. 272. Springer, Heidelberg (2001)
8. Davis, M., Logemann, G., Loveland, D.: A machine program for theorem-proving. Communications of the ACM 5, 394–397 (1962)
9. Eén, N., Biere, A.: Effective preprocessing in SAT through variable and clause elimination. In: Bacchus, F., Walsh, T. (eds.) SAT 2005. LNCS, vol. 3569, pp. 61–75. Springer, Heidelberg (2005)
10. Eén, N., Sörensson, N.: An extensible SAT-solver. In: Giunchiglia, E., Tacchella, A. (eds.) SAT 2003. LNCS, vol. 2919, pp. 502–518. Springer, Heidelberg (2004)
11. Goldreich, O.: Candidate one-way functions based on expander graphs. Electronic Colloquium on Computational Complexity (ECCC) 7(90) (2000)
12. Davis, M., Putnam, H.: A computing procedure for quantification theory. Journal of the ACM 7, 201–215 (1960)
13. Mossel, E., Shpilka, A., Trevisan, L.: On ϵ-biased generators in NC^0. Random Structures and Algorithms 29(1), 56–81 (2006)

A   MiniSat Experiment

Inverting Goldreich's function can be seen as the task of solving a constraint satisfaction problem with a planted solution. This suggests the use of a general-purpose SAT solver to solve the constraint satisfaction problem. We performed an experiment using MiniSat version 2.0 beta [10,9], which is one of the best publicly available SAT solvers. We always use the degree-five predicate $P_5(x) = x_1 \oplus x_2 \oplus x_3 \oplus (x_4 \wedge x_5)$. For each trial, we choose a new random graph of right-degree 5. MiniSat requires a boolean formula in conjuctive normal form as input, so we represent each constraint $P(x_{j_1}, x_{j_2}, x_{j_3}, x_{j_4}, x_{j_5}) = v_i$ by 16 clauses: one for each truth assignment to $x_{j_1}, \cdots, x_{j_5}$ that would violate the constraint.

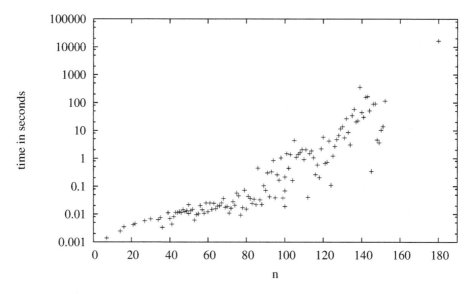

Fig. 1. Number of seconds taken by MiniSat to invert Goldreich's function for different values of n. We use the degree-five predicate $P_5(x) = x_1 \oplus x_2 \oplus x_3 \oplus (x_4 \wedge x_5)$ and a random bipartite graph of right-degree five.

We ran MiniSat on a Lenovo T61 laptop with 2GB of RAM and a 2.00GHz Intel T7300 Core Duo CPU. Fig. 1 plots the number of seconds taken to find a solution versus the input size n. The graph is plotted on a logarithmic scale. The time appears to grow exponentially in n.

Secret Sharing and Non-Shannon Information Inequalities[*]

Amos Beimel and Ilan Orlov

Dept. of Computer Science, Ben-Gurion University Be'er-Sheva, Israel

Abstract. The known secret-sharing schemes for most access structures are not efficient; even for a one-bit secret the length of the shares in the schemes is $2^{O(n)}$, where n is the number of participants in the access structure. It is a long standing open problem to improve these schemes or prove that they cannot be improved. The best known lower bound is by Csirmaz (J. Cryptology 97), who proved that there exist access structures with n participants such that the size of the share of at least one party is $n/\log n$ times the secret size. Csirmaz's proof uses Shannon information inequalities, which were the only information inequalities known when Csirmaz published his result. On the negative side, Csirmaz proved that by only using Shannon information inequalities one cannot prove a lower bound of $\omega(n)$ on the share size. In the last decade, a sequence of non-Shannon information inequalities were discovered. This raises the hope that these inequalities can help in improving the lower bounds beyond n. However, in this paper we show that all the inequalities known to date cannot prove a lower bound of $\omega(n)$ on the share size.

1 Introduction

A secret-sharing scheme is a mechanism for sharing data among a set of participants such that only pre-defined authorized subsets of participants can reconstruct the data, while any other subset has absolutely no information on the data. The collection of authorized subsets is called an access structure. For example, in a t-out-of-n threshold secret-sharing scheme, the access structure contains all subsets of size at least t. As an interesting "real-world" illustration of this situation: According to *Time Magazine* control of the nuclear weapon in Russia in the early 1990s depended upon a similar "two-out-of-tree" access mechanism, where the three parties were the President, the Defense Minister, and the Defense Ministry. Secret-sharing schemes, introduced by [40, 8, 30], are nowadays used in many cryptographic protocols, e.g., Byzantine agreement [38], secure multiparty computations [6, 14, 17], threshold cryptography [20], access control [36], and attribute-based encryption [27, 43].

An important issue in secret-sharing schemes is the size of the shares distributed to the participants. For most access structures, even the best known

---

[*] Partially supported by the Frankel Center for Computer Science at the Ben-Gurion University.

O. Reingold (Ed.): TCC 2009, LNCS 5444, pp. 539–557, 2009.

secret-sharing schemes (e.g., [7, 11, 21, 31, 42, 31]) are not efficient; the length of the shares for sharing an ℓ-bit secret is $\ell \cdot 2^{O(n)}$, where n is the number of participants in the access structure. The best lower bound was proved by Csirmaz [18]; he proved that for each n there exists an access structure with n participants such that any secret-sharing scheme with an ℓ-bit secret requires shares of length $\Omega(\ell n / \log n)$. There is a large gap between the upper bounds and the lower bounds. Closing this gap is a major open problem.

The entropy of a random variable, which was introduced by Shannon in the landmark paper [41], is a measure of the amount of uncertainty associated with the value of the random variable. Starting from the works of Karnin et al. [32] and Capocelli et al. [12], the entropy was used to prove lower bounds on the share size in secret sharing schemes [9, 22, 18, 19]. Specifically, Csirmaz's proof [18] uses only Shannon information inequalities, which were the only information inequalities known when Csirmaz published his result (this is true also for all the previous works mentioned above). On the negative side, Csirmaz proved that by using only Shannon information inequalities one cannot prove a lower bound of $\omega(n)$ on the share size. In the last decade, a sequence of non-Shannon information inequalities were discovered. This raises the hope that these inequalities can help in improving the lower bounds beyond n. However, in this paper we show that all the inequalities known to date cannot prove a lower bound of $\omega(n)$ on the share size.

1.1   Related Work

Threshold secret-sharing schemes, in which a subset is authorized iff its size is larger than some threshold, were independently introduced by Shamir [40] and Blakley [8] about thirty years ago. General secret sharing schemes were presented by Ito, Saito, and Nishizeki [30]; they presented a construction of a secret-sharing scheme for every monotone access structure. More efficient schemes for specific access structures were presented in, e.g., [7, 11, 21, 42, 31]. However, even these better constructions are not efficient and, for most access structure, the shares' size is exponential. Lower bounds for secret-sharing schemes were presented in [9, 22, 18, 19]; however, as stated above, there is a big gap between the upper and lower bounds. Super-polynomial lower bounds for *linear* secret-sharing schemes were presented in [1, 26].

In this work, we discuss using information inequalities for proving lower bounds on the share size in secret-sharing schemes. An information inequality is a linear inequality over the entropy of subsets of variables that holds for any random variables (for a formal definition see Section 2.1). For example, $H(X_1) + H(X_2) \geq H(X_1 X_2)$ is an information inequality. Many inequalities can be expressed as a linear combination of a single inequality involving the conditional mutual information, namely, $I(X; Y | Z) \geq 0$. Such inequalities are known as Shannon inequalities. It was an open problem for many years if there are information inequalities that are not implied by Shannon inequalities, i.e., if there are non-Shannon inequalities. The first non-Shannon inequality was given by Zhang and Yeung [47]. In the last decade, several additional non-Shannon inequalities were discovered [33, 46, 23, 44]. In particular, an interesting technique for deriving non-Shannon inequalities, called projection, was presented in [44]. Several papers have dealt with the characterization of information inequalities. Chan and Yeung [13] have characterized information inequalities using group-theoretic inequalities. Matúš [34] has proved that there are infinitely many independent information inequalities. Guille et al. [28] have given results concerning the structure of information inequalities and, more specially, results describing the minimal set of information inequalities when all the coefficient are 1 or −1, called Ingleton inequalities.

The information inequality of Zhang and Yeung [47] was used in several areas. It was used by Dougherty, Freiling, and Zeger [24] to prove bounds on the capacity of network coding, by Matúš [35] to prove that a function is not asymptotically entropic, and by Riis [39] to prove bounds on graph entropy of certain graphs. Furthermore, it was used by Beimel, Livne, and Padró [4] to prove lower bounds on the size of shares in secret-sharing schemes; they proved that there is a matroidial access structure – the Vamos access structure – that is not nearly ideal. We observe that this result can be proved using other information inequalities, e.g., the information inequalities of [23]. Furthermore, the information inequalities of [47, 23] can be used to prove that other matroidial access structures are not nearly ideal, e.g., the access structures induced by the matroids AG32r, F8, Q8 (for the definitions of these matroids see [37]).

This paper deals with limitations of the techniques for proving lower bounds on the size of shares in secret-sharing schemes, continuing the work of [3]. Beimel and Franklin [3] considered weakly-private secret-sharing schemes, in which any unauthorized set can never rule-out any secret (however, it might deduce, for example, that one secret is much less likely than other secrets). They show efficient constructions of weakly-private secret-sharing schemes (for large secret domains), implying that to prove lower bounds on the shares' size in secret-sharing schemes one must use the strong privacy requirement of secret-sharing schemes.

1.2   Our Results

In contrast to the success of applying the known information inequalities to proving lower bounds in several areas, we show that they cannot help in proving lower bounds of $\omega(n)$ on the share size in secret-sharing schemes. Let us elaborate on our proof. Csirmaz [18] in 1994 has proven his lower bound by translating the question of proving lower bounds on share size to proving that a certain linear programming instance does not have a small solution. Csirmaz constructed the linear program by using Shannon inequalities, which were the only information inequalities known in 1994. He proved a lower bound of $\Omega(n/\log n)$ times the secret size for an access structure with n parties. Furthermore, all previous lower bounds [32, 12, 9, 22] can be restated using Csirmaz's framework using Shannon inequalities. On the other hand, Csirmaz proved that for every access structure the linear program has a solution in which the objective function has value $O(n)$, implying that his framework cannot prove better lower bounds than $\Omega(n)$.

In the last decade, a sequence of non-Shannon information inequalities were discovered [47, 33, 46, 23, 44]. This gives hope that adding these inequalities to the linear program, one could prove better lower bounds on the share size. However, in this work we show that Csirmaz's solution to the linear program remains valid even after adding all the known information inequalities. That is, all the information inequalities known to date cannot prove lower bounds better than $\Omega(n)$ even if used simultaneously. Our proof that Csirmaz's solution remains valid after adding the new inequalities is much more involved than Csirmaz's proof for Shannon inequalities. We present a brute-force algorithm that checks if Csirmaz's solution remains valid given an information inequality.[1] We executed this algorithm, using a computer program, on all known information inequalities of [47, 33, 46, 23]. For [47, 46, 33, 44], which also give an infinite sequence of information inequalities, we manually executed the algorithm on a symbolic representation of the inequalities. The conclusion is that all the known information inequalities cannot help in proving better lower bounds than $\Omega(n)$.

We end the introduction with a few remarks. First, one cannot interpret our result as suggesting that information inequalities cannot help in improving the lower bounds. To the contrary, the conclusion of our paper is that new information inequalities should be sought. Hopefully, these new information inequalities would not be ruled-out by our algorithm. However, not failing the test in our algorithm is only the first step. Our algorithm only gives a necessary condition for an information inequality to be helpful in proving lower bounds of $\omega(n)$ on the share size. To use new inequalities, one has to prove that for some access structure the linear program with the new inequalities, and possibly with all the known inequalities, has only large solutions.

2   Preliminaries

In this section we review the relevant definitions from information theory and define secret-sharing schemes.

2.1   Basic Definitions from Information Theory and Information Inequalities

In this section, we review the basic concepts of Information Theory used in this paper. For a complete treatment of this subject see, e.g., [16]. All the logarithms here are of base 2.

The *entropy* of a random variable X is $H(X) \stackrel{\text{def}}{=} - \sum_{x, \Pr[X=x]>0} \Pr[X = x] \log \Pr[X = x]$. It can be proved that $0 \leq H(X) \leq \log |\text{supp}(X)|$, where $|\text{supp}(X)|$ is the size of the support of X (the number of values with probability greater than zero). The upper bound $|\text{supp}(X)|$ is obtained if and only if the distribution of X is uniform and the lower bound is obtained if and only if X is

---

[1] Our algorithm is highly inefficient. However, most known non-Shannon information inequalities have 4 or 5 variables, thus, executing the computer program returns an answer in a reasonable time (less than a minute).

deterministic. Given two random variables X and Y (possibly dependent), the *conditioned entropy* of X given Y is defined as $H(X|Y) \overset{\text{def}}{=} H(X,Y) - H(Y)$. From the definition of the conditional entropy, the following properties can be proved: $0 \leq H(X|Y) \leq H(X)$, where $H(X|Y) = H(X)$ if and only if X and Y are independent, and $H(X|Y) = 0$ if the value of Y completely determines the value of X. The *mutual information* between X and Y is defined as $I(X;Y) \overset{\text{def}}{=} H(X) - H(X|Y)$, and the *conditional mutual information* between X and Y given Z is defined as $I(X;Y|Z) \overset{\text{def}}{=} H(X|Z) - H(X|Y,Z)$. Entropies, conditional entropies, mutual information, and conditional mutual information are called *Shannon's information measures*.

Let $\{X_i\}_{i \in [m]}$ be a set of m jointly distributed random variables. For any subset I of $[m]$, let $X_I = (X_i)_{i \in I}$.

Definition 1 (Information Inequality). *An* information inequality over m variables is defined by 2^m constants $\{\alpha_A\}_{A \subseteq [m]}$, where $\alpha_A \in \mathbb{R}$, such that $\sum_{A \subseteq [m]} \alpha_A H(X_A) \geq 0$ for every m random variables $X_1, \ldots, X_m$.

For example, $H(X_1) + H(X_2) \geq H(X_1 X_2)$ is an information inequality. Many inequalities can be expressed as a linear combination of a single inequality involving the conditional mutual information, namely, $I(X_1; X_2 | X_3) \geq 0$ (this inequality can be stated as $H(X_1, X_3) + H(X_2, X_3) - H(X_1, X_2, X_3) - H(X_3) \geq 0$). Such inequalities are known as Shannon-type inequalities. Information inequalities that cannot be deduced from Shannon inequalities are called non-Shannon inequalities. For more background on information inequalities the reader may consult [45].

2.2   Secret Sharing

Definition 2 (Access Structure and Distribution Scheme). *Let* $P = \{p_1, \ldots, p_n\}$ *be a finite set of parties, and let* $p_0 \notin P$ *be a special party called the dealer. A collection* $\mathcal{A} \subseteq 2^P$ *is monotone if* $B \in \mathcal{A}$ *and* $B \subseteq C$ *imply that* $C \in \mathcal{A}$. *An* access structure *is a monotone collection* $\mathcal{A} \subseteq 2^P$ *of non-empty subsets of* P. *Sets in* $\mathcal{A}$ *are called* authorized, *and sets not in* $\mathcal{A}$ *are called* unauthorized.

A distribution scheme *$\Sigma = \langle \Pi, \mu \rangle$ with domain of secrets K is a pair, where μ is a probability distribution on some finite set R (the set of random strings) and Π is a mapping from $K \times R$ to a set of n-tuples $K_1 \times K_2 \times \cdots \times K_n$, where K_i is called the* share-domain *of p_i. A dealer distributes a secret $s \in K$ according to Σ by first sampling a string $r \in R$ according to μ, computing a vector of shares $\Pi(s, r) = (s_1, \ldots, s_n)$, and privately communicating each share s_i to party p_i.*

We next define secret-sharing schemes using the entropy function. It is convenient to view the secret as the share of the dealer p_0, and for every set $T \subseteq P \cup \{p_0\}$ to consider the vector of shares of T. Any probability distribution on the domain of secrets, together with the distribution scheme Σ, induces, for any $T \subseteq P \cup \{p_0\}$, a probability distribution on the vector of shares of the parties in T. We denote the random variable taking values according to this probability distribution on the vector of shares of T by S_T, and by S the random variable denoting the secret (i.e., $S = S_{\{p_0\}}$).

Definition 3 (Secret-Sharing Scheme). *We say that a distribution scheme is a secret-sharing scheme realizing an access structure $\mathcal{A}$ with respect to a given probability distribution on the secrets, denoted by a random variable S, if the following conditions hold.*

CORRECTNESS. *For every authorized set $T \in \mathcal{A}$, the shares of the parties in T determine the secret, i.e., $H(S|S_T) = 0$.*

PRIVACY. *For every unauthorized set $T \notin \mathcal{A}$, the shares of the parties in T do not disclose any information on the secret, that is, $H(S|S_T) = H(S)$.*

Remark 1. Although the above definition considers a specific distribution on the secrets, Blundo et al. [10] proved that its correctness and privacy are actually independent of this distribution: If a scheme realizes an access structure with respect to one distribution on the secrets, then it realizes the access structure with respect to any distribution with the same support. Furthermore, the above definition is equivalent to the definition of [15, 2, 5], where there is no probability distribution associated with the secrets and it is required that the probability of every vector of shares of an unauthorized set is the same given any secret.

Karnin et al. [32] have showed that for each non-redundant party (that is, a party that appears in at least one minimal authorized set) $H(S_i) \geq H(S)$, which implies that the size of the share of the party is at least the size of the secret.

Notation 1. *We use the following notation for two sets A and $\widehat{A}$. The set $\widehat{A}$ is a subset of $P \cup \{p_0\}$ and the set A is a subset of P, where $A = \widehat{A} \setminus \{p_0\}$, that is, if $p_0 \notin \widehat{A}$, then $A = \widehat{A}$, otherwise A is obtained by removing p_0 from $\widehat{A}$.*

3   Csirmaz Framework for Proving Lower Bounds and Its Limitations

3.1   Csirmaz Framework for Proving Lower Bounds

Csirmaz [18] has proved the best known lower bounds on the size of the shares in secret-sharing schemes. Towards this goal, he presented a framework for proving lower bounds and showed how to implement this framework to prove lower bounds for a specific access structure. The idea of the framework of Csirmaz is to construct a linear program such that lower bounds on the value of the objective function in this program imply lower bounds on the share size. Specifically, given an access structure $\mathcal{A}$ and a secret-sharing scheme realizing it, define the function $f(\widehat{A}) = H(S_{\widehat{A}})/H(S)$ for every $\widehat{A} \subseteq P \cup \{p_0\}$. The correctness and privacy of the secret-sharing scheme can be translated to constrains on the function f. Namely, (1) if $A \in \mathcal{A}$, then $f(A \cup \{p_0\}) = f(A)$, and (2) if $A \notin \mathcal{A}$, then $f(A \cup \{p_0\}) = f(A) + 1$. Proving lower bounds on the size of the shares is equivalent to proving that any n random variables $S_1, \ldots, S_n$ (i.e., shares) satisfying the above equalities imply that $\sum_{i=1}^{n} H(S_i)$ is large.

These constrains are translated to a linear program using known properties of the entropy function, namely, information inequalities. That is, we get a set of linear inequalities, where we want to minimize $\sum_{i=1}^{n} f(\{p_i\})$.

Csirmaz has constructed an access structure $\mathcal{A}$ that implies a linear program in which $\sum f(\{p_i\}) = \Omega(n^2/\log n)$, thus, for at least one party $f(\{p_i\}) = \Omega(n/\log n)$. This implies that in every secret-sharing scheme realizing $\mathcal{A}$ with an ℓ-bit secret, the share of at least one party is an $\Omega(\ell \cdot n/\log n)$-bit string. We next formally define and describe Csirmaz's framework.

Definition 4. *Given a secret sharing scheme over n parties, define the function $f : 2^{P \cup \{p_0\}} \to \mathbb{R}$ as follow: $f(\widehat{A}) = H(S_{\widehat{A}})/H(S)$ for every $\widehat{A} \subseteq P \cup \{p_0\}$.*

The properties of the entropy function implies that f is a polymatroid as defined below.

Definition 5. *Let Q be a finite set, and $g : 2^Q \to \mathbb{R}$ be a function assigning real numbers to subsets of Q. The system (Q, g) is a* polymatroid *if g satisfies the following conditions:*

non-negative: $g(A) \geq 0$ *for all $A \subseteq Q$ and $g(\emptyset) = 0$,*
monotone: *if $A \subseteq B \subseteq Q$, then $g(A) \leq g(B)$,*
submodular: $g(A) + g(B) \geq g(A \cup B) + g(A \cap B)$ *for every $A, B \subseteq Q$.*

Proposition 1 ([25]). *The function f defined in Definition 4 is a polymatroid.*

Combining Proposition 1 and the properties of secret-sharing scheme we get:

Proposition 2. *The function f defined in Definition 4 satisfies the following additional inequalities for every sets $A, B \subseteq P$:*

1. *If $A \subseteq B, A \notin \mathcal{A}$, and $B \in \mathcal{A}$, then $f(B) \geq f(A) + 1$,*
2. *If $A \in \mathcal{A}, B \in \mathcal{A}$, but $A \cap B \notin \mathcal{A}$, then $f(A) + f(B) \geq f(A \cap B) + f(A \cup B) + 1$.*

3.2 Limitation of Shannon Inequalities

Csirmaz [18] has proved that using his framework with only Shannon inequalities (which were the only information inequalities known when he published his result) one cannot prove lower bounds better than $\Omega(n)$. That is, his lower bound is the best possible up to a factor of $\log n$ using only Shannon inequalities.

In this section we explain how Csirmaz proved this limitation. Since Csirmaz proved his result in 1994, some non-Shannon information inequalities were discovered. In Section 6 we will show that these inequalities cannot prove better lower bounds than $\Omega(n)$ using Csirmaz's framework.

Theorem 1. *Given any access structure $\mathcal{A}$ on the n-element set P, there is a polymatroid $\widehat{g} : 2^{P \cup \{p_0\}} \to \mathbb{R}$ so that*

1. *For every $A \subseteq P$, $\widehat{g}(A \cup \{p_0\}) = \widehat{g}(A)$ if $A \in \mathcal{A}$ and $\widehat{g}(A \cup \{p_0\}) = \widehat{g}(A) + 1$ if $A \notin \mathcal{A}$.*
2. *$\widehat{g}$ satisfies the conditions of Proposition 2,*
3. *$\widehat{g}(\{p_i\}) \leq n$ for every $p_i \in P$.*

In order to prove this theorem, Csirmaz has defined a polymatroid $\widehat{g}$ that, on one hand, satisfies all the conditions and, on the other hand, $\widehat{g}(\{p_i\}) = n$. In other words, Csirmaz has shown that for every access structure the linear program has a small solution.

Definition 6 (The Csirmaz function). *Let* $n \in \mathbb{N}$. *Define the* the Csirmaz function $C_n : \{0, \ldots, n\} \to \mathbb{N}$ *as follows*

$$C_n(k) \stackrel{\text{def}}{=} n + (n-1) + \ldots + (n-k+1) = nk + \frac{k}{2} - \frac{k^2}{2}.$$

To prove Theorem 1, Csirmaz defined $g : 2^P \to \mathbb{N}$ as $g(A) \stackrel{\text{def}}{=} C_n(|A|)$. Next, he extended g to $\widehat{g} : 2^{P \cup \{p_0\}} \to \mathbb{N}$, where for every $A \subseteq P$ he defined $\widehat{g}(A) = g(A)$, and $\widehat{g}(A \cup \{p_0\}) = g(A)$ if $A \in \mathcal{A}$, and $\widehat{g}(A \cup \{p_0\}) = g(A) + 1$ if $A \notin \mathcal{A}$. It can be checked that $\widehat{g}$ satisfies the conditions of the theorem. The Csirmaz function is universal; it is used to construct a polymatroid for every access structure. We next prove that any such universal function is at least as large as the Csirmaz function. This lemma sheds some light why Csirmaz chose this function.

Lemma 1. *Let* $y_n : \{0, \ldots, n\} \to \mathbb{R}$ *be a function satisfying the following inequalities:*

1. *If* $A \subseteq B \subseteq Q$, *then* $y_n(|B|) \geq y_n(|A|) + 1$ *and* $y_n(0) = 0$,
2. *If* A *and* B *are subsets of* Q *such that* $A \nsubseteq B$ *and* $B \nsubseteq A$, *then* $y_n(|A|) + y_n(|B|) \geq y_n(|A \cap B|) + y_n(|A \cup B|) + 1$.

The Csirmaz Function $C_n(k)$ *is the minimal function that satisfies these requirements, i.e., for each* $1 \leq k \leq n, C_n(k) \leq y_n(k)$.

Proof. Let A, B be two sets of k elements each that are different in exactly one element. Thus, $|A \cap B| = k - 1$ and $|A \cup B| = k + 1$. From Item (2) in the lemma, for each $0 \leq k \leq n$

$$y_n(k) - y_n(k-1) \geq y_n(k+1) - y_n(k) + 1.$$

This implies that $y_n(k) - y_n(k-1) \geq y_n(n) - y_n(n-1) + n - k$ for every $0 \leq k \leq n$. By Item (1) in the lemma, $y_n(|B|) \geq y_n(|A|) + 1$. Thus, $y_n(n) - y_n(n-1) \geq 1$. Therefore,

$$y_n(k) - y_n(k-1) \geq n - k + 1. \tag{1}$$

By the requirement in the lemma $y_n(0) = 0$, thus, Inequality (1) with $k = 1$ implies $y_n(1) \geq n = C_n(1)$. By induction and by (1), $y_n(k) \geq y_n(k-1) + n - k + 1 \geq C_n(k-1) + n - k + 1 = C_n(k)$. □

4   When Can Information Inequalities Help?

In this section, we will define when information inequalities can help in improving lower bounds beyond $\Omega(n)$. We start with some notation; using this notation we will define two quantities for an information inequality, Δ and Λ. These quantities are used to define when an information inequality can help.

Notation 2. *Let $A_1, \ldots, A_m$ be m (not necessarily disjoint) sets. For $I \subseteq [m]$, denote $A_I = \bigcup_{i \in I} A_i$.*

Let $\sum_{I \subseteq [m]} \alpha_I H(X_I) \geq 0$ be an information inequality. Given an access structure $\mathcal{A}$, we fix some secret-sharing scheme realizing it. Therefore, the function $f(\widehat{A}) = H(S_{\widehat{A}})/H(S)$ where $\widehat{A} \subseteq P \cup \{p_0\}$ is well defined. Then, for every sets $\widehat{A}_1, \ldots, \widehat{A}_m \subseteq P \cup \{p_0\}$, the following inequality is valid $\sum_{I \subseteq [m]} \alpha_I f(\widehat{A}_I) \geq 0$. Recall that for every $1 \leq i \leq m$, we defined $A_i = \widehat{A}_i \setminus \{p_0\}$. Using this notation, $f(\widehat{A}_I) = f(A_I) + 1$ if $p_0 \in \widehat{A}_I$ and $A_I \notin \mathcal{A}$, otherwise, $f(\widehat{A}_I) = f(A_I)$.

Definition 7. *For an information inequality $\sum_{I \subseteq [m]} \alpha_I H(X_I) \geq 0$, an access structure $\mathcal{A}$, and sets $\widehat{A}_1, \ldots, \widehat{A}_m$, define Δ as $\Delta \stackrel{def}{=} -\sum_{I: p_0 \in \widehat{A}_I; A_I \notin \mathcal{A}} \alpha_I$.*

Claim 1. *Let $\widehat{A}_1, \ldots, \widehat{A}_m$ be m sets, $\sum_{I \subseteq [m]} \alpha_I H(X_I) \geq 0$ be an information inequality, and $\mathcal{A}$ be an access structure. Then, $\sum_{I \subseteq [m]} \alpha_I f(A_I) \geq \Delta$.*

Proof. Applying the rules $f(\widehat{A}_I) = f(A_I)$ if $p_0 \notin \widehat{A}_I$ or $A_I \in \mathcal{A}$, and $f(\widehat{A}_I) = f(A_I) + 1$ otherwise, the inequality $\sum_{I \subseteq [m]} \alpha_I f(\widehat{A}_I) \geq 0$ implies

$$\sum_{I \subseteq [m]} \alpha_I f(\widehat{A}_I) = \sum_{I: p_0 \notin \widehat{A}_I \vee A_I \in \mathcal{A}} \alpha_I f(A_I) + \sum_{I: p_0 \in \widehat{A}_I \wedge A_I \notin \mathcal{A}} \alpha_I (f(A_I) + 1)$$

$$= \sum_{I \subseteq [m]} \alpha_I f(A_I) - \Delta \geq 0. \qquad \square$$

Observe that Δ can be negative, positive, or equal to zero, but, as we will see later, the information inequality can be useful only when $\Delta > 0$.

Definition 8. *Let $\sum_{I \subseteq [m]} \alpha_I H(X_I) \geq 0$ be an information inequality. For sets $A_1, \ldots, A_m \subseteq P$ define Λ as $\Lambda \stackrel{def}{=} \sum_{I \subseteq [m]} \alpha_I \mathcal{C}_n(|A_I|)$.*

For every $I \subseteq [m]$, the size $|A_I|$ depends on some of the sizes of the intersections between the sets $A_1, \ldots, A_m$. Therefore, we define additional notation in order to represent these intersections. For an illustration of this notation see Fig. 1.

Notation 3. *Let $A_1, \ldots, A_m$ be m (not necessarily disjoint) sets. Denote $\delta_I \stackrel{def}{=} \bigcap_{i \in I} A_i \setminus \bigcup_{i \notin I} A_{\{i\}}$ and $t_I \stackrel{def}{=} |\delta_I|$ for $I \subseteq [m]$. In addition, for $\mathcal{I} \subseteq 2^{[m]}$, denote $\delta_{\mathcal{I}} \stackrel{def}{=} \bigcup_{I \in \mathcal{I}} \delta_I$.*

Observation 1. *$\delta_J \subseteq A_i$ if and only if $i \in J$, that is, $A_i = \bigcup_{i \in J} \delta_J$ and $A_I = \bigcup_{i \in I} A_i = \bigcup_{I \cap J \neq \emptyset} \delta_J$.*

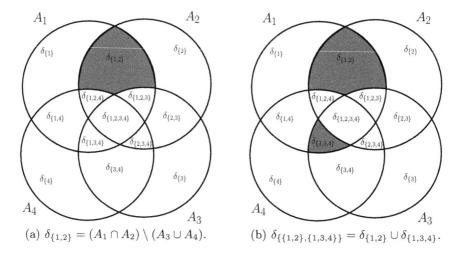

(a) $\delta_{\{1,2\}} = (A_1 \cap A_2) \setminus (A_3 \cup A_4)$.   (b) $\delta_{\{\{1,2\},\{1,3,4\}\}} = \delta_{\{1,2\}} \cup \delta_{\{1,3,4\}}$.

Fig. 1. An illustration of Notation 3 for $m = 4$. For clarity of the illustration, we assume that $\delta_{\{2,4\}} = \delta_{\{1,3\}} = \emptyset$.

Csirmaz has suggested a specific function defined in Definition 6 in order to show the limitations of Shannon information inequalities. We will prove in Lemma 4 that any information inequality remains valid after plugging in the Csirmaz function. That is, if $\sum_{I \subseteq [m]} \alpha_I H(X_I) \geq 0$ is an information inequality, then $\sum_{I \subseteq [m]} \alpha_I C_n(|A_I|) \geq 0$. So, our only hope is that Δ is "big" for some sets $\widehat{A}_1, \ldots, \widehat{A}_m \subseteq P \cup \{p_0\}$ and the corresponding sets $A_1, \ldots, A_m \subseteq P$, but, $\Lambda = \sum_{I \subseteq [m]} \alpha_I C_n(|A_I|)$ is negative (or "small"). If this condition does not hold, then the inequality cannot help.

Definition 9. *We say that an information inequality $\sum_{I \subseteq [m]} \alpha_I H(X_I) \geq 0$ can at most γ-help if $\Delta \leq \gamma \Lambda$ for every sets $\widehat{A}_1, \ldots, \widehat{A}_m \subseteq P \cup \{p_0\}$ and for every access structure $\mathcal{A}$, where $\Delta = -\sum_{I:p_0 \in \widehat{A}_I; A_I \notin \mathcal{A}} \alpha_I$ and $\Lambda = \sum_{I \subseteq [m]} \alpha_I C_n(|A_I|)$.*

Theorem 2. *Let $\gamma > 0$ be a constant. Consider a collection of information inequalities, where each information inequality in the collection can at most γ-help. Then, this collection of information inequalities cannot help improving the lower bounds beyond γn even if all inequalities are used simultaneously.*

Proof. Consider an access structure $\mathcal{A}$ and the "huge" linear program obtained for this access structure by applying each information inequality to every choice of subsets of the parties. We take the polymatroid $g(A_I) = \gamma C_n(|A_I|)$, and we get a solution that satisfies each inequality in the program, where $g(\{p_i\}) = \gamma n$. $\square$

When dealing with a finite collection of information inequalities, one can use a rougher notion than an information inequality that can at most γ-help.

Definition 10. *We say that an information inequality $\sum_{I \subseteq [m]} \alpha_I H(X_I) \geq 0$ cannot help (in improving the lower bounds beyond $\Omega(n)$) if for every sets $\widehat{A}_1, \ldots, \widehat{A}_m \subseteq P \cup \{p_0\}$ and for every access structure $\mathcal{A}$, if $\Delta > 0$ then $\Lambda > 0$.*

Observation 2. Let $\sum_{I \subseteq [m]} \alpha_I H(X_I) \geq 0$ be an information inequality that cannot help. Observe that $\Delta = -\sum_{I:p_0 \in \widehat{A}_I; A_I \notin \mathcal{A}} \alpha_I \geq -\sum_{I:p_0 \in \widehat{A}_I; A_I \notin \mathcal{A}; \alpha_I < 0} \alpha_I$. In addition, using Lemma 3 (proved later), if $\Lambda > 0$ then there exists a constant $\beta > 0$ that depends only on the coefficients of the information inequality (and, therefore, independent of the access structure and the number of parties in the access structure) such that $\Lambda \geq \beta$.[2] Thus, the information inequality can at most γ-help for some constant $\gamma > 0$. If we consider a *finite* collection of information inequalities, such that each inequality in the collection cannot help, then there is a constant $\gamma > 0$ such that each inequality in the collection can at most γ-help, and we can apply Theorem 2. Therefore, when dealing with a finite collection of information inequalities, we will check that each inequality in the collection cannot help; this is easier than calculating the maximal γ for each inequality.

5   Examples of Information Inequalities That Cannot Help

In this section, we demonstrate our method for proving that an information inequality cannot help by considering two example. First, we will demonstrate the calculations and the technique that we will use later on a simple Shannon inequality with two random variables. The fact that this inequality cannot help follows from Csirmaz's proof that using only Shannon inequalities one cannot prove better lower bounds. We reprove this result in order to supply a simple example of our method.

We consider the inequality $f(\widehat{A}_1) + f(\widehat{A}_2) \geq f(\widehat{A}_1 \cup \widehat{A}_2) + f(\widehat{A}_1 \cap \widehat{A}_2)$ for two sets $\widehat{A}_1, \widehat{A}_2 \subseteq P \cup \{p_0\}$. This inequality follows from the fact that the conditional mutual information is non-negative. We should calculate $\Lambda = C_n(|A_1|) + C_n(|A_2|) - C_n(|A_1 \cup A_2|) - C_n(|A_1 \cap A_2|)$. By Obseration 1, $|A_1| = t_1 + t_{1,2}$, $|A_2| = t_2 + t_{1,2}$, $|A_1 \cup A_2| = t_1 + t_{1,2} + t_2$, and $|A_1 \cap A_2| = t_{1,2}$.[3] Furthermore, $n = t_1 + t_{1,2} + t_2$. Therefore, for every $A_1, A_2 \subseteq P$

$$C_n(|A_1|) + C_n(|A_2|) - C_n(|A_1 \cup A_2|) - C_n(|A_1 \cap A_2|)$$
$$= (t_1 + t_{1,2}) \left[(t_1 + t_{1,2} + t_2) + \frac{1}{2} - \frac{(t_1 + t_{1,2})}{2} \right]$$
$$+ (t_2 + t_{1,2}) \left[(t_1 + t_{1,2} + t_2) + \frac{1}{2} - \frac{(t_2 + t_{1,2})}{2} \right]$$
$$- (t_1 + t_{1,2} + t_2) \left[(t_1 + t_{1,2} + t_2) + \frac{1}{2} - \frac{(t_1 + t_{1,2} + t_2)}{2} \right]$$
$$- t_{1,2} \left[(t_1 + t_{1,2} + t_2) + \frac{1}{2} - \frac{(t_{1,2})}{2} \right] \quad = \quad t_1 t_2.$$

---

[2] The value of β can be calculated by assigning $t_I = 1$ whenever $t_I > 0$.

[3] For simplicity of our notation, in the rest of the paper we sometimes write $t_{1,2}$ instead of $t_{\{1,2\}}$ (and similarly for other sets).

Assume that $p_0 \in \widehat{A}_1, \widehat{A}_2$. Thus, $p_0 \in \widehat{A}_1 \cup \widehat{A}_2, \widehat{A}_1 \cap \widehat{A}_2$. Before calculating Δ we have to decide which sets are in the access structure. If $A_1 \cup A_2 \notin \mathcal{A}$, then also $A_1, A_2, A_1 \cap A_2 \notin \mathcal{A}$. Thus, $f(A_1 \cup \{p_0\}) = f(A_1) + 1$, $f(A_2 \cup \{p_0\}) = f(A_2) + 1$, $f(A_1 \cup A_2 \cup \{p_0\}) = f(A_1 \cup A_2) + 1$, and $f(A_1 \cap A_2 \cup \{p_0\}) = f(A_1 \cap A_2) + 1$. Therefore, $\Delta = 0$ and the inequality cannot help using these selections. However, if $A_1, A_2 \in \mathcal{A}$, but $A_1 \cap A_2 \notin \mathcal{A}$, then $f(A_1 \cup \{p_0\}) = f(A_1)$, $f(A_2 \cup \{p_0\}) = f(A_2)$, $f(A_1 \cup A_2 \cup \{p_0\}) = f(A_1 \cup A_2)$, and $f((A_1 \cap A_2) \cup \{p_0\}) = f(A_1 \cap A_2) + 1$. Therefore, $\Delta = 1 > 0$ as needed. But the selection of $A_1, A_2 \in \mathcal{A}$ and $A_1 \cap A_2 \notin \mathcal{A}$ implies $A_1 \setminus (A_1 \cap A_2), A_2 \setminus (A_1 \cap A_2) \neq \emptyset$ which means that $t_1 > 0$ and $t_2 > 0$, thus, $\Lambda = t_1 \cdot t_2 \geq 1 > 0$ as well. In other words using these selections the inequality cannot help. Moreover, every other set of selections cannot help to achieve $\Delta > 0$ while $\Lambda = 0$.

To conclude, given an information inequality we want $\Delta > 0$ while $\Lambda = 0$. By different choices of which sets are in the access structure and which sets contain the dealer we get different values of Δ. We want choices that maximize Δ. However, notice that by choosing, for example, $A_1 \in \mathcal{A}$ while $A_2 \notin \mathcal{A}$, we must have that $A_1 \setminus A_2 \neq \emptyset$. Thus, the choices of which sets are in the access structure force that certain sets are non-empty, which might imply that $\Lambda > 0$.

5.1   The Zhang and Yeung Information Inequality Cannot Help

We next consider the Zhang and Yeung information inequality [47] – the first Non-Shannon inequality that was discovered – and prove that this inequality cannot help in proving lower bounds of $\omega(n)$.

Theorem 3 (The Zhang and Yeung Information Inequality [47, Theorem 3]). *For every four discrete random variables X_1, X_2, X_3, and X_4 the following inequality holds:*

$$3 \left[H(X_3 X_4) + H(X_2 X_4) + H(X_2 X_3) \right] + H(X_1 X_3) + H(X_1 X_2) - H(X_4)$$
$$- 2 \left[H(X_3) + H(X_2) \right] - H(X_1 X_4) - 4H(X_2 X_3 X_4) - H(X_1 X_2 X_3) \geq 0. \quad (2)$$

For every secret-sharing scheme and for every four sets $\widehat{A}_1, \widehat{A}_2, \widehat{A}_3, \widehat{A}_4 \subseteq P \cup \{p_0\}$ we can consider the random variables $X_i = S_{\widehat{A}_i}$ for $i = 1, \dots, 4$. Thus,

$$3 \left[f(\widehat{A}_3 \widehat{A}_4) + f(\widehat{A}_2 \widehat{A}_4) + f(\widehat{A}_2 \widehat{A}_3) \right] + f(\widehat{A}_1 \widehat{A}_3) + f(\widehat{A}_1 \widehat{A}_2) - f(\widehat{A}_4)$$
$$- 2 \left[f(\widehat{A}_3) - f(\widehat{A}_2) \right] - f(\widehat{A}_1 \widehat{A}_4) - 4f(\widehat{A}_2 \widehat{A}_3 \widehat{A}_4) - f(\widehat{A}_1 \widehat{A}_2 \widehat{A}_3) \geq 0. \quad (3)$$

By choosing which sets contain the dealer and which sets are in the access structure we get different values of Δ. We next apply the Csirmaz function on Inequality (3). We use the same process described above on each one of the terms of (3). After simplifications, we get the following polynomial Λ, which is a multivariate polynomial whose variables are $\{t_I : I \subseteq [m]\}$.

$$\overbrace{\frac{t_{1,2,3}+t_{1,2,3}^2}{2}} + t_1 t_{1,2,3} + t_{1,2} t_{1,2,3} + \overbrace{t_{1,2,4}+t_{1,2,4}^2} + 2t_1 t_{1,2,4} + 2t_{1,2} t_{1,2,4} + t_{1,2} t_{1,3}$$

$$+ \ t_{1,2,3} t_{1,3} + \overbrace{t_{1,3,4}+t_{1,3,4}^2} + 2t_1 t_{1,3,4} + 2t_{1,3} t_{1,3,4} + \overbrace{\frac{t_{1,4}+t_{1,4}^2}{2}} + t_1 t_{1,4} + 2t_{1,2} t_{1,4}$$

$$+ 2t_{1,2,4} t_{1,4} + 2t_{1,3} t_{1,4} + t_{1,3,4} t_{1,4} + t_{1,2,3} t_2 + 2t_{1,2,4} t_2 + t_{1,3} t_2 + 2t_{1,4} t_2 + t_{1,2} t_{2,3}$$

$$+ \ \overbrace{t_{2,3}+t_{2,3}^2} + t_1 t_{2,3} + t_{1,2,3} t_{2,3} + t_{1,3} t_{2,3} + 2t_2 t_{2,3} + \overbrace{\frac{t_{2,4}+t_{2,4}^2}{2}} + 2t_1 t_{2,4} + t_2 t_{2,4}$$

$$+ 2t_{1,2} t_{2,4} + 2t_{1,2,4} t_{2,4} + 2t_{1,4} t_{2,4} + t_{1,2} t_{2,3} + t_{1,2,3} t_{2,3} + 2t_{1,3,4} t_{2,3} + 2t_{1,4} t_{2,3} + 2t_2 t_{2,3}$$

$$+ 2t_{2,3} t_{2,3} + \overbrace{\frac{t_{3,4}+t_{3,4}^2}{2}} + 2t_1 t_{3,4} + 2t_{1,3} t_{3,4} + 2t_{1,3,4} t_{3,4} + 2t_{1,4} t_{3,4} + t_3 t_{3,4} + + t_1 t_4$$

$$+ 2t_{1,2} t_4 + 2t_{1,2,4} t_4 + 2t_{1,3} t_4 + 2t_{1,3,4} t_4 + t_{1,4} t_4 + t_2 t_4 + t_{2,4} t_4 + t_3 t_4 + t_{3,4} t_4 .$$

After applying the Csirmaz function we get a polynomial of degree 2 such that all its coefficients are non-negative. We are looking for the following situation: $\Lambda = 0$ while $\Delta > 0$. Since all coefficients are non-negative and $t_I \geq 0$ for every $I \subseteq [m]$, the value of Λ is zero if every monomial in Λ is zero. In particular, every term $\beta \cdot t_I$ or $\beta \cdot t_I^2$ in Λ has to be equal to zero. If the coefficient β is positive, then $t_I = 0$ must hold. Thus, $t_{1,2,3} = t_{1,2,4} = t_{1,3,4} = t_{1,4} = t_{2,3} = t_{2,4} = t_{3,4} = 0$. Let Λ' be the polynomial after setting these variables to be zero, that is,

$$\Lambda' = t_{1,2} t_{1,3} + t_{1,3} t_2 + t_{1,2} t_3 + 2t_2 t_3 + t_1 t_4 + 2t_{1,2} t_4 + 2t_{1,3} t_4 + t_2 t_4 + t_3 t_4 .$$

The polynomial Λ' should be zero, therefore, in the inequality above one of the variables (i.e., set size) in each monomial has to be zero (e.g., $t_{1,2} = 0$ or $t_{1,3} = 0$).

We use a brute-force algorithm for checking if it is possible that $\Delta > 0$ while $\Lambda = 0$. We have two decisions to make:

- For each $i \in \{1, \ldots, 4\}$ we should decide if $p_0 \in \widehat{A}_i$ or not.
- We have to decide which sets are in the access structure. Specifically, for each $I \subseteq [m]$ such that $\alpha_I \neq 0$ in the information inequality, we need to decide whether $A_I \notin \mathcal{A}$ or $A_I \in \mathcal{A}$. These decisions should be consistent with the constrains that some sets δ_J are of size zero.

Example 1. Assume that A_4 is the only minimal set in the in the access structure. Thus, the sets that are in the access structure are exactly those that include A_4. We add the dealer to $\widehat{A}_2$ and do not add it to any other set. After committing to these decisions we compute Δ as specified in Definition 7, $\Delta = -\sum_{2 \in I, 4 \notin I} \alpha_I = -(3+1-2-1) = -1 < 0$. Thus, these decisions cannot help.

Example 2. Assume that $A_{\{1,2\}}$ and $A_{\{2,3\}}$ are the only minimal sets in the in the access structure. This means that the sets that are in the access structure are exactly those that include $A_{\{1,2\}}$ or $A_{\{2,3\}}$. For example, $A_{\{1,2,3\}} \in \mathcal{A}$. We also add the dealer to every $\widehat{A}_i$, $1 \leq i \leq 4$. After committing to these two decisions we compute $\Delta = -\sum_{\{1,2\} \nsubseteq I \wedge \{1,3\} \nsubseteq I} \alpha_I = -(3+3+3-1-2-2-1-4) = 1 > 0$. Observe that $\Delta > 0$ as needed. But, $A_{\{1,2\}} \in \mathcal{A}$ while $A_{\{1,3\}} \notin \mathcal{A}$. This means that $A_{\{1,2\}} \setminus A_{\{1,3\}} = \delta_{\{\{2\},\{2,4\}\}} \neq \emptyset$. However, we have set $t_{2,4} = 0$, thus, $t_2 \neq 0$. In a similar way, $A_{\{2,3,4\}} \in \mathcal{A}$ while $A_{\{2,3\}} \notin \mathcal{A}$. This means that $A_{\{2,3,4\}} \setminus A_{\{2,3\}} = \delta_{\{\{4\},\{1,4\}\}} \neq \emptyset$. However, we have set $t_{1,4} = 0$, thus, $t_4 \neq 0$. Combining these two constraints we get $t_2 \cdot t_4 > 0$, which implies $\Lambda > 0$. Thus, as before, these decisions cannot help.

We have written a computer program that checks all the possibilities for including the dealer in the sets and for which sets are in the access structure. The computer program showed that for each possible combination either $\Delta \leq 0$ or $\Lambda > 0$ (or both). This means that the Csirmaz function is still a solution to the linear program and this inequality cannot help.

6   All Known Information Inequalities Cannot Help

In this section we describe an algorithm that checks if an information inequality cannot help. We executed this algorithm on all known information inequalities, except for two infinite collections of inequalities, and verified that they cannot help. Thereafter, we consider the two known infinite collections of information inequalities and show that they can at most γ-help for some constant $\gamma > 0$. Before presenting these results, we show how to compute the polynomial Λ efficiently and analyze its properties.

6.1   Properties of the Polynomial Λ

For every information inequality $\sum_{I \subseteq [m]} \alpha_I H(X_I) \geq 0$ and for every sets $A_1, \ldots, A_m$ we consider the quantity $\Lambda = \sum_{I \subseteq [m]} \alpha_I \mathcal{C}_n(|A_I|)$. By Obseration 1, $|A_I| = \sum_{I \cap J \neq \emptyset} t_J$. Thus, we consider $\Lambda = \sum_{I \subseteq [m]} \alpha_I \mathcal{C}_n(\sum_{I \cap J \neq \emptyset} t_J)$ as a polynomial in the variables $\{t_J\}_{J \subseteq [m]}$. We start with proving a property of information inequalities that we use in the analysis of our algorithm.

Lemma 2. *Let $\sum_I \alpha_I H(X_I) \geq 0$ be an information inequality. Then, for every $J \subseteq [m]$, $\sum_{I \cap J \neq \emptyset} \alpha_I \geq 0$.*

Proof. Define a random variable Y which is uniformly distributed in $\{0, 1\}$; in particular $H(Y) = 1$. Now define $X_1, \ldots, X_m$, where $X_j = Y$ iff $j \in J$ and $X_j = 0$ otherwise (that is, in the latter case X_j is a deterministic variable whose entropy is 0). This implies that $H(X_I) = 1$ iff $I \cap J \neq \emptyset$ and $H(X_I) = 0$ otherwise. Since the information inequality holds for every random variables, the lemma follows. □

Lemma 3. *For every information inequality the polynomial Λ is a multivariate polynomial with total degree 2. Furthermore, the coefficient of every monomial in Λ is non-negative and can be efficiently calculated from the information inequality (without applying the Csirmaz function).*

Proof. The fact that the polynomial Λ is a multivariate polynomial with total degree 2 can be deduced from the structure of the Csirmaz function (see Definition 6), that is, Λ is a sum of polynomials $\mathcal{C}_n(|A_I|) = \mathcal{C}_n(\sum_{I \cap J \neq \emptyset} t_J)$, where

$C_n(k)$ is polynomial of degree 2. Next, we compute the coefficients of Λ. Recall that $n = \sum_{I \subseteq [m]} t_I$.

$$
\Lambda = \sum_{I \subseteq [m]} \alpha_I C_n(|A_I|) = \sum_{I \subseteq [m]} \alpha_I \left[n|A_I| + \frac{|A_I|}{2} - \frac{|A_I|^2}{2} \right]
$$

$$
= \sum_{I \subseteq [m]} \alpha_I \left(\sum_{J:I \cap J \neq \emptyset} t_J + \sum_{J:I \cap J = \emptyset} t_J \right) \left(\sum_{J:I \cap J \neq \emptyset} t_J \right) + \sum_{I \subseteq [m]} \alpha_I \frac{\sum_{J:I \cap J \neq \emptyset} t_J}{2}
$$

$$
- \sum_{I \subseteq [m]} \alpha_I \frac{\left(\sum_{J:I \cap J \neq \emptyset} t_J \right)^2}{2}
$$

$$
= \sum_{I \subseteq [m]} \alpha_I \left(\frac{\sum_{J:I \cap J \neq \emptyset} t_J + \left(\sum_{J:I \cap J \neq \emptyset} t_J \right)^2}{2} + \sum_{J:I \cap J \neq \emptyset} t_J \cdot \sum_{J:I \cap J = \emptyset} t_J \right).
$$

We can now compute the coefficients of the monomials of the polynomial Λ:

1. βt_J: In this case $\beta = \frac{\sum_{I \cap J \neq \emptyset} \alpha_I}{2}$, i.e., the sum of the coefficients of sets that include δ_J. By Lemma 2 this sum is non negative.
2. βt_J^2: In this case $\beta = \frac{\sum_{I \cap J \neq \emptyset} \alpha_I}{2}$, again, this is the sum of the coefficients of sets that include δ_J.
3. $\beta t_J t_K$: In this case $\beta = \sum_{I : I \cap (J \cup K) \neq \emptyset} \alpha_I$. That is, β is the sum of coefficients of sets that include at least one of t_J and t_K, and by Lemma 2, $\beta \geq 0$. □

As all the coefficients in Λ are non-negative and all the values of t_I are non-negative, its value is always non-negative. That is,

Lemma 4. Let $\sum_{I \subseteq [m]} \alpha_I H(X_I) \geq 0$ be an information inequality. Then, for every sets $A_1, \ldots, A_m \subseteq P$, $\sum_{I \subseteq [m]} \alpha_I C_n(|A_I|) \geq 0$.

6.2 An Algorithm for Checking If an Information Inequality Cannot Help

We next present the algorithm that checks if an information inequality cannot help. The algorithm is a brute-force algorithm that checks, for each possible choice of adding the dealer or not adding the dealer to each set A_i and for each possible choice $A_I \in \mathcal{A}$ or $A_I \notin \mathcal{A}$ for each $I \subseteq [m]$, if $\Delta > 0$ while it is possible that $\Lambda = 0$. To check if Λ can equal 0 under some a specific choice, we check for each choice $t_I = 0$ and $t_I > 0$ for each $I \subseteq [m]$ if (1) $\Lambda = 0$ under this choice, and (2) this choice is consistent with the choice of sets that are in the access structure. The algorithm is formally described in Algorithm 1.

Input : An information inequality $\sum_{A \subseteq [m]} \alpha_A H(X_A) \geq 0$.
Output: "NO" if the information inequality cannot help, "YES" otherwise.

1 *Calculate the polynomial Λ using Lemma 3*;
2 **foreach** *monomial in Λ of the form βt_J where $\beta \neq 0$* **do** set $t_J = 0$;
3 Let Λ' be the resulting polynomial after setting these variables.;
4 **foreach** *choice of setting $A_I \in \mathcal{A}$ or $A_I \notin \mathcal{A}$ for each $\alpha_I \neq 0$ in the information inequality* **do**

> /* If there are q terms with non-zero coefficient in $\sum_{I \subseteq [m]} \alpha_I H(X_I) \geq 0$, there are 2^q combinations. */

5   **foreach** *choice of setting $p_0 \in \widehat{A}_i$ or $p_0 \notin \widehat{A}_i$ for every $1 \leq i \leq m$* **do**

> > /* There are 2^m combinations. */

6   Calculate $\Delta = -\sum_{I : p_0 \in \widehat{A}_I ; A_I \notin \mathcal{A}} \alpha_I$;
7   **if** $\Delta \leq 0$ **then** go to (5);

> > /* Check if it is possible that $\Lambda = 0$: */

8   **foreach** *choice of setting $t_I = 0$ or $t_I > 0$ for every $I \subseteq [m]$* **do**

> > > /* There are 2^{2^m} such combinations. */

9   **foreach** *monomial $\beta t_J t_K$ in Λ', where $\beta \neq 0$* **do**
10   **if** $t_J > 0$ and $t_K > 0$ *in the current explored combination* **then** go to (8);
11   **end**
12   **foreach** I, J *where $\alpha_I \neq 0$ and $\alpha_J \neq 0$ in the information inequality $\sum_{I \subseteq [m]} \alpha_I H(X_I) \geq 0$* **do**
13   **if** *in the current explored combination $A_I \in \mathcal{A}, A_J \notin \mathcal{A}$, and there is no $K \subseteq [m]$ such that $I \cap K \neq \emptyset, J \cap K = \emptyset$, and $t_K > 0$ in the current explored combination* **then** go to (8);
14   **end**
15   **return** *"YES"*
16   **end**
17   **end**
18 **end**
19 **return** *"NO"*

Algorithm 1. A brute-force algorithm that checks if an information inequality cannot help.

We have executed Algorithm 1 on the following non-Shannon inequalities:

- The first Non-Shannon inequality with four variables that was discovered by Zhang and Yeung in [47].
- The six Non-Shannon inequalities with four variables and anther one with five variables in [23].
- The five Non-Shannon inequalities with four variables in [44].
- The inequality of Ingleton in [29].[4]

For each of these inequalities the result is the same – the information inequality cannot help in proving lower bounds of $\omega(n)$.

---

[4] The inequality of Ingleton [29] holds only for linear-algebraic spaces.

Remark 2. The algorithm written above is not efficient. However, for our purpose – checking information inequalities with four or five variables – the algorithm is good enough. To be precise, the running of the computer program executing the algorithm takes less than a minute even for an information inequality of [23] that contains five variables. All other inequalities contain 4 variables and the running time is better.

Remark 3. Our algorithm gives a necessary condition for an information inequality to be helpful. We do not know of an information inequality that fulfills this necessary condition. We do have an example of a potential inequality that satisfies it: $H(X_1X_2) + H(X_1X_3) + H(X_2X_3) + H(X_4) \leq H(X_1X_2X_3) + H(X_1) + H(X_2) + H(X_3X_4)$. We stress that we do not know if this is an information inequality. It does satisfy Lemma 2 and some stronger conditions for being an information inequality.

6.3 Dealing with the Known Infinite Collections of Information Inequalities

There are a few examples for infinite sequences of Non-Shannon inequalities. The first infinite sequence of Non-Shannon inequalities was discovered by Zhang and Yeung in [47]; they show for every $n \in \mathbb{N}$ an information inequality with n variables. A sequence of Non-Shannon information inequalities generalizing the result of [47] appears in [33, 46]. Finally, an infinite sequence of Non-Shannon information inequalities with four variables was given in [44].

In [44] there is a symbolic inequality with four variables, where some of the coefficients are a function of a parameter s. This inequality is an information inequality for every assignment $s \in \mathbb{N}^+$. For example, for $s = 2$ it yields the Zhang and Yeung information inequality [47]. For this symbolic information inequality, we computed the symbolic polynomial Λ and proved that there is a constant $\gamma > 0$ such that for every $s \in \mathbb{N}^+$ the information inequality with parameter s can at most γ-help. We used a similar technique to deal with the infinite sequence presented in [46] that is more general than the infinite sequences presented in [47, 33]. For these sequences the result is that there is a constant $\gamma > 0$ such that every inequality in the sequence can at most γ-help.

Using Theorem 2 we conclude that all the known information inequalities cannot help in proving lower bounds of $\omega(n)$ on the size of the shares in secret-sharing schemes.

Theorem 4. *The information inequalities of [29, 47, 33, 46, 23, 44] cannot help in proving lower bounds of $\omega(n)$ even if they are used simultaneously.*

Acknowledgment. We thank the anonymous TCC referees for valuable comments.

References

[1] Babai, L., Gál, A., Wigderson, A.: Superpolynomial lower bounds for monotone span programs. Combinatorica 19(3), 301–319 (1999)
[2] Beimel, A., Chor, B.: Universally ideal secret sharing schemes. IEEE Trans. on Info. Theory 40(3), 786–794 (1994)

[3] Beimel, A., Franklin, M.: Weakly-private secret sharing schemes. In: Vadhan, S.P. (ed.) TCC 2007. LNCS, vol. 4392, pp. 253–272. Springer, Heidelberg (2007)

[4] Beimel, A., Livne, N., Padró, C.: Matroids can be far from ideal secret sharing. In: Canetti, R. (ed.) TCC 2008. LNCS, vol. 4948, pp. 194–212. Springer, Heidelberg (2008)

[5] Bellare, M., Rogaway, P.: Robust computational secret sharing and a unified account of classical secret-sharing goals. In: 14th CCS, pp. 172–184 (2007)

[6] Ben-Or, M., Goldwasser, S., Wigderson, A.: Completeness theorems for noncryptographic fault-tolerant distributed computations. In: 20th STOC, pp. 1–10 (1988)

[7] Benaloh, J.C., Leichter, J.: Generalized secret sharing and monotone functions. In: Goldwasser, S. (ed.) CRYPTO 1988. LNCS, vol. 403, pp. 27–35. Springer, Heidelberg (1990)

[8] Blakley, G.R.: Safeguarding cryptographic keys. In: Proc. of the 1979 AFIPS National Computer Conference, pp. 313–317 (1979)

[9] Blundo, C., De Santis, A., Gargano, L., Vaccaro, U.: On the information rate of secret sharing schemes. Theoretical Computer Science 154(2), 283–306 (1996)

[10] Blundo, C., De Santis, A., Vaccaro, U.: On secret sharing schemes. Inform. Process. Lett. 65(1), 25–32 (1998)

[11] Brickell, E.F.: Some ideal secret sharing schemes. Journal of Combin. Math. and Combin. Comput. 6, 105–113 (1989)

[12] Capocelli, R.M., De Santis, A., Gargano, L., Vaccaro, U.: On the size of shares for secret sharing schemes. J. of Cryptology 6(3), 157–168 (1993)

[13] Chan, T.H., Yeung, R.W.: On a relation between information inequalities and group theory. IEEE Trans. on Info. Theory 48(7), 1992–1995 (2002)

[14] Chaum, D., Crépeau, C., Damgård, I.: Multiparty unconditionally secure protocols. In: 20th STOC, pp. 11–19 (1988)

[15] Chor, B., Kushilevitz, E.: Secret sharing over infinite domains. J. of Cryptology 6(2), 87–96 (1993)

[16] Cover, T.M., Thomas, J.A.: Elements of Information Theory. John Wiley & Sons, Chichester (1991)

[17] Cramer, R., Damgård, I.B., Maurer, U.M.: General secure multi-party computation from any linear secret-sharing scheme. In: Preneel, B. (ed.) EUROCRYPT 2000. LNCS, vol. 1807, pp. 316–334. Springer, Heidelberg (2000)

[18] Csirmaz, L.: The size of a share must be large. In: De Santis, A. (ed.) EUROCRYPT 1994. LNCS, vol. 950, pp. 13–22. Springer, Heidelberg (1995)

[19] Csirmaz, L.: The dealer's random bits in perfect secret sharing schemes. Studia Sci. Math. Hungar. 32(3–4), 429–437 (1996)

[20] Desmedt, Y.G., Frankel, Y.: Shared generation of authenticators and signatures. In: Feigenbaum, J. (ed.) CRYPTO 1991. LNCS, vol. 576, pp. 457–469. Springer, Heidelberg (1992)

[21] van Dijk, M.: A linear construction of perfect secret sharing schemes. In: De Santis, A. (ed.) EUROCRYPT 1994. LNCS, vol. 950, pp. 23–34. Springer, Heidelberg (1995)

[22] van Dijk, M.: On the information rate of perfect secret sharing schemes. Designs, Codes and Cryptography 6, 143–169 (1995)

[23] Dougherty, R., Freiling, C., Zeger, K.: Six new non-Shannon information inequalities. In: ISIT 2006, pp. 233–236 (2006)

[24] Dougherty, R., Freiling, C., Zeger, K.: Networks, matroids, and non-Shannon information inequalities. IEEE Trans. on Info. Theory 53(6), 1949–1969 (2007)

[25] Fujishige, S.: Polymatroidal dependence structure of a set of random variables. Information and Control 39(1–3), 55–72 (1978)

[26] Gál, A.: A characterization of span program size and improved lower bounds for monotone span programs. Computational Complexity 10(4), 277–296 (2002)

[27] Goyal, V., Pandey, O., Sahai, A., Waters, B.: Attribute-based encryption for fine-grained access control of encrypted data. In: 13th CCS, pp. 89–98 (2006)

[28] Guille, L., Chan, T.H., Grant, A.: The minimal set of Ingleton inequalities. Technical Report 0802.2574, arxiv.org (2008), http://arxiv.org/abs/0802.2574

[29] Ingleton, A.W.: Conditions for representability and transversability of matroids. In: Proc. Fr. Br. Conf. 1970, pp. 62–67. Springer, Heidelberg (1971)

[30] Ito, M., Saito, A., Nishizeki, T.: Secret sharing schemes realizing general access structure. In: Globecom 1987, pp. 99–102 (1987)

[31] Karchmer, M., Wigderson, A.: On span programs. In: Proc. of the 8th IEEE Structure in Complexity Theory, pp. 102–111 (1993)

[32] Karnin, E.D., Greene, J.W., Hellman, M.E.: On secret sharing systems. IEEE Trans. on Info. Theory 29(1), 35–41 (1983)

[33] Makarychev, K., Makarychev, Y., Romashchenko, A., Vereshchagin, N.: A new class of non-Shannon type inequalities for entropies. Communications in Information and Systems 2(2), 147–166 (2002)

[34] Matúš, F.: Infinitely many information inequalities. In: IEEE International Symposium on Information Theory 2007, pp. 41–44 (2007)

[35] Matúš, F.: Two constructions on limits of entropy functions. IEEE Trans. on Info. Theory 53(1), 320–330 (2007)

[36] Naor, M., Wool, A.: Access control and signatures via quorum secret sharing. IEEE Transactions on Parallel and Distributed Systems 9(1), 909–922 (1998)

[37] Oxley, J.G.: Matroid Theory. Oxford University Press, Oxford (1992)

[38] Rabin, M.O.: Randomized Byzantine generals. In: Proc. of the 24th IEEE Symp. on Foundations of Computer Science, pp. 403–409 (1983)

[39] Riis, S.: Graph entropy, network coding and guessing games. Technical Report 0711.4175, arxiv.org (2007), http://arxiv.org/abs/0711.4175

[40] Shamir, A.: How to share a secret. Comm. of the ACM 22, 612–613 (1979)

[41] Shannon, C.E.: Communication theory of secrecy systems. Bell System Technical Journal 28(4), 656–715 (1949)

[42] Simmons, G.J., Jackson, W., Martin, K.M.: The geometry of shared secret schemes. Bulletin of the ICA 1, 71–88 (1991)

[43] Waters, B.: Ciphertext-policy attribute-based encryption: An expressive, efficient, and provably secure realization. Technical Report 2008/290, Cryptology ePrint Archive (2008), http://eprint.iacr.org/

[44] Xu, W., Wang, J., Sun, J.: A projection method for derivation of non-Shannon-type information inequalities. In: ISIT 2008, pp. 2116–2120 (2008)

[45] Yeung, R.W.: A First Course in Information Theory. Springer, Heidelberg (2006)

[46] Zhang, Z.: On a new non-Shannon type information inequality. Communications in Information and Systems 3(1), 47–60 (2003)

[47] Zhang, Z., Yeung, R.W.: On characterization of entropy function via information inequalities. IEEE Trans. on Info. Theory 44(4), 1440–1452 (1998)

Weak Verifiable Random Functions

Zvika Brakerski[1], Shafi Goldwasser[1,2,*],
Guy N. Rothblum[2,**], and Vinod Vaikuntanathan[2,3,***]

[1] Weizmann Institute of Science
[2] CSAIL, MIT
[3] IBM Research

Abstract. Verifiable random functions (VRFs), introduced by Micali, Rabin and Vadhan, are pseudorandom functions in which the owner of the seed produces a public-key that constitutes a commitment to all values of the function and can then produce, for any input x, a proof that the function has been evaluated correctly on x, preserving pseudorandomness for all other inputs. No public-key (even a falsely generated one) should allow for proving more than one value per input.

VRFs are both a natural and a useful primitive, and previous works have suggested a variety of constructions and applications. Still, there are many open questions in the study of VRFs, especially their relation to more widely studied cryptographic primitives and constructing them from a wide variety of cryptographic assumptions.

In this work we define a natural relaxation of VRFs that we call weak verifiable random functions, where pseudorandomness is required to hold only for randomly selected inputs. We conduct a study of weak VRFs, focusing on applications, constructions, and their relationship to other cryptographic primitives. We show:

- **Constructions.** We present constructions of weak VRFs based on a variety of assumptions, including general assumptions such as (enhanced) trapdoor permutations, as well as constructions based on specific number-theoretic assumptions such as the Diffie-Hellman assumption in bilinear groups.
- **Separations.** Verifiable random functions (both weak and standard) cannot be constructed from one-way permutations in a black-box manner. This constitutes the first result separating (standard) VRFs from any cryptographic primitive.
- **Applications.** Weak VRFs capture the essence of constructing non-interactive zero-knowledge proofs for all **NP** languages.

* Supported by NSF grants CCF-0514167, CCF-0635297, NSF-0729011, the Israel Science Foundation 700/08 and the Chais Family Fellows Program.
** Supported in part by NSF grants CCF-0635297, CNS-0430336, NSF-0729011, Israel Science Foundation 700/08 and a Symantec Graduate Fellowship.
*** Supported in part by NSF grants CCF-0635297 and Israel Science Foundation 700/08.

O. Reingold (Ed.): TCC 2009, LNCS 5444, pp. 558–576, 2009.

1   Introduction

Verifiable random functions (VRFs) were introduced by Micali, Rabin and Vadhan [1]. A VRF is a pseudorandom function (see Goldreich, Goldwasser and Micali [2]), that also enables a verifier to verify, given input x, output y and a proof, that the function has been computed correctly on x. The VRF's seed (or secret key) SK is associated with a public key PK. As usual, SK can be used to compute the function's output $y = F_{SK}(x)$ on input x, but it is also used to generate a *proof of correctness* $\pi = \Pi_{SK}(x)$. This proof can be used in conjunction with PK to convince a verifier that y is indeed the correct output on input x with respect to the public-key PK. Further, it is guaranteed that the verifier cannot accept two different values for an input x, even if PK is generated dishonestly. In other words, any string interpreted as a public-key of a VRF constitutes a commitment to at most one output per input.

VRFs are a natural primitive that combines the properties of pseudorandomness and verifiability; however, constructions of VRFs have been few and far between (see related work). In particular, unlike many central and natural cryptographic primitives, there are no known constructions of VRFs based on more *general assumptions*, such as the existence of a one-way function, or even the strong assumption of trapdoor permutations.

1.1   This Work

We propose a relaxation of VRFs, which we call *weak verifiable random functions* (or WVRF). Informally, a weak VRF is similar to a VRF, except that while VRFs require the output to be pseudo-random for *adversarially chosen* inputs, weak VRFs only require pseudorandomness to hold for *random inputs* (see Section 2 for the formal definition). Thus, weak VRFs are a natural relaxation of VRFs, analogous to the relaxation of *weak* pseudorandom functions (without verifiability), proposed by Naor and Reingold [3].

This work is a study of the power of weak VRFs. We present applications to building non-interactive zero-knowledge proofs by showing that the existence of non-interactive zero-knowledge proofs for all of **NP** in the common random string model is essentially equivalent to the existence of weak verifiable random functions in the *standard* model (i.e. without setup assumptions). Thus, we provide a new and conceptually simpler methodology for constructing and analyzing non-interactive zero-knowledge proof systems. We proceed by showing constructions of weak verifiable random functions from a variety of cryptographic assumptions (ones that are not known to imply standard VRFs). Finally, we present a black-box separation from other widely-studied cryptographic primitives (namely one-way permutations). These separations are also the first known separations of standard VRFs from any other cryptographic primitive. We proceed with an overview of each of these contributions.

Weak VRFs and NIZK. We begin by showing an intimate connection between weak VRFs and the study of non-interactive zero-knowledge proofs (NIZKs) for all **NP** languages (with an efficient-prover and in the common random string model). In one direction, we show that a weak VRF can be used to construct a NIZK for any **NP** language. This construction is based on the methodology of Feige, Lapidot and Shamir [4], and shows that weak VRFs can be used to implement their hidden-bits model. In a nutshell, to implement the hidden-bits model, we need a method for a prover (with a secret key) and a verifier (with a public key) to interpret a common random string as a sequence of bit commitments. Towards this end, we give the verifier a public key that guarantees that the commitments are binding, and the prover a secret key that allows non-interactive de-commitment. Weak VRFs are a natural solution to this problem. Modulo the technical details, we can implement the hidden-bits model by taking a random bit sequence x to be a commitment to the output of the weak VRF on input x. The verifier, given the weak VRF's public key, is guaranteed binding, and the pseudo-randomness (even though it only holds for *random* inputs) guarantees hiding. There is a subtle technical problem with the above construction, where the prover may choose a particularly badly formed public key, this is resolved using a certification technique due to Bellare and Yung [5].

Lemma 1. *If there is a family of weak verifiable random functions, then there are efficient-prover non-interactive zero-knowledge proofs in the common random string model for all of* **NP**.

In the other direction, we show that given a construction of (efficient-prover) NIZKs for all **NP** languages, and given also an injective one-way function, we can construct a weak VRF.[1] First recall that the existence of injective OWFs implies the existence of (ordinary) pseudorandom functions (PRFs) and non-interactive perfectly binding commitment schemes. The WVRF is as follows: the generator produces a seed for a PRF and uses it as secret key. The public key is a commitment to that seed. For a random input we use a part of the input as an input to the PRF. The weak VRF's output will be the output of the PRF on this input. The second part of the input is used as a CRS for a NIZK proof. The novelty of this construction is observing that since pseudorandomness should only hold when the adversary sees randomly distributed samples, the *input* can be used as an "honest" source of randomness (CRS in this case). Weak pseudorandomness follows, therefore, from the zero-knowledge property.

Lemma 2. *If there exist injective one-way functions and efficient-prover non-interactive zero-knowledge proofs in the common random string model for all of* **NP***, then there is a family of weak verifiable random functions.*

Thus, weak VRFs and efficient prover NIZK proof systems are essentially equivalent. We hope that weak VRFs, being a "clean" and natural primitive, will prove to be a useful tool or abstraction leading to new constructions of NIZK proof systems. See Section 3 for all details of the above relationship.

---

[1] In fact, we can replace the injective one-way function with a standard one-way function by either (a) allowing the prover and verifier to be non-uniform or (b) using derandomization assumptions (see Barak, Ong and Vadhan [6]).

Constructing Weak VRFs. We show efficient constructions of weak VRFs based on a variety of assumptions. We consider both general assumptions, such as the existence of enhanced trapdoor permutations, and specific number-theoretic assumptions such as the bilinear decisional Diffie-Hellman (BDDH) assumption. We note that none of the assumptions we use to construct weak VRFs are known to imply the existence of (strong) VRFs.

The first construction is a black-box construction using any (enhanced) trapdoor permutation, obtaining WVRFs with arbitrary long (polynomial) output length.[2] The second construction (with a single output bit) is based on the bilinear Diffie-Hellman assumption. In contrast, all known constructions of (strong) VRFs from bilinear maps make substantially stronger assumptions.

We note that these constructions are implicit in known constructions of non-interactive zero-knowledge proofs from trapdoor permutations in [4] and based on bilinear maps on elliptic curves in [7]. This is not surprising in light of the equivalence to NIZK proof systems, and in fact further reinforces our belief that weak VRFs are a natural primitive to focus on when constructing non-interactive zero-knowledge proofs.

See Section 4 for the full details of all the above constructions.

Black-Box Separations. Finally, we initiate a study of the relationship between weak VRFs and more extensively studied cryptographic primitives. We show that there are no black-box constructions of weak VRFs from one-way permutations (OWPs). We note that, given that both pseudorandom functions and signature schemes can be (black-box) constructed even from one-way functions [2,8,9,10], one might have hoped that it is possible to combine both the pseudorandomness and the verifiability properties and to construct weak VRFs (or even standard, strong VRFs) from one-way functions. However, while in signature schemes verifiability is guaranteed if the public-key has been selected properly, in VRFs we require verifiability even for adversarial ones. Indeed, our result shows that this cannot be done. We note that this is also the first separation result for (strong) VRFs, and thus sheds light on their complexity as well.

Formally, we show a black-box separation (see related work) between weak VRFs and OWPs. In fact we show this even for weak verifiable unpredictable functions, where pseudo-randomness is replaced with unpredictability.

Theorem 1 (informal). *There is no black-box construction of weak VRFs from OWPs.*

We prove this theorem by showing that any black-box construction of weak VRF fails with respect to some oracle implementing a OWP. From this result it follows immediately that VRFs also cannot be black-box reduced to OWPs.

This result differs from previous work on black-box separations (see related work below) in several ways. Unlike the seminal result of Impagliazzo and Rudich separating one-way permutations and key agreement protocols [11], we do not

---

[2] Note that we can always increase the output length of weak VRFs by concatenating multiple instances, but here this is done without increasing the key lengths.

show a relativizing separation. That is, we do not prove that there exists a oracle relative to which OWPs exist but weak VRFs do not. Moreover, our method is also different from the one introduced by Gertner, Malkin and Reingold [12] for proving a separation between trapdoor predicates and trapdoor functions. There, each construction is tailored with an oracle relative to which it is proved insecure. Informally speaking, we show that there exists an oracle $\mathcal{O}$ (that implements a OWP and encodes an **NP**-hard oracle) such that any construction that is correct (namely, both complete and sound) with respect to all one-way permutation oracles, fails to be pseudorandom with respect to $\mathcal{O}$ (see the proof intuition below for more details).

We remark that since we show that (enhanced) trapdoor permutations imply weak VRFs, it also follows from our separation that there is no black-box construction of such trapdoor permutations from one-way permutations. While this was already known as a corollary from [11], our proof is simpler (albeit somewhat more restricted).

Proof Intuition. Our adversary works by "reverse engineering" the oracle queries made by the secret and public key generator of the weak VRF. Essentially, what we want to do, is use the oracle (that can solve **NP**-hard problems) to find a "fake" secret-key corresponding to the given public key. This fake secret key will later be used (by the adversary) to predict the outputs of the weak VRF. The intuition is that these predictions (by the fake secret key) should be correct because they match the same public key as the one used with the "real" secret key. Note that this holds because we are assuming here strong "completeness" of the key generation algorithm: namely, it *always* (with probability 1) generates a valid secret-public key pair, and so even the fake secret key that we found "should" generate correct outputs. We need this completeness property (which all our constructions possess) to prove our separation result.

The problem we face is that the key-generator is itself an oracle circuit, and thus we cannot simply find a fake secret key that corresponds to the given public key and the given OWP oracle. We can, however, use our ability to solve **NP**-hard problems (via the given oracle) to find a fake secret key that matches the given public key and a (slightly) different oracle, by changing the oracle's answers on queries made by the key generator. Consider an oracle query that was made by the generator, and consider the case of changing the answer for this query in a way that does not affect the public key. This change may affect some values of the function, and so we are no longer guaranteed that the function's output on the real and fake secret key are the same. Note, however, that the function's output can only change when the verification algorithm makes an oracle query to a value whose output has changed. This follows from the weak VRF's soundness: the verification algorithm must make such an oracle query in order to tell the two cases apart. Therefore, if we take enough random samples, run the verification algorithm and "collect" its OWP oracle queries, we'd have a bank of "common" queries that holds all queries that affect a large portion of the values of the function. Then we can find, using the **NP**-hardness of the oracle, a secret key and simulated values to all other queries made by the generator (answers to the "common" queries are unchanged) that yield the same public-key. Changing these "uncommon" values, however, would not affect the value of the function almost anywhere, and thus we can use the fake secret key we found to predict the value of the function on a random point.

1.2   Related Work

Verifiable Random Functions. Bellare and Goldwasser [13] present a signature scheme based on combining a PRF and a NIZK proof system. While their scheme implies a PRF with verifiability properties, a falsely generated verification-key may enable the prover to make the verifier accept more than one output per input. Thus this construction falls short of the soundness requirement for a VRF.

Known constructions of VRFs are due to [1] based on strong RSA, Lysyanskaya [14] based on a strong version of the Diffie-Hellman assumption in bilinear groups, Dodis [15] based on the sum-free generalized DDH assumption, and Dodis and Yampolskiy [16] based on the bilinear Diffie-Hellman inversion assumption.[3]

Variants of VRFs have also been proposed and used for various applications, for instance the notion of of simulatable VRFs, introduced by Chase and Lysyanskaya [17], which were used to compile any single-theorem non-interactive zero-knowledge proof for a language L into a many-theorem non-interactive zero-knowledge proof for the same L. We stress that simulatable VRFs are defined in the public-parameters model and are incomparable to standard VRFs.

Non-Interactive Zero-Knowledge and Related Primitives. Non-interactive zero-knowledge proofs (NIZK)[4], introduced by Blum, Feldman and Micali [18], are proof systems where the prover P sends a single message to the verifier V to convince V of an (**NP**) statement, while conveying no more knowledge to the verifier except that the statement is true. NIZK proof systems have been immensely useful, including in the construction of non-malleable and chosen-ciphertext secure encryption schemes [19,20,21,22,23,24,25], signature schemes [13] and more. There have been a handful of constructions of NIZK proof systems from the time they were introduced, based on specific number theoretic assumptions and based on general assumptions (see below).

Specific number-theoretic assumptions that imply NIZKs include quadratic residuosity [18], the computational Diffie-Hellman assumption on (prime-order) bilinear groups due to Canetti, Halevi and Katz [7], and constructions due to Groth, Ostrovsky and Sahai based on the subgroup-decisional assumption on composite-order bilinear groups [26], and on the decisional linear assumption on (prime-order) bilinear groups [27].

---

[3] We note that both [1] and [16] construct VRFs with polynomial-size domains and later extend it to arbitrary domains via a tree-based construction, which impacts their efficiency.

[4] We stress that here we are dealing with computational NIZK *proofs*, as opposed to statistical NIZK *arguments*. In the latter, the soundness property holds only against cheating provers that are computationally bounded.

Many works have investigated constructions of NIZK proofs based on *general assumptions* – these include the construction of [4] based on (enhanced) trapdoor permutations (improved by Kilian and Petrank [28] for better efficiency), and more recently, the construction of [7] based on what they call verifiable trapdoor predicates. Both these constructions use primitives that have an explicit trapdoor structure which may not be inherent. Dwork and Naor [29] showed that NIZK proof systems can be constructed using a (strong) VRF (Dodis and Puniya [30] show an alternative construction by going through their notion of a verifiable random permutation). Our construction is from a *weak VRF* and is much more direct.

Goldwasser and Ostrovsky [31] proposed a new primitive called invariant signature schemes and showed that in the CRS model they are *equivalent* to non-interactive zero-knowledge proofs for all of **NP**. In [29], verifiable pseudorandom generators (VPRGs) are presented and shown to be equivalent to NIZK proofs in the CRS model. A VPRG is essentially a pseudorandom generator, such that the owner of the seed can post proofs of correctness for subsets of the generated bits, while maintaining hiding of the rest of them. It is shown that the existence of VPRGs in the CRS model is equivalent to the existence of NIZK in that model. Approximate VPRG is a variant of this notion, where the soundness requirement of the proof is relaxed. Approximate VPRGs in the *standard model* exist if and only if NIZK exists in the *CRS model*.

Black-Box Separations. A black-box reduction from primitive A to primitive B is essentially a construction of A that uses an oracle to B, such that the security of B implies the security of A. Most known reductions between cryptographic primitives are black-box. In [11], it was shown for the first time that it is possible to rule out the existence of black-box reductions between some primitives.[5] Such proofs of impossibility are referred to as *black-box separations*. Their result has been followed by many others, showing impossibility of various classes of black-box reductions between various cryptographic primitives and protocols. For a classification of black-box reductions (and separations), we refer the reader to the work of Reingold, Trevisan and Vadhan [32].

2 Preliminaries and Definitions

Verifiable Random Functions. We use the definition of verifiable random functions from Micali, Rabin and Vadhan [1]. A key feature of VRFs is their soundness property: soundness requires that no two distinct values can be proven to be $F_{SK}(x)$, for *any PK, and any x*, even ones that are adversarially chosen. This is a crucial difference between VRFs and other cryptographic primitives such as encryption and digital signatures, where the public/secret-keys are assumed to be chosen correctly. For a formal definition of VRFs, we refer the reader to [1].

---

[5] Specifically, [11] show that there is no *relativizing* reduction from secure key-agreement protocols to one-way permutations, thus ruling out black-box reductions where the proof of correctness is also black box. This has been improved by [32] to also rule out cases where the proof of correctness has "some" non-black-boxness.

Weak Verifiable Random Functions. Weak verifiable random functions maintain the key feature of VRFs, namely that even if the public-key PK is adversarially chosen, it is impossible for the adversary (even one who knows SK) to prove that $y = F_{SK}(x)$ and $y' = F_{SK}(x)$ for two different y and y'. However, in the case of weak VRFs, we relax this condition slightly by saying that *for every* PK, the completeness and soundness conditions hold *for most inputs* x (and not necessarily all the inputs, as in the case of standard VRFs). We stress that there are *no public-keys* PK for which the completeness and/or the soundness conditions fail on a large fraction of inputs.

The other major difference between the definitions of weak VRFs and standard VRFs is in the pseudorandomness condition: whereas in the case of VRFs, pseudorandomness holds against an adversary that can adaptively choose inputs x and obtain evaluations $F_{SK}(x)$, the weak VRF adversary gets evaluations of $F_{SK}(x)$ on *random values* x. This is in the spirit of weak PRFs presented in [3].

Definition 1 (Weak Verifiable Random Function). *A family of functions* $\mathcal{F} = \{f_s : \{0,1\}^{n(k)} \to \{0,1\}^{m(k)}\}_{s \in \{0,1\}^k}$ *is a family of* weak verifiable random functions *with security parameter* k *if there exist algorithms* (G, F, Π, V) *such that: the key-generation algorithm* $G(1^k)$ *is a PPT algorithm that outputs a pair of keys* (PK, SK); *the function-evaluator* $F_{SK}(x)$ *is a deterministic algorithm that outputs* $f_{SK}(x)$; *the Prover* $\Pi_{SK}(x)$ *is a deterministic algorithm that outputs a proof of correctness* π *and the Verifier* $V(PK, x, y, \pi)$ *is a PPT algorithm that either accepts or rejects a purported proof* π *of the statement "$y = F_{SK}(x)$".*

We require the following:

1. (Relaxed) Completeness: *for all* $(PK, SK) \leftarrow G(1^k)$ *and for all but a* 2^{-k} *fraction of* x's, *if* $y = F_{SK}(x)$ *and* $\pi = \Pi_{SK}(x)$, *then* $\Pr[V(PK, x, y, \pi) = \text{accept}] \geq 1 - 2^{-k}$. *The probability here is taken over the random coins of* V.
2. (Relaxed) Soundness: *for all* PK *and for all but a* 2^{-k} *fraction of* x's, *and for all* y_1, y_2, π_1, π_2 *such that* $y_1 \neq y_2$, $\Pr[V(PK, x, y_i, \pi_i) = \text{accept}] \leq 2^{-k}$ *for at least one* $i \in \{1, 2\}$.
3. Weak Randomness: *let* A *be a PPT algorithm, and let* $p(k)$ *be any polynomial. Then, the probability that* A *succeeds in the following experiment is at most* $\frac{1}{2} + \mathsf{negl}(k)$:

    $(PK, SK) \leftarrow G(1^k)$

    Choose $x_1, x_2, \ldots, x_{p(k)} \xleftarrow{R} \{0,1\}^{n(k)}$, $x^* \xleftarrow{R} \{0,1\}^{n(k)}$ *and* $b \xleftarrow{R} \{0,1\}$.

    If $b = 0$, *set* $y^* = F_{SK}(x^*)$, *otherwise set* $y^* \xleftarrow{R} \{0,1\}^{m(k)}$

    $b' \leftarrow A(1^k, PK, \{x_i, F_{SK}(x_i), \Pi_{SK}(x_i)\}_{i=1}^{p(k)}, x^*, y^*)$

    A *succeeds if* $b' = b$.

Weak verifiable unpredictable functions (VUF) are the same as the above, except that the weak randomness requirement is replaced by the weak unpredictability requirement below.

3′. Weak Unpredictability: *Consider the following experiment with the adversary* A, *and let* $p(k)$ *be any polynomial.*

$(PK, SK) \leftarrow G(1^k)$

Choose $x_1, x_2, \ldots, x_{p(k)} \leftarrow \{0,1\}^{n(k)}$, $x^* \leftarrow \{0,1\}^{n(k)}$.

$y^* \leftarrow A(1^k, PK, \{x_i, F_{SK}(x_i), \Pi_{SK}(x_i)\}_{i=1}^{p(k)}, x^*)$

A succeeds *if* $y^* = F_{SK}(x^*)$. *We require that the probability that A succeeds is at most* negl(k).

3 Weak Verifiable Random Functions and NIZK Proofs

In this section, we show that weak verifiable random functions and non-interactive zero-knowledge proofs are essentially equivalent. First, in Lemma 2, we construct a weak VRF given a non-interactive zero-knowledge (NIZK) proof system for all of **NP** (with an efficient prover, in the common random string model) and an injective one-way function. Secondly, in Lemma 1, we construct NIZK proof systems for every **NP** language, given any weak VRF.

Lemma 2 (restated). *If there exist injective one-way functions and efficient-prover non-interactive zero-knowledge proofs in the common random string model for all of* **NP** *, then there is a family of weak verifiable random functions.*

Proof. The construction is very similar to the construction of a signature scheme from (enhanced) trapdoor permutations, due to [13]: the difference is that in [13], the common random string for the NIZK proof system is part of the public-key of the resulting signature scheme, whereas in our case, it is part of the *input*. Informally speaking, the reason for this difference is that in a signature scheme, the public-key is completely trusted, whereas this is not the case for (both strong and weak) VRFs.

The key-generation algorithm picks s and s', two independent seeds for a pseudorandom function. The public-key PK for the WVRF is the commitment of the seed s, using randomness ρ. The secret-key SK is (s, ρ, s'). Namely, $PK = \text{COM}(s; \rho)$ and $SK = (s, \rho, s')$. The function $F_{SK}(r\|x)$ parses its input as r and x and outputs $f_s(x)$. The proof generator Π does the following: define the **NP** language

$$L = \{(PK, x, y) \mid \exists s, \rho \text{ such that } PK = \text{COM}(s; \rho) \text{ AND } y = f_s(x)\}$$

Π runs the prover algorithm for the NIZK proof system for the language L using r as the common random string. The randomness of the prover is $f_{s'}(x)$, and the output of the prover is the proof π. It is easy to see that given SK and x, the proof-generator Π is deterministic. The verifier V, given PK, x, y and π, runs the NIZK verifier on input the statement (PK, x, y) and the proof π and accepts if and only if the NIZK verifier accepts.

This construction assumes a pseudorandom function (which can be constructed from any one-way function [2,8]) and a non-interactive commitment scheme (which can be constructed from any injective one-way function, see Blum and Micali [33]).

Completeness of the WVRF follows from the perfect completeness of the NIZK proof system. Pseudorandomness follows via a standard hybrid argument, which we omit for lack of space.

Relaxed soundness follows from the perfect binding of the commitment scheme and the soundness of the NIZK proof system. Slightly more precisely, given any PK, there is at most one s such that $PK \in \text{COM}(s; \cdot)$ (where $\text{COM}(s; \cdot)$ denotes the set of all commitments of the string s). Thus, for all $y' \neq f_s(x)$, it follows that $(PK, x, y') \notin L$. By the soundness of the NIZK proof system, this means that with high probability over the input (that is, over r) the verifier will not accept any purported proof of the statement (PK, x, y') with high probability over its coin-tosses. □

Next, we show how to construct NIZK proofs for all of NP from any weak VRF. We do this by implementing the hidden-bit model of [4] using any weak VRF.

Lemma 1 (restated). *If there is a family of weak verifiable random functions, then there are efficient-prover non-interactive zero-knowledge proofs in the common random string model for all of* **NP**.

Informally, the idea for implementing the hidden-bits proof system is to let the prover P choose a pair of keys (PK, SK) for the weak verifiable random function, and let the hidden bits $(b_1, \ldots, b_m)$ (for some $m = \text{poly}(n)$) be defined as $b_i = F_{SK}(r_i)$, where $(r_1, \ldots, r_m)$ is the first part of the common random string. The prover can reveal any subset of the bits, simply by giving the verifier the proof $\Pi_{SK}(r_i)$ for the corresponding bits.

One potential problem is that the prover can select (PK, SK) depending on the common random string and potentially violate soundness. This is solved in the standard way of [4] by reducing the soundness error of the NIZK proof in the hidden-bit model.

A more subtle problem is that the prover may select (PK, SK) such that $F_{SK}(\cdot)$ is heavily unbalanced, thus introducing a bias into the distribution of the hidden bits. We handle this in a way that is similar to a certification procedure developed in [5].

We refer the reader to the full version [34] for a complete proof of this lemma.

4   Constructions of Weak Verifiable Random Functions

In this section, we show two efficient constructions of weak verifiable random functions (WVRF), as outlined in the introduction.

Construction from Trapdoor Permutations. For simplicity, we describe the construction from any (enhanced) *certified* trapdoor permutation, namely given a function f, it is possible in polynomial time to check that f indeed defines a one-to-one and onto function. This construction can be made to work with any (enhanced) trapdoor permutation, using a certification procedure of Bellare and Yung [5].

Let (f, f^{-1}) be an enhanced certified trapdoor permutation. Then, the construction of a WVRF (G, F, Π, V) is as follows: The key-generation algorithm G, on input 1^k, outputs $PK = f$, and $SK = f^{-1}$, where (f, f^{-1}) is a random trapdoor permutation together with its trapdoor. Let $f^{-i}(x)$ denote the result of f^{-1} applied i times to the input x. F_{SK} parses its input as (x, r), and outputs $(b_1, \ldots, b_\ell)$ where $b_i = \langle f^{-i}(x), r \rangle$. $\Pi_{SK}(x, r)$ outputs $f^{-(\ell+1)}(x)$. The verification algorithm V, given $PK, (x, r)$ and y, accepts if and only if $\langle f^i(\pi), r \rangle = b_{\ell-i+1}$ for all $1 \leq i \leq \ell$ and $f^{\ell+1}(\pi) = x$.

To sketch the proof of this construction, observe that perfect completeness is immediate. Soundness follows from the fact that f is a (certified) permutation. Pseudorandomness follows from the one-wayness of f, as well as the fact that we use the Goldreich-Levin hardcore bit.

Construction from the Computational Diffie-Hellman Assumption in Gap-DDH groups. Let $\mathbb{G}$ and $\mathbb{G}'$ be groups of prime order q, with a bilinear map $e : \mathbb{G} \times \mathbb{G} \to \mathbb{G}'$. Let g be a generator of $\mathbb{G}$. The WVRF (G, F, Π, V) is defined as follows: the key-generation algorithm $G(1^k)$ outputs $PK = g^a$ and $SK = a$, where a is a random element in $\mathbb{Z}_q$. $F_{SK}(r)$ uses r to sample a random element R in $\mathbb{G}$, [6] and outputs a hardcore bit of R^a (for example, the most significant bit of R^a). $\Pi_{SK}(r)$ simply outputs R^a. The verification algorithm, on input PK, x, y and π, accepts if $e(PK, x) = e(g, \pi)$ and y is the hardcore bit of π.

The fact that this is a weak VRF follows from the Diffie-Hellman assumption. The formal proof is omitted from this extended abstract.

5   Separations

In this section, we show a black-box separation between weak verifiable unpredictable functions (weak VUFs) and one-way permutations. Recall that both weak and standard VRFs are in particular also weak VUFs, and that weak VRFs can be constructed in a fully black-box manner from (enhanced) trapdoor permutations (eTDPs, see Section 4). This result, therefore, implies a separation between weak VRFs, standard VRFs and eTDPs and one-way permutations.

Technically, we show that there is no *semi black-box* reduction (a notion defined in [32], included below) from a weak VUF to a one-way permutation. In other words, we show that for every construction of a weak VUF from a one-way permutation, there is an oracle (which possibly depends on the construction) such that the construction fails with respect to the oracle.[7]

Definition 2 ([32]). *A tuple of oracle algorithms (G, F, Π, V) is a Semi-BB reduction from weak verifiable unpredictable functions to one-way permutations:*

- **Correctness.** *For every permutation $\mathcal{O}$, $(G^{\mathcal{O}}, F^{\mathcal{O}}, \Pi^{\mathcal{O}}, V^{\mathcal{O}})$ has (relaxed) completeness and soundness as in Definition 1.*
- **Security.** *For every permutation $\mathcal{O}$, if there exists a PPT oracle machine A such that $A^{\mathcal{O}}$ predicts $(G^{\mathcal{O}}, F^{\mathcal{O}}, \Pi^{\mathcal{O}}, V^{\mathcal{O}})$ in the sense of Definition 1, then there exists a PPT oracle machine S such that $S^{\mathcal{O}}$ inverts $\mathcal{O}$ with non-negligible probability.*

---

[6] In the case where $\mathbb{G}$ is a subgroup of an elliptic curve group, the sampling can be done efficiently. See the full version [34] for details.

[7] We note that our reduction does not preclude a relativizing reduction. Ruling out a relativizing reduction involves constructing an oracle relative to which no secure weak VUF exists. For more details on the different types of black-box reductions, see [32].

Using the definition above, we can formally state our claim. We show that the following holds.

Theorem 1 (formally stated). *There is no semi black-box reduction from a weak VUF to a one-way permutation. Namely, for every construction (G, F, Π, V) of a weak VUF, there is an oracle $\mathcal{O}$ such that $(G^{\mathcal{O}}, F^{\mathcal{O}}, \Pi^{\mathcal{O}}, V^{\mathcal{O}})$ is, in the terms of Definition 2, either incorrect or insecure.*

In the remaining of this section, we provide a sketch of the proof of Theorem 1 (Section 5.1, for the full proof, see the full version of this paper) and conclude with some remarks on limits and extensions of the proof (Section 5.2).

5.1 Proof Sketch of Theorem 1

The proof proceeds by contradiction. Fix (towards contradiction) some semi black-box reduction (see Definition 2) (G, F, Π, V). For any oracle $\mathcal{O}$ that implements a one-way permutation, $(G^{\mathcal{O}}, F^{\mathcal{O}}, \Pi^{\mathcal{O}}, V^{\mathcal{O}})$ is a weak VUF. For any such reduction, we show an oracle $\mathcal{O}$ and an adversary $A^{\mathcal{O}}$ that breaks the weak unpredictability of the defined weak VUF (w.r.t $\mathcal{O}$). Throughout the proof, let t_G (resp. t_V) denote the (polynomial in k) running times of $G^{\mathcal{O}}$ (resp. $V^{\mathcal{O}}$). For simplicity, we will assume that the verifier $V^{\mathcal{O}}$ is deterministic, throughout the rest of this proof.[8]

The oracle $\mathcal{O}$ is similar to the one presented in [32].[9] Roughly speaking, $\mathcal{O}$ both implements a one-way permutation (that is, no adversary with oracle access to $\mathcal{O}$ can compute x given $\mathcal{O}(x)$, for a random $x \in \{0, 1\}^n$), and is **NP**-hard (namely, with oracle access to $\mathcal{O}$, it is possible to decide every language in **NP**). A formal statement follows.

Proposition 1 (implicit in [32]). *There exists an oracle $\mathcal{O}$ which is (i) A length preserving permutation; (ii) One-way: there exists no PPT oracle machine A s.t. $A^{\mathcal{O}}$ inverts $\mathcal{O}$; and (iii) **NP**-hard: for any **NP** relation $\mathcal{R}$, there exists a polynomial-time oracle machine B that for any x where $\exists y.(x, y) \in \mathcal{R}$, $B^{\mathcal{O}}$ finds such a y, namely: $(x, B^{\mathcal{O}}(x)) \in \mathcal{R}$.*

We want to use the power of $\mathcal{O}$ to construct an adversary that predicts the weak VUF. Given a public-key PK, this can be done by finding a secret-key SK' such that (PK, SK') is a possible output of $G^{\mathcal{O}}(1^k)$ (this follows from completeness and soundness of the weak VUF). However, this requires finding a witness for an $\mathbf{NP}^{\mathcal{O}}$ relation, a task beyond the powers of our oracle. We thus relax the requirement. We present an **NP** relation that enables finding a secret-key SK' and an oracle $\mathcal{O}'$ such that (PK, SK') is a possible output of $G^{\mathcal{O}'}(1^k)$. Furthermore, $\mathcal{O}'$ is only a slight modification of $\mathcal{O}$: $\mathcal{O}'$ and $\mathcal{O}$ agree on

---

[8] However, see remark on handling probabilistic verifiers in Section 5.2.

[9] In [11], two oracles are used: a random oracle and a **PSPACE**-complete oracle. [32] show how this can be simplified into one oracle that is both a one-way permutation and is **PSPACE**-hard (the same argument holds for **NP**-hardness as well).

almost all inputs, and particularly on a set of "significant" inputs. We then show that such $SK', \mathcal{O}'$ can be used to predict the weak VUF. A detailed description follows.

We define an **NP**-relation $\mathcal{R}$ that will enable us to find SK' and a transcript of oracle query/answers (which will define $\mathcal{O}'$) that are consistent with PK and with a predefined query bank (a set of queries and answers from $\mathcal{O}$). The query bank will formally be represented by a set of queries $Q \subseteq \{0,1\}^*$ and a function $f_Q : Q \to \{0,1\}^*$ mapping them to answers. The input of $\mathcal{R}$, therefore, is formally denoted $z = (1^k, PK, Q, f_Q)$ (where 1^k is the security parameter). The corresponding witness consists of the new secret key SK', along with the rest of the information that enables simulating the generation of (PK, SK'): the randomness r that G uses, and the queries not in Q that G made, along with their respective answers. These are represented by $D \subseteq \{0,1\}^*$, $f_D : D \to \{0,1\}^*$ (the same way as Q). We require that $|r|, |D| \leq t_G(k)$. Formally, the witness for relation $\mathcal{R}$ is denoted $w = (r, SK', D, f_D)$ and $(z, w) \in R$ if (PK, SK') are produced by an execution of G with security parameter 1^k and randomness r, which makes oracle queries in $Q \cup D$, and gets answers according to f_Q, f_D. The verification procedure $\mathrm{Ver}_{\mathcal{R}}(z, w)$ for $\mathcal{R}$ simply simulates G for at most $t_G(k)$ steps and checks that (z, w) are consistent with the above.

Using an **NP**-hard oracle, we can compute a witness of $\mathcal{R}$ for any input, if such exists. We further note that if f_Q is consistent with $\mathcal{O}$, and if PK was in fact generated by $G^{\mathcal{O}}(1^k)$, then there always exists at least one witness for that input: the one that contains the actual random tape, secret key and oracle query/answers that were used in the generation of PK.

We are now ready to describe the adversary algorithm $A^{\mathcal{O}}$ (recall that A has oracle access to $\mathcal{O}$). For the remainder of the proof, fix the (PK, SK) generated by $G(1^k)$ for the unpredictability challenge.

The Adversary Algorithm. The adversary A operates in two stages. In the first stage, the *"exploration stage"*, the adversary receives a public-key PK for the weak VUF, as well as polynomially many evaluations of $F_{SK}^{\mathcal{O}}, \Pi_{SK}^{\mathcal{O}}$ on random inputs x_i. The adversary tries to learn the random-oracle queries that are "significant" in computing the function, and outputs a bank of oracle queries and answers. In the second stage, the *"conquering stage"*, the adversary (using the bank of queries) constructs a secret-key SK' and an (implicit) oracle $\mathcal{O}'$ for the same PK such that $F_{SK'}^{\mathcal{O}'}$ and $F_{SK}^{\mathcal{O}}$ coincide on most inputs. This enables the adversary to predict the value of $F_{SK}^{\mathcal{O}}$ on most inputs, in turn. The description of A follows.

The **exploration stage** of the adversary A.

INPUT: 1^k, PK and $\{x_i, y_i, \pi_i\}_{i=1}^{k^2 t_G(k)}$, where $(PK, SK) \leftarrow G^{\mathcal{O}}(1^k)$, $y_i = F_{SK}^{\mathcal{O}}(x_i)$ and $\pi_i = \Pi_{SK}^{\mathcal{O}}(x_i)$.

OUTPUT: A bank of queries consisting of a set of queries Q and a mapping $f_Q : Q \to \{0,1\}^*$ matching answer $f_Q(q) = \mathcal{O}(q)$ to every answer q.

ALGORITHM:

    1. Initialize the bank of queries $Q, f_Q = \emptyset$.

2. For $i = 1, \ldots, k^2 \cdot t_G(k)$ run $V^{\mathcal{O}}(PK, x_i, y_i, \pi_i)$. Save all the query-answer pairs made by V to the oracle $\mathcal{O}$ into the query bank (Q, f_Q). Output (Q, f_Q).

The **conquering stage** of the adversary A.

INPUT: PK, query bank (Q, f_Q) and a challenge $x^* \xleftarrow{R} \{0,1\}^{n(k)}$.
OUTPUT: $y^* \in \{0,1\}^{m(k)}$.
ALGORITHM:

1. Let $z = (1^k, PK, Q, f_Q)$ be an input for **NP** relation $\mathcal{R}$ described above, we can use $\mathcal{O}$ (which is **NP**-hard) to compute a witness $w = (r, SK', D, f_D)$ such that $(z, w) \in \mathcal{R}$ (as we mentioned, such witness must exist).

2. For all $q \in \{0,1\}^*$, define

$$\mathcal{O}'(q) = \begin{cases} f_D(q), & q \in D, \\ \mathcal{O}(q), & \text{otherwise.} \end{cases}$$

Note that $\mathcal{O}'(q)$ can be computed in polynomial time given access to $\mathcal{O}, D, f_D$. Using SK' and $\mathcal{O}'$, return $y^* = F_{SK'}^{\mathcal{O}'}(x^*)$.[10]

Analysis of the Adversary. Recall that we fixed PK, SK. We first define a notion of "frequent oracle queries" of the verification algorithm (with respect to PK and SK) and show that the bank of queries (Q, f_Q) that the adversary outputs in the exploration stage contains all the frequent oracle queries of the verification algorithm, with high probability.

We define the frequency $\mathsf{freq}(q)$ of a query q to the oracle $\mathcal{O}$ (with respect to PK and SK) to be the fraction of x's for which the verification algorithm, on input PK, x, $F_{SK}(x)$ and $\Pi_{SK}(x)$, makes the query q to the oracle $\mathcal{O}$ during its execution. More precisely,

$$\mathsf{freq}(q) = \Pr_{x \in \{0,1\}^{n(k)}} \left[V^{\mathcal{O}}(PK, x, F_{SK}^{\mathcal{O}}(x), \Pi_{SK}^{\mathcal{O}}(x)) \text{ makes query } q \text{ to } \mathcal{O} \right]$$

A query q is called α-frequent if $\mathsf{freq}(q) \geq 1/\alpha$. Let $\mathcal{F}_{\alpha(k)}$ be the set of all $\alpha(k)$-frequent queries. That is, $\mathcal{F}_{\alpha(k)} = \{q : \mathsf{freq}(q) \geq 1/\alpha(k)\}$. The following lemma states that the exploration stage succeeds in finding all frequent queries with very high probability.

Lemma 3 (exploration stage). *Let $\alpha(k) = k \cdot t_G(k)$. With probability at least $1 - \mathrm{poly}(k) \cdot e^{-k}$, at the end of the exploration stage of A, $Q \supseteq \mathcal{F}_{\alpha(k)}$.*

Proof. Consider an $\alpha(k)$-frequent query $q \in \mathcal{F}_{\alpha(k)}$. By definition, $\mathsf{freq}(q) \geq 1/\alpha(k)$. That is, for at least $1/\alpha(k)$ fraction of x's, $V^{\mathcal{O}}(PK, x, f_{SK}(x), \Pi_{SK}(x))$

---

[10] We remark that $\mathcal{O}'$ as defined is not necessarily a permutation, so $F_{SK'}^{\mathcal{O}'}(x^*)$ may not be well defined. In the full version [34] we show how this is fixed by defining a permutation $\mathcal{O}'$ s.t. $|\{q : \mathcal{O}(q) \neq \mathcal{O}'(q)\}| \leq 2|D|$. For the remaining of the analysis, we assume that $\mathcal{O}'$ is a permutation.

makes the query q to the oracle $\mathcal{O}$. Since the bank of queries Q contains all the oracle queries made by V on $k\alpha(k)$ random inputs x_i, the probability that q is not in the bank of queries is exponentially small. More precisely,

$$\Pr[q \notin Q] \le (1 - 1/\alpha(k))^{k\alpha(k)} \le e^{-k}$$

Union bounding over all queries in $\mathcal{F}_{\alpha(k)}$ shows that with probability all but $|\mathcal{F}_{\alpha(k)}| \cdot e^{-k}$, Q contains $\mathcal{F}_{kt_G(k)}$ (where the probability is over the randomness of x_i). Now, $|\mathcal{F}_{\alpha(k)}| \le \alpha(k) \cdot t_V(k) = k \cdot t_G(k) \cdot t_V(k)$ by simple counting. This completes the proof. □

The next lemma states that assuming the exploration stage completed properly, in the conquering stage A breaks the weak unpredictability of the VUF. We note that $\mathcal{O}'$ is one-way because it only differs from $\mathcal{O}$ on polynomially many inputs.

Lemma 4 (conquering stage). *Let Q, f_Q be an output of the exploration stage of A s.t. $Q \supseteq \mathcal{F}_{kt_G(k)}$. Then the conquering stage runs in $\mathrm{poly}(k)$ time and predicts $F_{SK}^{\mathcal{O}}(x^*)$ with probability at least $1 - 1/k$.*

Proof. We define an input $x \in \{0,1\}^{n(k)}$ to be *indifferent* (with respect to PK and SK) if the execution of the verification algorithm (with oracle access to $\mathcal{O}$), on input $(PK, x, F_{SK}^{\mathcal{O}}(x), \Pi_{SK}^{\mathcal{O}}(x))$ makes no oracle query $q \in D$ (recall that D is the set of queries computed in step 1 of the conquering stage). In other words, this execution of the verification algorithm is indifferent to whether it is given oracle access to $\mathcal{O}$ or $\mathcal{O}'$.

The following claims establish that all but a $1/k$ fraction of the inputs x are indifferent; and that for every indifferent input x, $F_{SK}^{\mathcal{O}}(x) = F_{SK'}^{\mathcal{O}'}(x)$. In other words, if x^* is an indifferent input, then the adversary (which outputs $F_{SK'}^{\mathcal{O}'}(x^*)$) succeeds in predicting $F_{SK}^{\mathcal{O}}(x^*)$. It follows that the adversary succeeds with probability $1 - \frac{1}{k}$.

Claim. Let $\mathcal{I}$ denote the set of indifferent inputs (with respect to $\mathcal{O}, PK$ and SK). Then, $\Pr_{x \in \{0,1\}^{n(k)}}[x \in \mathcal{I}] \ge 1 - 1/k$.

Proof: By definition of our **NP** relation $\mathcal{R}$, $Q \cap D = \emptyset$. Thus, if $Q \supseteq \mathcal{F}_{kt_G(k)}$ then $\mathcal{F}_{kt_G(k)} \cap D = \emptyset$. If we fix some query $q \in D$, then $q \notin \mathcal{F}_{kt_G(k)}$, meaning

$$\mathsf{freq}(q) = \Pr_{x \in \{0,1\}^{n(k)}} \left[V^{\mathcal{O}}(PK, x, F_{SK}^{\mathcal{O}}(x), \Pi_{SK}^{\mathcal{O}}(x)) \text{ makes query } q \right] \le 1/(kt_G(k)) .$$

Applying the union bound over all $|D| \le t_G(k)$ queries in D yields

$$\Pr_{x \in \{0,1\}^{n(k)}} \left[V^{\mathcal{O}}(PK, x, F_{SK}^{\mathcal{O}}(x), \Pi_{SK}^{\mathcal{O}}(x)) \text{ makes any query } q \in D \right] \le \frac{|D|}{kt_G(k)} \le \frac{1}{k} ,$$

and the claim follows. ∎

Claim. For all $x^* \in \mathcal{I}$, $F_{SK}^{\mathcal{O}}(x^*) = F_{SK'}^{\mathcal{O}'}(x^*)$.

Proof: For simplicity, assume that for every one-way permutation $\mathcal{O}$, the construction $(G^{\mathcal{O}}, F^{\mathcal{O}}, \Pi^{\mathcal{O}}, V^{\mathcal{O}})$ is correct for every input x.[11]

By the completeness of the weak VUF with respect to $\mathcal{O}$, we have that $V^{\mathcal{O}}(PK, x^*, F_{SK}^{\mathcal{O}}(x^*), \Pi_{SK}^{\mathcal{O}}(x^*))$ accepts. Since no queries in D are made during this computation, then clearly it would run in the exact same way with oracle access to $\mathcal{O}'$ rather than to $\mathcal{O}$. Thus,

$$V^{\mathcal{O}'}(PK, x^*, F_{SK}^{\mathcal{O}}(x^*), \Pi_{SK}^{\mathcal{O}}(x^*)) = V^{\mathcal{O}}(PK, x^*, F_{SK}^{\mathcal{O}}(x^*), \Pi_{SK}^{\mathcal{O}}(x^*)) = \mathsf{accept} .$$

Since $\mathcal{O}'$ is a OWP, $(G^{\mathcal{O}'}, F^{\mathcal{O}'}, \Pi^{\mathcal{O}'}, V^{\mathcal{O}'})$ is a weak VUF, which in particular, means that it satisfies the completeness and soundness properties. By its completeness, we get that $V^{\mathcal{O}'}(PK, x^*, F_{SK'}^{\mathcal{O}'}(x^*), \Pi_{SK'}^{\mathcal{O}'}(x^*))$ accepts. Soundness with respect to $\mathcal{O}'$ guarantees, therefore, that $F_{SK}^{\mathcal{O}}(x^*) = F_{SK'}^{\mathcal{O}'}(x^*)$.   □

Combining Lemmas 3, 4 (using the union bound) yields that A succeeds in predicting $F_{SK}^{\mathcal{O}}(x^*)$ with probability at least $1 - 1/k - \mathrm{poly}(k) \cdot e^{-k}$, contradictory to the alleged security of the reduction. Theorem 1 follows.   □

5.2   Additional Remarks

- **Handling Probabilistic Verifiers.** The analysis above disregarded the fact that the verifier V may not return the correct answer, with some small probability. Essentially, we handle this issue by using amplification by applying sequential repetition and then using a single random tape for all inputs. For details, we refer the reader to the full version [34].
- **On Requiring Perfect Completeness.** In the definition of VRF and weak VRF, we required that completeness holds for any (PK, SK) generated by G. When allowing *relaxed* completeness in Definition 1, the relaxation was over the *inputs* and not the keys. While this definition is frequently used, in some cases (e.g. [1]) the definition is so that the generator is allowed to output "bad" keys (ones that have no completeness for almost any input) with very small probability.

  While our proof does not cover such constructions, we notice that all known constructions (including that of [1]), can be presented as having perfect completeness in a **ZPP** sense. That is, where the generator is allowed to run for expected polynomial time rather than worst-case. Our construction can be slightly altered to work for such constructions as well.
- **Separating Trapdoor Permutations from OWPs.** As mentioned above, since there is a black-box reduction from weak VRFs to eTDPs, our separation also implies a Semi-BB separation of eTDPs from OWPs. The work of [11] implies a result that is stronger in two aspects: their separation (appended with a modification due to [32]) implies a ∀∃Semi-BB separation (see definition in [32]), and they show a separation from key-agreement. Our result, on the other hand, seems simpler and does not use heavy probability-theoretic machinery.

---

[11] This is as opposed to relaxed completeness and soundness as in Definition 1 which hold for almost all inputs.

- **Other Types of Black-box Separations.** Our result as presented does not rule out a relativizing reduction. To rule out a relativizing reduction, we must exhibit an oracle $\mathcal{O}$ relative to which no weak VUF exists. We show, essentially, that for every construction, there is an oracle that makes the construction fail as a weak VUF. Our adversary, however, works by generating a slightly different OWP, which is efficiently computable from the old one, and plugging it into the same construction. To get a separation, we require correctness (but not necessarily security) for the modified oracle. Therefore, while our separation only rules out Semi-BB reductions in the general case, it can also be interpreted as ruling out $\forall\exists$Semi-BB reductions if correctness holds for any OWP.
- **Inefficient Proof Generators.** We notice that while the adversary uses the code of oracle algorithms G, F, V, its use of Π is black-box only. Therefore, we additionally obtain that a semi-BB reduction is impossible even when the proof generator Π is allowed to be inefficient. [12]

References

1. Micali, S., Rabin, M., Vadhan, S.: Verifiable random functions. In: FOCS, p. 120 (1999)
2. Goldreich, O., Goldwasser, S., Micali, S.: How to construct random functions. J. ACM 33(4), 792–807 (1986)
3. Naor, M., Reingold, O.: Synthesizers and their application to the parallel construction of pseudo-random functions. J. Comput. Syst. Sci. 58(2) (1999)
4. Feige, U., Lapidot, D., Shamir, A.: Multiple noninteractive zero knowledge proofs under general assumptions. SIAM J. Comput. 29(1), 1–28 (1999)
5. Bellare, M., Yung, M.: Certifying permutations: Noninteractive zero-knowledge based on any trapdoor permutation. J. Cryptology 9(3), 149–166 (1996)
6. Barak, B., Ong, S.J., Vadhan, S.P.: Derandomization in cryptography. SIAM J. Comput. 37(2), 380–400 (2007)
7. Canetti, R., Halevi, S., Katz, J.: A forward-secure public-key encryption scheme. J. Cryptology 20(3), 265–294 (2007)
8. Håstad, J., Impagliazzo, R., Levin, L.A., Luby, M.: A pseudorandom generator from any one-way function. SIAM J. Comput. 28(4), 1364–1396 (1999)
9. Naor, M., Yung, M.: Universal one-way hash functions and their cryptographic applications. In: STOC, pp. 33–43 (1989)
10. Rompel, J.: One-way functions are necessary and sufficient for secure signatures. In: STOC, pp. 387–394 (1990)
11. Impagliazzo, R., Rudich, S.: Limits on the provable consequences of one-way permutations. In: STOC 1989: Proceedings of the twenty-first annual ACM symposium on Theory of computing, pp. 44–61. ACM, New York (1989)

---

[12] We note that there is a construction of a weak VRF from one-way permutations, using inefficient-prover non-interactive zero-knowledge for **NP** [4]. However, this construction is inherently *non-black-box*. Our separation result shows that this *must* be the case.

12. Gertner, Y., Malkin, T., Reingold, O.: On the impossibility of basing trapdoor functions on trapdoor predicates. In: FOCS 2001: Proceedings of the 42nd IEEE symposium on Foundations of Computer Science, Washington, DC, USA, p. 126. IEEE Computer Society, Los Alamitos (2001)

13. Bellare, M., Goldwasser, S.: New paradigms for digital signatures and message authentication based on non-interactive zero knowledge proofs. In: Brassard, G. (ed.) CRYPTO 1989. LNCS, vol. 435, pp. 194–211. Springer, Heidelberg (1990)

14. Lysyanskaya, A.: Unique signatures and verifiable random functions from the DH-DDH separation. In: Yung, M. (ed.) CRYPTO 2002. LNCS, vol. 2442, pp. 597–612. Springer, Heidelberg (2002)

15. Dodis, Y.: Efficient construction of (Distributed) verifiable random functions. In: Desmedt, Y.G. (ed.) PKC 2003. LNCS, vol. 2567, pp. 1–17. Springer, Heidelberg (2002)

16. Dodis, Y., Yampolskiy, A.: A verifiable random function with short proofs and keys. In: Vaudenay, S. (ed.) PKC 2005. LNCS, vol. 3386, pp. 416–431. Springer, Heidelberg (2005)

17. Chase, M., Lysyanskaya, A.: Simulatable vRFs with applications to multi-theorem NIZK. In: Menezes, A. (ed.) CRYPTO 2007. LNCS, vol. 4622, pp. 303–322. Springer, Heidelberg (2007)

18. Blum, M., Feldman, P., Micali, S.: Non-interactive zero-knowledge and its applications (extended abstract). In: STOC, pp. 103–112 (1988)

19. Blum, M., Feldman, P., Micali, S.: Proving security against chosen cyphertext attacks. In: Goldwasser, S. (ed.) CRYPTO 1988. LNCS, vol. 403, pp. 256–268. Springer, Heidelberg (1990)

20. Rackoff, C., Simon, D.R.: Cryptographic defense against traffic analysis. In: STOC, pp. 672–681 (1993)

21. Naor, M., Yung, M.: Public-key cryptosystems provably secure against chosen ciphertext attacks. In: STOC, pp. 427–437 (1990)

22. Dolev, D., Dwork, C., Naor, M.: Nonmalleable cryptography. SIAM J. Comput. 30(2), 391–437 (2000)

23. Sahai, A.: Non-malleable non-interactive zero knowledge and adaptive chosen-ciphertext security. In: FOCS, pp. 543–553 (1999)

24. Pass, R., Shelat, A., Vaikuntanathan, V.: Construction of a non-malleable encryption scheme from any semantically secure one. In: Dwork, C. (ed.) CRYPTO 2006. LNCS, vol. 4117, pp. 271–289. Springer, Heidelberg (2006)

25. Cramer, R., Hanaoka, G., Hofheinz, D., Imai, H., Kiltz, E., Pass, R., Shelat, A., Vaikuntanathan, V.: Bounded CCA2-secure encryption. In: Kurosawa, K. (ed.) ASIACRYPT 2007. LNCS, vol. 4833, pp. 502–518. Springer, Heidelberg (2007)

26. Groth, J., Ostrovsky, R., Sahai, A.: Perfect non-interactive zero knowledge for NP. In: Vaudenay, S. (ed.) EUROCRYPT 2006. LNCS, vol. 4004, pp. 339–358. Springer, Heidelberg (2006)

27. Groth, J., Ostrovsky, R., Sahai, A.: Non-interactive zaps and new techniques for NIZK. In: Dwork, C. (ed.) CRYPTO 2006. LNCS, vol. 4117, pp. 97–111. Springer, Heidelberg (2006)

28. Kilian, J., Petrank, E.: An efficient non-interactive zero-knowledge proof system for np with general assumptions. Journal of Cryptology 11, 1–27 (1998)

29. Dwork, C., Naor, M.: Zaps and their applications. SIAM J. Comput. 36(6), 1513–1543 (2007)

30. Dodis, Y., Puniya, P.: Feistel networks made public, and applications. In: Naor, M. (ed.) EUROCRYPT 2007. LNCS, vol. 4515, pp. 534–554. Springer, Heidelberg (2007)
31. Goldwasser, S., Ostrovsky, R.: Invariant signatures and non-interactive zero-knowledge proofs are equivalent. In: Brickell, E.F. (ed.) CRYPTO 1992. LNCS, vol. 740, pp. 228–245. Springer, Heidelberg (1993)
32. Reingold, O., Trevisan, L., Vadhan, S.P.: Notions of reducibility between cryptographic primitives. In: Naor, M. (ed.) TCC 2004. LNCS, vol. 2951, pp. 1–20. Springer, Heidelberg (2004)
33. Blum, M., Micali, S.: How to generate cryptographically strong sequences of pseudo random bits. In: FOCS, pp. 112–117 (1982)
34. Brakerski, Z., Goldwasser, S., Rothblum, G., Vaikuntanathan, V.: Weak verifiable random functions. MIT CSAIL Technical Report (2008)

Efficient Oblivious Pseudorandom Function with Applications to Adaptive OT and Secure Computation of Set Intersection

Stanisław Jarecki and Xiaomin Liu

University of California, Irvine

Abstract. An Oblivious Pseudorandom Function (OPRF) [15] is a two-party protocol between sender S and receiver R for securely computing a pseudorandom function $f_k(\cdot)$ on key k contributed by S and input x contributed by R, in such a way that receiver R learns only the value $f_k(x)$ while sender S learns nothing from the interaction. In other words, an OPRF protocol for PRF $f_k(\cdot)$ is a secure computation for functionality $\mathcal{F}_{\mathsf{OPRF}} : (k, x) \rightarrow (\perp, f_k(x))$.

We propose an OPRF protocol on committed inputs which requires only $O(1)$ modular exponentiations, and has a constant number of communication rounds (two in ROM). Our protocol is secure in the CRS model under the Composite Decisional Residuosity (CDR) assumption, while the PRF itself is secure on a polynomially-sized domain under the Decisional q-Diffie-Hellman Inversion assumption on a group of composite order, where q is the size of the PRF domain, and it has a useful feature that f_k is an injection for every k.

A practical OPRF protocol for an injective PRF, even limited to a polynomially-sized domain, is a versatile tool with many uses in secure protocol design. We show that our OPRF implies a new practical fully-simulatable adaptive (and committed) OT protocol secure without ROM. In another example, this oblivious PRF construction implies the first secure computation protocol of set intersection on committed data with computational cost of $O(N)$ exponentiations where N is the maximum size of both data sets.

1 Introduction

PRF and Oblivious PRF. A pseudorandom function (PRF) [17] is an efficiently computable keyed function $f_k(\cdot)$ whose values are indistinguishable, for a randomly chosen key k, from random elements in the function range. The *oblivious* PRF, or OPRF [15], is a protocol that allows the sender S, on input the key k, to let the receiver R compute the value $f_k(x)$ of a PRF $f_k(\cdot)$ on any input x of R's choice without releasing any other information to R, and do so *obliviously* in the sense that sender S learns nothing from the protocol, similarly as in oblivious transfer [28,14] or oblivious polynomial evaluation [24]. In other words, an OPRF protocol corresponding to a PRF function $f_k(\cdot)$ is a secure computation protocol for functionality $\mathcal{F}_{\mathsf{OPRF}} : (k, x) \rightarrow (\perp, f_k(x))$. To enforce consistency between several protocol instances it is helpful to extend the above fuctionality $\mathcal{F}_{\mathsf{OPRF}}$ to include verification whether k and x contributed by S and R correspond to some previously committed values. We call this extended functionality a *committed*

O. Reingold (Ed.): TCC 2009, LNCS 5444, pp. 577–594, 2009.

OPRF, and in parallel to public-key primitives we refer to the commitment to a PRF key k as a corresponding *public key pk*.

Examples of Oblivious PRF Applications: Keyword Search, Adaptive OT, SFE for Set Intersection. An oblivious PRF has numerous exciting applications. It was introduced as a primitive by Freedman et al. [15] with an application to privacy-preserving *Keyword Search*. This protocol problem is otherwise known as symmetrically-private Keyword PIR [12] or "Keyword OT", and it can be defined as a secure computation for the following functionality: Sender S contributes a set of (keyword,data) pairs $\{x_i, y_i\}_{i=1..N}$, where all x_i's are unique, receiver R contributes a keyword x, and the functionality gives data item y_i to R if $x_i = x$, or a symbol $\perp$ if all x_i's differ from x. The reduction of Keyword OT to Oblivious PRF is very simple provided that f_k is an injection for every k and that the keyword domain is polynomial-sized: The sender S picks two PRF keys k_1, k_2, publishes a set of pairs $\{(f_{k_1}(x_i), f_{k_2}(x_i) \oplus y_i)\}_{i=1..N}$ together with commitments pk_1, pk_2 and a proof of knowledge of the corresponding k_1, k_2. (The restriction that the keyword domain is polysized is needed for simulatability because it enables extraction of sender's inputs given keys k_1, k_2.) Then R computes $z_1 = frm[o] -- f_{k_1}(x)$ and $z_2 = f_{k_2}(x)$ via two instances of the oblivious PRF protocol, uses z_1 to check if there exists an x_i in S's set s.t. $x_i = x$, and if so then it uses z_2 to recover the corresponding y_i.

Essentially the same protocol is also a solution to the *Adaptive OT* problem, introduced by Naor and Pinkas [22]. A (fully simulatable) adaptive OT is a secure computation protocol for a reactive functionality where S contributes a sequence of N values $Y = \{y_i\}_{i=1..N}$, and then S and R can engage in any number of 1-out-of-N OT protocol instances on these values and any index i contributed by R. An adaptive OT can be implemented using oblivious (committed) PRF if S publishes a sequence $\{f_k(i) \oplus y_i\}_{i=1..N}$, a commitment pk and a ZK proof of knowledge of the corresponding k, and for each OT instance the two parties run an OPRF protocol on S's key k and R's adaptively chosen index i, which lets R compute $f_k(i)$ and thus retrieve y_i. (This protocol is related to the ROM-based adaptive OT scheme of Camenisch, Neven, and Shelat [8], as we explain in subsection below.)

Another application of oblivious PRF's was recently shown by Hazay and Lindell [19], who use an oblivious PRF to construct a very simple protocol for a set intersection function, where S and R contribute their respective sets of N data items, X and Y, and the protocol lets R compute the intersection $X \cap Y$ without revealing anything else. As in any secure function evaluation (SFE), if the protocol ensures that both parties enter previously committed inputs then both parties can compute the output (without guaranteeing fairness) if the SFE protocol is run in both directions. In Hazay-Lindell protocol, S picks a PRF key k, sends to R a commitment pk to k and a set of outputs of the PRF function $f_k(\cdot)$ on all its inputs, $X' = \{f_k(x)\}_{x \in X}$. Note that these PRF values are indistinguishable from n random values in f's range, and hence in particular X' cannot be efficiently correlated with X. Players S and R engage then in n OPRF instances on respective inputs k and y for all $y \in Y$, allowing R to compute the set $Y' = \{f_k(y)\}_{y \in Y}$. If f_k is an injection then R can conclude that $y \in X \cap Y$ iff $y \in X' \cap Y'$. Moreover, if we use committed OPRF and S proves that values $f_k(x)$ correspond to committed inputs x, the result is an SFE protocol for set intersection on committed data sets.

Previous Work on Oblivious PRF's. The first oblivious PRF construction was given by Naor and Reingold [25], based on the PRF construction given in the same paper, $f_k(x) = g^{\prod_{x_i=1} r_i}$, where key k is the sequence of T random elements $r_1, ..., r_T$ in $\mathbb{Z}_q$, q is the order of the group $\langle g \rangle$ in which the Decisional Diffie-Hellman assumption holds, and T is the bitlength of elements x in the PRF domain. This first oblivious PRF protocol required $O(t)$ rounds. Freedman et al. [15], based on the previous work of Naor and Pinkas [21,22], improved this to a constant-round protocol, secure under DDH. This protocol implements a "weak OPRF" notion (which is just as good in all the applications listed above), where the receiver is allowed to learn additional information about the PRF key as long as that information does not change the pseudorandomness of the PRF function on any new inputs. This protocol uses T parallel OT instances, one for each r_i value, which using best known OT techniques, e.g. [23,1], translates into $O(T)$ exponentiations. The OT-batching techniques of Ishai et al. [20] do not seem to reduce this overhead in the case of a single OPRF instance if the PRF domain size is smaller or equal to the security parameter. The Hazay-Lindell set intersection protocol implemented with this OPRF involves NT parallel OT instances, but it is an open problem to efficiently use the OT-batching techniques of [20] in this context because the case of malicious adversaries seems to require batching *committed* OT instances and proving that the committed OT inputs correspond to the PRF key.

The ROM-based Adaptive OT construction from any unique blind signature scheme given by Camenisch et al. [8] relies on a weak OPRF protocol for f_k defined as $f_k(m) = H(\mathsf{SIG}_k(m))$, which is a PRF in ROM if the signature scheme is CMA-secure, where the weak OPRF functionality on inputs (k, m) releases the signature $\mathsf{SIG}_k(m)$ in addition to the proper PRF value $H(\mathsf{SIG}_k(m))$. However, for the resulting protocol to be secure computation of this functionality we need a secure computation protocol for the blind signature functionality, $\mathcal{F}_{\mathsf{BlSg}}(k, m) \rightarrow (\perp, \mathsf{Sig}_k(m))$. While there exist efficient unique blind signature schemes, e.g. by Chaum [10] or Boldyreva [3], these schemes rely on non-standard "one more" type of security assumptions, and it is an open problem to extend such blind signature protocols to secure computation of $\mathcal{F}_{\mathsf{BlSg}}$, and to ensure that the computation proceeds on committed receiver's input m.

An Adaptive OT construction of Green and Hohenberger from blind IBE schemes [18] can also be seen as variant of an OPRF protocol. Assuming the ROM model one could define $f_k(ID) = H(sk_{ID})$ where k is the KDC's master private key and sk_{ID} is the IBE decryption key corresponding to identity ID. If this is a blind IBE scheme then we could use the blinded private-key retrieval protocol as an implementation of a weak OPRF which releases sk_{ID} in addition to the hash value $H(sk_{ID})$. However, efficient IBE's seem to require bilinear maps, which are not needed for the OPRF protocol we present, and moreover it remains an open problem to upgrade such construction to secure computation of the ideal OPRF functionality, and to further extend it to computation on committed inputs.

Our Result: Efficient (Committed) Oblivious PRF. We propose an OPRF construction which requires only $O(1)$ modular exponentiations, and has a constant number of communication rounds (two in ROM). The secure computation protocol for functionality $\mathcal{F}_{\mathsf{OPRF}} : (k, x) \rightarrow (\perp, f_k(x))$ for our PRF function family $f_k(\cdot)$ builds on the Camenisch-Shoup [9] version of Paillier encryption [26], and it is secure in the

CRS model under the Composite Decisional Residuosity (CDR) assumption. The PRF $f_k(\cdot)$ itself is a variant of the PRF construction of Dodis-Yampolskiy [13] based on the Boneh-Boyen signature [4], but moved to a group of a composite order, the safe RSA modulus on which the Camenisch-Shoup encryption operates. As far as we know, this OPRF construction is the first constant-round efficient OPRF which is actually a secure computation protocol for the ideal OPRF functionality $\mathcal{F}_{\mathsf{OPRF}}$.

This PRF has a useful feature that f_k is an injection for every key k, and it secure on a domain of bitstrings of length $|q|$ under the Decisional q-Diffie-Hellman Inversion (q-DHI) assumption on a group whose order is a safe RSA modulus. Consequently, by positive hardness results of [9,13] (see also related upper bounds on q-DHI hardness given by Jung Hee Cheon [11]), the domain of this PRF is polynomially-sized, but this restriction does not stop any of the OPRF applications we listed above. The one application it constrains is the secure computation of set intersection, where the inputs must be encoded in a polynomially-sized domain. Note that in many applications set intersection protocol might run on short inputs anyway, e.g. names, social security numbers, etc. However, extending the domain of this OPRF construction would allow secure computation of set intersection on larger input domains.

Another very useful feature of our OPRF construction is that it is an efficient *committed* OPRF, i.e. our secure computation protocol for extended version of the F_{OPRF} functionality which checks whether both parties contribute previously committed inputs into the protocol.

Technical Roadmap: We give a brief intuition for constructing an OPRF protocol for the Dodis-Yampolskiy PRF $f_k(x) = g^{1/(k+x)}$ in group $\langle g \rangle$ of composite order n, and an encryption scheme which is additively homomorphic on message domain $\mathbb{Z}_n$, like Paillier encryption. We assume that this encryption allows for shared decryption. Namely, one can form a "joint key" $pk = pk_s \cdot pk_r$ from two public keys pk_s, pk_r, and the corresponding private keys sk_s, sk_r allow for shared decryption of ciphertext encrypted under pk. This is true of the Camenisch-Shoup version [9] of Paillier encryption. Using expressions like $C_v^{(r)}$, $C_v^{(s)}$, and C_v, to denote ciphertexts which encrypt variable v under, respectively, pk_r, pk_s, or pk, we have that partial decryption of C_v under sk_r creates $C_v^{(s)}$, and a partial decryption of C_v under sk_s creates $C_v^{(r)}$.

The idea of our construction goes like this: Sender S and receiver R exchange public keys pk_s and pk_r and encrypt their inputs k and x under the joint key $pk = pk_s \cdot pk_r$ as C_k and C_x. Then by the homomorphism of the encryption $C_k \cdot C_x = C_\alpha$ where $\alpha = k + x$. Player R then randomizes the encrypted value α by picking random a in $\mathbb{Z}_n$ and computing $C_\beta = (C_\alpha)^a$, for $\beta = a \cdot \alpha$. R also encrypts a under pk, as C_a, partially decrypts C_β into $C_\beta^{(s)}$, and sends $(C_a, C_\beta^{(s)})$ to S. Sender S decrypts $C_\beta^{(s)}$ to get β and computes $C_\sigma = (C_a)^{1/\beta}$, for $\sigma = a/\beta = 1/\alpha = 1/(k+x)$. Finally, S picks an additive share σ_s of σ, computes $v_s = g^{\sigma_s}$, encrypts it as C_{σ_s} and computes $C_{\sigma_r} = C_\sigma / C_{\sigma_s}$, for $\sigma_r = \sigma - \sigma_s$. S also partially decrypts C_{σ_r} into $C_{\sigma_r}^{(r)}$ and sends $(v_s, C_{\sigma_r}^{(r)})$ to R. R decrypts $C_{\sigma_r}^{(r)}$ and uses σ_r to compute $v = v_s \cdot g^{\sigma_r} = g^{\sigma_s + \sigma_r} = g^{1/(k+x)}$.

Our actual protocol streamlines these operations, and uses solely keys pk_r, pk_s instead of the joint key pk, but the above sketch is an idea behind our construction.

Related Concurrent Work: We note that independently from our work, Belenkiy et al. [2] recently showed a different protocol for oblivious computation of the same function $f_k(x) = g^{1/(k+x)}$, also using Paillier encryption. Their protocol is somewhat similar to ours but it uses multiplicative rather than additive sharing of the crucial exponent value $\sigma = 1/(k+x)$, and it is about twice faster than our protocol as a result. Moreover, the protocol of [2] can work on groups with a 160-bit prime order unrelated to the RSA modulus n, instead of a composite order we need, leading to further speed-ups in the applications of this OPRF. While the initial version of this protocol published in [2] is not a secure computation of the OPRF functionality, it is not difficult to modify this protocol to secure computation in the CRS model using techniques similar to ours.

Organization: We introduce our notation, security assumptions, and important definitions in Subsection 2.1. In Subsection 2.2 we show the main tool in our efficient OPRF protocol construction, an additively homomorphic encryption scheme with verifiable encryption and decryption, and in Subsection 2.3 we show an efficient instantiation of such scheme, i.e. the Camenisch-Shoup encryption scheme. In Section 3 we present our main contribution, a construction of an OPRF protocol. Finally we show two applications of a committed OPRF, to Adaptive OT and to secure computation of the Set Intersection problem, in Sections 4 and 5 respectively.

2   Preliminaries and Tools

2.1   Notation, Definitions, and Security Assumptions

Notation. We use $a \leftarrow A$ to denote that a is the output of the (randomized) algorithm A and $a \leftarrow_R S$ if a is chosen uniformly at random from set S. If A is an algorithm then we will sometimes use $A(x)$ to denote a set of all possible outputs of A on input x. We use $P\{b\}$ to denote a zero-knowledge proof system that statement b holds, and $PoK\{a \mid \phi(a)\}$ to denote a zero-knowledge proof of knowledge of value a that satisfies a publicly computable relation ϕ. Finally, most numerical operations in the paper are group operations unless specifically noted otherwise.

Factoring Assumption (Definition). Let RSAGen denote an algorithm that picks safe RSA moduli with a given security parameter. Namely, RSAGen(1^κ) chooses two random primes p'_1, p'_2 s.t. $|p'_1| = |p'_2| = \kappa$ and $p_1 = 2p'_1 + 1$ and $p_2 = 2p'_2 + 1$ are also primes, and outputs $n = p_1 \cdot p_2$. We say that *factoring safe RSA moduli is hard* if for every efficient algorithm A the probability $\Pr[A(n) \in \{p_1, p_2\} \mid n \leftarrow \text{RSAGen}(1^\kappa)]$ is a negligible function of κ.

Decisional q-Diffie-Hellman Inversion (q-DHI) Problem (Definition). The *computational* q-DHI problem in a group with generator g and order n is to compute $g^{1/\alpha}$ given the tuple $(g, g^\alpha, \ldots, g^{(\alpha^q)})$, for random α in $\mathbb{Z}_n^*$. We define the hardness of the *decisional* version of this problem for any fixed constant q as follows: Let gGen be an algorithm which on input a security parameter κ picks a modulus n and a generator g of a multiplicative group $\mathbb{G}$ of order n. For example gGen can be a composition of RSAGen and an algorithm gGen' which on input n output by RSAGen finds the first prime p s.t. $n|p-1$, and sets g as any element of order n in $\mathbb{Z}_p^*$. We say that *the*

Decisional q-DHI Assumption holds on group (family) $\mathbb{G}$, if for every efficient algorithm $\mathcal{A}$ function $\epsilon_A(\kappa) = |\text{Real}_A(\kappa) - \text{Random}_A(\kappa)|$ is negligible, where:

$$\text{Real}_A(\kappa) = \Pr\left[\mathcal{A}(\mathsf{g}, \mathsf{g}^\alpha, \ldots, \mathsf{g}^{(\alpha^q)}, \mathsf{g}^{1/\alpha}) = 1 \mid (\mathsf{g}, n) \leftarrow \mathsf{gGen}(1^\kappa); \ \alpha \leftarrow \mathbb{Z}_n^*\right]$$

$$\text{Random}_A(\kappa) = \Pr\left[\mathcal{A}(\mathsf{g}, \mathsf{g}^\alpha, \ldots, \mathsf{g}^{(\alpha^q)}, \mathsf{h}) = 1 \ \middle| \ \begin{matrix}(\mathsf{g}, n) \leftarrow \mathsf{gGen}(1^\kappa); \ \alpha \leftarrow \mathbb{Z}_n^*; \\ \mathsf{h} \leftarrow \mathbb{G}\end{matrix}\right]$$

Pseudorandom Function (Definition). For notational simplicity we consider a version of the general PRF notion which is custom-made to fit the PRF implementation we consider in this paper. Namely, the PRF function f_k maps $|q|$-bit strings to elements of group $\mathbb{G}$. We say that *function (family)* f_k *defined by the key generation algorithm* KGen *is a PRF* if $|\text{Real}_A(\kappa) - \text{Random}_A(\kappa)|$ is a negligible function of κ, where:

$$\text{Real}_A(\kappa) \ = \ \Pr\left[\mathcal{A}^{f_k(\cdot|\neg x)}(v, \mathsf{st}) = 1 \ \middle| \ \begin{matrix}(k, \mathsf{pk}) \leftarrow \mathsf{KGen}(1^\kappa); \\ (x, \mathsf{st}) \leftarrow A^{f_k(\cdot)}(\mathsf{pk}); \ v \leftarrow f_k(x)\end{matrix}\right]$$

$$\text{Random}_A(\kappa) \ = \ \Pr\left[\mathcal{A}^{f_k(\cdot|\neg x)}(v, \mathsf{st}) = 1 \ \middle| \ \begin{matrix}(k, \mathsf{pk}) \leftarrow \mathsf{KGen}(1^\kappa); \\ (x, \mathsf{st}) \leftarrow A^{f_k(\cdot)}(n, \mathsf{g}, \mathsf{pk}); \ v \leftarrow \mathbb{G}\end{matrix}\right]$$

In the above experiments $f_k(\cdot|\neg x)$ denotes an oracle $f_k(\cdot)$ modified to output $\perp$ on x.

Dodis-Yampolskiy's PRF and Boneh-Boyen's Function in Composite-Order Groups. Our OPRF construction relies on a variant of the PRF scheme of Dodis-Yampolskiy [13], based on the Boneh-Boyen unpredictable function [4], with the sole modification being a substitution of a prime-order group with a group whose order is a safe RSA modulus. The Boneh-Boyen function [4] is $f_k(x) = \mathsf{g}^{1/(k+x)}$ where g generates a group $\mathbb{G}$ of prime order p and k is a random element in $\mathbb{Z}_p$. This function is unpredictable under the *computational* q-DHI assumption on $\mathbb{G}$. Dodis-Yampolskiy [13] considered the same function as a source of a verifiable pseudorandom function (VRF) in a group with a bilinear map, and showed that it is secure under a decisional version of the q-DHI assumption on the target group of the bilinear map. However, the same argument also shows that the decisional q-DHI assumption on group $\mathbb{G}$ itself implies that the Boneh-Boyen function is a PRF. Morevoer, the same arguments which were done by [4,13] for prime-order groups also imply that (1) the Boneh-Boyen function in a composite-order group remains a PRF under the decisional q-DHI assumption on such groups (and hardness of factoring) and (2) the same generic-group argument which motivated trust in the q-DHI assumption on the prime-order groups carries to composite-order groups as well.

Specifically, for security parameter κ we define the following PRF function family: The key generation algorithm picks $n \leftarrow \mathsf{RSAGen}(1^\kappa)$, $\mathsf{g} \leftarrow \mathsf{gGen}'(n)$, $k \leftarrow \mathbb{Z}_n^*$, and computes $\mathsf{pk} \leftarrow \mathsf{g}^k$. We define $f_k(x)$ for each $x \in \{0,1\}^{|q|}$ as follows:

$$f_k(x) = \begin{cases} \mathsf{g}^{1/(k+x)} & \text{if } gcd(k+x, n) = 1 \\ 1 & \text{otherwise} \end{cases}$$

Claim 1: The Decisional q-DHI Assumption holds on a generic group (family) with a safe RSA order.

Argument Sketch: This claim can be verified by inspecting the generic-group argument for the computational q-DHI problem on a prime-order group given in [4], because generic arguments for decisional rather than computational problems are identical, and all that this specific argument requires is that the group order has only large factors.

Claim 2: The function (family) f_k defined above is a PRF on the domain of q-domain strings if factoring safe RSA moduli is hard and if the Decisional q-DHI Assumption holds on group (family) $\mathbb{G}$.

Argument Sketch: Assuming decisional q-DHI holds in group $\mathbb{G}$ of order a safe RSA modulus n, the argument why $f_k(x)$ in $\mathbb{G}$ is a PRF follows from the argument of Theorem 1 in [13] and the fact that under the assumption of hardness of safe RSA moduli, efficient algorithms can encounter elements v s.t. $gcd(v, n) \neq 1$ only with negligible probability. The reduction given in the proof of Theorem 1 in [13] uses inverses $(\alpha - x_{i*} + x_i)$ in the exponent for all x_i in the domain of f_k, where x_{i*} is the value for which the adversary must distinguish $f_k(x_{i*})$ from a random element in $\mathbb{G}$. Note that by the factoring assumption the probability that in this game there appears i s.t. $(\alpha - x_{i*} + x_i)$ is *not* co-prime with n is negligible. Thus the argument for pseudorandomness of $f_k(x) = g^{1/(k+x)}$ shown in [13] for prime-order groups extends to groups whose order is hard to factor.

2.2 Additively Homomorphic Verifiable Encryption with Additional Properties

In order to construct an oblivious PRF, we need an additively homomorphic encryption scheme with *verifiable encryption* and *verifiable decryption*.

- Setup(κ) on security parameter κ outputs the public parameter par to be used in all subsequent algorithms. This par also defines the plaintext space $\mathcal{M}$, a finite additive group $\mathbb{Z}_n$ for some n.
  For notational simplicity, we omit explicit mention of par as the input to all the following algorithms.
- KGen outputs a random public/secret key pair (pk, sk). Parameters par also defines a relation KVal (for key validity) on all valid (pk, sk). We also require an efficient proof system

$$PoK\{sk \mid (pk, sk) \in \mathsf{KVal}\} \tag{1}$$

- $\mathsf{Enc}_{pk}(m)$ on public key pk and message m outputs a ciphertext C which is a random encryption of m under pk. We require an efficient realization of the following proof system given a public key pk and ciphertext C

$$PoK\{m \mid C \in \mathsf{Enc}_{pk}(m)\} \tag{2}$$

- $\mathsf{Dec}_{sk}(C)$ is a deterministic algorithm on secret key sk and ciphertext C that outputs a message m. We require an efficient realization of the following proof system given a public key pk and ciphertext C

$$P\{\exists\, sk, \text{ s.t. } m = \mathsf{Dec}_{sk}(C) \wedge (pk, sk) \in \mathsf{KVal}\} \tag{3}$$

Besides semantic security, we require the following properties of the above encryption scheme:

- **Additive Homomorphism:** We require that there is an efficient operation on ciphertexts, which for convenience we denote as a multiplication, s.t. $\mathsf{Enc}_{pk}(m_0) \cdot \mathsf{Enc}_{pk}(m_1) \in \mathsf{Enc}_{pk}(m_0 + m_1)$. We can also define exponentiation and division operations on ciphertexts, and by homomorphism of the encryption $(\mathsf{Enc}_{pk}(m))^a = \mathsf{Enc}_{pk}(a \cdot m)$ and $\mathsf{Enc}_{pk}(m_0)/\mathsf{Enc}_{pk}(m_1) \in \mathsf{Enc}_{pk}(m_0 - m_1)$, for any a and any m, m_0, m_1 in $\mathcal{M}$.
- **Verifiable Encryption:** We require an efficient realization of the following proof system, given public key pk, ciphertext C, and two elements g and y of some multiplicative group of order $|\mathcal{M}| = n$:

$$PoK\{m \mid C \in \mathsf{Enc}_{pk}(m) \wedge \mathsf{y} = \mathsf{g}^m)\} \tag{4}$$

In addition, we require an efficient realization of the following proof system, given public keys pk and pk' and ciphertexts C_1, C_2 and C', where C_1 and C_2 are supposed to be ciphertexts under public key pk.

$$PoK\{m \mid \exists \, m' \in \mathbb{Z}_n, \text{ s.t. } C' = \mathsf{Enc}_{pk'}(m') \wedge C_2 \in (C_1 \cdot \mathsf{Enc}_{pk}(m))^{m'}\} \tag{5}$$

- **Verifiable Decryption:** We require an efficient realization of the following proof system, given public keys pk and pk' and ciphertexts C, C_1 and C_2, where C is supposed to be ciphertext under public key pk and C_1 and C_2 are supposed to be ciphertexts under public key pk':

$$P\left\{ \begin{array}{l} \exists \, m, m' \in \mathbb{Z}_n, m'', sk \text{ s.t. } m = \mathsf{Dec}_{sk}(C) \wedge (pk, sk) \in \mathsf{KVal} \\ \wedge \, C_1 \in (C_2 \cdot \mathsf{Enc}_{pk'}(m'))^{m''} \wedge \mathsf{y} = \mathsf{g}^{m'} \wedge m \cdot m'' = 1 \bmod n \end{array} \right\} \tag{6}$$

2.3 Efficient Instantiation Using Camenisch-Shoup Encryption

The above encryption scheme can be efficiently instantiated with just the semantically secure version of Camenisch-Shoup Encryption [9].

- $\mathsf{Setup}(\kappa)$ generates public parameter $\mathsf{par} = (g, n)$, where n is safe RSA modulo, i.e. $n = p_1 \cdot p_2$, $p_1 = 2p_1' + 1$, $p_2 = 2p_2' + 1$, and p_1, p_2, p_1' and p_2' are all primes, and g is of order $p_1'p_2'$. Let $h = n + 1$ and $n' = p_1'p_2'$. The message space $\mathcal{M}$ is the additive group $\mathbb{Z}_n$.
- $\mathsf{KGen}(\mathsf{par})$ picks random x in $[0, \frac{n}{4}]$, computes $y \leftarrow g^x$, and sets $pk = y$ and $sk = x$.
- $\mathsf{Enc}_y(m)$ for $m \in \mathbb{Z}_n$, picks random $r \in [0, \frac{n}{4}]$, and outputs a ciphertext $C = (u, e) = (g^r, y^r h^m)$.
- $\mathsf{Dec}_x(u, e)$ computes $\hat{m} \leftarrow (e/u^x)^2$. If $\hat{m} \notin \langle h \rangle$ (i.e. if n does not divides $\hat{m} - 1$), it rejects the ciphertext. Otherwise, it sets $\hat{m}' \leftarrow \frac{\hat{m}-1}{n}$ (over integers), computes $\gamma \leftarrow \beta^{-1} \bmod n$, and outputs $m = \hat{m}'/2 \cdot \gamma \bmod n$

Semantic security of this encryption holds under the Composite Residuosity Assumption on $\mathbb{Z}_{n^2}^*$ [9]. It also satisfies all other properties listed in Section 2.2.

- This encryption scheme is additively homomorphic. The corresponding operation on the ciphertext is a pair-wise multiplication of the two components, i.e. if $C_1 = (u_1, e_1)$ and $C_2 = (u_2, e_2)$ then $C_1 \cdot C_2 = (u_1 \cdot u_2, e_1 \cdot e_2)$. Similarly $C_1/C_2 = (u_1/u_2, e_1/e_2)$ and $(C_1)^a = ((u_1)^a, (e_1)^a)$.

- All the proof systems listed in Section 2.2 can be realized for this encryption scheme by proof systems requiring $O(1)$ exponentiations from each party. We defer description of these proofs to the full version of this paper, but all these proof systems are very similar to the verifiable encryption proof system given by Camenisch-Shoup in [9]. Such HVZK proofs can be converted to ZK proof systems using known compilation techniques, or into Non-Interactive ZK using the Fiat-Shamir heuristic in the Random Oracle Model. We note that the proof systems resulting from the last compilation are as efficient as the underlying HVZK proof systems, and they remain zero-knowledge and simulation sound under parallel composition.

3   Construction of an OPRF Protocol

We show the construction of a secure computation protocol of the ideal OPRF functionality for the PRF defined in the previous section, using the additively homomorphic encryption scheme with verifiable encryption, decryption, and other useful properties listed in Section 2.2. This protocol is illustrated in Figure 1, with all the proof systems denoted non-interactively for notational simplicity.

Theorem 1. *Assuming hardness of factoring of safe RSA moduli, a semantically secure encryption scheme on $\mathbb{Z}_n$ which satisfies the properties listed in Section 2.2, and assuming that each proof (of knowledge) system in Figure 1 is zero-knowledge and (strong) simulation-sound, the protocol in Figure 1 is a secure computation protocol for functionality $\mathcal{F}_{\mathsf{OPRF}}$.*

Proof. Constructing an ideal-world sender SIM_s *from a malicious real sender* S^*:
First, we show the construction of the ideal-world sender SIM_s which interacts with the real world sender S^* and the ideal functionality $\mathcal{F}_{\mathsf{OPRF}}$. SIM_s proceeds as follows:

- If S^* succeeds in the proof π_1, then SIM_s runs the extractor algorithm for π_1 with S^* to extract k, s.t. $\mathsf{pk} = \mathsf{g}^k$.
- Then SIM_s simulates the real-world receiver R as follows:
  1. $(pk_r, sk_r) \leftarrow \mathsf{KGen}$
  2. $a \leftarrow_R \mathbb{Z}_n^*, C_a^{(r)} \leftarrow \mathsf{Enc}_{pk_r}(a)$
  3. $\beta \leftarrow_R \mathbb{Z}_n^*, C_\beta^{(s)} \leftarrow \mathsf{Enc}_{pk_s}(\beta)$
  4. $C_\beta^{(s)} \leftarrow \mathsf{Enc}_{pk_s}(\beta)$
  5. Send $(pk_r, C_a^{(r)}, C_\beta^{(s)})$ and simulate the proof π_2.
- If the proof π_3 verifies, then SIM_s sends k to $\mathcal{F}_{\mathsf{OPRF}}$. Note that $\mathcal{F}_{\mathsf{OPRF}}$ on SIM_s's input k and ideal-world receiver $\bar{\mathsf{R}}$'s input x outputs $f_k(x)$ to $\bar{\mathsf{R}}$.

Let $\mathcal{Z}$ be a distinguisher that controls the sender S^*, feeds the input of the receiver R, and also sees the output of R. Now we argue that $\mathcal{Z}$'s view in the real world (S^*'s view + R's output) and its view in the idea world (S^*'s view + ideal receiver $\bar{\mathsf{R}}$'s output) are indistinguishable. This is done by showing a series of games $\mathsf{Game}_0, \ldots, \mathsf{Game}_5$, each interacting with $\mathcal{Z}$, where each Game_{i+1} modifies Game_i slightly, and arguing that $\mathcal{Z}$'s views in Game_i and Game_{i+1} are indistinguishable, where Game_0 runs S^* together with the real receiver R's protocol, while Game_5 runs the above simulator SIM_s (with

Common input: $(n, \mathsf{g}, \mathsf{pk}, \mathsf{par})$

S's private input: k, s.t. $\mathsf{g}^k = \mathsf{pk}$ R's private input: x

$(pk_s, sk_s) \leftarrow \mathsf{KGen}, C_k^{(s)} \leftarrow \mathsf{Enc}_{pk_s}(k)$

$\pi_1 \leftarrow PoK \left\{ k \left| \begin{array}{l} C_k^{(s)} \in \mathsf{Enc}_{pk_s}(k) \\ \mathsf{pk} = \mathsf{g}^k \end{array} \right. \right\}$

$\xrightarrow{\quad pk_s, C_k^{(s)}, \pi_1 \quad}$

If π_1 verifies, then $(pk_r, sk_r) \leftarrow \mathsf{KGen}$
$a \leftarrow_R \mathbb{Z}_n^*, \; C_a^{(r)} \leftarrow \mathsf{Enc}_{pk_r}(a)$
$C_\beta^{(s)} \leftarrow \left(C_k^{(s)} \cdot \mathsf{Enc}_{pk_s}(x) \right)^a$

$\pi_2 \leftarrow PoK \left\{ x \left| \begin{array}{l} \exists\, a, \text{ s.t. } C_a^{(r)} \in \mathsf{Enc}_{pk_r}(a) \\ C_\beta^{(s)} \in \left(C_k^{(s)} \cdot \mathsf{Enc}_{pk_s}(x) \right)^a \end{array} \right. \right\}$

$\xleftarrow{\quad pk_r, C_a^{(r)}, C_\beta^{(s)}, \pi_2 \quad}$

If π_2 verifies, then $\beta \leftarrow \mathsf{Dec}_{sk_s}(C_\beta^{(s)})$
If $gcd(n, \beta) \neq 1$, send $\perp$ to R and abort
$\gamma \leftarrow (\beta)^{-1} \bmod n, \; \sigma_s \leftarrow_R \mathbb{Z}_n, \; v_s \leftarrow \mathsf{g}^{\sigma_s}$
$C_{\sigma_r}^{(r)} \leftarrow (C_a^{(r)})^\gamma \cdot \mathsf{Enc}_{pk_r}(-\sigma_s)$

$\pi_3 \leftarrow P \left\{ \begin{array}{l} \exists\, \beta, \sigma_s, \gamma, sk_s, \text{ s.t.} \\ \beta = \mathsf{Dec}_{sk_s}(C_\beta^{(s)}) \\ (pk_s, sk_s) \in \mathsf{KVal} \\ C_{\sigma_r}^{(r)} \in (C_a^{(r)})^\gamma \cdot \mathsf{Enc}_{pk_r}(-\sigma_s) \\ v_s = \mathsf{g}^{\sigma_s}, \; \beta \cdot \gamma = 1 \bmod n \end{array} \right\}$

$\xrightarrow{\quad v_s, C_{\sigma_r}^{(r)}, \pi_3 \quad}$

Output $\perp$ if receiving $\perp$ from S, or if π_3 fails.
$\sigma_r \leftarrow \mathsf{Dec}_{sk_r}(C_{\sigma_r}^{(r)}), \; v_r \leftarrow \mathsf{g}^{\sigma_r}$, output $v_s \cdot v_r$

Fig. 1. Construction of an OPRF Protocol

oracle access to S^*), which simulates an ideal world sender, the ideal functionality $\mathcal{F}_{\mathsf{OPRF}}$, and the ideal receiver $\bar{\mathsf{R}}$.

Game$_1$: Same as Game$_0$ except that instead of proving π_2, R simulates it. By zero-knowledge of the π_2 proofs system, $\mathcal{Z}$'s views in Game$_0$ and Game$_1$ are indistinguishable.

Game$_2$: Same as Game$_1$ except that (a.) If S^* succeeds in the proof π_1, then Game$_2$ runs the extractor algorithm for π_1 with S^* to extract k, s.t. $\mathsf{pk} = \mathsf{g}^k$; and (b) If the proof π_3 verifies, then Game$_2$ outputs $f_k(x) = \mathsf{g}^{1/(k+x)}$ as its final output (or $\perp$ if $gcd(k+x, n) \neq 1$). Note that Game$_2$ knows both k and x. For (a), by strong simulation soundness of the proof system π_1, Game$_2$ extracts k with non-negligible probability. For (b), by simulation soundness of proof system π_3, $\mathcal{Z}$'s views in Game$_1$ and Game$_2$ are indistinguishable.

Game$_3$: Same as Game$_2$ except that as long as $gcd(k + x, n) = 1$, Game$_3$ does the following:

1. $(pk_r, sk_r) \leftarrow \mathsf{KGen}$
2. $\beta \leftarrow_R \mathbb{Z}_n^*, \; C_\beta^{(s)} \leftarrow \mathsf{Enc}_{pk_s}(\beta)$
3. $a \leftarrow \beta/(k + x), \; C_a^{(r)} \leftarrow \mathsf{Enc}_{pk_r}(a)$
4. Simulate the proof π_2.

Note that the probability that $gcd(k + x, n) \neq 1$ is negligible if factoring safe RSA moduli is hard, and if $gcd(k + x, n) = 1$ then the tuple $(pk_r, C_a^{(r)}, C_\beta^{(s)})$ is distributed identically in Game$_2$ and Game$_3$, so $\mathcal{Z}$'s view of these two games are indistinguishable.

Game$_4$: Same as Game$_3$ except that in line 3 above, value a is replaced by random $a' \in \mathbb{Z}_n^*$. We claim that by semantic security of the encryption scheme, $\mathcal{Z}$'s view in Game$_3$ and Game$_4$ are indistinguishable. A reduction Red can be constructed as follows. Getting the challenger's public key pk^*, it sets $pk_r = pk^*$ and follows Game$_3$ except that when C_a is to be computed, it sends (a, a') to the challenger and gets back the challenge C^*. It sets $C_a^{(r)} \leftarrow C^*$, and continue following Game$_3$. If $\mathcal{Z}$ distinguishes Game$_3$ and Game$_4$, then Red breaks the semantic security of the encryption scheme.

Game$_5$: Game$_5$ is the ideal world game between SIM$_s$ (with access to S^*), $\mathcal{F}_{\mathsf{OPRF}}$, and the ideal world receiver $\bar{\mathsf{R}}$. Instead of computing $v = f_k(x)$ as the last step of Game$_4$ (this modification is shown in the description of Game$_2$ Step b), in Game$_5$, let SIM$_s$ follow the protocol in Game$_4$ until $f_k(x)$ is to be computed, then send k to $\mathcal{F}_{\mathsf{OPRF}}$. $\mathcal{F}_{\mathsf{OPRF}}$ computes $v = f_k(x)$ on k given by SIM$_s$ and x given by $\bar{\mathsf{R}}$, and sends v to $\bar{\mathsf{R}}$. It is easy to see that $\mathcal{Z}$'s view in Game$_4$ and Game$_5$ are identical.

Constructing an ideal-world receiver SIM$_r$ *from a malicious real-world receiver* R*:
We first describe the construction of SIM$_r$:

- SIM$_r$ picks $(pk_s, sk_s) \leftarrow_R$ KGen, random k' in $\mathcal{M}$, sets $C_k^{(s)} \leftarrow \mathsf{Enc}_{pk_s}(k')$, sends pk_s and $C_k^{(s)}$ to R*, and simulate the proof π_1.
- If the proof π_2 verifies, SIM$_r$ runs the extractor algorithm of π_2 with R* to extract x and sends it to $\mathcal{F}_{\mathsf{OPRF}}$.
- Getting $v = f_k(x)$ from $\mathcal{F}_{\mathsf{OPRF}}$, which computes v on ideal-world sender $\bar{\mathsf{S}}$'s input k and SIM$_r$'s input x, SIM$_r$ does the following:
  1. If $f_k(x) = 1$, then SIM$_r$ sends $\perp$ to R* and aborts.
  2. $\sigma_r \leftarrow_R \mathbb{Z}_n$
  3. $v_s \leftarrow v/g^{\sigma_r}$
  4. $C_{\sigma_r}^{(r)} \leftarrow \mathsf{Enc}_{pk_r}(\sigma_r)$
  5. send $(v_s, C_{\sigma_r}^{(r)})$ and simulate the proof π_3.

Let $\mathcal{Z}$ be a distinguisher that controls the receiver R*, feeds the input of the sender S, and also sees the output of S. We still show a series of games to argue that the environment $\mathcal{Z}$'s view in the real protocol and its view in the ideal world protocol is indistinguishable. Let Game$_0$ be the protocol executed by the real world sender, and let Game$_4$ be the ideal world game.

Game$_1$: Game$_1$ is the same as Game$_0$ except that S instead of proving π_1 and π_3, it simulates the two proofs. By zero-knowledge (simulatability) of these proof systems, $\mathcal{Z}$'s view in Game$_0$ and its view in Game$_1$ are indistinguishable.

Game$_2$: Game$_2$ is the same as Game$_1$ except that if the proof π_2 verifies, Game$_2$ extracts x from R* (The probability that Game$_2$ extracts x is non-negligible because of simulation soundness of the proof system π_2). $\mathcal{Z}$'s views in Game$_1$ and Game$_2$ are indistinguishable.

Game$_3$: Game$_3$ is the same as Game$_2$ except it does the following after extracting x:
1. $v = f_k(x)$; if $v = 1$, send $\bot$ to R* and abort.
2. $\sigma_r \leftarrow_R \mathbb{Z}_n$, $v_s \leftarrow v/g^{\sigma_r}$
3. $C_{\sigma_r}^{(r)} \leftarrow \text{Enc}_{pk_r}(\sigma_r)$
4. send $(v_s, C_{\sigma_r}^{(r)})$ and simulate the proof π_3.

Note that $f_k(x) = 1$ if and only if $gcd(k + x, n) \neq 1$, and if $gcd(k + x, n) \neq 1$, then $\beta = a(k + x)$ for any a is not co-prime with n, i.e. $gcd(\beta, n) \neq 1$, so the real sender (as well as Game$_2$) in this case will also send $\bot$ to R* and abort. If $v \neq 1$, the pair $(v_s, C_{\sigma_r}^{(r)})$ distributes identically in Game$_2$ and Game$_3$. Therefore, $\mathcal{Z}$'s view in Game$_2$ and Game$_3$ are identical.

Game$_4$: In Game$_4$, we let the simulator SIM$_r$ to follow the protocol in Game$_3$ except that when v is to be computed, SIM$_r$ sends the extracted x to the ideal functionality $\mathcal{F}_{\text{OPRF}}$ which also gets input k from the ideal world sender $\bar{S}$, and gets the value $v = f_k(x)$ from the functionality $\mathcal{F}_{\text{OPRF}}$ instead. Then SIM$_r$ continues following the protocol in Game$_3$. It is easy to see that $\mathcal{Z}$'s view in Game$_2$ and Game$_3$ are identical, and Game$_3$ is in fact the ideal world game among ideal sender S, the ideal functionality $\mathcal{F}_{\text{OPRF}}$ and the ideal receiver SIM$_r$ who has oracle access to R* and simulates R* in the ideal world.

Extension to secure computation of $\mathcal{F}_{\text{OPRF}}$ on *committed* inputs. Note that in our implementation of the OPRF functionality the sender's input k is already committed in the key pk $= g^k$, but only a slight modification to the protocol is needed if we extend the OPRF functionality so that the receiver's input x is committed as well. The receiver can commit to x using Pedersen commitment [27] in group $\langle g \rangle$ of order n, i.e. $\text{Com}_x = g^x h^{r_x}$ for random $h \in \langle g \rangle$ specified in the CRS (or picked by the sender) and random $r_x \in \mathbb{Z}_n$. The proof system π_2 in receiver's first step must then be extended to a proof of knowledge of not just x but also r_x s.t. $\text{Com}_x = g^x h^{r_x}$. Using the hiding and binding properties of Pedersen commitment it is not difficult to extend our proof of security of the basic OPRF protocol in Figure 1 to this case. Finally, such as proof system is easy to realize because the order of group $\langle g \rangle$ is the same n which acts on the plaintext in encryption $\text{Enc}_{pk}(x)$. Using techniques for proving equality between exponents in groups of different orders, e.g. [5,7], one can extend it to accommodate commitments to x using more standard groups of smaller (and prime) order.

Parallel composition and re-using sender's first message. This OPRF scheme can be composed sequentially because it is a secure computation protocol. This directly implies security of the adaptive OT protocol based on such OPRF, as explained in Section 4 below. However, by inspection of the proof, using standard hybrid arguments one can argue that this OPRF scheme can also be composed in parallel, provided that the ZK proof systems it uses remain zero-knowledge and simulation-sound under parallel composition. The resulting protocol is a secure computation of the "parallel OPRF" functionality, where the receiver enters a sequence of data items $(x_1, \ldots, x_t)$ and gets a sequence of PRF values $f_k(x_1), \ldots, f_k(x_n)$ in return. The secure computation for this parallel OPRF functionality enables constant-round secure computation protocol for set intersection, as we explain in Section 5.

It is also easy to see that one can reduce the round complexity and/or the sender's computation's time in both of these applications if the sender re-uses the same pk_s, sk_s, and $C_k^{(s)}$ values in each instance of the OPRF subprotocol. One can think of this as a case of sender's re-using the same randomness when executing the first step of the protocol. Since we cannot prevent a malicious sender from doing so anyway, such modified protocol remains secure against a malicious sender. For malicious receiver, one can see by inspection of the proof of Theorem 1 that this re-use does not change anything in the security proof. Note moreover that in the ROM model one can realize the proof π_1 non-interactively as a tuple of a few group elements. Therefore in ROM the sender's first message can be included as a common input to the protocol, which results in a 2-round OPRF protocol in ROM.

4 Adaptive Oblivious Transfer from an OPRF Scheme

In this section, we show a construction of t-out-of-N OT protocol from an OPRF protocol, where the sender takes N messages $(m_1, \ldots, m_N)$ as input, while receiver takes t indices $(i_1, \ldots, i_t)$ as input. At the end of the protocol, receiver gets $\{m_{i_j}\}_{j=1,\ldots,t}$, while sender gets nothing. An OT protocol is adaptive if receiver can query on indices $i_1, \ldots, i_t$ adaptively one after another. Using an OPRF protocol, we show below the construction of an adaptive OT protocol and in Figure 2.

- Let (g, n) be the common input.
- Sender picks random $k \in \mathbb{Z}_n$, sets $\mathsf{pk} = g^k$, and then computes $c_i = m_i \cdot f_k(i)$ for every $i = 1, \ldots, N$. It sends pk, $\{c_i\}_{i=1,\ldots,N}$ to the receiver.
- For $j = 1, \ldots, t$:
  - Sender and receiver interacts in an OPRF with sender's input k and receiver's input i_j;
  - Receiver gets $v_{i_j} = f_k(i_j)$ and recovers m_{i_j} by computing c_{i_j}/v_{i_j}.

As we point out at the end of Section 3, the OPRF protocol of that section takes only two-rounds in ROM if the sender re-uses the first message of the OPRF protocol in each instance. Therefore the combination of these two protocols results in an optimal two-round adaptive OT.

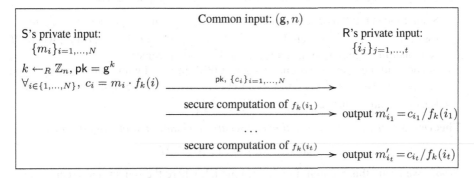

Fig. 2. Adaptive OT Protocol from an OPRF Protocol

Theorem 2. *The above construction of OT is a secure computation of the ideal t-out-of-N OT functionality*

$$\mathcal{F}_{\mathsf{OT}}\left(\{m_i\}_{i=1,\ldots,N}, \{i_j\}_{j=1,\ldots,t}\right) = \left(\perp, \{m_{i_j}\}_{j=1,\ldots,t}\right)$$

Proof. We argue this theorem in the hybrid model using the secure computation of OPRF as a blackbox.

Constructing ideal world sender SIM_s *from a malicious sender* S^* *in the real world:* Getting k from S^*, SIM_s computes $v_i = f_k(i)$ for each $i = 1, \ldots, N$, and sends $\{\bar{m}_i = c_i/v_i\}_{i=1,\ldots,N}$ to $\mathcal{F}_{\mathsf{OT}}$, which on $\{\bar{m}_i\}_{i=1,\ldots,N}$ from SIM_s and $\{i_j\}_{j=1,\ldots,t}$ from the ideal world receiver R, outputs $\{\bar{m}_{i_j}\}_{j=1,\ldots,t}$ to the receiver. In the real world, what the receiver R outputs is $\{c_{i_j}/f_k(i_j)\}_{j=1,\ldots,t}$, which is also $\{\bar{m}_{i_j}\}_{j=1,\ldots,t}$. S^* learns nothing either interacting with the real world receiver R via the ideal functionality $\mathcal{F}_{\mathsf{OPRF}}$ or interacting with SIM_s in the ideal world. Therefore, the environment $\mathcal{Z}$'s views in the real world and ideal world are indistinguishable.

Constructing ideal world receiver SIM_r *from a malicious receiver* R^* *in the real world:* SIM_r first sends to R^* a random pk' in $\langle g \rangle$ as well as a tuple of $\{r_i\}_{i=1,\ldots,N}$ where each r_i is random in the range of $\mathcal{F}_{\mathsf{OPRF}}$. On getting each i_j for $j = 1, \ldots, t$, SIM_r sends $\{i_j\}_{j=1,\ldots,t}$ to $\mathcal{F}_{\mathsf{OT}}$ and gets back $\{m_{i_j}\}_{j=1,\ldots,t}$. Then SIM_r sends $\{r_{i_j}/m_{i_j}\}_{j=1,\ldots,t}$ to R^*. As $\mathcal{F}_{\mathsf{OPRF}}$ is a pseudorandom function, R^*'s view in the real protocol ($\{c_i\}_{i=1,\ldots,N}$, $\{f_k(i_j)\}_{j=1,\ldots,t}$) is indistinguishable from ($\{r_i\}_{i=1,\ldots,n}$, $\{r_{i_j}/m_{i_j}\}_{j=1,\ldots,t}$) for each r_i chosen at random in the range of $\mathcal{F}_{\mathsf{OPRF}}$. Since both the real world sender S and ideal world sender $\bar{S}$ output a $\perp$, the environment $\mathcal{Z}$'s views in the real world and ideal world are indistinguishable.

5   Secure Computation of Set Intersection from an OPRF Scheme

Using the Hazay-Lindell construction [19] one can easily convert a secure computation protocol for OPRF functionality into secure computation of the set intersection problem. Let $M_s = \{m_i^{(s)}\}_{i=1,\ldots,N}$ and $M_r = \{m_i^{(r)}\}_{j=1,\ldots,N}$ denote, respectively, the sender's and the receiver's data sets. The secure set intersection protocol should allow the receiver to compute $M_s \cap M_r$ while the sender gets nothing from the interaction. Using a secure computation protocol for parallel OPRF functionality for the PRF from Section 3), the construction goes as follows, on common inputs (g, n):

- Sender picks random $k \in \mathbb{Z}_n$ and sets $\mathsf{pk} = g^k$. Then it computes $v_i^{(s)} = f_k(m_i^{(s)})$ for every $i = 1, \ldots, N$, sets $V_s = \{v_i^{(s)}\}_{i=1,\ldots,N}$ and sets $V_s' = \Pi(V_s)$, where Π is a random permutation. It sends pk and V_s' to the receiver.
- Sender and receiver interact in a parallel OPRF protocol with sender's input k and receiver's inputs $(m_1^{(r)}, \ldots, m_N^{(r)})$. Let $(v_1^{(r)}, \ldots, v_N^{(r)})$ be receiver's outputs;
- Receiver outputs the set $\left\{ m_j^{(r)} \text{ s.t. } v_j^{(r)} \in V_s' \right\}$

Theorem 3. *The above construction is a secure computation protocol for functionality*

$$\mathcal{F}_{\mathsf{SI}}(M_s, M_r) = (\perp, M_s \cap M_r)$$

Proof. We argue this theorem in a hybrid model, where the sender and receiver communicate using the ideal functionality $\mathcal{F}_{\mathsf{OPRF}}$ when they invoke the OPRF protocol.

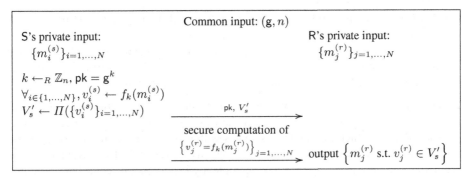

Fig. 3. Computing Set Intersection from an OPRF protocol

Constructing ideal world sender SIM_s *from a malicious sender* S^* *in the real world:*
SIM_s first gets pk and the set $V'_s = \{v_i^{(s)}\}_{i=1,\ldots,N}$ from S^*. Then when getting the
key k from S^*, SIM_s tries every possible input in the range of the PRF to reconstruct
the set $\bar{M}_s = \{\bar{m}_i^{(s)}\}_{i=1,\ldots,N}$ from $V'_s = \{v_i^{(s)}\}_{i=1,\ldots,N}$, and sends $\bar{M}_s$ to $\mathcal{F}_{\mathsf{SI}}$, which
computes $\bar{M}_s \cap M_r$ on SIM_s's input $\bar{M}_s$ and the ideal world receiver $\bar{\mathsf{R}}$'s input M_r, and
sends $\bar{M}_s \cap M_r$ to $\bar{\mathsf{R}}$. The real world receiver R in the hybrid model gets $v_i^{(s)} = f_k(m_i^{(s)})$
for every $\bar{m}_i^{(s)} \in \bar{M}_s$ as well as $v_j^{(r)} = f_k(m_j^{(r)})$ for every $j = 1,\ldots,N$. So R's output
is also $\bar{M}_s \cap M_r$. S^* learns nothing either interacting with the real world receiver R via
the ideal functionality $\mathcal{F}_{\mathsf{OPRF}}$ or interacting with SIM_s in the ideal world. Therefore,
the environment $\mathcal{Z}$'s views in the real world and ideal world are indistinguishable.

Constructing ideal world receiver SIM_r *from a malicious receiver* R^* *in the real world:*
SIM_r first sends to R^* a random pk' in $\langle g \rangle$ as well as a tuple of $\{r_i\}_{i=1,\ldots,N}$ where each
r_i is random in the range of the pseudorandom function. On getting $M_r = \{m_i^{(r)}\}_{j=1,\ldots,N}$
from the receiver R^*, it sends M_r to the ideal functionality $\mathcal{F}_{\mathsf{SI}}$, which computes $M_s \cap$
M_r on the ideal world sender S's input M_s and SIM_r's input M_r, and sends $M_s \cap M_r$
to SIM_r. For each $m_j^{(r)} \in M_r$, if $m_j^{(r)} \in M_s \cap M_r$, then SIM_r sets $\bar{v}_j^{(r)} = r_i$ for some
(previously not picked) i; otherwise it sets $\bar{v}_j^{(r)}$ at random in the range of the pseudoran-
dom function. SIM_r sends $\{\bar{v}_j^{(r)}\}_{j=1,\ldots,N}$ to R^*. Because of $\mathcal{F}_{\mathsf{OPRF}}$ is a pseudorandom
function, R^*'s view in the real protocol $(\{\Pi(v_i^{(s)}\}_{i=1,\ldots,N}), \{\bar{v}_i^{(r)}\}_{i=1,\ldots,N})$ is indistinguishable from $(\{r_i\}_{i=1,\ldots,n}, \{\bar{v}_i^{(r)}\}_{i=1,\ldots,N})$ in the simulated game above. Since both
the real world sender S and ideal world sender $\bar{\mathsf{S}}$ output a $\perp$, the environment $\mathcal{Z}$'s views
in the real world and ideal world are indistinguishable.

Extension to computing on committed inputs: As in the OPRF protocol in Section
3, it is easy to extend this protocol so both parties execute on committed inputs. The
receiver can be forced to execute on committed inputs if we replace the basic OPRF
functionality by the committed OPRF, as sketched as the end of Section 3. And if
$\{\mathsf{Com}_i\}_{i=1,\ldots,N}$ are Pedersen commitments on the sender's inputs in the same group
$\langle g \rangle$ of order n, then it is easy to extend the sender's first message with a ZK proof of
knowledge of values m_i, for each i, s.t. $v_i^{(s)} = f_k(m_i) = g^{1/(k+m_i)}$ where $pk = g^k$ and
m_i is committed in Com_i. Note that since we use the same commitment for sender's and

receiver's inputs, both parties can securely (but not fairly) compute the set intersection if they run the same protocol twice, with the roles reversed. Another use of computing on committed data is for the sender to be able to verify that the receiver holds some authorization on the values it enters into the computation, e.g. in the form of a signature, or an unlinkable credential [6], on the commitments to these values.

Extension to transfer of associated data, or "Index OT": In addition to letting the receiver discover any item $m_i^{(s)}$ which it shares with the sender, the receiver can also get some data d_i which the sender associates with this item. If the receiver has only a single item then this protocol problem is known as "index OT". The advantage of the protocol given here is that it has $O(N)$ complexity if both parties contribute N items. Assume that each d_i is an element of group $\langle g \rangle$. The sender generates two PRF keys k_1 and k_2 and publishes $(f_{k_1}(m_i^{(s)}), f_{k_2}(m_i^{(s)}) * d_i)$ for each i. For each receiver's item $m_j^{(r)}$ the two parties engage in *two* instances of the OPRF protocol, on two sender's keys k_1 and k_2. The receiver then uses $f_{k_1}(m_j^{(r)})$ to decide if there exists $m_i^{(s)} = m_j^{(r)}$, and $f_{k_2}(m_j^{(r)})$ to retrieve the data d_i associated with $m_i^{(s)}$.

5.1   Efficiency Estimation of the Set Intersection Protocol

It is interesting to compare the set intersection protocol resulting from the oblivious PRF protocol of Figure 1 with the FNP protocol by Friedman et al. [16], both in the honest-but-curious and the malicious models. We compare only straightforward implementations of both schemes, and summarize this comparison in table 4. Clearly, it is possible that either algorithm can be further optimized.

Let N_s, N_r denote the number of entries in, respectively, the sender's and the receiver's data sets, and let k be the number of bits needed for representing each entry. Assume that the FNP protocol is implemented using ElGamal encryption over a 160-bit subgroup of $\mathbb{Z}_{p'}^*$ for a 1024-bit prime p'. Note that multiplications in the OPRF protocol in Figure 1 are either modulo a 2048-bit Paillier modulus n^2 or modulo the first prime p s.t. n divides $p - 1$. Therefore if m is the cost of a single multiplication (or squaring) modulo a 1024-bit modulus then we can assume that these two different mults in our scheme cost about 3m and m, respectively.

In the honest-but-curious model the main bottleneck of the FNP protocol is the oblivious evaluation of the receiver's encrypted polynomial, with N_r coefficients, on N_s points in the sender's database. Using Horner's rule, this would take $2N_sN_r$ of k-bit exponentiations, i.e. $3kN_sN_r$m. In the malicious model, using standard commitments and zero-knowledge proofs of arithmetic relationships between committed values, it seems that each k-bit exponentiation would have to be replaced by at least two 160-bit exponentiations, so the total cost would grow by at least a factor of $320/k$.

In the set intersection protocol of Figure 3, using the OPRF of Figure 1, the sender's costs in the honest-but-curious model consist of N_s evaluations of the PRF (N_s fixed-base 1000-bit exponentiations, contributing $500N_s$m), N_r Paillier decryptions (N_r 1000-bit exponentiations mod n^2, contributing $4500N_r$m), and N_r computations of v_s values ($500N_r$m) and $C_{\delta_r}^{(r)}$ values (this is a mixture of variable and fixed-base exponentiations which we estimate to amount to under $13000N_r$m), for the total cost of $18000N_r$m. For the receiver we have N_r computations of $C_a^{(r)}$ and $C_\beta^{(s)}$ values

	FNP [16]	our protocol
honest-but-curious model	$3kN_sN_r$m	$(500N_s + 32000N_r)$m
malicious model	$\geq 960N_sN_r$m	$(1000N_s + 64000N_r)$m

Fig. 4. Comparison of Efficiency between the FNP protocol [16] and the one proposed here

and N_r Paillier decryptions and computations of v_r values. This is a mixture of variable and fixed-base exponentiations modulo n^2 and p, for the estimated total cost of $14000N_r$m. Thus we estimate the total cost of this protocol in the passive model as $(500N_s + 32000N_r)$m. In the malicious model this cost grows by only a factor of 2.

Summarizing these estimations, it seems that if N_s and N_r are comparable then the new protocol is faster than FNP in the honest-but-curious model for N_s on the order of $11000/k$, while in the malicious model the new protocol should be faster for $N_s \geq 67$.

References

1. Aiello, W., Ishai, Y., Reingold, O.: Priced oblivious transfer: How to sell digital goods. In: Pfitzmann, B. (ed.) EUROCRYPT 2001. LNCS, vol. 2045, p. 119. Springer, Heidelberg (2001)
2. Belenkiy, M., Camenisch, J., Chase, M., Kohlweiss, M., Lysyanskaya, A., Shacham, H.: Delegatable anonymous credentials. Cryptology ePrint Archive, Report 2008/428 (2008)
3. Boldyreva, A.: Threshold signatures, multisignatures and blind signatures based on the gap-diffie-hellman-group signature scheme. In: Desmedt, Y.G. (ed.) PKC 2003. LNCS, vol. 2567, pp. 31–46. Springer, Heidelberg (2002)
4. Boneh, D., Boyen, X.: Short signatures without random oracles. In: Cachin, C., Camenisch, J.L. (eds.) EUROCRYPT 2004. LNCS, vol. 3027, pp. 56–73. Springer, Heidelberg (2004)
5. Boudot, F., Traoré, J.: Efficient publicly verifiable secret sharing schemes with fast or delayed recovery. In: Varadharajan, V., Mu, Y. (eds.) ICICS 1999. LNCS, vol. 1726, pp. 87–102. Springer, Heidelberg (1999)
6. Camenisch, J.L., Lysyanskaya, A.: A signature scheme with efficient protocols. In: Cimato, S., Galdi, C., Persiano, G. (eds.) SCN 2002. LNCS, vol. 2576, pp. 268–289. Springer, Heidelberg (2003)
7. Camenisch, J.L., Michels, M.: Separability and efficiency for generic group signature schemes (Extended abstract). In: Wiener, M. (ed.) CRYPTO 1999. LNCS, vol. 1666, p. 413. Springer, Heidelberg (1999)
8. Camenisch, J.L., Neven, G., Shelat, A.: Simulatable adaptive oblivious transfer. In: Naor, M. (ed.) EUROCRYPT 2007. LNCS, vol. 4515, pp. 573–590. Springer, Heidelberg (2007)
9. Camenisch, J.L., Shoup, V.: Practical verifiable encryption and decryption of discrete logarithms. In: Boneh, D. (ed.) CRYPTO 2003. LNCS, vol. 2729, pp. 126–144. Springer, Heidelberg (2003)
10. Chaum, D.: Blind signatures for untraceable payments. In: CRYPTO (1982)
11. Cheon, J.H.: Security analysis of the strong diffie-hellman problem. In: Vaudenay, S. (ed.) EUROCRYPT 2006. LNCS, vol. 4004, pp. 1–11. Springer, Heidelberg (2006)
12. Chor, B., Gilboa, N., Naor, M.: Private information retrieval by keywords. Cryptology ePrint Archive, 1998/003 (1998)
13. Dodis, Y., Yampolskiy, A.: A verifiable random function with short proofs and keys. In: Vaudenay, S. (ed.) PKC 2005. LNCS, vol. 3386, pp. 416–431. Springer, Heidelberg (2005)

14. Even, S., Goldreich, O., Lempel, A.: A randomized protocol for signing contracts. Communications of ACM 28(6) (1985)

15. Freedman, M.J., Ishai, Y., Pinkas, B., Reingold, O.: Keyword search and oblivious pseudorandom functions. In: Kilian, J. (ed.) TCC 2005. LNCS, vol. 3378, pp. 303–324. Springer, Heidelberg (2005)

16. Freedman, M.J., Nissim, K., Pinkas, B.: Efficient private matching and set intersection. In: Cachin, C., Camenisch, J.L. (eds.) EUROCRYPT 2004. LNCS, vol. 3027, pp. 1–19. Springer, Heidelberg (2004)

17. Goldreich, O., Goldwasser, S., Micali, S.: How to construct random functions. J. ACM, 33(4) (1986)

18. Green, M., Hohenberger, S.: Blind identity-based encryption and simulatable oblivious transfer. In: Kurosawa, K. (ed.) ASIACRYPT 2007. LNCS, vol. 4833, pp. 265–282. Springer, Heidelberg (2007)

19. Hazay, C., Lindell, Y.: Efficient protocols for set intersection and pattern matching with security against malicious and covert adversaries. In: Canetti, R. (ed.) TCC 2008. LNCS, vol. 4948, pp. 155–175. Springer, Heidelberg (2008)

20. Ishai, Y., Kilian, J., Nissim, K., Petrank, E.: Extending oblivious transfers efficiently. In: Boneh, D. (ed.) CRYPTO 2003. LNCS, vol. 2729, pp. 145–161. Springer, Heidelberg (2003)

21. Naor, M., Pinkas, B.: Oblivious transfer and polynomial evaluation. In: STOC (1999)

22. Naor, M., Pinkas, B.: Oblivious transfer with adaptive queries. In: Wiener, M. (ed.) CRYPTO 1999. LNCS, vol. 1666, p. 573. Springer, Heidelberg (1999)

23. Naor, M., Pinkas, B.: Efficient oblivious transfer protocols. In: SODA (2001)

24. Naor, M., Pinkas, B.: Oblivious polynomial evaluation. SIAM J. Comput. 35(5) (2006)

25. Naor, M., Reingold, O.: Number-theoretic constructions of efficient pseudo-random functions. J. ACM 51(2) (2004)

26. Paillier, P.: Public-key cryptosystems based on composite degree residuosity classes. In: Stern, J. (ed.) EUROCRYPT 1999. LNCS, vol. 1592, p. 223. Springer, Heidelberg (1999)

27. Pedersen, T.P.: Non-interactive and information-theoretic secure verifiable secret sharing. In: Feigenbaum, J. (ed.) CRYPTO 1991. LNCS, vol. 576, pp. 129–140. Springer, Heidelberg (1992)

28. Rabin, M.O.: How to exchange secrets by oblivious transfer. Technical report, Harvard University (1981)

Towards a Theory of Extractable Functions

Ran Canetti[1,*] and Ronny Ramzi Dakdouk[2,**]

[1] Tel Aviv University, Tel Aviv, Israel
canetti@tau.ac.il
[2] Yale University, New Haven, CT
dakdouk@cs.yale.edu

Abstract. Extractable functions are functions where any adversary that outputs a point in the range of the function is guaranteed to "know" a corresponding preimage. Here, knowledge is captured by the existence of an efficient extractor that recovers the preimage from *the internal state of the adversary*. Extractability of functions was defined by the authors (ICALP'08) in the context of perfectly one-way functions. It can be regarded as an abstraction from specific knowledge assumptions, such as the Knowledge of Exponent assumption (Hada and Tanaka, Crypto 1998).

We initiate a more general study of extractable functions. We explore two different approaches. The first approach is aimed at understanding the concept of extractability in of itself; in particular we demonstrate that a weak notion of extraction implies a strong one, and make rigorous the intuition that extraction and obfuscation are complementary notions.

In the second approach, we study the possibility of constructing cryptographic primitives from simpler or weaker ones while maintaining extractability. Results are generally positive. Specifically, we show that several cryptographic reductions are either "knowledge-preserving" or can be modified to be so. Examples include reductions from extractable weak one-way functions to extractable strong ones, from extractable pseudorandom generators to extractable pseudorandom functions, and from extractable one-way functions to extractable commitments. Other questions, such as constructing extractable pseudorandom generators from extractable one way functions, remain open.

1 Introduction

Extractability plays a central role in cryptographic protocol design and analysis. In its basic form, it relates to two-party protocols where one of the parties (a "prover") has secret input, and tries to convince the other party (a "verifier") that it holds the secret. The idea is to argue that if the verifier accepts the interaction, then the prover indeed "knows" the secret. More concretely, extractability makes the following requirement: Given access to the internals of *any* (potentially malicious) prover, it is possible to explicitly and efficiently compute the secret value as long as the verifier accepts an interaction. (Many variants of this notion exist, of course. See e.g. [12].)

---

* Supported by NSF grant CFF-0635297 and US-Israel Binational Science Foundation Grant 2006317, a European Union Marie Curie grant, and the Check Point Institute for Information Security.
** Supported by NSF grant #0331548.

The notion of **extractable functions** extends the concept of extractability to the more basic setting of computing a function. Here the task of "convincing a verifier" is replaced by "outputting a value in the range of the function". More specifically, any machine that generates a point in the range "knows" a corresponding preimage in the sense that a preimage is efficiently recoverable given the internal state of the machine.

Extractable functions were coined in [8] for the specific goal of defining extractable perfectly one-way (EPOW) functions.[1] These functions were demonstrated to have some interesting applications, such as new ways to realize Random Oracles and new three-round Zero-Knowledge arguments based on weaker assumptions than previously known. Furthermore, it was demonstrated that extractable functions can be viewed as an abstraction from specific knowledge assumptions, such as the Knowledge of Exponent (KE) assumption [16,3] or the Proof of Knowledge (POK) assumption [20], in much the same way as the notion of one-way function is an abstraction of the Discrete Log (DL) assumption.

This work attempts to initiate a more general study of extractable functions. Specifically, we address two goals: First, we try to understand exactly what extraction means and how different notions of extraction (and lack of it) are related. Second, we study the possibility of constructing complex primitives from simpler ones while preserving extractability. We note that the latter approach may help in basing cryptographic protocols that use or require specific knowledge assumptions, on a general computational notion, which in turn may be concretely realized by alternative assumptions.

Before discussing this work in more detail, we provide a high level overview of the two versions of knowledge extraction defined in [8]: interactive and noninteractive extraction.

Noninteractive extraction. Noninteractive extraction is an abstraction of specific knowledge assumptions as mentioned in the previous paragraph. Informally, there is a family of functions and the adversary gets a description of a specific function from the family, and tries to output a point in the range of this function. This function family is considered noninteractively extractable if whenever the adversary generates a point in the range, it knows a corresponding preimage. In other words, for every such adversary there is a corresponding extractor that computes a preimage from the private input of the adversary. One extreme example of extractable functions is the identity function where the output itself reveals the input. Obviously, such functions are of lesser interest to cryptographic applications than functions with computational hardness properties. On another extreme, if the function is a one-way permutation, then it is easy to output a valid image without knowing a preimage; specifically, output a random string in the range. In this work, we concentrate on functions that enjoy both properties, namely, extractability and computational hardness.

Unlike proofs of knowledge [15,2], this notion of extraction does not *require* efficient verification. In other words, the range of the function is not necessarily efficiently verifiable. Therefore, it may not be possible to decide if the adversary generates a point in the range (and consequently, knows a preimage). However, this notion guarantees the implication: If the adversary generates an image, it knows a preimage. We mention that the construction in [8] has a range that is efficiently verifiable in the presence of some auxiliary information (about the function itself).

---

[1] Informally, a probabilistic function is perfectly one-way if it hides all partial information about the input [7].

Extraction can be studied with or without auxiliary information. We would like to consider extraction in the presence of auxiliary information as this is a more useful and meaningful notion. Auxiliary information can be either dependent or independent [14] (here, the dependence is on the specific function under study). We remark that dependent auxiliary information is inseparable from independent auxiliary information when extraction is required for a single function, f. This is so because it is not possible to prevent an adversary with access to auxiliary information from receiving dependent auxiliary information, e.g., $f(x)$. Moreover, the notion of a single extractable function with auxiliary information is not realizable for one-way functions. Specifically, by the one-wayness assumption, there is no extractor for the adversary that receives $f(x)$, for a uniform x, and simply outputs it. Consequently, we relax the requirement to extraction for a family of functions, i.e., a function is chosen uniformly from the family. Indeed, the KE assumption is already formulated in terms of function families.

In this work, we focus on extraction with independent auxiliary information only. Formulating and realizing extraction with dependent auxiliary information is tricky. For instance, it is possible that $f(x)$ is hidden in the input in some clever way such that it is easy to recover $f(x)$ but not x. For example, the input of the adversary may look like $(r_1, u_1), \ldots, (r_n, u_n)$, where the ith bit of $f(x)$ is $\langle r_i, u_i \rangle$. [8] addresses this issue by restricting the dependency of auxiliary information so that only a sequence of images under f can be part of dependent auxiliary information. Moreover, Zheng and Seberry [25] follow the same approach under the notion of "sole-samplability".

Interactive extraction. This notion is geared towards probabilistic functions, and can work for single functions as well as families. In interactive extraction, the adversary engages in a 3-round game with a challenger. The objective of the game is to show that the adversary is capable of computing a function, f, on some point, x, that he chooses, but using random coins for f that the challenger chooses. In other words, the goal is to show that the adversary is capable of computing a "large" fraction of the possible images of x under f (recall f is probabilistic). In more detail, the adversary, A, sends, in the first round, a point, $y_0 = f(x, r_0)$, where x and r_0 are chosen by A. The challenger responds with random coins, r_1, in the second round and A has to send back $y_1 = f(x, r_1)$. In this setting, consistency means that y_0 and y_1 have a common preimage x. Interactive extraction means if the adversary is able to answer consistently, then it knows a common preimage. As in the noninteractive case, this form of knowledge is captured computationally by the existence of an extractor that recovers a preimage from the private input of the adversary. We emphasize that no verification of consistency is assumed to occur. The knowledge requirement states that *if the adversary is consistent*, it must know a preimage.

Unlike noninteractive extraction, interactive extraction is required to work for any function. In other words, the function is fixed once and for all, and any auxiliary information is allowed to depend on this function. Intuitively, this is realizable because the challenge in the second round forces the adversary to compute an image "online".

In interactive extraction, we focus mainly on *probabilistic* functions because for deterministic functions, this notion is equivalent to noninteractive extraction. (To use the 3-round game of interactive extraction on a deterministic function, f, view f as a probabilistic function that simply ignores the random coins, i.e. $f(x, r) = f(x)$ for any x and r.)

It is worth mentioning that noninteractive extraction can be viewed as a two-round interactive extraction analogous to the three-round extraction discussed above. Specifically, in the first round the challenger sends a random function from the family and the adversary responds with a point in the range of this function. That is, there is a fixed function, g, the challenger sends a random r, and the adversary responds with $g(x, r) = f_r(x)$.

1.1   Our Work

We approach extractable functions from two different angles.

First, we attempt to address the question: What makes a function extractable? Moreover, if a function is extractable with noticeable success, does this mean that it is extractable in a strong sense? Towards answering these questions, we show that every function satisfies either a "mild" form of obfuscation [1] or a "mild" form of extraction. In other words, lack of extractability can be viewed as inability to "reverse engineering" or obfuscatability. This is indeed what one might naïvely expect - a function is either extractable or obfuscatable, and we show that this naïve thinking is correct to some extent. We then address the second question posed at the beginning of this paragraph. We find out that for a large class of functions, notably, POW functions with auxiliary information, the answer to this question is positive.

Second, we try to construct complex extractable primitives from simpler ones. In general, extractable functions exist, e.g., the identity function. However, extractable functions are more useful in cryptographic applications if they satisfy certain hardness assumptions. Thus, in the second line of work, we address the question: Is it possible to build primitives with complex hardness properties from weaker hardness assumptions while maintaining extractable properties? For instance, suppose we have an extractable weak one-way function, can we build an extractable strong one-way function? Results indicate that answers to such questions are mostly positive.

On the first line of work. We discuss interactive extraction before noninteractive extraction.

On interactive extraction versus obfuscation. This line of work starts with an observation that extraction and obfuscation complement each other in a natural way. In other words, if a function is not extractable, then this lack of extractability is some form of obfuscation. Specifically, we call a function weakly (and interactively) extractable if for any adversary that is consistent in the interactive game with noticeable probability, there is a corresponding extractor that recovers a preimage with noticeable success. Moreover, the obfuscation mentioned previously relates to inability to "reverse engineer" an obfuscated program that produces images under the function. In other words, there is an obfuscated code that receives r as input and computes $f(x, r)$ for some x "hidden" in the obfuscated code. In more detail, we call f weakly obfuscatable if the following holds. There is an obfuscator that produces a program capable of correctly computing the function $f_x(r) = f(x, r)$ with noticeable probability, where x is chosen according to some well-spread distribution and then "hidden" in the program. Also, the program is considered obfuscated in the sense that it is hard to recover x from the obfuscated program, when x is drawn from the well-spread distribution mentioned above. The corresponding theorem can be stated in words as:

Theorem 1: Every family of probabilistic functions is either weakly extractable or weakly obfuscatable.

We emphasize that Theorem 1 is a general observation on any family of functions and does not assume anything about the family, not even that it is efficiently computable. Informally, this theorem can be argued for as follows. Suppose a function, f, is not weakly extractable. Then, there is an adversary A that answers consistently in the 3-round game of interactive extraction, and yet there is no extractor that recovers a preimage x. We use A to construct an obfuscation for the function f_x. The obfuscation simply contains the description of A and a corresponding private input that causes A to answer consistently. To compute $f_x(r)$, simulate A, send r in the second round of the extraction game, and output the response of A. Functionality of this obfuscation follows from consistency of A while the hiding property follows directly from the assumption that no extractor is able to recover x. We point out that finding an obfuscation of f_x may not be efficient, however, the obfuscation itself is efficient because A is.

Amplifying knowledge extraction. Theorem 1 is not entirely satisfactory because extraction is guaranteed to occur only noticeably often. So, we address the issue of amplifying extraction. We show how to do so under a necessary (for the class of injective functions) and sufficient assumption on the function. Specifically, we assume what we call "weak verification". Weak verification is a notion introduced to show that some form of verification is necessary and sufficient for knowledge amplification. Moreover, it is implied by common verification notions such as public verification for probabilistic functions [7]. Informally, weak verification means for any adversary A that outputs images in the range of f, there is a corresponding verifier, V, which given some x and the private input of A, decides whether the output of A is a valid image of x under f. In other words, V has to decide whether there exists an r such that $f(x, r) = A(z, r_A)$, where z and r_A are the auxiliary information and random coins for A. Moreover, V is allowed to fail with some arbitrary small, yet noticeable probability. We use the term "extraction (respectively, verification) with vanishing but noticeable error" to mean that for every polynomial, p, there is an extractor (respectively, verifier) that fails no more than $\frac{1}{p}$ fraction of the time. The corresponding theorem can be stated in words as follows.

Theorem 2: Every weakly-verifiable family of probabilistic functions is either weakly obfuscatable or extractable with vanishing but noticeable error. Moreover, if an injective family of functions is extractable with vanishing but noticeable error, then it is weakly verifiable.

At a very high level, the proof of Theorem 2 uses a variant of Impagliazzo's hard-core lemma [19] to amplify weak extraction to extraction with vanishing but noticeable error. Informally, we use the lemma to construct a family, $\mathbb{U}$, of machines that take the input of A and attempt to extract a preimage, x, from it. This family has the property that when all its members fail, no machine can succeed *noticeably*. We then construct a family of distributions on the input of A, one distribution for each input length n, such that any member of $\mathbb{U}$ succeeds only negligibly often (as n increases). Consequently, if $\mathbb{U}$ is not a family of extractors with vanishing but noticeable error, then the distributions just mentioned have a noticeable weight in proportion to the original one. Using Theorem 1 on A and the new distributions imply the existence of an extractor with noticeable success. However, this contradicts the amplification lemma.

Interactively-extractable POW functions. An important corollary to Theorem 2 is that every POW function with auxiliary information is interactively extractable (see Corollary 2 for a more formal presentation). This supersedes the corresponding transformation of [8] from POW with auxiliary information to extractable POW function. Moreover, the current result is more efficient in that the challenger needs to send a single challenge instead of n.

Towards negligible error. We can obtain negligible failure probability if we relax the notion of extraction so that it applies only to "reliably-consistent adversaries". Intuitively, an adversary is reliably consistent if its consistency is noticeable. In other words, disregarding input on which the adversary is consistent only negligibly often, there is a fixed polynomial, p, such that $\frac{1}{p}$ is a lower bound on the probability of consistency (here, the probability is taken over the random challenge). The corresponding theorem can be stated as follows:

Theorem 3: Every weakly-verifiable family of probabilistic functions is either weakly obfuscatable or extractable with negligible error for adversaries that are reliably consistent.

Moreover, if an efficiently computable and verifiable family of functions is extractable with negligible error, then every corresponding adversary is reliably consistent.

The proof this theorem is very similar to the previous one but it uses a stronger amplification lemma in the uniform model. Informally, the lemma states that there is a family of polynomial-time machine, $\mathbb{U}$, such that no machine can succeed in inverting a function where all members of $\mathbb{U}$ fail. (Contrast this lemma with the previous one, where the guarantee is that no machine can succeed *noticeably* where $\mathbb{U}$ fails.)

On noninteractive extraction versus obfuscation. Results similar to those for interactive extraction hold in this case. However, they are weaker in the sense that functions seem to be more likely to satisfy a weaker notion of obfuscation. Informally, the obfuscated program receives a function description, k, as input and outputs $f_k(x)$ for some x hidden in the program that may depend on k. Moreover, it is hard to recover x from the obfuscated code. The results and proofs are similar. Two issues are worth highlighting. First, following the discussion at the beginning of this introduction, the function is not fixed in advance. Rather, it is sampled from a well-spread distribution and given to the adversary. Second, a corollary to these results states that injective functions that are extractable with vanishing but noticeable error are extractable with negligible error.

On the second line of research: Constructing extractable functions. Taking another approach towards a theory of extractable functions, we study knowledge-preserving reductions among cryptographic primitives. In other words, we address the question: given a noninteractively extractable cryptographic primitive, is it possible to construct another primitive while maintaining extraction? We attempt to answer this question by reviewing the literature on cryptographic reductions and investigating whether these reductions maintain extraction. Here, we focus solely on noninteractive extraction because deterministic one-way functions are not interactively extractable (Corollary 1). The results are positive: Most reductions maintain extractability or can be modified to do so. The following is a list of reductions that preserve extractability.

1. *Extractable weak one-way functions* $\implies$ *extractable strong one-way functions.* (This is the standard reduction [24,12].)
2. *Extractable pseudorandom generators* $\implies$ *extractable pseudorandom functions.* This reduction uses the construction of [13]. We assume, in addition to the extractable pseudorandom generator, G_1, another pseudorandom generator, G_2 that is not necessarily extractable but remains pseudorandom in the presence of G_1, i.e., $G_1(x), G_2(x)$ is pseudorandom when x is uniform.
3. *Extractable one-way functions* $\implies$ *extractable* $1-1$ *trapdoor functions.* This construction assumes, in addition, the existence of a $1-1$ trapdoor function that remains one-way in the presence of the extractable function.
4. *Extractable one-way functions* $\implies$ *extractable public-key encryption.* This reduction, assumes, in addition, a trapdoor permutation. Here, extractable public-key encryption is against passive adversaries and it means that it is hard to generate a ciphertext without knowledge of the plaintext and *without seeing another ciphertext*. On the other hand, extractability against active adversaries, that is adversaries that can see other ciphertext is known in the literature as plaintext-aware encryption [5,18,4,11]. We mention that this notion requires extraction with dependent auxiliary information and is left for future work.
5. *Extractable one-way functions* $\implies$ *extractable* 2-*round commitments.* Extractable commitments means if the sender commits correctly (i.e., the commitment can be opened) then it knows the message at the commit stage. This reduction uses either the construction of [6] or of [21]. We note that [23] independently constructs extractable 2-round commitments from plaintext-aware encryption.

The main reduction missing from this list is from one-way functions to pseudorandom generators. Even though we give a reduction from the KE and DDH assumptions to extractable pseudorandom generators, constructing such generators from extractable one-way functions remains open. In this work, we take a step towards this goal by giving a reduction from a "strongly" extractable one-way function, where extraction is required to hold even when $f(x)$ is represented unambiguously in a different way. Refer to Section 4 for a detailed presentation of all results regarding knowledge-preserving reductions.

Organization. We present the first approach in the context of interactive extraction in Section 3 (the corresponding results on noninteractive extraction can be found in the full version of the paper), and the second line of research in Section 4. Formal definitions of extractable functions appear in Section 2. Due to space limitation, formal proofs appear only in the full version of the paper.

2   Preliminaries

We define here interactive and noninteractive extraction. Note that these definitions require negligible extraction error. In Section 3, we study weaker forms of extraction, where the extractor succeeds noticeably or fails with vanishing but noticeable probability.

Definition 1 (Noninteractive extraction). *A randomized family ensemble,* $\mathbb{F} = \{\{F_k\}_{k \in K_n}\}_{n \in \mathbb{N}}$, *is called* ***noninteractively extractable*** *if for any PPT A, any* *well-spread distribution, K_n, on the function description, any distribution, $\mathbb{ZR} = \{ZR_n\}_{n\in\mathbb{N}}$, on auxiliary information and the private input of A, there is polynomial-time machines, $\mathcal{K}$, such that:*

$$Pr[(z, r_A) \leftarrow ZR_n, \ k \leftarrow K_n, \ y = A(k, z, r_A), \ x = \mathcal{K}(k, z, r_A):$$

$$\exists r, f_k(x, r) = y \ or \ \forall x', r', y \neq f_k(x', r')] > 1 - \mu(n).$$

Definition 2 (Interactive Extraction). *A randomized family ensemble, $\mathbb{F} = \{\{F_k\}_{k\in K_n}\}_{n\in\mathbb{N}}$, is called **interactively extractable** if for any PPT A, any distribution, $\mathbb{ZR} = \{ZR_n\}_{n\in\mathbb{N}}$, on auxiliary information and the private input of A, there is polynomial-time machines, $\mathcal{K}$, such that for any $k \in K_n$:*

$$Pr[(z, r_A) \leftarrow ZR_n, \ r_1 \leftarrow R_n, \ (y_0, s) = A(z, r_A), \ y_1 = A(s, r_1), \ x = \mathcal{K}(z, r_A):$$

$$\exists r_0, f_k(x, r_0) = y_0 \ or \ (\forall x', \ (\forall r_0, \ y_0 \neq f_k(x', r_0)) \ or \ y_1 \neq f_k(x', r_1))] > 1 - \mu(n).$$

3   On Obfuscation Versus Interactive Extraction

We present the three theorems mentioned in the introduction concerning the connection between obfuscation and interactive extraction with different extraction rates. Recall, the first theorem says that every function is either weakly extractable or weakly obfuscatable. The second theorem builds on the first one to imply that every weakly verifiable function is either weakly obfuscatable or extractable with vanishing but noticeable error. The final theorem states that negligible-error extraction can be achieved if and only if certain conditions on the adversary are met. These conditions, termed "reliable consistency" in the introduction, are discussed and formalized in Section 3.2.

The statement that any function is either extractable or obfuscatable is to some degree intuitive. After all, these two notions are complementary in some way. For instance, suppose there is an obfuscated program that hides a license key inside it and is able to compute a new hash of the key. If we look at such a program from an extractability point of view, this means that there is a machine that simulates this program and computes the functionality mentioned above. Moreover, no extractor can recover the license key by the assumption that the obfuscated program hides it. Going in the reverse direction, it seems intuitive that the existence of an extractor for every adversary implies the absence of an obfuscation of such a functionality.

In the next theorem, we formalize and show that the intuition mentioned in the previous paragraph is sound. In more detail, statement 1 of this theorem (the obfuscation clause) states that there is a well-spread distribution, $\mathbb{X}$, on the input (think of this as the license key of the previous example) and an obfuscator, G_n, that takes a license key, x, and produces an obfuscated program, $g(x)$. In turn, $g(x)$ takes an input r and produces a new image of x using r as random coins for the function, i.e., $g(x)(r) = f(x, r)$. Moreover, $g(x)$ is required to be one-way in x but not required to succeed in computing this functionality more than noticeably often. In the theorem, we use the terminology $g(x)(\perp)$ to refer to a fixed hash of x available in the clear in the obfuscated program. On the other hand, statement 2 (the extraction clause) says that any adversary, A, with any distribution on its input, z, r_A (z is auxiliary information and r_A is the random coins for A), that is consistent in the 3-round game discussed in the introduction, there is a corresponding extractor that recovers a preimage. In more detail, A is supposed to produce, with noticeable success, an image, y_0 in the first round and then again y_1 in the third round, such that there is a preimage common to both y_0 and y_1. Moreover, the extractor is supposed to succeed only noticeably often.

Theorem 1. *Let* $\mathbb{F} = \{f_n\}_{n \in \mathbb{N}}$ *be any randomized family of functions and* $\mathbb{R} = \{R_n\}_{n \in \mathbb{N}}$ *be any distribution on the randomness domain of* $\mathbb{F}$. *Then, exactly one of the following two statements should hold:*

1. *There is a well-spread distribution* $\mathbb{X}$ *on the input domain of* $\mathbb{F}$, *a probabilistic function,* $\mathbb{G} = \{G_n\}$ *such that for any nonuniform polynomial-time machine,* A: *(Obfuscation)*

$$Pr[x \leftarrow X_n,\ g(x) \leftarrow G_n(x),\ x' = A(g(x)) : \exists r',\ g(x)(\bot) = f_n(x', r')] \le \mu(n).^2$$

   (Functionality)

$$Pr[x \leftarrow X_n,\ g(x) \leftarrow G_n(x),\ r \leftarrow R_n : \exists r',\ g(x)(r) = f_n(x, r)\ and\ g(x)(\bot) \\ = f_n(x, r')],$$

   is nonnegligible in n. *Moreover,* $g(x)(r)$ *is efficiently computable, for any* r.
2. *For any probabilistic polynomial-time machine (PPT),* A, *any infinite subset of security parameters,* $\mathbb{N}'$, *any distribution,* $\mathbb{ZR} = \{ZR_n\}_{n \in \mathbb{N}'}$, *on auxiliary information and the private input of* A, *if:* *(Consistency)*

$$Pr[(z, r_A) \leftarrow ZR_n,\ r_1 \leftarrow R_n,\ (y_0, s) = A(z, r_A),\ y_1 = A(s, r_1) :$$

$$\exists x',\ r_0,\ y_0 = f_n(x', r_0))\ and\ y_1 = f_n(x', r_1))], \tag{1}$$

   is nonnegligible in n, *then there exists a nonuniform polynomial-time machine,* $\mathbb{K}$, *such that:* *(Extraction)*

$$Pr[(z, r_A) \leftarrow ZR_n,\ (y_0, s) = A(z, r_A),\ x = K(z, r_A) : \exists r_0,\ y_0 = f_n(x, r_0)], \tag{2}$$

   is nonnegligible in n.

We emphasize that the previous theorem holds for any function. That is, it does not assume anything about the function, not even that it is efficiently computable. At a high level, the proof proceeds as follows. If f is not extractable, we take an adversary that violates this property and construct from it a distribution on the input to f (for clarity, refer to this as the license distribution) and an obfuscation on this distribution such that the obfuscation hides the license but is able to compute new images of it. In more detail, the license distribution is the distribution induced by A on preimages of its consistent output. For instance, if A *always* outputs $f_n(0, r_0)$ in the first round and $f_n(0, r_1)$ in the third round (in this case there is a straightforward extractor), then the induced distribution always samples 0. Moreover, the corresponding obfuscation

---

[2] Here and in the rest of the paper, μ denotes a negligible function.

is simply the input of A that causes A to output valid images of the license. Observe that the license distribution is well-spread because otherwise the nonuniform extractor can invert with noticeable probability. Therefore, using this license distribution with the corresponding obfuscation, statement 1 follows from the negation of statement 2. The other direction is easier to see and has been referred to in the second paragraph of this section.

Corollary 1. *Any deterministic one-way function is not even weakly extractable. That is, any deterministic one-way function satisfies statement 1 of Theorem 1. Moreover, this remains true if the function is not efficiently computable.*

3.1 Amplifying Extraction

Theorem 1 says each function has a weakly extractable or weakly obfuscatable property. Next, we investigate conditions that allow for amplifying knowledge extraction in the interactive setting. In particular, the goal in this section is to reach a vanishing but noticeable extraction error. Recall from the introduction, this term means that for every polynomial, p, there is an extractor that may depend on p and fails at most $\frac{1}{p}$ of the time. In Section 3.2, we address extraction with negligible error.

Not surprisingly, functions that admit such a property require more than the negation of statement 1 of Theorem 1. Recall that Theorem 1 holds for any function, in particular, not efficiently-computable functions. However, to decrease the extraction error, efficient verification is needed. For the purpose of amplifying extraction, common notions of verification (e.g., Definition 3) are sufficient. However, a weaker but contrived form of verification is also sufficient, and, in the case of injective functions (i.e., for all y, there is no more than one x such that $y = f_n(x, r)$ for some r), is also necessary. Thus, we use this notion in the following theorem for the purpose of achieving a characterization instead of an implication. Informally, weak verification means that there is a verifier tailored for every adversary, A. It receives x and the input of A and determines whether the output of A is a valid image of x. Moreover, the verifier is allowed to fail, when A is consistent, with noticeable probability.

Definition 3 (Efficient Verification, [7])
A function family , $\mathbb{F} = \{f_n\}_{n \in \mathbb{N}}$, satisfies efficient verification if there exists a deterministic polynomial time algorithm, $V_{\mathbb{F}}$ such that:

$$\forall n \in \mathbb{N}, \ x \in \{0,1\}^n, \ y \in range(f_n), \ V_{\mathbb{F}}(x, y) = 1 \ \textit{iff} \ \exists r, y = f_n(x, r).$$

Definition 4 (Weak Verification)
A function family , $\mathbb{F} = \{f_n\}_{n \in \mathbb{N}}$, satisfies weak verification if for every PPT, A (with input z, r_A), any distribution, $\mathbb{ZR} = \{ZR_n\}_{n \in \mathbb{N}'}$, on auxiliary information and the private input of A, and any polynomial p, there exists a nonuniform polynomial-time machine, $V_{A,ZR,p}$, such that for sufficiently large $n \in \mathbb{N}'$:

$$Pr[(z, r_A) \leftarrow ZR_n, \ r_1 \leftarrow R_n, \ (y_0, s) = A(z, r_A), \ y_1 = A(s, r_1) :$$

$$(\exists x, r_0, \ V_{A,ZR,p}(x, z, r_A) \neq 1 \ and \ f_n(x, r_0) = y_0 \ or \ \exists x, V_{A,ZR,p}(x, z, r_A) = 1$$
$$and \ \forall r_0, f_n(x, r_0) \neq y_0)$$

$$and \ (\exists x, r_0, \ f_n(x, r_0) = y_0 \ and \ f_n(x, r_1) = y_1)] < \frac{1}{p(n)}.$$

Theorem 2. *Let* $\mathbb{F} = \{f_n\}_{n \in \mathbb{N}}$ *be any randomized function family that is weakly extractable (satisfies statement 2 of Theorem 1). If $\mathbb{F}$ is weakly verifiable (as in Definition 4), then for any PPT A, any distribution, $\mathbb{ZR} = \{ZR_n\}_{n \in \mathbb{N}'}$, on auxiliary information and the private input of A, there exists a family of nonuniform polynomial-time machines, $\mathbb{U} = \{U_i\}_{i \in \mathbb{N}}$ such that for any polynomial p, there is an index i_p where for all $i \geq i_p$ and sufficiently large $n \in \mathbb{N}'$:*

$$Pr[(z, r_A) \leftarrow ZR_n, \ r_1 \leftarrow R_n, \ (y_0, s) = A(z, r_A), \ y_1 = A(s, r_1), \ x = U_i(z, r_A):$$

$$(\exists r_0, f_n(x, r_0) = y_0 \ or \ (\forall x', \ (\forall r_0, \ y_0 \neq f_n(x', r_0)) \ or \ y_1 \neq f_n(x', r_1))] > 1 - \frac{1}{p(n)}.$$
$$(3)$$

Moreover, this implication is an equivalence for injective functions.

The proof uses, in an essential way, an amplification lemma which is a version of Impagliazzo's hard-core lemma [19] applied to this setting. At a very high level, this lemma asserts the existence of a family of machines, $\mathbb{U}$, such that "no machine can succeed noticeably where all of these machines fail". Using this lemma, we then claim that for every polynomial, p, there is a member $U_{i_p} \in \mathbb{U}$ that fails in extracting a preimage with a probability at most $\frac{1}{p}$. If this were not to be the case, then this means that there is some polynomial p, where every machine in $\mathbb{U}$ fails with probability at least $\frac{1}{p}$. This implies that there is a noticeable fraction of the domain where A is consistent yet all members of $\mathbb{U}$ fail. Lets restrict the distribution on the input of A to those on which such an event occurs. We then apply Theorem 1, in particular, statement 2, to obtain an extractor with noticeable success contradicting the lemma.

The following corollary is one of the main applications of this result.

Corollary 2. *Every POW function with auxiliary information that is collision resistant and has public randomness is extractable with vanishing but noticeable error in the interactive setting (as in Theorem 2).*

3.2   Towards Extraction with Negligible Error

The previous section underscores the conditions that are necessary (at least for injective functions) and sufficient for extraction with vanishing but noticeable error. Here, we address the question of obtaining extraction with negligible error. As before, we show necessary and sufficient conditions to achieve this objective. However, unlike the previous results, the conditions are on the adversary itself and not on the function under study. Moreover, as we discuss later on, this result is in the uniform setting only.

Conditions for extraction with negligible error. As we mentioned in the introduction, extraction with negligible error requires "reliable consistency" on the behalf of the adversary. Informally, we show that negligible extraction error is possible for a particular adversary, A, if it can answer challenges consistently with probability bounded from below the inverse of some fixed polynomial. Informally, it may be the case that A answers consistently with noticeable probability. Yet, depending on its input, its corresponding consistency probability (taken over the random coins of the challenger) can be arbitrary small though still noticeable. In such a scenario, extraction can not achieve negligible error because as answers are less likely to be consistent, extraction requires more effort and time to find a preimage. On the other hand, if for almost all of its input, A answers consistently with a probability bounded from below by an inverse polynomial, this bound can be translated into an upper bound on the running time of the extractor.

We elaborate on these conditions through a toy example. Suppose there is a function, f and an adversary A with the following properties. A outputs a consistent pair (y_0, y_1) with probability $\frac{1}{n^i}$ for every element in the ith $\frac{2^n}{n}$ fraction of the input domain for A. Here, the probability is taken over random coins sent by the challenger in round 2. Formally, we have for every n, and every $(z, r_A) \in \left[\frac{i2^n}{n}, \frac{(i+1)2^n}{n}\right]$:

$$Pr[r_1 \leftarrow R_n, \ (y_0, s) = A(z, r_A), \ y_1 = A(s, r_1) : \exists x, r_0, \ f_n(x, r_0) = y_0$$

$$\text{and } f_n(x, r_1) = y_1] = \frac{1}{n^i}.$$

Now, it may be the case that extraction depends on how successful A is in answering challenges. If this is so, then extraction is proportional to consistency. In other words, as A becomes less consistent (that is, as its input is chosen from the upper fraction of the domain), extraction requires more time to achieve the same success rate. In such a scenario, it turns out that overwhelming success requires super-polynomial time. In other words, noticeable extraction error is unavoidable.

In the previous example, we assume that A has a noticeable success in every fraction of the input domain. Also, we assume that A can not do any better. In other words, A can not amplify its success rate. However, there are cases where A can indeed amplify its success, e.g., A may provide wrong answers intentionally even though it can easily compute the correct ones. In such a scenario, extraction with negligible error is possible. As an example, consider an adversary, A, that provides wrong answers intentionally. A receives x as input, computes i such that $x \in \left[\frac{i2^n}{n}, \frac{(i+1)2^n}{n}\right]$, and gives the correct answer only if $r_1 \in [0, \frac{2^n}{n^i}]$. Even though A satisfies the previous condition, an extractor can easily recover x by reading it from the input. So, we need a meaningful way to separate the notion of "truthful" failure from "intentional" failure. In the next theorem, we capture the notion of intentional failure through the existence of another machine A' that behaves similarly to A, yet it amplifies its consistency.

Uniform Setting. The proof of Theorem 2 uses a diagonalization technique to show that no machine can succeed "substantially" where the family $\mathbb{U}$ fails. The diagonalization is over machines that succeeds noticeably over inputs of some length n. This technique works because this set of machines is enumerable. (Specifically, there are at most n machines that each succeeds exclusively with probability $\frac{1}{n}$ and so on.) However, this technique fails when we try to use it to achieve negligible error in polynomial time. Two factors seem to prevent this technique from working. First, the set of nonuniform polynomial-time machines is not enumerable and so we can not diagonalize over this set (as we discuss later on, we use the enumeration of uniform machines to prove this result in the uniform setting). Second, if we instead consider machines that succeed exclusively, as in the previous theorem, we need to take into account those that succeed with negligible probability, yet the probability is not "very negligible", say, $\frac{1}{n^{\log n}}$.

However, this causes $\mathbb{U}$ to be slightly super-polynomial. Consequently, the next theorem applies to the uniform setting only. It uses a uniform version of Theorem 1 which can be found in the full version of the paper.

In words, reliable consistency in the next theorem refers to a new machine, A', that replaces an adversary, A, with the purpose of undoing any intentional failure on behalf of A. The conditions on A' are as follows: (1) the output of A' is equivalent to A in the first round, (2) the consistency of A' is not any worse than that of A, and (3) there is a fixed polynomial, $p_{A'}$, such that almost all inputs to A' cause it to be either consistent negligibly or with probability at least $\frac{1}{p_{A'}}$. If there is such an A' then extraction with negligible extraction error is possible. Moreover, the converse is also true for efficiently computable and verifiable functions.

Theorem 3. *Let* $\mathbb{F} = \{f_n\}_{n \in \mathbb{N}}$ *be any randomized function family that satisfies the uniform version of statement 2 of Theorem 1 and is weakly verifiable (as in Definition 4, except with respect to uniform deterministic machines).*

Let A be any PPT and $\mathbb{ZR} = \{ZR_n\}_{n \in \mathbb{N}'}$ *be any distribution on auxiliary information and the private input of A. If there is another PPT, A', satisfying the following three conditions of reliable consistency:*

1. $A'(z, r_A) = A(z, r_A)$ for all z, r_A.
2.

$$Pr[(z, r_A) \leftarrow ZR_n, \; r_1 \leftarrow R_n, \; (y_0, s) = A'(z, r_A), \; y_1 = A'(s, r_1) :$$
$$\exists x', \; r_0, \; y_0 = f_n(x', r_0)) \text{ and } y_1 = f_n(x', r_1))]$$
$$\geq Pr[(z, r_A) \leftarrow ZR_n, \; r_1 \leftarrow R_n, \; (y_0, s) = A(z, r_A), \; y_1 = A(s, r_1) :$$
$$\exists x', \; r_0, \; y_0 = f_n(x', r_0)) \text{ and } y_1 = f_n(x', r_1))] - \mu(n)$$

3. There exists a polynomial $p_{A'}$, such that for any polynomial $q > p_{A'}$:

$$Pr[(z, r_A) \leftarrow ZR_n :$$

$$\tfrac{1}{q(n)} \leq Pr[r_1 \leftarrow R_n, (y_0, s) = A'(z, r_A), y_1 = A'(s, r_1, a_{A'}) : \exists x', r_0, y_0 = f_n(x', r_0) \text{ and }$$
$$y_1 = f_n(x', r_1)] \leq \tfrac{1}{p_{A'}(n)}] \leq \mu(n)$$

then there is a deterministic polynomial-time machine, $\mathcal{K}$ such that for $n \in \mathbb{N}'$:

$$Pr[(z, r_A) \leftarrow ZR_n, \; r_1 \leftarrow R_n, \; (y_0, s) = A(z, r_A), \; y_1 = A(s, r_1), \; x = \mathcal{K}(z, r_A) :$$

$$\exists r_0, f_n(x, r_0) = y_0 \text{ or } (\forall x'(\forall r_0, \; y_0 \neq f_n(x', r_0)) \text{ or } y_1 \neq f_n(x', r_1))] > 1 - \mu(n). \tag{4}$$

Moreover, if $\mathbb{F}$ is efficiently computable and verifiable (as in Definition 3), then the converse is also true.

The proof is similar to that of Theorem 2. There are two points worth highlighting. The proof uses a uniform version of the amplification lemma. Informally, this lemma provides a family of machines, $\mathbb{U}$, such that any machine can not succeed even negligibly where this family fails. At a high level, each $U_i \in \mathbb{U}$ contains the first i machines in an enumeration of uniform polynomial-time machine. This ensures that every polynomial-time machine is eventually included in the family. We claim that there is a member of this family that achieves negligible extraction error. If this were not to be the case, then for every member U_i there is a polynomial p_i such that U_i fails with probability at least $\frac{1}{p_i}$. Note that p_i may increase as i increases. However, by the third condition on A', consistency of A' is bounded from below by the inverse of a fixed polynomial which is independent of p_i. This is important because when we restrict the input distribution to where A' is consistent and $\mathbb{U}$ fails, A' remains consistent with noticeable probability. Consequently, we can apply Theorem 1 to get an extractor with noticeable success contradicting the lemma.

Corollary 3. *Any deterministic and efficiently-verifiable (i.e., given x and y, it is easy to decide whether $f(x) = y$) function is extractable with negligible error if and only if it is weakly extractable in the uniform setting.*

4   Knowledge-Preserving Reductions

In Section 3, we investigate the relationships among different notions of extraction. We address questions regarding the possibility that functions satisfy some extractability properties, such as weak extraction, extraction with noticeable error, or extraction with negligible error. Results in this line of work show equivalence among some notions of extraction, e.g., extraction with noticeable error is equivalent to extraction with nonnegligible success for deterministic and efficiently verifiable functions (Corollary3).

Here, we take a different approach. Specifically, we investigate building extractable functions with additional hardness properties from extractable functions with simpler computational assumptions. In particular, we revisit the literature on reductions among primitives to see if these reductions or variations of preserve noninteractive extraction.

The results are mostly positive. In particular, reductions from weak one-way functions to strong one-way functions, from one-way functions to 2-round commitments and public-key encryption scheme (assuming in addition a trapdoor permutation) are knowledge preserving or can be easily modified to be so. Moreover, extractable pseudorandom generators imply extractable pseudorandom functions and extractable 2-round commitments. One important open question is whether extractable one-way functions imply extractable pseudorandom generators. In pursuit of answering this question, we show that the HILL construction [17] is not knowledge preserving. On the other hand, an extractable pseudorandom generator can be constructed from the KE and the DDH assumptions.

Next, we provide a detailed presentation of these results. They address noninteractive extraction with negligible error only. Interactive extraction is primarily useful for probabilistic functions because by Corollary 1, deterministic one-way functions and pseudorandom generators are not interactively extractable. As for probabilistic functions, [8] provides a transformation from POW functions to interactively-extractable POW functions. Moreover, every POW function with auxiliary information and public randomness is interactively extractable (Corollary 2).

From extractable weak one-way to extractable strong one-way functions. The standard reduction from weak one-way functions to strong one-way functions [24,12] is knowledge preserving. Specifically, let $\mathbb{F} = \{\{f_k\}_{k \in K_n}\}_{n \in \mathbb{N}}$ be a family of weak functions with $\frac{1}{p}$ as a lower bound on the failure probability of all polynomial-time machines. Furthermore, suppose that $\mathbb{F}$ is extractable with negligible error with respect to some well-spread distribution, $\mathbb{K}$, on the function description. Then, the family, $\mathbb{G} = \{\{g_k\}_{k \in K_n}\}_{n \in \mathbb{N}}$, where $g_k(x_1, ..., x_{np(n)}) = f_k(x_1), ..., f_k(x_{np(n)})$, is also extractable with respect to $\mathbb{K}$.

Let A be any adversary that receives k, z, r_A as input (where z and r_A are auxiliary information and random coins of A, respectively) and outputs y in the range of G_k. Let B be a machine that receives k, z, r_A, i as input and outputs y_i, where i is uniform and $A(k, z, r_A) = y_1, ..., y_{np(n)}$. Note that B outputs a valid image under f_k with at least the same probability as A outputs a valid image under g_k. Therefore, there is a corresponding extractor, $\mathcal{K}_B$, for B. Let $\mathcal{K}_A$ be an extractor for A that runs $\mathcal{K}_B$ on k, z, r_A, i for $i = 1$ to $np(n)$. Except with negligible probability, if A outputs a valid image, $\mathcal{K}_B$ computes the correct images for all $f_k(x_i)$. Thus, $\mathcal{K}_A$ is a negligible-error extractor for A.

From extractable one-way functions to extractable pseudorandom generators. First, we point out that the HILL construction [17] of pseudorandom generator from even injective one-way functions is not knowledge preserving. Specifically, the family, $\mathbb{G}$, is not extractable, where $G_k(x, h) = h(f_k(x)), h, p(x)$, f_k is an extractable, $1 - 1$ one-way function, h is a hash function, and p is a hardcore predicate for f_k. This is so because the adversary, that receives and outputs a random string, succeeds with noticeable probability in producing a valid image under G_k. On the other hand, no extractor can recover a preimage because G_k is pseudorandom.

Constructing extractable pseudorandom generators from extractable one-way functions remains open. The obstacle seems to be that somehow, $f_k(x)$, should be easy to compute from the output of the generator so that it is possible to use the original extractor to recover x. Consequently, for G to be a pseudorandom generator, it should also be easy to compute $f_k(x)$ from a random string, for some x. However, the range of f may be distinguishable from uniform, e.g., the first n bits may always be 0. So, it is not clear how to put $f_k(x)$ in the output without compromising pseudorandomness.

A point worth mentioning here is that it is possible to construct extractable pseudorandom generators from a stronger knowledge requirement on the one-way function. The original knowledge assumptions states that any adversary that outputs $f_k(x)$ *as a sequence of bits* "knows" x. Consider the following stronger version. Informally, if an adversary outputs $f_k(x)$ specified in another representation, it should still know x. In particular, the type of representation, $\mathcal{R}$, we are interested in is a randomized representation of strings, where $\mathcal{R}(y, r)$ is indistinguishable from uniform and every $\mathcal{R}(y, r)$ has a unique preimage (except with negligible probability). We give a concrete example: Let π be a one-way permutation and b be a corresponding hardcore predicate. Then, $\mathcal{R}(y, r_1, ..., r_{|y|}) = \pi(r_1), ..., \pi(r_{|y|}), y \oplus b(r_1), ..., b(r_{|y|})$. Note that $\mathcal{R}$ is pseudorandom and unambiguous, in that there is a single y as a valid preimage of any output. Now, if f_k is extractable with respect to this representation, then the following construction is an extractable family of pseudorandom generators.

$$G_k(x, r_1, ..., r_{|f_k(x)|}) = \mathcal{R}(f_k(x), r_1, ..., r_{|f_k(x)|}), G'(x) \oplus r_1, ..., r_{|f_k(x)|},$$

where G' is another pseudorandom generator with a suitable expansion factor that remains pseudorandom in the presence of f (but G' is not assumed to be extractable). In other words, $f(x), G'(x)$ is assumed to be indistinguishable from $f(x), U_{|G'(x)|}$ (in this section, U_l denotes a uniform variable over strings of length l).[3]

Finally, we mention that the knowledge of exponent assumption [16] (with the DDH assumption) imply the existence of extractable pseudorandom generators, specifically, $G_{g,g^a}(x) = g^x, g^{ax}$, where g is a generator for the group for which these assumptions apply.

From extractable pseudorandom generators to extractable pseudorandom functions.
The notion of extractable pseudorandom functions is slightly different from the notions considered so far. Informally, a pseudorandom function is extractable if any adversary that computes $f_k(x, r)$, for any r that a challenger chooses, has a corresponding extractor that recovers x.

Formally, for any PPT A, any well-spread distribution, K_n, on the function description, any distribution, $\mathbb{ZR} = \{ZR_n\}_{n \in \mathbb{N}'}$, on auxiliary information and the private input of A, there is polynomial-time machines, $\mathcal{K}$, such that:

$$Pr[(z, r_A) \leftarrow ZR_n, \ k \leftarrow K_n, \ x = \mathcal{K}(k, z, r_A):$$

$$\exists r, f_k(x, r) \neq A(k, z, r_A, r) \text{ and } \exists x', \forall r', f_k(x', r') = A(k, z, r_A, r')] \leq \mu(n).$$

The construction of extractable pseudorandom functions uses the construction of [13] on all input, except 0. On input 0, the output is exactly that of the extractable generator in order to allow for successful extraction. Formally, let G^1 be any injective and extractable pseudorandom generator with a $2n^2$ (or more) expansion factor. Let b a hardcore bit for G^1 and $G_k^2(x_1, \ldots, x_n) = G_k^1(b(x_1), \ldots, b(x_n))$, where $|x_1| = \cdots = |x_n| = n$. W.l.o.g. assume G^2 has a $2n$ expansion factor, otherwise, trim the output to a suitable length. Let $\mathbb{F}'$ be the family of pseudorandom functions obtained by applying the construction of [13] on G^2. Then, the extractable family of pseudorandom functions, $\mathbb{F} = \{\{f_k\}_{k \in K_n}\}_{n \in \mathbb{N}}$, is defined as follows:

$$f_k((x_1, \ldots, x_n), r) = \begin{cases} G_k^1(x_1), \ldots, G_k^1(x_n) & \text{if } r = 0 \\ f_k'((x_1, \ldots, x_n), r) & \text{otherwise} \end{cases}$$

Let A be any PPT that receives k, z, r_A, r and outputs $f_k(x_1, \ldots, x_n, r)$ for some $x_1, \ldots, x_n$. Let B be a machine that receives k, z, r_A, i (where i is uniform), computes $A(k, z, r_A, 0) = G_k^1(x_1), \ldots, G_k^1(x_n)$ and outputs $G^1(x_i)$. Since G^1 is extractable, there is a machine, $\mathcal{K}_B$ that recovers the corresponding x_i on input k, z, r_A, i. Then, the extractor, $\mathcal{K}_A$, for A and $\mathbb{F}$, simulates $\mathcal{K}_B$ on input k, z, r_A, i, for $i = 1, \ldots, n$, and outputs $x_1, \ldots, x_n$.

From extractable one-way functions to extractable public-key encryption. Before we discuss extractable public-key encryption, we briefly mention that private-key encryption with a "strong" extraction property (that is, plaintext-aware [5]) can be easily constructed from standard computational assumptions *without knowledge assumptions*. However, we emphasize that not all private-key encryption are extractable, e.g., a random string is a valid ciphertext under $E_{sk}(m, r) = r, m \oplus f_{sk}(r)$ [12], where f_{sk} is

---

[3] Note that the machine that outputs a random string as a possible representation of $f_k(x)$ under $\mathcal{R}$ does not succeed considerably better than the machine that output a random string as a possible $f_k(x)$.

a pseudorandom function. However, the previous construction can be easily modified to become extractable. Specifically, $E_{sk=(sk_1,sk_2)}(m,r) = r, m \oplus f_{sk_1}(r), f_{sk_2}(m,r)$ has the property that without knowledge of sk, it is hard to find a *new* ciphertext even if the adversary sees encryption of multiple messages.

Extractable one-way functions can be used with a trapdoor permutation to construct public-key encryption schemes with the property that any adversary that computes a ciphertext *without seeing another ciphertext* "knows" the corresponding plaintext. This notion is similar to plaintext-aware encryption [5,18,4,11]. Informally, the latter notion says that no adversary, with access to ciphertext of messages it may not know, can produce a ciphertext without knowing the corresponding plaintext. In this work we focus on extraction with independent auxiliary information only. So, we leave the study of constructing plaintext-aware encryption from extractable functions to future work as it requires extraction with dependent auxiliary information [8]. We note that [8] constructs plaintext-aware encryption from extractable POW functions with dependent auxiliary information.

Let $\mathbb{F} = \{\{f_k\}_{k \in K_n}\}_{n \in \mathbb{N}}$ and $\Pi = \{\{\pi_{pk}\}_{pk \in PK_n}\}_{n \in \mathbb{N}}$ be families of extractable one-way functions and trapdoor permutations, respectively. Moreover, suppose that $\mathbb{F}$ and Π remain one-way with respect to each other, specifically, for a uniform r, k, pk, $f_k(r), \pi_{pk}(r)$ is one-way. Let b be a hardcore predicate for the function $g_{k,pk}(r) = f_k(r), \pi_{pk}(r)$. Note that g is extractable and injective. Let $E_{k,pk}(m, (r_1, \ldots, r_n)) = g_{k,pk}(r_1), \ldots, g_{k,pk}(r_n), m \oplus b(r_1), \ldots, b(r_n)$. It can be show that for any adversary that computes a valid ciphertext, without seeing another ciphertext, there is an extractor that recovers $r_1, \ldots, r_n$ and consequently, m.

From extractable one-way functions to extractable $1 - 1$ trapdoor functions. Observe that g, as defined above, is an extractable $1 - 1$ trapdoor function if $\mathbb{F}$ and Π remain one-way with respect to each other. Moreover, the same result holds when Π is a family of $1 - 1$ trapdoor functions.

Extractable commitments. Informally, an extractable commitments guarantee *at the commit stage* that the sender knows the secret if the commitment is valid (that is, it can be opened). Even though in a stand-alone protocol, this additional property may seem irrelevant (because the sender reveals the secret in the decommit stage and nothing happens between these two stages), it is one of several important properties that come into play in more complex protocols with stronger security requirement. Thus, extractable commitments in the CRS model were introduced and studied in [22,9,10] as part of zero-knowledge proofs and universally-composable commitments.

We show that known commitments constructions from injective one-way function [6] and from pseudorandom generators [21] can be easily modified into 2-round extractable commitments if the underlying primitives are extractable. We note that Ventre and Visconti [23], independently construct 2-round extractable commitments from plaintext-aware encryption schemes (with additional assumptions).

Extractable commitments from $1 - 1$ extractable, one-way functions. Let $\mathbb{F}$ be a family of injective and extractable one-way functions. The 2-commitment starts with the receiver sending a random function description, k, and the sender responds with $f_k(u_1), \ldots, f_k(u_n), m \oplus b(u_1), \ldots, b(u_n)$, where b is a hardcore bit for f_k. Note that it is essential for the hiding property that the family, $\mathbb{F}$ be one-way with respect to *any* function in the family.

Extractable commitments from extractable pseudorandom generators. We modify the 2-round commitment scheme of [21] to make it extractable. In the first round, the receiver sends random strings $r_1, \ldots, r_n$ and the description, k, for the pseudorandom generator. In the second round, the senders responds with $g_k(u_1) \oplus r_1^{m_1}, \ldots, g_k(u_n) \oplus r_n^{m_n}$, where $r_i^{m_i} = r_i$ if $m_i = 0$ and $r_i^{m_i} = 0^{3n}$, otherwise. As in the previous construction, every function in the family is assumed to be pseudorandom.

References

1. Barak, B., Goldreich, O., Impagliazzo, R., Rudich, S., Sahai, A., Vadhan, S.P., Yang, K.: On the (Im)possibility of obfuscating programs. In: Kilian, J. (ed.) CRYPTO 2001. LNCS, vol. 2139, p. 1. Springer, Heidelberg (2001)
2. Bellare, M., Goldreich, O.: On defining proofs of knowledge. In: Brickell, E.F. (ed.) CRYPTO 1992. LNCS, vol. 740, pp. 390–420. Springer, Heidelberg (1993)
3. Bellare, M., Palacio, A.: The knowledge-of-exponent assumptions and 3-round zero-knowledge protocols. In: Franklin, M. (ed.) CRYPTO 2004. LNCS, vol. 3152, pp. 273–289. Springer, Heidelberg (2004)
4. Bellare, M., Palacio, A.: Towards plaintext-aware public-key encryption without random oracles. In: Lee, P.J. (ed.) ASIACRYPT 2004. LNCS, vol. 3329, pp. 48–62. Springer, Heidelberg (2004)
5. Bellare, M., Rogaway, P.: Optimal asymmetric encryption. In: De Santis, A. (ed.) EUROCRYPT 1994. LNCS, vol. 950, pp. 92–111. Springer, Heidelberg (1995)
6. Blum, M.: Coin flipping by phone. In: IEEE Computer conference (1982)
7. Canetti, R.: Towards realizing random oracles: Hash functions that hide all partial information. In: Kaliski Jr., B.S. (ed.) CRYPTO 1997. LNCS, vol. 1294, pp. 455–469. Springer, Heidelberg (1997)
8. Canetti, R., Dakdouk, R.R.: Extractable perfectly one-way functions. In: Aceto, L., Damgård, I., Goldberg, L.A., Halldórsson, M.M., Ingólfsdóttir, A., Walukiewicz, I. (eds.) ICALP 2008, Part II. LNCS, vol. 5126, pp. 449–460. Springer, Heidelberg (2008)
9. Canetti, R., Fischlin, M.: Universally composable commitments. In: Kilian, J. (ed.) CRYPTO 2001. LNCS, vol. 2139, p. 19. Springer, Heidelberg (2001)
10. Di Crescenzo, G.: Equivocable and extractable commitment schemes. In: Cimato, S., Galdi, C., Persiano, G. (eds.) SCN 2002. LNCS, vol. 2576, pp. 74–87. Springer, Heidelberg (2003)
11. Dent, A.W.: The cramer-shoup encryption scheme is plaintext aware in the standard model. In: Vaudenay, S. (ed.) EUROCRYPT 2006. LNCS, vol. 4004, pp. 289–307. Springer, Heidelberg (2006)
12. Goldreich, O.: Foundations of Cryptography. Cambridge University Press, Cambridge (2001)
13. Goldreich, O., Goldwasser, S., Micali, S.: How to construct random functions. Journal of the ACM 33 (1986)
14. Goldwasser, S., Kalai, Y.T.: On the impossibility of obfuscation with auxiliary input. In: FOCS (2005)
15. Goldwasser, S., Micali, S., Rackoff, C.: The knowledge complexity of interactive proof-systems. In: STOC (1985)
16. Hada, S., Tanaka, T.: On the existence of 3-round zero-knowledge protocols. In: Krawczyk, H. (ed.) CRYPTO 1998. LNCS, vol. 1462, p. 408. Springer, Heidelberg (1998)
17. Hastad, J., Levin, L., Impagliazzo, R., Luby, M.: Construction of a pseudorandom generator from any one-way function. SIAM Journal on Computing (1999)
18. Herzog, J.C., Liskov, M., Micali, S.: Plaintext awareness via key registration. In: Boneh, D. (ed.) CRYPTO 2003. LNCS, vol. 2729, pp. 548–564. Springer, Heidelberg (2003)
19. Impagliazzo, R.: Hard-core distributions for somewhat hard problems. In: FOCS (1995)

20. Lepinski, M.: On the existence of 3-round zero-knowledge proofs. M.S. Thesis (2002)
21. Naor, M.: Bit commitments using pseudorandom generators. Journal of Cryptology (1991)
22. De Santis, A., Di Crescenzo, G., Persiano, G.: Necessary and sufficient assumptions for non-interactive zero-knowledge proofs of knowledge for all NP relations. In: Welzl, E., Montanari, U., Rolim, J.D.P. (eds.) ICALP 2000. LNCS, vol. 1853, p. 451. Springer, Heidelberg (2000)
23. Ventre, C., Visconti, I.: Message-aware commitment schemes (unpublished manuscript, 2008)
24. Yao, A.C.: Theory and application of trapdoor functions. In: FOCS (1982)
25. Zheng, Y., Seberry, J.: Immunizing public key cryptosystems against chosen ciphertext attacks. Journal on Selected Areas in Communication (1993)

Author Index